MW01357985

FINE PARTICLES

SURFACTANT SCIENCE SERIES

1. Nonionic Surfactants, *edited by Martin J. Schick* (see also Volumes 19, 23, and 60)
2. Solvent Properties of Surfactant Solutions, *edited by Kozo Shinoda* (see Volume 55)
3. Surfactant Biodegradation, *R. D. Swisher* (see Volume 18)
4. Cationic Surfactants, *edited by Eric Jungermann* (see also Volumes 34, 37, and 53)
5. Detergency: Theory and Test Methods (in three parts), *edited by W. G. Cutler and R. C. Davis* (see also Volume 20)
6. Emulsions and Emulsion Technology (in three parts), *edited by Kenneth J. Lissant*
7. Anionic Surfactants (in two parts), *edited by Warner M. Linfield* (see Volume 56)
8. Anionic Surfactants: Chemical Analysis, *edited by John Cross*
9. Stabilization of Colloidal Dispersions by Polymer Adsorption, *Tatsuo Sato and Richard Ruch*
10. Anionic Surfactants: Biochemistry, Toxicology, Dermatology, *edited by Christian Gloxhuber* (see Volume 43)
11. Anionic Surfactants: Physical Chemistry of Surfactant Action, *edited by E. H. Lucassen-Reynders*
12. Amphoteric Surfactants, *edited by B. R. Bluestein and Clifford L. Hilton* (see Volume 59)
13. Demulsification: Industrial Applications, *Kenneth J. Lissant*
14. Surfactants in Textile Processing, *Arved Datyner*
15. Electrical Phenomena at Interfaces: Fundamentals, Measurements, and Applications, *edited by Ayao Kitahara and Akira Watanabe*
16. Surfactants in Cosmetics, *edited by Martin M. Rieger* (see Volume 68)
17. Interfacial Phenomena: Equilibrium and Dynamic Effects, *Clarence A. Miller and P. Neogi*
18. Surfactant Biodegradation: Second Edition, Revised and Expanded, *R. D. Swisher*
19. Nonionic Surfactants: Chemical Analysis, *edited by John Cross*
20. Detergency: Theory and Technology, *edited by W. Gale Cutler and Erik Kissa*
21. Interfacial Phenomena in Apolar Media, *edited by Hans-Friedrich Eicke and Geoffrey D. Parfitt*
22. Surfactant Solutions: New Methods of Investigation, *edited by Raoul Zana*
23. Nonionic Surfactants: Physical Chemistry, *edited by Martin J. Schick*
24. Microemulsion Systems, *edited by Henri L. Rosano and Marc Clausse*
25. Biosurfactants and Biotechnology, *edited by Naim Kosaric, W. L. Cairns, and Neil C. C. Gray*
26. Surfactants in Emerging Technologies, *edited by Milton J. Rosen*
27. Reagents in Mineral Technology, *edited by P. Somasundaran and Brij M. Moudgil*
28. Surfactants in Chemical/Process Engineering, *edited by Darsh T. Wasan, Martin E. Ginn, and Dinesh O. Shah*
29. Thin Liquid Films, *edited by I. B. Ivanov*
30. Microemulsions and Related Systems: Formulation, Solvency, and Physical Properties, *edited by Maurice Bourrel and Robert S. Schechter*
31. Crystallization and Polymorphism of Fats and Fatty Acids, *edited by Nissim Garti and Kiyotaka Sato*

32. Interfacial Phenomena in Coal Technology, *edited by Gregory D. Botsaris and Yuli M. Glazman*
33. Surfactant-Based Separation Processes, *edited by John F. Scamehorn and Jeffrey H. Harwell*
34. Cationic Surfactants: Organic Chemistry, *edited by James M. Richmond*
35. Alkylene Oxides and Their Polymers, *F. E. Bailey, Jr., and Joseph V. Koleske*
36. Interfacial Phenomena in Petroleum Recovery, *edited by Norman R. Morrow*
37. Cationic Surfactants: Physical Chemistry, *edited by Donn N. Rubingh and Paul M. Holland*
38. Kinetics and Catalysis in Microheterogeneous Systems, *edited by M. Grätzel and K. Kalyanasundaram*
39. Interfacial Phenomena in Biological Systems, *edited by Max Bender*
40. Analysis of Surfactants, *Thomas M. Schmitt*
41. Light Scattering by Liquid Surfaces and Complementary Techniques, *edited by Dominique Langevin*
42. Polymeric Surfactants, *Irja Piirma*
43. Anionic Surfactants: Biochemistry, Toxicology, Dermatology. Second Edition, Revised and Expanded, *edited by Christian Gloxhuber and Klaus Künstler*
44. Organized Solutions: Surfactants in Science and Technology, *edited by Stig E. Friberg and Björn Lindman*
45. Defoaming: Theory and Industrial Applications, *edited by P. R. Garrett*
46. Mixed Surfactant Systems, *edited by Keizo Ogino and Masahiko Abe*
47. Coagulation and Flocculation: Theory and Applications, *edited by Bohuslav Dobiáš*
48. Biosurfactants: Production • Properties • Applications, *edited by Naim Kosaric*
49. Wettability, *edited by John C. Berg*
50. Fluorinated Surfactants: Synthesis • Properties • Applications, *Erik Kissa*
51. Surface and Colloid Chemistry in Advanced Ceramics Processing, *edited by Robert J. Pugh and Lennart Bergström*
52. Technological Applications of Dispersions, *edited by Robert B. McKay*
53. Cationic Surfactants: Analytical and Biological Evaluation, *edited by John Cross and Edward J. Singer*
54. Surfactants in Agrochemicals, *Tharwat F. Tadros*
55. Solubilization in Surfactant Aggregates, *edited by Sherril D. Christian and John F. Scamehorn*
56. Anionic Surfactants: Organic Chemistry, *edited by Helmut W. Stache*
57. Foams: Theory, Measurements, and Applications, *edited by Robert K. Prud'homme and Saad A. Khan*
58. The Preparation of Dispersions in Liquids, *H. N. Stein*
59. Amphoteric Surfactants: Second Edition, *edited by Eric G. Lomax*
60. Nonionic Surfactants: Polyoxyalkylene Block Copolymers, *edited by Vaughn M. Nace*
61. Emulsions and Emulsion Stability, *edited by Johan Sjöblom*
62. Vesicles, *edited by Morton Rosoff*
63. Applied Surface Thermodynamics, *edited by A. W. Neumann and Jan K. Spelt*
64. Surfactants in Solution, *edited by Arun K. Chattopadhyay and K. L. Mittal*
65. Detergents in the Environment, *edited by Milan Johann Schwuger*

66. Industrial Applications of Microemulsions, *edited by Conxita Solans and Hironobu Kunieda*
67. Liquid Detergents, *edited by Kuo-Yann Lai*
68. Surfactants in Cosmetics: Second Edition, Revised and Expanded, *edited by Martin M. Rieger and Linda D. Rhein*
69. Enzymes in Detergency, *edited by Jan H. van Ee, Onno Misset, and Erik J. Baas*
70. Structure–Performance Relationships in Surfactants, *edited by Kunio Esumi and Minoru Ueno*
71. Powdered Detergents, *edited by Michael S. Showell*
72. Nonionic Surfactants: Organic Chemistry, *edited by Nico M. van Os*
73. Anionic Surfactants: Analytical Chemistry, Second Edition, Revised and Expanded, *edited by John Cross*
74. Novel Surfactants: Preparation, Applications, and Biodegradability, *edited by Krister Holmberg*
75. Biopolymers at Interfaces, *edited by Martin Malmsten*
76. Electrical Phenomena at Interfaces: Fundamentals, Measurements, and Applications, Second Edition, Revised and Expanded, *edited by Hiroyuki Ohshima and Kunio Furusawa*
77. Polymer-Surfactant Systems, *edited by Jan C. T. Kwak*
78. Surfaces of Nanoparticles and Porous Materials, *edited by James A. Schwarz and Cristian I. Contescu*
79. Surface Chemistry and Electrochemistry of Membranes, *edited by Torben Smith Sørensen*
80. Interfacial Phenomena in Chromatography, *edited by Emile Pefferkorn*
81. Solid–Liquid Dispersions, *Bohuslav Dobiáš, Xueping Qiu, and Wolfgang von Rybinski*
82. Handbook of Detergents, *editor in chief: Uri Zoller*
    Part A: Properties, *edited by Guy Broze*
83. Modern Characterization Methods of Surfactant Systems, *edited by Bernard P. Binks*
84. Dispersions: Characterization, Testing, and Measurement, *Erik Kissa*
85. Interfacial Forces and Fields: Theory and Applications, *edited by Jyh-Ping Hsu*
86. Silicone Surfactants, *edited by Randal M. Hill*
87. Surface Characterization Methods: Principles, Techniques, and Applications, *edited by Andrew J. Milling*
88. Interfacial Dynamics, *edited by Nikola Kallay*
89. Computational Methods in Surface and Colloid Science, *edited by Małgorzata Borówko*
90. Adsorption on Silica Surfaces, *edited by Eugène Papirer*
91. Nonionic Surfactants: Alkyl Polyglucosides, *edited by Dieter Balzer and Harald Lüders*
92. Fine Particles: Synthesis, Characterization, and Mechanisms of Growth, *edited by Tadao Sugimoto*

ADDITIONAL VOLUMES IN PREPARATION

Analysis of Surfactants: Second Edition, Revised and Expanded, *Thomas M. Schmitt*

Physical Chemistry of Polyelectrolytes, *edited by Tsetska Radeva*

Fluorinated Surfactants and Repellents: Second Edition, Revised and Expanded, *Erik Kissa*

Thermal Behavior of Dispersed Systems, *edited by Nissim Garti*

Surface Characteristics of Fibers and Textiles, *edited by Christopher M. Pastore and Paul Kiekens*

Liquid Interfaces in Chemical, Biological, and Pharmaceutical Applications, *edited by Alexander G. Volkov*

FINE PARTICLES

Synthesis, Characterization, and Mechanisms of Growth

edited by

Tadao Sugimoto

Institute for Advanced Materials Processing
Tohoku University
Sendai, Japan

MARCEL DEKKER, INC. NEW YORK • BASEL

Library of Congress Cataloging-in-Publication Data

Fine particles: synthesis, characterization, and mechanisms of growth / edited by Tadao Sugimoto.
p. cm. — (Surfactant science series; 92)
ISBN 0-8427-0001-5 (alk. paper)
1. Nanoparticles. 2. Nanostructure materials. I. Sugimoto, Tadao. II. Series.

TA418.78.F57 2000
660′.2945—dc21

00-031587

This book is printed on acid-free paper.

Headquarters
Marcel Dekker, Inc.
270 Madison Avenue, New York, NY 10016
tel: 212-696-9000; fax: 212-685-4540

Eastern Hemisphere Distribution
Marcel Dekker AG
Hutgasse 4, Postfach 812, CH-4001 Basel, Switzerland
tel: 41-61-261-8482; fax: 41-61-261-8896

World Wide Web
http://www.dekker.com

The publisher offers discounts on this book when ordered in bulk quantities. For more information, write to Special Sales/Professional Marketing at the headquarters address above.

Current printing (last digit):
10 9 8 7 6 5 4 3 2 1

PRINTED IN THE UNITED STATES OF AMERICA

Preface

Studies of the syntheses, mechanisms of formation, and properties of fine particles represent some of the essential aspects of colloid science and engineering. Although scientists have dealt with these kinds of materials for much longer than a century, a dramatic surge in general interest has occurred only recently. The reason for this appeal is the recognition of the importance of finely dispersed matter in nature and in countless applications, including ceramics, catalysts, electronics, magnetics, pigments, cosmetics, and medical diagnostics.

Since the properties of these particulate materials are basically determined by their mean size, size distribution, external shape, internal structure, and chemical composition, the science in the mechanistic study of particle formation and the fundamental technology in their synthesis and characteristic control may constitute the background for the essential development of colloid science and pertinent industries. Scientists have now learned how to form ''monodispersed'' fine particles of different shapes of simple or mixed chemical compositions, and, as a result, it is now possible to design many powders of exact and reproducible characteristics for a variety of uses. These achievements are especially important in the manufacture of high-quality products requiring stringent specification of properties.

Although many books have been published on various aspects of the science and technology of finely dispersed matter, no comprehensive text is available that covers both the fundamental mechanisms and practical procedures in the preparation and characterization of fine particles. The purpose of this book is to rectify this situation and offer a systematically organized review of the studies in this area of materials, including recent remarkable developments. This book covers the science and technology related to the preparation of fine particles in liquids and gases, with special emphasis on monodispersed particles varying in modal sizes from several nanometers to several micrometers. The chapters deal with inorganic and organic materials according to their chemical composition, including different metal compounds, such as (hydrous) oxides, chalcogenides (sulfides, selenides, tellurides), silver halides, sulfates, phosphates, apatites, carbonates, and nitrides. Other chapters describe silica, metals, carbon nanotubes, and polymer latexes. The last two chapters are devoted to the surface modification of inorganic and organic particles and to particles of specific functions, including magnetic particles, luminous particles, and fine composite particles prepared mainly by mechanical processes. Each chapter for a chemical species consists of subsections for different synthetic methods of fine particles of the specific chemical species, in which concrete procedures of each method, analyses of the growth processes, characterization of the products, and their resulting physical or chemical properties are delineated on the basis of their causality. Wherever possible, information is offered on the mechanisms of formation and specific characteristics of each family of compounds, with indications of problems that still need to be resolved. Strong emphasis is given to extensive referencing of the relevant literature.

In view of the large variety of materials described in this volume and the numerous methods for the preparation of the fine particles of individual materials, no one author could adequately cover the entire subject matter. For this reason I asked a number of specialists in the field to contribute chapters on specific topics. It is gratifying that so many have been willing to undertake the tasks requested of them, which makes this publication truly representative of the present state of the science and engineering of finely dispersed matter. I deeply appreciate their invaluable contributions.

The objective of this volume is to cover the science and technology of particle synthesis in a comprehensive and up-to-date way by frontline contributors so that it will be useful to specialists who want to find information on the latest topics and their underlying backgrounds, with easy access to original works through the abundant references. Because the text is arranged plainly and systematically from the fundamentals to the highest levels related to particle synthesis, this book is also expected to be widely used as a textbook or reference for graduate or undergraduate students and general researchers in a wide variety of fields.

Tadao Sugimoto

Contents

Contributors

David J. Elliot, Ph.D. Research Associate, CSIRO, Molecular Science, South Clayton, Victoria, Australia

Fernand Fiévet, Sc.D. Professor, Department of Chemistry, Université Paris 7–Denis Diderot, Paris, France

D. Neil Furlong, B.Sc.(Hons), Ph.D. Professor and Pro-Vice Chancellor, Research and Development Division, Royal Melbourne Institute of Technology, Melbourne, Australia

Herbert Giesche, Ph.D. Associate Professor, New York State College of Ceramics, Alfred University, Alfred, New York

Franz Grieser, Ph.D. Reader in Physical Chemistry, Advanced Mineral Products Research Center, School of Chemistry, University of Melbourne, Parkville, Victoria, Australia

Karen Grieve, B.Sc.(Hons) School of Chemistry, University of Melbourne, Parkville, Victoria, Australia

Hirotaka Honda, Ph.D. Associate Professor, Faculty of Industrial Science and Technology, Science University of Tokyo, Hokkaido, Japan

Dorothee Ingert, Ph.D. Scientist, Laboratory SRSI, Université Pierre et Marie Curie, Paris, France

Tatsuo Ishikawa, Ph.D. Professor, School of Chemistry, Osaka University of Education, Osaka, Japan

Saburo Iwama, Dr.Eng. Professor, Department of Applied Electronics, Daido Institute of Technology, Nagoya, Japan

Haruma Kawaguchi, Ph.D. Professor, Department of Applied Chemistry, Faculty of Science and Technology, Keio University, Yokohama, Japan

Keisaku Kimura, D.Sc. Professor, Department of Material Science, Himeji Institute of Technology, Akou-gun, Hyogo, Japan

Masumi Koishi, D.Sc. Professor, Faculty of Industrial Science and Technology, Science University of Tokyo, Hokkaido, Japan

Kijiro Kon-no, D.Sc. Professor, Department of Industrial Chemistry, Institute of Interface Science, Science University of Tokyo, Tokyo, Japan

Takashi Kyotani, Ph.D. Associate Professor, Institute for Chemical Reaction Science, Tohoku University, Sendai, Japan

Egon Matijević, Ph.D. Victor K. LaMer Professor, Department of Chemistry and Center for Advanced Materials Processing, Clarkson University, Potsdam, New York

Laurence Motte, Ph.D. Assistant Professor, Laboratory SRSI, Université Pierre et Marie Curie, Paris, France

Masaaki Oda, Ph.D. Manager, Nano-Particle Department, Vacuum Metallurgical Company Ltd., Sanbu-cho, Sanbu-gun, Chiba, Japan

Takashi Ogihara, Ph.D. Associate Professor, Department of Material Science and Engineering, Fukui University, Fukui, Japan

K. Osseo-Asare, Ph.D. Professor, Department of Materials Science and Engineering, Pennsylvania State University, University Park, Pennsylvania

Masataka Ozaki, Ph.D. Professor, Department of Environmental Science, Yokohama City University, Yokohama, Japan

Richard E. Partch, Ph.D. Senior Professor, Department of Chemistry, Clarkson University, Potsdam, New York

Marie-Paule Pileni, Ph.D. Professor, Department of Physical Chemistry, Université Pierre et Marie Curie, Paris, France

Yahachi Saito, Ph.D. Professor, Department of Electrical and Electronic Engineering, Faculty of Engineering, Mie University, Tsu, Japan

Ronald S. Sapieszko, Ph.D. Senior Research Scientist, Aveka, Inc., Woodbury, Minnesota

Mamoru Senna, Ph.D. Professor, Department of Applied Chemistry, Faculty of Science and Technology, Keio University, Yokohama, Japan

Tadao Sugimoto, Ph.D. Professor, Institute for Advanced Materials Processing, Tohoku University, Sendai, Japan

Akira Tomita, Ph.D. Professor, Institute for Chemical Reaction Science, Tohoku University, Sendai, Japan

Naoki Toshima, Dr.Eng. Professor, Department of Materials Science and Engineering, Science University of Tokyo at Yamaguchi, Onoda, Japan

Kohji Yoshinaga, Dr.Eng. Professor, Faculty of Engineering, Department of Applied Chemistry, Kyushu Institute of Technology, Kitakyushu, Japan

FINE PARTICLES

1

Metal Oxides

EGON MATIJEVIĆ, RONALD S. SAPIESZKO, TAKASHI OGIHARA,
TADAO SUGIMOTO, KIJIRO KON-NO, RICHARD E. PARTCH,
and MASAAKI ODA

1.1 FORCED HYDROLYSIS IN HOMOGENEOUS SOLUTIONS

EGON MATIJEVIĆ
Clarkson University, Potsdam, New York

RONALD S. SAPIESZKO
Aveka, Inc., Woodbury, Minnesota

1.1.1 Introduction

Precipitation of metal (hydrous) oxides in aqueous solutions is a process that continuously proceeds in nature and is carried out for countless reasons in laboratories and industry. For example, such compounds are the result of corrosion processes, and they are used as pigments, catalysts, etc., as well as in gravimetric analysis. In practice, it is very easy to produce a precipitate of a metal (hydrous) oxide: One needs only to add a sufficient amount of a base to a metal salt solution! However, achieving dispersions of these materials as uniform colloidal particles is a much more demanding task. The reason for the difficulties lies in the sensitivity of the properties of the products to many experimental parameters. A small change in temperature, pH, concentration of reactant solutions, methods of mixing, etc. may result in an entirely different composition, shape, size, or uniformity of the resulting particles. Indeed, in the past, very few monodispersed metal oxide colloids were reported in the literature, the oldest of which may be the formation of nearly spherical hematite (α-Fe_2O_3) particles by Steele (1) and of akageneite (β-FeOOH) by Heller and associates (2,3), which they used to study the schiller layers. In later years, Furusawa and Hachisu described the preparation of rather uniform colloidal tungsten oxide (4,5).

Subsequent to these early studies, several techniques have been developed to produce well-dispersed metal (hydrous) oxides of different chemical compositions (single or biphasic), consisting of uniform particles in a variety of shapes, including spheres. These efforts are certainly justified in view of the ever-increasing recognition of the importance of such materials in numerous applications in various areas of modern technology and medicine.

Direct interactions of aqueous metal salt solutions with bases failed, as a rule, to yield uniform particles; thus, it was necessary to design some other approaches to achieve such dispersions. The fundamental ideas were based on the fact that seldom, if ever, can a metal ion in an aqueous solution be directly converted to an oxide. Instead, the initial steps involve the formation of metal hydroxides or oxyhydroxides. Either these products remain stable as precipitated, or they convert to oxides on aging by processes that may take different lengths of time but that can be accelerated by various dehydration methods. Consequently, metal ion hydrolysis is the essential initial reaction in the formation of metal (hydrous) oxide particles in aqueous solutions.

To produce uniform colloids by precipitation, it is necessary to control several stages in the process, including nucleation and particle growth. In principle, a short burst of nuclei, followed by the capture of constituent solute species, should yield monodispersed particles. This appealing concept was originally proposed by LaMer to describe the formation of monodispersed sulfur hydrosols (6), and it was generally accepted by the scientific community active in the field. More recently, it has been recognized in numerous cases that the formation of monodispersed colloids is a more involved process; that is, the nuclei grow to nanosized particles, which then aggregate in a controlled manner into the final uniform colloids (7). Recently, a model that accounts for such a mechanism was developed (8), and is described later in this chapter.

Regardless of the steps in the precipitation, the essential initial chemical reaction is the metal ion hydrolysis. Thus, in order to obtain uniform particles, different approaches are needed to manipulate the hydrolysis reaction, rather than to directly add strong bases to metal salt solutions. Two broadly successful methods have been:

1. The introduction of hydroxide ions into the metal salt solution in a kinetically controlled manner
2. The deprotonation of hydrated metal ions

1.1.1.1 Controlled Addition of Hydroxide Ions

As mentioned earlier, it has proven difficult to precipitate uniform metal (hydrous) oxides by direct mixing of solutions of metal salts and bases, especially of concentrated ones. Instead, it is possible to control the generation of hydroxide ions in situ, which can be achieved if certain organic compounds, such as formamide or urea, are slowly decomposed in aqueous solutions of metal salts. The rate of these processes depends on the pH, temperature, concentration of the reactants, and the presence or absence of metal complexing agents. In particular, precipitations caused by reactions in solutions of urea have been widely employed, although the products consist primarily of metal carbonates or basic metal carbonates. For this reason, this method is described in more detail in Chapter 7.1 of this volume.

1.1.1.2 Deprotonation of Hydrated Metal Ions (Forced Hydrolysis)

The observation that the pH of salt solutions, especially of those containing polyvalent metal ions, decreases on heating has lead to a different, yet very elegant, technique to generate well-defined metal (hydrous) oxides. Obviously, the increased acidity of such solutions must be due to the release of protons from the hydrated cations, which in turn change to hydroxide complexes. This process was termed by the senior author as ''forced hydrolysis'' (7,9–11). At appropriate temperatures and initial values of the pH, which differ for each system, the so induced hydrolysis may proceed until solid metal hydroxides or hydrous oxides are precipitated. Since the generated solute intermediates are precursors to the formation of the solids, the control of the deprotonation of hydrated metal ions in aqueous solutions is critical to the nature of the resulting dispersions. Under specific conditions, the final particles may indeed appear uniform in size and shape. In order to yield desired colloids, the temperature of these processes will depend on the hydrolyzability of the cation; in most cases, especially those involving multivalent metal ions, temperatures below 100°C suffice for the hydrolysis reactions to proceed at reasonable rates. When the deprotonation is more difficult to achieve (e.g., with most divalent cations), it is necessary to use higher temperatures, i.e., hydrothermal processing, or somewhat higher initial pH values, which can be adjusted with a mild base. Thus, in principle, all that is needed to produce monodispersed metal (hydrous) oxides by forced hydrolysis is to age certain salt solutions at elevated temperatures.

However, other parameters, such as the salt concentration, ionic strength, and especially the natures of anions in the reacting solution, play essential roles in determining the properties of the precipitated solids. The effects of anions are related to their tendency to be incorporated in the solute complexes formed on aging, which in turn differ with each cation. These anion-containing solutes often act as precursors to the solid-phase formation, affecting the properties of the final products. Various phenomena are illustrated and discussed in the text that follows.

It is noteworthy that a rather convenient approach to produce monodispersed metal (hydrous) oxides is to hydrolyze metal alkoxides. As described elsewhere in this volume, this process is rather efficient, but it does involve the use of expensive chemicals and organic solvents. Furthermore, the hydrolysis rates of some metal alkoxides are so rapid as to preclude the preparation of uniform particles.

This chapter summarizes the present state of the art of the forced hydrolysis approach by considering specific cations, particularly those of greatest practical and theoretical interest, using aqueous solutions of common salts. In addition to being economical in the manufacture of different products, the described procedure can also help in the development of a better understanding of different processes, such as corrosion of metals or formation of minerals, to mention a few. It should be emphasized that the focus of this chapter is on dispersions of narrow particle size distributions, normally designated as ''monodispersed'' systems. While a number of general reviews have been published on monodispersed colloids (7,9–21), this chapter specifically addresses the problems related to metal (hydrous) oxides.

1.1.2 Methods of Preparation

1.1.2.1 Laboratory Techniques

It took almost a century from the time Faraday obtained the first monodispersed colloid, his well-known gold sol (22), until the works of Steele (1) and Heller (2,3) mentioned earlier. Therefore, it is even more surprising that all one needs to obtain a number of such compounds is a constant-temperature oven or bath, because many of the ''monodispersed'' metal (hydrous) oxides are generated by simply heating the respective salt solutions at moderately elevated temperatures, usually $<100°C$. However, uniform particles precipitate only under a narrow set of conditions, which have to be ascertained experimentally by systematically varying the critical parameters. Depending on the hydrolyzability of the cation, the process may take several minutes to several hours or longer (9–11).

While precipitation in homogenous solutions is an exceedingly simple method, usually rather low concentrations of electrolytes must be used if well-dispersed uniform particles are to be achieved. This requirement is based on the need to keep the ionic strength of the system below a critical value in order to prevent the coagulation of the precipitates, which consist, almost without exception, of charged particles. In some instances, concentrations of the reactants can be increased, if stabilizers are added into the systems, although the latter may affect the particle uniformity and/ or shape.

The forced hydrolysis approach can be employed for the generation of monodispersed simple or internally composite metal (hydrous) oxides. In the latter case, the resulting particles are mostly internally inhomogeneous.

Another very useful technique that alleviates some of the aforementioned difficulties is the controlled double-jet precipitation (CDJP) process, which can be extended to a triple jet design. The technique was originally developed to produce uniform large particles (23–25), particularly of silver halides for photographic applications. Recently, it was demonstrated that the procedure may be used for the synthesis of a variety of uniform dispersions, including those of metal oxides of different modal sizes, ranging from several nanometers to several micrometers (26–29).

A schematic representation of a laboratory apparatus for CDJP is given in Figure 1.1.1. In principle, the reacting solutions are introduced into a constant temperature chamber at desired flow rates by means of peristaltic pumps. The predetermined volume of solutions in the reactor may contain stabilizing, reducing, or other agents, or it may be used to control the reaction pH.

This procedure offers several advantages over some other techniques, because it is fast and it can be employed with much higher reactant concentrations.

1.1.2.2 Scaling Up Processes

In order to produce larger quantities of monodispersed metal (hydrous) oxides, the laboratory techniques must be scaled up. In doing so, it is necessary to recognize that it is not possible to deviate much from the optimum conditions established by systematic studies on the small scale. For example, increasing reactant concentrations to generate more material, or raising the temperature to speed up the process, may yield products different than expected based on the preliminary experiments. Instead, it is necessary to come up with engineering designs that utilize the parameters that yielded desired dispersions.

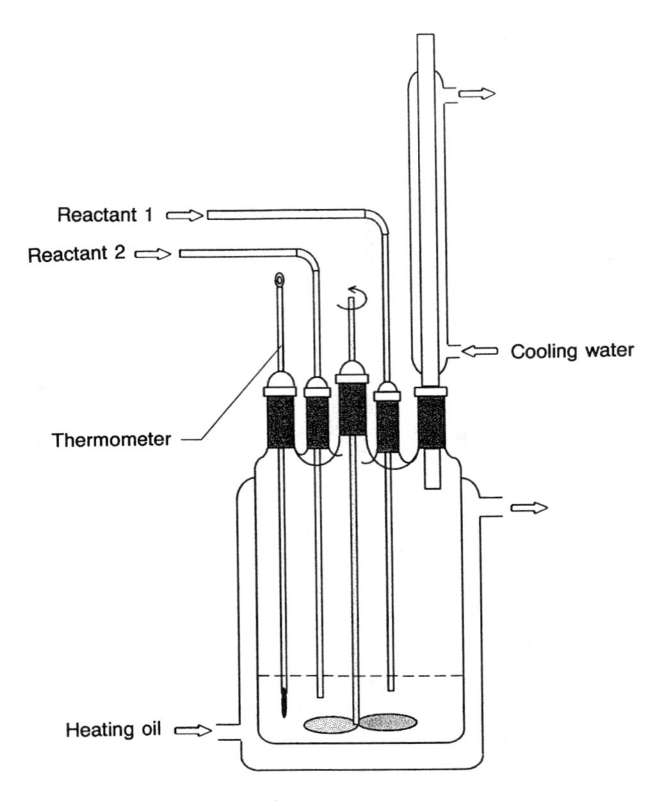

Fig. 1.1.1 Schematic presentation of the controlled double-jet precipitation (CDJP) setup.

It was demonstrated in a laboratory setup that uniform iron oxide (hematite) particles could be continuously produced in a flow system (30). The concept was then extended to a plug-flow type reactor, illustrated schematically in Figure 1.1.2 (31). The major part of the reactor is a coil of tubing submerged in a constant-temperature water or oil bath. The length and the diameter of the tubing and the flow rate of the reactant solutions determine the residence time necessary to reproduce optimum conditions, as identified for a given system. The desired laminar plug flow is achieved by inserts in the tubing. This setup was used to produce in 1 kilogram per day quantities of different monodispersed colloids of various compositions, including yttrium basic carbonate, silica, aluminum hydrous oxide, and barium titanate (31,32).

The CDJP process can also be readily scaled up. For example, it was demonstrated that large batches (e.g., 500 g) of uniform indium hydroxide could be obtained in a bench-size apparatus (26).

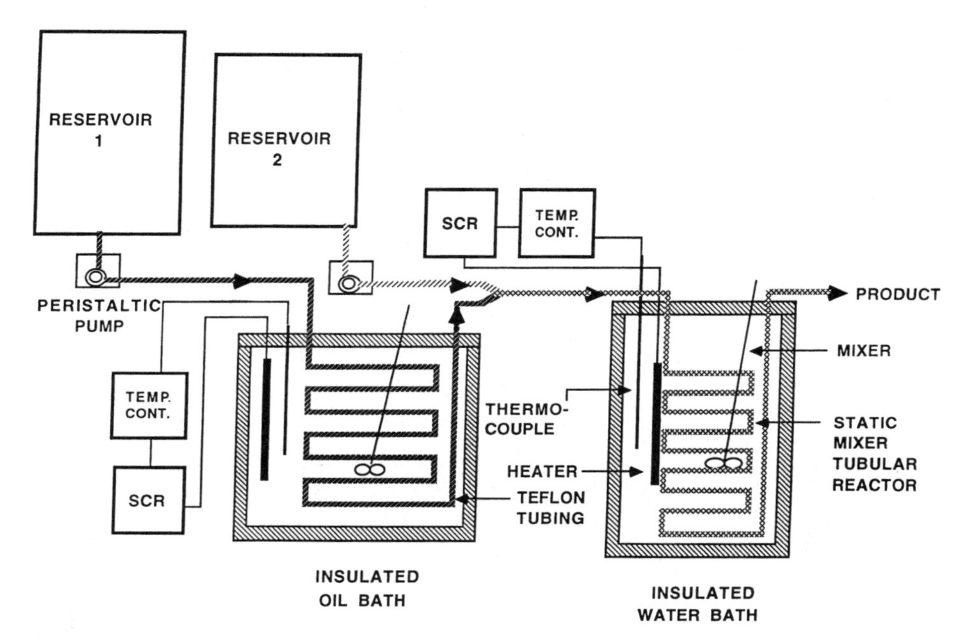

Fig. 1.1.2 Schematic presentation of the plug-flow type of a reactor for the continuous preparation of monodispersed colloids by precipitation from homogeneous solutions. (From Ref. 31.)

1.1.3 Mechanisms of Particle Formation

In discussing the mechanisms of the formation of monodispersed colloids by precipitation in homogeneous solutions, it is necessary to consider both the chemical and physical aspects of the processes involved. The former require information on the composition of all species in solution, and especially of those that directly lead to the solid phase formation, while the latter deal with the nucleation, particle growth, and/or aggregation stages of the systems under investigation. In both instances, the kinetics of these processes play an essential role in defining the properties of the final products.

1.1.3.1 Physical Aspects

The first condition necessary to form a solid in a solution is to exceed its solubility. In ionic solutions, the products of the concentrations (activities) of the actual reactants must be higher than required by the solubility product of the resulting compound at a given temperature.

However, the thermodynamic supersaturation condition does not suffice to predict or control various properties of a precipitate, because the solid-phase formation proceeds through several stages, each of which can affect the composition, size, and shape of the final particles.

A scheme representing the sequence of events in a supersaturated solution is shown in Figure 1.1.3. The essential steps are the initial formation of the solid phase,

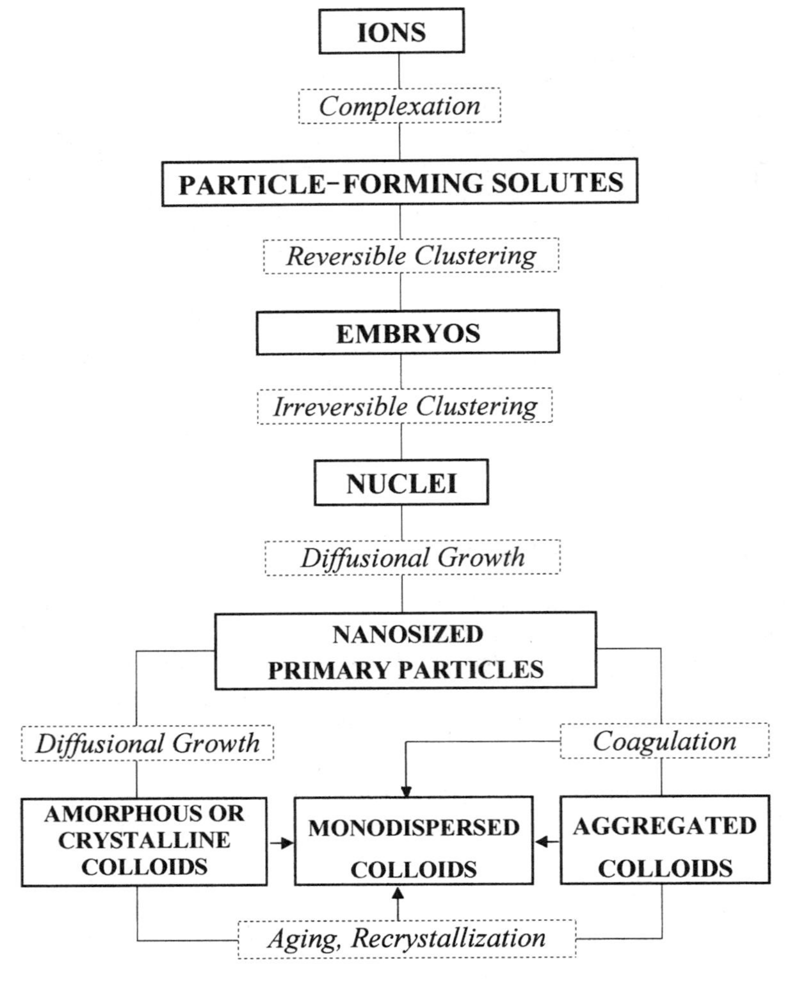

Fig. 1.1.3 Stages in the formation of colloidal particles by precipitation in homogeneous solutions.

i.e., of nuclei, followed by their growth. The latter may continue by diffusion of the constituent species to and accretion onto these nuclei until larger (colloid) particles are the end product. Alternately, the nuclei may undergo limited growth to nanosized precursors. Under certain conditions, the process can be arrested at this stage, or this finely dispersed matter can aggregate to form colloid dispersions. This controlled aggregation process is much more common than originally believed.

It should be recognized that a seemingly completed precipitation process can be subject to secondary reactions, designated in the overall scheme as "aging," which may involve dehydration, crystallization, recrystallization, ripening, etc. of the original particles.

Under certain conditions both the diffusional growth and the aggregative growth pathways may lead to "monodispersed" systems. As mentioned earlier, the mechanism by which the former process could yield uniform particles was proposed by Victor K. LaMer (6), whose original scheme, reproduced numerous times, is shown in Figure 1.1.4. This kind of process is most likely operational in the formation of nanosized particles or larger amorphous spheres.

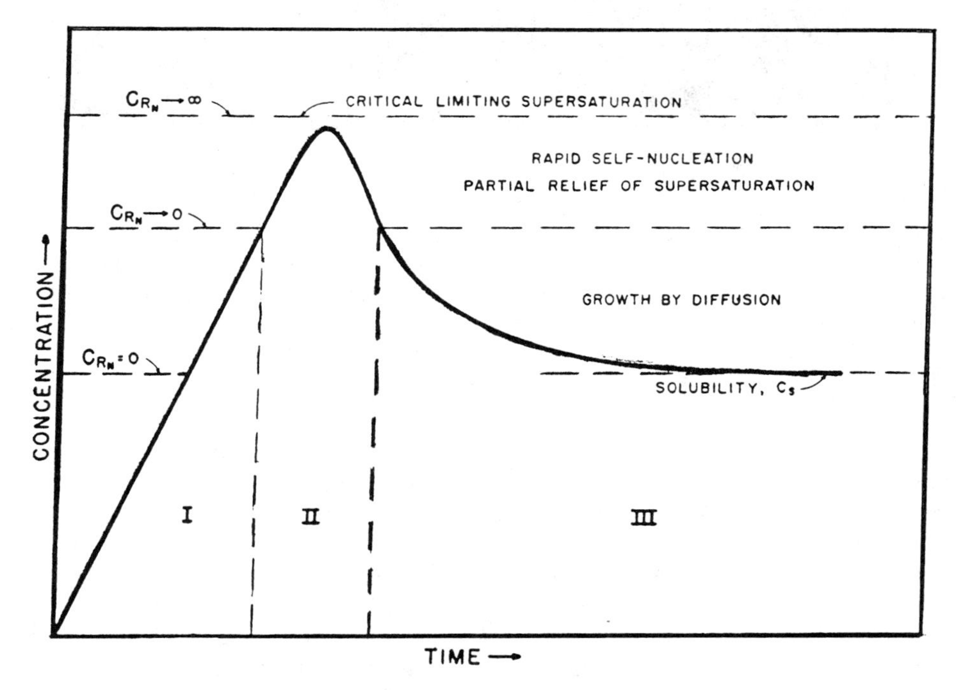

Fig. 1.1.4 The original reaction scheme for the formation of monodispersed colloids as proposed by LaMer.

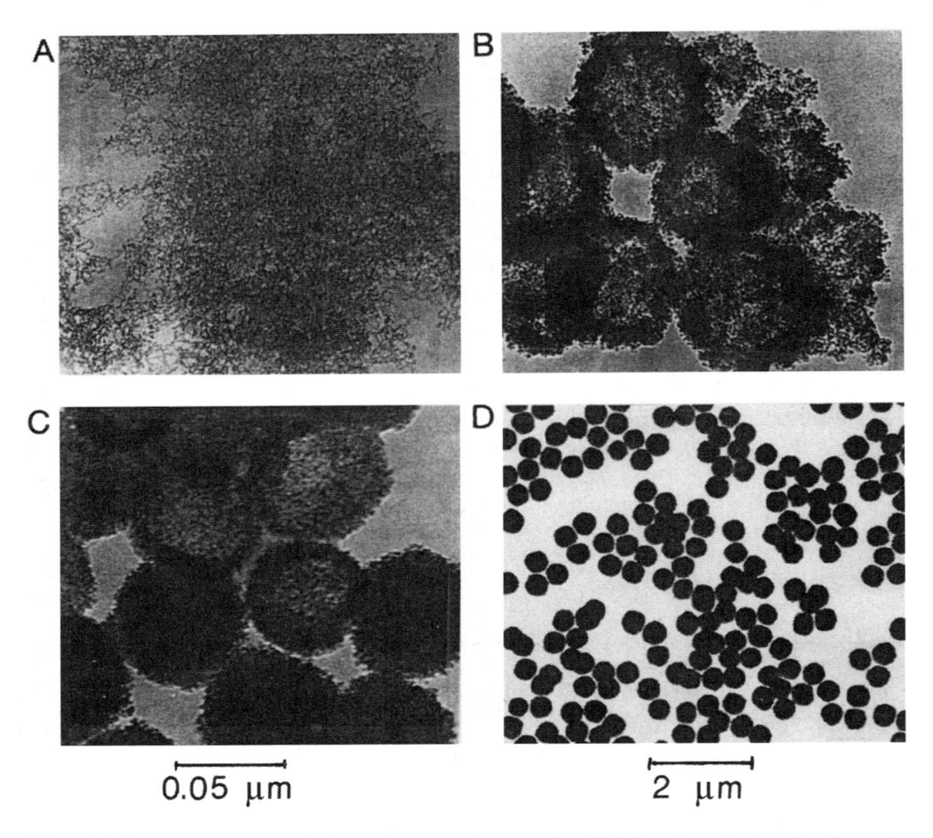

Fig. 1.1.5 (A–C) Transmission electron micrographs (TEM) taken during the formation of CeO_2 particles by forced hydrolysis of an acidic (4.0×10^{-2} mol dm^{-3} H_2SO_4) solution of $Ce(SO_4)_2$ (1.0×10^{-3} mol dm^{-3}), heated at 90°C for up to 6 h. (D) TEM of this dispersion at the completion of aging after 48 h. (From Ref. 33.)

In contrast, much experimental evidence has become available in the past few years clearly demonstrating that uniform colloids in a variety of shapes, including spheres, are generated by the aggregation of much smaller (nanosize) precursors (26,27,33). The time-resolved electron micrographs in Figure 1.1.5 illustrate such a sequence of events leading to uniform spheres of CeO_2 (33). Extensive studies with colloidal gold particles, which also show their composite nature (Figure 1.1.6), have further substantiated the aggregation mechanism (in contrast to a possible dissolution and reprecipitation process). The size, as determined by x-ray analysis, of the clearly discernible subunits in the lower part of this figure is identical with the size of the nanosized precursors of gold prepared separately (21,34).

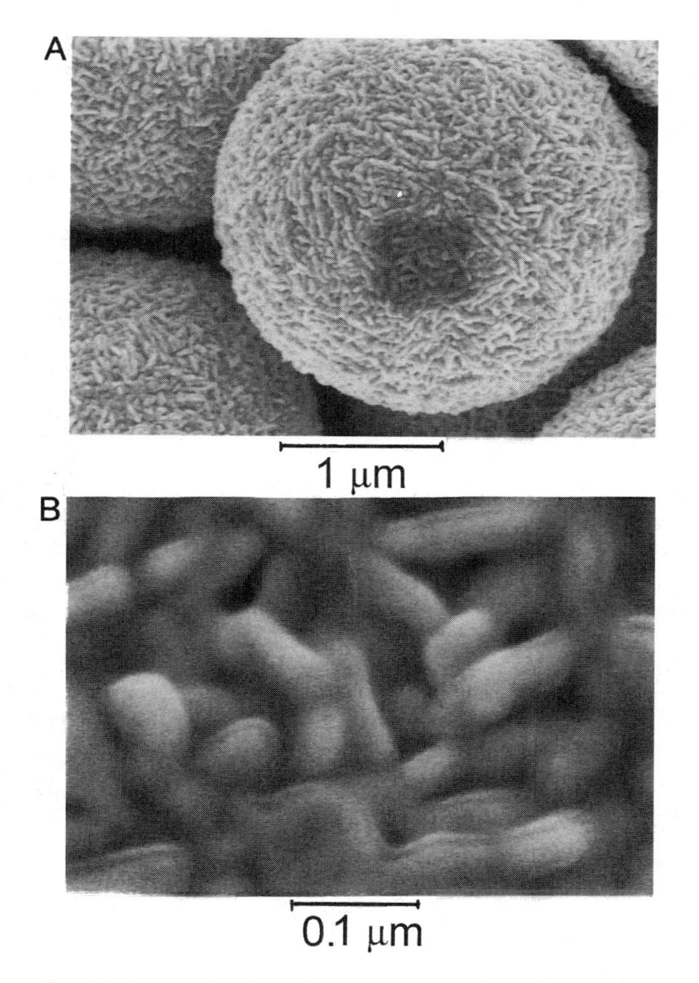

Fig. 1.1.6 (A) Field emission micrograph of a gold particle obtained by reducing a $HAuCl_4$ solution with iso-ascorbic acid in the presence of gum arabic. (B) Enlarged area of the same particle. (From Ref. 34.)

Recently, a model has been developed that explains the size selection mechanism, which is based on experimentally accessible parameters, yet shows that particle uniformity can be achieved by aggregation (8). This mechanism assumes that the nuclei formed in a supersaturated solution rapidly grow to primary nanoparticles (singlets), which then aggregate into larger colloidal particles. Certain conditions must be met for the final products to be uniform in size. Thus, electrostatic repulsion of nanosize precursors must be mitigated or eliminated in the course of the process, e.g., by changing the pH in a dispersion of a metal hydrous oxide to reach the isoelectric point, or by increasing the ionic strength to screen the charge. The growing aggregates must be at a sufficiently low number concentration and the diffusion constant of the singlets should be much larger than that of the aggregates. Finally, no continuous formation of nanoparticles should occur; that is, once formed, their concentration must decay by aggregation. The dominance of the irreversible singlet capture in the growth process can then result in the uniformity of the final dispersion. The model was successfully tested on the formation of uniform gold spheres (8).

1.1.3.2 Chemical Aspects

Any precipitation process in a homogeneous solution depends on the composition and the concentration of all solutes. Some of the latter are directly involved in the solid-phase formation, while others may indirectly affect the final products by their contribution to the ionic strength, control of the pH, etc. Obviously, in order to elucidate the mechanism of the precipitation process in a given homogeneous solution, it is necessary to establish the speciation of all complexes, especially those that affect the supersaturation condition preceding nucleation.

In the formation of metal (hydrous) oxides, hydrolyzed metal ions are the primary constituent species. In the "forced hydrolysis" method, the latter are generated by deprotonation of the coordinated water of the hydrated cation at elevated temperatures, according to:

$$xM(H_2O)_n{}^{z+} \rightleftharpoons M_x(OH)_y(H_2O)_{(nx-y)}^{(xz-y)+} + yH^+ \qquad (1)$$

Since the hydrolysis of each metal ion depends on the concentration, pH, and temperature of the solution, both the composition and the rate of formation of the resulting solutes are affected by all these parameters. In turn, the natures of the final metal (hydrous) oxide particles are determined by the speciation of the hydrolyzed intermediates.

In the presence of anions that are readily incorporated in such solutes, it is necessary to take into consideration additional complexes, which may be essential in defining the chemistry and morphology of the precipitates, especially in the formation of uniform particles.

The quantitative evaluation of all complexes in a given solution, along with changes in their concentrations in the course of the precipitation process, is an arduous and often elusive task, and only a limited number of such systems have been fully analyzed.

The formation of monodisperse amorphous, spherical particles of chromium hydroxide is offered here as an example. These dispersions are readily obtained by simply heating chrome alum solutions (35). It was established experimentally that the sulfate ion was essential in order to obtain uniform dispersions; thus, when chromium nitrate solutions were aged, monodispersed particles were produced only if a sulfate salt was introduced into the reaction.

The composition of all solutes under the same experimental conditions in the presence and in the absence of sulfate ions was determined by the radiotracer paper

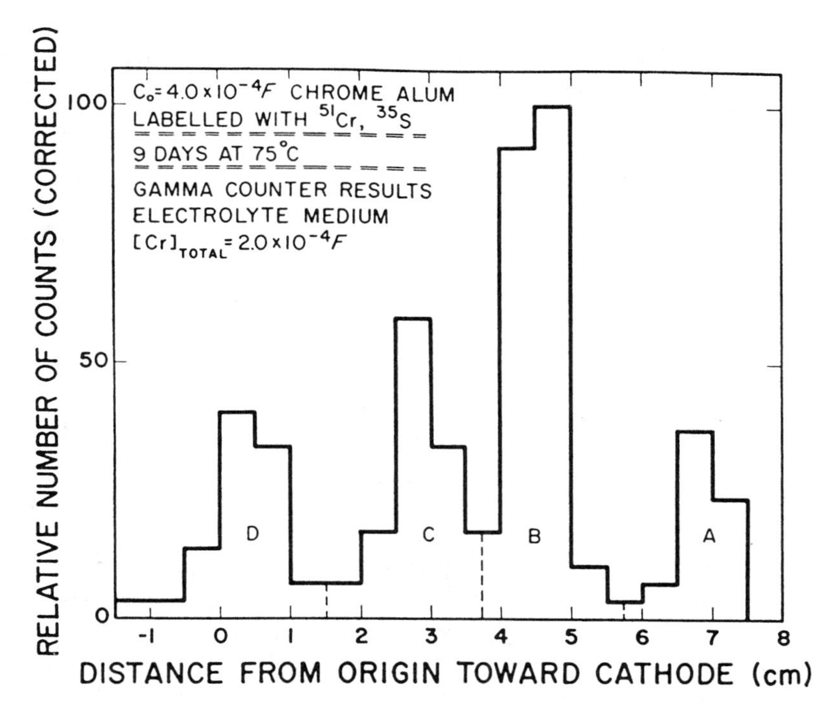

Fig. 1.1.7 Complex solutes present in solutions of chrome alum, which on aging yield monodispersed amphorous spherical chromium hydroxide particles. A labeled solution initially 4.0×10^{-4} mol dm^{-3} in chrome alum was heated at 75°C for 9 days. After cooling and filtering off the particles, the filtrate was subjected to paper electrophoresis. The corresponding solutes are: (A) Cr^{3+}, (B) $Cr_2(OH)_2SO_4^{2+}$, (C) $Cr(OH)_2^+$, (D) $Cr(OH)SO_4$ and $Cr(OH)_3$. (From Ref. 36.)

electrophoresis technique (36). The change in the concentration of these complexes was then followed during particle growth. Figure 1.1.7 shows that the aged solutions contained only solute complexes of 0, +1, +2, and +3 charge, which were identified by analyzing their content using radioactive isotopes ^{51}Cr and ^{35}S. It was further established that the essential species in the formation of the particles were $Cr(OH)_3^0$ (solute) and $[Cr_2(OH)_2SO_4]^{2+}$, which led to the following proposed mechanism of the particle growth:

$$m[Cr_2(OH)_2SO_4]^{2+} + mSO_4^{2-} \text{ and } nCr(OH)_3(aq) \tag{2}$$

$$\downarrow$$

$$[Cr(OH)(SO_4)]_m \text{ (polymer)} + nCr(OH)_3(aq) + 2mH_2O \tag{3}$$

$$\downarrow$$

$$[Cr(OH)_3]_{m+n} \text{ (solid)} + mSO_4^{2-} + 2mH^+ \tag{4}$$

Table 1.1.1 Molar Concentrations of Various Complexes Present at Different Temperatures in a Solution Initially 0.18 mol dm^{-3} in $Fe(NO_3)_3$ and 0.53 mol dm^{-3} in Na_2SO_4 (Total Concentrations) at pH 1.7

	Temperature		
Species	25°C	55°C	80°C
Fe^{3+}	1.78×10^{-2}	8.89×10^{-3}	3.30×10^{-3}
$FeOH^{2+}$	1.07×10^{-3}	1.60×10^{-3}	1.46×10^{-3}
$Fe(OH)_2^{+}$	8.90×10^{-5}	6.40×10^{-4}	5.23×10^{-3}
$Fe_2(OH)_2^{4+}$	4.80×10^{-4}	4.00×10^{-4}	1.10×10^{-4}
$FeSO_4^{+}$	1.58×10^{-1}	1.66×10^{-1}	1.69×10^{-1}
$FeHSO_4^{2+}$	2.57×10^{-3}	2.25×10^{-3}	8.00×10^{-4}
SO_4^{2-}	1.06×10^{-1}	7.60×10^{-2}	5.00×10^{-2}
HSO_4^{-}	2.91×10^{-2}	6.32×10^{-2}	1.22×10^{-1}
$NaSO_4^{-}$	2.35×10^{-1}	2.23×10^{-1}	1.87×10^{-1}

taking into account that the final particles consisted essentially of only $Cr(OH)_3$ and that the pH during the precipitation process decreased. The suggested mechanism was substantiated by additional data described in the original paper (36).

In another example, the analysis of solutions in which either monoclinic or hexagonal monodispersed ferric basic sulfates (mineral group of jarosite) were precipitated (37) showed the coexistence of a large number of solute complexes listed in Table 1.1.1 (38). A significant dependence of speciation was noted as a result of relatively small changes in the temperature, which was also reflected in the properties of the resulting particles. Thus, a hexagonal solid precipitated at 25°C, due to the dominance of $FeOH^{2+}$ and $FeSO_4^{+}$ solutes, according to

$$FeOH^{2+} + 2FeSO_4^{+} + 6H_2O \rightarrow Fe(SO_4)_2(OH)_5 \cdot 2H_2O(hex) + 4H^{+} \quad (5)$$

(with analogous stoichiomentry applicable to $Fe_2(OH)_2^{4+}$).

At 80°C, the increasing amount of $Fe(OH)_2^{+}$ explains the simultaneous appearance of monoclinic particles, according to:

$$3Fe(OH)_2^{+} + FeSO_4^{+} + 4H_2O \rightarrow Fe_4(SO_4)(OH)_{10}(monocl) + 4H^{+} \quad (6)$$

The kind of analyses just exemplified must be carried out for each precipitating system, because in every case the composition of the solution will differ depending on the hydrolyzability of the cation, complexation with anions, concentration of reactants, temperature, etc. Such projects are exceedingly time-consuming, which explains the paucity of well-documented published cases.

1.1.4 Survey of Published Works

In this chapter, the results of published works are summarized with special emphasis on metal (hydrous) oxides, which have been most widely studied. To assist the

Table 1.1.2 Literature Survey on Uniform Metal (Hydrous) Oxides

Metal	References
Al	32, 39–44
Sc	45
Ti	46–48
V	49, 50
Cr	35–36, 51–58
Fe	1–3, 30, 37, 38, 59–103
Co	104
Ni	105
Cu	27, 106–109
Zn	28, 110, 111
Ga	112, 113
Ge	114
Zr	115–124
In	26, 125–129
Sn	130, 131
La, Ce, and RE[a]	33, 132, 133
Hf	134, 135
W	4, 5, 136–138
Bi	139
Th	140

[a] Rare earths.

reader, Table 1.1.2 lists the pertinent references on each metal (hydrous) oxide as they are cited throughout this chapter. It should be again emphasized that only published studies reporting uniform particles prepared by the forced hydrolysis process are reviewed, while disregarding metal alkoxides as starting reactants.

It would be impractical to consider each metal separately; thus, the review is divided into three sections as follows:

1. Aluminum, chromium, and zirconium (hydrous) oxides
2. Iron(III) (hydrous) oxides
3. Other metal (hydrous) oxides

1.1.4.1 Aluminum, Chromium, and Zirconium (Hydrous) Oxides

These three elements are considered together, because of the amount of information available, the similarity of the natures of the products, and the importance in various

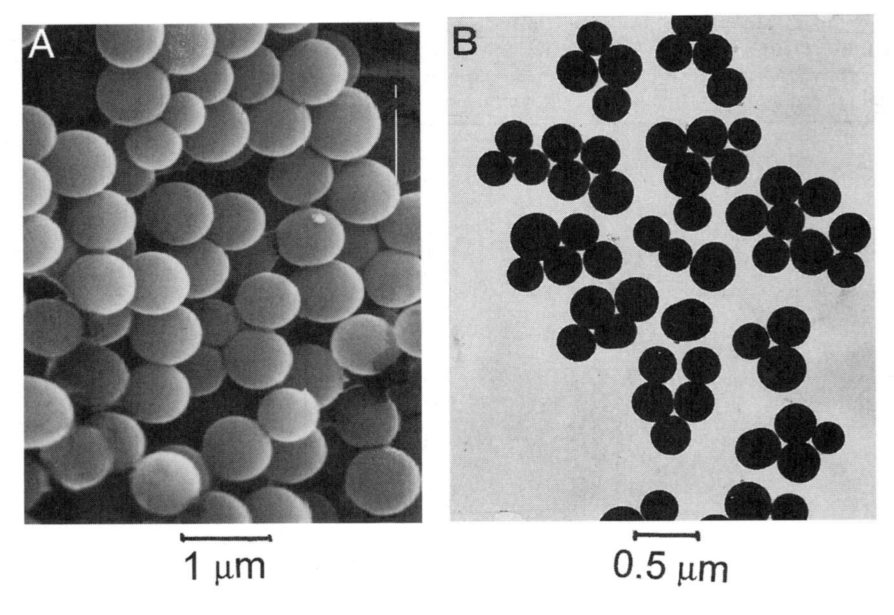

Fig. 1.1.8 Scanning electron micrographs (SEM) of (A) aluminum (hydrous) oxide particles, obtained by aging at 100°C for 72 h a solution containing 1×10^{-3} mol dm^{-3} $Al(ClO_4)_3$ and 1×10^{-3} mol dm^{-3} $AlNH_4(SO_4)_2$. (B) TEM of chromium (hydrous) oxide particles, obtained by aging at 75°C for 24 h a 4.0×10^{-4} mol dm^{-3} solution of $CrK(SO_4)_2$.

applications. As a rule, forced hydrolysis using the respective metal salt solutions yielded spherical amorphous particles. Furthermore, to obtain uniform dispersions, anions (especially the sulfate ion) play an important role in all cases.

Figure 1.1.8 illustrates the aluminum (39) and chromium (hydrous) oxide (55) particles prepared under conditions given in the legend. Some details of the mechanisms of formation for the chromium system are given in Section 1.1.3.2.

In principle, the three metals considered here readily hydrolyze, yielding polymeric solute complexes, which are precursors to the solid-phase formation. In all cases, a certain amount of anions (e.g., sulfate ion) is usually incorporated in the original particles, but can be removed by the exchange with hydroxide ions when treated with a base. Figure 1.1.9 shows size distributions, obtained by light scattering in situ, of aluminum hydrous oxide particles as prepared and after the sulfate ions were removed (12). While the uniformity and morphology were retained, the particle size decreased by the purification process. The presence of the sulfate ion also had an effect on the electrokinetic behavior, as demonstrated in Figure 1.1.10, which shows the mobility data of the same aluminum hydrous oxide particles containing the sulfate ion and after this anion was leached out. The isoelectric points (i.e., p)

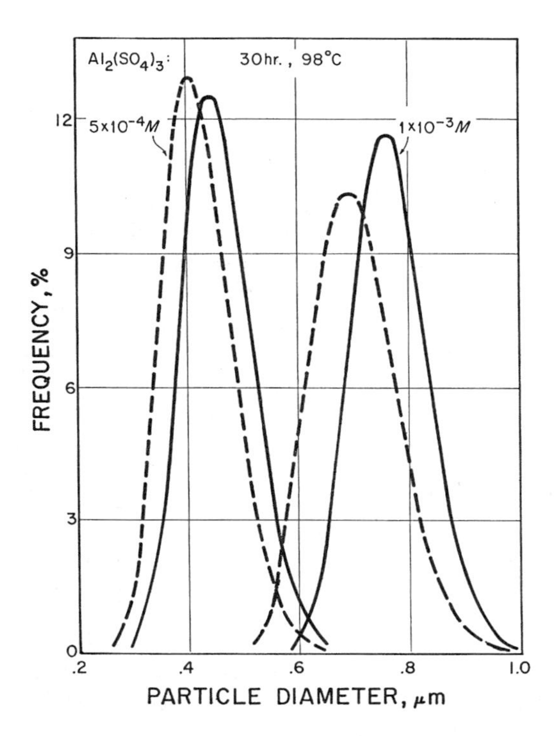

Fig. 1.1.9 Particle size distribution curves obtained by light scattering for two different aluminum hydrous oxide sols prepared by aging at 98°C for 30 h solutions containing 5×10^{-4} and 1×10^{-3} mol dm^{-3} $Al_2(SO_4)_3$, respectively (solid lines). Dashed lines show the corresponding size distributions of the same sols after all sulfate ions in the particles had been exchanged for hydroxyl groups.

at pH 7.2 and 9.3 correspond to the lowest and the highest values reported in the literature, with the latter being characteristic of pure aluminum hydroxide (141).

It is noteworthy that reversing the charge on such particles from positive to negative by increasing the pH with a base does not affect the size distribution of the particles, as documented in the example of chromium hydroxide (51).

The effect of anions needs special consideration. For example, if aqueous dispersions of purified aluminum hydrous oxide particles (i.e., freed of sulfate ions) are reheated, crystallization into boehmite takes place with simultaneous change in the morphology (Figure 1.1.11A) (39). On the other hand, heating aluminum chloride or perchlorate solutions at higher temperatures results directly in uniform (pseudo)-boehmite particles of unique shapes, as illustrated in Figure 1.1.11B (40).

Forced hydrolysis of zirconium sulfate solutions yielded reasonably uniform spheroidal particles in the presence of formamide (118). Smaller zirconia (~80 nm)

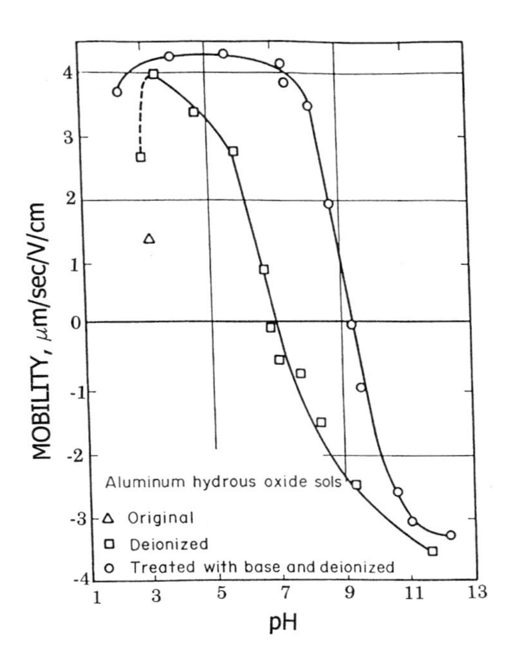

Fig. 1.1.10 The electrophoretic mobilities of aluminum hydrous oxide sols prepared by aging at 98°C for 24 h a 1×10^{-3} mol dm^{-3} solution of $Al_2(SO_4)_3$ and treated as follows: △, original sol at room temperature; □, sol freed of the original electrolyte by centrifugation and redispersed in dilute NaOH solutions; ○, pH of the sol adjusted to 9.7 by NaOH, followed by deionization, and redispersion at various pH values.

particles were prepared by aging $ZrO(NO_3)_2$ solutions. It was shown that the latter were formed by the aggregation process of crystallites ~3 nm in size (117).

1.1.4.2 Iron(III) (Hydrous) Oxides

Despite the fact that the hydrolysis of the ferric ion is exceedingly sensitive to various experimental parameters (temperature, pH, etc.), hematite (α-Fe_2O_3) and akageneite (β-FeOOH) were apparently the first reasonably uniform colloidal metal (hydrous) oxides dispersions reported in the literature, as already indicated in the introduction. Since then, this family of compounds has been the most extensively investigated, with specific emphases on particle uniformity, composition, and morphology.

The nature of the products is strongly dependent on the anion of the ferric salt used in the forced hydrolysis processes. For this reason, this review first considers results obtained with ferric chloride, followed by the studies involving ferric sulfate solutions.

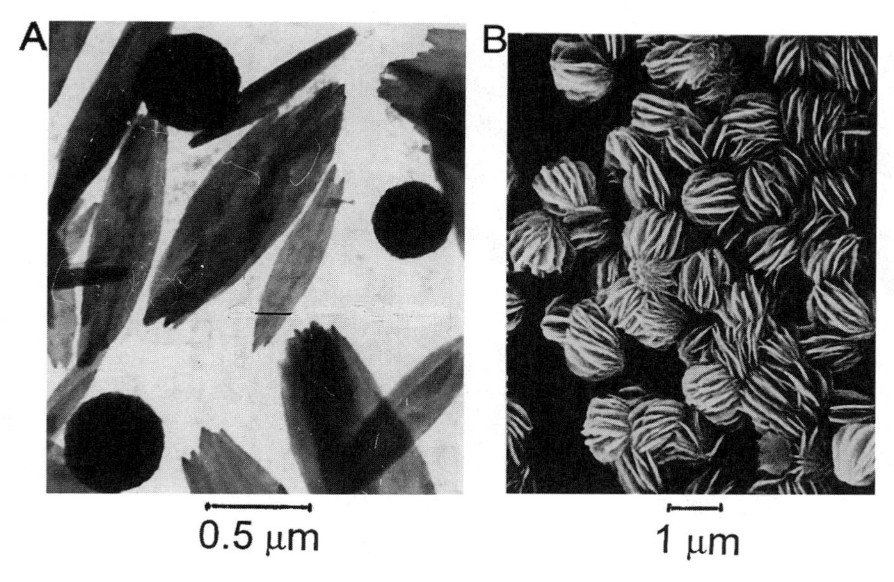

Fig. 1.1.11 (A) TEM of a sol prepared by aging at 98°C for 24 h a 1×10^{-3} mol dm^{-3} solution of $Al_2(SO_4)_3$, followed by adjustment of the pH to 9.7 with NaOH, deionization, repeptization at pH ~8, and additional aging at 98°C for 24 h. (From Ref. 39.) (B) SEM of pseudoboehmite particles prepared by aging at 125°C for 12 h a 5×10^{-3} mol dm^{-3} solution of $Al(ClO_4)_3$ of pH 3.9 at 125°C for 12 h. (From Ref. 40.)

Hydrolysis of $FeCl_3$ Solutions. Aging of ferric chloride solutions yields, as a rule, either colloidal akageneite (β-FeOOH) or hematite (α-Fe_2O_3). However, the two forms are closely related in the formation of the precipitates (95,142).

One special, somewhat surprising, aspect of the hydrolysis of ferric chloride solutions is the variety of morphologies that have been obtained by altering the experimental parameters, often over a small range of conditions, as originally reported by the senior author (65). Thus, uniform dispersions of hematite particles of spherical (67), ellipsoidal (59,71,77), rod-like (65), cubic (69,70), platelet-type (82), and other shapes (65,88) have been produced. Figure 1.1.12(A–C) illustrates several such dispersions prepared under the conditions given in the legend. An example of ellipsoidal hematite particles is shown in Figure 1.1.19. In contrast, β-FeOOH normally precipitates in rod-like form, as illustrated in Figure 1.1.12D.

At this time it is impossible to predict a priori a specific morphology from a certain set of experimental parameters. Instead, Figure 1.1.13 tabulates the conditions that have been used to produce hematite particles of different shapes, starting with acidified $FeCl_3$ solutions (65).

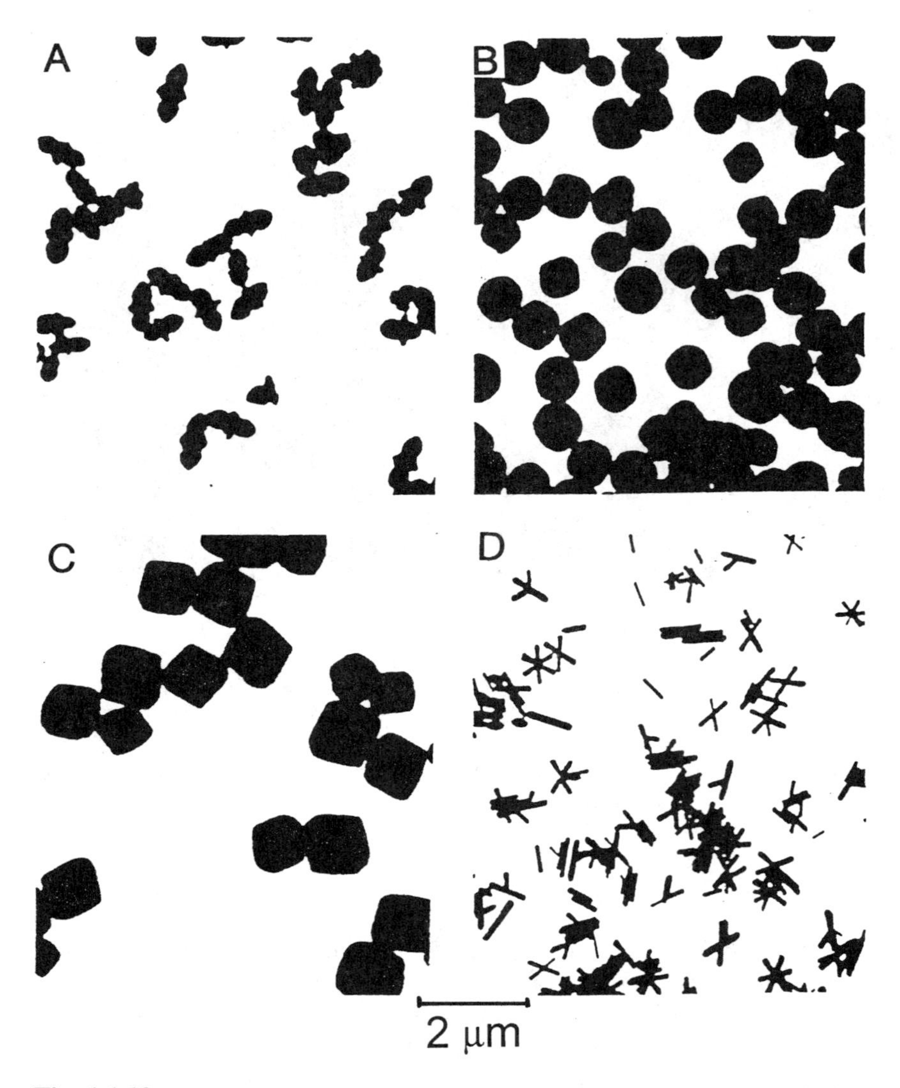

Fig. 1.1.12 TEM of particles obtained in solutions of $FeCl_3$ + HCl under the following conditions:

	Fe^{3+} (mol dm^{-3})	Cl^- (mol dm^{-3})	Initial pH	Final pH	Temp. of aging (°C)	Time of aging
(a)	0.018	0.104	1.3	1.1	100	24 h
(b)	0.315	0.995	2.0	1.0	100	9 days
(c)	0.09	0.28	1.65	0.88	100	24 h
(d)	0.09	0.28	1.65	0.70	150	6 h

(A), (B), (C) Hematite, α-Fe_2O_3; (D) Akageneite, β-FeOOH.

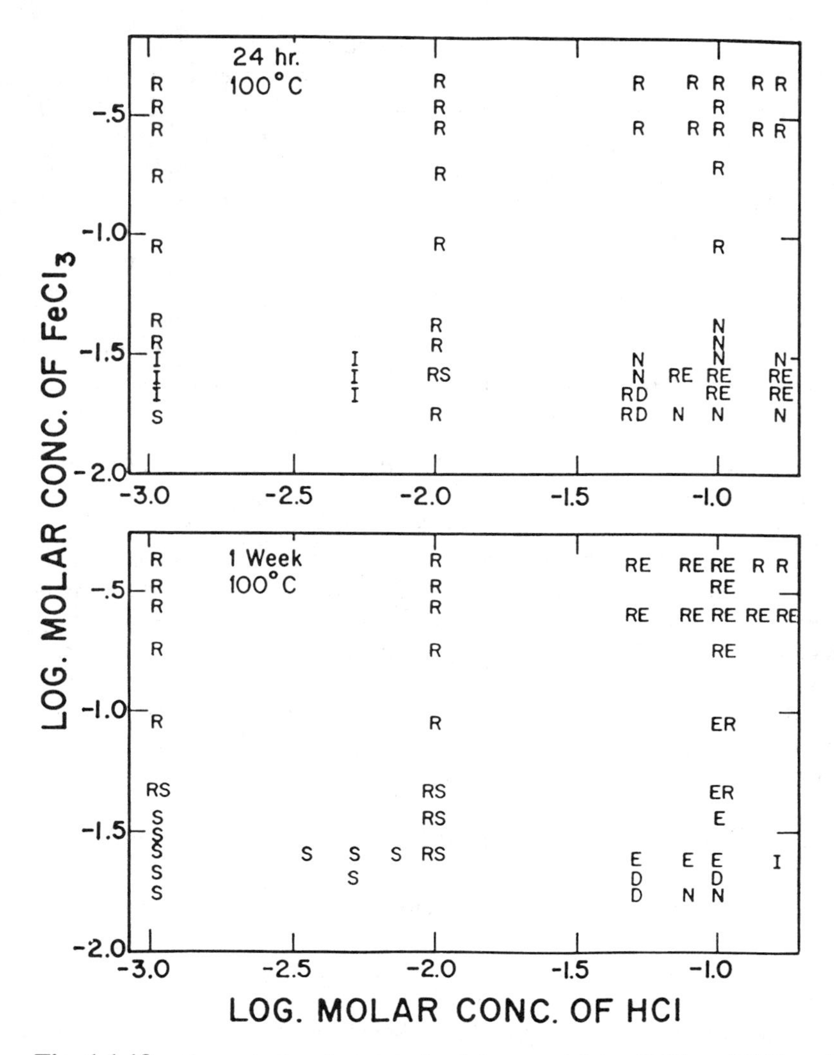

Fig. 1.1.13 Concentration domains of solutions containing $FeCl_3$ and HCl aged at 100°C for 24 h (upper) and for 1 week (lower). N, no particle formation. Particle shapes: D, double ellipsoids; E, ellipsoidal; I, irregular of varying sizes; R, rod-like; S, spherical. Pairing of symbols indicates a mixture of corresponding particles in the suspension. Particle composition: R, β-FeOOH; all other particles, α-Fe_2O_3. (From Ref. 65.)

Once particles have been produced, it is possible to elucidate reasons for the observed variations in their habits. Specifically, different geometries of hematite particles were explained by the complex mechanisms of their formation, which consist of first precipitating akageneite precursors, subsequently undergoing controlled aggregation and structural transformation into hematite (95). Indeed, it was indicated that the morphological properties of the precursors were responsible for different shapes of the final products.

There are some other conditions that may affect the appearance of hematite. For example, ellipsoidal particles of various anisometries were obtained by the addition of small amounts of phosphate ions into the aging $FeCl_3$ solutions (71). A possible explanation for the effect of this anion was suggested in Refs. 100 and 143.

Substituting $Fe(NO_3)_3$ or $Fe(ClO_4)_3$ for $FeCl_3$, had no significant effects on the natures of the resulting hematite or akageneite particles (65).

Hydrolysis of $Fe_2(SO_4)_3$ Solutions. The forced hydrolysis of ferric sulfate solutions produced some of the most uniform crystalline colloid dispersions, all of the alunite family (37). The essential constituent in all these solids is the sulfate ion. Depending on conditions, either hexagonal or monoclinic particles are obtained as discussed in Section 1.1.3.2 (37). Examples of such ferric basic sulfate dispersions are illustrated by electron micrographs in Figures 1.1.14 and 1.1.15. The uniformity and the crystal habit of the hexagonal particles is clearly seen in Figure 1.1.14, while the scanning electron microscopy (SEM) in Figure 1.1.15A shows indeed a huge number of identical colloidal crystals of the same composition. The reason for the coexistence of monoclinic and hexagonal particles (Figure 1.1.15B) is also discussed in Section 1.1.3.2.

1.1.4.3 Other Metal (Hydrous) Oxides

Some comments are offered here regarding titanium, copper, zinc, and indium (hydrous) oxides.

Hydrolysis of aqueous titanium solutions at elevated temperatures yield normally amorphous spherical particles (47), which on calcination at moderate temperatures crystallize to anatase while retaining the spherical shape. To convert these particles to rutile requires much higher temperatures, a process usually accompanied by sintering. The exception seems to be a study carried out with highly acidic $TiCl_4$ solutions in the presence of Na_2SO_4, which on very slow aging (37 days!) produced spherical rutile particles (46).

Cu(I) oxide particles of different shapes were obtained by reductive hydrolysis of Cu(II) tartrate solutions in the presence of glucose (107,109) using a process originally described by Andreasen (106). More recently, an extensive study dealt with the precipitation of different copper (hydrous) oxide and oxide particles using controlled double-jet precipitation (CDJP). Thus, depending on experimental conditions, hexagonal platelets or rods of $Cu_2(OH)_3NO_3$ and ellipsoids or needles of CuO

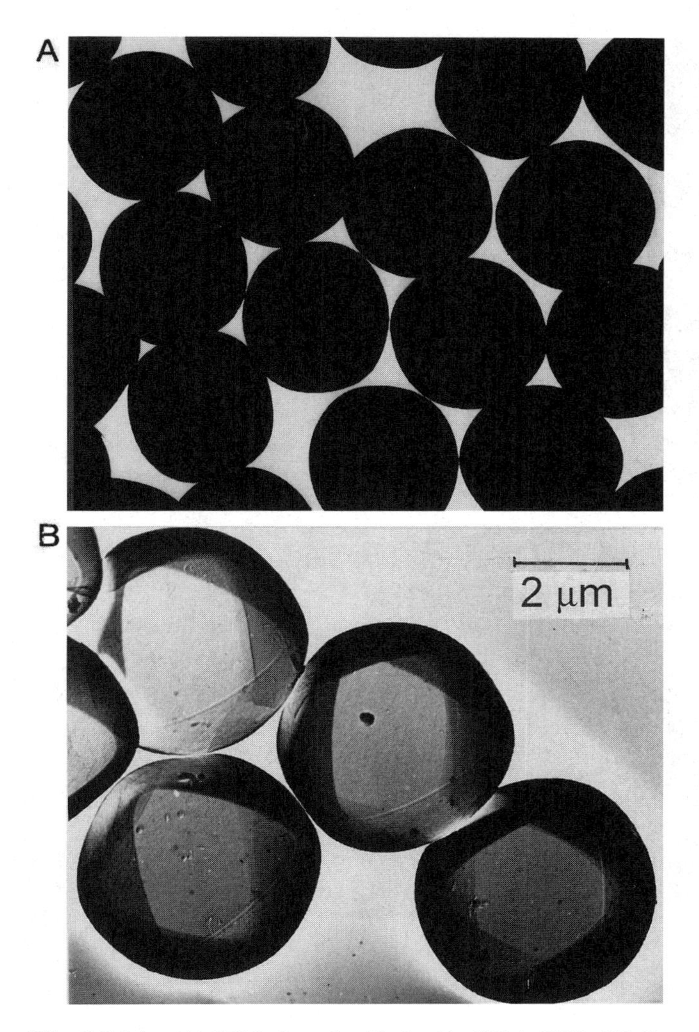

Fig. 1.1.14 (A) TEM of carphosiderite, $Fe_3(SO_4)_2(OH)_5 \cdot 2H_2O$, particles obtained by aging at 80°C for 0.5 h a solution containing 0.18 mol dm^{-3} $Fe(NO_3)_3$ and 0.53 mol dm^{-3} Na_2SO_4. (B) TEM of the carbon replica of the same particles. (From Ref. 37.)

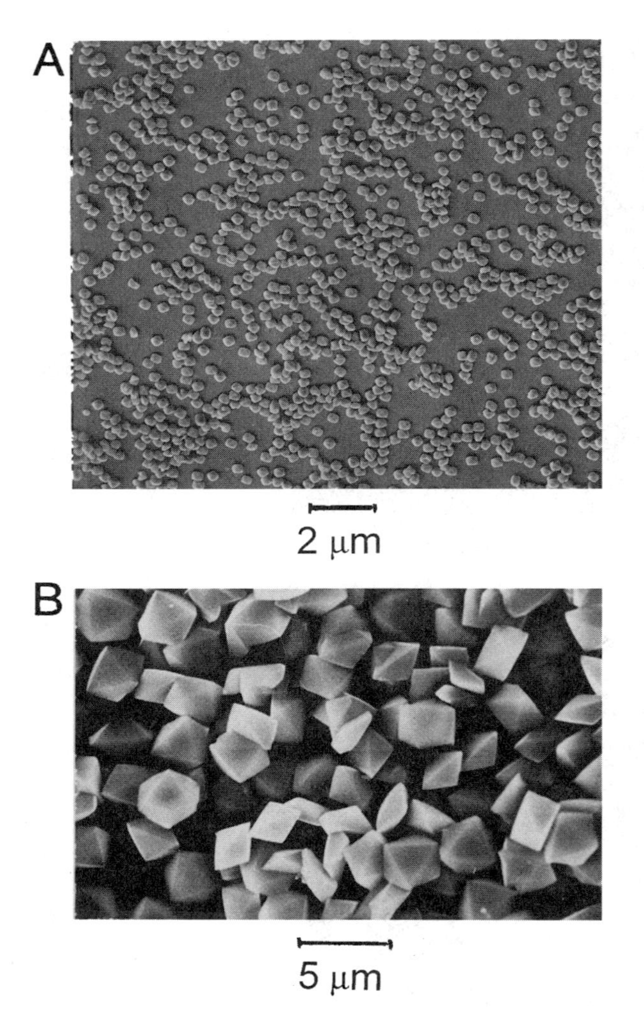

Fig. 1.1.15 (A) SEM of smaller carphosiderite particles. (B) SEM of a mixture of hexagonal and monoclinic alunite particles, obtained by aging at 100°C for 3 h a solution containing 8.8×10^{-2} mol dm^{-3} $Fe_2(SO_4)_3$.

Fig. 1.1.16 (A) SEM of intertwined zincite (ZnO) particles, obtained by aging at 90°C for 3 h solutions containing 5.0×10^{-3} mol dm^{-3} $Zn(NO_3)_2$, and 1.9×10^{-2} mol dm^{-3} NH_4OH at pH 8.8; (B) ZnO particles obtained by aging at 150°C for 2 h solutions containing 4.0×10^{-2} mol dm^{-3} $Zn(NO_3)_2$ and 1.2×10^{-3} mol dm^{-3} KOH at pH 13.3. (From Ref. 110.)

were generated. The same work also provided direct evidence for the aggregation mechanism of such uniform particles (27).

Since Zn^{2+} ion does not hydrolyze as readily as cations of higher charge, bases need to be added in order to initiate precipitation by forced hydrolysis. Particles of different morphologies, including spheroids and ellipsoids, formed in the presence of different weak bases on heating solutions of zinc salts (Figure 1.1.16) (110). Dispersions of narrow size distributions resulted under specific reactant concentrations as shown in Figure 1.1.17 (110). X-ray diffraction analysis showed all particles of different shapes to be composed of zincite, ZnO.

Spherical amorphous particles of zinc oxide were prepared by hydrolyzing zinc acetate solutions in the presence of diethylene glycol (111), or zinc nitrate solutions containing triethanolamine (28).

Hydrolysis of $In(NO_3)_3$ solutions, containing different amounts of Na_2SO_4, yields indium hydroxide particles of various shapes (125,126). Some of these particles are displayed in Figure 1.1.18, all of which show distinct x-ray diffraction patterns, although the spectra could not be related to any reported indium compound (125,126). On calcination at 1000°C, these solids are converted to In_2O_3.

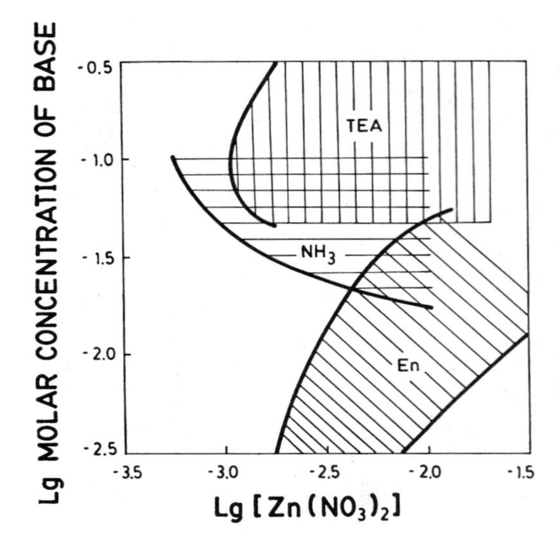

Fig. 1.1.17 Composition domains of well-defined particles formed by aging at 90°C for 1 h solutions containing zinc nitrate and different bases. TEA; triethanolamine; En; ethylenediamine. (From Ref. 110.)

No specific descriptions of (hydrous) oxides of other metals are offered here. However, the references given in Table 1.1.2 can direct the reader to the published studies on relevant systems.

1.1.5 Coated Particles

For many reasons it may be advantageous to cover one kind of particles with a layer of a different chemical composition. In doing so it is possible to alter optical, magnetic, electric, catalytic, adsorptive, and other properties of the original dispersion. Furthermore, one may produce particles of a given morphology, if the latter cannot be achieved directly, by selecting the core of the desired shape and encasing it with a shell of the material of interest.

In principle, the coating may be produced in two ways, i.e., by (1) interacting the cores with preformed smaller particles of a different composition (essentially by heterocoagulation), and (2) by forming the shell directly by chemical reaction on preformed particles dispersed in salt solutions yielding the coating. In both instances it is essential that conditions be optimized in order to avoid having mixed dispersions of coated and constituent particles.

In this chapter, only those examples are offered in which shells of metal (hydrous) oxides on various cores were formed by ''forced hydrolysis.'' Systems

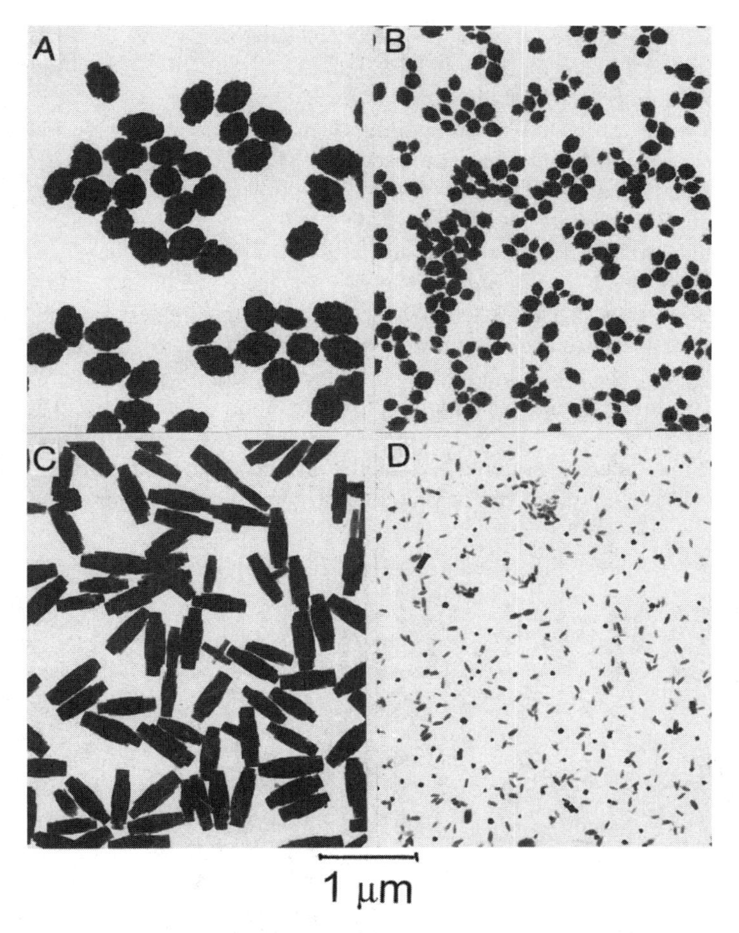

Fig. 1.1.18 TEM of indium hydroxide particles obtained by aging at 85°C for 2 h solutions containing 7×10^{-3} mol dm^{-3} $In(NO_3)_3$ and different concentrations of Na_2SO_4: (A) 5×10^{-4} mol dm^{-3}, pH 2.6; (B) 5×10^{-4} mol dm^{-3}, pH 3.0; (C) 1×10^{-4} mol dm^{-3}, pH 2.7; and (D) 1×10^{-4} mol dm^{-3}, pH 3.0. (From Ref. 125.)

in which urea was used to produce the coatings are described in Chapter 7.1. It should be noted that shells can be rather rapidly generated by the hydrolysis of different alkoxides in the presence of cores, but such cases are not discussed here, nor are those in which coatings are produced by adsorption of small particles.

In a number of studies, ellipsoidal hematite (α-Fe_2O_3) (71) was the core material of preference, because the coating materials appeared, as a rule, as spheres when precipitated alone. This morphological difference made it possible to clearly distinguish between dispersions consisting of only coated particles and those in which excess shell material precipitated as separate entities.

Figure 1.1.19 illustrates such hematite particles coated with chromium hydroxide in which the latter was produced by the hydrolysis of a chrome alum solution in the presence of α-Fe_2O_3 cores on aging the dispersion at 85°C. For greater efficiency the chromium salt solution was either preheated or preheated with KOH (but avoiding precipitation of chromium hydroxide) (144).

Figure 1.1.20 shows the differential thermal analysis (DTA) data for the cores, of chromium hydrous oxides particles prepared in the absence of hematite, and of coated particles. It is obvious that the latter behave as the coating material, when alone. This example clearly indicates the possibility of having the surface site characteristics of chromium hydrous oxide induced onto ellipsoidal iron oxide particles. The latter morphology cannot be achieved by direct precipitation of the same chromium compound.

Analogously, hematite particles could be coated with zirconium hydrous oxide by aging zirconium sulfate solutions in the presence of cores, which required the

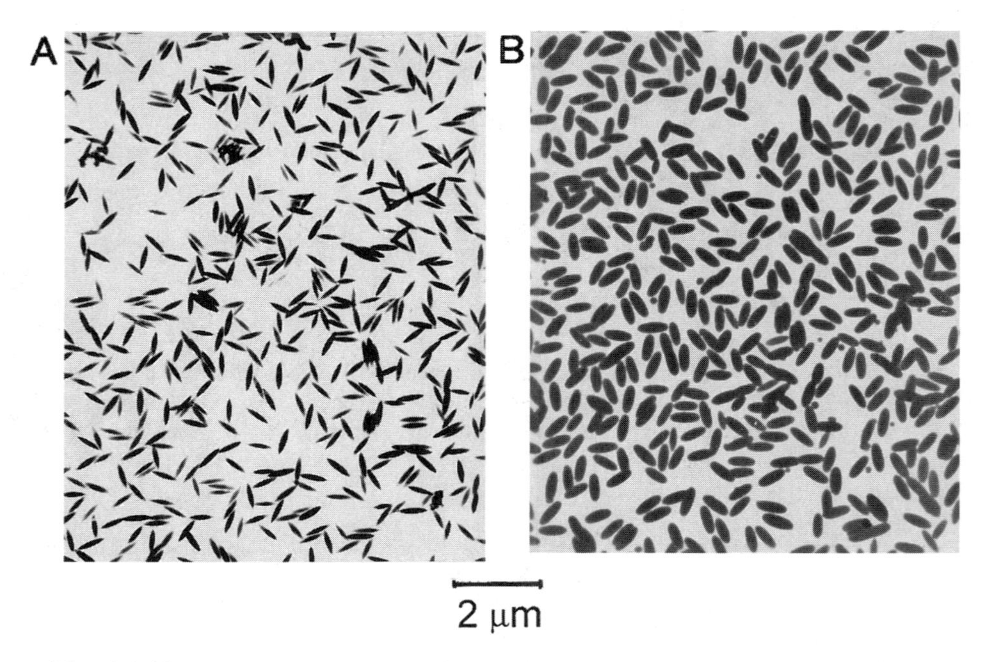

Fig. 1.1.19 (A) TEM of ellipsoidal hematite (α-Fe_2O_3) particles and (B) of the same particles coated by chromium hydrous oxide obtained by aging at 85°C for 6 h 40 cm^3 of a dispersion containing 30 mg dm^{-3} of the cores in the presence of 5.0×10^{-4} mol dm^{-3} chrome alum. (From Ref. 144.)

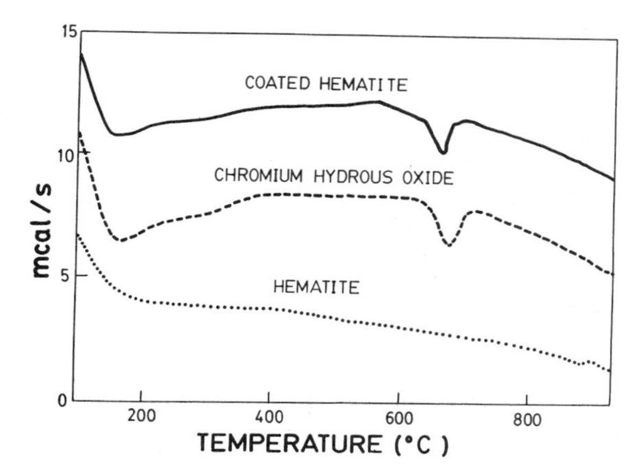

Fig. 1.1.20 Differential thermal analysis data for chromium hydrous oxide, hematite, and hematite particles coated with chromium hydrous oxide. Data refer to dispersions as illustrated in Fig. 1.1.19.

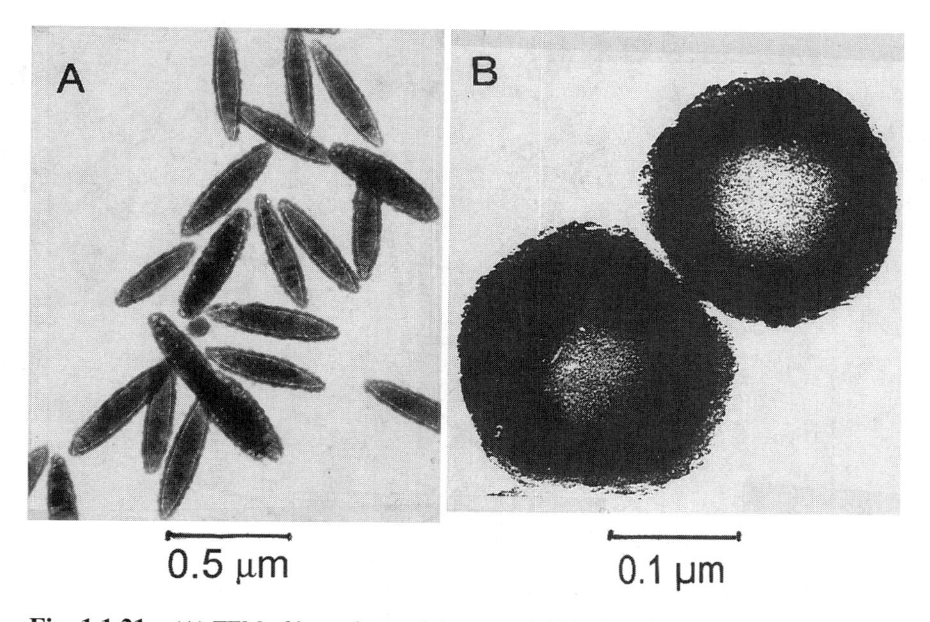

Fig. 1.1.21 (A) TEM of hematite particles coated with zirconium hydrous oxide by aging at 70°C for 2 h an aqueous dispersion containing 600 mg dm^{-3} Fe_2O_3 cores, 5×10^{-3} mol dm^{-3} zirconium sulfate, 5 vol% formamide, and 0.5 wt% polyvinylpyrrolidone. (From Ref. 118.) (B) TEM of hollow zirconium oxide particles obtained by calcining at 800°C for 3 h a sample of polystyrene latex particles coated with zirconium hydrous oxide using the procedure described in A. (From Ref. 145.)

addition of formamide to control the rate of the hydrolysis process. Furthermore, uniform shells were obtained if polyvinylpyrrolidone (PVP) was present in the reaction mixture (118). The uniformity of the shell is clearly evident in Figure 1.1.21A.

Using a similar procedure, it was also possible to coat cationic and anionic latexes with zirconium hydrous oxide (145). On careful heating such powders in air at temperatures >500°C the core polymer was vaporized and the shell was calcined to yield hollow spheres of ZrO_2, as illustrated in Figure 1.1.21B (145).

Silica has been used both as a core and shell material. For example, monodispersed silica spheres were coated with titania by decomposing titanyl sulfate, $TiOSO_4$, in acidic solutions at 90°C (146). The particles so produced showed good hiding power, to be useful as paper whiteners (147). Due to the uniformity of the cores and shells, the optical properties of such dispersions were predictable and reproducible, as shown in Figure 1.1.22, which compares the scattering coefficient,

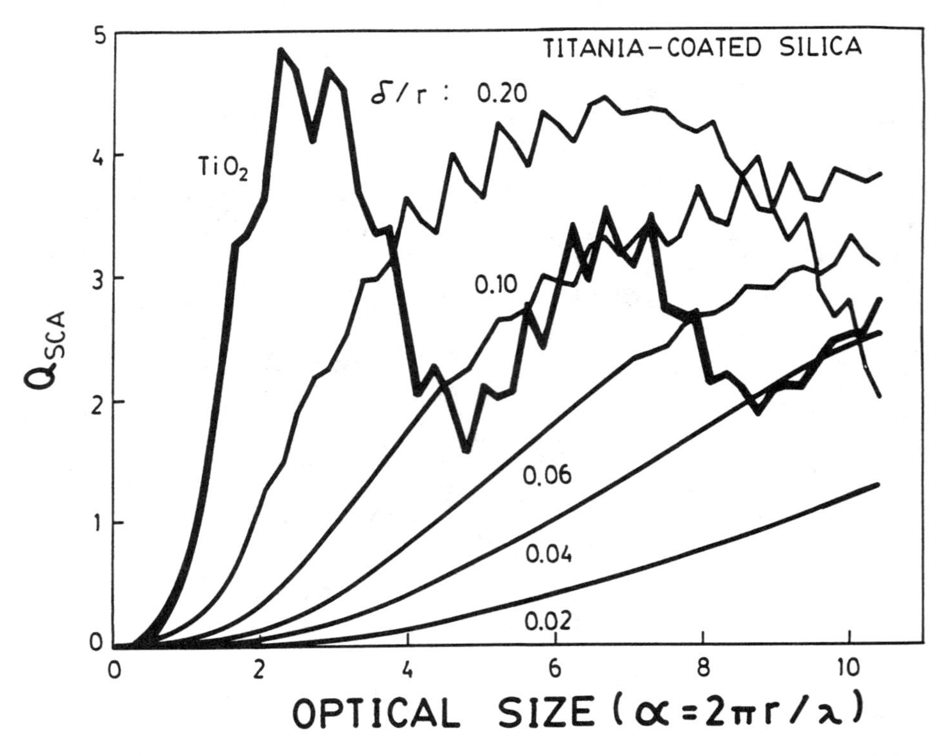

Fig. 1.1.22 Scattering coefficient (Q_{SCA}) versus the optical size ($2\pi r/\lambda$) of silica particles coated with layers of titania of different thickness, incorporated in a cellulose matrix. r, Radius of coated particles; δ, thickness of the titania shell; λ, wavelength in the medium.

Q_{SCA}, of pure titania particles and silica particles coated with shells of titania of different thicknesses, δ. It is quite apparent that the Q_{SCA} of coated particles is much less sensitive to the particle size, *r*, than the pure titania.

In contrast, needle-type boehmite particles were covered with silica by decomposition of sodium silicate to produce isotropic birefringent dispersions (148).

Finally, rod-like particles of indium hydrous oxide, dispersed in solutions of $HfOCl_2$ and Na_2SO_4, were coated with hafnium hydrous oxide by simply aging the entire system for 24 h at room temperature (149).

REFERENCES TO SECTION 1.1

1. FA Steele. J Colloid Sci 9:166, 1954.
2. H Zocher, W Heller. Z Anorg Allg Chem 186:75, 1930.
3. JHL Watson, W Heller, T Schuster. Proc Eur Reg Conf Electron Microscopy Delft, 1960, vol. 1, p 229.
4. K Furusawa, S Hachisu. J Colloid Interface Sci 15:115, 1968.
5. K Furusawa, S Hachisu. Sci Light 12:157, 1963.
6. VK LaMer. Ind Eng Chem 44:1270, 1952.
7. E Matijević. Chem Mater 5:412, 1993.
8. V Privman, DV Goia, J Park, E Matijević. J Colloid Interface Sci, 213:36, 1999.
9. E Matijević. Langmuir 10:8, 1994.
10. E Matijević. Annu Rev Mater Sci 15:483, 1985.
11. E Matijević. In: DJ Wedlock, ed. Controlled Particle, Droplet and Bubble Formation. London: Butterworth-Heinemann, 1994, pp 39–59.
12. E Matijević. Progr Colloid Polym Sci 61:24, 1976.
13. JTG Overbeek. Adv Colloid Interface Sci 15:251, 1982.
14. M Harita, B Delmon. J Chim Phys 83:859, 1986.
15. T Sugimoto. Adv Colloid Interface Sci 28:65, 1987.
16. E Matijević. Control of Powder Morphology. In: LH Hench, JK West, eds. Chemical Processing of Advanced Materials. New York: Wiley, 1992, pp 513–527.
17. E Matijević. In: E. Pelizzetti, ed. Fine Particles Science and Technology. NATO ASI, Series 3. Dordrecht: Kluwer, 1996, vol 12, pp 1–16.
18. E Matijević. Progr Colloid Polym Sci 10:38, 1996.
19. E Matijević. Curr Opin Colloid Interface Sci 1:176, 1996.
20. E Matijević. J Eur Ceram Soc 18:1357, 1998.
21. DV Goia, E Matijević. New J Chem 22:1131, 1998.
22. M Faraday. Phil Trans R Soc 147:145, 1857.
23. CR Berry. In: TJ James, ed. The Theory of the Photographic Process, 4th ed. New York: Macmillan, 1977, p 88.
24. J Stávek, T Hamslik, V Zapletal. Mater Lett 9:90, 1990.
25. J Stávek, M Šipek, J Nývlt. Proc Int Symp on Preparation of Functional Materials and Industrial Crystallization, Osaka, Japan, 1989, p 17.
26. L Wang, LA Pérez-Maqueda, E Matijević. Colloid Polym Sci 276:847, 1998.
27. SH Lee, Y-S Her, E Matijević. J Colloid Interface Sci 186:193, 1997.
28. Q Zhong, E Matijević. J Mater Chem 6:443, 1996.
29. Y-S Her, E Matijević, MC Chon. J Mater Res 10:3106, 1995.

30. N Kallay, I Fischer, E Matijević. Colloids Surf 13:145, 1985.
31. Y-S Her, E Matijević, WR Wilcox. Powder Technol 61:173, 1990.
32. Y-S Her, S-H Lee, E Matijević. J Mater Res 11:156, 1996.
33. WP Hsu, L Rönnquist, E Matijević. Langmuir 4:31, 1988.
34. DV Goia, E Matijević. Colloids Surf 146:139, 1999.
35. R Demchak, E Matijević J Colloid Interface Sci 31:257, 1969.
36. A Bell, E Matijević. J Inorg Nucl Chem 37:907, 1975.
37. E Matijević, RS Sapieszko, JB Melville. J Colloid Interface Sci 50:567, 1975.
38. RS Sapieszko, RC Patel, E Matijević. J Phys Chem 81:1061, 1977.
39. R Brace, E Matijević. J Inorg Nucl Chem 35:3691, 1973.
40. WB Scott, E Matijević. J Colloid Interface Sci 66:447, 1978.
41. S Hamada, Y Kudo, S Hasegawa. Colloid Polym Sci 269:290, 1991.
42. F Montino, G Spoto. US patent 4,902,494, 1990.
43. SK Milonjić. Mater Sci Forum 214:197, 1996.
44. S Hamada, Y Kudo, T Kinoshita A Kanaya. Shikizai Kyokaishi 70:163, 1997.
45. S Hamada, Y Kudo, K Masuzawa. Shikizai Kyokaishi 67:688, 1994.
46. E Matijević, M Budnik, L Meites. J Colloid Interface Sci 61:302, 1977.
47. A Kato, Y Takeshita, Y Katatae. In: IA Aksay, GL McVay, DR Ulrich, eds. Processing and Science of Advanced Ceramics. Mater Res Soc, 1989, p 13.
48. M Yokota, M Naka. Funtai Oyobi Funmatsu Yakin 39:235, 1992.
49. JB Donnet. Compt Rend Acad Sci France 227:508, 1948.
50. JHL Watson, W Heller, W Wojtowicz. Science 109:274, 1949.
51. E Matijević, AD Lindsay, S Kratohvil, ME Jones, RI Larson, NW Cayey. J Colloid Interface Sci 36:273, 1971.
52. RI Larson, EF Fullam, AD Lindsay, E Matijević. Am Inst Chem Eng J 19:602, 1973.
53. AC Zettlemoyer, M Siddiq, FJ Micale. J Colloid Interface Sci 66:173, 1978.
54. A Bell, E Matijević. In Al Smith, ed. Particle Growth in Suspensions. New York: Academic Press, 1973, pp 179–193.
55. A Bell, E Matijević. J Phys Chem 78:2621, 1974.
56. R Sprycha, E Matijević. Colloids Surf 47:195, 1990.
57. R Sprycha, J Jablonski, E Matijević. Colloids Surf 67:101, 1992.
58. R Sprycha, J Jablonski, J Szczypa. Croat Chem Acta 71:1155, 1998.
59. JHL Watson, RR Cardell, W Heller. J Phys Chem 66:1757, 1962.
60. TG Spiro, SE Allerton, J Renner, A Terzis, R Bills, P Saltman. J Am Chem Soc 88: 2721, 1966.
61. RHH Wolf, M Wrischer, J Šipalo-Žuljević. Kolloid Z Z Polym 215:57, 1967.
62. TG Spiro, L Pape, P Saltman. J Am Chem Soc 89:5555, 1967.
63. RJ Atkinson, AM Posner, JP Quirk. J Inorg Nucl Chem 30:2371, 1968.
64. JT Kenney, WP Townsend, JA Emerson. J Colloid Interface Sci 42:589, 1973.
65. E Matijević, P Scheiner. J Colloid Interface Sci 63:509, 1978.
66. T Sugimoto, E Matijević. J Colloid Interface Sci 74:227, 1980.
67. RS Sapieszko, E Matijević. J Colloid Interface Sci 74:405, 1980.
68. Y Maeda, S Hachisu. Colloids Surf 6:7, 1983.
69. S Hamada, E Matijević. J Colloid Interface Sci 84:274, 1981.
70. S Hamada, E Matijević. J Chem Soc Faraday Trans I 78:2147, 1982.
71. M Ozaki, S Kratohvil, E Matijević. J Colloid Interface Sci 102:146, 1984.

72. S Hamada, T Hanami, Y Kudo. Nippon Kagaku Kaishi 6:1065, 1984.
73. JHA van der Woude, PL de Bruyn. Colloids Surf 12:179, 1984.
74. M Ozaki, E Matijević. J Colloid Interface Sci 107:199, 1985.
75. S Hamada, S Niizeki, Y Kudo. Bull Chem Soc Jpn 59:3443, 1986.
76. NHG Penners, LK Koopal. Colloids Surf 19:337, 1986.
77. E Matijević, S Cimaš. Colloid Polym Sci 265:155, 1987.
78. NHG Penners, LK Koopal. Colloids Surf 28:67, 1987.
79. T Ishikawa, E Matijević. Langmuir 4:26, 1988.
80. S Hamada. Prog Batteries Sol Cells 7:96, 1988.
81. S Hamada, Y Kudo, T Matsumoto. Bull Chem Soc Jpn 62:1017.
82. M Ozaki, N Ookoshi, E Matijević. J Colloid Interface Sci 137:546, 1990.
83. K Kandori, Y Kawashima, T Ishikawa. J Chem Soc Faraday Trans 87:2241, 1991.
84. G Wang, Z Chen, Y Zhang, S Yin. Wuli Huaxue Xuebao 7:699, 1991.
85. NJ Reeves, S Mann. J Chem Soc Faraday Trans 87:3875, 1991.
86. U Schwertmann, RM Cornell. In: Iron Oxides in the Laboratory—Preparation and Characterization. New York: VCH Publ Inc., 1991.
87. K Kandori, Y Kawashima, T Ishikawa. J Colloid Interface Sci 152:284, 1992.
88. MP Morales, T Gonzalez-Carreno, CJ Serna. J Mater Res 7:2538, 1992.
89. T Sugimoto, K Sakata. J Colloid Interface Sci 152:587, 1992.
90. K Kandori, Y Kawashima, T Ishikawa. J Mater Sci Lett 12:288, 1993.
91. T Sugimoto, MM Khan, A Muramatsu. Colloids Surf A 70:167, 1993.
92. T Sugimoto, MM Khan, A Muramatsu, H Itoh. Colloids Surf A 79:233, 1993.
93. T Sugimoto, K Sakata, A Muramatsu. J Colloid Interface Sci 159:372, 1993.
94. S Kan, Y Zhang, X Peng, S Li, L Xiao. Gaodeng Xuexiao Huaxue Xuebao 14:1013, 1993.
95. JK Bailey, CJ Brinker, ML Mecartney. J Colloid Interface Sci 157:1, 1993.
96. K Kandori, S Tamura, T Ishikawa. Colloid Polym Sci 272:812, 1994.
97. D Shindo, GS Park, Y Waseda, T Sugimoto. J Colloid Interface Sci 168:478, 1994.
98. K Kandori, I Horii, A Yasukawa, T Ishikawa. J Mater Sci 30:2145, 1995.
99. T Nakamura, H Kurokawa. J Mater Sci 30:4710, 1995.
100. M Ocaña, MP Morales, CJ Serna. J Colloid Interface Sci 171:85, 1995.
101. S Kan, X Zhang, S Yu, D Li, L Xiao, G Zou, T Li, W Dong, Y Lu. J Colloid Interface Sci 191:503, 1997.
102. K Kandori, A Yasukawa, T Ishikawa. J Colloid Interface Sci 180:446, 1996.
103. K Kandori, N Ohkoshi, A Yasukawa, T Ishikawa. J Mater Res 13:1698, 1998.
104. T Sugimoto, E Matijević. J Inorg Nucl Chem 41:165, 1979.
105. L Durand-Keklikian, I Haq, E Matijević. Colloids Surf A 92:267, 1994.
106. AHM Andreasen. Kolloid-Z 104:181, 1943.
107. P McFadyen, E Matijević. J Colloid Interface Sci 44:95, 1973.
108. S Hamada, Y Kudo, I Ishiyama. Shikizai Kyokaishi 69:658, 1996.
109. P McFadyen, E Matijević. J Inorg Nucl Chem 35:1883, 1973.
110. A Chittofrati, E Matijević. Colloids Surf 48:65, 1990.
111. D Jezequel, J Guenot, N Jouini, F Fievet. J Mater Res 10:77, 1995.
112. S Hamada, K Bando, Y Kudo. Nippon Kagaku Kaishi 6:1068, 1984.
113. S Hamada, K Bando, Y Kudo. Bull Chem Soc Jpn 59:2063, 1986.
114. S Hamada. In: E Pelizzetti, ed. Fine Particles Science and Technology. Dordrecht: Kluwer, 1996, pp 97–107.

115. A Bleier, RM Cannon. Am Ceram Soc Bull 61:336, 1982.
116. MA Blesa, AJG Maroto, SI Passaggio, NE Figliolia, G Rigotti. J Mater Sci 20:4601, 1985.
117. A Bleier, RM Cannon. In: CJ Brinker, DE Clark, DR Ulrich, eds. Better Ceramics Through Chemistry II. Pittsburg: Mater Res Soc 1986, pp 71–78.
118. A Garg, E Matijević. J Colloid Interface Sci 126:243, 1988.
119. G Rinn, H Schmidt. In: GL Messing, ER Fuller, H Hausner, eds. Ceramic Powder Science II. Westerville, OH: Am Ceram Soc, 1988, p 23.
120. RR Wusirika. US patent 4,719,091, 1988.
121. H Qui, C Feng, J Guo. Wuji Cailiao Xuebao 8:156, 1993. Chem Abstr 120:248617s.
122. YT Moon, DK Kim, CH Kim. J Am Ceram Soc 78:1103, 1995.
123. E Kato, M Hirano, Y Kobayashi, K Asoh, M Mori, M Nakata. J Am Ceram Soc 79: 972, 1996.
124. LA Perez-Maqueda, E Matijević. J Mater Res 12:3280, 1997.
125. K Yura, KC Fredrikson, E Matijević. Colloids Surf 50:281, 1990; 53:411, 1991.
126. S Hamada, Y Kudo, K Minigawa. Bull Chem Soc Jpn 63:102, 1990.
127. S Hamada, Y Kudo, T Kobayashi. Colloids Surf A 79:227, 1993.
128. PG Harrison, JK McGiveron, CC Harrison. J Sol-Gel Sci Technol 2:295, 1994.
129. LA Perez-Maqueda, L Wang, E Matijević. Langmuir 14:4397, 1998.
130. M Ocaña, CJ Serna, E Matijević. Mater Lett 12:32, 1991.
131. M Ocaña, CJ Serna, E Matijević. Colloid Polym Sci 273:681, 1995.
132. E Matijević. US patent 5,015,452, 1991.
133. V Briois, CE Williams, H Dexpert, F Villain, B Cabane, F Deneuve, C Magnier. J Mater Sci 28:5019, 1993.
134. M Ocaña, D Hoffman, E Matijević. J Mater Chem 1:87, 1991.
135. M Ocaña, M Andres, M Martinez, CJ Serna, E Matijević. J Colloid Interface Sci 163: 262, 1994.
136. JHL Watson, W Heller, W Wojtowicz. J Chem Phys 16:979, 999, 1948.
137. W Heller, W Wojtowicz, JHL Watson. J Chem Phys 16:998, 1948.
138. ABD Brown, SM Clarke, AR Rennie. Colloid Polym Sci 276:549, 1998.
139. E Han, Z Wu, Z Chen, Y Chang, L Zhu. Wuli Huaxue Xuebao 9:94, 1993.
140. NB Milić, E Matijević. J Colloid Interface Sci 85:306, 1982.
141. GA Parks. Chem Rev 65:177, 1965.
142. MA Blesa, E Matijević. Adv Colloid Interface Sci 29:173, 1989.
143. M Ocaña, R Rodriguez-Clemente CJ Serna. Adv Mater 7:212, 1995.
144. A Garg, E Matijević. Langmuir 4:38, 1988.
145. N Kawahashi, C Persson, E Matijević. J Mater Chem 1:577, 1991.
146. WP Hsu, R Yu, E Matijević. J Colloid Interface Sci 156:56, 1993.
147. E Matijević, WP Hsu, MR Kuehnle. US patent 5,248,556, 1993.
148. AP Philipse, A-M Nechifor, C Pathmamanoharan. Langmuir 4:38, 1988.
149. K Yura, E Matijević. J Colloid Interface Sci 155:328, 1993.

1.2 HYDROLYSIS OF METAL ALKOXIDES IN HOMOGENEOUS SOLUTIONS

TAKASHI OGIHARA
Fukui University, Fukui, Japan

1.2.1 Introduction

It is well understood that the potential of ceramic materials is strongly dependent on powder characteristics such as particle size, size distribution, and morphology. Powder preparation techniques through the gas and liquid phases are an area of considerable interest and increasingly important in the development of advanced ceramics. The chemical synthesis of nonagglomerated powders, which have submicrometer size with narrow size distribution, is available for the production of ceramic components with uniform and fine-grained microstructure at lower sintering temperatures and shorter sintering times. In addition, ceramic powders for which the particle morphology and chemical homogeneity are precisely controlled are required. It has been also suggested that monodispersed ceramic powders are preferable to raw materials for enhancing the densification and controlling the grain growth. These powders have been prepared through the following processes: (1) sol-gel method (1–3), (2) spray pyrolysis using organometallic compound (4–6), (3) coprecipitation of inorganic salt solution (7,8), (4) thermal decomposition of dilute inorganic salts using urea or formamide (9,10), and (5) hydrothermal reaction of inorganic salts (11–13). Alkoxide-derived spherical, monodisperse powders have been noted as the best precursors for the ceramic materials. The alkoxide route offers significant advantages over conventionally produced powders; for example, this leads to high-purity and homogeneous powders at lower processing temperature than solid-state reaction, spray pyrolysis, and hydrothermal reaction. In this chapter, the controlled hydrolysis, emulsion process, industrial processing techniques, and formation mechanism for alkoxide-derived monodispersed ceramic powders are introduced.

1.2.2 Controlled Hydrolysis of Metal Alkoxide in Alcohol Solution

Nucleophilic metal alkoxide reacts rapidly with electrophilic water to generate the several hydrolysis products with complex intermediate species containing some residual alkoxy groups. The concentration ratio of the starting reagents and the nature of alkoxide and the solvent influence the morphology of hydrolysis products. Overall, the formation reaction for particles consists of hydrolysis of metal alkoxide and condensation of the hydrolysis species. The hydrolysis reaction of metal alkoxide is shown in Eq. (1). The hydrolysis species generated may differ with the hydrolysis conditions.

$$M(OR)_n + xH_2O \rightarrow M(OR)_{n-x}(OH)_x + xROH \quad (1)$$

This product undergoes condensation to produce the polymer species.

$$M(OR)_{n-x}(OH)_x + M(OR)_{n-x}(OH)_x \rightarrow M_2(OR)_{2(n-x)-1}(OH)_{2x-1}O + H_2O \quad (2)$$

$$M(OR)_{n-x}(OH)_x + M(OR)_{n-x}(OH)_x \rightarrow M_2(OH)_{2(n-x)-1}(OR)_{2x-1}O + ROH \quad (3)$$

Condensation proceeds through either water elimination [Eq. (2)] or alcohol elimination [Eq. (3)].

The preparation of monodisperse particles requires a short nucleation period, which produces homogeneous nucleation and growth throughout the solution. Several conditions are necessary to successfully synthesize monodisperse particles from metal alkoxide in alcohol solution: (1) The reagent concentration and molar ratio of alkoxide to water must be proper to promote homogeneous nucleation; (2) reagents must be completely mixed before nucleation; (3) insoluble impurities must be removed from the reagent to prevent heterogeneous nucleation; and (4) the colloid stability in the solution must be maintained in the stage of initial reaction stage by using repulsive steric or electrostatic forces.

Monodisperse oxide powders such as SiO_2 (14), TiO_2 (15–19), ZrO_2 (20,21), Ta_2O_5 (22), and PZT (23,24) have been synthesized by the controlled hydrolysis of metal alkoxide in a dilute alcohol solution. Figure 1.2.1 shows SEM photographs of typical single- and multicomponent oxide powders derived from metal alkoxide. The physical and chemical properties of as-prepared powders are listed in Table 1.2.1. As-prepared particles, which have submicrometer size with spherical morphology, are amorphous and hydrated, but crystallize to oxide at elevated temperature. The chemical composition of multicomponent oxide powders such as $Y_2O_3-ZrO_2$ and PZT was in agreement with starting precursor composition. The crystallization also resulted in particles that retained spherical morphology. Particle characteristics such as particle size, size distribution, and state of agglomeration were dependent on the synthesis factors such as solution concentration, reaction temperature, and reaction time. Figure 1.2.2 shows the concentration range of metal alkoxide and water for the formation of monodispersed TiO_2 particles. The optimum concentrations for preparing monodispersed particles were 0.01 to 0.2 mol/dm^3 metal alkoxide and 0.05 to 0.5 mol/dm^3 water. At higher water and metal alkoxide concentrations, the particles aggregated due to rapid, uncontrolled hydrolysis. On the other hand, at lower water and metal alkoxide concentrations, polydisperse particles were formed due to slow hydrolysis. The water concentration, rather than metal alkoxide concentration, strongly affected the particle size distribution and state of aggregation. Monodisperse particles were given at the temperature from 25 to 50°C. However, the

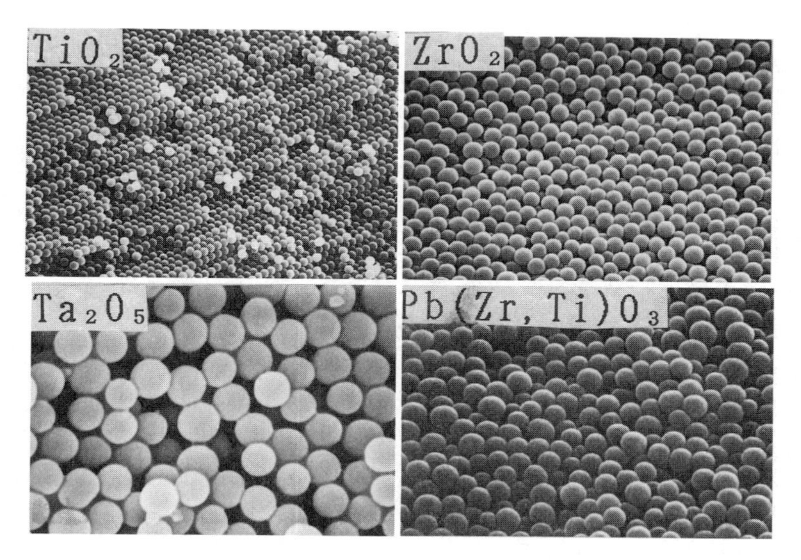

Fig. 1.2.1 SEM photographs of typical monodisperse oxide fine particles.

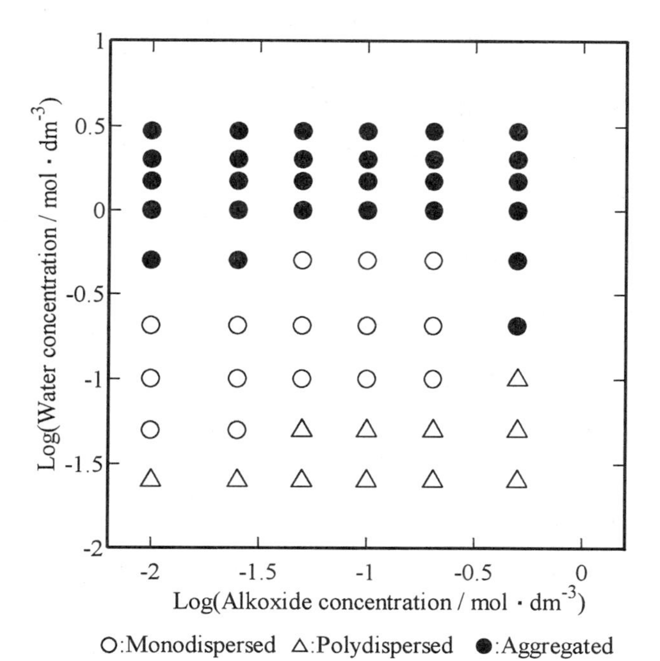

Fig. 1.2.2 Concentration range of metal alkoxide and water for the formation of monodisperse TiO_2 particles.

Table 1.2.1 Chemical and Physical Properties of Monodisperse Oxide Powders

	TiO_2	ZrO_2	3mol% Y_2O_3-ZrO_2	Ta_2O_5	Nb_2O_5	Pb(Zr, Ti)O_3
Precursor	$Ti(OC_2H_5)_4$	$Zr(OC_4H_9)_4$	$Zr(OC_4H_9)_4$ Y(iso-$OC_3H_7)_3$	$Ta(OC_4H_9)_5$	$Nb(OC_4H_9)_5$	Pb(iso-$OC_3H_7)_2$ $Zr(OC_4H_9)_4$ $Ti(OC_2H_5)_4$
Morphology	Spherical	Spherical	Spherical	Spherical	Spherical	Spherical
Particle size (μm)	0.54	0.64	0.67	0.68	0.51	0.75
Particle size distribution (σ_g)	1.06	1.08	1.10	1.10	1.16	1.17
Crystal phase	Amorphous ↓ 400°C Anatase ↓ 800°C Rutile	Amorphous ↓ 400°C Tetragonal ↓ 1000°C Monoclinic	Amorphous ↓ 1000°C Tetragonal	Amorphous ↓ 740°C β-Ta_2O_5	Amorphous ↓740°C β-Nb_2O_5	Amorphous ↓ 600°C Tetragonal
State of agglomeration	Unagglomerated	Unagglomerated	Unagglomerated	Unagglomerated	Unagglomerated	Unagglomerated
Composition (atom. conc.%)			Zr 97.10 Y 2.90			Pb 47.32 Zr 26.15 Ti 26.53
Yield (%)	70	60	67	73	70	75
Voltaile	C, H	C, H	C, H	C, H	C, H	C, H
Weight loss (%)	12	10	14	185	12	10
Iso electric point (IEP)	9.0	10.5	10.3	10.5	9.8	10.5

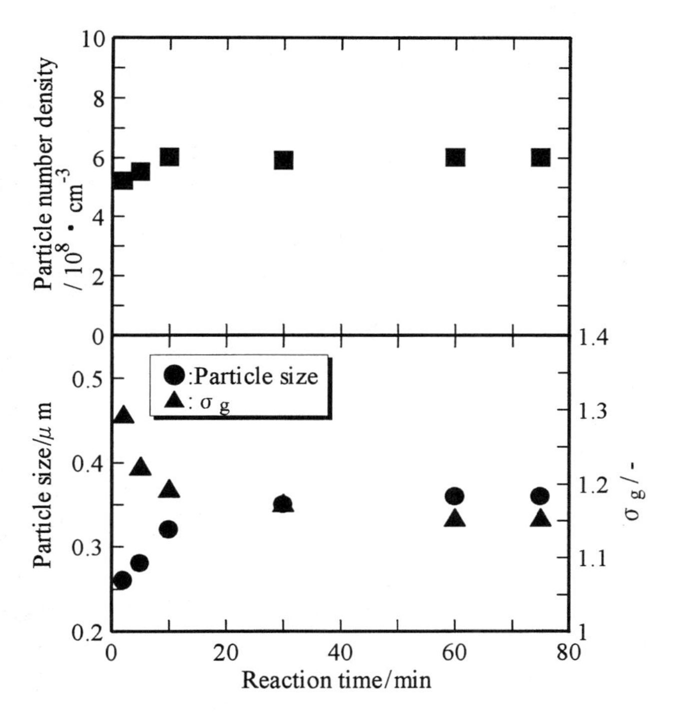

Fig. 1.2.3 Growth curves of monodisperse TiO_2 particles.

particles were aggregated at more than 50°C because the higher hydrolysis rate made it impossible to control the nucleation and growth. Figure 1.2.3 shows the growth curves of monodispersed TiO_2 particles examined up to 75 min. The growth of TiO_2 particles was initially rapid, followed by slow growth. After 2 min, the particle size was 0.26 μm and particle size distribution was broad. After 10 min, the value of particle size distribution (σ_g) was <1.2. Then the particle size became nearly constant because of the decrease of metal alkoxide concentration in mother liquid. The particle number density became almost constant after 2 min. Coalescence of nuclei is expected to occur before then. However, the particles grow alone without the coalescence of particles after at least 2 min.

The growth mechanism of alkoxide-derived monodispersed particles using seed ZrO_2 particles (25) was examined by Nielsen's chronomal analysis (26). Chronomal analysis has been used to establish the growth mechanism of systems in which the particles have uniform size and shape and the number of particles in a unit volume remains nearly constant. This treatment permits distinguishing between diffusion controlled growth and surface reaction controlled growth, i.e., polynuclear

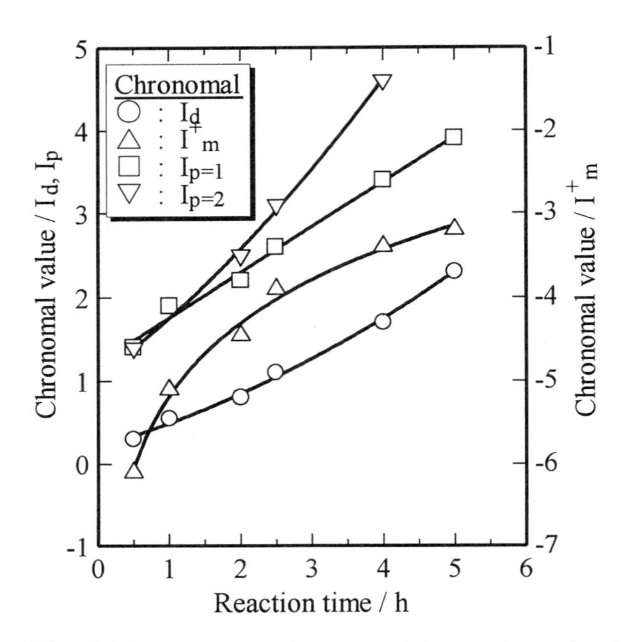

Fig. 1.2.4 Chronomal plots for particle growth as a function of reaction time. (From Ref. 25. Courtesy of M. Mecklenborg, The American Ceramic Society Inc., Westerville, Ohio.)

layer and mononuclear layer. The relation between chronomal value and reaction time for each growth mechanism is shown in Figure 1.2.4. A linear relationship can be obtained only for the first-order polynuclear layer growth ($I_{P=1}$). Similar plots were also made for the other growth mechanism, but their plots were not liners. From chronomal analysis, the rate-determining step of surface nucleation on ZrO_2 seed particles was polynuclear-layer growth of first order. nakanishi et al. (27) reported that the growth of alkoxide-derived SiO_2 seed particles was the fourth-order polynuclear-layer growth ($I_{P=4}$). This finding indicates that ZrO_2 particles grow not by diffusion from the solution, but surface reaction with polynuclear-layer growth as shown in Figure 1.2.5; i.e., each layer is the result of intergrowth of surface nuclei on seed particles. The rate constant of surface nucleation determined by chronomal equation of the first-order polynuclear layer growth was 6.1×10^{-6} (cm/s).

1.2.3 Hydrolysis of Metal Alkoxide in Emulsion

To date, it has been difficult to produce monodisperse spherical powders from metal alkoxides with low valence, such as Al and Fe, because the hydrolysis and condensation of metal alkoxides in alcohol are too rapid to allow control of particle size, size distribution, and morphology. Many previous studies reported (28–30) that alkoxide-

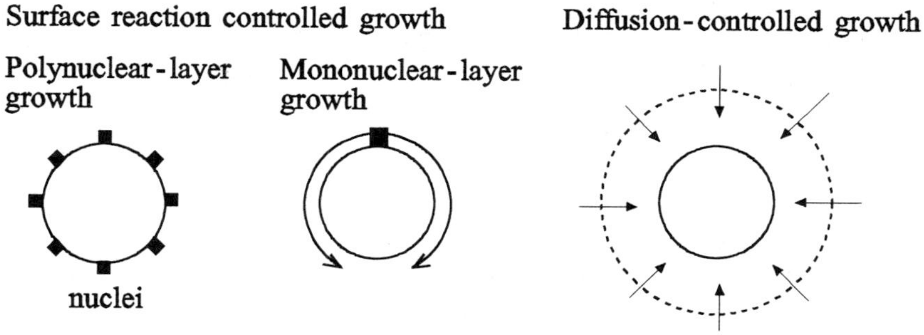

Surface reaction controlled growth
Polynuclear-layer growth
Mononuclear-layer growth
nuclei
Diffusion-controlled growth

Fig. 1.2.5 Schematic diagram of particle growth mechanism in solution. (From T. Sugimoto, Hyomen Surface, 29, 978 (1991). Courtesy of T. Konno, Koshinsya, Tokyo.)

derived powders became gels with various textures and structures. The formation of hydroxide sols is given by the hydrolysis of metal alkoxides using several acidic catalysts. It is well known that the peptization and crystal structure of sol is greatly influenced by the type of acid and hydrolysis temperature. However, controlling the particle composition and state of agglomeration is more difficult.

Recently, a novel emulsion route, in which emulsion droplets can act as a separate reaction chamber and reduce their size to particle size, has become available for producing spherical and nonaggregated multicomponent ceramic powders from nitrate solution (31–34) and metal alkoxide (35–41) with high precipitating kinetics. The emulsion route is a unique approach for the processing of spherical monodispersed oxide particles. A nonaqueous emulsion composed of metal alkoxide droplets dispersed in a polar, inert second phase is formed. After forming the emulsion, the water was added to the emulsion to convert to alkoxide droplets to oxide particles. The hydrolysis occurred at the interface of alkoxide droplets, followed by the diffusion of partially hydrolyzed alkoxide within alkoxide droplets, leading to the condensation of another alkoxide. This process occurs faster than the formation of a solid shell. In order to form the emulsion, the solvent must be immiscible and unreactive with metal alkoxide. The Hildebrand parameter (δ) (42) is used as a guide to select the solvent. The metal alkoxide is soluble in a solvent with low δ, such as alcohol. The immiscible solvents are selected as those with large δ, such as dipolar aprotic solvents. Table 1.2.2 shows the effect of solvent on the formation of aluminum alkoxide emulsion.

Acetonitrile was the most efficient of all dipolar aprotic solvent for the formation of monodisperse particles. Figure 1.2.6 shows the relation between solubility curves of metal alkoxide and particle size in octanol/acetonitrile solution. The solubility of metal alkoxide in octanol gradually decreased with increasing acetonitrile concentration. Metal alkoxide was precipitated to form an emulsion when the acetonitrile was added at more than 40 vol%. The particle size increased with increasing

Table 1.2.2 Effect of Solvent on the Formation of Emulsion from Aluminum Alkoxide

Organic solvent	Solubility parameter (δ ($MPA^{1/2}$)	Reactivity	Hydrogen bonding capacity	Solution	As-prepared particles
Esters					
Propylene carbonate	27.2	No reaction	Moderate	Emulsion	Monodisperse
Nitriles					
Acetonitrile	24.3	No reaction	Moderate	Emulsion	Monodisperse
Propionitrile			Poor	Emulsion	Monodisperse
Sulfur compounds					
Dimethyl sulfoxide	24.5	No reaction	Moderate	Emulsion	Monodisperse
Amides					
Dimethylformamide	24.8	No reaction	Moderate	Emulsion	Monodisperse
Nitrocompounds					
Nitromethane	26.0	Reaction	Strong	Clear solution	—
Ketones					
Acetone	20.2	Reaction	Strong	Clear solution	Monodisperse
Methyl ethyl ketone	19.0	Reaction	Moderate	Clear solution	Aggregation
Methyl isobutyl ketone	17.2	Reaction	Moderate	Clear solution	Aggregation
Acetylacetone			Strong	Clear solution	—

Note. $\Delta\delta$, butanol 23.3, octanol 21.1.

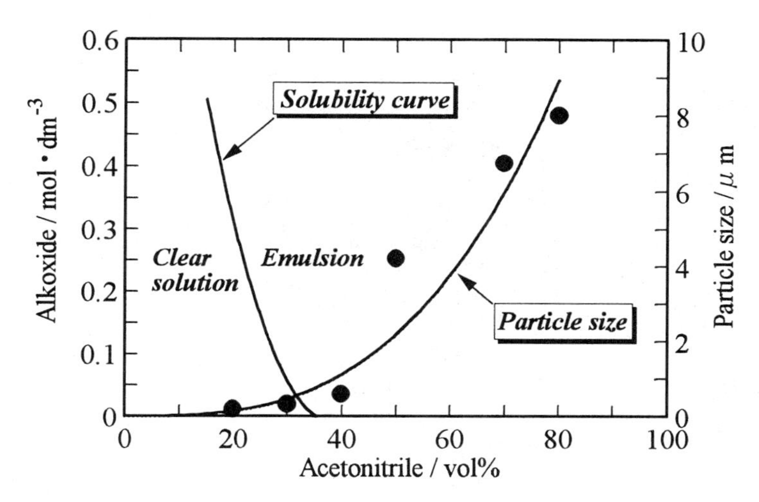

Fig. 1.2.6 Relation between the solubility curve of metal alkoxide and particle size in octanol/acetonitrile.

acetonitrile concentration. Thus, the droplet size of metal alkoxide increases with acetonitrile concentration. The emulsion process made it possible to control the particle size from submicrometer to micrometer. Figure 1.2.7 shows SEM photographs of alumina particles obtained at various acetonitrile concentrations. Monodispersed spherical particles with submicrometer size were formed at 40 vol%. Thus, the addition of 40 vol% acetonitrile to an alkoxide–octanol solution resulted in alkoxide emulsion droplets with a narrow size distribution. Large, polydispersed particles with micrometer size were obtained at 70 vol%. Thus, alkoxide droplets with large size and broad size distribution were formed in emulsion with increasing acetonitrile. However, aggregated particles with irregular morphology were formed at more than 90 vol% because the gelation was occurred in emulsion state.

The formation of alumina particles from emulsion was observed in situ using He–Ne laser photo scattering. Figure 1.2.8 shows the change of particle size distribution as a function of reaction time. The change of particle size distribution suggested that the formation mechanism of monodispersed particles in the octanol/acetonitrile solution was different from homogeneous nucleation and growth of particles by the hydrolysis of alkoxide in alcohol. Bimodal size distribution corresponding to particles and alkoxide droplets was observed at initial reaction (5 s). The size distribution corresponding to alkoxide droplets disappeared at 900 s. This means that all unreacted alkoxide droplets in emulsion convert to the particles at 900 s. The size distribution corresponding to particles became narrow with reaction time, while the particle size gradually was decreased until 900 s. After 900 s, the particle size remained

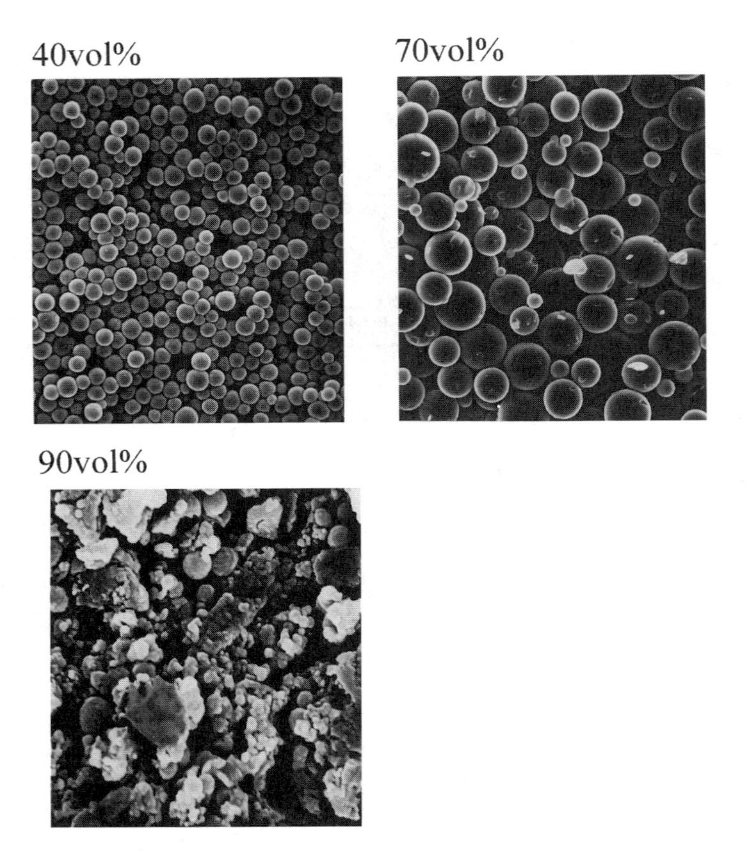

Fig. 1.2.7 SEM photographs of alumina particles obtained at acetonitrile concentration indicated.

a constant. Therefore, it was found that particle growth did not occur during the reaction.

In this system, various monodisperse oxide particles containing multiple cations can be also easily synthesized. Generally, it is difficult to control the hydrolysis of metal alkoxide mixtures because the individual alkoxides hydrolyze at very different rate. The difference in hydrolysis rate makes it impossible to obtain ceramic powders that have the same stoichiometry as the starting solution. An alkoxide emulsion containing multications was prepared for overcoming this difficulty. To demonstrate the possibility of emulsion technique for preparing a multicomponent oxide, mullite $3Al_2O_3{\cdot}2SiO_2$ (43), $BaTiO_3$, $SrTiO_3$, $PbTiO_3$ (44), and yttrium alumi-

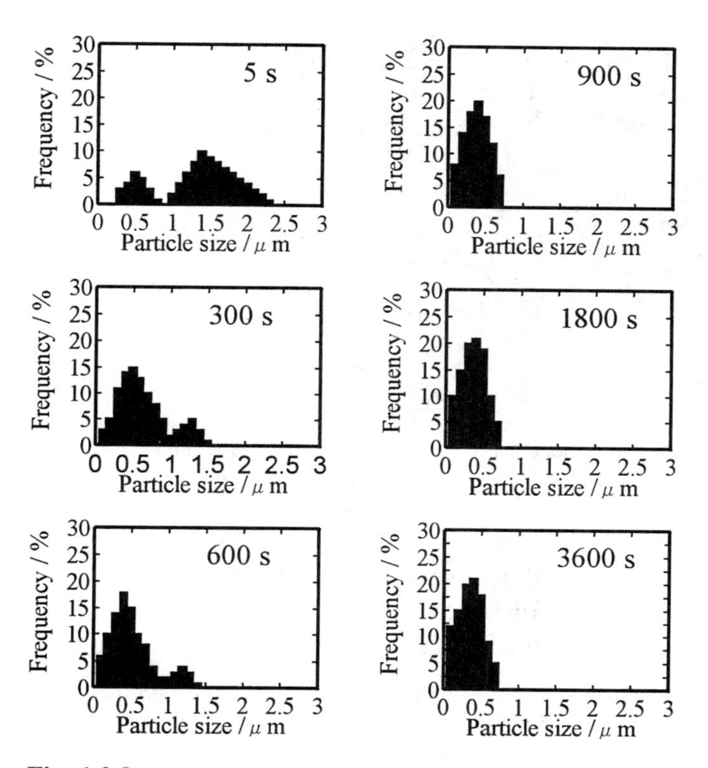

Fig. 1.2.8 Relation between reaction time and particle size distribution.

num garnet (YAG) (45) particles have been synthesized by the hydrolysis of alkoxide emulsion in the combination of alcohol and dipolar aprotic solvent. The hydrolysis rate of metal alkoxide used for the synthesis of these powders was too rapid to control the nucleation in the dilute alcohol, leading to particle agglomeration.

Figure 1.2.9 shows SEM photographs of mullite and $BaTiO_3$ particles prepared from alkoxide emulsion at 40 wt% acetonitrile. Monodisperse particles containing multications were easily given under the same synthesis condition as aluminum alkoxide emulsion. Chemical composition of as-prepared powders determined by inductively coupled plasma (ICP) analysis is listed in Table 1.2.3. ICP analysis revealed that the chemical compositions of as-prepared powders were in good agreement with starting solution compositions. Thus, the composition of emulsion droplets is homogeneous at the molecular level. These powders were crystallized at a lower temperature than the solid-state reaction. After the crystallization, the stoichiometric composition of these powders was retained.

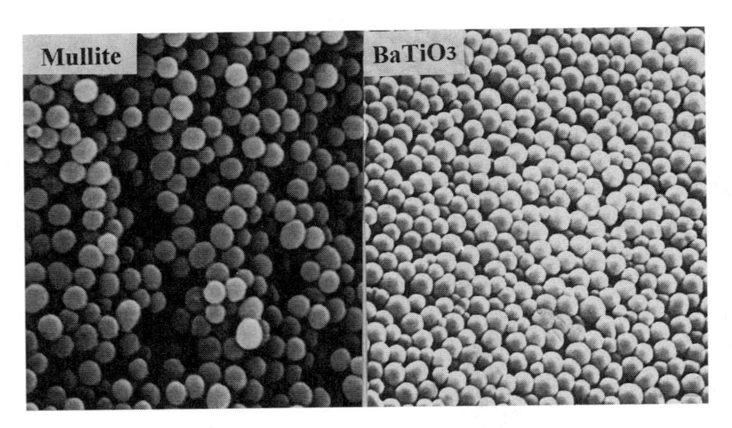

Fig. 1.2.9 SEM photographs of monodisperse particles obtained from emulsion.

1.2.4 Preparation of Monodisperse Oxide Powders by Hydrolysis of Metal Alkoxide Using Continuous Tube-Type Reactor

Generally, alkoxide-derived monodisperse oxide particles have been produced by batch processes on a beaker scale. However, on an industrial scale, the batch process is not suitable. Therefore, a continuous process is required for mass production. The stirred tank reactors (46) used in industrial process usually lead to the formation of spherical, oxide powders with a broad particle size distribution, because the residence time distribution in reactor is broad. It is necessary to design a novel apparatus for a continuous production system of monodispersed, spherical oxide particles. So far, the continuous production system of monodisperse particles by the forced hydrolysis

Table 1.2.3 Chemical Composition of Monodisperse Oxide Powders Determined by ICP Analysis

	As-prepared powders (mol%)						
Solution composition	BaO	SrO	PbO	TiO_2	Al_2O_3	SiO_2	Y_2O_3
$BaTiO_3$ (50:50)	49.5	—	—	50.5	—	—	—
$SrTiO_3$ (50:50)	—	49.8	—	50.2	—	—	—
$PbTiO_3$ (50:50)	—	—	49.5	50.5	—	—	—
Mullite (60:40)	—	—	—	—	61.0	39.0	—
YAG (62.5:37.5)	—	—	—	—	62.0	—	38.0

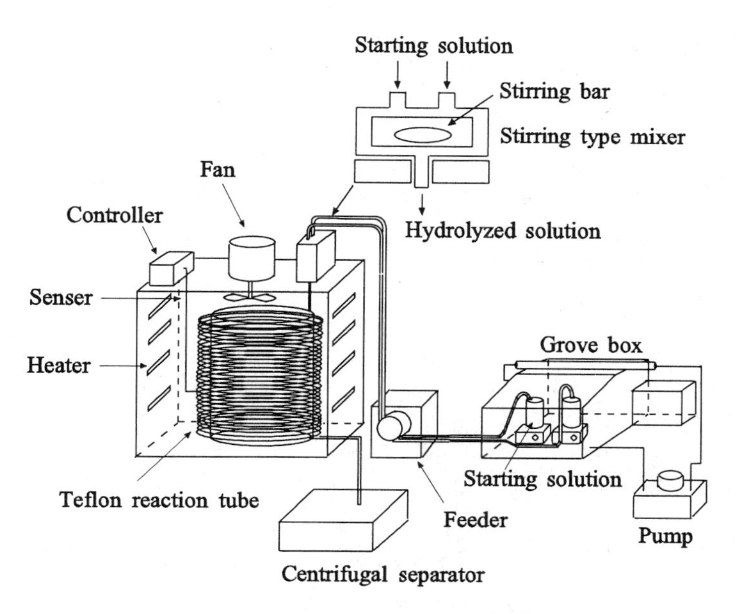

Fig. 1.2.10 Schematic diagram of CTTR system. (From T. Ogihara et al., J Soc Powder Technol Jpn 31, 620 (1994). Courtesy of H. Masuda, The Society of Powder Technology, Kyoto, Japan.)

of metal salts (47,48) and the controlled hydrolysis of metal alkoxide (49) has been developed. Ring et al. (50,51) offered continuous flow reactors using a glass-bead-packed-bed and static mixer. These mixers led to the formation of particles with broad size distribution because the residence time distribution in the reactor became broad. A schematic diagram of a continuous tube-type reactor (CTTR) is shown in Figure 1.2.10. This system was developed by modifying the stirred tank reactors. The CTTR system consisted of starting solution reservoirs (alkoxide and water solutions), a feeder, stirring type mixer, and Teflon reaction tube (52,53). Alkoxide and water solutions were introduced into the stirring type mixer by a feeder at equal rates. The starting solutions were rapidly mixed by rotating the stirring bar with outside magnet. After mixing, the hydrolyzed solution was introduced into a Teflon reaction tube (diameter 40 mm $\times$ 250 m). The Teflon reaction tube was wound on cylinders and heated in the thermostatted box, in which the temperature difference between entrance and exit was always kept within 2°C. The reaction time in which the solution was passed though the tube was controlled by the supply rate of feeder. An ideal plug flow was formed in the Teflon reaction tube. The particles passed through the solution were collected by using centrifugal separator and then dried at 100°C.

Several monodisperse, spherical oxide particles were produced from the hydrolysis of metal alkoxide in ethanol or emulsion solution using the CTTR system.

The effects of experimental factors such as reactant concentration, reaction time and temperature, flow rate, and type of mixer on the formation of monodispersed particle were investigated. Typical scanning electron microscopy (SEM) photographs and particle size distributions of Fe_2O_3 powders prepared from the CTTR and batch process are shown in Figure 1.2.11. Both the CTTR and batch process led to spherical Fe_2O_3 particles without agglomeration. The particle sizes of Fe_2O_3 powders prepared by CTTR and batch process were 0.43 and 0.4 μm, respectively, while the values of σ_g for the particles given by CTTR and batch process were 1.12 and 1.18, respectively. It was found that the particle characteris-

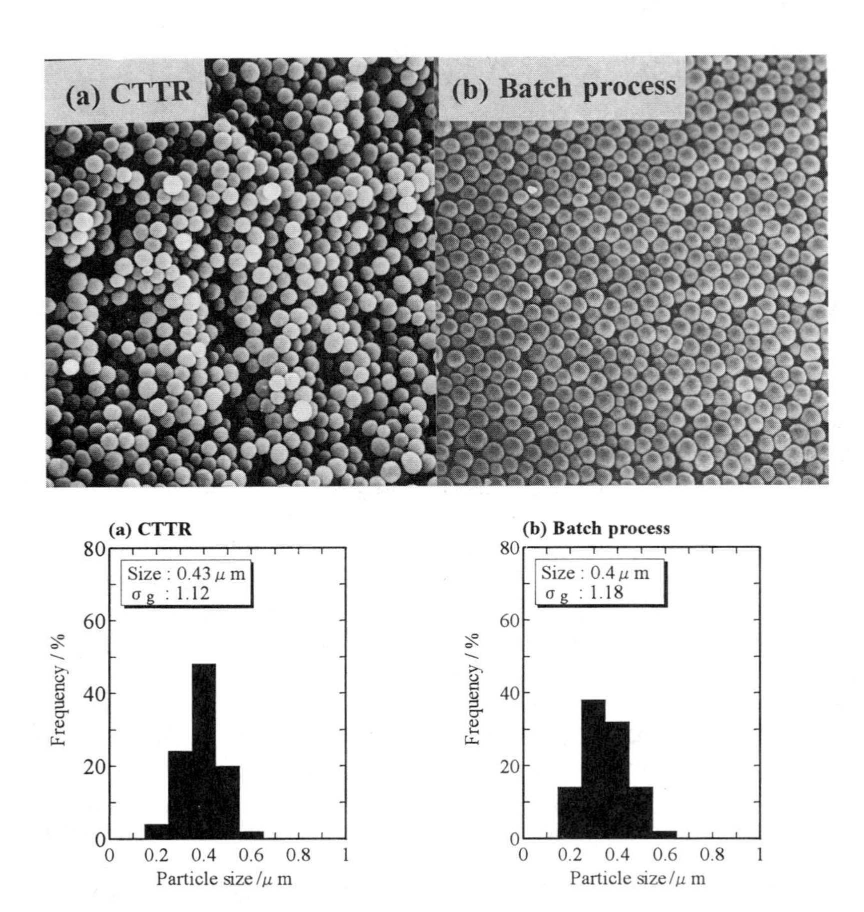

Fig. 1.2.11 SEM photographs and particle size distribution of monodisperse Fe_2O_3 particles produced by CTTR and batch process. (From T. Ogihara et al., J Soc Powder Technol Jpn 31, 620 (1994). Courtesy of H. Masuda, The Society of Powder Technology, Kyoto, Japan.)

tics of monodispersed Fe_2O_3 powders produced by CTTR were comparable to those obtained by batch process.

Monodisperse Fe_2O_3 particles were produced under the conditions of 0.01–0.2 mol/dm^3 alkoxide solution and 0.05–1.5 mol/dm^3 water solution by CTTR. The relations between the concentration of starting solution and resultant particle characteristics (the particle size, σ_g, yield, and particle number density) are shown in Figure 1.2.12. The particle size increased with increasing alkoxide concentration in the range where monodisperse particles were formed, because the size of alkoxide droplets increased with alkoxide concentration. The yield and particle number density also increased with increasing alkoxide concentration. On the other hand, the particle size decreased with increasing water concentration. However, the yield and particle number density increased with increasing water concentration. When the water concentration was high, the diffusion of water to alkoxide droplets interface was promoted to form a lot of particles in a short time.

Figure 1.2.13 shows the relation between flow rate and σ_g of particles and also the change of σ_g when the static mixer is used. The value of σ_g gradually decreased to form monodisperse particles with increasing the flow rate. When the mixing effect of the stirring type mixer and static mixer was compared, it was found that the particle size distribution of particles obtained by using the stirring type mixer was narrower than that obtained by the static mixer at an equal flow rate. To obtain monodisperse particles, it is important to uniformly mix metal alkoxide with the water. It is difficult to mix alkoxide with the water in the octanol because the octanol is hydrophobic and of higher viscosity. The stirring type mixer used in this work enabled the uniform mixing of metal alkoxide and the water compared with the static mixer. Thus, the shear stress produced by the stirring type mixer is stronger than that given by the static mixer.

The particle characteristics such as particle size, σ_g, yield, and particle number density were independent on the reaction time and temperature examined in the formation of Fe_2O_3 particles from an emulsion state. However, the particle characteristics are generally influenced by the reaction time and temperature in the particle formation from the hydrolysis in alcohol solution, because the reaction time and temperature promote the hydrolysis reaction.

To estimate the production efficiency of the CTTR system, monodisperse Fe_2O_3 particles were continuously produced through 10 h under the constant operation condition. The changes of average size, σ_g, yield, and particle number density of monodisperse Fe_2O_3 particles are shown in Figure 1.2.14. The particle size of Fe_2O_3 particles remained constant through the continuous production. The dispersions of σ_g, yield, and particle number density were observed for 3 h, but they became constant after 3 h. This result suggests that it is possible to produce monodisperse particles by using the CTTR system with a high reproducibility. Using a larger volume of stirring type mixer and wider diameter of Teflon reaction tube, a large amount of monodispersed powders can be produced.

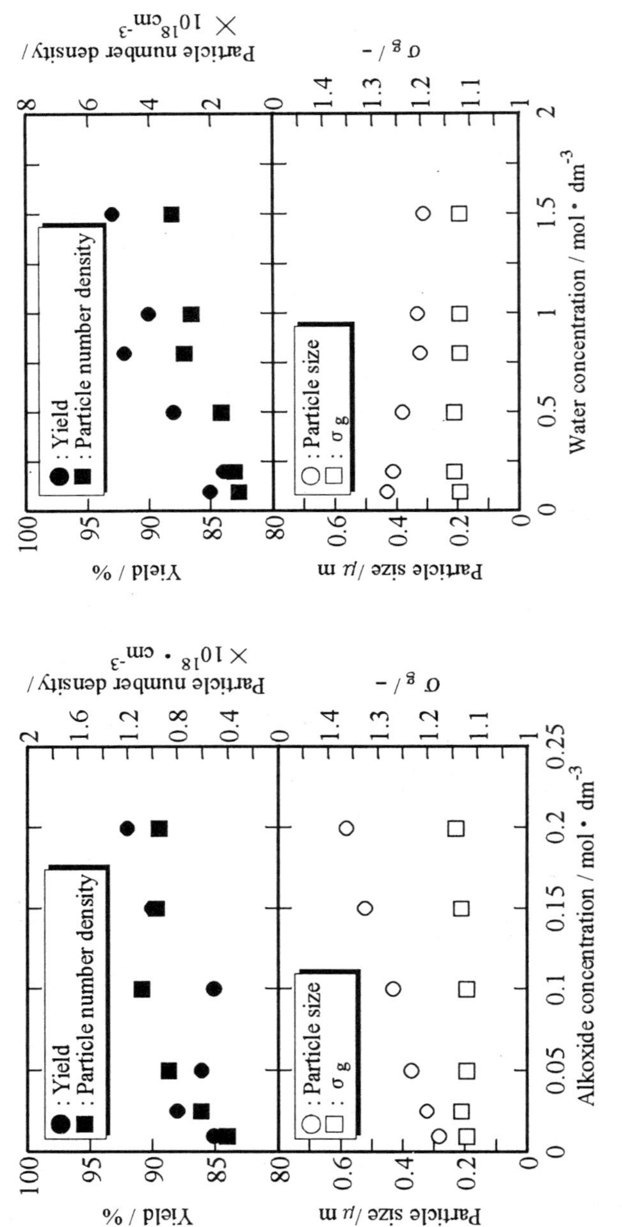

Fig. 1.2.12 Relation between concentration and particle characteristics. (From T. Ogihara et al., J Soc Powder Technol Jpn 31, 620 (1994). Courtesy of H. Masuda, The Society of Powder Technology, Kyoto, Japan.)

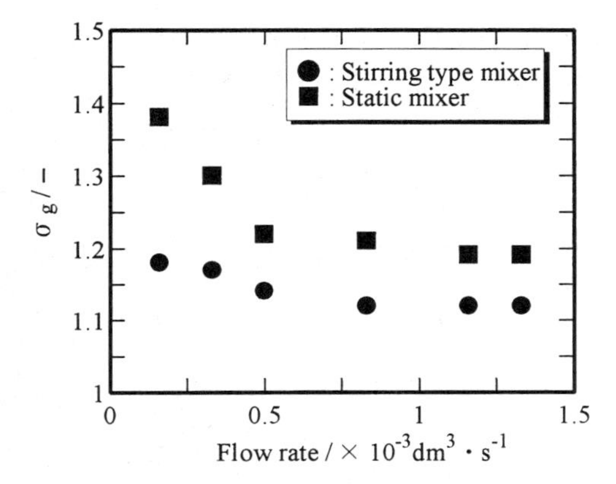

Fig. 1.2.13 The effect of mixing on the formation of monodisperse particles. (From T. Ogihara et al., J Soc Powder Technol Jpn 31, 620 (1994). Courtesy of H. Masuda, The Society of Powder Technology, Kyoto, Japan.)

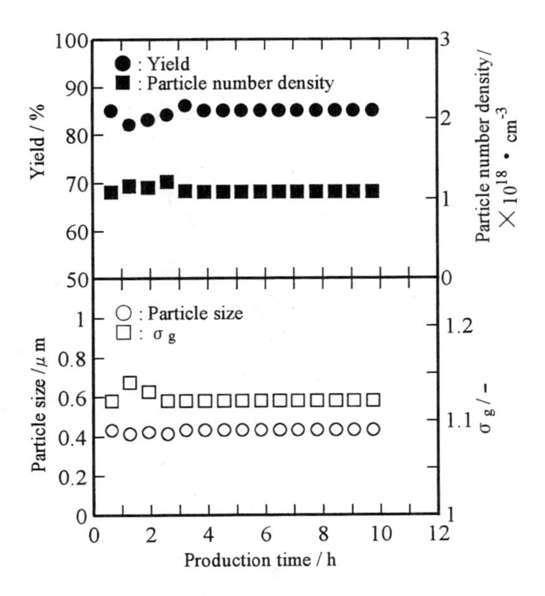

Fig. 1.2.14 Variance of particle characteristics through continuous production. (From T. Ogihara et al., J Soc Powder Technol Jpn 31, 620 (1994). Courtesy of H. Masuda, The Society of Powder Technology, Kyoto, Japan.)

1.2.5 Preparation of Monodisperse, Spherical Oxide Particles by Hydrolysis of Metal Alkoxide Using a Couette–Taylor Vortex Flow Reactor

An alternative novel continuous reactor system was developed for industrial production of monodispersed oxide powders via alkoxide route. Couette–Taylor vortex (C-T) flow was used for uniform mixing of solution. Each of the Taylor vortices can be worked as a well-mixed batch reaction vessel with same residence time (54). Figure 1.2.15 shows a schematic diagram of a C-T flow and reactor system. This system (55) consists of starting solution reservoirs (alkoxide and water solution), feeder, static mixer, and C-T flow reactor consisting of two coaxial vertical cylinders (diameter 50 mm × 1500 mm). Two coaxial cylinders with rotation of the inner cylinder form the annular flow. The annular space formed at between inner and outer cylinder is 12 mm wide and 1500 mm long. The ratio of the governing inertia and viscosity forces for the annular flow is given by the dimensionless Taylor (Ta) number. The Taylor number is equal to the Reynolds number in cylindrical coordinates. The Taylor number is given by Eq. (4):

$$\mathrm{Ta} = \frac{\omega b^{3/2} R_i^{\ 1/2}}{\nu} \tag{4}$$

where ω is the angular speed of inner cylinder, ν the viscosity of fluid, b the annular

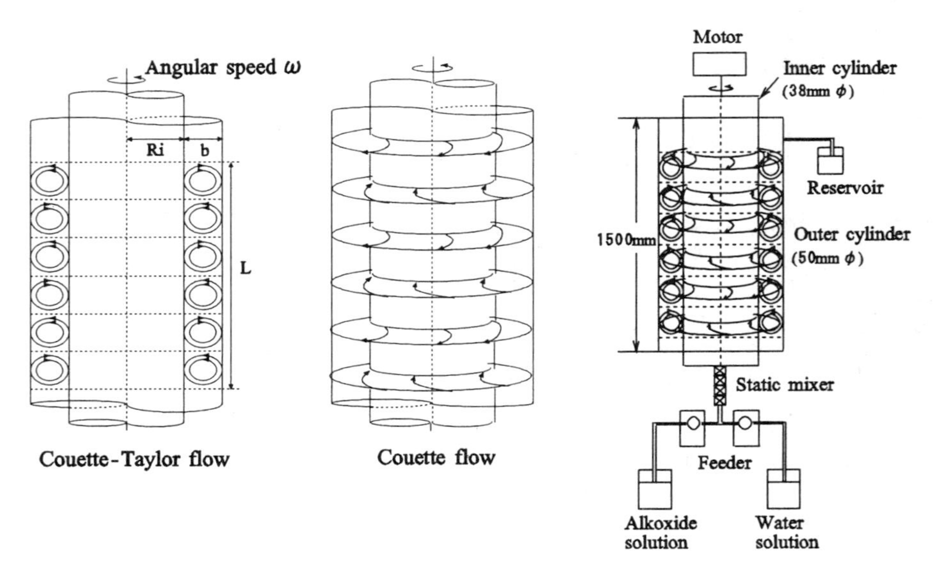

Fig. 1.2.15 Schematic diagram of Couette–Tayler vortex flow and reactor system. (From Ref. 55. Courtesy of Y. Kokubo, The Ceramic Society of Japan, Tokyo.)

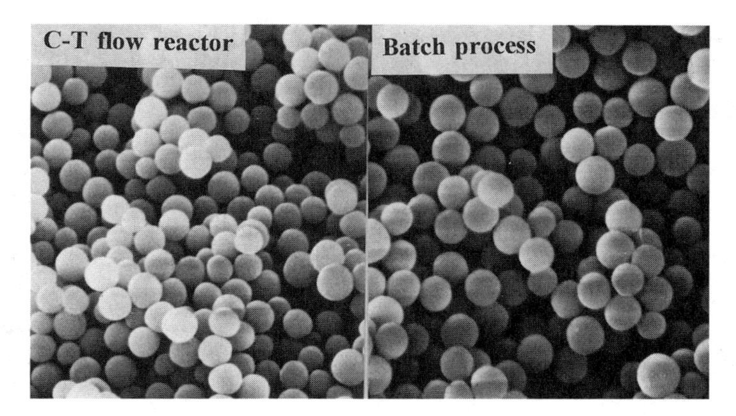

Fig. 1.2.16 SEM photographs of silica particles produced by C-T flow reactor and batch process.

gap width, and R_i the inner cylinder radius. The transition occurs from Couette flow to C-T flow, when Ta exceeds a critical value. Furthermore, a series of counterrotating cellular vortex flow pairs oriented axially along the cylinder is formed between the inner and outer cylinders.

Alkoxide and water solutions were introduced into the static mixer by a feeder with equal rates. After mixing, the hydrolyzed solution was introduced into the C-T flow reactor and injected from the reactor bottom. The residence time of C-T flow was 1 h. Ethyl silicate was continuously hydrolyzed with ammonia in ethanol to prepare monodisperse silica particles using this system. Figure 1.2.16 shows SEM photographs of monodisperse silica particles produced by a C-T flow reactor and batch process. The resultant particle characteristics were comparable to that obtained by using batch process. The effect of Taylor number on the particle size and σ_g of silica particles is shown in Figure 1.2.17. Couette flow was developed at Ta less than 30, and then polydisperse silica particles were formed. The diffusion of alkoxide and water occurs with Couette flow, leading to the heterogeneous concentration region in the system. C-T flow developed at Ta more than 30, and then monodisperse silica particles were formed. Thus, the diffusion of alkoxide and water among the vortex is successfully controlled by the formation of C-T flow. The particle size decreased with increasing Ta. The value of σ_g was kept to constant. The greater the Taylor number is, the more nuclei are generated, because the alkoxide and water are locally well mixed. However, silica particles were aggregated at Ta more than 160. This suggests that the agglomeration of nuclei occur due to vigorously mixing in vortex. The change of particle size, σ_g, yield, and particle number density of monodisperse silica particles through the continuous production is shown in Figure 1.2.18. The resultant particle characteristics remained constant up to 5 h. Monodis-

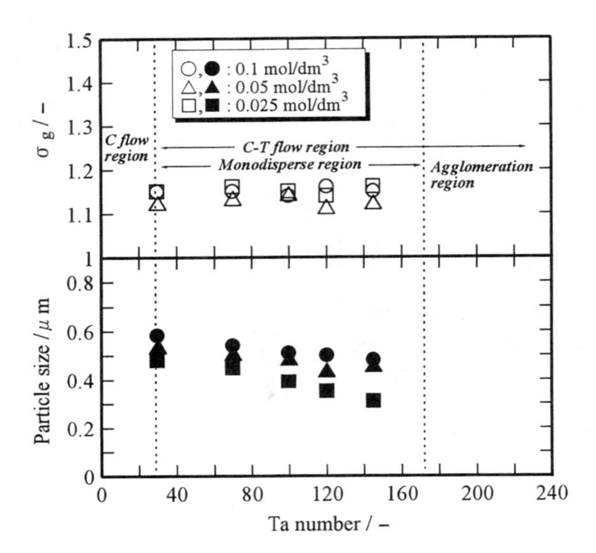

Fig. 1.2.17 Relation between Taylor number and particle size of monodispersed silica particles.

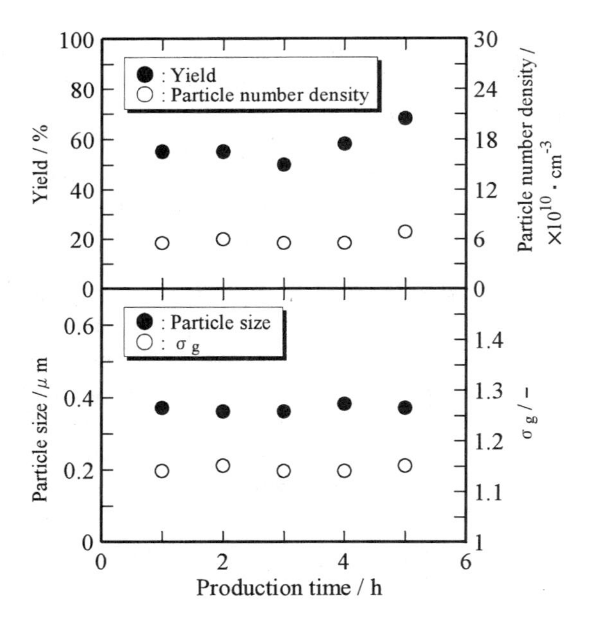

Fig. 1.2.18 Variance of particle characteristics through continuous production. (From Ref. 55. Courtesy of Y. Kokubo, The Ceramic Society of Japan, Tokyo.)

persed silica particles could be continuously produced for 5 h by using this reactor with reproducibility.

1.2.6 Summary

Monodisperse spherical oxide particles were prepared by the hydrolysis of metal alkoxide in homogeneous alcohol in an emulsion state. The formation mechanism from homogeneous alcohol and emulsion state was discussed by chronomal analysis and in situ observation using laser photo scattering. Two types of continuous systems for the industrial production of monodispersed oxide powders were also offered.

1. Monodisperse oxide particles were only formed at the narrow concentration range in alcohol solution. Monodisperse particles, which have submicrometer order with narrow size distribution, were amorphous and hydrated. The spherical morphology was retained after the crystallization. The rate-determining step of particle growth was polynuclear layer growth of first order from Nielsen's chronomal analysis.
2. An emulsion process led to the formation of monodispersed particles from metal alkoxide with low valence. Alkoxide droplets in alcohol solution were given by using solvent with a larger Hildebrand parameter. Each alkoxide droplet acts as a microreactor, allowing control of particle size, size distribution, morphology, and chemical composition. In situ observation using laser photo scattering demonstrated that the particle formation mechanism was different with the nucleation and growth mechanism.
3. CTTR using a Teflon tube was developed for industrial production of monodispersed particles. The important steps in CTTR system are (a) uniform and rapid mixing with stirring type mixer and (b) ideal plug flow in Teflon reaction tube. The quality of monodisperse particles produced by CTTR was comparable to those obtained by batch process. The CTTR system demonstrated that monodisperse particles could be continuously produced for 10 h with reproducibility.
4. A novel reactor using C-T flow was designed as an alternative continuous reactor system to CTTR. Taylor number influenced the mixing effect of solution. Couette flow led to the formation of polydisperse silica particles because of the difference of concentration by the diffusion of alkoxide and water. Monodispersed particles were obtained in the region of C-T flow. The particle size decreased with increasing Taylor number in the region of C-T flow. A higher Taylor number lead to the formation of agglomerated particle due to the collision of nuclei in the vortex. Monodisperse silica particles can be produced with reproducibility by this system.

REFERENCES TO SECTION 1.2

1. KS Mazdyasni. Ceram Intel 8:42, 1982.
2. BE Yoldas. Am Ceram Soc Bull 54:289, 1975.

3. PH Colomban. Ceram Intel 15:23, 1989.
4. H Ishizawa, O Sakurai, N Mizutani, M Kato. Yogyo-Kyokai-Shi 93:382, 1985.
5. H Ishizawa, O Sakurai, N Mizutani, M Kato. Am Ceram Soc Bull 65:1399, 1986.
6. N Mizutani, TQ Liu. In: GL Messing, S. Hirano, H Hausner, eds. Ceramic powder science III. Westerville, OH: American Ceramic Society, 1991, pp 59–73.
7. JR Moyer, AR Prunier Jr, NN Hughes, R Winter. In: CJ Brinker, DE Clark, DR Ulrich, eds. Better Ceramics Through Chemistry II. Mat Res Soc Symp Proc 73. Pittsburgh, PA: Material Research Society, 1986, pp 117–122.
8. RW Schwartz, DJ Eichorst, D Payne. In: CJ Brinker, DE Clark, DR Ulrich, eds. Better Ceramics Through Chemistry II. Mat Res Soc Symp Proc 73. Pittsburgh, PA: Material Research Society, 1986, pp 123–128.
9. MD Sacks, TY Tseng, SY Lee. Am Ceram Soc Bull 63:301, 1984.
10. JE Blendell, HK Bowen, RL Coble. Am Ceram Soc Bull 63:797, 1984.
11. KC Beal. In: GL Messing, KS Mazdiyasni, JW McCauley, RA Harber, eds. Ceramic Powder Science, Advances in Ceramics vol 21. Westerville, OH: American Ceramic Society, 1986, pp 33–41.
12. E Tani, M Yoshimura, S Somiya. J Am Ceram Soc 64:C-181, 1981.
13. M Suzuki, S Udaira, H Masuya, H Tamura. In: GL Messing, ER Fuller Jr, H Hausner, eds. Ceramic Powder Science II, A, Ceramic Transactions, Vol. 1, Westerville, OH: American Ceramic Society, 1988, pp 163–170.
14. W Stöber, A Fink, E Bohn. J Colloid Interface Sci 26:62, 1968.
15. EA Barringer, HK Bowen. J Am Ceram Soc 65:C-199, 1982.
16. EA Barringer, HK Bowen. Langmuir 1:414, 1985.
17. B Fegley Jr, P White, HK Bowen. J Am Ceram Soc, 67:C-113, 1984.
18. T Ikemoto, K Uematsu, N Mizurani, M Kato. Yogyo-Kyokai-Shi 93:261, 1985.
19. JH Jean, TA Ring. Langmuir 2:251, 1986.
20. T Ogihara, N Mizutani, M Kato. Ceram Intel 13:35.
21. K Uchiyama, T Ogihara, T Ikemoto, N Mizutani, M Kato. J Mater Sci 22:4343, 1987.
22. T Ogihara, T Ikemoto, N Mizutani, M Kato, Y Mitarai. J Mater Sci 21:2771, 1986.
23. T Ogihara, H Kaneko, N Mizutani, M Kato. J Mater Sci Lett 7:867, 1988.
24. H Hirashima, E Onishi, M Nakagawa. J Non-Cryst Solids 121:404, 1990.
25. T Ogihara, N Mizutani, M Kato. J Am Ceram Soc 72:421, 1989.
26. AE Nielsen. Kinetics of Precipitation. Oxford: Pergamon Press, 1964, pp 29–85.
27. K Nakanishi, Y Takamiya. J Ceram Soc Jpn 96:719, 1988.
28. BE Yoldas. J Appl Chem Biotechnol 23:803, 1973.
29. BE Yoldas. J Mater Sci 21:1087, 1986.
30. C Sanchez, JL Livage, M Henry, F Babonneau. J Non-Cryst Solids 100:65, 1988.
31. K Richardson, M Akinc. Ceram Intel 13:253, 1987.
32. A Celikaya, M Akinc. In: GL Messing, ER Fuller, H Hausner, eds. Ceramic Powder Science II, A, Ceramic Transactions, vol 1. Westerville, OH: American Ceramic Society, 1988, pp 110–118.
33. M Akinc, A Celikaya. In: GL Messing, KS Mazdiyasni, JW McCauley, RA Haer, eds. Ceramic Powder Science, Advances in Ceramics, vol. 21. Westerville, OH: American Ceramic Society, 1987, pp 57–68.
34. I Sevinc, Y Sarikara, M Akinc. Ceram Intel 17:1, 1991.
35. T Kanai, WE Rhine, HK Bowen. In: GL Messing, ER Fuller, H Hausner, eds. Ceramic Powder Science II, A, Ceramic Transactions. Westerville, OH: American Ceramic Society, 1988, pp 119–126.

36. M Ikeda, SK Lee, K Shinozaki, N Mizutani. J Ceram Soc Jpn 100:183, 1991.
37. AB Hardy, G Gowda, TJ Mcmahon, RE Riman, WE Rhine, HK Bowen. In: JD Mackenzie, DL Ulrich, eds. Ultrastructure Processing of Advances Ceramics. New York: Wiley Interscience, 1987, pp 407–426.
38. AB Hardy, WE Rhine, HK Bowen. J Am Ceram Soc 76: 97, 1993.
39. WT Minehan, GL Messing. J Non-Cryst Solids 121:375, 1990.
40. T Ogihara, N Nakajima, T Yanagawa, N Ogata, K Yoshida. J Am Ceram Soc 74:2263, 1991.
41. T Ogihara, T Yanagawa, N Ogata, K Yoshida. J Ceram Soc Jpn 101:315, 1993.
42. AE Barton. CRC Handbook of Solubility Parameters and Other Cohesion Parameters. Boca Raton, FL: CRC Press, 1983, pp 249–277.
43. T Ogihara, T Yanagawa, N Ogata, K Yoshida, M Iguchi, N Nagata, K Ogawa. J Ceram Soc Jpn 102:778, 1994.
44. T Ogihara, T Yanagawa, N Ogata, K Yoshida, M Iguchi, N Nagata, K Ogawa. J Soc Powder Technol Jpn 31:795, 1994.
45. T Ogihara, K Wada, T Yoshida, T Yanagawa, N Ogata, K Yoshida, N Matsushita. Ceram Intel 19:159, 1993.
46. AV Zyl, PM Smit, AI Kingon. J Mater Sci 78:217, 1986.
47. E Matijevic, WP Hsu. Colloids Surf 118:506, 1987.
48. N Kallay, I Fischer. Colloids Surf 13:145, 1985.
49. T Ogihara, M Ikeda, M Kato, N Mizutani. J Am Ceram Soc 72:1598, 1989.
50. TA Ring. Chem Eng Sci 39:1731, 1984.
51. JH Jean, DM Goy, TA Ring. Am Ceram Soc Bull 65:1574.
52. T Ogihara, M Iizuka, T Yanagawa, N Ogata, K Yoshida. J Mater Sci 27:55, 1992.
53. T Ogihara, M Yabuchi, T Yanagawa, N Ogata, K Yoshida, N Nagata, K Ogawa, U Maeda. Adv Powder Technol 8:73, 1997.
54. K Kataoka, T Takigawa. A I Chem Eng J 27:504, 1981.
55. T Ogihara, G Matsuda, T Yanagawa, N Ogata, K Fujita, M Nomura. J Ceram Soc Jpn 103:151, 1995.

1.3 PHASE TRANSFORMATION FROM SOLID PRECURSORS

TADAO SUGIMOTO
Institute for Advanced Materials Processing, Tohoku University, Sendai, Japan

The most typical system in this category may be the ''gel-sol system'' recently developed for the preparation of general monodisperse particles in large quantities, in which highly viscous condensed gels are used as a solid precursor. The solid precursor in the form of a gel protects the product particles against coagulation by fixing them on the gel network and, at the same time, works as a reservoir of metal ions to be dissolved and release metal ions by degrees. If a solid precursor itself does not form a gel structure, some substances such as lyophilic polymers, surfactants, etc. are used as a subsidiary additive to form a gel-like structure. This idea is based on an earlier finding of selective formation of monodisperse particles on a precursory gel-like solid precipitated from a homogeneous dilute solution.

1.3.1 Dilute Systems

1.3.1.1 Cobalto-Cobaltic Oxide

Sugimoto and Matijević (1) prepared monodisperse cubic particles of cobalto-cobaltic oxide (Co_3O_4; 0.1–0.2 μm) with a spinel structure by hydrolysis of cobalt(II) with partial oxidation of Co^{2+} ions over 90°C through 100°C for several hours, starting from an aqueous solution of 10^{-2} mol dm^{-3} Co(II) acetate at pH 7.3. A green cobalt hydroxide gel precipitated at first and then the final particles formed theron. The oxidation of Co^{2+} ions was caused by oxygen dissolved in the solution, and thus it was greatly pronounced by bubbling oxygen or air. The counterions, acetate, worked as a pH buffer and also as a component of the precursor complexes to the Co_3O_4 particles (2). Too high pH over 8 gave no precipitation of Co_3O_4 on the $Co(OH)_2$ gel, so that a specific complex such as $CoAc^+$ may be responsible for the precipitation of the Co_3O_4 solid. The $Co(OH)_2$ gel served as a gel network to hold each Co_3O_4 particle to prevent coagulation and as a reservoir of Co^{2+} ions. This observation led to the idea of the gel-sol method later on. Figure 1.3.1 shows pH-dependent yields of Co_3O_4 and $Co(OH)_2{\cdot}nH_2O$ prepared under different atmospheres such as oxygen, air, and nitrogen. A transmission electron micrograph (TEM) of a carbon replica of the Co_3O_4 particles prepared in air is shown in Figure 1.3.2, wherein a considerable amount of amorphous $Co(OH)_2$ gel is found to remain.

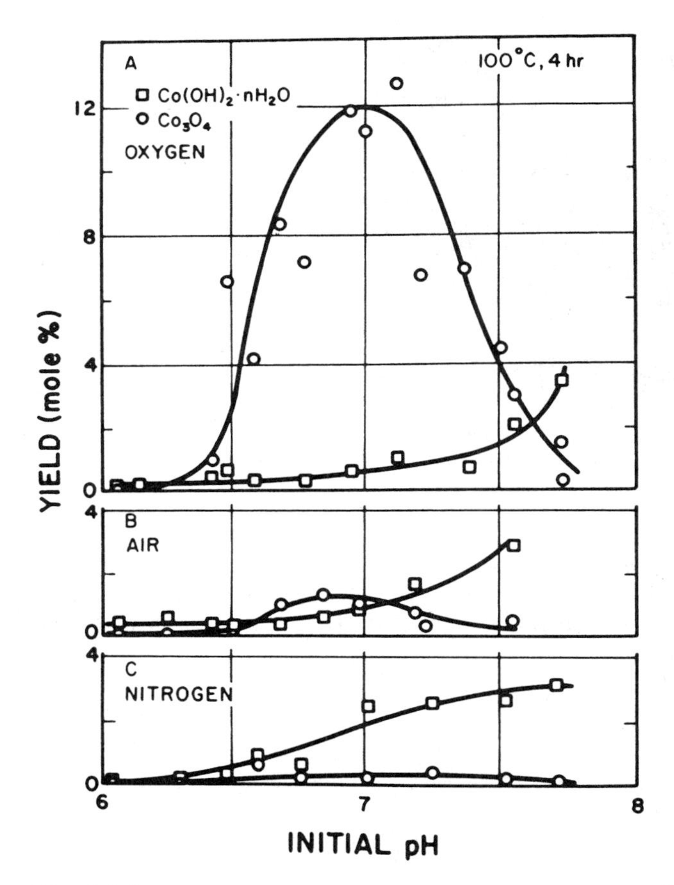

Fig. 1.3.1 Yields of Co_3O_4 and $Co(OH)_2 \cdot nH_2O$ in 1×10^{-2} mol dm^{-3} Co(II) acetate solution aged for 4 h at 100°C in oxygen (A), air (B), and nitrogen (C) as a function of initial pH. (From Ref. 1.)

1.3.1.2 Magnetite and Ferrites

Sugimoto and Matijević (3) prepared uniform spherical particles of magnetite (Fe_3O_4) as well by partial oxidation of ferrous hydroxide gel with nitrate. The uniform magnetite particles were obtained at a slight excess of Fe^{2+}, and the mean size critically depended on the excess concentration of Fe^{2+} or pH.

Figure 1.3.3 shows TEM images for the time evolution of the solid phase, revealing the formation process of the uniform magnetite particles. First, the $Fe(OH)_2$ gel precipitated on mixing ferrous sulfate with potassium hydroxide. Then the ferrous hydroxide gel shortly turned into irregular platelets holding extremely small magne-

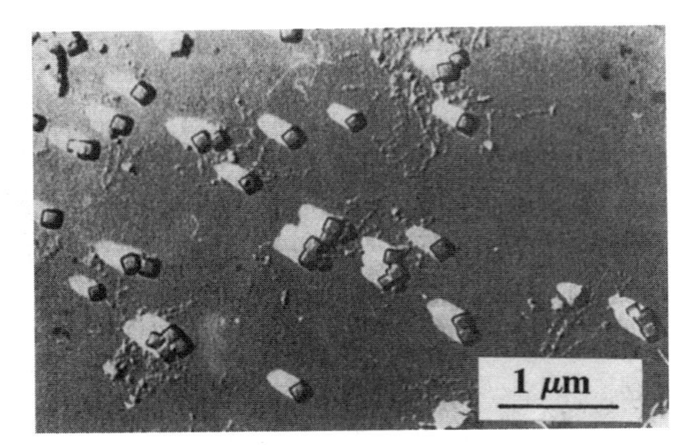

Fig. 1.3.2 TEM image of a carbon replica of monodispersed Co_3O_4 particles obtained on aging for 4 h at 100°C in 10^{-2} mol dm^{-3} Co(II) acetate solutions in air. (From Ref. 1.)

tite particles (<0.1 μm) thereon by partial oxidation on introduction of potassium nitrate as a mild oxidizing agent of ferrous ions at 90°C. The platelets of a dark green color are thought to be a partly oxidized $Fe(OH)_2$ called green rust (4). The phase transformation from the pure $Fe(OH)_2$ to the green rust finished within 15 min at 90°C, as revealed by radiochemical analysis (3). In this early stage, the fine magnetite particles increased in number whereas no appreciable growth took place without coagulation among them, owing to the gel-like substrate of the partly oxidized ferrous hydroxide. In the course of dissolution of the substrate with an accumulation of the primary particles of magnetite, they suddenly started coagulation to form clusters as the nuclei of the secondary particles consisting of a limited number of primary particles. The secondary nuclei promptly gathered the neighboring primary particles within the individual attraction fields, presumably by the magnetic attraction in addition to van der Waals forces at pH close to the isoelectric point. The residual gel substrate might prevent random coagulation among the isolated secondary particles to yield uniform spherical magnetite particles.

The coagulation of the primary particles appears to be against the general rules for the formation of monodisperse particles. However, if the clustering of the primary particles is regarded as the nucleation of the secondary particles, this system is still in compliance with the rules. In this case, the selective growth of the secondary particles seems to be due to their own increasing attractive forces of magnetism and van der Waals potential with their growth. As a consequence, the role of the remaining gel network is essential for separation of the growing secondary particles with a proper interparticle distance necessary for keeping them out of each potential field of attraction. Here it seems noteworthy that the nucleation and succeeding growth of the secondary particles are not necessarily synchronized among the individual secondary particles, unlike usual monodisperse systems, but the uniformity is

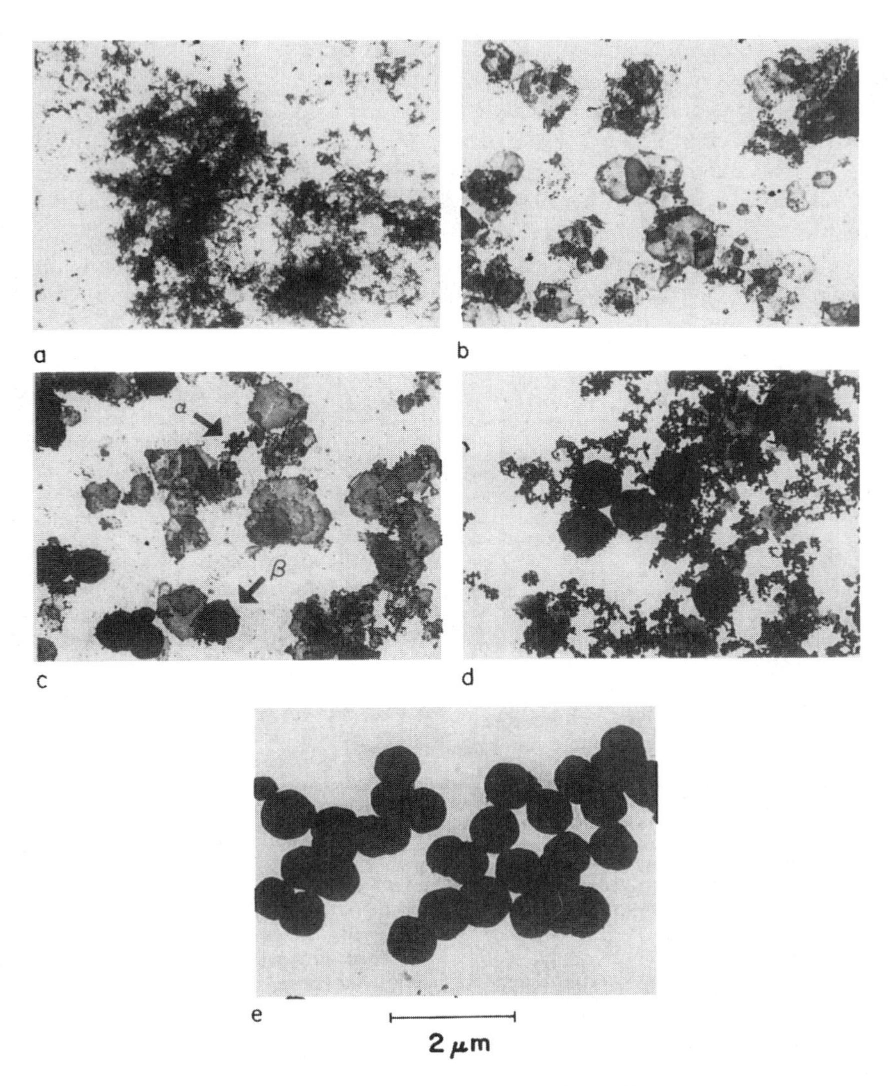

Fig. 1.3.3 TEM images for the time evolution of the solid phase in a system consisting of 2.5×10^{-2} mol dm^{-3} $Fe(OH)_2$, 5.0×10^{-3} mol dm^{-3} excess concentration of $FeSO_4$ aged at 90°C for (a) 0, (b) 15, (c) 30, (d) 45, and (e) 120 min. The arrows α and β in (c) indicate aggregates of primary particles in the relatively early and later stages, respectively. (From Ref. 3.)

achieved because the final size of each secondary particle is predetermined by the range of the attractive force to gather the surrounding primary particles.

The mean size of the magnetite particles was strongly dependent on the excess concentration of the ferrous ions or pH, ranging from 0.03 to 1.1 μm. The maximum particle size was found at pH ~6.7, and the size was dramatically lowered on either side of this pH. Since the specific pH is close to the isoelectric point of magnetite, the strong pH dependence of the particle size seems to be due to the drastic change in the repulsive force of the electric double layer about the isoelectric point between the growing secondary particles and the neighboring primary particles. Figure 1.3.4

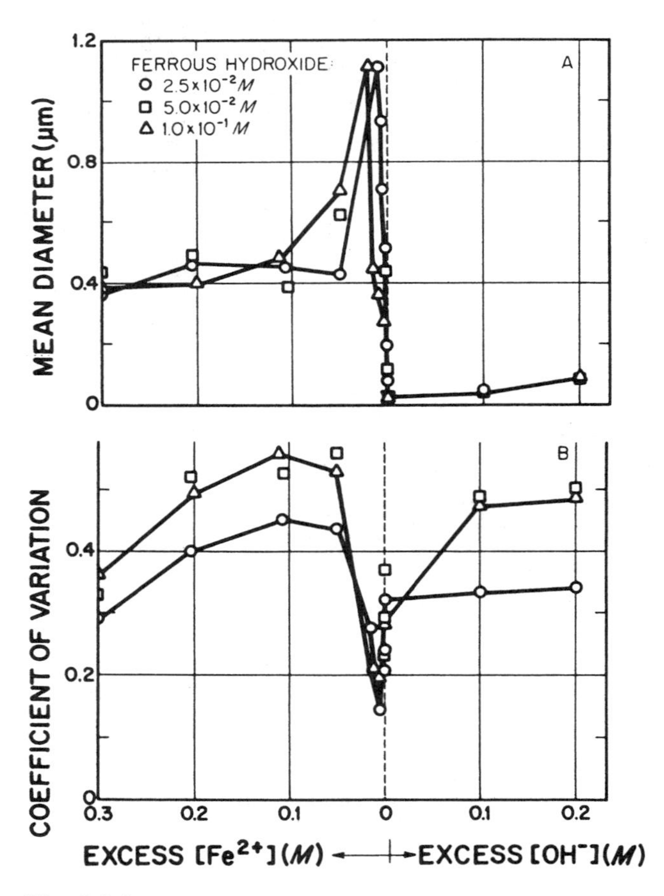

Fig. 1.3.4 Mean diameter of Fe_3O_4 particles (A) and coefficient of variation of their size distribution (B) as a function of excess concentrations of Fe^{2+} and OH^- at different contents of $Fe(OH)_2$ gel ($[KNO_3]$ = 0.2 *M;* temp. = 90°C; time = 4 h). (From Ref. 3.)

shows the effects of an excess Fe^{2+} ions and excess OH^- ions on the mean diameter and coefficient of variation of the size distribution with different ferrous hydroxide contents. Figure 1.3.5 shows scanning electron micrographs of spherical Fe_3O_4 particles prepared with different excess concentrations of $FeSO_4$.

It is no wonder that the particles are spherical but crystalline, if one considers the formation mechanism. The rather smooth surface of the spherical magnetite may be due to the rapid contact recrystallization of the constituent primary particles (5), forming the rigid polycrystalline structure. However, it must be noted that polycrystalline spheres are also prepared by normal deposition of monomeric solute, as shown in the formation of the uniform spherical polycrystalline particles of metal sulfides in Chapters 3.1–3.3. Thus, while we may be able to predict the final particle shape and structure from the formation mechanism, it is risky to conclude the formation mechanism only from characterization of the product. As a rule, scrupulous analyses are needed for concluding the growth mechanism in a particle system.

In a similar manner, uniform spherical particles of nickel ferrite ($NiFe_2O_4$) (6), cobalt ferrite ($CoFe_2O_4$) (7), and cobalt-nickel ferrite ($Co_xNi_{1-x}Fe_2O_4$) (8) were prepared as well.

1.3.1.3 Hematite

Precipitation of ferric hydroxide gel was also observed in the preparation of spindle-like hematite (α-Fe_2O_3) particles in a dilute ferric chloride solution in the presence of phosphate (9). In this case, however, the positive role of the gel was not definite since similar uniform hematite paricles were obtained as well in homogeneous systems in the presence of the same anions (9). Also, Hamada and Matijević (10) prepared uniform particles of pseudocubic hematite by hydrolysis of ferric chloride in aqueous solutions of alcohol (10–50%) at 100°C for several days. In this reaction, it was observed that acicular crystals of β-FeOOH precipitated first, and then they dissolved with formation of the pseudocubic particles of hematite. The intermediate β-FeOOH appears to work as a reservoir of the solute to maintain an ideal supersaturation for the nucleation and growth of the hematite. Since the β-FeOOH as an intermediate and the pseudocubic shape are not peculiar to the alcohol/water medium (11), alcohol may favor the uniform particles formation as a poorer solvent in terms of the dielectric constant to give rise to a single burst of nucleation of hematite, apart from the subsequent growth process. In fact, the concentration of ferric ions in the supernatant solution was as low as 5% of the initial concentration of ferric chloride (1.9×10^{-2} mol dm^{-3}) after the precipitation of β-FeOOH, while it was 50% in an alcohol-free aqueous solution of ferric chloride with a comparable initial concentration (2.0×10^{-2} mol dm^{-3}) (12). Since this system also starts from a dilute homogeneous solution, it can apparently be classified into the homogeneous systems, but it involves a nature of the heterogeneous systems in itself. The growth mechanism in such a system has been studied in detail by Sugimoto and Muramatsu (12). Particularly, they proved by a seeding analysis that hematite particles in such

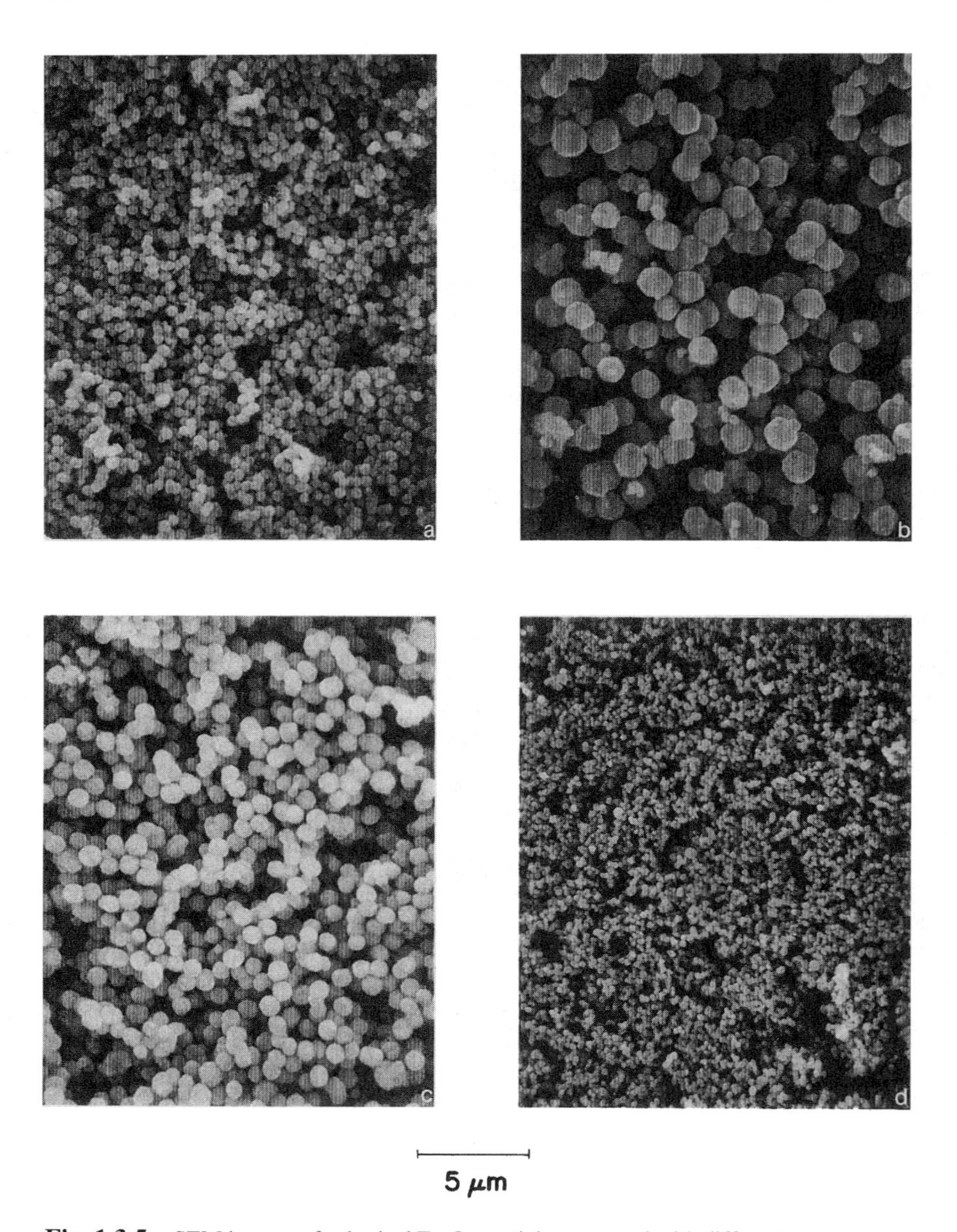

Fig. 1.3.5 SEM images of spherical Fe_3O_4 particles prepared with different excess concentrations of Fe^{2+}: (a) 3.0×10^{-1}, (b) 1.0×10^{-2}, (c) 5.0×10^{-3}, (d) 1.0×10^{-3} mol dm^{-3}, where $[Fe(OH)_2] = 2.5 \times 10^{-2}$ *M*, $[KNO_3] = 0.2$ *M*, temp. = 90°C, and time = 4 h. (From Ref. 3.)

a system are grown by deposition of monomeric species, and not by aggregation of preformed primary particles of hematite. The seeding analysis is based on a principle that if the growth of particles proceeds through aggregation of preformed primary particles, the overall reaction is not accelerated by addition of seeds, since seed particles do not enhance the formation of primary particles in the solution phase. On the other hand, if the particles are grown by deposition of monomeric species, the overall reaction must be accelerated by increase of the deposition area, as long as the rate-determining step is the deposition of the monomeric species. Figure 1.3.6 shows the dramatic effect of seeds on the reaction rate of the entire process (12).

1.3.2 Condensed Systems

In most cases, monodisperse particles are synthesized in dilute systems such as 10^{-4} to 10^{-2} mol dm^{-3} in concentration in order to overcome the essential problem of coagulation. The low productivity may be the most serious problem for general monodispersed particles to be used as industrial products despite their ideally controlled properties. To resolve this fundamental problem, a new general process named the ''gel-sol method'' has been invented (13–28). This method is based on the idea that if we make use of an extremely condensed precursor gel as a matrix of the subsequently generated product particles as well as a reservoir of the metal (and hydorxide) ions, it may be possible to prevent the coagulation of the particles by fixing them in the gel matrix even at a high concentration of electrolyte and keep them growing without renucleation at a moderate supersaturation by the constant release of the metal (and hydoxide) ions.

1.3.2.1 Hematite

Based on this expectation, the synthesis of monodisperse hematite (α-Fe_2O_3) particles as one of the most popular metal oxides was attempted first, and its exceedingly uniform pseudocubic particles were successfully obtained from a highly condensed ferric hydroxide gel (13,14). In a typical procedure, 5.4 mol dm^{-3} NaOH was added to the same volume of 2.0 mol dm^{-3} $FeCl_3$ in 10 min at room temperature under agitation, and the resulting highly viscous $Fe(OH)_3$ gel (pH ~2.0) was aged in an oven preheated at 100°C for 8 days. Through this simple procedure, monodisperse pseudocubic hematite particles, 1.6 μm in edge length, were produced with yield of nearly 100%.

As shown by the transmission electron micrographs in Figure 1.3.7, the initial $Fe(OH)_3$ gel (a) turned into very fine needle-like β-FeOOH crystals in a few hours at 100°C (b). After 1 day, very small but uniform pseudocubic particles were observed with the slightly grown β-FeOOH (c). They are α-Fe_2O_3 particles in their early stage of growth. With the progress of aging, the α-Fe_2O_3 particles continued to grow at the expense of the β-FeOOH particles until finally the β-FeOOH particles totally disappeared after 8 days. Hence, the α-Fe_2O_3 particles were formed through a two-step phase transformation of $Fe(OH)_3 \rightarrow \beta$-$FeOOH \rightarrow \alpha$-$Fe_2O_3$ by a dissolu-

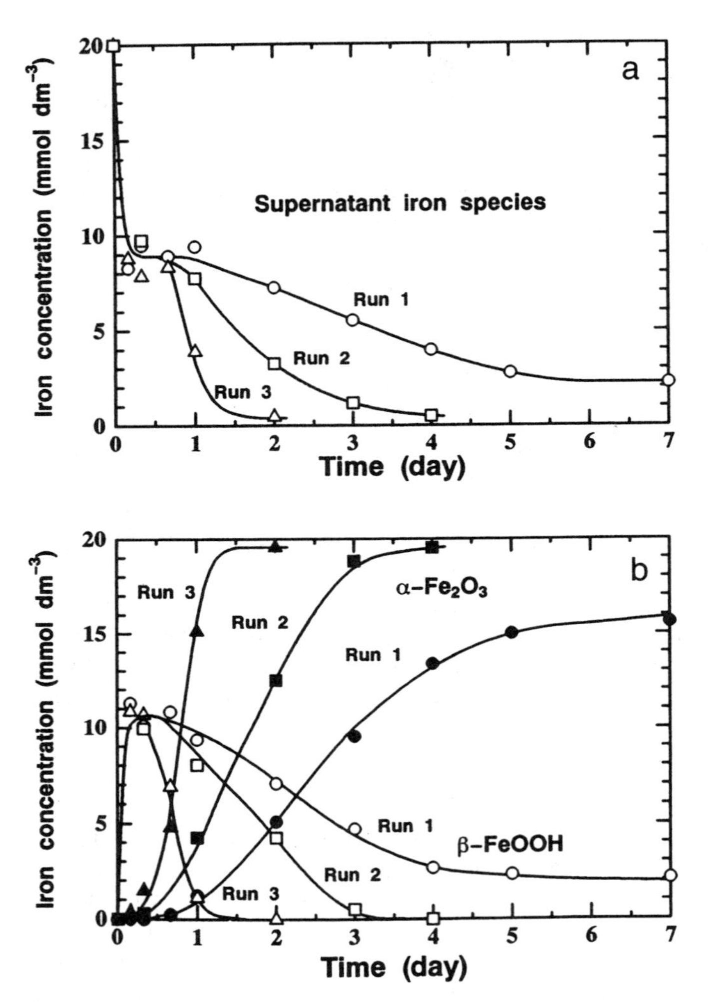

Fig. 1.3.6 (a) Changes of the supernatant iron concentrations in runs 1, 2, and 3 with different contents of the α-Fe_2O_3 seeds: 0 (○); 1.50×10^{-7} (□); 1.50×10^{-6} (△) mol dm^{-3}. (b) Changes of the concentrations of the β-FeOOH (○, □, △) and α-Fe_2O_3 (●, ■, ▲), where the circles, squares, and triangles correspond to runs 1, 2, and 3, respectively. (From Ref. 12.)

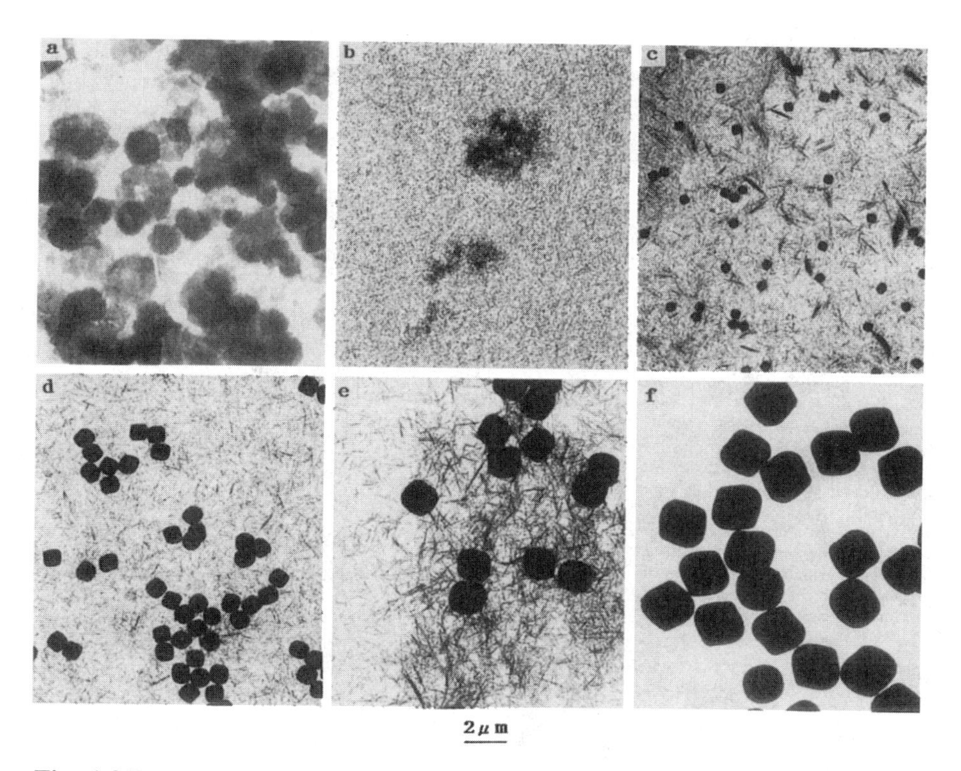

Fig. 1.3.7 Phase transformation and growth of the pseudocubic hematite particles with aging time under the standard conditions (temp. = 100°C, excess [Fe^{3+}] = 0.1 mol dm^{-3}, and [$Fe(OH)_3$] = 0.9 mol dm^{-3}): (a) 0 h, (b) 6 h, (c) 1 day, (d) 2 days, (e) 4 days, and (f) 8 days. (From Ref. 14.)

tion–recrystallization process, as clearly verified by the seeding analysis (14). In this phase transformation process, the acicular β-FeOOH also formed viscous network, preventing the coagulation of α-Fe_2O_3 particles.

On the other hand, it was found that the initial pH or excess concentration of Fe^{3+} ions had a strong effect on the final size of the product, as shown in Figure 1.3.8 (13). The drastic enhancement of nucleation of the hematite with increasing initial pH or decreasing excess concentration ferric ions suggests a dramatic increase of some specific ferric hydoxide complex as a precursor such as $(HO)_2FeOFe(OH)_2$ for the formation of hematite particles with increasing pH (29), since the supersaturation for the formation of hematite in terms of solubility product is identical for all pH due to the presence of $Fe(OH)_3$ gel.

Also, Figure 1.3.9 shows the evolution of the supernatant iron species, composition of the solid phase, and pH, measured at room temperature. The sharp drop

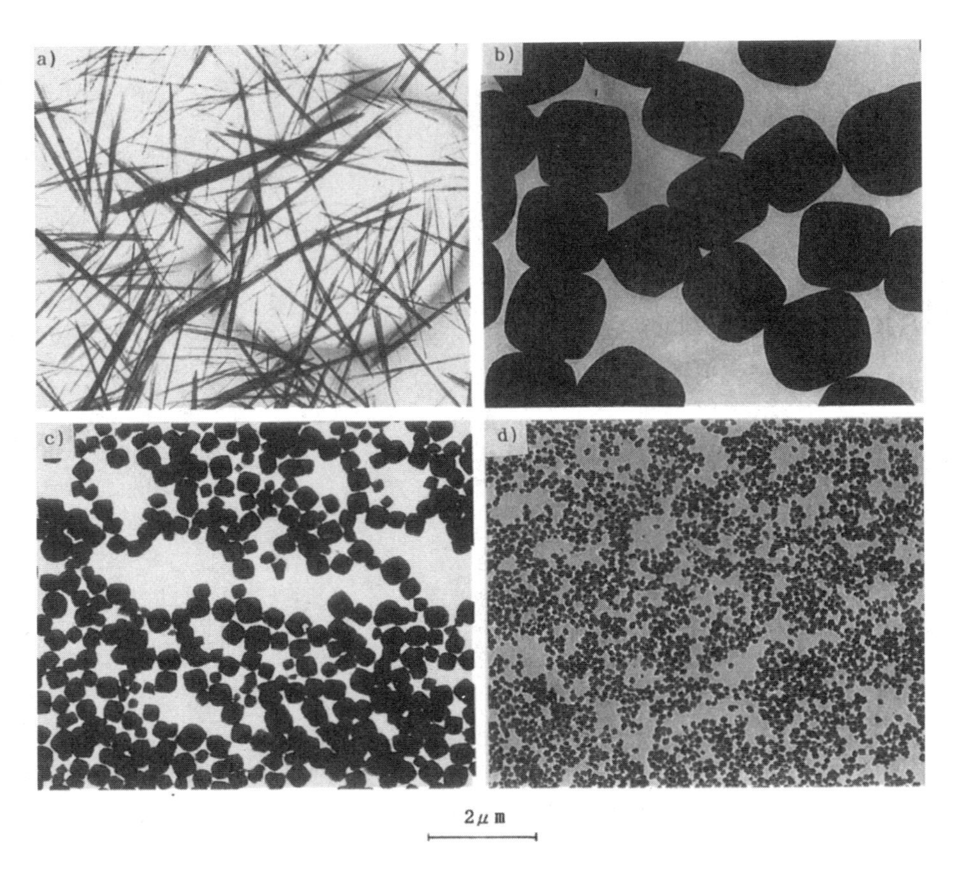

Fig. 1.3.8 TEM images of the particles of (a) β-FeOOH and (b)–(d) α-Fe_2O_3 prepared by aging gels or a $FeCl_3$ solution of x mol dm^{-3} Fe^{3+} + $(1 - x)$ mol dm^{-3} $Fe(OH)_3$ at 100°C for 8 days: x = (a) 1.00, (b) 0.10, (c) 0.05, or (d) 0.00. (From Ref. 13.)

of pH from 2.0 to 1.0 with the transformation of $Fe(OH)_3$ to β-FeOOH served to cease the nucleation of hematite, since it efficiently lowered the supersaturation of the precursor complex. The nucleation stage was found to be limited within a time range from the start of mixing of NaOH with the $FeCl_3$ solution to the end of the formation of β-FeOOH after ~3 h of aging at 100°C. Hence, the system fulfills all requirements for monodisperse particles formation; i.e., inhibition of random coagulation, separation between nucleation and growth stages, and reserve of monomers.

Size control of the hematite particles must be performed during this nucleation period by controlling the temperature or pH, or by adding very fine seeds of α-Fe_2O_3.

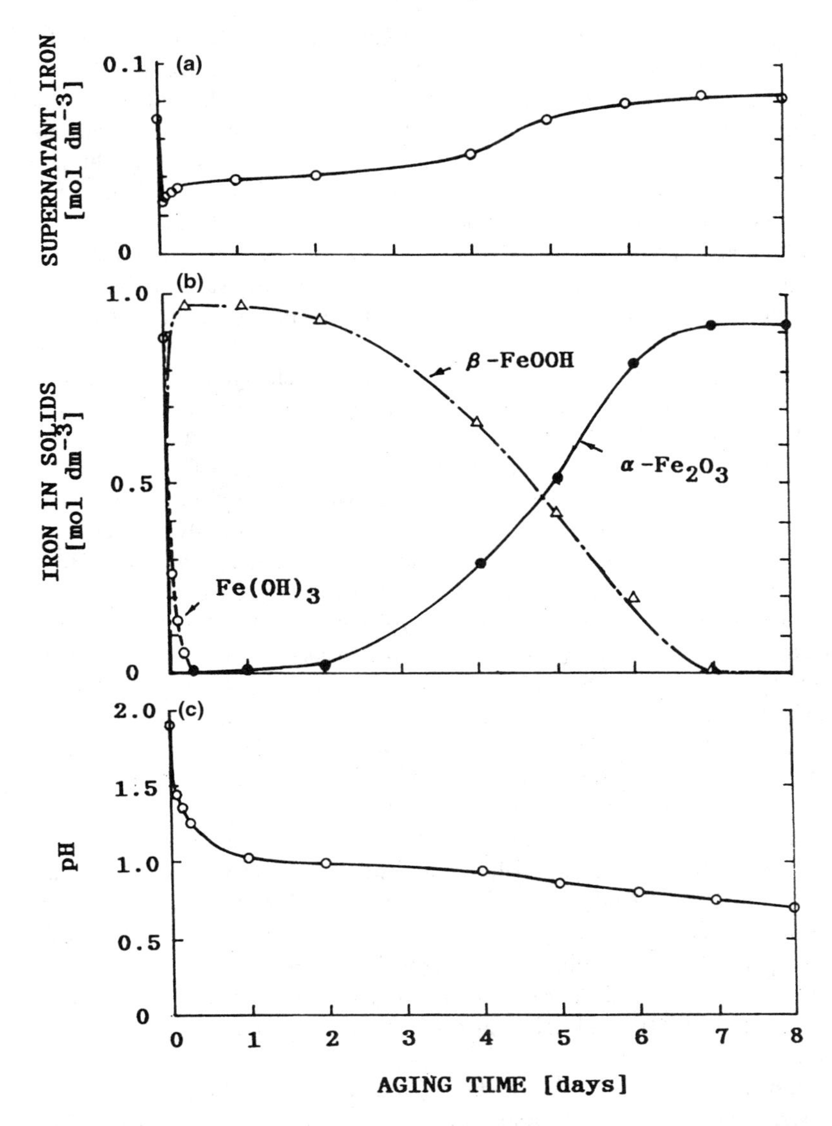

Fig. 1.3.9 Changes of the composition of iron species and pH with aging time, where all measurements were conducted at room temperature: (a) supernatant iron, (b) iron in solids, and (c) pH. (From Ref. 14.)

For example, the mean size can continuously be reduced from a few micrometers to ~0.3 μm without degrading the uniformity by varying the temperature from below room temperature to 100°C during the addition of 5.4 *N* NaOH to 2.0 *M* $FeCl_3$ (15). The same purpose can be achieved to some extent by raising the pH through decreasing the excess concentration of ferric ions, but in this case some size distribution broadening occurs due to an enhanced growth with the nucleation. Probably, the best way to control the particle size is the addition of ultrafine seeds to the $Fe(OH)_3$ gel under otherwise the standard conditions, because the simultaneous growth of the generated nuclei is not accelerated in this procedure. If we use very fine seeds of ~3 nm in this method, it is possible to extend the lower limit of the final size to 0.03 μm without degrading the monodispersity (15).

The shape of the hematite particles is dramatically changed from the pseudocube to an ellipsoid or peanut-like shape by addition of sulfate or phosphate ions to the $Fe(OH)_3$ gel (15–18). In contrast to these anisometric uniform particles prepared in acidic conditions (pH <2), monodisperse platelet-type α-Fe_2O_3 particles were obtained by aging highly condensed β-FeOOH particles suspended in a strong alkaline medium with 7.5 *N* NaOH at 70°C for 8 days (20). If we simply age $Fe(OH)_3$ gel in ordinary alkaline conditions, we have to raise the aging temperature to 150°C or higher to convert fairly stable α-FeOOH (goethite) particles generated as an intermediate into α-Fe_2O_3. In this case, only polydisperse tabular hematite particles are produced, since the high temperature raises the probability of renucleation of α-Fe_2O_3 during their growth. Scanning electron micrographs of the pseudocubic particles prepared under the standard conditions, ellipsoidal particles with 10^{-2} mol dm^{-3} SO_4^{2-}, peanut-type particles with 3×10^{-2} mol dm^{-3} SO_4^{2-}, and platelet particles in the strong alkaline condition are shown in Figure 1.3.10. The surface planes of the pseudocubic partilces were the {012} faces, the long axis of the ellipsoids and peanuts coincided with the *c*-axis of the hexagonal system, and the basal and side planes of the platelets were the {001} and {012} faces, respectively, as identified by the OPML-XRD (18). The anisotropic growth of these characteristic particles was explained by the specific adsorption of the anions or the organic additives to some particular faces, inhibiting the growth of these faces: i.e., Cl^- to the {012} faces; SO_4^{2-}, PO_4^{3-}, and the organic additives to the faces parallel to the c-axis, and OH^- to the {001} and {012} faces. Thus, these characteristic forms are reaction-controlled steady forms.

The anions such as Cl^-, SO_4^{2-}, PO_4^{3-}, and OH^- have a strong effect not only on the shape, but also on the internal structure. Figure 1.3.11 shows transmission electron micrographs of ultrathin sections of pseudocubic and peanut-type particles of hematite prepared with a ultramicrotome (21–23). Obviously, the pseudocubic and peanut-type particles are polycrystals consisting of much smaller subcrystals of a definite orientation. From the overall definite particle shape and the regular orientation of the subcrystals, it seems reasonable to consider that the overall growth mode of these polycrystalline particles is basically identical to ordinary monocrystalline particles in which two-dimensional surface nuclei are repeatedly generated epitaxi-

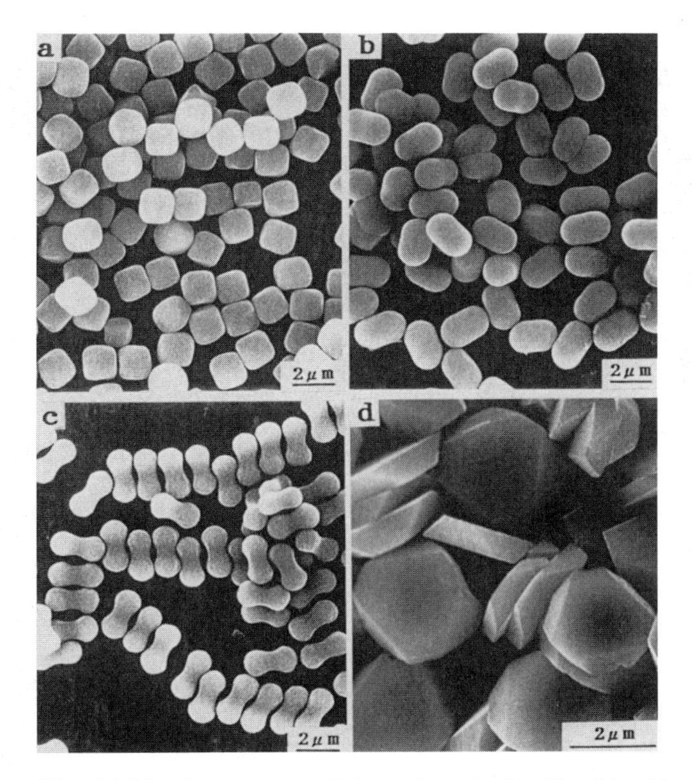

Fig. 1.3.10 SEM images of (a) pseudocubic, (b) ellipsoidal, (c) peanut-type, and (d) platelet-type hematite particles. The particles of (a), (b), and (c) were prepared under the same conditions as those of the particles in Fig. 1.3.7 but with 10^{-2} and 3.0×10^{-2} mol dm^{-3} Na_2SO_4 for (b) and (c), respectively. The platelet particles in (d) were prepared by aging a β-FeOOH suspension (~0.9 mol dm^{-3}) at 70°C for 8 days in a medium of 2 mol dm^{-3} NaCl and 7.5 mol dm^{-3} NaOH. (From Refs. 16 and 20.)

ally on the growing surfaces and laterally developed reflecting the lattice structure of the substrate crystal planes, but there must be some specific reason for the formation of the internal discontinuity. The cause of the internal discontinuity has been explained in terms of the blockage of mutual fusion of the developed surface nuclei by the strong adsorption of Cl^- and/or SO_4^{2-} ions to the side faces of the surface nuclei, leading to the internal grain boundaries as a result of the repetition of this process. On the other hand, the platelet particles were found to be single crystals from the close observation of their ultrathin sections (24). In addition, there were significant differences between the species of anions including Cl^-, SO_4^{2-}, and PO_4^{3-} in their effect on the internal structure, and considerable amounts of the added anions were found to be incorporated in the hematite particles (15,18).

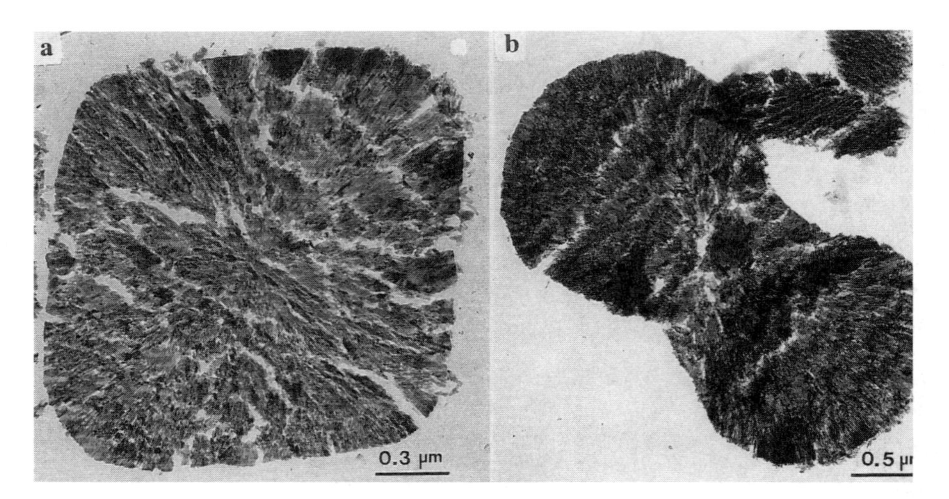

Fig. 1.3.11 Electron micrographs of the ultrathin sections of (a) pseudocubic and (b) peanut-type hematite particles. (From Ref. 21.)

As mentioned earlier, the hematite particles are grown by direct deposition of monomeric species through a dissolution–recrystallization mechanism, as shown by the seeding analysis in which the total reaction is dramatically accelerated by the presence of seeds (14). In this sense, the growth mechanism of hematite particles in the gel-sol system is exactly the same as that in a dilute homogeneous system. This fact suggests at the same time that the dissolution of β-FeOOH particles preceding to the deposition of the monomeric species onto hematite is not the rate-determining step of the entire process. In general, if the overall reaction is accelerated by addition of the seeds, the growth of the particles is controlled by the deposition of the monomeric species, and anyone of the preceding processes for the release of monomers from their reservoir, such as hydrolysis of metal alkoxides or metal ions, dissociation of metal complexes, dissolution of solid precursors, etc., is not the rate-determining step of the total reaction. This method can be applied to any kind of particulate systems, including homogeneous systems. Here it should be noted that if the overall reaction rate is limited by the release rate of the monomers from their reservoir, addition of seeds has no effect on the overall reaction rate, regardless of the growth mechanism. In other words, even if addition of seeds has no influence on the overall reaction rate, it does not necessarily mean the aggregative growth mechanism.

The seeding technique can also be used for controlling the size of the monodispersed hematite particles. If we combine seeding technique with a shape controller such as sulfate or phosphate ions, the size and shape with different aspect ratios of

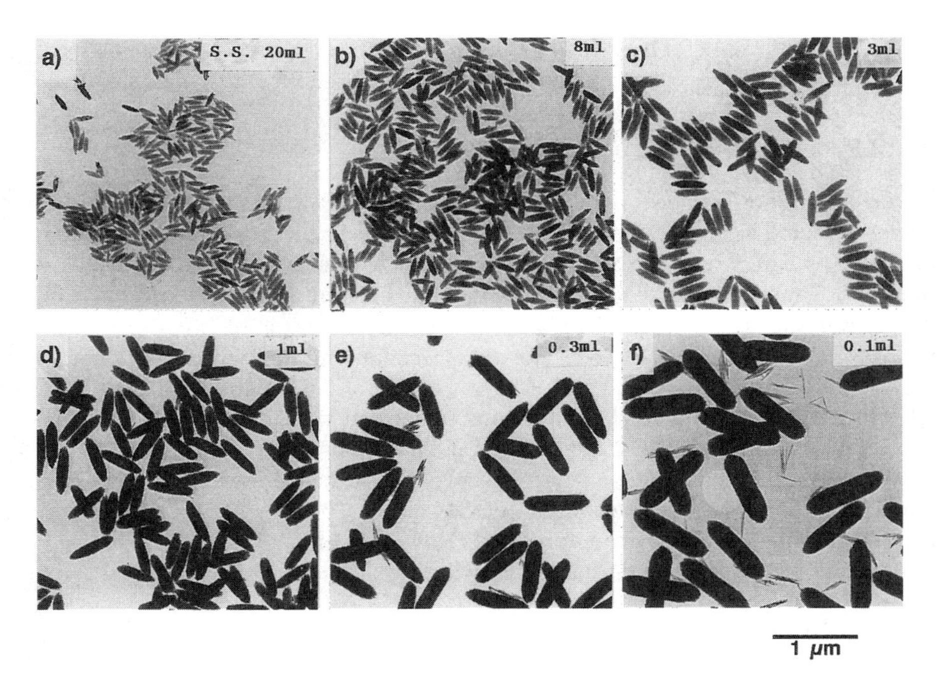

Fig. 1.3.12 TEM images of ellipsoidal hematite particles of different sizes prepared by the seeding technique at 100°C with different contents of seed suspension (S.S.) in the presence of a fixed concentration of sulfate ions at 3.0×10^{-2} mol dm^{-3}. The aging times were fixed at 2 days for (a), (b), and (c) and 3 days for (d), (e), and (f). (From Ref. 15.)

ellipsoid can be controlled independently in a wide range. Figure 1.3.12 shows an example of size control of monodispersed ellipsoidal hematite particles at a constant aspect ratio with 3.0×10^{-2} mol dm^{-3} Na_2SO_4 by variation of the amount of added seeds of 8.3 nm in mean diameter (15).

Shape control of hematite particles is also possible with organic additives such as organic phosphates (30); citrate (31); oxalate (32); dihydroxybenzene (19); dihydroxynaphthalene (19), ethylenediamine tetraacetic acid (EDTA) (19), and nitrilotriacetic acid (NTA) (19). These shape controllers produce ellipsoidal particles, except for citrate, which yields rod-like particles. Most of them are strongly adsorbed to the crystal planes parallel to the *c*-axis of the hexagonal crystal, yielding the anisotmetric particles elongated along the *c*-axis. Interestingly, only hydroquinone (*p*-dihydroxybenzene) and catechol (*o*-dihydroxybenzene) of dihydroxybenzenes are effective shape controllers, while resorcinol (*m*-dihydroxybenzene) is ineffective. This is because dihydroquinones that are originally not adsorbed to hematite can be adsorbed when they are oxydized to a quinone form by ferric ions and polymerized, which is possible only for the *p*- and *o*-dihydroxybenzenes. Most of the polymerized species of hydroquinone was found to be incorporated into the hematite particles.

1.3.2.2 Cobalt Ferrite

As an application of the use of β-FeOOH as a solid precursor, cobalt ferrite ($CoFe_2O_4$) particles of ~50 nm diameter with a narrow size distribution have been produced by aging a mixed solid precursors, β-FeOOH and β-$Co(OH)_2$, at a molar ratio of 2:1 in alkaline media (33). A concentrated NaOH solution (15 cm^3; 4 mol dm^{-3}) was added to 40 cm^3 of a highly condensed suspension of β-FeOOH (1.25 mol dm^{-3}) and β-$Co(OH)_2$ (0.625 mol dm^{-3}), containing small amounts of $FeCl_3$ (4.0×10^{-2} mol dm^{-3}) and $CoCl_2$ (2.0×10^{-2} mol dm^{-3}), to adjust the pH to 13, which was then aged at 40°C for 30 min to nucleate the cobalt ferrite. After the nucleation stage, the pH was lowered to 11 and the temperature was raised to 80°C, followed by aging at this temperature for a week to grow the ferrite particles. The arrangement of pH and the use of β-FeOOH and β-$Co(OH)_2$ of low solubilities contribute to the formation of uniform particles by preventing the renucleation and coagulation of the particles during the growth stage.

1.3.2.3 Titania

Spindle-like uniform particles of titania (TiO_2; anatase) have been prepared by the gel-sol method (26,27). Triethanolamine and titanium tetraisopropoxide were mixed in nitrogen atmosphere at a molar ratio 2:1, and then doubly distilled water was added to make a stock solution containing 0.50 mol dm^{-3} titanium ions. To the stock solution an equal volume of 2.0 mol dm^{-3} NH_3 was added at room temperature, and aged at 100°C for 24 h in a Pyrex culture tube (first aging), followed by aging in an autoclave at 140°C for 3 days (second aging). The triethanolamine was used as a stabilizer of titanium tetraisopropoxide to prevent hydrolysis at room temperature. The first aging was for the gel formation of titanium hydroxide, and the possibility of the nucleation of titania during the first aging was negligibly small because the final size and shape of the product were virtually unaffected by the difference in length of the first aging ranging from 8 to 24 h after the rigid gel had been formed. The nucleation of the TiO_2 seemed to be limited within the early stage of the second aging, including the time for the elevation of temperature to 140°C, when the concentration of titanium ions in the supernatant solution of the gel was drastically lowered at the same time due to the enhanced hydrolysis of the remaining titanium alkoxide complex of triethanolamine and incorporation into the gel network (see Fig. 1.3.13). After the rapid drop of the supernatant concentration of titanium ions, the slope of its reduction became very small, corresponding to the steady state in the growth stage of the titania with the dissolution of the titanium hydroxide gel. The electron micrographs in Figure 1.3.14 display the time evolution of the solid phase in the second aging.

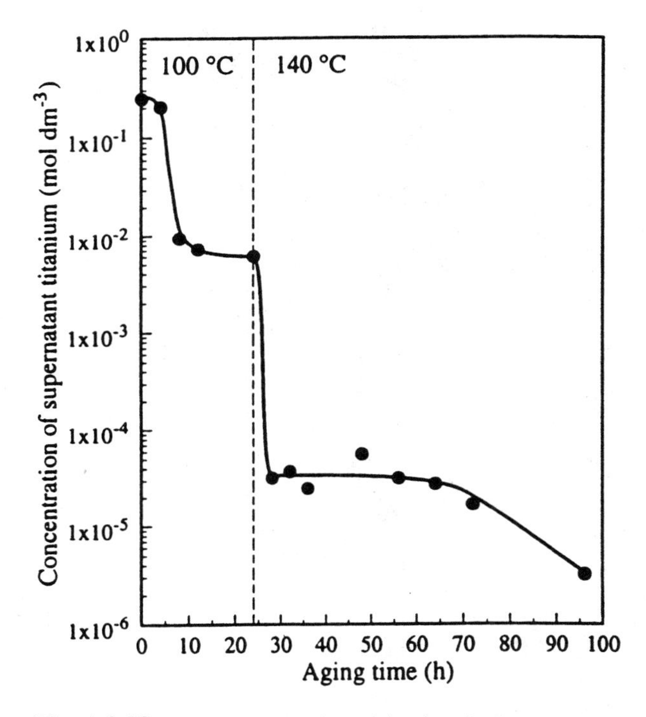

Fig. 1.3.13 The concentration of titanium in the supernatant solution as a function of aging time under the standard conditions. (From Ref. 27.)

When seed particles of titania were introduced into the gel-sol system before the start of the first aging, the reaction completed only in 1 day, in contrast to 3 days in the standard system without seeds. This result clearly demonstrates that the rate-determining step of the total growth reaction is not the dissolution process of the hydroxide gel, and that the titania particles are grown by the deposition of the solute and not by aggregation of preformed primary particles. Also, when the first aging for gelation was skipped, a drastic aggregation of the product was observed, showing the important role of the gel network as an anticoagulant.

1.3.2.4 Zirconia

Nanosized crystalline spheres of zirconia (ZrO_2) ranging from 2.5 to 15 nm in mean diameter with a narrow size distribution have been prepared by a method similar to that for the preparation of the uniform anatase titania (34). In a typical system, zirconium(IV) *n*-propoxide is mixed with triethanolamine at a molar ratio of 1 : 3 in a dry box filled with dry air to form a stable compound of Zr^{4+} against the rapid

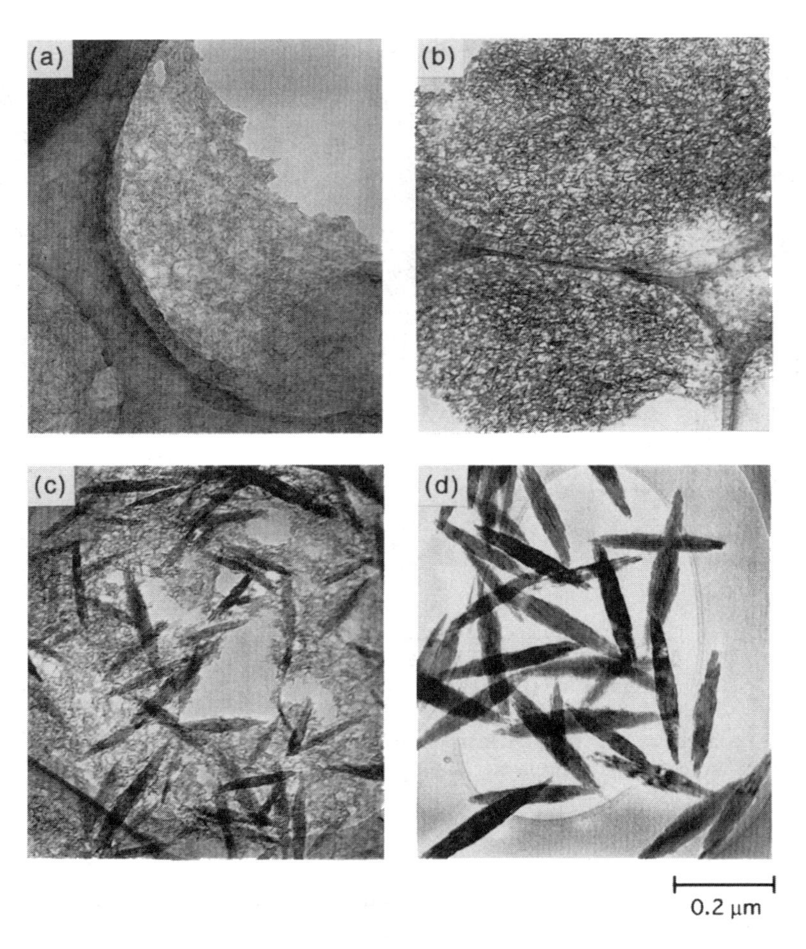

Fig. 1.3.14 Time evolution of the solid phase for the formation of monodisperse spindly TiO_2 particles (anatase) in the second aging of the gel-sol process: (a) after the first aging for 24 h at 100°C, but before the second aging at 140°C; (b) 1 day; (c) 2 days; (d) 3 days. (From Ref. 27.)

hydrolysis of zirconium alkoxide at room temperature. After aging the solution at room temperature for 24 h, doubly distilled water and NH_3 solution are added to the solution in order to make a solution of 0.5 mol dm^{-3} in Zr^{4+} and 1.0 mol dm^{-3} in ammonia (pH 10.7). In this stage no reaction takes place. The resulting solution is transferred into an autoclave and then aged at 200°C for 3 days without stirring in a preheated oil bath to nucleate and grow the zirconia particles. Thus obtained 15-nm particles are shown in an ordinary TEM and high-resolution TEM (HRTEM)

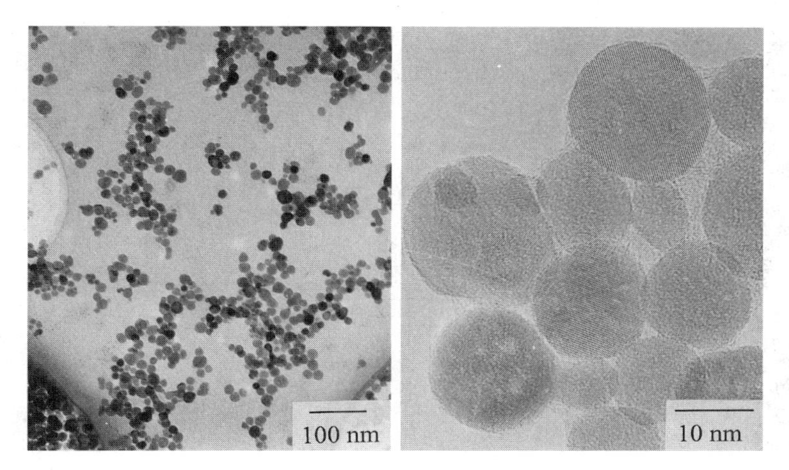

Fig. 1.3.15 Ordinary TEM and HRTEM of the tetragonal spheres of zirconia prepared by the gel-sol method under the standard conditions. (From Ref. 34.)

in Figure 1.3.15. In the HRTEM, one may find the lattice structure of each nanosphere, revealing that the spherical particles are all single crystals (tetragonal). If the 1.0 mol dm^{-3} ammonia is replaced by 0.81 mol dm^{-3} acetic acid and pH is adjusted to 7.1, 2.5-nm monoclinic zirconia particles are obtained. Since the stable crystal structure of large dimensions at low temperatures around 200°C is monoclinic, the formation of tetragonal nanoparticles may suggest that the specific surface energy of the tetragonal crystal of zirconia is lower than that of the monoclinic one. If this is the case, the restoration of monoclinic structure for the nanoparticles grown in the presence of acetate seems to be due to the reduction of the high specific surface energy of the monoclinic particles by adsorption of acetate. Both of them, whether doped or undoped with yttria, showed excellent sintering performances: low sintering temperatures and high sintering densities.

1.3.2.5 Cupric Oxide

Monodisperse leaflet-like cupric oxide (CuO) particles are prepared by addition of 0.80 mol dm^{-3} NaOH to the same volume of 0.40 mol dm^{-3} $Cu(NO_3)_2$ at 40°C and subsequent aging at the same temperature for 6 h with stirring. Figure 1.3.16 shows a transmission electron micrograph of thus prepared leaflet-like CuO particles, which are grown by direct deposition of the monomeric solute furnished through dissolution of a condensed needle-like cupric hydroxide precursor, as has been verified by Fourier transform infrared (FTIR) and seeding analyses (28). Figure 1.3.17 exhibits the effect of seeding in the standard system on the reaction rate for the formation of CuO particles from $Cu(OH)_2$ gel, with 0.20 mol dm^{-3} CuO seeds

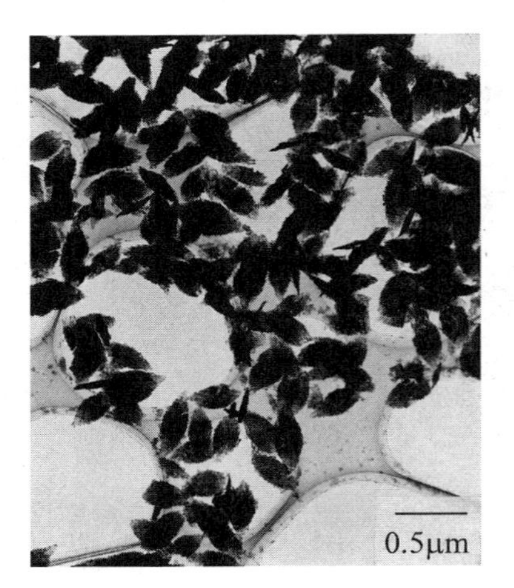

Fig. 1.3.16 Monodisperse leaflet-like CuO particles prepared by phase transformation from 0.20 mol dm^{-3} $Cu(OH)_2$ precursor. (From Ref. 28.)

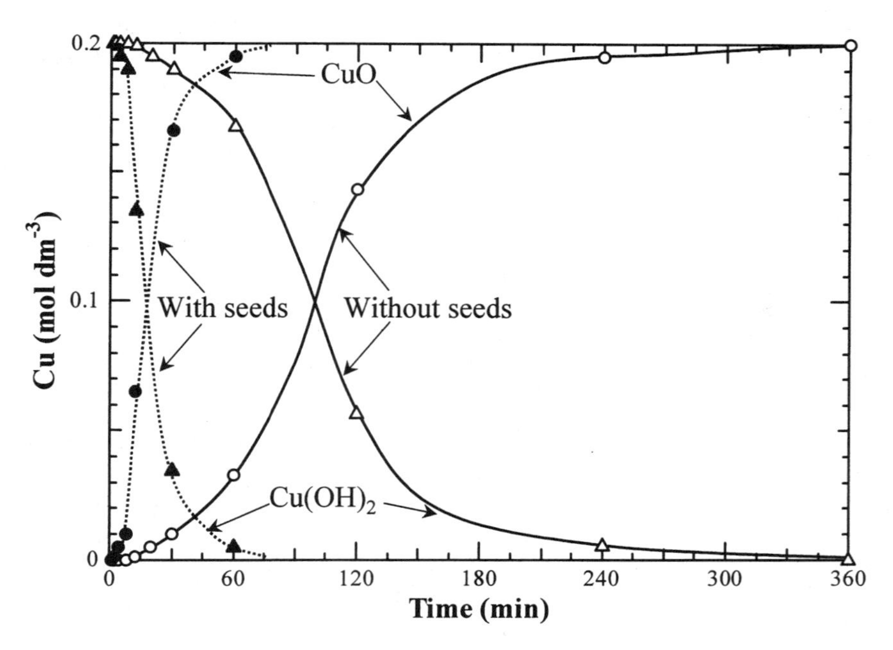

Fig. 1.3.17 The seed effect on the reaction rate of phase transformation from 0.20 mol dm^{-3} $Cu(OH)_2$ in the presence or absence of 0.20 mol dm^{-3} CuO seeds. (From Ref. 28.)

prepared by the standard process. The composition and shape of the copper oxide particles are identical to those prepared earlier by Lee et al. in a dilute controlled double-jet system in which cupric nitrate and sodium hydroxide solutions were introduced simultaneously into a large volume of water, followed by aging of the precipitate (35). The growth mechanism of the CuO particles in this system is probably the same as that in the gel-sol system. In fact, the dramatic effect of seeds in acceleration of the total reaction was observed in a dilute system as well, in which 2.86×10^{-2} mol dm^{-3} NaOH was added to the same volume of 1.43×10^{-2} mol dm^{-3} $Cu(NO_3)_2$ and aged at 40°C for 1 h, in the absence or presence of 7.15×10^{-3} or 1.43×10^{-2} mol dm^{-3} CuO seeds (28). The final concentrations of starting materials in this dilute system with no seeds are equivalent to those in the double-jet system.

1.3.2.6 Cuprous Oxide

Uniform spherical particles of cuprous oxide (Cu_2O; polycrystals; 0.27 μm in mean diameter) were prepared by aging a 0.5 mol dm^{-3} CuO suspension containing 0.5

Fig. 1.3.18 SEM images of CuO precursor particles (a) and uniform polycrystalline spheres of Cu_2O (b) prepared therefrom by phase transformation. (From Ref. 36.)

mol dm^{-3} N_2H_4 (hydrazine) and 3 wt% deionized gelatin for 3 h at 30°C under constant agitation (36). The initial pH was adjusted to 9.3 ± 0.1 at 30°C. Figure 1.3.18 shows SEM images of the starting material CuO and the product Cu_2O. Figure 1.3.19 shows TEM images revealing the phase transformation process from CuO to Cu_2O. Use of amorphous $Cu(OH)_2$ as a reservoir of copper ions in place of the crystalline CuO yielded polydispersed particles due to the too high solubility of $Cu(OH)_2$, causing renucleation of Cu_2O during its growth. Thus, the choice of solid reservoirs in terms of solubility and dissolution rate is generally important for controlling the supersaturation in the formation of uniform particles by the gel-sol method, like the choice of chelating agents in the formation of uniform metal sulfide particles in chelate systems. The role of the gelatin is the same as those in the CdS system in Chapter 3.3. On the other hand, Figure 1.3.20 shows the changes of the concentration of Cu^{2+} ions, pH, and total concentration of soluble Cu species in the solution phase in a standard system but without gelatin. The initial increase in pH after the introduction of hydrazine is due to the dissolution of the CuO solid caused

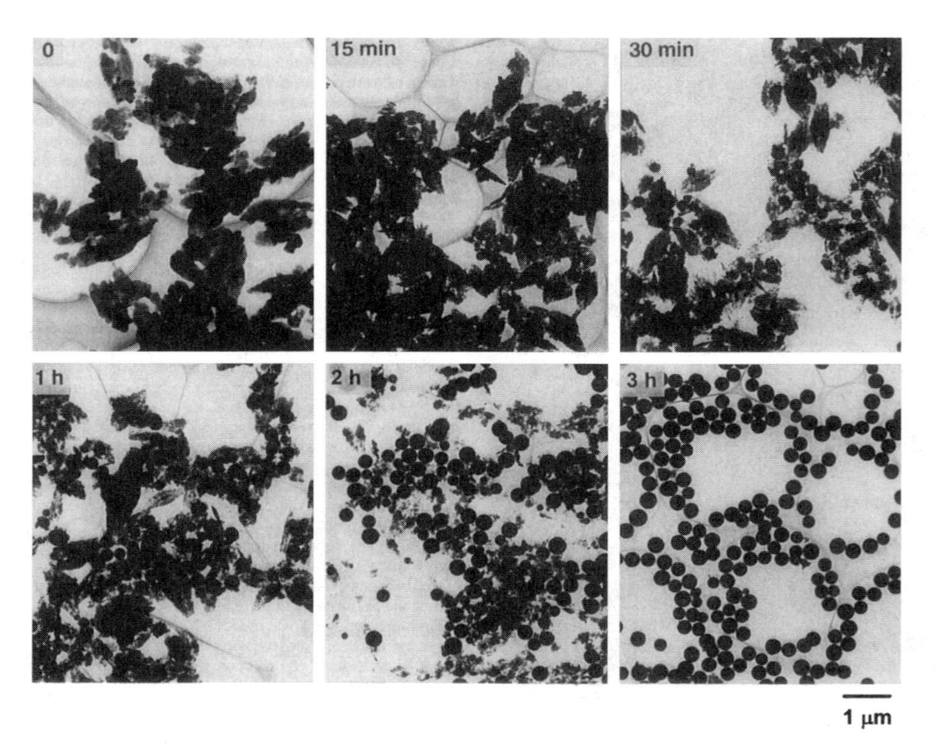

Fig. 1.3.19 TEM images showing the phase-transformation process from CuO to Cu_2O through dissolution of CuO. (From Ref. 36.)

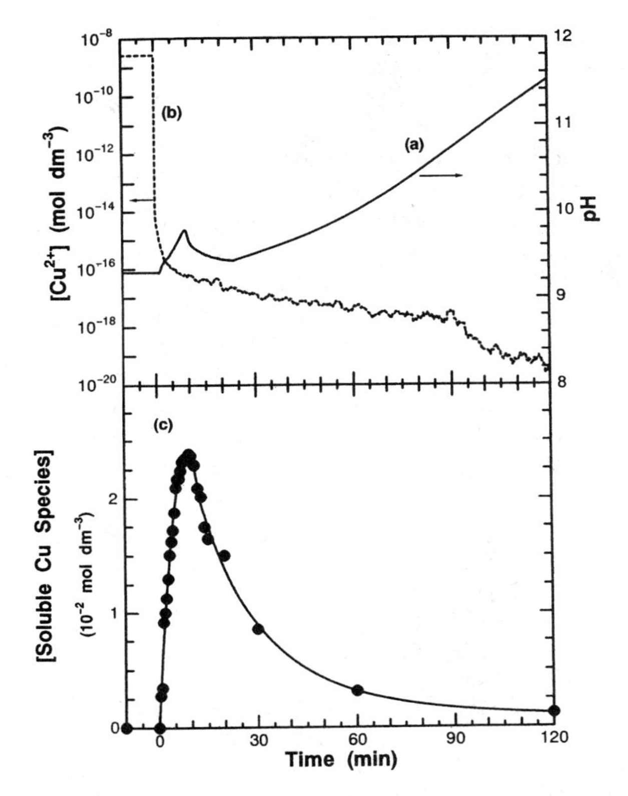

Fig. 1.3.20 Changes in pH (a), concentration of free Cu^{2+} (b), and total concentration of soluble Cu species (c) with aging time in the absence of gelatin under otherwise standard conditions. (From Ref. 36.)

by the sharp drop of $[Cu^{2+}]$ with complexation of N_2H_4 with Cu^{2+} in the solution phase initially in equilibrium with CuO. The subsequent drop of pH after a peak corresponds to the reduction of Cu^{2+} to Cu^+, probably through electron transfer from the coordinated N_2H_4 ligands, leading to the nucleation and growth of the Cu_2O particles, as revealed by the corresponding drop of the concentration of the soluble Cu species. The final reincrease of pH is due to the formation of ammonia with the oxidation of N_2H_4 of a basicity lower than NH_3. It was found from quantitative analysis of the by-products that 38% of N_2H_4 was consumed by Eq. (1) and 15% by Eq. (2).

$$N_2H_4 + CuO \rightarrow \tfrac{1}{2}Cu_2O + \tfrac{1}{2}N_2 + NH_3 + \tfrac{1}{2}H_2O \tag{1}$$

$$N_2H_4 + 4CuO \rightarrow 2Cu_2O + N_2 + 2H_2O \tag{2}$$

In this system, strict control of the initial pH is essential for forming the uniform Cu_2O particles. When the initial pH was below 9.0, the rate-determining step for the growth of Cu_2O shifted from the dissolution process of CuO to the deposition process of Cu_2O monomers due to the lowered reducing activity of hydrazine, so that the size distribution of the product became broad due to the lifted supersaturation above the critical level, leading to renucleation of Cu_2O. Since it is unlikely that the slight pH difference strongly affects the selectivity in aggregation of primary particles to the growing secondary particles, and since there was no characteristic particle group to be regarded as primary particles, there seems to be no possibility of the aggregative growth mechanism. In contrast to the formation of the monocrystalline Cu_2O particles of different crystal habits such as cubes, cuboctahedra, and octahedra produced by reduction of copper(II) tartrate with glucose (37), the nonepitaxial growth of the spherical polycrystalline Cu_2O particles may be brought about by the adsorbed N_2H_4 on the surface of the Cu_2O particles, blocking the rearrangement of the Cu_2O molecules for the epitaxial nucleation on the surfaces.

REFERENCES TO SECTION 1.3

1. T Sugimoto, E Matijević. J Inorg Nucl Chem 41:165, 1979.
2. K Bridger, RD Patel, E Matijević. Colloid Surf 13:145, 1985.
3. T Sugimoto, E Matijević. J Colloid Interface Sci 74:227, 1980.
4. W Feitknecht. Z Elektrochem 63:34, 1959.
5. T Sugimoto, G Yamaguchi. J Phys Chem 80:1579, 1976; T Sugimoto, G Yamaguchi. J Crystal Growth 34:253, 1976.
6. AE Reganzzoni, E Matijević. Corrosion 38:212, 1982.
7. H Tamura, E Matijević. J Colloid Interface Sci 90:100, 1982.
8. AE Reganzzoni, E Matijević. Colloids Surf 6:189, 1983.
9. M Ozaki, S Kratohvil, E Matijević. J Colloid Interface Sci 102:146, 1984.
10. S Hamada, E Matijević. J Chem Soc Faraday Trans I 78:2147, 1982.
11. E Matijević, P Scheiner. J Colloid Interface Sci 63:509, 1978.
12. T Sugimoto, A Muramatsu. J Colloid Interface Sci 184:626, 1996.
13. T Sugimoto, K Sakata. J Colloid Interface Sci 152:587, 1992.
14. T Sugimoto, K Sakata, A Muramatsu. J Colloid Interface Sci 159:372, 1993.
15. T Sugimoto, Y Wang, A Muramatsu. Colloids Surfaces A Physicochem Eng Aspects 134:265, 1998.
16. T Sugimoto, MM Kahn, A Muramatsu. Colloids Surfaces A Physicochem Eng Aspects 70:167, 1993.
17. T Sugimoto, MM Kahn, A Muramatsu. Colloids Surfaces A Physicochem Eng Aspects 79:233, 1993.
18. T Sugimoto, A Muramatsu, K Sakata, D Shindo. J Colloid Interface Sci 158:420, 1993.
19. T Sugimoto, H Itoh, T Mochida. J Colloid Interface Sci 205:42, 1998.
20. T Sugimoto, S Waki, H Itoh, A Muramatsu. Colloids Surfaces A Physicochem Eng Aspects 109:155, 1996.
21. D Shindo, S Aita, G Park, T Sugimoto. Mater Trans JIM 34:1226, 1993.
22. G Park, D Shindo, Y Waseda, T Sugimoto. J Colloid Interface Sci 177:198, 1996.

23. D Shindo, G Park, Y Waseda, T Sugimoto. J Colloid Interface Sci 168:478, 1994.
24. D Shindo, B Lee, Y Waseda, A Muramatsu, T Sugimoto. Mater Trans JIM 34:580, 1993.
25. T Sugimoto, H Itoh, H Miyake. J Colloid Interface Sci 188:101, 1997.
26. T Sugimoto, M Okada, H Itoh. J Colloid Interface Sci 193:140, 1997.
27. T Sugimoto, M Okada, H Itoh. J Disp Sci Tech 19:143, 1998.
28. A Muramatsu, T Sugimoto. Proc 51st Symp Colloid Interface Chem, Chem Soc Jpn., Chiba, 1998. p 144.
29. T Sugimoto, Y Suzuki, T Kato, A Muramatsu. Proc 51st Symp Colloid Interface Chem, Chem Soc Jpn, Chiba, 1998, p 242.
30. S Matsumoto. Shikizai 57:602, 1984; S Matsumoto. Kesshoh Seichoh Gakkaishi 16: 191, 1989.
31. U Schwertmann, WR Fischer, H Papendorf. Trans 9th Int Congr Soil Sci 1:645, 1968.
32. WR Fischer, U Schwertmann. Clays Clay Miner 23:33, 1975.
33. T Sugimoto, Y Shimotsuma, H Itoh. Powder Tech 96:85, 1998.
34. T Sugimoto, S Waki, A Muramatsu. Proc 51st Symp Colloid Interface Chem, Chem Soc Jpn, Chiba, 1998, p 245.
35. S Lee, Y Her, E Matijević. J Colloid Interface Sci 186:193, 1997.
36. A Muramatsu, T Sugimoto. J Colloid Interface Sci 189:167, 1997.
37. P McFadyen, E Matijević. J Colloid Interface Sci 44:95, 1973.

1.4 REACTION IN MICROEMULSIONS

KIJIRO KON-NO
Science University of Tokyo, Tokyo, Japan

Monodisperse and spherical Fe_3O_4 particles of 3–5 nm in diameter were synthesized by Kon-no and coworkers (1–3) and López-Quintela and Rivas (4,5). The particles can be produced easily by mixing of Aerosol OT solution containing aqueous $FeCl_3$ and polyoxyethylene(6) nonylphenyl ether solution containing aqueous $FeCl_2$ and ammonia, or by addition of aqueous $FeCl_2$ into Aerosol OT solution containing aqueous $FeCl_3$ and ammonia (1–3). The particles showed that a typical superparamagnetic behavior due to very small size and the magnetic property measurements yielded a coercive of 12 Oe and a saturation magnetization of 4100 G, respectively. The dispersion of the particles was significantly stable for long time and hence showed a fluid magnetic behavior after the solvent was evaporated (1–3). Microcrystalline YBaCuO, $BaFe_{12}O_{19}$, and Fe_2O_3 particles were synthesized by Shah and coworkers (6–8). YBaCuO particles were produced by calcination of the corresponding oxalate particles, which were achieved by the precipitation of Y, Ba, and Cu nitrates with ammonium oxalate in cetyltrimethyl ammonium bromide-1-butanol–*n*-octane systems (6). The particles were a high-density $YBa_2Cu_3O_{7-x}$ with narrow size distribution and showed a strong Meissner signal and superconducting transition temperature of 93 K. Same particles were also synthesized in order microemulsions by López-Quintela and Rivas (4). $BaFe_{12}O_{19}$ particles were also synthesized by the calcination at 950°C for 12 h of barium-iron carbonate precipitates obtained by mixing two W/O microemulsions of aqueous $(NH_4)_2CO_3$ and a mixture of aqueous barium nitrate and ferric nitrate (7). The precursor carbonates were fairly monodisperse particles with a size range of 5–15 nm, and by calcination they changed to hexagonal barium ferrite with a size range of 50–100 nm. The magnetic property measurements of the powder yielded an instrinsic coercivity of 5397 Oe and a saturation magnetization, lower than the values theoretically calculated (9,10). The discrepancy between experimental and calculated values may be due to reverse domain nucleation and poor crystallization of the particles. On the other hand, Fe_2O_3 particles produced by the calcination of $Fe(OH)_3$, precipitated by adding NH_4OH into microemulsified aqueous $Fe(NO_3)_3$, by sorbitane monooleate in the 2-ethylhexanol–hydrocarbon system, were found to consist of spherical α-Fe_2O_3, somewhat elongated γ-Fe_2O_3, amorphous Fe_2O_3 fine particles depending on the concentration of the precursor $Fe(NO_3)_3$ (8). The α- and γ-Fe_2O_3 particles were found at a particle size of 30 nm. In 0.312% $Fe(NO_3)_3$ solution, amorphous Fe_2O_3 particles of 5 nm in size were produced and were crystallized to γ-Fe_2O_3 at 290°C; further structural phase transition from γ- to α-Fe_2O_3 was observed at 400°C. This transition was irreversible. $LaNiO_3$ particles can be prepared by the calcination of the corresponding oxalate precursors achieved by the precipitation of lanthanum and nickel nitrates with oxalic acid in polyoxyethylene(5) nonylphenyl ether–petroleum ether systems (11). The precursors were spherical and uniform particles with an average size of about 20 nm in diameter. Spherical and single-phase $LaNiO_3$ powder of about 2 μm was formed when the nanoprecursor of 1 : 1 molar ratio of La to Ni was calcinated at 800°C for 20 h. However, the powder was an aggregates consisting of fine particles of about 0.2 μm in diameter and exhibited a paramagnetic susceptibility of about 5.6×10^{-6} cgs units/g at 20°C.

Unlike the synthesis of metal oxide particles by calcination of precursors mentioned earlier, other metal oxide particles such as SiO_2 and ZrO_2 have been synthesized by the hydrolysis of the corresponding metal alcoxides in nonionic and ionic reversed micelle systems. Since spherical and uniform SiO_2 particles by the ammonia-catalyzed hydrolysis of tetraethylorthosilicate (TEOS) in nonionic and ionic reversed micelles were synthesized by Kon-no and coworkers (12), the kinetics of TEOS hydrolysis, the growth mechanism of the particles, and the factors affecting the particle size and their dispersion stability have been examined by many workers. The formation of SiO_2 particles takes 1 weeks after TEOS is added into the reaction systems. Recently, by using Fourier transform infrared spectroscopy, transmission electron microscopy, and light scattering techniques, the kinetics of TEOS hydrolysis and growth mechanism of SiO_2 particles in polyoxyethylene(4) nonylphenyl ether–heptane systems were studied in detail by Chang and Fogler (13). In Fourier Transform infrared spectra measured over the TEOS hydrolysis period, the absorption intensity of the Si-O-C stretching band of TEOS located at 967 cm^{-1}, i.e., the concentration of TEOS, decreased exponentially with reaction time, whereas that of ethanol produced by the reaction increased approximately four times greater than the decrease in initial TEOS concentration and that of water decreased approximately twofold. This result indicates the overall reaction holds stoichiometrically in the TEOS hydrolysis process. The apparent rate constant k_h, for TEOS hydrolysis with different concentrations of water, ammonia, and surfactant was determined from the slopes of straight lines for the plots of absorption intensity of TEOS band at 967 cm^{-1} versus reaction time. Consequently, the values of k_h obtained increased approximately linearly with the aqueous ammonia concentration, indicating that the rate of TEOS hydrolysis in the systems is approximately first order with respect to the concentration of aqueous ammonia. A similar reaction order could be also estimated in respect with TEOS concentration in *n*-heptane solutions of surfactants and other polyoxyethylene alkylphenyl ethers–*n*-hepatane systems (14), but that with respect to water concentration was approximately zero order, suggesting that the water does not significantly affect the rate of hydrolysis. However, the reaction with respect to the surfactant concentration was approximately one-half order, suggesting that the role of surfactant in the rate of TEOS hydrolysis is significantly complicated. Probably, the role of surfactant in the hydrolysis may be to expedite the molecular contact

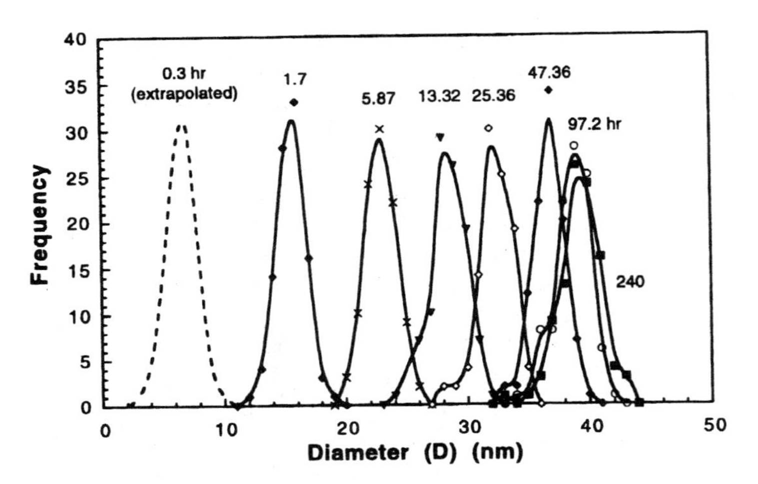

Fig. 1.4.1 Size distribution of silica particles grown in a microemulsion during TEOS hydrolysis under conditions of 0.126 *M* surfactant, 0.0357 *M* TEOS, 0.104 *M* NH_3, and 0.235 *M* H_2O. Figures above each distribution profiles express reaction time (h). The dashed profile is an extrapolated distribution achievable at the earliest reaction time by assuming a size-independent particle growth.

between the hydroxy ion and TEOS, which diffuses into the water droplets from the bulk oil phase. In addition, increasing the surfactant concentration may also enlarge the area of the micelle droplet surface and may favor a higher rate of mass transfer of TEOS into the droplets. The growth mechanism of SiO_2 particles was examined by transmission electron microscopy and light scattering techniques. When the changes of SiO_2 particle size with reaction time are measured by transmission electron microscopy, as seen in Figure 1.4.1, the SiO_2 size distribution shape and standard deviation are found to be preserved over the reaction period. This result indicates that the growth of SiO_2 particles follows a reaction-controlled behavior in the sense in which individual particle sizes grow at the same velocity. On the other hand, for diffusion-controlled growth, neither the particle-size distribution shape nor the standard deviation is preserved (15). This size-independent growth indicates that the exchange of hydrolyzed silica reacting species among micelle droplets is sufficiently fast compared with the condensation of reacting species onto the growing particles. As a result, the reacting species maintain an even distribution over all droplets irrespective of the size of the growing particles therein, and condense onto silica-particle surfaces at a size-dependent rate. The fast intermicellar exchange of hydrolyzed silica species can be attributed to the fusion–fission process of micelle droplets. The integrated area under the dashed profile, shown in Figure 1.4.1, indicates that the minimum time period for nucleating these silica particles is about

15–20 min, and during the period first 97% of silica particles are nucleated and the other 3% would be most likely nucleated at a later reaction time. When the apparent rate constant k_c for SiO_2 particle growth is estimated from the change of average particle size in electron micrographs and of scattered light intensity from a microemulsion during TEOS hydrolysis, its value determined to be 0.031 and 0.034 h^{-1}, respectively. These k_c values agreed with k_c value of 0.031 h^{-1} and k_h value of 0.031 h^{-1} estimated by Fourier transform infrared spectroscopy. The fact that the calculated k_h and k_c values are the same demonstrates that the growth of SiO_2 particles is controlled by the extremely slow hydrolysis of TEOS. On the other hand, Osseo-Asare and Arriagada suggested the mechanism of SiO_2 particle growth by the evaluation of the number of particles produced in a given volume of microemulsion, and assumed that the conversion of TEOS is completed (16). The number of particles per unit volume of microemulsion is calculated from the terminal size of particles produced, and their values obtained at different Wo (= $[H_2O]/[\text{surfactant}]$) are presented in Table 1.4.1. It is found that each particle is formed by about 10^6 aggregates at low Wo, while fewer aggregates are involved per particle at larger Wo values. This result suggests that particle growth dose not occur only by addition of TEOS monomer to existing nuclei, in which case the number of resulting particles would correspond to the number of nuclei initially formed. However, the forming of only one stable nucleus out of 10^6 aggregates seems too low in probability, considering that, on the average, there are two or more hydrolyzed TEOS molecules per aggregates as deduced from N_t values in Table 1.4.1. Alternatively, therefore, it is more likely that besides monomer addition to existing particles, particle growth also involves aggregation of nuclei particles during micelle–micelle collision. This may be also confirmed by the fact that the size of final particles is larger than that of micelle droplets.

Table 1.4.1 Parameters of Reversed Micelles and SiO_2 Particles

	Micelle		Particle			
Wo	N[a]	Na_a[b] (10^{18})	D_p[c] (nm)	N_p[d] (10^{13})	N_a/N_p (10^4)	N_t[e]
0.7	30	30.1	71.6	2.49	120.0	2.7
1.0	40	15.8	61.4	3.95	41.0	5.1
1.3	54	9.0	—	—	—	9.1
1.6	63	6.3	49.5	7.54	8.4	12.9
2.3	89	3.0	48.5	8.02	3.7	26.6

[a] N, aggregation number of micelle.
[b] N_a, number of aggregates in 5.06 mL of microemulsion.
[c] D_p, final diameter of particle obtained after 42 days.
[d] Assumed particle density = 1.82 g mL^{-1} with an SiO_2 content of 92.7 wt%.
[e] Number of TEOS molecules per aggregate.

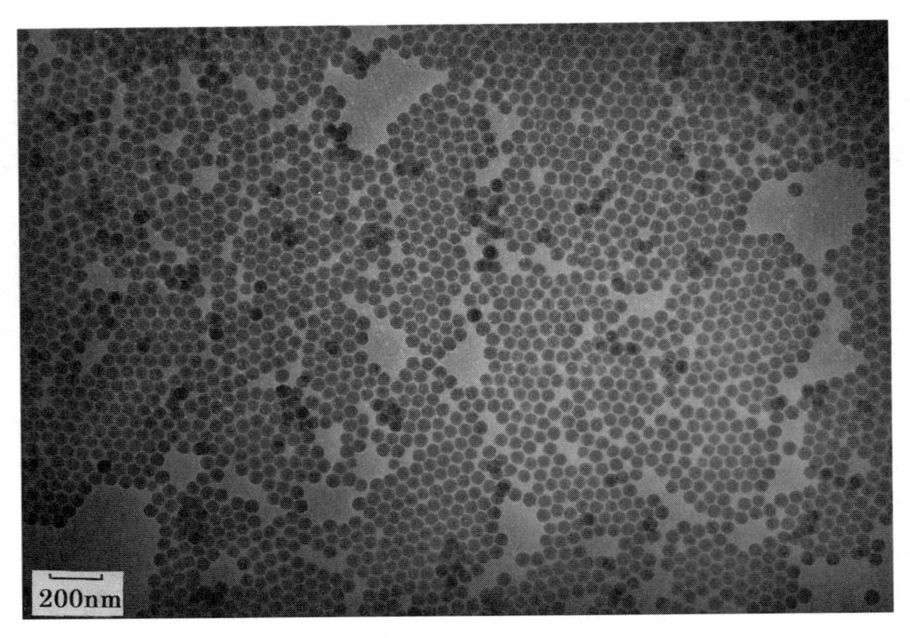

Fig. 1.4.2 Transmission electron micrograph of SiO_2 particles produced at Wo = 5 in 0.10 mol kg^{-1} polyoxyethylene(6) nonylphenyl ether–cyclohexane system. [TEOS] 0.10 mol kg^{-1}; $[NH_3]$/[surfactant] = 0.5.

Generally, SiO_2 particles produced in nonionic surfactant microemulsions are largely spherical and monodisperse, and their size lies in the range of 40–70 nm in diameter as showed in Figure 1.4.2. The particle size and size distribution are affected significantly by micelle components such as water, reactants, surfactants, catalysts, and so on. If the formation of SiO_2 particles occurs by the condensation of hydrolyzed silica species onto the surface of nuclei, the number of the species per micelle droplet is larger at large Wo. Thus, the formation of a large number of nuclei at large Wo is expected, which would then lead to smaller particles. In other words, this is explained by the dilution effect of hydrolyzed species in micellar water droplets Indeed, as shown in Figure 1.4.3, the size of SiO_2 particle produced in the polyoxyethylene(6) nonylphenyl ether–cyclohexane system depended little on Wo values below Wo = 4, but above Wo = 4 they decreased rapidly, and again further decreased from Wo = 9 (12). The values of Wo = 4 and Wo = 9 correspond to the transitions from reversed micelles containing water bound to the oxyethylene groups of surfactants to swollen micelles holding water interacted with the hydrated polar groups and from swollen micelles to W/O microemulsions containing bulk-like water, respectively (Y Kurokawa, K Kawai, K Kon-no, unpublished results).

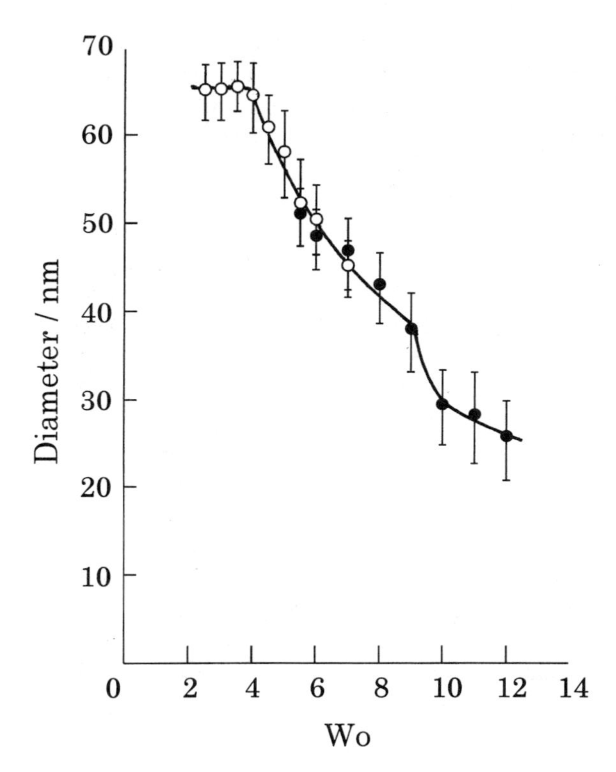

Fig. 1.4.3 Change of size of SiO_2 particles with Wo in 0.10 mol kg^{-1} polyoxyethylene(6) nonylphenyl ether–cyclohexane system. [TEOS] 0.10 mol kg^{-1}; ○: [NH_3]/[surfactant] = 0.5. ●: [NH_3]/[surfactant] = 2.25.

This suggests that the particle size not only is controlled by the micellar droplet size, but also is restricted by the states of water droplets in the interior of micelles. In polyoxyethylene(5) nonylphenyl ether in cyclohexane, however, the size was slowly decreased in the range of Wo = 0.7–2.3 (16). Such gradual decrease was also observed in polyoxyethylene(9) octylphenyl ether and polyoxyethylene(12) decylphenyl ether in heptane, whereas in polyoxyethylene(9) dodecylphenyl ether system the size remained in 48 nm in the range of 0.174–0.73 *M* H_2O and then decreased significantly to 36 nm at 1.1 *M* H_2O (14). In addition, when SiO_2 particles are synthesizd at vicinity of the limiting amount of aqueous ammonia in polyoxyethylene(5) nonylphenyl ether–cyclohexane system, the resulting particles exhibit a bimodal size distribution (17). This is due to the formation of new silica nuclei in a water-rich phase, which is separated from the single microemulsion phase by ethanol produced during particle formation process. In the polyoxyethylene(6) nonylphenyl ether–cyclohexane system, however, the particle size was independent of the concentration of ammonia (12). This fact indicates that ammonia plays a role in the creation of spherical particles, because no particles form in the absence of ammonia. As expected from the dilution effect mentioned, indeed, the particle size increased with increasing the concentration of surfactant. A similar decrease in particle size was also observed in polyoxyethylene(9) octylphenyl ether and polyoxyethylene(12) decylphenyl ether systems, but in the range of 0.09–0.22 *M* polyoxyethylene(9) dodecylphenyl ether system the particle size remain almost a constant (14). The magnitude of particle size by the kind of surfactant was the order of polyoxyethylene(9) dodecylphenyl ether > polyoxyethylene(9) octylphenyl ether > polyoxyethylene(12) decylphenyl ether in the range of 0.04–0.26 *M* surfactants in heptane. In solvent effect on particle size, small particles were produced when heptane is replaced by cyclohexane in polyoxyethylene(9) dodecylphenyl ether (14). This may be due to the increase of compartmentalization of the hydrolyzed species, because the cyclohexane molecule, having a shorter length or a higher polarity, deeply penetrates into surfactant films. On the other hand, the SiO_2 dispersion systems after hydrolysis reaction are generally stable. If SiO_2 particles are synthesized at relatively lower water-to-TEOS ratio than the stoichiometric ratio of the overall SiO_2 formation reaction, SiO_2 particles became unstable and large flocs form. This result indicates that the residual water film around the particles was necessary to stabilize SiO_2 particles in microemulsions. Residual water acted as an intermediate medium that hydrogen bonded both surfactant molecules and the surface of SiO_2 particles. As a result, the surfactants attached themselves to the surface of SiO_2 particles and sterically stabilized them from flocculating (17). Recently, Chang, Liu, and Asher have developed a synthetic method for preparing a large SiO_2 particles and SiO_2–CdS nanocomposite particles of ~40–300 nm in microemulsions of Triton N-101/cyclohexane (18). The pure large SiO_2 particles are produced by further reactions of TEOS with SiO_2 particles used as seed. In this system, the particles of 77 ± 4 nm used as seed can be grown to 103 ± 5 to 147 ± 6 nm with different amounts of TEOS added. For SiO_2–CdS composite particles, the morphologically different particles such as homogeneous dispersion of CdS quantum dots (~2.5 nm in diameter), large patches of CdS, cores of CdS, shells of CdS, and so on are produced by controlling the simultaneous coprecipitation of cadmium nitrate and ammonium sulfide in micelle droplets. When these composite particles have the CdS inclusions etched out with strong acid, the SiO_2 particles have craters on their surface, and spherical voids 2.4 nm in diameter within the particles are obtained (19). Such CdS composite particles or void particles may be useful materials for nonlinear optics or as a novel catalyst support medium.

The synthesis of SiO_2 particles by ammonia-catalyzed hydrolysis of TEOS has been also carried out in Aerosol OT reversed micelle systems. The particles precipitated in this systems are spheres, but they have generally a broad size distribution comparing with that of nonionic reversed micelles; compared to the normalized standard deviation of <10% in polyoxyethylene(5) nonylphenyl ether, that in Aero-

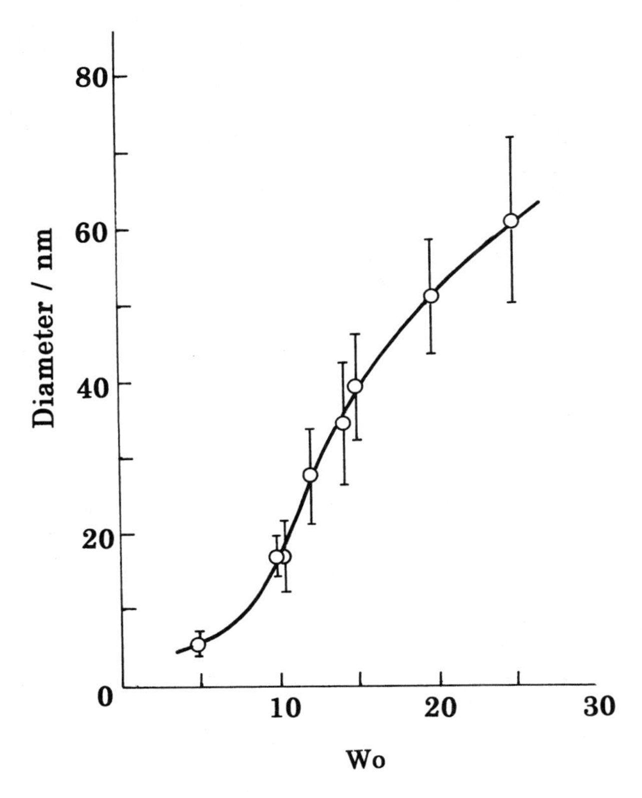

Fig. 1.4.4 Change of size of SiO_2 particles with Wo in 0.10 mol kg^{-1} Aerosol OT–cyclohexane system. [TEOS] 0.25 mol kg^{-1}; $[NH_3]$/[surfactant] = 1.0.

sol OT is 20% (20). In decane and isooctane (20,21), no particles are produced in region below about Wo = 4, irrespective of the presence of reversed micelles, but in the region above that Wo, spherical particles with significant size distribution were produced. The particle size in isooctane shown in Figure 1.4.4 increased rapidly with Wo in a swollen micelle region of Wo = 5–10, and in the W/O microemulsion region above Wo = 10 they saturated with increasing Wo (21), whereas the particle size in decane increased gradually up to Wo = 10. No particles are formed under conditions below Wo = 4 due to the unavailability for TEOS hydrolysis reaction of water droplets, since the droplets in that region bind tightly to the ionic polar groups of surfactant. A broad size distribution at higher Wo region is due to the destruction of the protective surfactant film by the hydrolysis of Aerosol OT with ammonia during the reaction. The destruction of protictive film by the hydrolysis of Aerosol OT led also to unstable dispersion systems (20). In toluene systems, however, the particles were formed even in the region of Wo < 4 and their size increased with increasing Wo, within the range Wo $= 1.8$–6.9 (22). Addition of benzyl alcohol affected the particle size produced at Wo $= 6$ in an Aerosol OT–decane system (23), below 1.5 molar ratio of benzyl alcohol to Aerosol OT; the size of particles produced after 18.5 h increased with increasing molar ratio, whereas the particles after 815 h were coagulated. These effects of benzyl alcohol on particle size are attributed to the inhibition of nucleation by fast rearrangement of hydrolyzed silica species. On the other hand, the nitrogen adsorption isotherms of particles produced in isooctane showed that the particles have microporous and specific surface area of about 100–300 $m^2\ g^{-1}$ (24).

Unlike TEOS hydrolysis, SiO_2 particles have been also prepared by hydrolysis of Na_2SiO_2 and Na_4SiO_2 in nonionic reversed micelle systems. Spherical and polydisperse particles of 31.8 nm mean diameter were produced in polyoxyethylene(9.5) octylphenyl ether–hexanol–cyclohexane systems (25), but more uniform and dense particles were precipitated by hydrochloric acid-catalyzed hydrolysis in a mixture of polyoxyethylene(5) nonylphenyl ether and polyoxyethylene(9) nonylphenyl ether in cyclohexane systems at pH 11 (26). The uniform particle formation at higher pH is attributed to the charge repulsion by OH^- adsorbed on particle surface. The particles of specific surface area of 347 $m^2\ g^{-1}$ can be obtained by calcination of particles produced at pH 2.

Besides SiO_2 particles, GeO_2, ZrO_2, and TiO_2 particles have been also synthesized by the hydrolysis of the corresponding metal alkoxides and metal chloride in ionic and nonionic reversed micelle systems. In contrast to SiO_2 particles, GeO_2 particles can be produced by the hydrolysis of germanium tetraethoxide with water droplets in the absence of ammonia. Therefore the reaction system is available to examine the relationship between the size or morphology of particles and the states of water droplets or micelles reported by Kon-No and coworkers (27). Figure 1.4.5 shows scanning electron micrographs of GeO_2 particles produced at Wo $= 3$, 8, and 35 in the Aerosol OT–cyclohexane system (28). Monodisperse and polyhedron particles were formed in both the reversed micelles of Wo $= 3$ and swollen micelles of Wo $= 8$, but in W/O microemulsions of Wo $= 35$ they became relatively uniform cubic particles. In the region of Wo $= 4$ and 5, which changes from reversed micelles to swollen micelles, the resulting particles were a mixtures of small and large particles (~40:60%) formed in both the micelles. The sizes of particles formed in each micelle are presented in Table 1.4.2. Particle size increased with increasing Wo in both the reversed and swollen micelle systems, whereas in W/O microemulsion systems the trend was the opposite. These results clearly indicate that the size and morphology of particles are controlled by the size of water droplets or the states of water in each micelle; namely, polyhedron particles are formed by hydrolysis with the water bound tightly to the ionic polar groups of surfactants in reversed micelles and with the water held together with the hydrated polar groups by hydrogen bonding in swollen micelles, and cubic particles are produced with bulk-like water in W/O microemulsions. X-ray diffraction measurements showed that small and

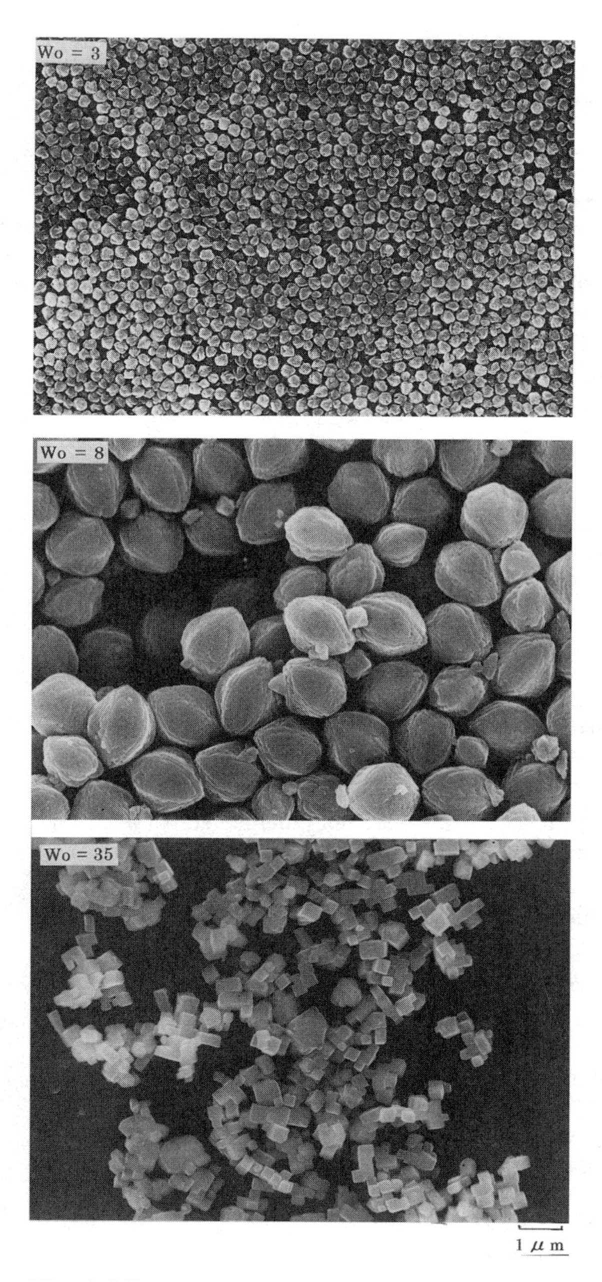

Fig. 1.4.5 Transmission electron micrographs of GeO_2 particles produced at various Wo values in 0.10 mol kg^{-1} Aerosol OT–cyclohexane system. $[Ge(OC_2H_5)_4]$ 0.02 mol kg^{-1}.

Table 1.4.2 Change of the Size[a] of GeO_2 Particles with Wo in Aerosol OT–Cyclohexane Systems

Wo	Micellar state[b]	Polyhedron (nm)		Cubic (nm)
		Small	Large	
2	RMs	80 ± 9	—	—
3	RMs	272 ± 21	—	—
4	RMs	623 ± 130	972 ± 96	—
5	SMs	643 ± 140	1003 ± 140	—
6	SMs	—	112 ± 73	—
8	SMs	—	1650 ± 110	—
10	SMs	—	1750 ± 110	—
13	MEs	—	—	993 ± 127
15	MEs	—	—	1021 ± 84
25	MEs	—	—	479 ± 53
35	MEs	—	—	265 ± 39

[a] The length of axis for polyhedron particles and of a side for cubic particles.
[b] RMs, reversed micelles; SMs, swollen micelles; MEs, W/O microemulsions.

moderate sized particles produced at Wo = 3 and 35 transformed from α-quartz type crystal to rutile type crystals at around 970 and 1050°C, whereas large particles formed at Wo = 8 were fused above 1100°C. These results were confirmed by differential thermal analysis (DTA) and Fourier transform infrared measurements. On the other hand, the dispersed TiO_2 particles were not precipitated by water and ammonia-catalyzed hydrolysis of titanium tetraisoproxide in both the Aerosol OT– and polyoxyethylene(6) nonylphenyl ether–cyclohexane systems (29). Hydrolysis by sulfuric acid droplets in a mixture of these surfactants, however, induced the formation of spherical and amorphous TiO_2 precursors in the range of 30–240 nm in diameter, as seen in Figure 1.4.6.

Infrared spectra suggested that a sulfate ion coordinates to two titanium atoms as a bidentate in particles. The maximum particle size was found at Aerosol OT mole fraction of 0.35 in the mixtures. The particle size increased linearly with increasing the concentration of sulfuric acid at any Wo, but with increasing Wo the effect was the opposite at any sulfuric acid concentration. These effects on the particle size can be explained qualitatively in relation with the extent of number of sulfate ions per micelle droplet. These precursor particles yield amorphous and nanosized TiO_2 particles, reduced by 15% in volume by washing of ammonia water. The TiO_2 particles transformed from amorphous to anatase form at 400°C and from anatase form to rutile form about at 800°C. In Triton X-100–*n*-hexanol–cyclohexane systems, however, spherical and amorphous titanium hydroxide precursor were precipitated by hydrolysis of $TiCl_4$ (30). When the precursor particles were calcinated,

Fig. 1.4.6 Transmission electron micrographs of TiO_2 precursors produced in 0.20 mol kg^{-1} mixture of Aerosol OT and polyoxyethylene(6) nonylphenyl ether–cyclohexane. Aerosol OT mole fraction 0.65; $[Ti(O\text{-}i\text{-}C_3H_7)_4] = 0.01$; Wo = 2.5; $[H_2SO_4]/[\text{surfactant}] = 0.05$.

they were transformed completely to anatase phase at 700°C and to rutile phase at 1000°C. From transmission electron micrographs, it is found that both the anatase and rutile particles are aggregates consisting of 15–30 nm and 15–30 nm sizes, respectively. In ultraviolet (UV) transmittance measurements of a TiO_2 suspension in a 1 : 1 mixture of water and polyethylene glycol, the particles calcinated at lower temperature adsorbed more ultraviolet radiation, and the amount of absorption decreased with an increase in calcination temperature. From a comparison of these results with specific surface area of particles, it is found that the amount of UV absorption increases with a decrease in the particle size. For catalytic activity for the photodegradation of phenol, only the anatase form showed significant degradation of phenol, whereas the rutile form is totally inactive. More small TiO_2 particles, e.g., 9 ± 0.5 nm in diameter, were yielded by the hydrolysis of $TiCl_4$ in regions of Wo = 4.5–13.5 in cetyldimethylbenzylammnomium chloride–benzene (31). When the preparation of ZrO_2 particles was tried by hydrolysis of zirconium tetra-*n*-butoxide with sulfuric acid droplets in a polyoxyethylene(6) nonylphenyl ether (NP-6)–cyclohexane system, the resulting spherical particles were suggested to be zirconium sulfate complexes such as $[Zr_2O(SO_4)_4]^{2-}$, from Fourier transform infrared measurements (32). The particle size increased from 10 to 75 nm as Wo increased from 2 to 40 at a given concentration of sulfuric acid, and the size distribution became significantly broad in W/O microemulsions of Wo = 28–40. The particle size also increased with increasing the concentration of sulfuric acid at a given Wo value. The amorphous particles obtained by washing with ammonia water were transformed to the monoclinic form of ZrO_2 by calcination for 2 h at 800°C.

REFERENCES TO SECTION 1.4

1. K Kon-no, M Gobe, K Kandori, A Kitahara. Int Conf Surface and Colloid Science, Jerusalem, 1981.
2. M Gobe, K Kon-no, K Kandori, A Kitahara. J Colloid Interface Sci 93:291, 1983.
3. M Gobe, K Kon-no, K Kandori, A Kitahara. Shikizai 57:380, 1984.
4. MA López-Quintela, J Rivas. J Colloid Interface Sci 158:446, 1993.
5. L Liz, MA López-Quintela, J Mira, J Rivas. J Mater Sci 29:3797, 1994.
6. P Ayyub, AN Maitra, DO Shah. Phys C 168:571, 1990.
7. V Pillai, P Kumar, MS Multani, DO Shah. Colloid Surf A 80:695, 1993.
8. P Ayyub, MS Multani, M Barma, VR Palker, DO Shah. Phys C 21:2229, 1988.
9. K Haneda, H Kojima. J Appl Phys 44:3760, 1973.
10. BT Shirk, WR Buessem. J Appl Phys 40:1294, 1969.
11. LM Gan, HSO Chan, LH Zhang, CH Chew, BH Loo. Mater Chem Phys 37:263, 1994.
12. K Kon-no, M Yanagi, A Kitahara. 6th Int Conf Surface and Colloid Science, Hakone, Japan, 1988.
13. C Chang, HS Fogler. AIChE J 42:3151, 1996.
14. C Chang, HS Fogler. Langmuir 13:3295, 1997.
15. JA Dirksen, TA Ring. Chem Eng Sci 46:2389 1991.
16. K Osseo-Asare, FJ Arriagada. Colloid Surf 50:321, 1990.
17. FJ Arriagada, K Osseo-Asare. Colloid Surf 69:105, 1992.
18. SY Chang, L Liu, SA Asher. J Am Chem Soc 116:6739, 1994.
19. SY Chang, L Liu, SA Asher. J Am Chem Soc 116:6745, 1994.
20. FJ Arriagada, K Osseo-Asare. J Colloid Interface Sci 170:8, 1995.
21. K Kon-no, M Yanagi, K Kandori, A Kitahara. 2nd Adv Meet Colloid and Interface Chemistry, Tokyo, 1987, p 1.
22. P Spiard, JE Mark, A Guyot. Polym Bull 24:173, 1990.
23. FJ Arriagada, K Osseo-Asare. J Dis Sci Tech 15:59, 1994.
24. H Yamaguchi, I Ishikawa, S Kondo. Colloid Surf 37:71, 1989.
25. W Wang, X Fu, J Tang, L Jiang. Colloid Surf A 81:81, 1993.
26. LM Gan, K Zhang, CH Chew. Colloid Surf A 110:199, 1996.
27. K Kawai, K Hamada, K Kon-no. Bull Chem Soc Jpn 65:2715, 1992.
28. K Kon-no. In E Pelizzetti, ed. Fine Particles Science and Technology. Dordrecht, Netherlands: Kluwer, 1996, p 431.
29. D Kaneko, T Kawai, K Kon-no. Shikizai 71:224, 1998.
30. V Chhabra, V Pillai, BK Mishra, A Morrone, DO Shah. Langmuir 11:307, 1995.
31. E Joselevich, I Willner. J Phys Chem 98:7628, 1994.
32. K Kawai, A Fujino, K Kon-no. Colloid Surf A 109:245, 1996.

1.5 SYNTHESIS OF MONODISPERSED COLLOIDS BY CHEMICAL REACTIONS IN AEROSOLS

EGON MATIJEVIĆ and RICHARD E. PARTCH
Clarkson University, Potsdam, New York

1.5.1 Introduction

Most of the techniques described in this volume deal with chemical reactions involving bulk liquids or solutions. Among these, the most common and versatile are various precipitation processes, which have resulted in a large number of uniform dispersions consisting of particles of simple or mixed (internally or externally) chemical compositions in a variety of shapes (1–3).

Despite the successes of these chemical procedures, they are still fraught with difficulties in terms of the predictability of the properties of the final products. Thus, except in a very few cases, it is impossible to predict conditions that would yield particles of a given shape. Even obtaining a desired particle size by a chemical process that can produce a monodisperse system must be established experimentally. When dealing with finely dispersed matter of internally mixed composition additional problems are encountered, because the molar ratio of constituents in the solid phase usually differs from those in solutions in which the precipitates are formed (4,5).

In this chapter a process is described that can alleviate some of the difficulties described above. The technique is based on chemical reactions in aerosols (6). Specifically, droplets of a reactant flowing in an inert carrier gas are contacted with the vapor of a coreactant, resulting—as a rule—in spherical solid particles. If the latter are internally chemically mixed, their composition is determined by the contents of the reactants in the original droplets, since each of the latter acts as a separate ''reaction container.''

It will be also demonstrated that the same technique is applicable to organic dispersions. Furthermore, it is shown that the method can be extended to making coated particles in a continuous process in which cores are formed first, followed by encasing them in layers of different compositions and thicknesses.

It should be noted that the procedure described in this chapter is strictly limited to the interactions of *droplets* with surrounding gases, with much emphasis on the conditions that would yield powders of narrow size distributions. However, no attempt is made to describe the literature on particle formation from a single levitated droplet (7). The technique used in these studies also differs from those in which the vapor of a compound is chemically or thermally decomposed or reacted with a gas to yield solid particles of a different composition (8–11).

There are other aerosol methods which can yield uniform powders, such as by dispersing aqueous *dispersions* of particles (e.g. of latex) and evaporating the water (12). In this case each droplet should contain only one particle, a task not easily accomplished. Alternatively, it is possible to nebulize *solutions* of electrolytes or other substances, which on removal of the liquid result in solid particles, dispersed in the carrier gas (13,14). This process has been expanded to include sintering of resulting solid aerosols in a continuous process to produce powders for various applications (15–18).

Finally, it is possible to produce aerosols by vaporization of solids and subsequent condensation, which under certain conditions may yield uniform spherical particles as shown on examples of NaCl (19–23), AgCl (24–26), V_2O_5 (27), etc. It is quite apparent that all these techniques are based on physical changes of the matter that do not involve chemical reactions, while the emphasis in this chapter is on using the described aerosol technique to produce inorganic materials, in particular metal oxides and polymers, *by chemical processes.*

A series of reviews on expanded uses of the aerosol technology appeared in a special issue of *Aerosol Science and Technology* (28) and in a recent comprehensive text (29).

1.5.2 Description of the Aerosol Method

The essential steps in the method described in this chapter consist of:

1. Generation of droplets containing one or more reactive liquids
2. Use of evaporation and nucleation phenomena to narrow the size distribution of the droplets
3. Exposure of the droplets to a coreactant vapor
4. Reaction of the liquids in the droplets with the surrounding vapor
5. Collection of the solid aerosol product

The first requirement of this technique is to produce precursor reactant droplets. It is also necessary that the chemical process proceeds relatively rapidly to prevent the coalescence of the aerosol before the reaction is completed.

There are many liquids capable of rapidly generating a solid product, when their droplets dispersed in an inert gas are contacted with a vapor coreactant. Such compounds include metal alkoxides, halides, and oxyhalides that react with water, hydrogen sulfide, etc., and organic monomers that can undergo either addition or condensation polymerization in the presence of appropriate initiators. The resulting powders consist of spherical particles, the size distribution of which depends on the method and conditions of droplet generation.

1.5.3 Generation and Interactions of Aerosols

Obviously, in order to obtain uniform final particles of a given size, the original droplets must be of a narrow size distribution and of predetermined modal diameters.

In principle one may proceed in two directions, either to break up liquid jets, such as by ultrasonic energy sources, or by condensation of vapors. The techniques based on the latter process have been shown more adaptable to obtaining ''monodispersed'' liquid aerosols. Sinclair and LaMer were first to construct a generator that met this requirement, in which a nuclei-laden gas was brought into contact with a hot liquid, whereupon the gas became saturated with the vapor (30). On subsequent cooling, the aerosol was formed by condensation onto nuclei. Although the described apparatus was capable of yielding uniform droplets, the reproducibility of the resulting aerosols was poor and the output was unstable on prolonged operation.

Several attempts were made to design equipment based on the Sinclair–LaMer concept, of which the most successful was the so called ''falling film'' generator (31,32). The essential parts of the latter are schematically shown in Figure 1.5.1 (31). In principle, the reacting liquid runs down on the inner walls of a tube kept at a constant temperature by a thermostatted liquid circulating through a jacket surrounding the tube. The vapor is picked up by the carrier gas containing nuclei, while the liquid that did not vaporize is recirculated. The gas with the vapor is cooled in a condenser to produce the droplets. The properties of the resulting aerosols depend on a number of parameters, including temperature (which controls the vapor pressure of the liquid), flow rate of the carrier gas, and the concentration of nuclei (33–35). It was established that a sharpening of the size distribution was achieved by the re-evaporation of the resulting aerosol and its recondensation (32). The same apparatus can be used for the preparation of internally chemically composite droplets, by cocondensation of a mixture of reactant vapors.

A variety of materials (NaCl, NaF, AgCl) can serve as nucleating agents, consisting of extremely small particles, obtained by heating the powders at appropriate temperatures, and swept by the carrier gas before entering the falling film generator (20,36).

Several types of manifolds, capable of maintaining the laminar flow of the aerosol, have been used to contact the droplets with a vapor in order to produce various powders (6,37,38). Figure 1.5.2 illustrates two such chambers, the operations of which are described in the legends. Of particular interest is design B, because it allows one to study the rate of chemical interactions in the droplets on exposure to the vapor (37).

Another generator that was successfully used in the formation of polymer colloids, particles of mixed composition, and coated particles is shown in Figure 1.5.3f, in which the falling film tube is replaced by a boiler (38). The complete apparatus described in Figure 1.5.3 makes it possible to produce cores and shells in a continuous process (39).

These generators, and some others described in the literature, such as the Berglund apparatus (40), result in droplets of a rather narrow size distribution (31,32,39,41), but their major difficulty is in the extremely low quantities of products. While these designs are most useful in establishing the optimum conditions for the preparation of uniform colloids, they are impractical for mass production. To achieve

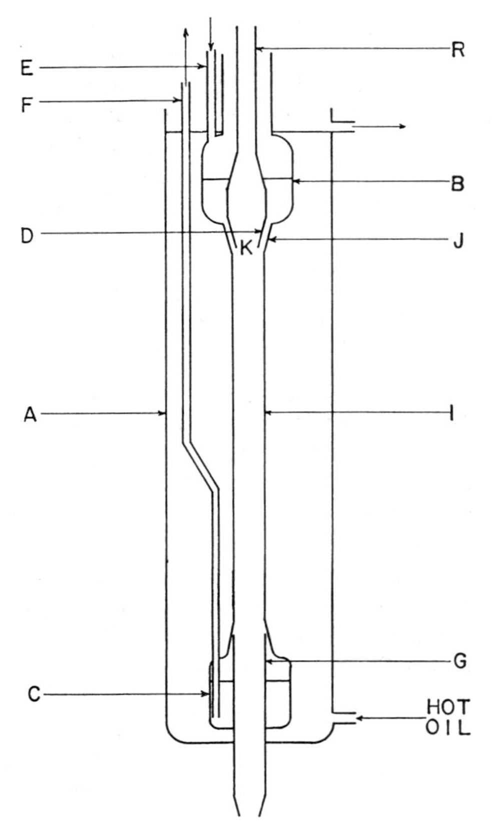

Fig. 1.5.1 Schematic diagram of the falling film aerosol generator. (A) Constant-temperature oil bath; (B) and (C) reservoirs containing the reactant liquid; (D) and (J) glass joints controlling the flow of the moving film; (E) and (F) tubes to the variable-speed pump; (G) exit tube; (I) boiler tube; (K) carrier gas (laden with nuclei) entrance. (From Ref. 31.)

the latter, the information obtained by these model systems must be used in the design of the scale-up equipment.

An alternate way to produce droplets is by dispersing liquids using different mechanical devices, such as rotating disks (42–44), or air-assist or piezo (45–49) ultrasonic nozzles (50,51). These generators can yield much larger amounts of liquid aerosols and are useful in the preparation of internally chemically composite particles, or for the production of colloids from solutions rather than from pure or miscible liquids. One such generator is illustrated in Figure 1.5.4, which was utilized in the synthesis of mixed silicates (52) and organic/inorganic composite particles by the

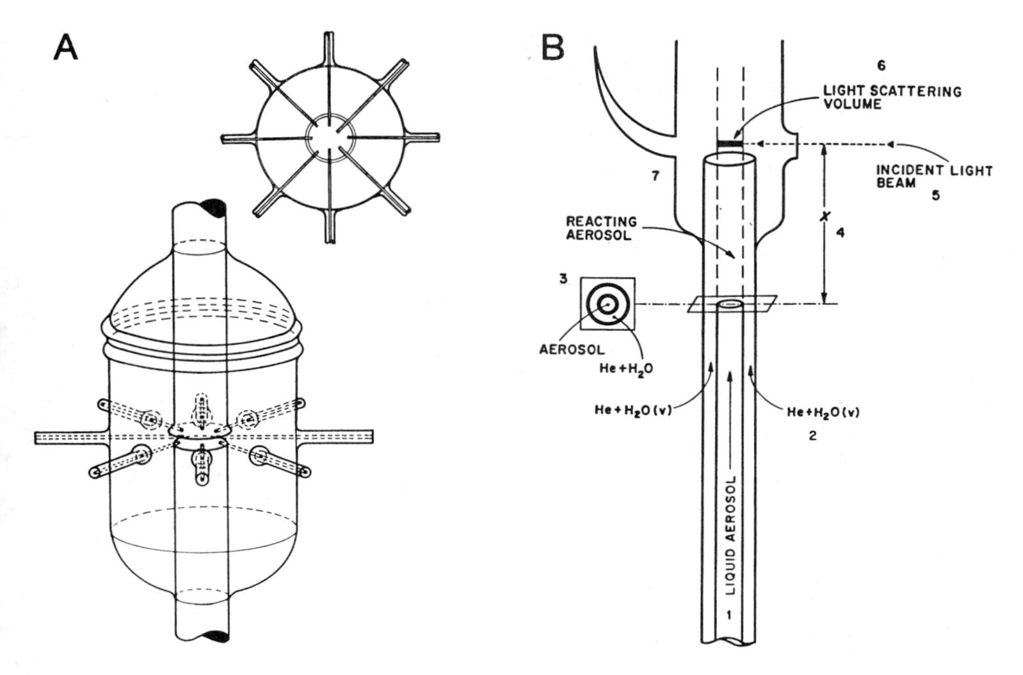

Fig. 1.5.2 Schematic diagrams of aerosol-vapor mixing manifolds. A: The aerosol passes through the center tube while the reactant vapor (e.g., water vapor) is introduced through the capillaries (6). B: (1) Inner tube for the liquid aerosol flow, (2) outer tube for the humidified gas (e.g. helium), (3) cross section of the chamber, (4) variable reaction distance (X), (5) light beam, (6) illuminated light scattering volume, (7) light scattering cell. (From Ref. 37.)

aerosol process (53). A similar apparatus was constructed by Ocaña et al. (54). The disadvantage of these kinds of equipment is in aerosol uniformity; as a rule, only polydispersed droplets are achieved (55). Consequently, such generators are convenient when a narrow particle size distribution is not essential in the product applications.

1.5.4 Preparation of Metal Oxides

Various metal alkoxides are ideal starting materials for the preparation of metal (hydrous) oxides by the described aerosol techniques, because many of these compounds are in the liquid state at room temperature, easily vaporized, and exceedingly reactive with water vapor. Additional advantage is the purity of the resulting powders, because the only products of the chemical reactions are the metal (hydrous) oxide and alcohol. The particles are, therefore, free of impurities, such as various ions, normally present in solids prepared from different salts.

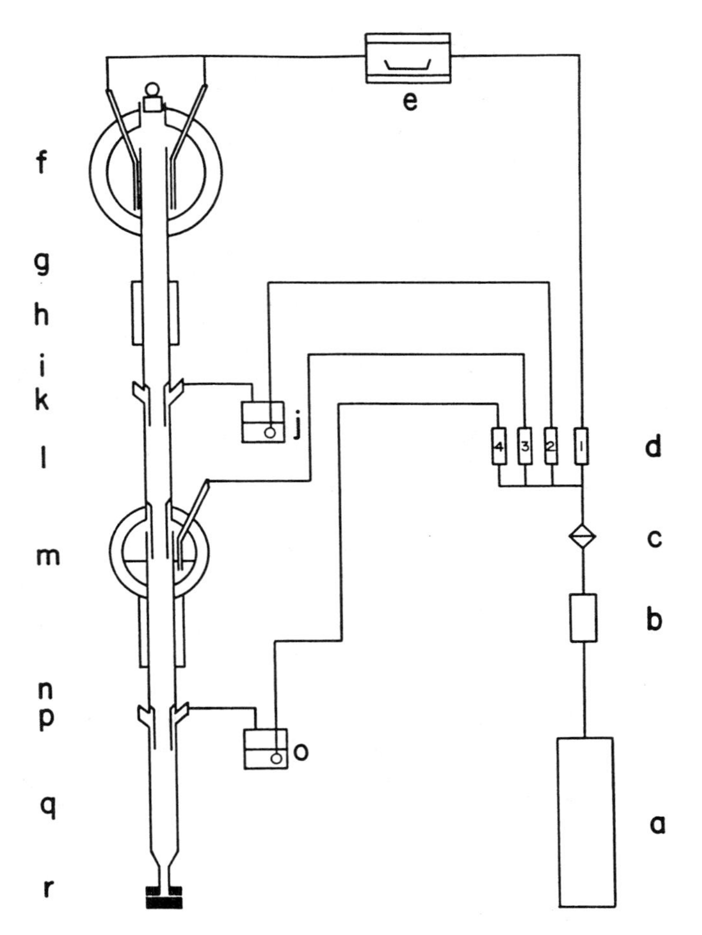

Fig. 1.5.3 Aerosol generator assembly for the production of composite or coated particles: (a) carrier gas (e.g., helium) tank, (b) drying column containing silica gel and molecular sieve, (c) Millipore membrane of 0.1 μm pore size, (d) flow meters, (e) nuclei (e.g., NaCl, AgCl) generator, (f) and (m) boilers containing different reactant liquids, (g) and (n) condensation chambers, (h) heater, (i) chamber for recondensation of droplets, (q) and (e) coreactant containers, (k) and (p) vapor injection ports, (l) and (g) reaction chambers, (r) powder collector. (From Ref. 39.)

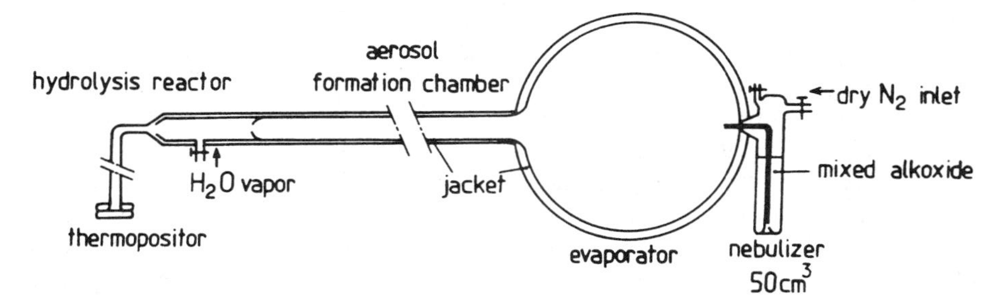

Fig. 1.5.4 Schematic diagram of the apparatus for the generation of the liquid aerosol by nebulization and subsequent hydrolysis. (From Ref. 52.)

The first material obtained by the hydrolysis of metal alkoxides aerosols was titanium dioxide. Using the falling film generator, droplets either of titanium ethoxide, $Ti(OEt)_4$, or of titanium isopropoxide, $Ti(i\text{-}OPr)_4$, were interacted with water vapor to yield uniform amorphous spheres of TiO_2 (56,57). Depending on the temperature of the generator (74.5–99.5°C) and the flow rate of the carrier gas, the diameter of the particles could be varied over the range of 0.1 to 0.6 μm. Figure 1.5.5 exempli-

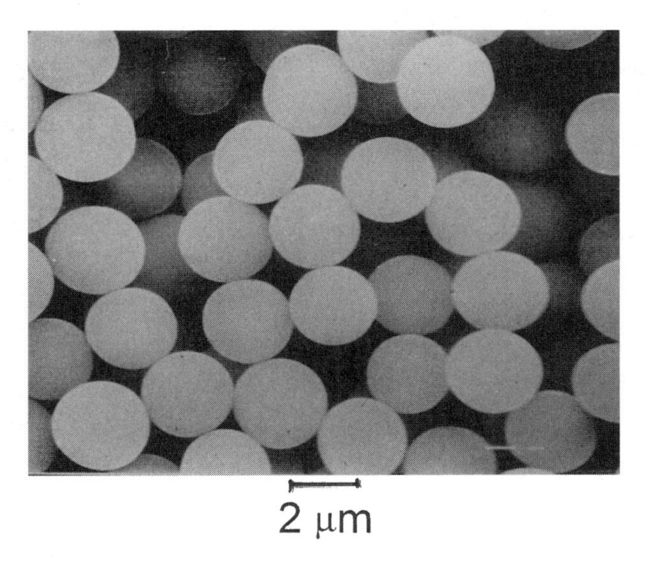

Fig. 1.5.5 Scanning electron micrograph of titanium dioxide particles prepared by the hydrolysis of a titanium (IV) ethoxide aerosol. Helium flow rate 800 cm^3 min^{-1}; generator temperature 96.5°C; nuclei (AgCl) oven temperature 620°C.

fies a powder so prepared. The amorphous particles calcined at 450°C were converted to essentially pure anatase, while retaining their shape and dispersity. Such powders of titanium oxide could be readily suspended in water and their isoelectric point indicated the high purity of the solids.

The kinetics of the hydrolysis of $Ti(OEt)_4$ droplets was systematically studied using the manifold displayed in Figure 1.5.2b (37). Since the original aerosols and the resulting solids were sufficiently uniform, the light-scattering polarization ratio method (20,37) was used to determine the particle size distribution in situ in course of the hydrolysis process. Figure 1.5.6 shows the polarization ratios measured at different times (0–340 ms), which were then evaluated in terms of the resulting size

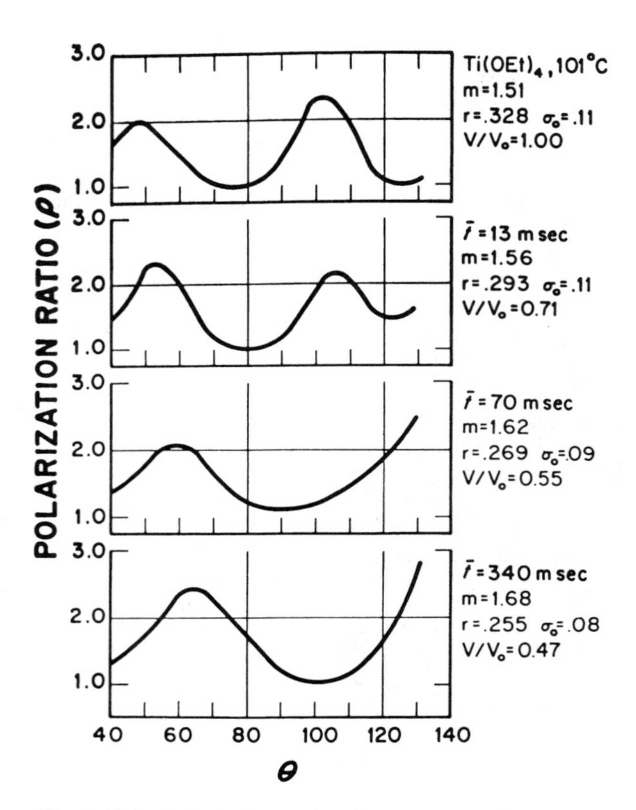

Fig. 1.5.6 Polarization ratio of the scattered light against the angle for the unreacted titanium (IV) ethoxide, $Ti(OEt)_4$, aerosol generated at 101°C, and the same aerosol reacted with water vapor at three distances (*X*) corresponding to 13, 39, and 340 ms reaction time. On the right-hand side are given calculated modal radii and refractive indices from the light scattering data. The bottom values are for the solid titania. (From Ref. 37.)

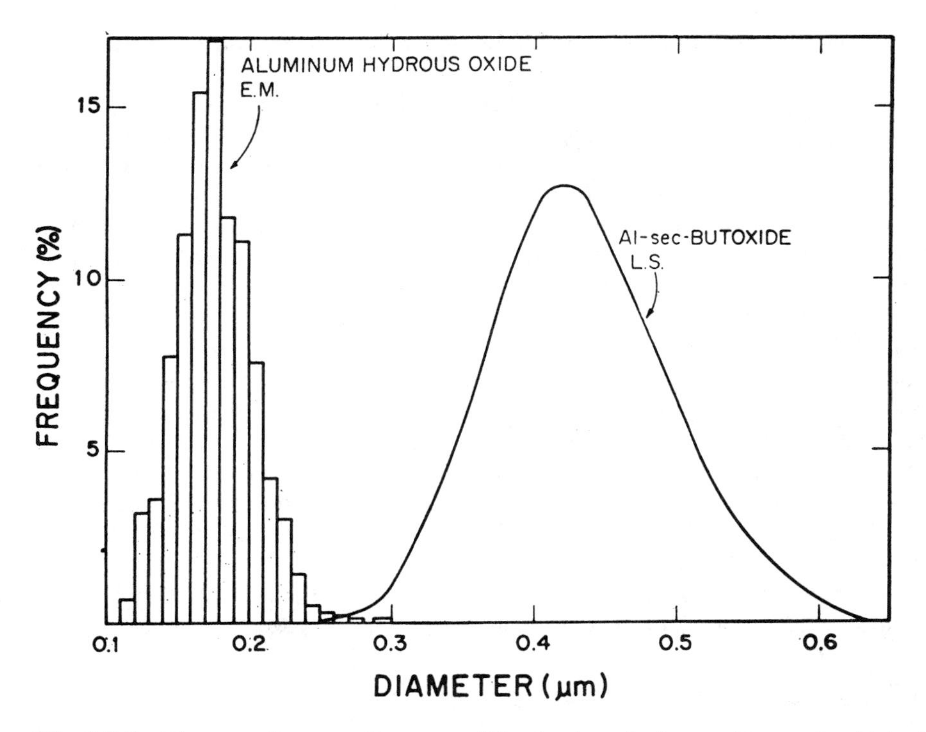

Fig. 1.5.7 Electron microscopy histogram of the aluminum hydrous oxide powder obtained from a liquid aerosol of aluminum *sec*-butoxide, Al(*sec*-OBu)$_3$, the droplet size distribution of which, obtained from light scattering, is given by the full line. Helium carrier gas flow rate was 1000 cm^3 min^{-1} and the temperatures of the boiler 122°C, of the reheater 130°C, and of the AgCl nuclei furnace 610°C. (From Ref. 58.)

distribution parameters and refractive indices. The bottom curve is for the final solid particles, indicating the process to be indeed very fast; the conversion of the liquid droplets into titania was completed in ~300 ms.

The same technique was used to prepare aluminum (hydrous) oxide powders (58,59). The modal diameters of the latter increased from 0.2 to 0.6 μm when the boiler temperature was raised from 122 to 150°C. Figure 1.5.7 compares the size distribution of alkoxide droplets as determined by light scattering and the histogram of the resulting aluminum oxide particles evaluated from electron micrographs. The ratio of modal diameters corresponds to the difference in densities of these materials. Later it was demonstrated that reasonably uniform aluminum (hydrous) oxide powders could be manufactured in larger quantities by hydrolysis of Al(sec-OBu)$_3$ droplets using a turbulent-flow aerosol process (60).

The hydrolysis of metal alkoxide aerosols was then applied for the preparation of additional oxides (SiO_2 ZrO_2) (61).

Other liquids can be used in the described aerosol technique, as long as the required vapor pressures can be achieved over a reasonable range of temperatures. For example, TiO_2 was produced from $TiCl_4$ droplets (56); due to the higher vapor pressure characteristics of this liquid, the generator was kept at lower temperatures (0 to 40°C). The resulting particles were less uniform than those generated from Ti(IV) alkoxides and they contained a small amount of Cl, as one would expect considering the composition of the starting reactant.

Analogously, colloidal tin(IV) oxide with particles of 1–2 μm in diameter was obtained by the interaction of $SnCl_4$ droplets with water vapor, which required considerably lower temperatures (−15 to −30°C) (62). To promote the rate of hydrolysis at these temperatures, a small amount of ammonia was added into the gas stream. The original amorphous particles started to crystallize into cassiterite on calcination at 400°C, and they retained their sphericity at treatments up to 1000°C.

Titania was also produced with $Ti(i\text{-}OPr)_4$ by spray hydrolysis; i.e., the aerosol was obtained by nebulizing the liquid (63), but the particles so obtained were polydisperse.

These methods should be clearly distinguished from the formation of metal oxides by the pyrolysis of the corresponding alkoxide vapors (64–66).

1.5.5 Preparation of Polymer Latexes

Both evaporation/condensation and nebulization equipments have been employed to generate droplets of reactive organic monomers, which could undergo the polymerization to powders when exposed to an initiator vapor.

For example, spheres of narrow size distribution of poly(*p-tert*-butylstyrene) were obtained by exposing the corresponding monomer droplets to the trifluoromethanesulfonic acid initiator vapor (Fig. 1.5.8). The polymer particles ranged in diameter from 1 to 3 μm, and their uniformity was sensitive to the monomer-to-initiator mass ratio. Under certain conditions the normally smooth spheres appeared connected through polymer whiskers (Fig. 1.5.9) (67). Using styrene monomer and adjusting the boiler temperature, uniform polystyrene particles up 10 μm in diameter and spheres of 20 μm of broader size distribution could be prepared (67).

The same initiator in contact with droplets of ''divinylbenzene'' (which was a mixture of *ortho, meta,* and *para* isomers) yielded divinyl/ethyl-vinylbenzene copolymer particles of great uniformity with diameters as large as 30 μm (38).

The aerosol technique can also be used to produce polymer colloids by addition polymerization. Thus, when droplets of toluene-2,4-diisocyanate (TDI) or 1,6-hexamethylene diisocyanate (HDI) were brought into contact with ethylenediamine (EDA) vapor (in the apparatus shown in Fig. 1.5.3) spherical polyurea particles with modal diameters of 1–3 μm were formed. The entire process, i.e., the formation of droplets and the polymerization, was carried out at moderate temperatures (<80°C)

Fig. 1.5.8 Scanning electron micrograph of poly(*p*-*ter*-butylstyrene) particles obtained by polymerization of monomer droplets in the aerosol phase. The monomer and initiator flow rates were 1.2 $dm^3\ min^{-1}$ and 40 $cm^3\ min^{-1}$ and the boiler and initiator reservoir temperatures were 50°C and 25°C, respectively. The initiator was injected into the flowing aerosol at two positions. The modal diameter of these particles is 1.8 μm. (From Ref. 67.)

(68). Since this process depends on the stoichiometry of the reactants, sufficient amounts of EDA must be present to produce fully solidified polymer particles. Incomplete reactions yielded a polyurethane shell, which on the removal of unreacted liquid in the core by evaporation resulted in hollow particles (68). It would appear that the solid encapsulating polymer inhibits the diffusion of EDA into the rest of the original droplet.

The polydivinybenzene colloids prepared by the aerosol technique were carbonized to yield uniform porous spheres of carbon of relative high specific surface areas (69).

Submicrometer-size droplets of methyldichlorosilane were produced at cryoscopic temperatures and reacted with ammonia to give high-molecular-weight solid silazane derivatives in a rapid reaction. On calcination these precursors ended as spherical mixed silicon nitrides of α- and β-Si_3N_4 (70).

Finally the aerosol process could be used to coat particles, as exemplified in the system of titania cores and polyurea shells (39). In this case the titania particles were generated from Ti(IV) ethoxide as described earlier, then contacted with dried

Fig. 1.5.9 Soft-shell poly(*p-ter*-butylstyrene) particles connected by polymer threads. The monomer and initiator flow rates were 2.7 $dm^3\ min^{-1}$ and 9 $cm^3\ min^{-1}$ and the boiler and initiator reservoir temperatures were 50°C and 92°C, respectively. (From Ref. 67.)

vapors of either HDI or TDI in a helium stream, resulting in their condensation on the titania cores. The solids so wetted were exposed to the EDA vapor, producing the polyurea shell. By careful control of the experimental conditions (the temperature of boilers and manifolds, flow rate, etc.), uniform polymer coatings of different thicknesses could be obtained as displayed in Figure 1.5.10.

1.5.6 Internally Composite Systems

The reaction of aerosol droplets with gases can also be used to prepare particles of internally mixed composition, such as consisting of different metal oxides. In principle, these powders can be obtained by first cocondensing vapor of the two or more volatile metal compounds (preferably alkoxides), or by nebulizing liquids of mixed composition. In both cases the droplets are then reacted with vapors.

Mixed titania/alumina collodial spheres were prepared in an apparatus consisting of two falling film generators in series upstream of the manifold. In the first generator, $Ti(OEt)_4$ was vaporized at temperatures ranging between 78 and 101°C, and the droplets, obtained by condensing this vapor, were then introduced into the

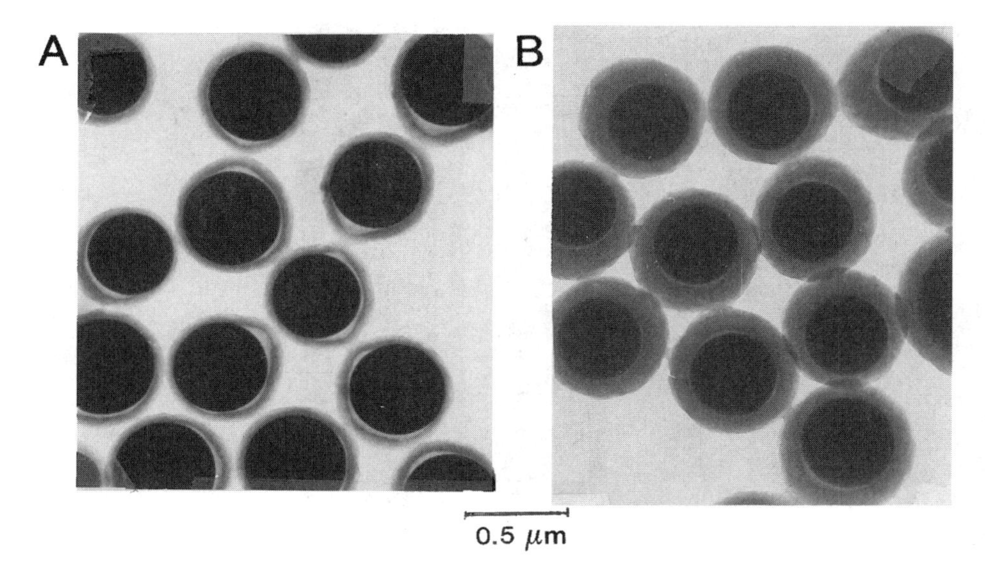

Fig. 1.5.10 Transmission electron micrographs of two different thicknesses of polyurea coating on titania core particles. In each case the titania was obtained under the same conditions as those in Fig. 1.5.3. HDI was kept at 80°C at flow rates of (A) 0.19 $dm^3\ min^{-1}$ and (B) 0.42 $dm^3\ min^{-1}$. The flow rate and temperature of ethylenediamine were held constant at 44.3 $cm^3\ min^{-1}$ and 25°C, respectively. (From Ref. 39.)

second generator kept at 110–132°C, in which they revaporized and were mixed with the vapor of Al(*sec*-OBu)$_3$ (71). The final mixture was cooled to produce droplets of mixed composition. By a careful control of varius experimental conditions, rather uniform spheres were obtained as shown in Figure 1.5.11. The molar ratio of the titania to alumina in the particles could be varied from 6.8 : 1 to 0.03 : 1 by adjusting the reaction parameters (temperature, flow rate, etc.). However, in all cases the ESCA analysis of the powders, and the isoelectric point of their dispersions in water, showed the surface to consist mostly of titania. Consequently, the internal composition of particles varied from the center to the periphery.

The freshly prepared powders were amorphous. On calcination at 750°C, the sample containing the large excess of titania ([Ti]/[Al] = 6 : 1) gave x-ray diffraction patterns with very strong lines characteristic of rutile and very weak lines of γ-alumina.

Mixed silica/titania particles were also prepared in a modified version of the apparatus illustrated in Figure 1.5.3 using $Ti(OEt)_4$ and $SiCl_4$ as reactants (72). The generated droplets of $Ti(OEt)_4$ absorb $SiCl_4$ vapor in the first manifold and cause a metathesis of the chloride and ethoxide groups, creating particles having a liquid core encapsulated in solid $TiCl_3(OEt)$ (73). This aerosol on exposure to water vapor

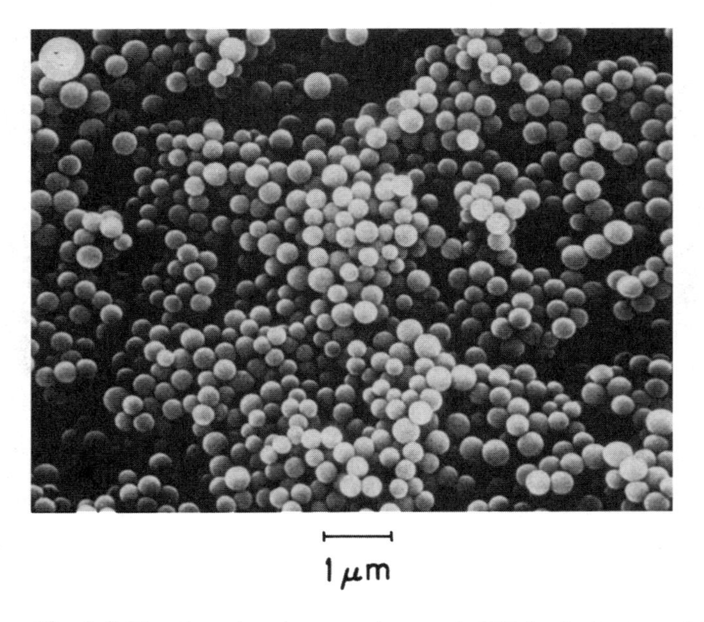

Fig. 1.5.11 Scanning electron micrograph (SEM) of mixed metal hydrous oxide particles generated from mixed $Ti(OEt)_4$ and $Al(sec\text{-}OBu)_3$ vapors at flow rate of 1.51 $dm^3\ min^{-1}$ and boiler temperatures of 75°C and 125°C. (From Ref. 71.)

in the second manifold yielded solid particles of both oxides with TiO_2-enriched surface (72).

It is expected that the preparation of particles of mixed composition by the aerosol technique should be easier and with greater throughput per unit time, if the droplets are generated by nebulization. Silica/titania particles were obtained by spraying the mixed solutions of $Ti(OEt)_4$ and tetraethyl orthosilicate (TEOS) by means of an ultrasonic nozzle, and subsequent hydrolysis with water vapor (74). The rather uniform aerosols could be achieved by separating the larger droplets from the smaller ones through their collision with the wall of a bent tube (75). Again the compositions of the resulting solids differed from those expected from the original liquids; the surface contained more titania than the bulk.

The spray method was used to produce alumina pigments doped with Cr, Mn, and Co. In these experiments, $Al(sec\text{-}OBu)_3$ was mixed with solutions of the corresponding metal nitrates in *sec*-butoxide, the resulting liquids were nebulized, and then the droplets were hydrolyzed (76). The major purpose of these studies was to obtain inorganic pigments and to evaluate their color properties by altering the amount of dopants in the aluminum oxide matrix. For the same reason, the vanadium doped zirconia powders were prepared by this aerosol technique from mixtures of Zr(IV) *n*-propoxide and vanadium oxychloride (41).

The nebulization was also employed to generate composite powders for specific applications, such as in ceramics, by hydrolyzing with water vapor droplets containing $Al(sec\text{-}OBu)_3$ and silicon methoxide in the atomic ratio Al/Si = 3. This ratio of alkoxides was chosen in order to produce mullite, which was achieved by calcination of the resulting amorphous particles at rather high temperatures (up to 1400°C) (52). In another approach a mixed Al-Mg-Si ethoxide was first synthesized, and then nebulized and hydrolyzed as usual (77). Depending on the experimental conditions, the powders calcined at 500°C exhibited structures of pure cordierite, or mixed with forsterite. In all of these described cases the nebulization yielded spherical but polydisperse particles.

In summary, the described aerosol technique offers a new way to produce exceedingly pure and morphologically predictable colloids. The products should especially find applications in areas of high value materials. While the principles of the procedure have been well established, it is now necessary to concentrate efforts on equipment designs that will yield large amounts of uniform droplets.

REFERENCES TO SECTION 1.5

1. This volume, Chapters 1.1–1.4, 2.1, 2.2, 3.1–3.3, 4.1–4.3, 5.1, 5.2, 6.1, 6.2, 7.1, 7.2, 9.2, 11.1, 11.2, and 13.1.
2. E Matijević. Langmuir 10:8, 1994.
3. E Matijević. Chem Mater 5:412, 1993.
4. F Ribot, S Kratohvil, E Matijević. J Mater Res 4:1123, 1989.
5. WP Hsu, G Wang, E Matijević. Colloids Surf 61:255, 1991.
6. D McRae, E Matijević, EJ Davis. J Colloid Interface Sci 53:411, 1975.
7. EJ Davis, MF Buehler. MRS Bull 15:26, 1990.
8. IN Tang, KH Fung. J Aerosol Sci 20:609, 1989.
9. JD Casey, JS Haggerty. J Mater Sci 22:4307, 1987.
10. Y-H Zhang, CK Chan, JF Porter, W Guo. J Mater Res 13:2602, 1998.
11. F Kirbir, H Komiyama. Adv Ceram Mater 3:511, 1988.
12. DP Bhanti, SK Dua, P Kotrappa, NS Pimpale. J Aerosol Sci 9:261, 1978.
13. B Binek, B Dohnalová. Staub-Reinhalt Luft 27:30, 1967.
14. CD Hendricks, S Babil. J Phys E 5:905, 1972.
15. CL Salisbury, G Tuncel, JM Ondov. Aerosol Sci Technol 15:156, 1991.
16. K Okada, A Tamaka, S Hayashi. J Mater Res 9:1709, 1994.
17. MV Cabañas, JM Gongález-Calbet, M Vallet-Regí. J Mater Res 9:712, 1994.
18. Y-L Wang, ZQ Tan, Y Zhu, AR Moodenbaugh, M Suenaga. J Mater Res 7:3175, 1992.
19. M Vallet-Regí, V Ragel, J Román, JL Martínez, M Labeau, JM González-Calbet. J Mater Res 8:138, 1993.
20. E Matijević, WF Espenscheid, M Kerker. J Colloid Sci 18:91, 1989.
21. WF Espenscheid, E Matijević, M Kerker. J Phys Chem 68:2831, 1964.
22. K Spurny, V Hampl. Collection Czech Chem Commun. 30:507, 1965.
23. S Kitani, S Ouchi. J Colloid Interface Sci 23:200, 1967.

24. HG Scheibel, J Porstendörfer. J Aerosol Sci 14:113, 1983.
25. WF Espenscheid, E Willis, E Matijević, M Kerker. J Colloid Sci 20:501, 1965.
26. E Matijević, M Kerker, KF Schulz. Disc Faraday Soc 30:178, 1960.
27. RT Jacobsen, M Kerker, E Matijević. J Phys Chem 71:514, 1967.
28. PK Hopke, ed. Aerosol Sci Technol 19(4), 1993.
29. TT Kodas, MJ Hampden-Smith. Aerosol Processing of Materials. New York: Wiley-VCH, 1999.
30. D Sinclair, VK LaMer. Chem Rev 44:245, 1949.
31. G Nicolaon, DD Cooke, M Kerker, E Matijević. J Colloid Interface Sci 34:534, 1970.
32. G Nicolaon, DD Cooke, EJ Davis, M Kerker, E Matijević. J Colloid Interface Sci 35: 490, 1971.
33. JC Barrett, CF Clement. J Aerosol Sci 19:223, 1988.
34. F Family, P Meakin. Phys Rev A 40:3836, 1989.
35. TH Tsang, SM Cook, ME Marra. Aerosol Sci Technol 12:386, 1990.
36. C-M Huang, M Kerker, E Matijević, DD Cooke. J Colloid Interface Sci 33:244, 1970.
37. BJ Ingebrethsen, E Matijević. J Colloid Interface Sci 100:1, 1984.
38. K Nakamura, RE Partch, E Matijević. J Colloid Interface Sci 99:118, 1984.
39. FC Mayville, RE Partch, E Matijević. J Colloid Interface Sci 120:135, 1987.
40. RN Berglund, BYH Liu. Environ Sci Technol 7:147, 1973.
41. M Kerker, E Matijević, WF Espenscheid, WA Farone, S Kitani. J Colloid Sci 19:213, 1964.
42. C Rodes, J Smith, R Crouse, G Ramachandran. Aerosol Sci Technol 13:220, 1990.
43. C Roth, R Köbrîch. J Aerosol Sci 19:939, 1988.
44. H Toivonen, MR Bailey. Aerosol Sci Technol 11:196, 1989.
45. D Baskin, J Wolfenstine, EJ Lavernia. J Mater Res 9:362, 1994.
46. A Hickey, PR Byron. J Pharm Sci 76:338, 1987.
47. J Rubio, JL Oteo, M Villegas, P Duran. J Mater Sci 32:643, 1997.
48. XF Dai, AH Lefebvre, J Rollbuhler. Am Soc Mech Eng Report 88-GT-7, 1988.
49. K-J Choi, B Delcorio. Rev Sci Instrum 61:1689, 1990.
50. M Langlet, E Senet, JL DesChanvres, G DeLabouglise, F Weiss, JC Joubert. J Less-Comm Mater 151:399, 1989.
51. A Pebler, RG Charles. MRS Bull 24:1069, 1989.
52. R Salmon, E Matijević. Ceram Int 16:157, 1990.
53. L Durand-Keklikian, R Partch. Colloids Surf 41:327, 1989.
54. M Ocaña, J Sanz, T Gonzales-Carreño, CJ Serna. J Am Ceram Soc 76:2081, 1993.
55. JS Chin, AH LeFebvre. Int J Turbo Jet Engines 3:293, 1986.
56. M Visca, E Matijević. J Colloid Interface Sci 68:308, 1979.
57. E Matijević, M Visca. US patent 4,241,042; Fr 2429184; GB 2023115, 2070579; JP 55/023090, 1980.
58. BJ Ingebrethsen, E Matijević. J Aerosol Sci 11:271, 1980.
59. A Sood, R Township, RA Marra. US patent 4,678, 657, 1987.
60. TT Kodas, A Sood, SE Pratsinis. Powder Technol 50:47, 1987.
61. M Ocaña, V Fornes, CJ Serna. Ceram Int 18:99, 1992.
62. M Ocaña, E Matijević. J Aerosol Sci 21:811, 1990.
63. S Gablenz, D Voltzke, H Pabicht, J Neumann-Zdralek. J Mater Sci Lett 17:537, 1998.
64. K Okuyama, Y Kousaka, N Tohge. AIChE J 32:2010, 1986.

65. K Okuyama, JT Jeung, Y Kousaka, HV Neguyen, JJ Wu. Chem Eng Sci 44:1369, 1989.
66. QH Powell, G Fotou, T Kodas, B Anderson, Y Guo. J Mater Res 12:552, 1997.
67. R Partch, E Matijević, AW Hodgson, BE Aiken. J Polym Sci Polym Chem Ed 21:961, 1983.
68. RE Partch, K Nakamura, KJ Wolfe, E Matijević. J Colloid Interface Sci 105:560, 1985.
69. S Gangolli, R Partch, E Matijević. Colloids Surf 41:339, 1989.
70. CJ Zimmermann, RE Partch, E Matijević. J Aerosol Sci 22:881, 1991.
71. BJ Ingebrethsen, E Matijević, RE Partch. J Colloid Interface Sci 95:228, 1983.
72. A Balboa, RE Partch, E Matijević. Colloids Surf 27:123, 1987.
73. DC Bradley, DAW Hill. J Chem Soc 2101, 1963.
74. E Matijević, Q Zhong, RE Partch. Aerosol Sci Technol 22:162, 1995.
75. V Burkholz. Droplet Separation. Weinheim: VCH, 1989.
76. M Ocaña, M Martinez-Gallego. Colloid Polym Sci 275:1010, 1997.
77. P Tartaj, CJ Serna, J Soria, M Ocaña. J Mater Res 13:413, 1998.

1.6 REACTION IN GAS PHASES

MASAAKI ODA
Vacuum Metallurgical Company Ltd., Sanbu-cho, Sanbu-gun, Chiba, Japan

1.6.1 Introduction

In recent years, much attention has been given to investigate ultrafine particles (UFPs). Physical and chemical properties of UFPs, which may be different from those of bulk materials, has been one of the most interesting subjects. Thus a great deal of effort has been given to prepare UFPs with uniform particle size or, in a more practical sense, with narrower size distribution. Formation methods of UFP through the process of condensation of evaporated atoms or molecules are called the gas evaporation methods (1,2). Among the known physical and chemical methods used in preparing UFPs, the gas evaporation method has been considered to be advantageous in preparing the particles with fewer impurities and better crystallized structures, since the particles are grown in a pure inert gas atmosphere.

The gas evaporation methods are classified depending on the heating method used, such as resistance heating, induction heating, arc heating, and laser heating. There are several ways to synthesize oxide UFPs by the gas evaporation method. One method produces UFPs by heating and evaporating metals and semi-metals in an oxygen atmosphere causing oxidation. When a piece of metal is heated in a mixture of inert gas containing oxygen and evaporated, metal oxide UFPs are obtained. To heat the metal, resistance or arc heating can be used. The former has problems due to a reaction between the molten metal and the heating device and oxidation of the heating device itself. In the latter, the electrode itself is evaporated, so it does not have the problems that occur in the resistance heating. A second method produces metal UFPs by evaporating metals in an inert gas atmosphere and transfers them to the region where oxidation occurs. This method is developed for monodisperse oxide UFPs in cases where the original metal can be evaporated by an induction heating and oxidation reaction can generate enough heat to recrystalize oxide UFPs such as Fe. A third method involves oxide melting followed by evaporation. Laser is the only tool to evaporate oxide itself to form UFPs (3).

In this chapter, the first and the second methods are explained.

1.6.2 Formation of Oxide UFPs by Arc Discharge Heating (4)

1.6.2.1 Formation Process

A schematic diagram of the oxide UFP formation apparatus is shown in Figure 1.6.1. The apparatus consists of a formation chamber, consumable positive and negative

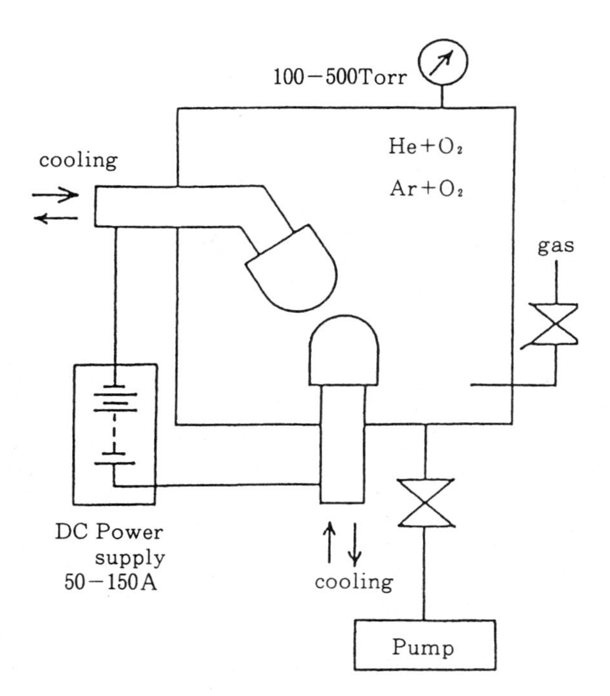

Fig. 1.6.1 Schematic diagram of oxide UFP formation apparatus. (From Ref. 4.)

electrodes, DC power supply, a gas supply and a vacuum pumping system. Both electrodes are made of the materials, which are evaporated in cases where the material can be shaped easily, such as Al or Zr. The evaporated materials are placed on a water-cooled Cu hearth as the positive electrode against the negative electrode, which is made of carbon in cases where the material cannot be shaped easily, such as Si or Mg. DC arc discharge is started between the electrodes to evaporate the materials in the atmosphere of He or Ar gas mixed with O_2 gas. Evaporated metal atoms are collided with atmospheric gas and cooled and reacted condensing into particles. Formed particles are deposited on a inner wall of the formation chamber. The atmospheric gas pressure is controlled between 150 torr and 500 torr. The electric current is controlled between 50 A and 200 A.

1.6.2.2 Identification of the Particles

TEM pictures of the UFPs formed under the condition of He (120 torr) mixed with O_2 (30 torr) and 50 A are shown in Figure 1.6.2. The characters of the formed UFPs are summarized in Table 1.6.1.

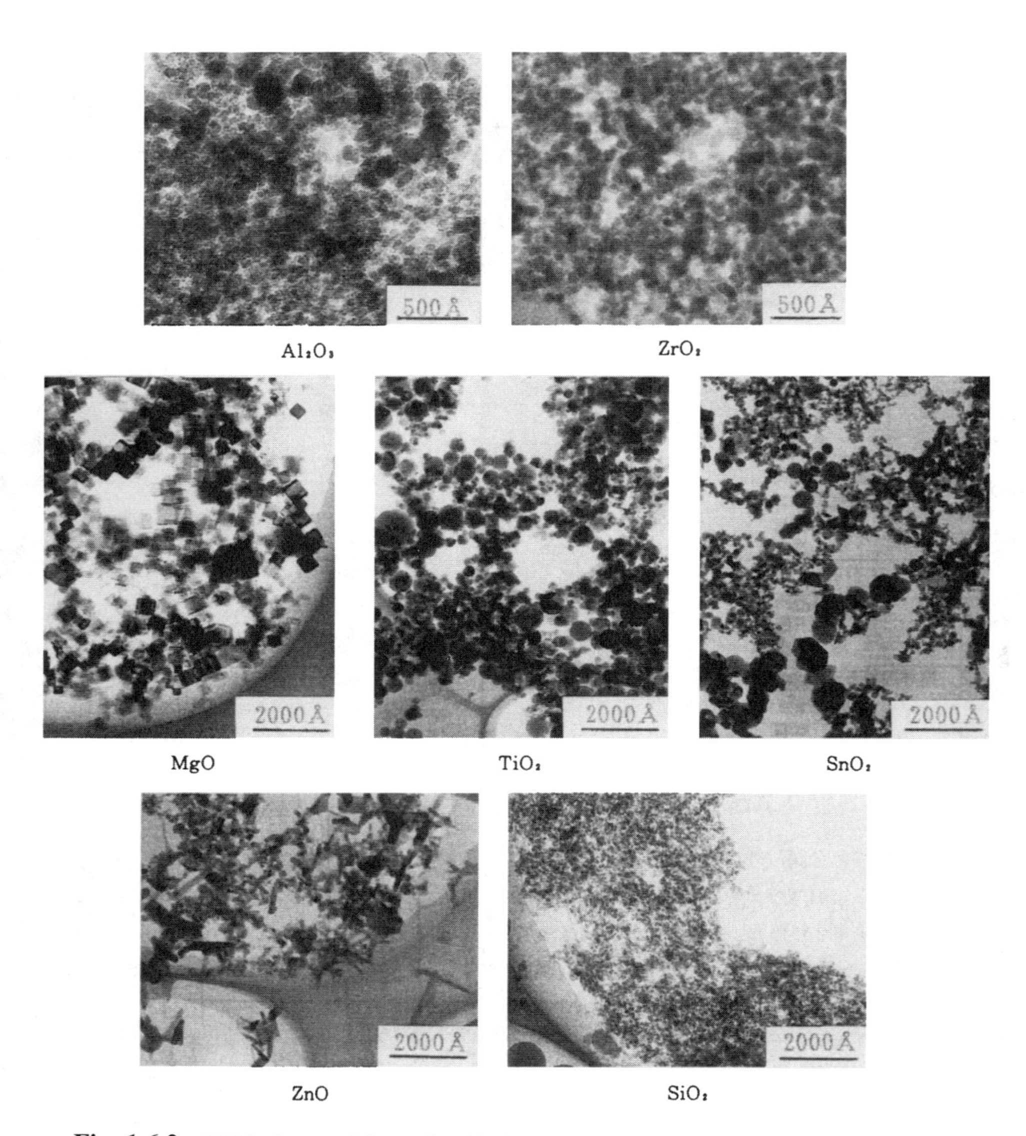

Fig. 1.6.2 TEM picture of formed oxide UFPs. (From Ref. 4.)

Table 1.6.1 Characters of Formed Oxide UFPs

Kind of UFP	Average diameter (Å)	Morphology	Crystal phase
Al_2O_3	100	Spherical	Gamma
ZrO_2	80	Spherical	Tetragonal
MgO	300	Cube	Cubic
TiO_2	300	Spherical	Anatase
SnO_2	300	Spherical	Tetragonal
ZnO	300	Elongated	Zincite (hexagonal)
SiO_2	100	Spherical	Amorphous

Source. From Ref. 4.

1.6.2.3 Size Control

Size dependence of Al_2O_3 particles versus the kinds of atmospheric gas and the arc discharge electric currents are shown in Figure 1.6.3. When He (120 torr) mixed with O_2 (30 torr) gas is used, the average particle sizes in the condition of 50 A and 150 A are 100 Å and 150 Å, respectively. When Ar (120 torr) mixed with O_2

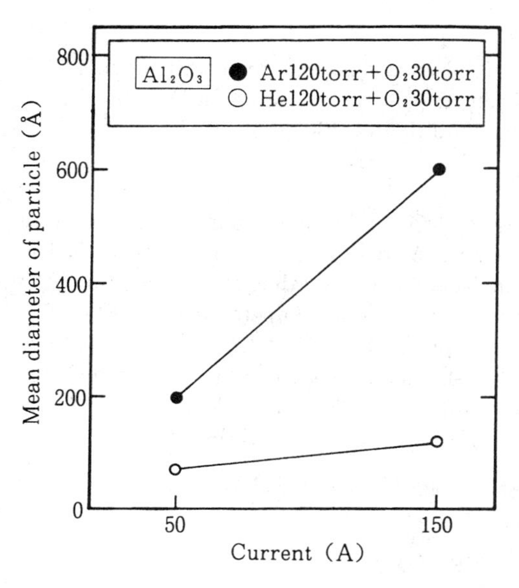

Fig. 1.6.3 Size dependence of Al_2O_3 particles versus arc discharge current. (From Ref. 4.)

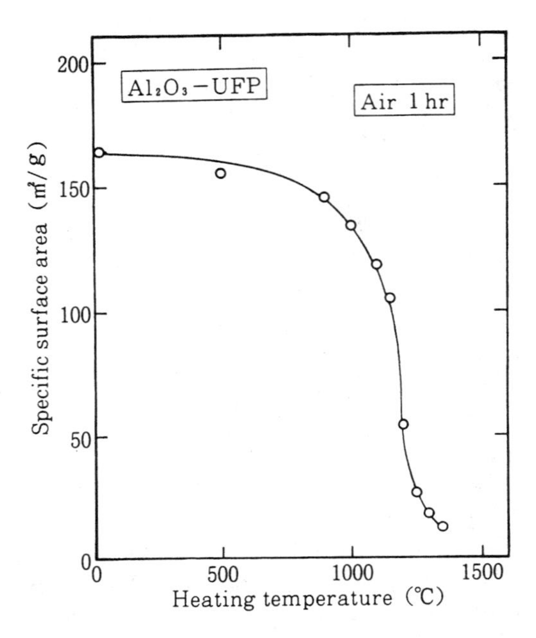

Fig. 1.6.4 Specific surface area dependence of Al_2O_3 versus heat-treated temperature. (From Ref. 4.)

(30 torr) gas is used, the average particle sizes in the condition of 50 A and 150 A are 200 Å and 600 Å, respectively. When Ar (470 torr) mixed with O_2 (30 torr) gas is used, the average particle size in the condition of 200 A is 1000 Å.

1.6.2.4 Properties of Al_2O_3 UFP

Dependence of the specific surface area of Al_2O_3 against the heat treated temperature is shown in Figure 1.6.4. The specific surface area of the original particles is 164 m^2/g. The specific surface area is gradually reduced according to the increase of the treatment temperature, but remaining at 134 m^2/g at the temperature of 1000°C. The specific surface area starts to reduce sharply at 1200°C and down to 10 m^2/g at 1350°C. X-ray diffraction patterns of heat-treated Al_2O_3 are shown in Figure 1.6.5. The particles heat-treated at temperatures up to 1150°C have γ-Al_2O_3 crystal structure, but they are transformed into α-Al_2O_3 structure over 1200°C.

1.6.3 Formation of Monodispersed Oxide UFP by Induction Heating (5)

1.6.3.1 General Description of the Method

It is known that grown particles easily form agglomerates. Individual fine particles in the agglomerate are thought to no longer possess the physical and chemical proper-

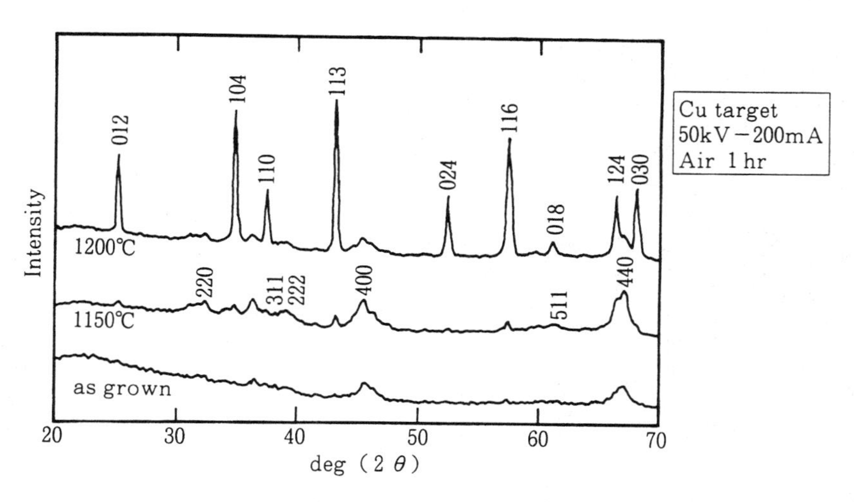

Fig. 1.6.5 X-ray diffraction patterns of heat-treated Al_2O_3. (From Ref. 4.)

ties of single fine particles. Therefore, much investigation has been made to prepare monodispersed fine particles with the gas evaporation method. The formation of monodispersed single-crystal Fe_3O_4 particles with an average diameter of 240 Å, and a distribution of full width at half maximum (FWHM) of 60 Å is presented.

In the gas evaporation method using induction heating as an evaporation source, atoms evaporated from a crucible collide with gas atoms and condense to form a fine particle. Further growth of the particle takes place with the coalescence of the particles. It was shown that the particle size and dispersion state of the particles strongly depends on the distance from the evaporation source (6). The particles near the evaporation source are small and in a monodispersed state. For example, monodispersed iron particles of about 200 Å in diameter can be obtained near the edge of the vapor flame with He gas pressure of 2.4 torr. Particles as small as 20 Å in diameter are obtainable by sampling at the inner part of the vapor flame. When the sampling is made outside of the vapor flame, particles larger than 200 Å can be obtained. Coalescence growth of a single particle stops just outside of the vapor flame region. The sizes of the particles in this region are determined primarily by the inert gas pressure and the temperature of the evaporation source. In the region farther from the evaporation source, the particles in monodispersed state form a necklace-shaped chain, whose length becomes longer as the distance from the evaporation source increases. Therefore, monodispersed UFPs are obtainable by selectively sampling the particles at the region where the particles are in desired size.

1.6.3.2 Formation Process

A schematic diagram of the formation apparatus is shown in Figure 1.6.6. The apparatus consists of three vacuum chambers, namely, the formation, the reaction,

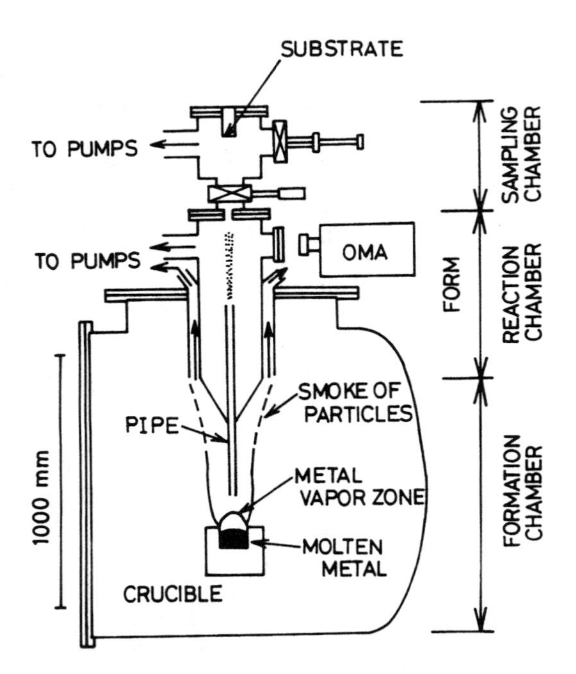

Fig. 1.6.6 Schematic diagram of monodispersed oxide UFP formation apparatus. (From Ref. 5.)

and the sampling chamber, each of which is differentially evacuated. The typical pressures of the chambers are 1, 10^{-2}, and $10^{-4}{}_{\mathrm{torr}}$ or lower for the formation, reaction, and sampling chambers, respectively.

In the formation chamber, iron metal chunks of 99.99% purity are placed in a ZrO_2 crucible 130 mm in diameter and evaporated by induction heating in the He gas atmosphere of 2.4 torr. The temperature of the evaporation source is measured with an optical pyrometer and is 1730°C. Selective collection of the particles is carried out by utilizing He gas flow caused by the difference of the pressures between the formation and the reaction chamber. The inlet of the pipe connecting the chambers is placed at the position where the particles are of desired size and in the desired dispersion state, i.e., either in a monodispersed state or in a chained form with desired length. The particles at the inlet are drawn in the pipe and transported to the reaction chamber together with He gas. Fluid dynamical calculation estimates the flying velocity of the particles about 100 m/s at the end of the connecting pipe in the reaction chamber.

Solid–gas reaction only on the surface or on the entire particle may occur while the particles fly through the reaction chamber by introducing reactive gas in the chamber. In the present study, O_2 gas at 0.013 torr is introduced into the chamber in order to form iron oxide particles. Thermal radiation from the particles in the reaction chamber during the oxidation is measured with an optical multichannel analyzer in order to estimate the temperature of the particles at oxidation reaction, which appears to be at about 1000°C.

The particles continue to fly into the sampling chamber through an orifice between the reaction and the sampling chamber. The pressure of the sampling chamber is 9.5×10^{-4} torr. The particles are collected in a form that is convenient for characterization or application. For example, the particles are collected on a microgrid for transmission electron microscope (TEM) observation and on a polyimid-film for Mössbauer and x-ray diffraction studies. A standard passivation treatment, namely, slow introduction of O_2 gas followed by the introduction of dry air to the chamber, is made.

1.6.3.3 Identification of the Particles

Particle size distribution and dispersion state are observed by using a TEM. Figure 1.6.7 shows that the particles are in mono-dispersed state. Observation was made for the sample of 249 particles under 6×10^5 magnification. The mean particle size and the FWHM of the particle size distribution are determined to be 240 Å and 60 Å, respectively. It should be noted that the distribution of the present sample is noticeably smaller than that of the particles prepared by a conventional gas evapora-

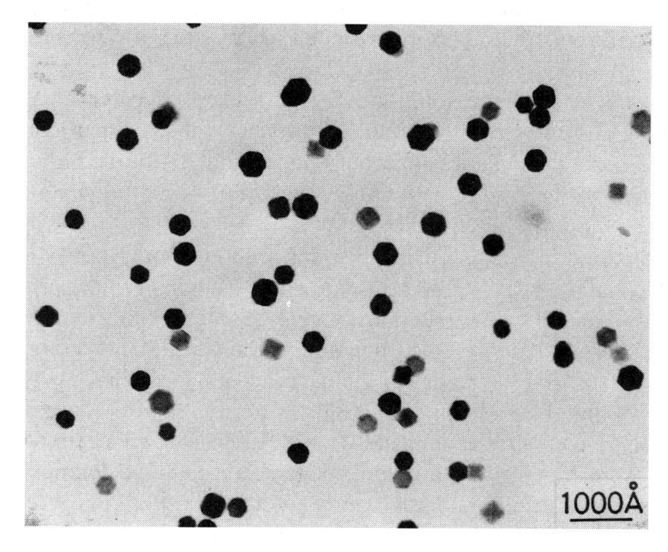

Fig. 1.6.7 TEM picture of monodispersed Fe_3O_4 UFP.

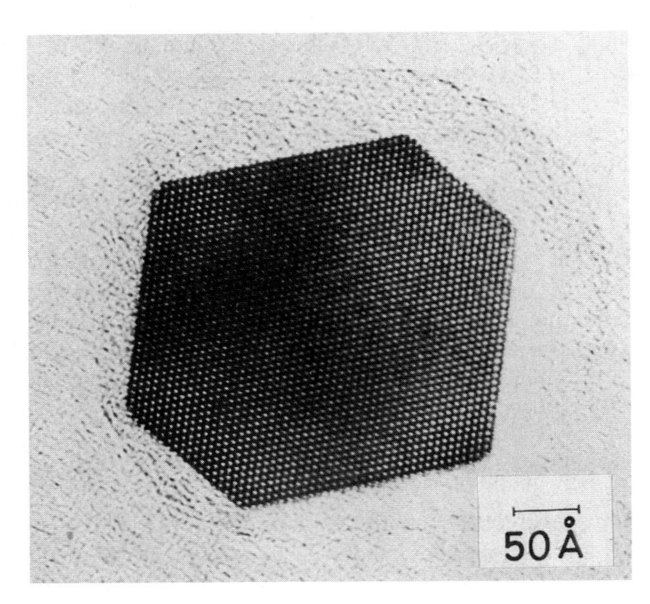

Fig. 1.6.8 High-resolution TEM image of a single Fe_3O_4 UFP. (From Ref. 5.)

tion method, here about 100 Å for the particles of which the average size is 250 Å. Electron diffraction patterns of the particles indicate that the particles are Fe_3O_4. It is noted that weak diffraction spots that might be assigned to FeO(OH) are seen in addition to Fe_3O_4 rings.

Figure 1.6.8 shows a typical high-resolution TEM image of the particle. A lattice image of the particle is clearly seen. Also the figure shows that the particle has good crystal habit, and contains no lattice defects. Detailed analysis of the high-resolution TEM images, such as the determination of crystal habit, is the subject of further study.

X-ray powder diffraction was recorded using a conventional x-ray powder diffractometer with Cu-Kα radiation. Polyimide film on which sample particles are deposited is glued on a glass sample holder with vacuum grease. Figure 1.6.9 shows the recorded diffraction pattern. An analysis of the pattern is made by comparing the lattice parameters and diffraction intensities of the particles and those of known iron compounds, and shows that the particles are Fe_3O_4.

Further identification of the particles is made with ^{57}Fe Mössbauer spectroscopy. Mössbauer spectra were recorded with a conventional constant acceleration spectrometer with ^{57}Co in Rh matrix as a γ-ray source. Velocity calibration was made using a 5-μm α-Fe foil at 293 K. Figure 1.6.10 shows the Mössbauer spectra of the sample recorded at 293 K and 4.2 K. Spectra were fitted with theoretical

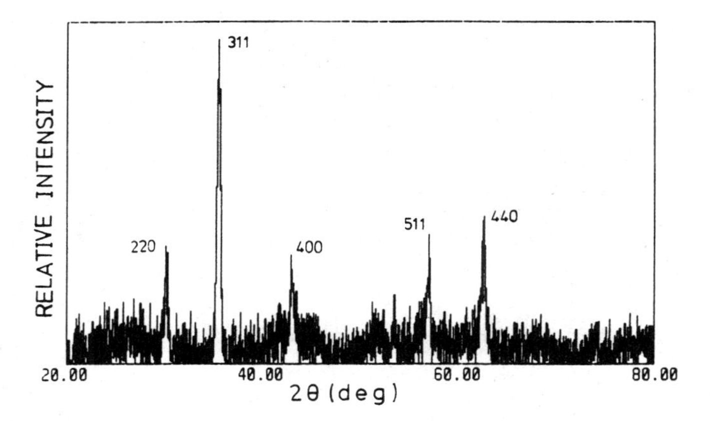

Fig. 1.6.9 X-ray powder diffraction pattern of monodispersed Fe_3O_4 particles at room temperature. (From Ref. 5.)

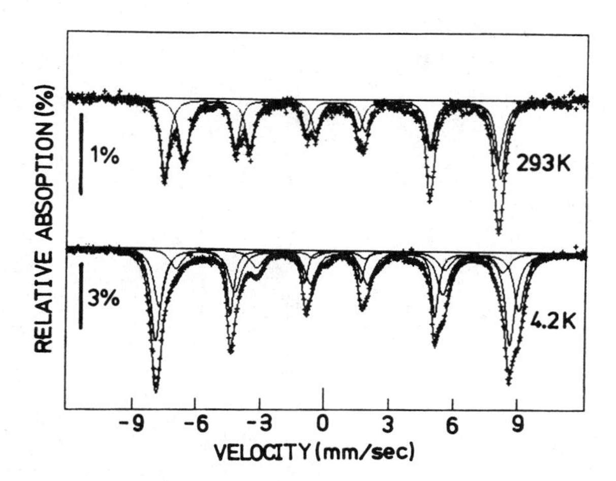

Fig. 1.6.10 Mössbauer spectra of monodispersed Fe_3O_4 particles recorded at 293 K and 4.2 K. (From Ref. 5.)

spectra using a conventional least squares fitting program. Hyperfine parameters obtained for the both spectra are consistent with those of Fe_3O_4 (7). The rearrangement of the Mössbauer absorption lines due to the Verway transition (8), which is peculiar to Fe_3O_4, takes place in the present particles, although the Verway transition temperature is not determined. Mössbauer results also support that the present particles are Fe_3O_4.

Note that both the x-ray diffraction and Mössbauer characterization do not reveal the presence of phases other than Fe_3O_4 with the present statistics of the respective data sets. The presence of a very small amount of FeO(OH) is suggested in the electron diffraction pattern. It is not known whether FeO(OH) exists on the surface of the particles and/or as an independent particle to date.

REFERENCES TO SECTION 1.6

1. See, for example, A Tsasaki, N Saegusa, M Oda. IEEE Trans Magn MAG-19:1731, 1983, and references therein.
2. K Kimoto, Y Kamiya, M Nonomiya, R Uyeda. Jpn J Appl Phys 2:702, 1963.
3. See, for example, Proc 2nd Symp Advanced Photon Processing and Measurement Technologies, Manufacturing Science and Technology Center, Tokyo, 1998, p 28.
4. M Tuneizumi, E Fuchita, M Oda. New Ceram 8:79, 1990.
5. M Oda, N Saegusa. Jpn J Appl Phys 24(9):L702, 1985.
6. E Fuchita, M Oda, S Kashu. Proc 7th Int Conf Vacuum Metal, Iron and Steel Institute of Japan, Tokyo, 1982, p 973.
7. See, for example, RS Hargrove, W Kudig. Solid State Commun 8:303, 1970, and references therein.
8. EJW Verway, PW Haayman. Physica 8:979, 1941.

2

Silica

HERBERT GIESCHE and K. OSSEO-ASARE

2.1 HYDROLYSIS OF SILICON ALKOXIDES IN HOMOGENEOUS SOLUTIONS

HERBERT GIESCHE
Alfred University, Alfred, New York

2.1.1 Introduction

In 1956 Gerhard Kolbe (1) published the first results that showed that spherical silica particles could be precipitated from tetraethoxysilane in alcohol solutions when ammonia was present as the catalyzing base. Several years later, in 1968, Stöber, Fink, and Bohn (2) continued in this research area and published the frequently cited ''original'' article for the preparation of monodispersed silica particles form alkoxide solutions. Stöber et al. improved the precipitation process and described the formation of exceptionally monodispersed silica particles. The final particle size could be controlled over a wide range from about 50 nm to 1 1/2 μm. Variations of the particle size could be achieved by different means, e.g., temperature, water and ammonia concentration, type of alcohol (solvent), TEOS (tetraethoxysilane) concentration, or mixing conditions.

The process did not draw ''too'' much attention until about 20 years later when several research groups started to apply those monodispersed silica particulates in various studies and the particle formation and growth process was studied in more detail. Van Helden, de Kruif, Vrij, Philipse, van Blaaderen, et al. (3–15) studied optical properties and the particle growth process by light scattering. Other methods were applied by Bogush and Zukoshi et al. (16–21). Van der Werff et al. (22,23) used highly concentrated suspensions of the silica particles for rheological studies. Different research groups utilized the highly monodispersed particulates to form ordered sphere packing structures for porosity studies (24–27) or the sintering behavior of the silica particles was analyzed and compared with theoretical models (28–35). The particle formation and growth mechanism was focus of several studies, which are discussed in more detail in a later section. Numerous other applications have been demonstrated for the ''Stöber'' silica particles, and those are described in Section 2.1.5. Main advantages of the Stöber silica particles are that the particles are exceedingly uniform in size, the size can be easily controlled, and the overall process proceeds at relative high concentrations and thus yields larger amounts of powders at comparable low cost. Several companies have adopted the production process; for example, the materials are used in special chromatography applications.

2.1.2 Experimental Procedure

With slight variations, the Stöber silica precipitation process proceeds from the same chemicals. The starting material is TEOS, tetraethoxysilane, $Si(OC_2H_5)_4$; the solvent is an alcohol (preferably ethanol); water is added; and ammonia acts as the catalyst to initialiate the hydrolysis and condensation reaction. In a very schematic way the reaction could be described as follows:

Hydrolysis:

$$Si(OC_2H_5)_4 \xrightarrow{+H_2O \quad -HOC_2H_5} Si(OC_2H_5)_3OH$$

$$\xrightarrow{+H_2O \quad -HOC_2H_5} Si(OC_2H_5)_2(OH)_2 \text{ etc.}$$

Condensation:

$$Si(OC_2H_5)_3OH \xrightarrow{+Si(OC_2H_5)_3OH \quad -H_2O} (H_5C_2O)_3Si\text{-}O\text{-}Si(OC_2H_5)_3 \text{ etc.}$$

$$\rightarrow SiO_2 \text{ particle}$$

The formation of silica particles is only in part influenced by the purity of the chemical constituents. However, pure chemicals and here especially the purity of the TEOS will help to get reproducible results as well as to achieve narrow particle size distributions.

2.1.2.1 The One-Step (Stöber) Process

The original process was described by Stöber, Fink, and Bohn (2). A suitable alkoxysilane is reacted in the corresponding alcohol, water, and ammonia mixture. Usually the reaction is performed at room temperature, but higher or lower temperatures can also be applied, if so desired. Stöber et al. described the influence of different alkoxides and alcohols as well as the water and ammonia concentration on the resulting particle size. A more specific example is provided next.

Tetraethyl orthosilicate should be distilled immediately before use, since it quite often contains higher oligomers, which might influence the particle formation and especially the monodispersity of the resulting particle size distribution. Prior to the distillation, CaO should be added in order to capture HCl impurities, which often remain from the synthesis of TEOS. All other chemicals should be of analytical or reagent grade and can be used without further purification. Some experiments may require a higher NH_3 concentration in relation to the permitted water content. In order to achieve this, part of the regular ethanol can be replaced by ethanol that has been saturated with ammonia. The latter is prepared by bubbling NH_3 through the ethanol at 195 K (−78°C). The ammonia concentrations should be checked in the final solutions (prior to the precipitation) by titration with HCl. The water and TEOS concentrations are calculated from the weight fractions of the different portions.

The precipitation proceeds then as follows. Mixture A, alcohol, water, and ammonia, and mixture B, TEOS in alcohol (1:1 volume ratio), are brought to the

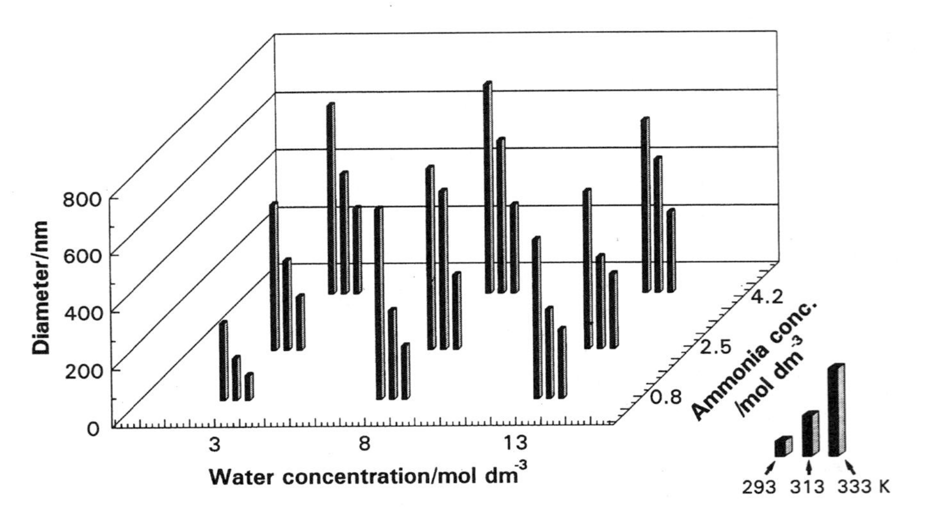

Fig. 2.1.1 Mean particle size at different reaction conditions. (From Ref. 37.)

desired temperatures in closed containers. Then mixture A is transferred into a round-bottom flask, equipped with a paddle stirrer and a condenser, and placed in a constant temperature bath. Thereafter mixture B is added under vigorous stirring (1000 rpm). The mixing is stopped after 15 s. In spite of the constant-temperature bath, the temperature in the center of the reaction vessel will increase slightly by up to 2 K during the first few minutes after the addition of TEOS, which is due to the exothermic character of the reaction. Using a mixture of 0.2 mol dm^{-3} TEOS, 8 mol/L water, and 1 mol dm^{-3} ammonia will produce about 0.5- to 0.8-μm particles at room temperature in ethanol as the solvent.

The influence of different concentrations and reaction temperatures is shown in Figure 2.1.1. In general, particles will become larger at lower reaction temperatures. Additional results for temperature changes between −20°C and +60°C and the influence of the solvent (alcohol) have been published elsewhere (36). When tetramethoxysilane is used instead of tetraethoxysilane the resulting particles are smaller. The same trend is observed when ethanol is replaced by methanol as the solvent. Longer and more branched alkoxides as well as higher alcohols as solvents will produce correspondingly larger particles. The tetraethoxysilane concentration has an obvious influence on the particle size; higher concentrations will form larger particles. However, in most cases the concentrations is kept around 0.2 mol dm^{-3} since larger concentrations often lead to nonspherical particulates and smaller concentrations obviously will result in a lower amount of product. Other effects from the addition of acetone or salts have also been described elsewhere (18,19). Depending on the concentrations and the temperature, the reaction may be completed in just a few minutes or it may take several hours. The specific example given in the preceding paragraph results in a relative fast reaction, since the ammonia concentration of 1 mol dm^{-3} is quite high and the reaction is also not limited by a small (insufficient) amount of water. The reaction will be completed in less than 1 h. More specific information with respect to the reaction rate is provided in Chapter 3.

Bogush et al. (17) summarized their experiments by fitting the experimental particle size data with the following equation:

$$d = A\,[H_2O]^2 \exp(-B\,[H_2O]^{1/2}) \tag{1}$$

where

$$A = [TEOS]^{1/2}\,(82 - 151\,[NH_3] + 1200\,[NH_3]^2 - 366\,[NH_3]^3)$$

and

$$B = 1.05 - 0.523\,[NH_3] - 0.128\,[NH_3]^2$$

The equation is valid for 298 K (25°C) and the following concentration ranges given in mol/dm^3: TEOS 0.1–0.5, water 0.5–17, and ammonia 0.5–3. The calculated value of d represents the mean particle diameter given in nanometers.

Reaction Rate. The reaction rate can be defined in different ways. One can distinguish between the hydrolysis and the condensation reaction, or one could simply relate the reaction rate to the particle growth rate. Depending on the technique applied, the various relations are determined. This section now focuses on the reaction rate as determined from the particle growth.

The reaction can either be stopped by addition of trimethylchlorosilane [also called ‘‘endcapping’’; see additional literature for further details (37,39–42)] or the particle growth can be determined in situ by light-scattering techniques (37). In general, the growth follows a first-order reaction with respect to the TEOS concentration. The reaction rates vary over several orders of magnitude depending on the reaction conditions; see Figure 2.1.2.

As can be seen from Figure 2.1.2, the water and ammonia concentrations have a significant influence on the reaction rate, and the corresponding reaction orders with respect to the water and ammonia concentrations can also be determined. The following overall equation can then relate the different reaction parameters to the growth rate of the particles. The equation is given for the TEOS, water, ammonia, ethanol system in the temperature range between 20 and 60°C.

$$\begin{aligned} d[SiO_2\ \text{particle}]/dt &= k_{TEOS}[TEOS] \\ &= k'\left\{\exp\left(\frac{-E_a}{RT}\right)\right\}[H_2O]^{1.18}[NH_3]^{0.97}[TEOS] \\ &= 2.36\ s^{-1}(\text{mol dm}^{-3})^{-2.15} \\ &\qquad \left\{\exp\left(\frac{-3256\ K}{T}\right)\right\}[H_2O]^{1.18}[NH_3]^{0.97}[TEOS] \end{aligned}$$

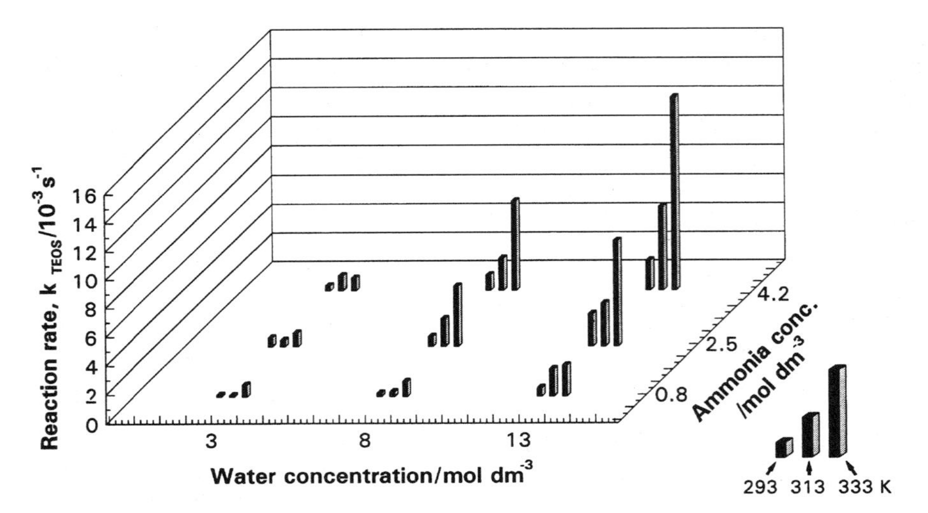

Fig. 2.1.2 Reaction rate of the Stöber silica particle growth reaction at a constant TEOS concentration of 0.2 mol/L; each data triplet indicates the reaction temperature of 293, 313, and 333 K, respectively. (From Ref. 37.)

The results correspond well with data published in other papers on similar reaction systems; see Figure 2.1.3.

2.1.2.2 Controlled Growth Reaction

After initial seed particles have been prepared according to the procedure described in Section 2.1.2.1, a careful addition of further TEOS, ammonia, and water can lead to a controlled growth of the existing seed particles without the occurrence of a second particle nucleation process. In practice this can be achieved by the addition of smaller quantities of TEOS to the seed solution. A stepwise process is described in further detail by Bogush et al. (20) or Giesche (38). It can also be done in a quasi-continuous setup as shown in Figure 2.1.4. The previously prepared seed suspension is gently stirred in the flask at about 100 to 200 rpm. Faster stirring will result in too much loss of ammonia into the gas phase, and less stirring does not provide the necessary mixing effect in the system. TEOS is premixed with ethanol in a 1 : 1 ratio and added by means of a peristaltic pump on the left side, whereas an ammonia, water, and ethanol mixture is introduced on the right-hand side of the flask. The rate of addition can be adjusted by the peristaltic pumps. It is important that the TEOS does not prereact too much with ammonia in the gas phase, which could cause a secondary nucleation effect. Thus, the TEOS outlet is carefully adjusted to obtain free falling droplets through the condenser setup. In addition, the ammonia present in the gas phase is kept away from the TEOS orifice by rinsing the TEOS

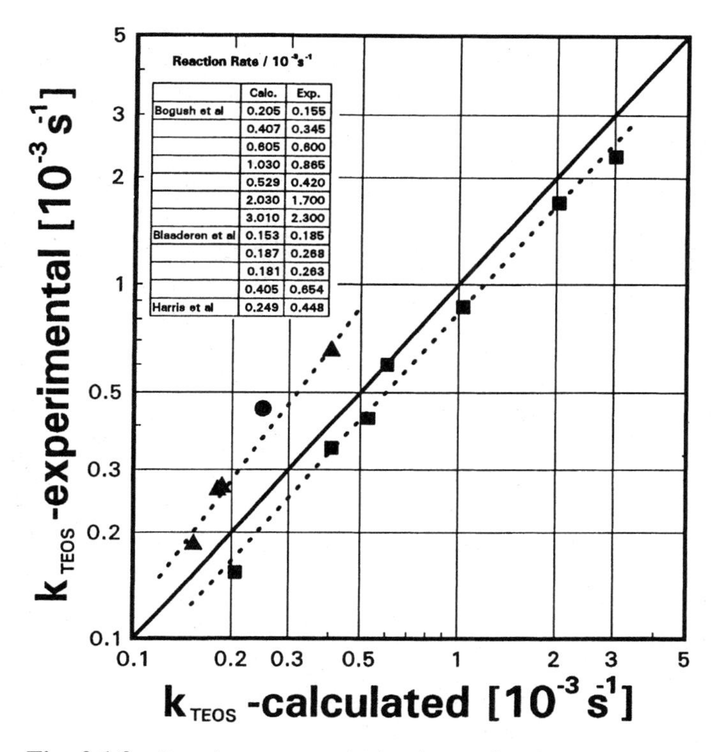

Reaction Rate / $10^{-3}s^{-1}$

	Calc.	Exp.
Bogush et al	0.205	0.155
	0.407	0.345
	0.605	0.600
	1.030	0.865
	0.529	0.420
	2.030	1.700
	3.010	2.300
Blaaderen et al	0.153	0.185
	0.187	0.268
	0.181	0.263
	0.405	0.654
Harris et al	0.249	0.448

Fig. 2.1.3 Reaction rate as calculated according to Eq. (2) compared with experimental values by Bogush and Zukoski (19), van Blaaderen et al. (13), Harris et al. (43), or Matsoukas and Gulari (44,45). (From Ref. 37.)

outlet with dry air. Large quantities of precisely controlled particle sizes can be produced by this process.

Faster rates of addition can be applied at smaller particle sizes. The larger the particles grow, the more slowly the TEOS has to be added. It is very easy to grow particles that are less than 0.2 μm in size, but the growth of particles larger than 1.5 μm has to be done under very carefully controlled conditions. The following general guidelines can be used to avoid a secondary nucleation. The ammonia concentration should be kept between 0.5 and 1 mol dm^{-3} and the water concentration should be kept at about 5 to 8 mol dm^{-3}. The TEOS concentration can be slightly increased during the process to values of 1 mol/L or even higher in some cases. The amount of existing silica in the suspension can be doubled within 4 h as long as the particle size is less than 0.5 μm, but for particle sizes of 1 μm the same addition should be done over a period of 8 h. Even larger particle sizes require correspondingly longer addition times. For larger particles it is also important to keep the solution

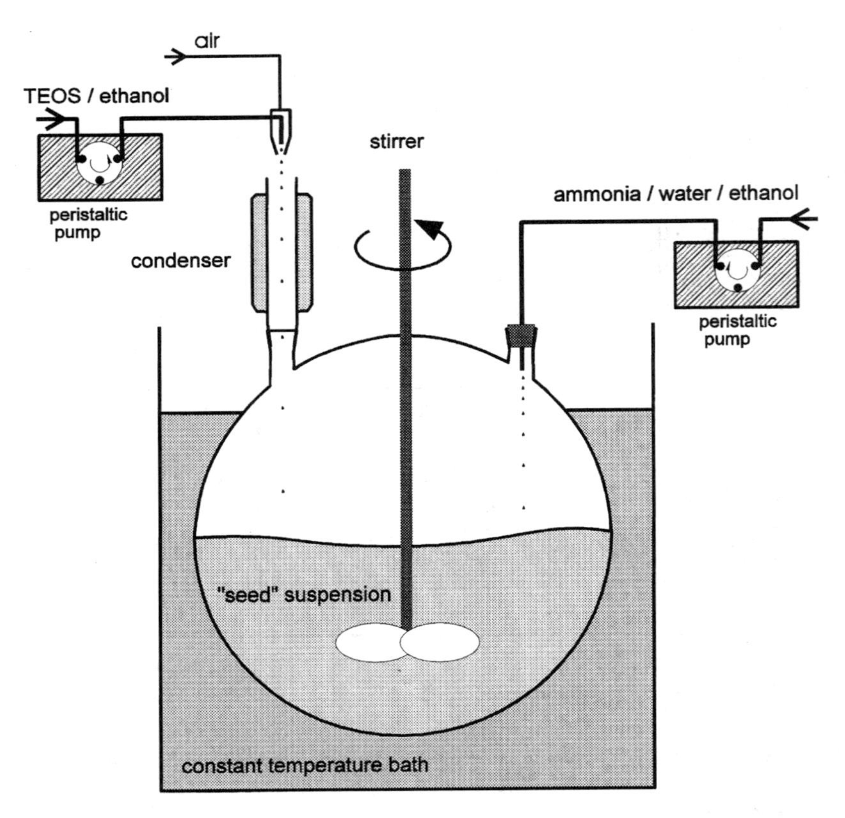

Fig. 2.1.4 Schematic setup for a controlled growth procedure. (From Ref. 38.)

at a high enough particle concentration in order to keep the interparticle distance small. However, one has to consider at the same time that particles will start to agglomerate when the suspension becomes too crowded and doublet or triplet particles occur. The latter unwanted particles can still be removed from the suspension by a simple sedimentation process if they are noticed in an early stage.

Several similar synthesis processes have been described in the literature that lead to particles of up to 10 μm (38,46–49). The formation of larger particles is based in all those processes on the hydrolysis of alkoxysilanes and the deposition of silica onto previously prepared seed particles.

2.1.2.3 Continuous Reactor

The previously described processes are batch type processes, which have certain limitations with respect to their scale-up possibilities. A continuous production pro-

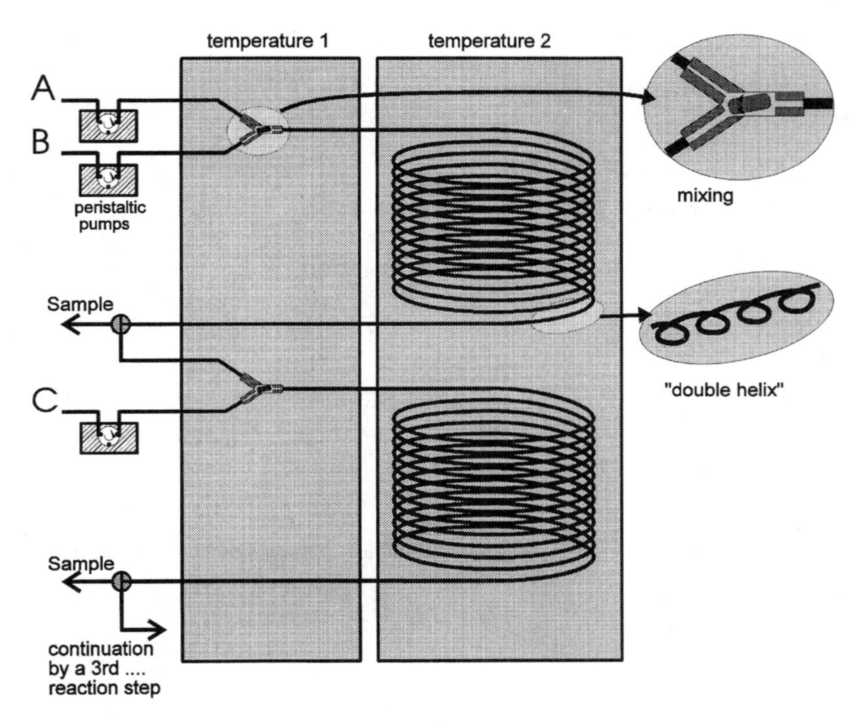

Fig. 2.1.5 Schematic drawing of a continuous reactor system. (From Ref. 38.)

cess has several important advantages with respect to a larger production process and the economics of the process (38). Figure 2.1.5 shows the schematic drawing of a continuous production setup. It primarily consists of three sections: a pumping, a mixing, and a reaction zone. The flow of the two components A (ammonia/water/ethanol) and B (TEOS/ethanol), is adjusted by two peristaltic pumps. After thoroughly mixing the two liquids at room temperature or at 273 K (0°C), the reaction proceeds at 313 K (40°C), set by a second temperature bath. The length of time at which the reaction proceeds at T_1 and T_2 can be adjusted by the length and size of the tubing as well as by the pumping speed.

In some cases a second (or further) reaction step(s) can be added in order to adjust the particle size, to increase the mass fraction, or to modify the surface of the silica particles prepared in the first step. The process proceeds in a similar way as already described for the first continuous preparation step; however, mixture A is now substituted by the previously prepared silica dispersion, and mixture B is replaced either by a further amount of TEOS or by various organotrialkoxysilanes (mixture C).

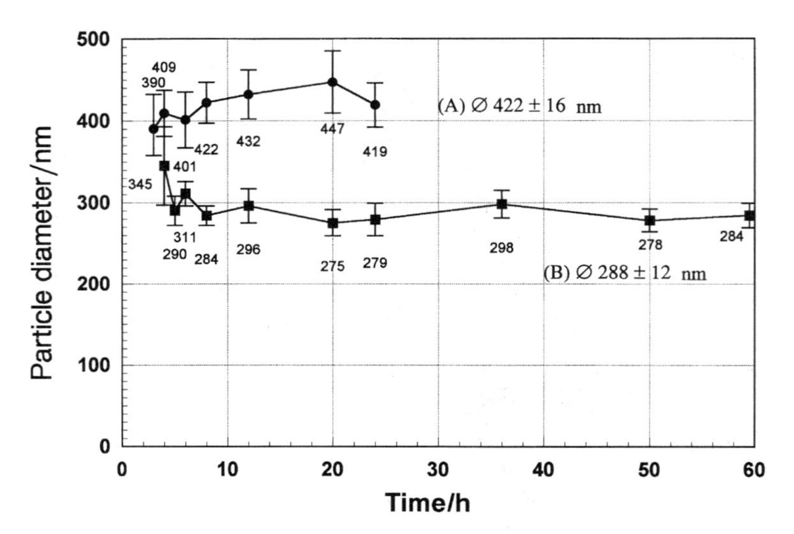

Fig. 2.1.6 Particle size variations during the continuous synthesis process; reaction conditions are described in the text. (From Ref. 38.)

Figure 2.1.6 shows the results of such a continuous synthesis process. It shows the variation of the mean particle size during the experiment. The error bars indicate the standard deviation of the particle size distribution of each sample based on the transmission electron micrographs (number distribution). The experiment was performed under the following conditions: (A) ammonia, water, and TEOS concentrations were 0.8, 8.0, and 0.2 mol dm^{-3}; $T_1 = 273$ K, $T_2 = 313$ K; total flow rate was 2.8 cm^3 min^{-1}; 100 m reaction tube of 3 mm diameter residence time $\approx$ 4 h; and (B) ammonia, water, and TEOS concentrations were 1.5, 8.0, and 0.2 mol dm^{-3}; $T_1 = 273$ K, $T_2 = 313$ K; total flow rate was 8 cm^3 min^{-1}; 50 m reaction tube of 6 mm diameter, residence time $\approx$ 3 h. Further details and other examples are described elsewhere (38). Unger et al. (50) also described a slightly modified continuous reaction setup in another publication.

2.1.2.4 Porous Silica Beads

An interesting modification of the Stöber silica process has been described by Unger et al. (50). By using a mixture of TEOS and an alkyltriethoxysilane they were able to synthesize monodispersed porous silica particles. The porosity is created by the alkyl groups, which act like space holder. After calcination/burnout of the organics, a well-defined porosity is left behind in the silica particles. The materials are used for very fast high-pressure liquid chromatography.

2.1.3 Characterization

Numerous techniques have been applied for the characterization of Stöber silica particles. The primary characterization is with respect to particle size, and mostly transmission electron microscopy has been used to determine the size distribution as well as shape and any kind of aggregation behavior. Figure 2.1.7 shows a typical example. As is obvious from the micrograph, the Stöber silica particles attract a great deal of attention due to their extreme uniformity. The spread (standard distribution) of the particle size distribution (number) can be as small as $\pm 1\%$. For particle sizes below 50 nm the particle size distribution becomes wider and the particle shape is not as perfectly spherical as for all larger particles. Recently, high-resolution transmission electron microscopy (TEM) has also revealed the microporous substructure within the particles (see Fig. 2.1.8) (51), which is further discussed in the section about particle formation mechanisms.

Frequently, particle size characterization is done by light scattering. However, this is often done as a check for the light-scattering technique or the instrument rather than for particle size determination. In addition, small-angle x-ray scattering (SAXS) or neutron scattering has been used, and these techniques indicated a similar nanometer-sized substructure within the particles, providing important information

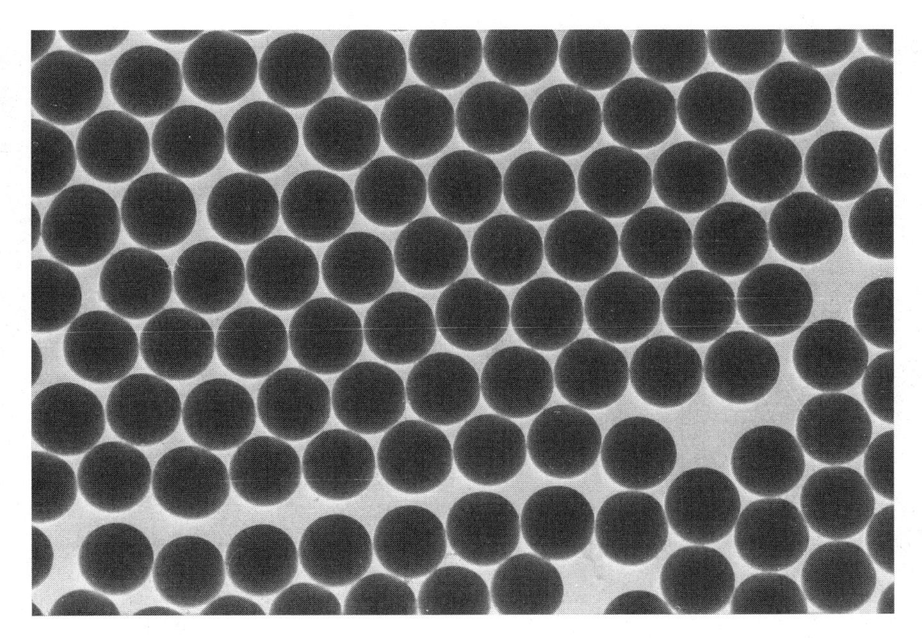

Fig. 2.1.7 Transmission electron micrograph of Stöber silica particles.

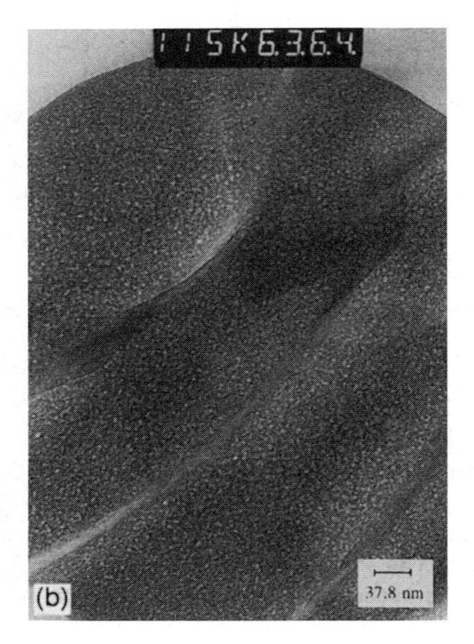

Fig. 2.1.8 Transmission electron micrograph showing the internal structure of Stöber silica particles. (From Ref. 51.)

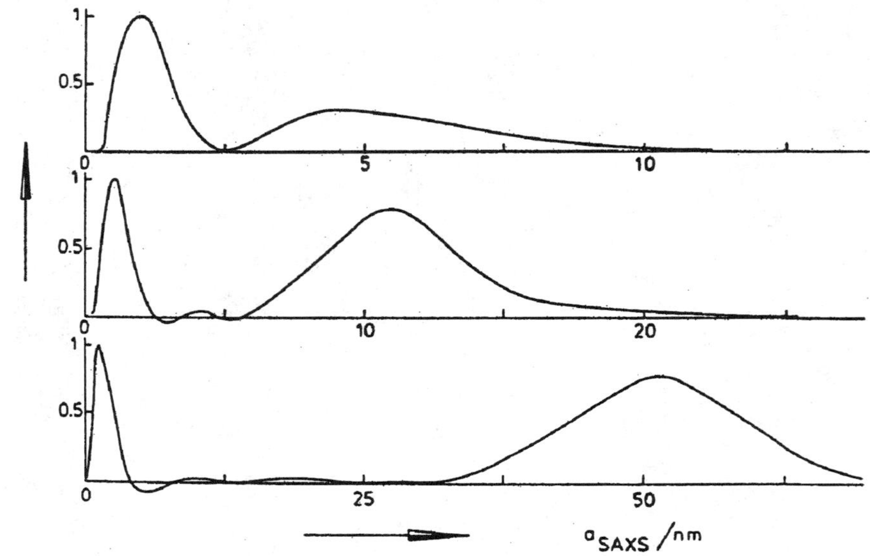

Fig. 2.1.9 Small-angle x-ray scattering (SAXS) of Stöber silica particles. (From Ref. 14.)

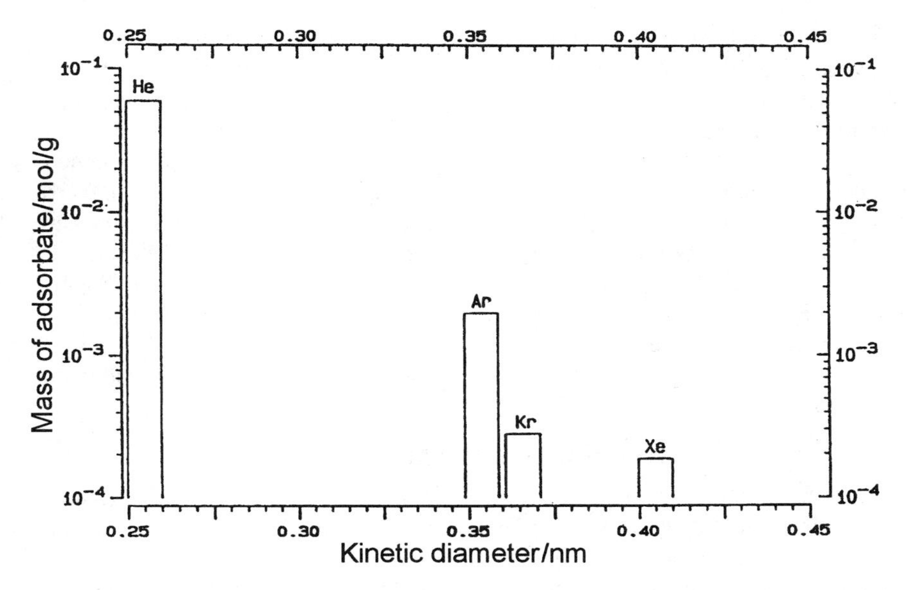

Fig. 2.1.10 Adsorption of gases on/in Stöber silica particles. (Courtesy of Horst Reichert, University of Mainz, Germany.)

about the particle growth mechanisms. Figure 2.1.9 shows the SAXS result published by van Helden et al. (14). This is discussed in more detail in Section 2.1.4.

Surface area analysis by gas adsorption is another technique frequently used to characterize powders. However, in the case of Stöber silica the results are not very reliable. The geometrical particle size is only vaguely related to the measured surface area. In many cases the adsorption analysis cannot be performed at all, since it takes an extremely long time to reach the adsorption equilibrium for the individual data points. It has been noted that the size of the adsorbate molecule plays a critical role. Small molecules like helium can penetrate the micropore structure within the particles and a higher surface area is calculated, whereas larger molecules like xenon cannot penetrate the pores and a lower surface area is determined. Nitrogen, argon, and krypton are right at the cutoff line, and in some cases the molecules can penetrate the structure whereas in other cases they can not (see Fig. 2.1.10). This also depends to a major degree on the pretreatment conditions of the powder. A calcination at temperatures of about 1000°C will densify the particles and the internal pores will disappear.

Stöber silica particles also show a low density of the powder as precipitated. All reported literature values are at or below a density of 2.0 g cm^{-3}, and van Helden et al. (14,15) reported values of as low as 1.61 g cm^{-3}. These results are in accordance with the previously discussed microporosity and TEM substructure in the particles.

Only at calcination temperatures above 800°C does the density increase to the literature value of amorphous silica of 2.2 to 2.25 g cm^{-3}. The exact microstructure within the Stöber silica particles depends very much on the specific precipitation conditions, which are discussed in more detail in section 2.1.4.

Thermogravimetric analysis in combination with other spectroscopy methods indicated that the original particles contain larger amounts of ammonia, unreacted ethoxy groups, and a larger amount of hydroxyl groups in the interior of the particles.

^{29}Si MAS-NMR experiments by van Blaaderen et al. (11), Labrosse et al. (51), Humbert (52), and Davis et al. (53) have indicated the same porous microstructure within the Stöber silica particles as observed by TEM and the surface area analysis. The publications reported high values for the Q^3 and the Q^2 species, which are an indication of a very open internal structure or molecular network. Q^4 values of approximately 65%, Q^3 of 30%, and Q^2 of about 5% were reported.

2.1.4 Formation/Reaction Mechanisms

Several growth and formation mechanisms have been proposed for the formation of monodispersed Stöber silica particles. Silica in general is an extremely well-studied system, and there are numerous publications with respect to the hydrolysis and condensation reaction. At present there are two major formation mechanisms that have been used to explain the formation of Stöber silica particles.

First, there is the monomer addition growth model as presented by Matsoukas and Gulari (44,45), in analogy to a ''LaMer'' precipitation diagram (54), which describes the nucleation by exceeding the supersaturation limit and the growth by condensation of monomeric silicic acid on the surface of the existing particles (nuclei). It is mainly a chemical type of explanation, focusing on the hydrolysis and condensation rate and the solubility of silicic acid.

Second, nucleation and growth of Stöber silica particles is modeled by a controlled aggregation mechanism of subparticles, a few nanometers in size, as for example presented by Bogush and Zukoski (19). Colloidal stability, nuclei size, surface charge, and diffusion and aggregation characteristics are the important parameters in this model.

Earlier, the typical LaMer type nucleation and growth mechanism had been proposed and several initial results agreed with that mechanism quite well. It explained the very uniform particle size distribution and it also provided a good explanation for the seeded growth process. Moreover, the observed reaction could be interpreted by a simple first-order reaction kinetic, and activation energy and reaction rate constants could be determined. At present the aggregation and growth mechanism is favored by more and more research groups, and it is mainly used at this point to explain the formation of the Stöber silica particles.

LaMer, Monomer Addition Growth Model. Most of the recent publications (13,18,37,43–45) concerning the ''Stöber'' silica precipitation describe a first-order hydrolysis of TEOS as the rate-limiting process in the silica particle precipitation. The second reaction step, the condensation reaction, was found to be faster by at least a factor of 3 (13,18,19). In addition, the appearance of an induction period would favor a LaMer type reaction mechanism, since it will take a certain time for the hydrolysis of TEOS to exceed the equilibrium concentration or the critical nucleation concentration of silicic acid in the solution, which is then followed by the precipitation of silica. Yet the latter hypothesis seemed somehow incompatible with observations by Van Blaaderen et al. (13) and Philipse (6). They observed an induction period in seeded growth experiments. At first this seems to be contradictory to the LaMer type reaction mechanism, but actually it can be explained under specific assumptions/observations, further details are provided in a publication by Flemming (55).

Aggregation Growth Model. On the other hand, an equal number of papers have been published that support an aggregation growth mechanism. One of the most convincing results is the fine structure within the particles as observed by Labrosse and Burneau (50) with high-resolution transmission electron microscopy (see Fig. 2.1.8). This observation is also in close agreement with the detection of ultramicropores and the high surface area of the samples unless they are calcined at higher temperatures or the SAXS experiments by van Helden et al. (36) showing nanometer-sized subunits, pores, or particles, within the silica particles (see Fig. 2.1.8).

The aggregation growth model assumes that nanometer-sized silica particles (nuclei) are formed continuously during the hydrolysis and condensation of TEOS. These nuclei are colloidally unstable, and they will coagulate and form larger units. Above a critical aggregate size, they are stable. Further growth is accomplished by addition of subsequent primary nuclei to the existing ''particle'' surface of the agglomerates. The number and size of these units will be determined not only by the reaction kinetics, but also by different parameters effecting the dispersion stability, like the ionic strength of the solution, temperature, charges on the particle surface, pH, and solvent properties (like viscosity, dielectric constant, etc.).

Bogush and Zukoski (18–21) explained several of their observations with this model. For example the final particle size increased from 343 nm to 710 nm diameter when the precipitation mixture contained 10^{-2} mol dm^{-3} NaCl, even as the reaction rate remained nearly constant. The lower colloidal stability in the system containing the higher NaCl concentration (higher ionic strength) explained why a smaller number of aggregate units were formed at the beginning and consequently particles grew larger in size.

The formation of primary nanometer-sized subunits, as an intermediate step in the growth reaction, can also account for the observed induction period that was observed in the seeded growth experiments (13,38,56,57). The condensation reaction can proceed parallel to the hydrolysis reaction from the very first beginning, and still it will take a certain time (induction time) to produce the primary subunits, which then may agglomerate to form the new seed nuclei or, in the case of seeded growth, adhere to the surface of already existing particles.

Both models, the monomer addition and the aggregation growth model have convincing arguments. Yet the aggregation mechanism is more favored by most research groups at this point. Nevertheless, one has to be very careful with any general statement, since the reaction mechanism might be different under different reaction conditions. As mentioned in a previous publication (37,38), the Stöber silica particles might very well be first formed by an aggregation mechanism from nanometer-sized subunits and at a later point in the reaction the mechanism might change to a monomer addition mechanism. This might happen when the concentration of reactive silica has decreased and the overall reaction rate is much slower than at the beginning of the precipitation. A slower reaction rate should certainly favor the monomer addition mechanism.

2.1.5 Applications

2.1.5.1 Chemically Modified Silica Surface

Chemically modified nonporous and porous silica spheres are applied in chromatography and as filler or inserts in composite materials. In its original state the surface of Stöber silica is covered with hydroxyl groups, about 8 μmol OH/m^2. The chemical nature of the surface can then be dramatically changed by a reaction with organosilanes or by grafting polymers onto the surface (58–60). Silanization is a very convenient method to introduce functional organic groups, and one simple method to link amino, mercapto, or alkyl groups to the surface is described by Bradley et al. (61). Polymers can also be grafted onto the surface by a direct esterification process (62) or through urethane linkage (63). Numerous other methods are described in the literature; see Unger et al. (64) for further examples.

2.1.5.2 Chromatography

Nonporous silica spheres have been extensively used as support materials in fast high-performance liquid chromatography (HPLC). Especially biopolymers can be successfully separated by means of reversed-phase (64), hydrophobic interaction (65), ion exchange (66), or affinity chromatography (67). Due to the small particle size and the absence of intraparticle porosity, the separations are extremely fast and the biological activity can be retained to a higher degree when compared with traditional HPLC separation techniques. One example is shown in Figure 2.1.11, and further examples are provided in numerous publications by K. K. Unger et al. Most recently, Stöber silica particles have been also prepared with a porous particle structure, and those powders can be used like standard HPLC material, utilizing the much faster separation times that can be achieved with the micrometer-sized particles.

2.1.5.3 Dyes and Pigments

Other studies have shown how monodispersed silica particles can be used as carrier material for pigments and dyes (68–73). Various reactive dyes, containing a sulfonic acid group, can be attached to the surface or incorporated into the growing silica particles (71). The resulting pigment particles have better color properties due to the con-

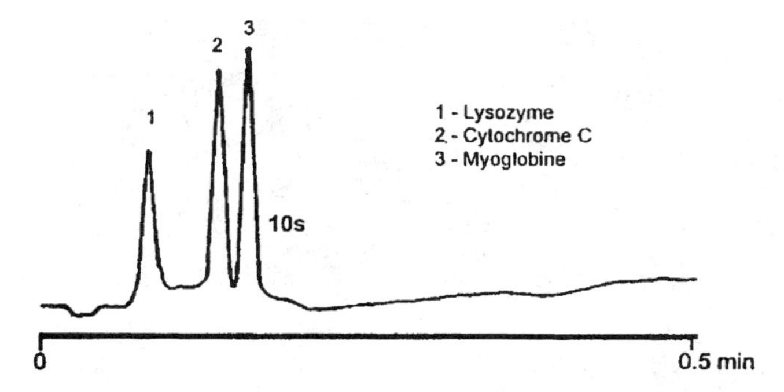

Fig. 2.1.11 HPLC example of fast separation of LCM proteins. (From Ref. 50.)

trolled particle morphology. In another example, the organic dye molecules are attached through aluminum acting as an intermediate coupling agent (72) or fluorescent dyes are attached through an aminopropylsilane–thioisocyanate coupling mechanism (12). All these materials can be used as direct pigments, or can be used for diagnostic application when the silica particle is mainly used as a carrier and the attached dye molecule helps to detect the presence of the particles in specific tests.

2.1.5.4 Rheology studies

Several research groups have studied the rheological behavior of hard-sphere colloidal dispersions. Stöber silica was used in those studies as a model system. In order to achieve the "hard-sphere" behavior, the silica particles were surface modified with an octadecyl chain by boiling them in octadecyl alcohol, described in more detail by van Helden et al. (15). The particles are then suspended in an organic liquid, like cyclohexane, and do not show any long-range electrostatic repulsion but rather are stabilized due to the steric effects of the octadecyl chains on the particle surface. The viscoelastic properties of concentrated dispersions above a volume concentration of 50% indicated interesting shear thickening/thinning phenomena, and at even higher concentrations the formation of various ordered structures has been observed. Further details of those model studies are described in the corresponding publications (3,22,23,73–79).

2.1.5.5 Ordered Packing and Controlled Pore Structure

The formation of ordered sphere-packing structures was observed in certain rheological experiments as just described. Due to the extremely uniform size of the particles, an ordered dense packing structure will develop during sedimentation of the Stöber silica particles (see Fig. 2.1.12) when the dispersion is either sterically or electrostatically stabilized. The gemstone opal is essentially based on this principle (80–88). A transmission electron replica picture is shown in Figure 2.1.13. The uniform

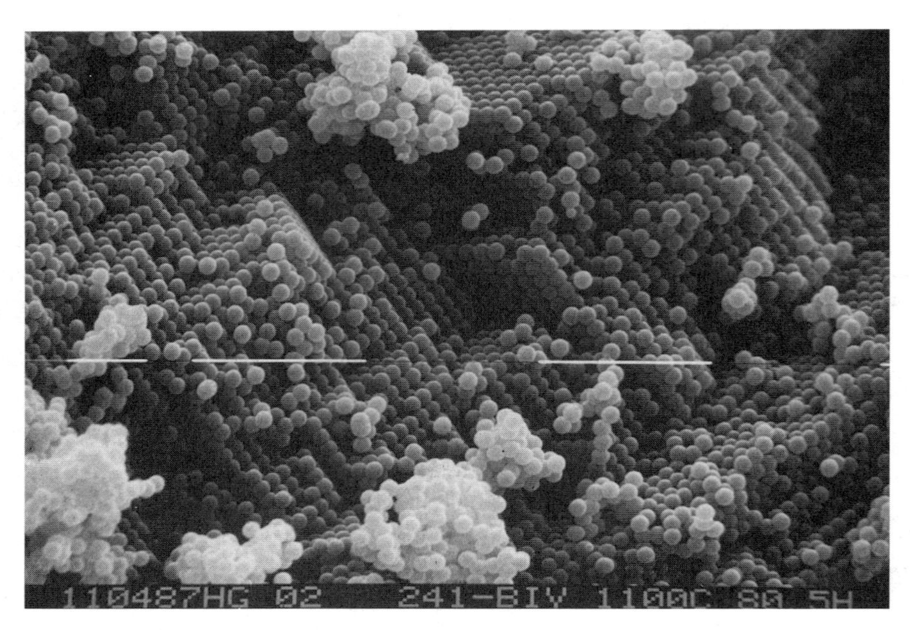

Fig. 2.1.12 Scanning electron micrograph, showing the ordered arrangement of silica particles after a slow sedimentation process. (From Ref. 26.)

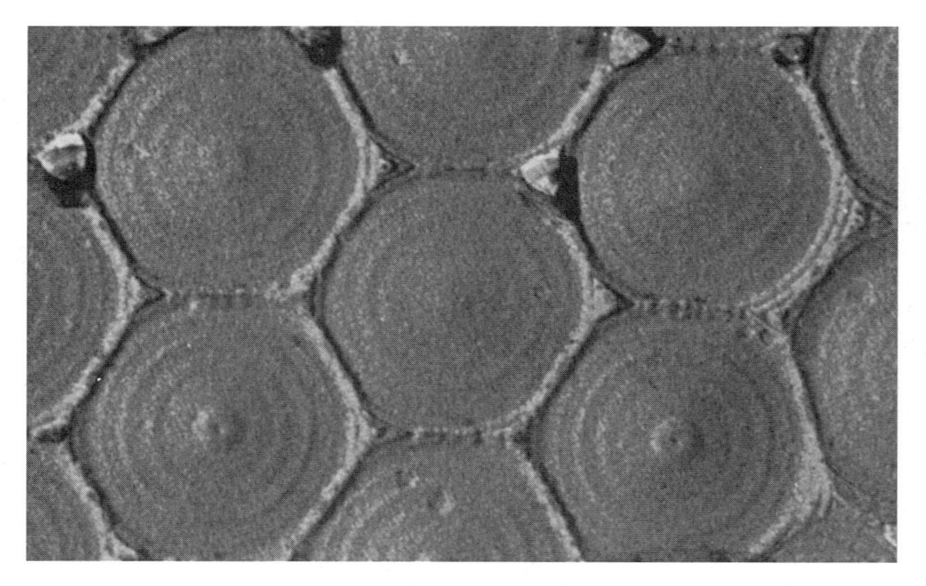

Fig. 2.1.13 Transmission electron micrograph of carbon replica of natural opal. (From Ref. 82.)

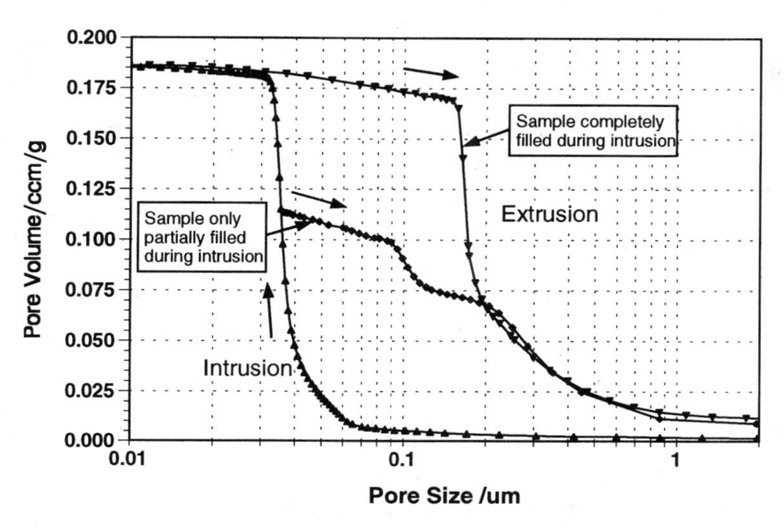

Fig. 2.1.14 Mercury porosimetry data for ordered silica packing structures.

particle arrangement or pore structure can be seen. This is used to test various porestructure analysis techniques, primarily nitrogen adsorption and mercury porosimetry. Systematic studies have been performed by several groups, and good correlations between the predicted and the experimental results were reported (24–27). Especially in the case of mercury porosimetry, the analysis technique has led to new insights into the actual intrusion and extrusion behavior of mercury into and out of the pore structure. A two-step extrusion curve was observed when the pore system was only partially filed with mercury, whereas the ''standard'' one-step extrusion curve was noticed when the pore system was completely filled during the preceding intrusion run (see Fig. 2.1.14). Further details of those effects are described in the corresponding publications (25–27).

2.1.5.6 Sintering Studies

The sintering of model systems has always been of major interest, and silica is one of the most frequently studied materials. Numerous studies have investigated sol-gel–derived materials, but some have also investigated the monodispersed Stöber silica as a model system. The latter allowed testing several of the proposed sintering models, and in general a good agreement was observed between the theoretical time-shrinkage or densification behaviour and the experiments (28–35,89). Yet when the particle size was varied by more than an order of magnitude, the results indicated that there was a major discrepancy between the theoretical prediction and the actual measured data. Usually viscous flow is assumed to be the principal mechanism for sintering or densification of silica. This mechanism predicts that the time to sinter a sample to a specific density (sintering stage) should be directly proportional to the particle size. However, some experiments, using Stöber silica particles of 0.09, 0.2, 0.43, and 1.0 μm in diameter, indicated that the sintering time was proportional to the square of the particle size in those tests (29). The question remains of whether there is a second effect superimposed, e.g., a surface-accumulated impurity, or whether viscous flow sintering is not the correct sintering mechanism in this situation.

REFERENCES TO SECTION 2.1

1. G Kolbe. Das komplexchemische Verhalten der Kieselsäure: Dissertation, Friedrich-Schiller Universität Jena, 1956.
2. W Stöber, A Fink, E Bohn: J Colloid Interface Sci 26:62–69, 1968.
3. CG de Kruif, EMF van Iersel, A Vrij. J Chem Phys 83(9):4717–4725, 1985.
4. JW Jansen, CG de Kruif, A Vrij. J Colloid Interface Sci 114(2):471–480, 1986.
5. PW Rouw, CG de Kruif. Phys Rev A 39(10):5399–5408, 1989.
6. AP Philipse. Colloid Polym Sci 266:1174–1180, 1988.
7. AP Philipse, A Vrij. J Chem Phys 87(10):5634–5643, 1987.
8. AP Philipse, A Vrij. J Colloid Interface Sci 128(1):121–136, 1989.
9. AP Philipse. J Mater Sci Lett 8:1371–1373, 1989.
10. AP Philipse, BC Bonekamp, HJ Veringa. J Am Ceram Soc 73(9):2720–2727 1990.
11. A van Blaaderen, APM Kentgens. J Non-Cryst Solids 149:161–178, 1992.
12. A van Blaaderen, A Vrij. Langmuir 8:2921–2931, 1992.
13. A van Blaaderen, J van Geest, A Vrij. J Colloid Interface Sci 154(2):481–501, 1992.
14. AK van Helden, A Vrij. J Colloid Interface Sci 76(2):418–433, 1980.
15. AK van Helden, JW Jansen, A Vrij. J Colloid Interface Sci 81(2):354–368, 1981.
16. GH Bogush, GL Dickstein, P Lee, KC Zukoski, CF Zukoski IV. In: CJ Brinker, De Clark, DR Ulrich, eds. Mater Res Soc Symp Proc vol 121, Better Ceramics Through Chemistry III. Pittsburgh, PA: Materials Research Society, 1988, pp 57–65.
17. GH Bogush, MA Tracy, CF Zukoski IV. Non-Cryst Solids 104:95–106, 1988.
18. GH Bogush CF Zukoski IV. J Colloid Interface Sci. 142(1):1–18, 1991.
19. GH Bogush, CF Zukoski IV. J Colloid Interface Sci 142(1):19–34, 1991.
20. GH Bogush, CJ Brinker, PD Majors, DM Smith. In: BJJ Zelinski, CJ Brinker, DE Clark, DR Ulrich, eds. Materials Research Society Symp Proc vol 180, Better Ceramics Through Chemistry IV. Pittsburgh, PA: Materials Research Society, 1990, pp 491–494.
21. CF Zukoski, MK Chow, GH Bogush, JL Look. In: BJJ Zelinski, CJ Brinker, DE Clark, DR Ulrich, eds. Materials Research Society Symp Proc vol 180, Better Ceramics Through Chemistry IV. Pittsburgh, PA; Materials Research Society, 1990, pp 131–140.
22. JC van der Werff, CG de Kruif. J Rheol 33(3):421–454, 1989.
23. JC van der Werff, CG de Kruif, C Blom, J Mellema. Phys Rev A 39(2):795–807, 1989.
24. S Bukowiecki, B Straube, KK Unger. In: JM Haynes, P. Rossi-Doria, eds. Principles and Applications of Pore Structural Characterization. Bristol: Arrowsmith, 1985, pp 43–55.
25. H Giesche, KK Unger, U Müller, U Esser. Colloids Surf 37:93–113, 1989.
26. H Giesche. In: S. Komarneni, DM Smith, JS Beck, eds. Materials Research Society Symp Proc vol 371, Advances in Porous Materials. Pittsburgh, PA: Materials Research Society, 1995, pp 501–510.
27. H Giesche. In: B McEnaney, TJ Mays, J Rouquerol, F Rodriguez-Reinoso, KSW Sing, KK Unger, eds. Characterization of Porous Solids IV Cambridge, UK: Royal Society of Chemistry, 1997, pp 171–179.

28. H Suzuki, S Takagi, H Morimitsu, Hirano. J Ceram Soc Jpn 100:284–287, 1992.
29. H Giesche, KK Unger. In: H Hausner, GL Messing, S Hirano, eds. Ceramic Powder Processing Science, 2nd Int Conf Berchtesgaden 1988, Proc. Köln: Deutsche Keramische Gesellschaft, 1989, pp 755–764.
30. MD Sacks, TS Yeh, SD Vora. In: H Hausner, GL Messing, S Hirano, eds. Ceramic Powder Processing Science, Proc Second Int Conf Berchtesgarden FRG, October 1988. Köln: Deutsche Keramische Gesellschaft, 1989, pp 693–704.
31. MD Sacks, T-Y Tseng. J Am Ceram Soc 67(8):526–532, 1984.
32. MD Sacks, T-Y Tseng. J Am Ceram Soc 67(8):532–537, 1984.
33. MD Sacks, SD Vora. J Am Ceram Soc 71(4):245–249, 1988.
34. MD Sacks, GW Scheiffele, N Bozkurt, R Raghunathan. In: S Hirano, GL Messing, H Hausner, eds. Ceramic Transactions, Ceramic Powder Science IV. Westerville, OH: American Ceramic Society, 1991, pp 437–455.
35. T Shimohira, A Makishima, K Kotani, M Wakakuwa. In: S Somiya, S Saito, eds. Proc Int Symp Factors in Densification and Sintering of Oxide and Non-Oxide Ceramics. Tokyo: Tokyo Institute of Technology, 1978, pp 119–127.
36. CG Tan, BD Bowen, N Epstein. J Colloid Interface Sci 118:290–293, 1987.
37. H Giesche. J Eur Ceram Soc 14:189–204, 1994.
38. H Giesche. J Eur Ceram Soc 14:205–214, 1994.
39. HP Calhoun, CR Masson. In: M Gielen, ed. Reviews on Silicon, Germanium, Tin and Lead Compounds. Tel-Aviv: Freund, 1981, p 153.
40. G Garzo, D Hoebbel, ZJ Ecsery, Ujszaszi. J Chromatogr 167:321–336, 1978.
41. MF Bechtold, RD Vest, L Plambeck Jr. J Am Chem Soc 90:4590–4598, 1968.
42. P Schubert. Der Einfluss der Silicatquelle auf die Synthese von ZSM-5—Neue Wege zur Synthese und Charakterisierung. Dissertation, Universität Mainz, 1985.
43. MT Harris, OA Basaran, CH Byers. In: GL Messing, S-I Hirano, H Hausner, eds. Ceramic Transactions, vol 12, Ceramic Powder Science III. Westerville, OH: American Ceramic Society, 1990, pp 119–127.
44. T Matsoukas, E Gulari. J Colloid Interface Sci 124:252–261, 1988.
45. T Matsoukas, E Gulari. J Colloid Interface Sci 132:13–21, 1989.
46. KK Unger, H Giesche, JN Kinkel. DE patent 3534143 A1, 25 September 1985; 3616133 A1, 14 May 1986; US patent 4,911,903, March 27, 1990; 4,775,520, October 4, 1988.
47. ES Kovats, L Jelinek, C Erbacher. European patent 0.574.642 A1.
48. TJ Barder, PD Dubois. US patent 4.983.369, Jan. 8, 1991.
49. S Coenen, CG de Kruif. J Colloid Interface Sci 124:104–111, 1988.
50. C Kaiser, M Hanson, H Giesche, J Kinkel, KK Unger. In: E. Pelizzetti, ed. Fine Particle Science and Technology: from Micro to Nanoparticles. Dordrecht: Kluwer, 1995, pp 71–84.
51. A Burneau, A Labrosse. J Non-Cryst Solids 221:107–124, 1997.
52. B Humbert. J Non-Cryst Solids 191:29–37, 1995.
53. PJ Davis, R Deshpande, DM Smith, CJ Brinker, RA Assink. J Non-Cryst Solids 167: 295–306, 1994.
54. VK LaMer, RH Dinegar. J Am Chem Soc 72:4847–4854, 1950.
55. BA Fleming. J Colloid Interface Sci 110:40–64, 1986.
56. S Coenen, CG De Kruif. J Colloid Interface Sci 124(1):104–110, 1988.
57. K Yoshizawa, Y Sugoh, Y Ochi. Sci Ceram 14:125–131, 1988.
58. KC Vrancken, L de Coster, P van der Voort, PJ Grobert. J Colloid Interface Sci 170: 71, 1995.

59. A van Blaaderen, J Vrij. J Colloid Interface Sci 156:1, 1992.
60. JN Kinkel. Darstellung and Charakterisierung von Siliziumdioxidträgermaterialien zur Trennung von Biopolymeren durch Hochdruckflüssigchromatographie. Dissertation, Universität Mainz, Germany, 1984.
61. RD Badley, WT Ford, FJ McEnroe, RA Assink. Langmuir 6:792, 1990.
62. H Ben Ouda, H Hommel, AP Legrand, H Balard, E Papirer. J Colloid Interface Sci 122:441, 1988.
63. K Bridger, B Vincent. Eur Polym J 16:1017, 1980.
64. KK Unger, G Jilge, JN Kinkel, MTW Hearn. J Chromatogr 359:61–72, 1986.
65. R Janzen, KK Unger, H Giesche, JN Kinkel, MTW Hearn. J Chromatogr 397:91–97, 1987.
66. G Jilge, KK Unger, U Esser, H-J Schäfer, G Rathgeber, W Müller. J Chromatogr 476: 135–145, 1989.
67. AI Liapis, B Anspach, ME Findley, J Davies, MTW Hearn, KK Unger. Biotechnol Bioeng 34:467–477, 1989.
68. L Horner, H Ziegler. Z Naturforsch 42b:643–660, 1987.
69. S Kaneko, H Saitoh, Y Maejima, M Nakamura. Anal Lett 22(6):1631–1641, 1989.
70. FM Winnik, B Keoshkerian. US patent 4,877,451, 1989.
71. H Giesche, E Matijevic. Dyes Pigments 17:323–340, 1991.
72. WP Hsu, R Yu, E Matijevic. Dyes Pigments 19:179–201, 1992.
74. BJ Ackerson. J Rheol 34(4):553–590, 1990.
75. DR Jones, B Leary, DV Boger. J Colloid Interface Sci 147(2):479–495, 1991.
76. DAR Jones. Depletion flocculation of sterically stabilized particles. PhD thesis, Bristol, 1988.
77. L Marshall, CF Zukoski IV. In: IA Aksay, GL McVay, DR Ulrich, eds. Mater Res Soc Symp vol 155, Processing Science of Advanced Ceramics. Pittsburgh, PA: Materials Research Society, 1989, pp 65–72.
78. L Marshall, CF Zukoski IV. J Phys Chem 94(3):1164–1171, 1990.
79. PN Pusey, W van Megen. Nature 320:340–342, 1986.
80. WB Russel. In: GL Messing, S Hirano, H Hausner, eds. Ceramic Transactions vol 12, Ceramic Powder Science III. Westerville OH: American Ceramic Society, 1990, pp 361–373.
81. PJ Darragh, JV Sanders. Gems Gemol 11(10):291–298, 1965.
82. PJ Darragh, AJ Gaskin, BC Terrell, JV Sanders. Nature 209:13–16, 1966.
83. PJ Darragh, AJ Gaskin, JV Sanders. Sci Am 234:84–95, 1976.
84. JB Jones, JV Sanders, ER Segnit. Nature 204:990–991, 1964.
86. JV Sanders. Nature 204:1151, 1964.
87. JV Sanders. Acta Cryst A 24:427–434, 1968.
88. JV Sanders. Acta Cryst A 32:334–338, 1976.
89. JV Sanders. Philos Mag A 42(6):705–720, 1980.
90. TN Zaslavskaya, LD Fedorovich. Izv Akad Nauk SSSR Neorg Mater 25(7):1152–1154, 1989.
91. DW Johnson Jr, EM Rabinovich, JB MacChesney, EM Vogel. J Am Ceram Soc 66(10): 688–693, 1983.

2.2 HYDROLYSIS OF SILICON ALKOXIDES IN MICROEMULSIONS

K. OSSEO-ASARE
Pennsylvania State University, University Park, Pennsylvania

2.2.1 Introduction

The techniques generally available for the solution synthesis of nanoparticles may be broadly classified into those that rely on microreactors, and those that exploit growth modifiers (1–7). In microreactor techniques, crystal growth is constrained by forcing particle formation to occur in small isolated volumes. On the other hand, in the case of techniques that rely on growth modifiers, chemical additives such as organic solvents, organic polymers, etc. are introduced into the reaction medium to modify the rates of particle nucleation, growth, or aggregation. Microemulsions, as reaction media, are unique in that they offer both microreactor and growth-modifier capabilities (7–16). In the water-in-oil (w/o) variety of these compartmentalized fluids, fine microdrops of aqueous phase (2–30 nm) are trapped within aggregates of surface-active molecules dispersed in an external oil phase. The dispersed water pools constitute spatially separated nanoreactors that allow the solubilized reacting species to be distributed at the molecular level. At the same time, the surfactant-stabilized microcavities (reverse micelles) provide a cage-like effect that limits particle nucleation, growth, and aggregation.

This chapter focuses on silica synthesis via the microemulsion-mediated alkoxide sol-gel process. The discussion begins with a brief introduction to the general principles underlying microemulsion-mediated silica synthesis. This is followed by a consideration of the main microemulsion characteristics believed to control particle formation. Included here is the influence of reactants and reaction products on the stability of the single-phase water-in-oil microemulsion region. This is an important issue since microemulsion-mediated synthesis relies on the availability of surfactant/oil/water formulations that give stable microemulsions. Next is presented a survey of the available experimental results, with emphasis on synthesis protocols and particle characteristics. The kinetics of alkoxide hydrolysis in the microemulsion environment is then examined and its relationship to silica-particle formation mechanisms is discussed. Finally, some brief comments are offered concerning future directions of the microemulsion-based alkoxide sol-gel process for silica.

2.2.2 Synthesis of Silica Nanoparticles

2.2.2.1 General Principles

Alkoxide-based silicas are typically derived from tetramethoxysilane [$Si(OCH_3)_4$, TMOS] and tetraethoxysilane [$Si(OCH_2CH_3)_4$, TEOS] as precursors (17). The silica

Table 2.2.1 Silica Synthesis in Microemulsions

Material	Microemulsion system	Reactants	Comments
SiO_2 nanoparticles	AOT/isooctane/water/ammonia (R = 6–18)	$TEOS/H_2O$ (0.7–3.6 M NH_3); water/TEOS molar ratio = 0.25–4	Polydisperse, 15–70 nm porous particles; specific surface area = 100–300 $m^2\ g^{-1}$ (20)
SiO_2 nanoparticles (surface functionalized)	AOT/toluene/water/ammonia (R = 1–6)	TEOS + $MPTMS/H_2O$, NH_3; water/TEOS molar ratio = 2.8–17	Spherical, 20–70 nm particles; vinyl groups in surface via MPTMS (21)
SiO_2 nanoparticles (surface functionalized)	POELE/cyclohexane/water/ammonia (R = 5.5)	TEOS + $MPTMS/H_2O$, NH_3	Spherical particles, d_p ~28 nm (21)
SiO_2 nanoparticles (surface functionalized)	AOT/toluene/water/ammonia (R = 2.1–10)	TEOS + $MPTMS/H_2O$ + NH_3 (30 wt%); water/TEOS molar ratio = 6–21	Particle size 28–113 nm (22)
SiO_2 nanoparticles (surface functionalized)	Isopropanol/toluene/water/ammonia	TEOS + $MPTMS/H_2O$ + NH_3 (30 wt%)	Highly unstable particle dispersion (22)
SiO_2 nanoparticles	NP-4/heptane/water/ammonia ([NP-4] = 0.06–0.25 M; $[H_2O]$ = 0.12–0.58 M; $[NH_3]/[H_2O]$ = 0.086–2.0)	$TEOS/H_2O$ + NH_3 (29 wt%); [TEOS] = 0.0357 M in μE–TEOS system; h = $[H_2O]/[TEOS]$ = 3.36–16.2	Silica particle size 26–43 nm, depending on water and surfactant concentrations; particle size distribution approximately constant during growth period (23)
SiO_2 nanoparticles	NIS/heptane/water/ammonia (NIS = NP-4, NP-5, or DP-6; 0.04–0.26 M NIS, 0.174–1.10 M H_2O, 0.075 M NH_3)	$TEOS/H_2O$ + NH_3 (29 wt%); [TEOS] = 0.018 M in μE–TEOS mixture	Monodisperse spherical nanoparticles, d_p = 28–50 nm; particle size increased in the order: NP-5 > NP-4 > DP-6 (24)

SiO_2 nanoparticles	NP-5/oil/water/ammonia (oil = heptane, cyclohexane, or heptane/cyclohexane (50/50 v/o); 0.04–0.23 *M* NP-5, 0.174 *M* H_2O, 0.075 *M* NH_3)	$TEOS/H_2O + NH_3$ (29 wt%); [TEOS] = 0.018 *M* in μE–TEOS mixture	Monodisperse spherical nanoparticles, d_p = 30–75 nm; particle size increased in the order: heptane > heptane/cyclohexane > cyclohexane (24)
SiO_2 nanoparticles	NP-5/cyclohexane/water ammonia ([NP-5] = 0.09–0.3 *M; R* = 0.7–2.3)	$TEOS/H_2O + NH_3$ (29.6 wt%); [TEOS] = 0.027 *M* in μE–TEOS mixture; $h = [H_2O]/[TEOS] = 7.8$	Spherical monodisperse particles, d_p = 50–70 nm (25)
SiO_2 nanoparticles	NP-5/cyclohexane/water/ammonia ([NP-5] = 0.056–0.277 *M; R* = 0.5–3.5)	$TEOS/H_2O + NH_3$ (29.6 wt%); [TEOS] = 0.023 *M* in μE–TEOS mixture; $h = [H_2O]/[TEOS] = 7.8$	Spherical monodisperse particles, d_p = 35–68 nm, depending on *R* and time; minimum in d_p at $\sim R$ = 1.5–2 (26)
SiO_2 nanoparticles	NP-5/cyclohexane/water/ammonia ([NP-5] = 0.05–0.3 *M; R* = 0.5–6.8)	$TEOS/H_2O + NH_3$ (1.6–29.6 wt%); [TEOS] = 0.0082–0.102 *M* in μE–TEOS mixture; $h = [H_2O]/[TEOS]$ = 1.6–19.9	Spherical monodisperse particles, d_p = 32–76 nm; minimum in d_p at $\sim R$ = 2–3 (27,28)
SiO_2 nanoparticles	NP-5/cyclohexane/water/ammonia (*R* = 0.7–5.4)	$TEOS/H_2O + NH_3$; [TEOS] = 0.025 *M; h* = $[H_2O]/[TEOS]$ = 7.8	Effect of *R* on the time evolution of particle size was investigated (29)

(continued)

Table 2.2.1 Continued

Material	Microemulsion system	Reactants	Comments
SiO_2 nanoparticles	NP-5/cyclohexane/water/ammonia ([NP-5] = 0.05–0.3 *M; R* = 2.29–6.8)	$TEOS/H_2O + NH_3$ (29.6 wt%); [TEOS] = 0.0082–0.102 *M* in μE–TEOS mixture; h = $[H_2O]/[TEOS]$ = 1.6–19.9)	Spherical particles, d_p = 30–70 nm; bimodal size distribution; phase separation during synthesis reaction (30)
SiO_2 nanoparticles	AOT/decane/water/ammonia ([AOT] = 0.09–0.24 *M R* = 2.0–9.5)	$TEOS/H_2O + NH_3$ (13.9 wt%); [TEOS] = 0.044 *M; h* = $[H_2O]/[TEOS]$ = 18.5	No particles observed (TEM) for $R < 4$; d_p = 10–60 nm; d_p increased with *R* (31)
SiO_2 nanoparticles	AOT/decane/benzyl alcohol (BA)/water/ammonia (*R* = 6.8, BA/AOT molar ratio = 0–2.5)	$TEOS/H_2O + NH_3$ (13.9 wt%); [TEOS] = 0.044 *M; h* = $[H_2O]/[TEOS]$ = 18.5	Microemulsions with BA/AOT > 1.5 became unstable during synthesis reaction; nearly spherical nanoparticles; maximum in particle size at BA/AOT = 1.5 (32)
Bioactive molecules/SiO_2 nanoparticles	Triton X-100/cyclohexane/hexanol/water/ammonia	$TMOS/H_2O + NH_3$ (1.05 *M*); bioactive molecules: isothiocyanate-dextran (FITC-dextran, mol. mass = 19.6 kD), [^{125}I]tyraminylinulin (mol. mass 5 kD), horseradish peroxidase (HRP, mol. mass 40 kD)	72 h Conditioning (refrigerator), vacuum evaporation, wash with ammonia buffer (pH ~9), particle characterization via dynamic light scattering, TEM; monodisperse particles ~31 nm; entrapment efficiency determined (33)
PuO_2/SiO_2 nanoparticles	AOT/hexane/CMPO/water/HNO_3	$TEOS/H_2O$, HNO_3, PuO_2	SiO_2-encapsulated PuO_2 nanoparticles recovered from organic phase by precipitation with base (hydrazine or NH_3) (34)

CdS/SiO_2 nanoparticles	Triton N-101/ cyclohexane/ hexanol/water/ ammonia	TEOS/H_2O, NH_3, Cd $(NO_3)_2$, $(NH_4)_2S$	CdS core, shell, homogeneously dispersed, or surface bonded quantum dots prepared, depending on sequencing of CdS precipitation and TEOS hydrolysis; presence of hexanol strengthened microemulsion against destabilization by ethanol reaction product (35–37)
CdS/SiO_2 nanoparticles	Igepal CO-520/ cyclohexane/ water/ammonia	TEOS/H_2O, NH_3, $Cd(NO_3)_2$, $(NH_4)_2S$	CdS core, shell, homogeneously dispersed, or surface-bonded quantum dots prepared, depending on sequencing of CdS precipitation and TEOS hydrolysis (35–37)
CdS/SiO_2 nanoparticles	Igepal CO-520/ cyclohexane/ water/ammonia	TEOS/H_2O, NH_3, $Cd(NO_3)_2$, Na_2S	CdS core; optical properties investigated (38)
CdS/SiO_2 nanoparticles	AOT/isooctane/water/ ammonia	TEOs/H_2O, NH_3, $Cd(NO_3)_2$, Na_2S	CdS core; optical properties investigated (38)
Ag/SiO_2 nanoparticles	Igepal CO-520/ cyclohexane/ water/ammonia	TEOS/H_2O, NH_3, $Ag(NO_3)_2$, N_2H_4	Ag core; optical properties investigated (38)
Ag/SiO_2	AOT/isooctane/ cyclohexane/ water/ammonia	TEOS; shH_2O, NH_3, $Ag(NO_3)_2$, N_2H_4	Ag core; optical properties investigated (38)
Mesoporous polymers (SiO_2 template)	Triton N-101/ cyclohexane/ hexanol/water/ ammonia	TEOS/H_2O, NH_3; HF; DVB, EDMA	Dissolution of silica from polymer/SiO_2 nanocomposites yields nanoporous polymer materials (39)

(continued)

Table 2.2.1 Continued

Material	Microemulsion system	Reactants	Comments
SiO_2 gel	SDS/pentanol/water	$TEOS/H_2O$ (pH 2, HCl)	Approximately triangular μE stability region (corner compositions: 100% C_5OH; 35% SDS, 45% C_5OH, 20% water; 5% SDS, 25% C_5OH, 70% water); 9 wt% TEOS gives transparent silica gel over nearly entire μE stability region; increase in TEOS shifts the transparent gel region toward the water-rich corner (40)
SiO_2 gel	SDS/pentanol/water	$TEOS/H_2O$ (pH 2, HCl)	Phase diagrams were determined for SDS/pentanol/water system with and without ethanol or TEOS addition (41)
SiO_2 gel	SDS/pentanol/water	$TEOS/H_2O$ (pH 2, HNO_3)	Viscoelectric properties were investigated. Gelation time: 30–35 h (42)
SiO_2 gel	Ethanol/C_n OH (n = 4,5,6)/H_2O	$TEOS/H_2O$ (pH 1)	Phase diagrams were determined with and without TEOS (43)
SiO_2 gel	CTAB/decanol/decane/formamide (nonaqueous μE)	$TEOS/H_2O$ (pH 1, HNO_3); TEOS/μE = 4:1 (w/w); $TEOS/H_2O$ = 0.5 (molar ratio)	Both decanol- and formamide-continuous μEs yielded gels (44)
SiO_2 gel	AOT/decanol/glycerol (nonaqueous μE)	$TEOS/H_2O$ (pH 1, HNO_3)	Gels obtained only in glycerol-rich region of phase diagram (44)
SiO_2 gel	AOT/decane/glycerol (nonaqueous μE)	$TEOS/H_2O$ (pH 1, HNO_3)	No gels were obtained (44)

SiO_2 gel	CTAB/decanol/decane/formamide (nonaqueous μE; decanol/decane = 75/25 w/w)	TEOS/H_2O (pH 1); water/TEOS molar ratio = 2	Viscoelastic properties of silica gels were investigated; gels based on formamide/(decane + decanol) molar ratio ~2.5–3 gave the highest elasticity (45)
SiO_2 gel	CTAB/decanol/decane/formamide (nonaqueous μE)	TEOS/H_2O (pH 1, HNO_3); μE/H_2O/TEOS = 5/1/5 (w/w)	Condensation rate monitored with ^{29}Si-NMR (46)
SiO_2 gel	SDS/octanol/toluene/ethylene glycol [nonaqueous μE; weight ratio: SDS/octanol, toluene (50/50 w/w)/ethylene glycol = 13.1/29.8/29.0]	TEOS/H_2O (pH 2, HCl)	Reaction rate monitored with ^{29}Si-NMR (47)
$Cu(NO_3)_2$/SiO_2 gel	TEGDE/cyclohexane/water (25% TEGDE, 40% cyclohexane, 35% aqueous solution)	TEOS/H_2O (pH 1.25 HNO_3) + $Cu(NO_3)_2$ (48%); TEOS/μE = 1:1 (w:w)	$Cu(NO_3)_2$-encapsulated silica gel product (48)
Laser dye/SiO_2 gel	CTAB/decanol/decane/formamide (nonaqueous μE)	TEOS/H_2O (pH 1, HNO;3; 10^{-2} *M* laser dye)	Silica gels doped with laser dyes (rhodamine B, rhodamine 6G) gave fluorescence quantum yields indicating promise as candidate solid-state laser dye materials (49)
Laser dye/SiO_2 gel	AOT/decanol/glycerol (nonaqueous μE)	TEOS/H_2O (pH 1, HNO_3; 10^{-2} *M* laser dye)	Silica gels doped with laser dyes (rhodamine B, rhodamine 6G) were synthesized (49)

(continued)

Table 2.2.1 Continued

Material	Microemulsion system	Reactants	Comments
Laser dye/SiO_2 gel	CTAB/decanol/decane/formamide (nonaqueous μE)	TEOS/H_2O (pH 1, HNO_3; coumarin 120 or coumarin 311)	Preparation of silica gels doped with coumarin 120 and coumarin 311 (50)
SiO_2 gel	DDAB/toluene/water (48.7% DDAB, 19.5% decane, 31.8% aqueous silica sol)	TMOS (partially hydrolyzed)/H_2O (+ 0.4–10 wt% HF)	Bicontinuous μEs used as templates for microporous silica gels; monodisperse pores (2 nm pore radius); large specific surface area (~103 m^2/g) (51)
Au/SiO_2 gel	DDAB/decane/water (1–5% DDAB)	TEOS (or partially hydrolyzed TEOS)/H_2O (+ TBAOH); $AuCl_3$ + $LiBH_4$ ($[Au(III)]/[BH_4^-]$ = 1:3); h = 1–4	Mesoporous silica gel, pore diameter = 5–30 nm; surface areas 4–340 m^2/g; Au nanoclusters, d_p = 5.6–7.1 nm (52)
Gelatin/SiO_2 gel	AOT/isooctane/gelatin/water ;[R = 30, 14% w/v gelatin (referred to the aqueous pseudophase)]	TMOS/H_2O (0.1 M HCl or 0.1 M NaOH); water/TMOS molar ratio = 4	Product: silica-gelatin nanocomposites, sl30 nm silica particles (53)
Chitosan/SiO_2 gel	AOT/isooctane/chitosan/water	TMOS/H_2O (0.1 M HCl or 0.1 M NaOH)	Silica–chitosan nanocomposites (53)
SiO_2 gel	AOT/isooctane/water	TMOS/H_2O (0.1 M HCl or 0.1 M NaOH)	Silica nanoparticles (~30 nm) aggregate to form porous gel structure (53)

formation process can be conveniently viewed in terms of a hydrolysis step, Eq. (1), that generates silanol groups (-SiOH), followed by polycondensation reactions, Eq. (2) and (3), that form silicon–oxygen–silicon bonds:

$$Si(OR)_4 + xH_2O = Si(OR)_{4-x}(OH)_x + xROH \quad (1)$$

$$\equiv Si\text{-}OH + HO\text{-}Si\equiv \; = \; \equiv Si\text{-}O\text{-}Si\equiv + H_2O \quad (2)$$

$$\equiv Si\text{-}OH + RO\text{-}Si\equiv \; = \; \equiv Si\text{-}O\text{-}Si\equiv + ROH \quad (3)$$

These reactions are responsive to both acid and base catalysis, and can be manipulated to give a variety of silica products, e.g., discrete particles, monolithic gels, films, and fibers. This technique of materials synthesis via alkoxide hydrolysis has become known as sol-gel processing (17). It should be noted, however, that under certain conditions, gelation may be confined only to the interior of discrete particles (base-catalyzed systems), while the sol may consist of polymeric networks rather than individual particles (acid-catalyzed systems).

In 1968, Stober et al. (18) reported that, under basic conditions, the hydrolytic reaction of tetraethoxysilane (TEOS) in alcoholic solutions can be controlled to produce monodisperse spherical particles of amorphous silica. Details of this silicon alkoxide sol-gel process, based on homogeneous alcoholic solutions, are presented in Chapter 2.1. The first attempt to extend the alkoxide sol-gel process to microemulsion systems was reported by Yanagi et al. in 1986 (19). Since then, additional contributions have appeared (20–53), as summarized in Table 2.2.1. In the microemulsion-mediated sol-gel process, the microheterogeneous nature (i.e., the polar–nonpolar character) of the microemulsion fluid phase permits the simultaneous solubilization of the relatively hydrophobic alkoxide precursor and the reactant water molecules. The alkoxide molecules encounter water molecules in the polar domains of the microemulsions, and, as illustrated schematically in Figure 2.2.1, the resulting hydrolysis and condensation reactions can lead to the formation of nanosize silica particles.

2.2.2.2 Microemulsion Behavior

Two main microemulsion microstructures have been identified: droplet and bicontinuous microemulsions (54–58). In the droplet type, the microemulsion phase consists of solubilized micelles: reverse micelles for w/o systems and normal micelles for the o/w counterparts. In w/o microemulsions, spherical water drops are coated by a monomolecular film of surfactant, while in w/o microemulsions, the dispersed phase is oil. In contrast, bicontinuous microemulsions occur as a continuous network of aqueous domains enmeshed in a continuous network of oil, with the surfactant molecules occupying the oil/water boundaries. Microemulsion-based materials synthesis relies on the availability of surfactant/oil/aqueous phase formulations that give stable microemulsions (54–58). As can be seen from Table 2.2.1, a variety of surfactants have been used, as further detailed in Table 2.2.2 (16). Also, various oils have been utilized, including straight-chain alkanes (e.g., *n*-decane, *n*-hexane),

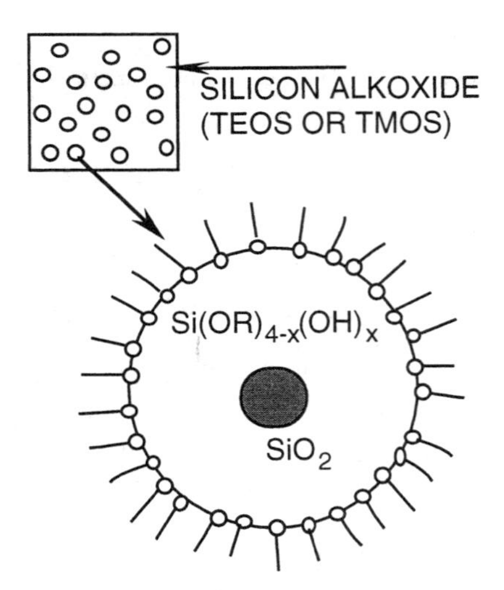

Fig. 2.2.1 Silica synthesis by alkoxide hydrolysis in microemulsion water pools.

cyclohexane, and toluene. It is important to recognize the multiple roles played by water molecules in the microemulsion sol-gel process; water is needed to: (1) form a stable microemulsion before and after alkoxide addition (this requires that the surfactant polar groups be adequately hydrated and that sufficient water be available to constitute a water pool with bulk-water-like properties), (2) hydrolyze alkoxide molecules [Eq. (1)], and (3) solvate the alcohol molecules produced by the alkoxide hydrolysis reaction [Eq. (3)]. Table 2.2.1 further reveals that, in some cases, bulk water in the microemulsions can be replaced with a polar nonaqueous solvent such as formamide (44–46,49,50), glycerol (44), and ethylene glycol (47).

For a given surfactant/oil system, the water molecules first added go to hydrate the surfactant polar groups. This hydration of the polar groups enhances surfactant aggregation via dipole–dipole interactions. With increasing water addition, unbound or ''free'' water molecules appear within the polar domains of the surfactant aggregates. Figure 2.2.2 illustrates this situation for the NP-5/cyclohexane/water and NP-5/cyclohexane/water/ammonia systems (25). The fluorescence spectral data [based on the probe ruthenium tris(bipyridyl), $Ru(Bpy)_3$] indicate that with increase in the water-to-surfactant molar ratio (R) above about 1.2, there is a red shift in the wavelength giving the maximum intensity in the emission spectrum. Above $R \approx 1–1.2$, the maximum intensity is observed at a wavelength similar to that of the probe in water. Corresponding results for the anionic surfactant AOT are presented in Figure 2.2.3 (31); the fluorophore was 1,3,6,8-pyrenetetrasulfonic acid, sodium salt (PTS). Figure 2.2.3 shows that the fluorescence intensity of PTS in the microemulsion system approaches that in bulk water as R exceeds about 8. This trend is reflective

Table 2.2.2 Surfactants Used in Microemulsion-Mediated Silica Synthesis

Type of surfactant	Name of surfactant	Chemical formula
Anionic	Aerosol OT (AOT), sodium bis(2-ethylhexyl) sulfosuccinate	$ROOC{-}CH_2$ / $ROOC{-}CHSO_3^-\ Na^+$ (the two $ROOC$ groups bonded through $CH_2{-}CH$); $R = CH_2CH(C_2H_5)(CH_2)_3CH_3$
	Sodium dodecyl sulfate (SDS)	$C_{12}H_{25}OSO_3^-Na^+$
Cationic	Didodecyldimethylammonium bromide (DDAB)	$[N(R_1)_2(R_2)_2]^+Br^-$; $R_1 = C_{12}H_{25}$, $R_2 = CH_3$
	Cetyltrimethylammonium bromide (CTAB)	$[N(R_1)(R_2)_3]^+Br^-$; $R_1 = C_{16}H_{33}$, $R_2 = CH_3$
Nonionic	Polyoxyethylene (n) dodecyl ether ($C_{12}EO_n$) or polyethylene glycol ether; n = 4 (tetraoxyethylene dodecylether, $C_{12}EO_4$) or tetraethyleneglycol dodecyl ether (TEGDE) or polyoxyethylene(4) lauryl ether (POELE)	$C_{12}H_{25}{-}(OCH_2{-}CH_2)_nOH$
	Polyoxyethylene (n) nonylphenyl ether [NP-4, n = 4 (Triton N-42); NP-5, n = 5 (Triton N-57, Igepal CO-520); NP-6, n = 6 (Triton N-60); NP-9, n = 9–10 (Triton N-101)]	$C_9H_{19}{-}C_6H_4{-}(OCH_2{-}CH_2)_nOH$
	Polyoxyethylene(n) octylphenyl ether [OP-1, n = 1 (Triton X-15); OP-3, n = 3 (Triton X-35); OP-5, n = 5 (Triton X-45); OP-10, n = 9–10 (Triton X-100)]	$C_8H_{17}{-}C_6H_4{-}(OCH_2{-}CH_2)_nOH$

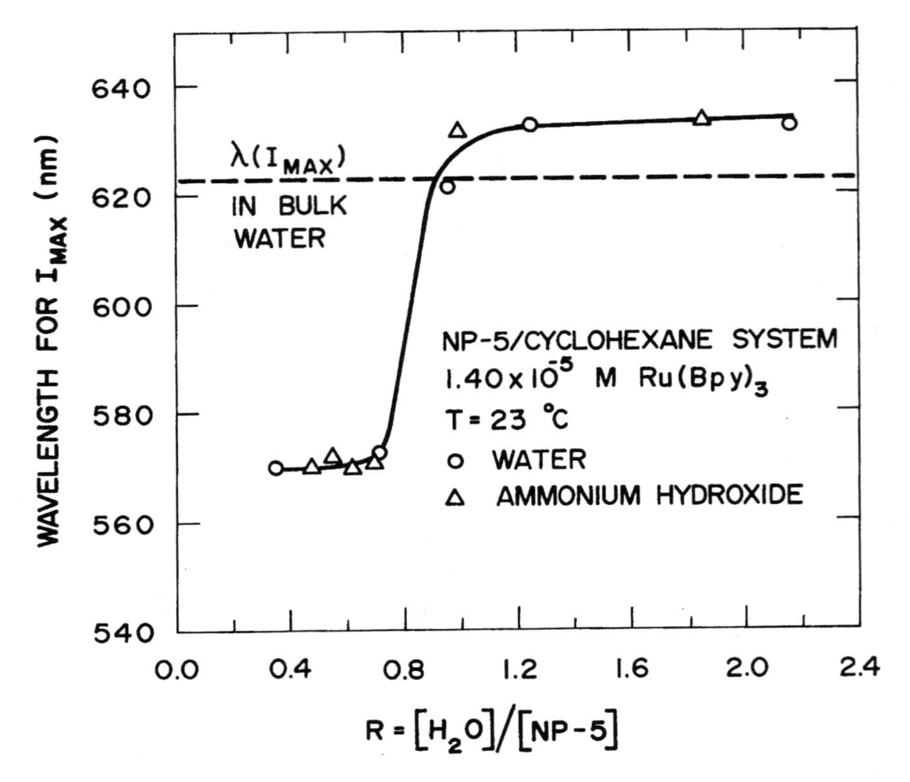

Fig. 2.2.2 Effect of the water-to-surfactant molar ratio (R) on the wavelength of maximum $Ru(Bpy)_3$ fluorescence intensity in the NP-5/cyclohexane/water and NP-5/cyclohexane/water/ammonia microemulsion systems; $\lambda_{ex} = 460$ nm. (From Ref. 25.)

of the fact that ''free'' water molecules become available in AOT-based systems when R exceeds 10–12 (31).

For a given surfactant, the ability to form a single-phase w/o microemulsion is a function of the type of oil, nature of the electrolyte, solution composition, and temperature (54–58). When microemulsions are used as reaction media, the added reactants and the reaction products can also influence the phase stability. Figure 2.2.4 illustrates the effects of temperature and ammonia concentration on the phase behavior of the NP-5/cyclohexane/water system (27). In the absence of ammonia, the central region bounded by the two curves represents the single-phase microemulsion region. Above the upper curve (the solubilization limit), a water-in-oil microemulsion coexists with an aqueous phase, while below the lower curve (the solubility limit), an oil-in-water water microemulsion coexists with an oil phase. It can be seen that introducing ammonia into the system results in a shift of the solubilization

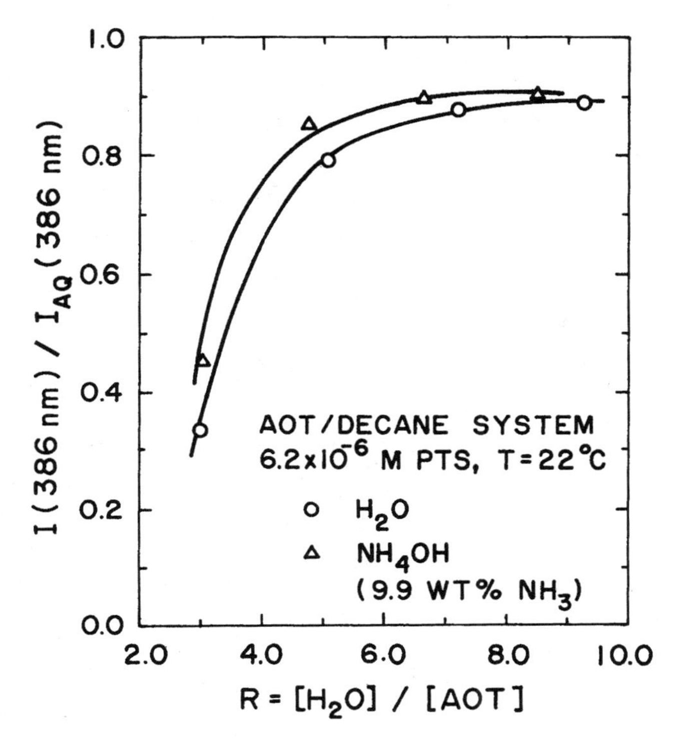

Fig. 2.2.3 Effect of the water-to-surfactant molar ratio (R) on the ratio of fluorescence intensities of PTS at 386 nm in the AOT/decane/water or AOT/decane/water/ammonia microemulsion (I) and in aqueous solution (I_{AQ}). (From Ref. 31.)

curve to lower temperatures. Figure 2.2.5 shows the effect of ethanol addition on the phase behavior (30). With increasing amounts of the alcohol, both the solubilization and solubility limits shift toward higher temperatures. The work of Friberg et al. (41) further illustrates the effect of ethanol on microemulsion phase stability. In the case of the system sodium dodecyl sulfate (SDS)/pentanol/water, it was found that a higher ethanol content led to an increase in the amount of water needed to form a stable microemulsion. Thus, the added ethanol destabilized the microemulsion. A possible cause of this effect is the competition between ethanol and the surfactant (SDS) molecules for water molecules.

Microemulsions are dynamic systems in which droplets continually collide, coalesce, and reform in the nanosecond to millisecond time scale. These droplet interactions result in a continuous exchange of solubilizates. The composition of the microemulsion phase determines the exchange rate through its effect on the elasticity of the surfactant film surrounding the aqueous microdomains. Compared with nonionic surfactant-based microemulsions, AOT reverse micelles have a more rigid

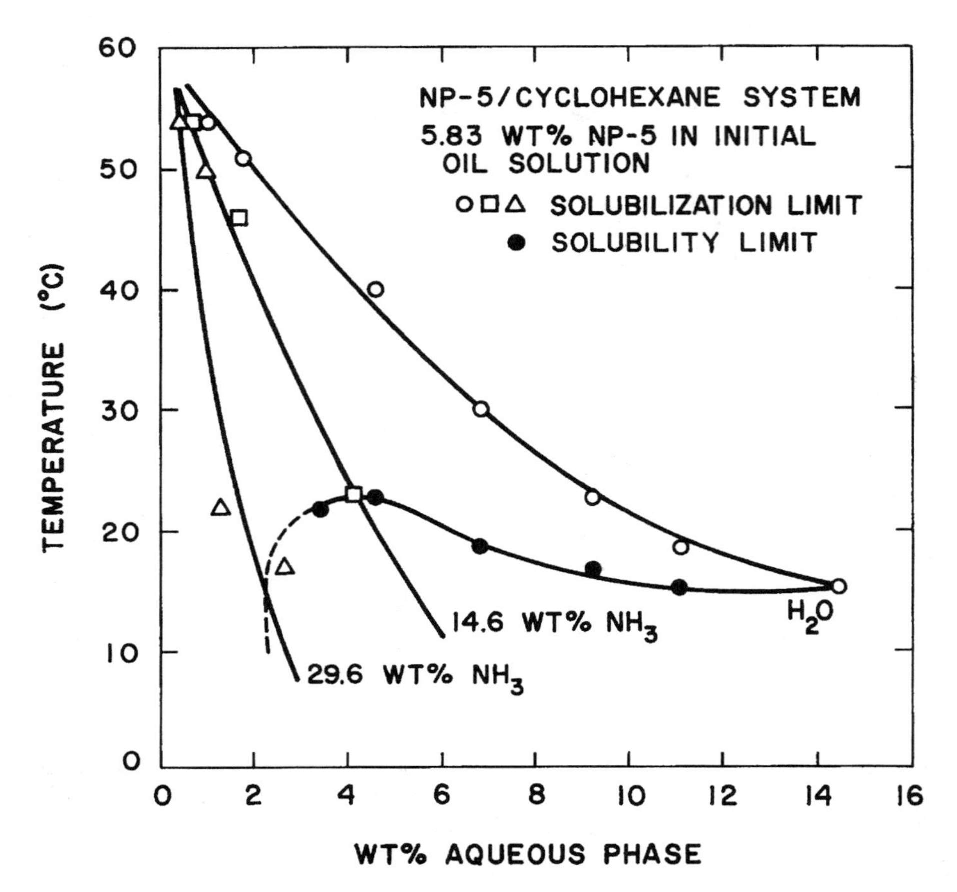

Fig. 2.2.4 Solubilization diagram for H_2O and aqueous NH_3 solutions in the NP-5/cyclohexane system. (From Ref. 27.)

interface, and consequently, the exchange rate constant (k_{ex}) tends to be lower (e.g., by an order of magnitude) than those for nonionic surfactants (59–63).

2.2.2.3 Synthesis Protocols

Three main methods are available for microemulsion-mediated materials synthesis, i.e., microemulsion-plus-trigger, microemulsion-plus-microemulsion, and microemulsion-plus-reactant (14,16). It can be seen from Table 2.2.1 that silica has been prepared by the microemulsion-plus-reactant (MPR) protocol exclusively; that is, the oil-soluble alkoxide is added to the microemulsion. Referring to Table 2.2.1, four

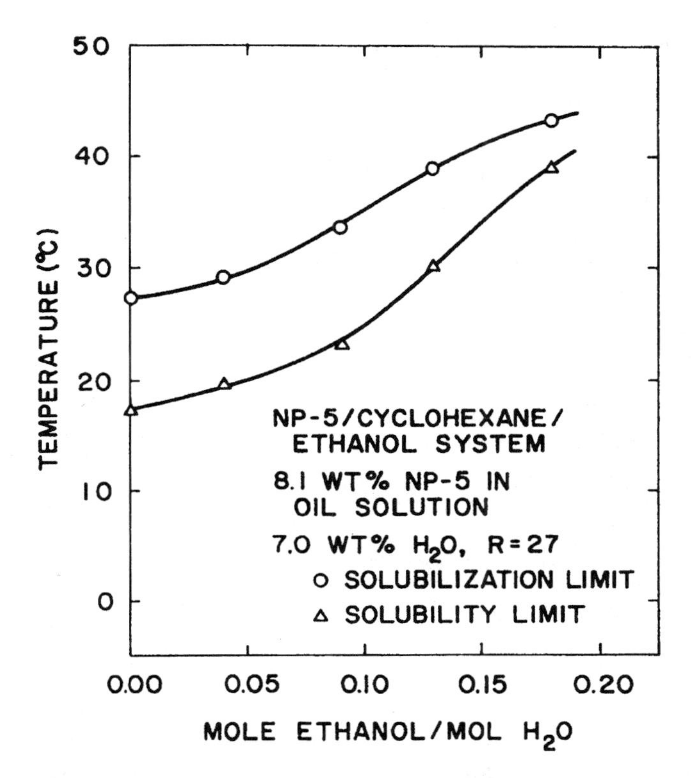

Fig. 2.2.5 Effect of ethanol on the one-phase water-in-oil microemulsion region in the NP-5/cyclohexane/water system. (From Ref. 30.)

main types of silica products may be identified: pure silica nanoparticles, modified nanoparticles, pure silica gels, and modified gels. To prepare these different materials, a number of variations of the basic MPR method have been developed, as illustrated in Figures 2.2.6–2.2.11.

Silica Nanoparticles. The base-catalyzed hydrolysis of silicon alkoxides in microemulsions produces nanoparticles (20–39). Aqueous ammonia has been used primarily as the base, with AOT and nonionic polyoxyethylene ethers as the main surfactants. Figure 2.2.6 presents a flow diagram for the synthesis of pure silica (23–32); the microemulsion is first prepared and then the alkoxide is added. As can be seen from Table 2.2.1, the microemulsions include the systems AOT/isooctane/water/ammonia, AOT/toluene/water/ammonia, NP-5/cyclohexane/water/ammonia, and NP-4/heptane/water/ammonia. Typical reaction times are 1–5 days. Various modified silica nanoparticles have also been prepared, including hydropho-

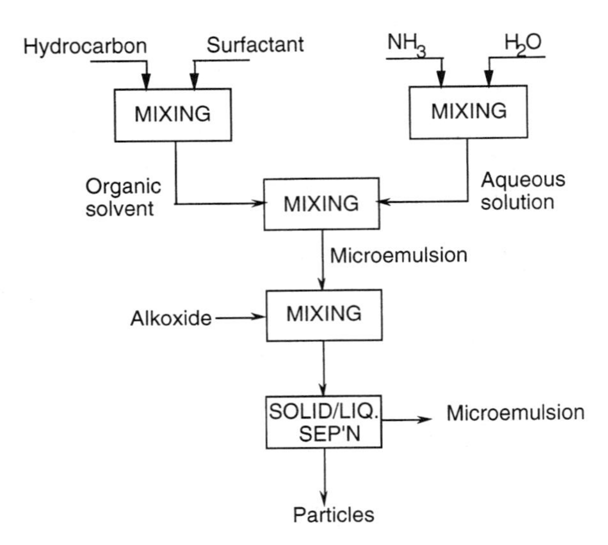

Fig. 2.2.6 Silica particle preparation via the base-catalyzed microemulsion-plus-reactant (MPR) method. (From Refs. 23–32.)

bic silica (21,22), as well as silica doped with bioactive molecules (33), metal oxides (34), metal sulfides (35–38), and metals (38,52). A flow diagram for preparing functionalized (hydrophobic) silica nanoparticles in AOT/toluene/water/ammonia microemulsions, according to the method of Espiard et al. (21,22), is shown in Figure 2.2.7. Trialkoxypropylsilanes of the type $(RO)_3Si(CH_2)_3X$ served as the functionalizing reagents. Most of the investigations were conducted with methacryloyl propyltrimethoxysilane (MPTMS), but other compounds were also considered, where the functional group X included the amino group, chlorine atom, and thiol group. To achieve surface modification, two approaches were used: direct copolymerization of TEOS and the coupling agent (the solid arrow for MPTMS in Figure 2.2.7) and the core-shell polycondensation of the coupling agent onto the preformed silica particles (the dashed arrow for MPTMS in Figure 2.2.7).

Jain et al. (33) used the microemulsion system Triton X-100/cyclohexane/hexanol/water/ammonia to prepare silica nanoparticles with entrapped bioactive macromolecules: fluorescein isothiocyanate-dextran (FITC-Dx) (mol. mass 19.6 kD), [^{125}I]tyraminylinulin (mol. mass 5 kD), and horseradish peroxidase (HRP) (mol. mass 40 kD). The biomolecules were first solubilized in the microemulsion, and the alkoxide (TMOS) was then added. To ensure small particle sizes, the reaction was conducted under ice-cold temperatures (in a refrigerator for 72 h).

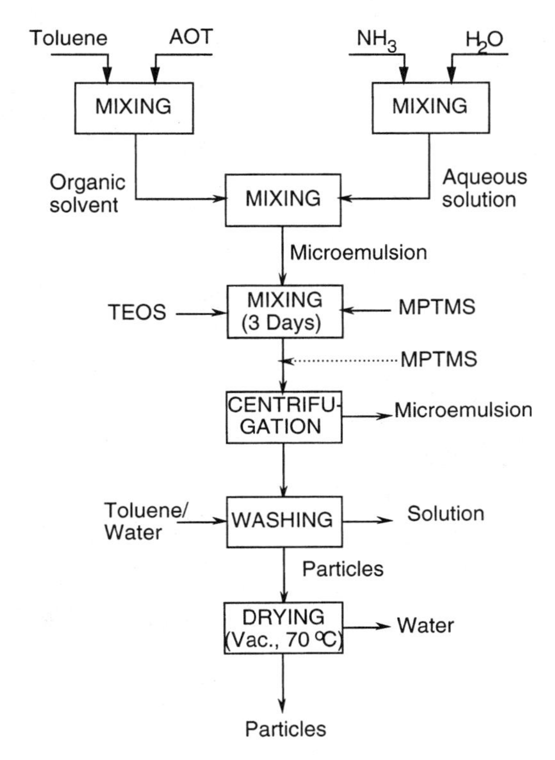

Fig. 2.2.7 Preparation of functionalized silica nanoparticles. (From Refs. 21,22.)

An elegant method for treating hazardous materials has been reported by Chaiko (34) in which reverse micellar-catalyzed solvent extraction was coupled with microemulsion-mediated materials synthesis. Polymeric plutonium (presumably nanoparticles of PuO_2) initially in an aqueous nitric acid medium was transferred into an organic phase via a liquid–liquid extraction process. The extractant octylphenyl-*n,n*-diisobutylcarbamoylmethylphosphine oxide (CMPO) was used and the initial organic phase was the AOT/hexane/CMPO reverse micellar system. Organic/aqueous contact resulted in the cosolubilization of some water and nitric acid with the plutonium nanoparticles in the organic phase. After liquid–liquid separation, TEOS was introduced into the organic phase, followed by base [e.g., hydrazine (N_2H_4) or ammonia] addition. The resulting hydrolysis reactions with the solubilized water (catalyzed by the solubilized nitric acid) resulted in silica-encapsulated PuO_2 particles, which, following solid/liquid separation, can be processed into monolithic silicate waste forms for long-term storage.

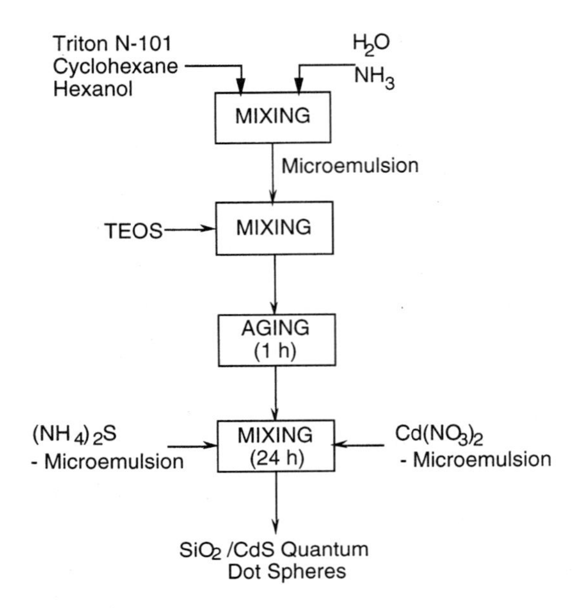

Fig. 2.2.8 Preparation of CdS (dispersed quantum dot)/SiO_2 nanocomposite particles. (From Refs. 35–37.)

Chang et al. (35–37) used controlled precipitation techniques to prepare CdS/SiO_2 nanocomposites in the Igepal CO 520/cyclohexane/water/ammonia and Triton N-101/cyclohexane/hexanol/water/ammonia microemulsions. By manipulating the sequence and duration of reagent addition, CdS/SiO_2 nanocomposites of widely different morphologies were produced, including CdS quantum dots homogeneously dispersed in silica (''raisin bread''), CdS-core/SiO_2, CdS-shell/SiO_2, CdS surface patch/SiO_2, CdS sandwich, surface-bound CdS quantum dots (''freckles''), and CdS-welded silica spheres. Figure 2.2.8 presents a flow diagram for the preparation of CdS quantum dot/SiO_2 spheres. The other types of composites were made by adopting slight modifications of this flowscheme. For example, to prepare the CdS-core particles, CdS was first prepared by combining a Cd(II)-containing microemulsion and an S(II)-containing microemulsion. The alkoxide was then added and the resulting mixture aged for an appropriate length of time (e.g., 20 h). Adair et al. (38) used similar techniques to prepare CdS/SiO_2 and Ag/SiO_2 nanocomposites.

The work of Mallouk et al. (39) offers an interesting extension of the microemulsion sol-gel technique. In this case, microemulsion-derived silica nanoparticles were used as templates for preparing ordered mesoporous polymers with tailored pore sizes. Utilizing the Triton N-101/cyclohexane/hexanol/water/ammonia microemulsion, monodisperse silica nanoparticles were first synthesized. The silica product was recovered from the microemulsion phase by a combination of centrifugation and ethanol washes. The dried particles (80°C) were pressed into tabular pellets, the pores within which were subsequently filled with divinylbenzene (DVB), ethyleneglycol dimethacrylate (EDMA), or their mixtures. Polymerization of these monomers was followed by an aqueous hydrofluoric acid treatment for silica dissolution.

Silica Gels. The acid-catalyzed alkoxide sol-gel process produces gels (17). Friberg and coworkers (40–50) pioneered the extension of this process to silica synthesis in microemulsions; both aqueous and nonaqueous microemulsions were used. For aqueous microemulsions, experiments were conducted mostly with the SDS/pentanol/water/acid system. A representative flow diagram is shown in Figure 2.2.9. The nonaqueous microemulsion systems utilized included CTAB/decanol/decane/formamide and AOT/decane/glycerol (44–46,49,50). The experimental approach followed the sequence: nonaqueous microemulsion preparation, water addition, and then TEOS addition.

Cussler and coworkers (51) were interested in using bicontinuous microemulsions (based on didodecyldimethylammonium bromide, DDAB) as templates for producing high surface area silica gels. Their method is illustrated in Figure 2.2.10. First, a sol derived from partially hydrolyzed TMOS was prepared by adding the alkoxide to a dilute aqueous solution of H_2SO_4. The methanol reaction product acted as a cosolvent, transforming the initially biphase system into a homogeneous

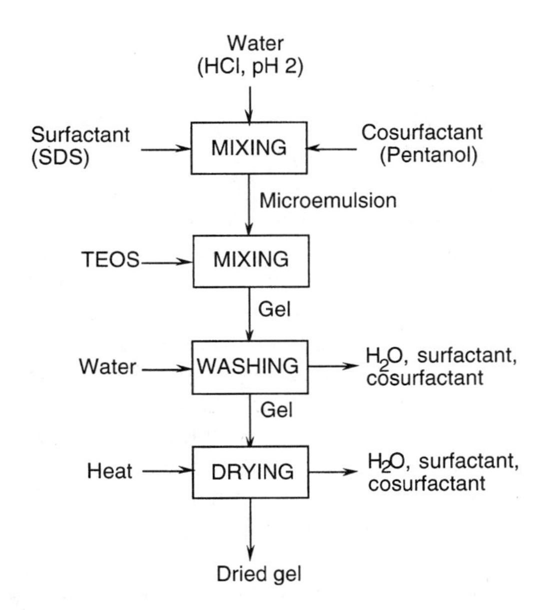

Fig. 2.2.9 Silica gel preparation via the microemulsion-plus-reactant method. (From Ref. 40.)

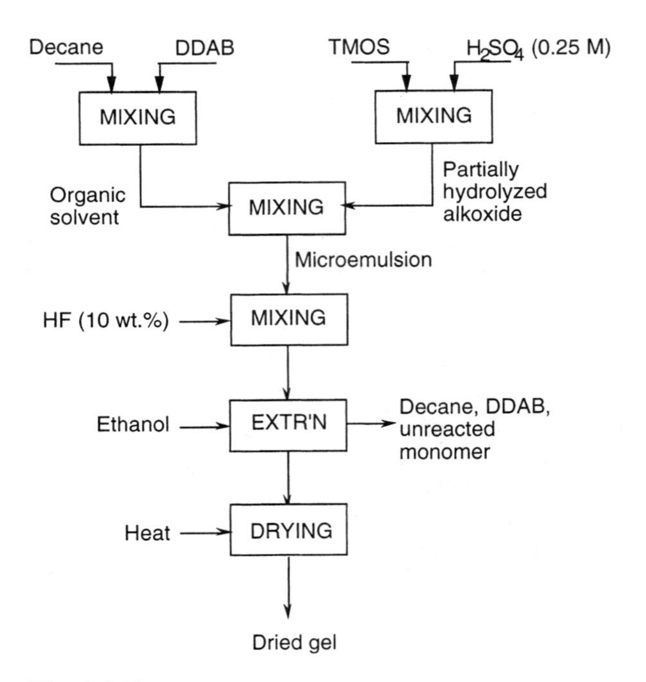

Fig. 2.2.10 Silica gel preparation using partially hydrolyzed alkoxide. (From Ref. 51.)

solution. The silica-containing solution was then introduced into a DDAB/decane solution to form the microemulsion. Aqueous HF, a polymerization catalyst, was next added. The subsequent reaction converted the previously clear bicontinuous microemulsion into a transparent solid, without phase separation.

Martino et al. (52) developed a heterogeneous catalyst preparation method in which the traditional sequence of steps was reversed: i.e., metal clusters were first produced in a reverse micellar phase and then encapsulated in silica gel via microemulsion-mediated silicon alkoxide hydrolysis. Gold particles were prepared in a water-free DDAB/toluene reverse micellar phase, following the method of Wilcoxon et al. (64). Dry $AuCl_3$ powder was first dissolved in the DDAB/toluene solution and then the alkoxide [TEOS or partially hydrolyzed TEOS, i.e., poly(diethoxysilane, mol. wt. = 610 g/gmol)] was added. Following this, a reducing agent, lithium borohydride ($LiBH_4$) dissolved in tetrahydrofuran (THF) was introduced. An aqueous solution of tetrabutylammonium hydroxide was subsequently added to initiate gelation.

The work of Watzke and Dieschbourg (53) emphasized the preparation of silica–biopolymer nanocomposites. In this case, silica was synthesized within the aqueous microdomains of a microemulsion organogel. As summarized in Figure

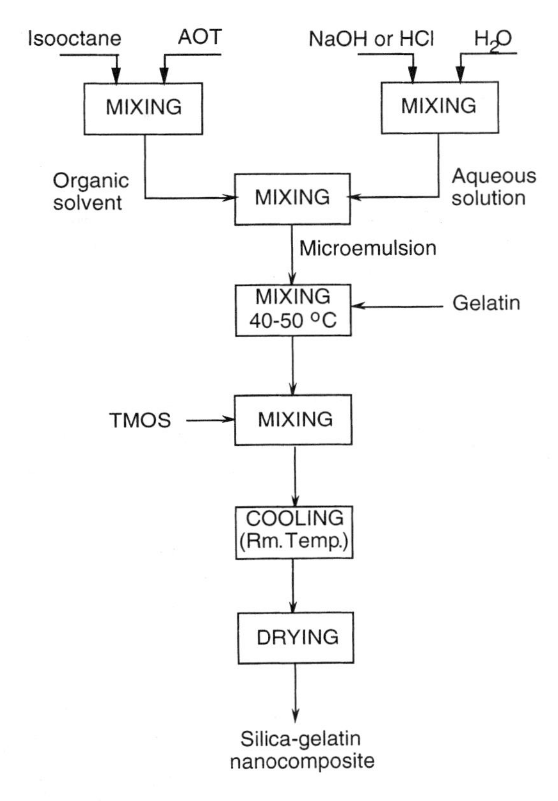

Fig. 2.2.11 Preparation of silica-gelatin nanocomposites. (From Ref. 53.)

2.2.11, to prepare silica–gelatin nanocomposites, a microemulsion based on AOT/isooctane/water/gelatin was first formed, and the resulting gelatin organogel was melted by heating. The alkoxide (TMOS) was then added and the solution gelled by cooling to room temperature. Two extreme views of the microstructure of an organogel are presented in the literature (15). In one case, cylindrical units of hydrated gelatin are covered by surfactant monolayers, and in the other case, reverse micelles are interconnected by gelatin. Presumably, the alkoxide hydrolysis occurs within the respective aqueous microdomains. In addition to gelatin, other biopolymers were used, including chitosan and sodium alginate.

2.2.2.4 Silica Product Characterization

The materials resulting from the microemulsion-mediated sol-gel process have been characterized primarily for particle size and size distribution. Transmission electron microscopy (TEM) is the main technique that has been used for this. Other properties that have been determined include density, specific surface area, porosity, nature of silanol groups, and degree of siloxane bond formation.

Pure Silica Nanoparticles. Based on experiments conducted with the TEOS/AOT/isooctane/water/ammonia system, and water-to-surfactant molar ratios in the range 5.7–15.8, Yamauchi et al. (20) prepared silica particles with diameters of 14–71 nm. Also using AOT, but with a different oil, i.e., toluene, Espiard et al. (21,22) similarly obtained small spherical silica particles with diameters in the range of 20–70 nm when the water-to-surfactant molar ratio was varied from 1 to 6. According to Yamauchi et al. (20), particles given an isooctane/acetone–water/acetone wash, followed by vacuum drying (443 K, 2 h) showed complete elimination of the organic moieties of the AOT surfactant (as determined by FTIR spectra and TGA-DTA analysis). Similar observations were made by Espiard et al. (21,22). However, Yamauchi et al. (20) were unable to achieve complete removal of the associated sodium ions; the final material typically contained 2% residual sodium.

Additional characterization of their silica particles was performed by Yamauchi et al. (20). Specific surface areas in the range of 90–290 $m^2 g^{-1}$ were obtained with the nitrogen BET technique. It was concluded from nitrogen and water adsorption data that the silica particles possessed internal pores and estimates of the micropore volumes were obtained as 0.05–0.13 $cm^3 g^{-1}$. The presence of porosity was also deduced from TEM observation of ~1 nm black spots within the spherical particles; these features were attributed to primary oligomeric particles. In the silicon alkoxide literature, polymers ranging from dimers to octamers are typically classified as oligomeric (17). The observed porosity was attributed to the entrapment of water inside the polymeric network of primary oligomers of silicic acid. Infrared spectra revealed three absorption bands at 3740, 3670, and 3500 cm^{-1}, attributable to free, inner, and surface-hydrogen-bonded silanol groups, respectively. The ammonia concentration in the initial aqueous pseudophase had a significant effect on the relative distribution of the silanol groups. Samples prepared using relatively low ammonia concentrations tended to have the highest content of free silanols. In contrast, surface hydrogen-bonded silanols predominated in samples generated under relatively high ammonia concentrations.

The experimental data of Espiard et al. (21,22), based on the AOT/toluene/water/ammonia system, showed an increase in particle size with increase in R. This observation led the authors to conclude that the droplet size of the microemulsion water pool was a key determinant of particle size. The effect of R on particle size was also investigated for silica nanoparticles synthesized in AOT/decane/water/ammonia microemulsions (31). No particles were observed below about $R = 4$. However, as R increased from 5 to 9.5, the particle size also increased, in agreement with the observations of Espiard et al. (21,22). As noted previously (see Figure 2.2.3), in this microemulsion system, free water pools do not become available until the water-to-surfactant molar ratio exceeds about 10. It can be concluded, therefore, that below $R = 4$, water molecules are strongly bound to the surfactant polar groups as well as the sodium counterions and are not available for TEOS hydrolysis. The increase in particle size with R (and, therefore, with increase in droplet size) suggests that particle nucleation in the AOT system is primarily an intramicellar process where the water core plays the role of a microreactor. AOT reverse microemulsions are characterized by a relatively rigid oil/water interfacial surfactant layer, so that the rates of intermicellar exchange in this system are the lowest among most reverse microemulsion systems (61,62).

It is noteworthy that the final particle sizes obtained with AOT microemulsions were significantly greater than the corresponding water pool diameters (21,22,31). This suggests that further particle growth relied on intermicellar matter exchange. The important role of the exchange process in particle synthesis was demonstrated with experiments in which AOT microemulsions were formulated with different amounts of benzyl alcohol (BA) (32), a reagent that is known to significantly increase the rate of inter-micellar communication in AOT reverse microemulsions (62,63). It was found that for BA/AOT ratios below 1.5, both the final particle size and the rate of growth increased markedly with increase in the BA/AOT ratio. On the other hand, above a BA/AOT ratio of 1.5, the growth rate decreased again and was similar to that obtained in the absence of the additive. This latter effect is attributable to the formation of an excess aqueous phase in the course of the reaction (phase instability), so that the growing particles (being located in this aqueous phase) were less accessible to the unreacted TEOS mainly solubilized in the coexisting w/o microemulsion (32).

The outcome of the synthesis reaction is highly dependent on the nature of the surfactant. Thus, with the same water-to-alkoxide molar ratio of 17, and approximately the same water-to-surfactant molar ratio ($R = 5.5$–6), Espiard et al. (21) found that AOT- and POELE-based microemulsions gave widely different particle sizes, i.e., 65 nm and 28 nm, respectively. Systematic investigations exploring the relationships between microemulsion composition and the particle formation process have been reported by Chang and Fogler (23,24) and Arriagada and Osseo-Asare (25–30) for NP-type microemulsions. Using the NP-4/heptane/water/ammonia system, Chang and Fogler (23) observed that, for a given surfactant concentration, increase in the amount of water led to a decrease in particle size. On the other hand, when the water concentration was kept constant, the particle size went through a minimum (at $R = [H_2O]/[NP\text{-}4] = 1.9$) with increase in surfactant concentration. The effects of surfactant molecular structure and type of oil were also investigated (24). The particle size of the silica particles followed the order: NP-5 $>$ NP-4 $>$ DP-6. The particle size was also found to be sensitive to the type of oil, with the size decreasing in the order: heptane $>$ heptane/cyclohexane (50/50 v/o) $>$ cyclohexane. The conductivity of the microemulsion followed the order: NP-5 $>$ NP-4 $\approx$ DP-6. The fact that the NP-5 microemulsion produced the largest particles was therefore attributed to the relatively high rate of intermicellar matter exchange (as suggested by the high conductivity). In the case of the oil type, the observed trend was rationalized in terms of the greater ability of the molecules of the more polar oil (cyclohexane) to more closely approach the surfactant polar groups. The resulting decrease in the droplet size correlates with the decrease in the particle size.

The work of Arriagada and Osseo-Asare (25–30) was based on the NP-5/cyclohexane/water/ammonia water system. Particles in the range of 50–70 nm were produced with standard deviations below 8.5% around the mean diameters. Figure 2.2.12 (25) presents a typical TEM micrograph of the resulting silica particles. It was observed in certain experiments that, even though particle synthesis was initiated in a clear one-phase microemulsion, the fluid phase became unstable during the reaction and phase separation occurred (30). The continuing nucleation and growth

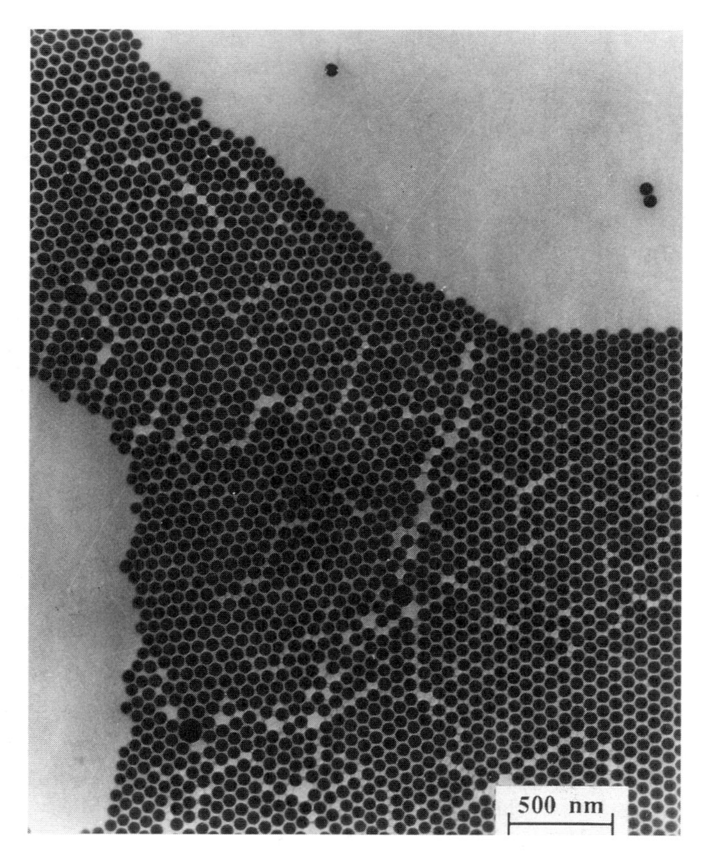

Fig. 2.2.12 TEM micrograph of SiO_2 particles obtained by TEOS hydrolysis in the NP-5/cyclohexane/water/ammonia microemulsion system. (From Ref. 25.)

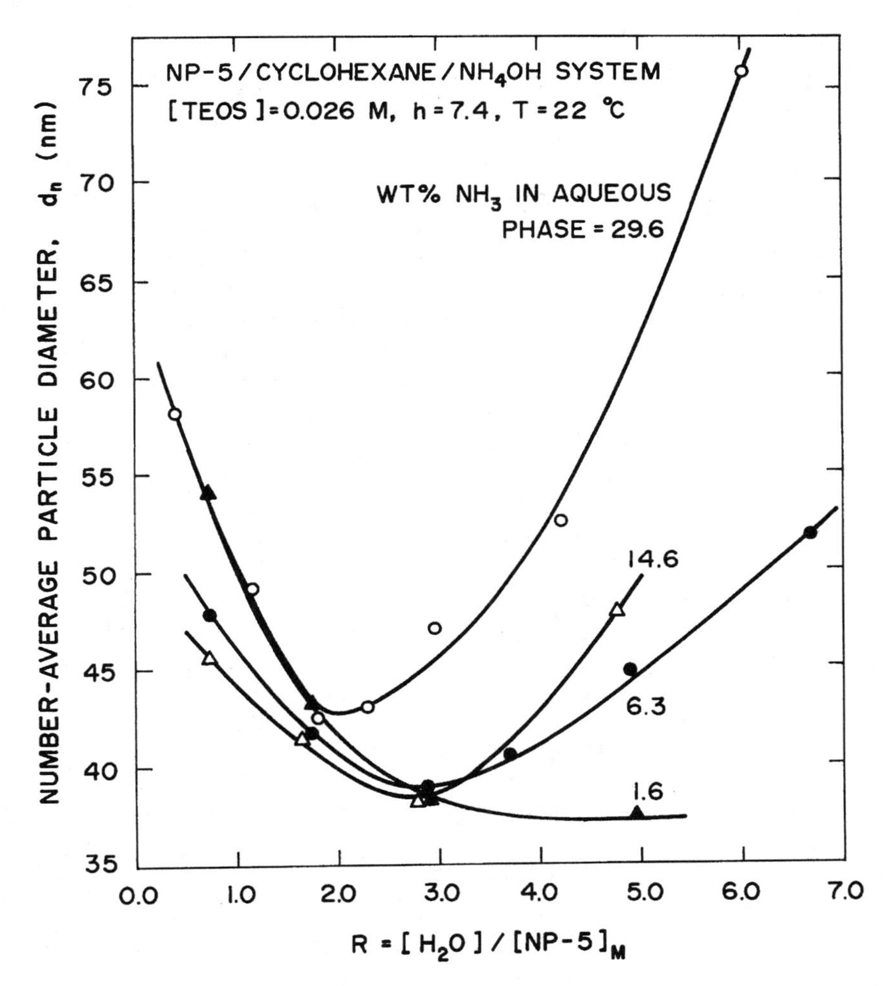

Fig. 2.2.13 Effect of the water-to-surfactant molar ratio (R) on the mean diameter of SiO_2 particles prepared with different ammonia concentrations. (From Ref. 27.)

of silica in the resulting biphase system produced a bimodal size distribution. Referring to the phase diagrams presented in Figures 2.2.4 and 2.2.5, the phase separation may be traced to microemulsion destabilization via H_2O depletion and ethanol release (30). Similar observations were made by Chang and Fogler (23). Figure 2.2.13 (27) shows particle size versus R data for several constant ammonia concentrations. For concentrated ammonium hydroxide (29.6%), the particle size goes through a minimum as R increases. A similar but less dramatic trend is observed for intermediate ammonia concentrations (i.e., 6.3 and 14.6%). In the most dilute ammonia solution utilized (1.6% NH_3), no minimum is observed within the R range investigated, and the particle size decreases continuously.

These observations result from a complex interplay of several factors that are sensitive to the water content in the microemulsion. With increase in R, the surfactant aggregation number (N), the concentration of "free water," and the intermicellar communication rate (k_{ex}) all increase (59–63). The particle size data (27) have been analyzed in terms of a statistical nucleation model based on the reverse micellar populations, the partition of TEOS molecules between the reverse micellar pseudophase and the bulk oil phase, and the Poisson distribution of TEOS molecules and hydroxyl ions among the reverse micelles (28). The decrease in particle size with R is attributable to the formation of an increasing number of nuclei. The subsequent rise in particle size at relatively high R values may be the result of particle aggregation. The increase in k_{ex} that accompanies an increase in R (59,60) is expected to facilitate this. It is likely that at higher ammonia concentrations, larger particles are observed at high R values because a high fraction of the nuclei redissolves. It is known that silica is relatively unstable in highly basic solutions, dissolving to give anionic species such as $SiO(OH)_3^-$ (17,65,66).

Chang et al. (35–37) used both NP-5 (Igepal CO 520) and NP-9 (Triton N-101). Utilizing the Igepal microemulsion with $R = 5$ and $h = 4.8$, 41.3-nm SiO_2 particles were prepared with a 4.2% relative standard deviation of the particle size. For a given h, increase in R decreased the particle size; also, for a given R, an increase in h decreased the particle size. It was observed, further, that increase in TEOS concentration increased the polydispersity of the product particles. This effect was attributed to the corresponding increase in concentration of the ethanol reaction product and the associated phase destabilization (Fig. 2.2.5). This ethanol-induced phase instability was minimized when the NP-5 microemulsion was replaced with the Triton N-101 system. Apparently, the presence of hexanol in the formulation made this microemulsion more tolerant of ethanol. It was therefore possible to make larger particles in the Triton-101 microemulsions. A seeded growth technique was used to produce relatively large particles (up to 150 nm).

Modified Silica Nanoparticles. As already noted above, silica nanoparticles have been modified by surface functionalization (21,22) and by doping with bioactive molecules (33), metal oxides (34), metal sulfides (35–38), and metals (38,52). In the case of surface-modified silicas, properties of interest include particle size, particle density, the C/Si ratio, and the degree of surface functionalization. In the work of Espiard et al. (21,22), the density of the silica product was obtained as 1.7 g cm^{-3}, irrespective of the sample, a value that is significantly lower than the value of 2.29 g cm^{-3} for pure silica. Also, elemental analysis gave a higher carbon-to-silicon weight ratio than that expected for complete replacement of the alkoxy groups by siloxane bonds. It was concluded, therefore, that each alkoxysilane molecule retains at least one alkoxy group. It was found, in general, that the concentration of reverse micelles was about 130 times higher than the silica particle concentration. This suggested to the authors that either only a certain number of water pools serve as locales for silica nucleation, or the previously formed particles aggregate to yield a smaller number of final particles. The concentration of surface double bonds was not significantly affected by the synthesis route (i.e., co-polycondensation vs. core-shell polycondensation); values in the range of 4.4–7.6 vinyl groups/nm^2 were obtained. This result reflects the more rapid condensation of $Si(OH)_4$ relative to $RSi(OH)_3$. The relative reaction rates of TEOS and MPTMS were measured via ^{29}Si-NMR and it was found that TEOS reacted faster than MPTMS (22).

The enzyme-containing nanosize silica particles prepared by Jain et al. (33) were characterized for their particle size (dynamic light scattering and TEM), entrapment efficiency, in vitro leaching capacity, and enzyme activity. For all three biomolecules, the entrapment efficiency was 80–90% and the entrapped molecules were found to be stable towards leaching (up to 45 days). The enzymatic activity of the entrapped molecules was lower than that of the corresponding free molecules, a result that was attributed to diffusional constraints.

Chang et al. (35–37) presented TEM micrographs that demonstrate that controlled precipitation, coupled with selective dissolution, can be used successfully to prepare CdS/SiO_2 nanocomposites of widely different morphologies. A typical ''raisin bread'' nanocomposite containing 3 mol% CdS was characterized by ~100 nm particle size and ~2.5 nm CdS quantum dots. Core-shell particles were prepared with 6 nm CdS cores and an overall particle size of ~39 nm. Specific surface areas as high as 183 m^2/g were reported. In general, elemental analysis of the CdS/SiO_2 composites revealed a high level of CdS incorporation, although less than 50% incorporation was observed for some preparations.

Pure Silica Gels. In the experiments of Friberg et al. (40–50) involving acid-catalyzed hydrolysis of TEOS, both aqueous (40–43,48) and nonaqueous (44–47,49,50) microemulsions were utilized. Using SDS/pentanol/water microemulsions, and relatively low TEOS concentrations, it was possible to prepare transparent silica gels over the entire microemulsion stability region. It was found, however, that with increasing concentrations of TEOS, the microemulsion stability region that could support the transparent silica gel shrank toward the water-rich corner of the ternary SDS/pentanol/water phase diagram. In the case of the nonaqueous microemulsion systems, the AOT/decane/glycerol system gave no gels (44). With the CTAB/decanol/decane/formamide system (44), a minimum formamide/CTAB composition was needed to achieve gelation of the microemulsion. Increase in surfactant concentration or the $C_{10}OH/C_{10}$ ratio decreased the gelation time. The xerogels experienced severe cracking when decane was added to the $C_{10}OH$-based microemulsions. The origins of this effect are not clear. On the basis of ^{29}Si-NMR spectral data, it was concluded that the condensation followed the sequence: monomer → dimer → linear trimer → cyclic tetramer.

Watzke and Dieschbourg (53) also synthesized pure silica gels under acidic conditions. The microemulsion underwent a phase separation, and this was attributed to the released methanol. Apparently, silica synthesis occured in the lower aqueous phase, i.e., under essentially homogeneous conditions. The presence of a relatively high content of silanol groups was indicated by MAS ^{29}Si-NMR; the relative species distributions followed the order: $Q^3 > Q^4 > Q^2$ (where the superscript gives the number of siloxane bonds in the oligomer). It is well known that in the homogeneous silica sol-gel process, acid catalysis yields polymeric network structures, whereas base catalysis results in dense colloidal particles (17). It was concluded by Watzke and Dieschbourg (53), however, that in the case of microemulsions and organogels, silica gel formation via alkoxide precursors involves an initial formation of discrete colloidal particles, followed by gelation. More detailed investigations are needed in order to support this claim.

Viscoelastic properties of microemulsion silica gels were investigated by Friberg et al. (42). In a study based on the SDS/pentanol/water system, it was found that, for reaction times under 32 h, both the eleastic and loss moduli were very low and were highly frequency dependent. In contrast, longer reaction times (e.g., 37 h) gave comparatively high moduli and the frequency dependence was weak. Gelation time was ~33 h. Cussler et al. (51) investigated the effects of drying techniques on the properties of their silica gels. The as-produced silica gel was treated with ethanol to remove the surfactant, decane, and the residual alkoxide. Vacuum or supercritical drying of the microemulsion-derived silica gels yielded specific surface areas that were greater than those obtained with conventional vacuum-dried gels.

Modified Silica Gels. The Cu-doped silica gels of Jones et al. (48), when examined by UV-visible spectroscopy, revealed that the metal was retained in the gels as the Cu^{2+} ion. The presence of crystalline copper nitrate was indicated by x-ray diffraction, and SEM micrographs revealed the copper nitrate as 1–15 μm crystals located both inside and on the surface of the silica network. Friberg et al. (49) reported that gels doped with rhodamine B or rhodamine 6G gave fluorescence quantum yields that indicated that the microemulsion gel method is a promising technique for preparing solid-state laser dye materials. The results with relatively polar dyes (coumarins) were not as encouraging, however (50).

Transparent silica gels were obtained in the gold-cluster-containing reverse micellar systems investigated by Martino et al. (52). Based on small-angle x-ray scattering (SAXS) investigations, it was concluded that introduction of TEOS into DDAB/toluene solutions partially destroyed the DDAB reverse micelles. Gold clusters with particle sizes in the range of 5.6–7.1 nm were synthesized. This cluster size was about twice as large as that obtained in the absence of TEOS (64), and this difference was attributed to TEOS/DDAB interaction. The gelation time (t_g) increased with increase in the water-to-alkoxide molar ratio (h); with TEOS as the alkoxide precursor, t_g was 1.5 and 24 h respectively for $h = 1$ and 4. As noted by Martino et al. (52), this trend is opposite to that observed in conventional sol-gel processing in alcoholic solutions. This discrepancy was rationalized (52) in terms of the microheterogeneity of the microemulsion phase. Increasing h simultaneously increases R (the water-to-surfactant molar ratio). In general, reverse micelles increase in size as R increases, and this leads to a decrease in the net surface area of water pools. The result is a decrease in the effective water-to-alkoxide molar ratio, since alkoxide hydrolysis occurs only at the surfactant interfaces within the reverse micelles. The as-prepared gels were dried with supercritical CO_2 to produce aerogels or dried under ambient conditions to prepare xerogels. In general, the materials derived directly from TEOS had lower specific surface areas, compared with those prepared with the partially hydrolyzed alkoxide. For the aerogels based on the prehydrolyzed alkoxide, increase in h increased the surface area while increase in surfactant concentration had the opposite effect. Also, comparable surface areas were obtained for the xerogels. In the case of the TEOS-based aerogels, the surface area versus h trends showed maxima, while the surface area versus surfactant trends exhibited minima. The corresponding TEOS-based xerogels had negligible surface areas.

According to Watzke and Dieschbourg (53), alkoxide hydrolysis in the AOT/gelatin organogel resulted in a white solid mass; this is in contrast to the transparent microemulsion silica gels of Friberg et al. (40). Scanning electron microscopy indicated a marked effect of pH on the microstructure of the silica-gelatin nanocomposite gels. Porous materials were obtained with both basic and neutral pH. However, low pH yielded a porous network of filamentous material, whereas reaction at higher pH gave a porous tubular structure. In both cases, the pore size distribution was broad. Also, both the filaments and the tube walls consisted of small silica particles. Pure silica gels prepared under neutral conditions did not have this tubular appearance. SEM revealed a highly porous network of silica particles. Thus, the tubular appearance reflects the incorporation of the silica particles into the gelatin network. Magic-angle spinning ^{29}Si-NMR (MAS ^{29}Si-NMR) was used to examine the degree of crosslinking among the Si atoms of the silica network. The NMR investigations revealed that for neutral conditions, both pure silica gels and silica–gelatin composites contained primarily Q^4 and Q^3 species, indicating a relatively high degree of cross-linking. Also, both materials contained the same proportions of Q^4 and Q^3 species. It was concluded, therefore, that the gel formation mechanism was identical for both pure silica and the nanocomposite. That is, the gels are a result of particle aggregation followed by gelation.

2.2.3 Silica Formation Kinetics and Mechanisms

2.2.3.1 General Trends

Systematic investigations into the kinetic aspects of silica formation in microemulsions have been undertaken by only a few researchers (23,24,29). Various forms of useful qualitative and semi-quantitative kinetic information are available, however. For example, size exclusion chromatography (SEC) was used by Espiard et al. (22) to monitor TEOS concentration as a function of time for different R values (R = 4, 6, 10). An initial period of rapid reaction (approximately first 6 h) was observed, followed by a much decreased rate. The reaction took about 3 days to reach completion, and the final conversion was typically 80–90%. Both the initial rate and the final conversion increased with *R*. The inability to achieve complete conversion was attributed to immobilization of a portion of the water molecules through their interaction with the surfactant polar groups. Friberg and Yang (40) monitored the consumption of water by using Karl Fischer titration. The water content was found to decrease rapidly in the first hour. Using ^{13}C-NMR, the generation of ethanol and the consumption of alkoxide were followed for 3, 10, and 50 min. Friberg et al. also investigated the kinetics of the gelation reaction with ^{29}Si-NMR (46,47) and viscoelastic property measurements (42,45).

2.2.3.2 Steps in the Particle Formation Process

Following alkoxide addition to the microemulsion phase, the formation of silica particles proceeds via a series of steps, which include: (1) association of TEOS molecules with the reverse micelles, (2) TEOS hydrolysis and formation of monomers, (3) nucleation, (4) particle growth, (5) nuclei dissolution, (6) intermicellar exchange of monomers, (7) ionization of monomers, and (8) particle surface ionization (28). It is likely that step 1, the partition of an alkoxide molecule between the reverse micellar pseudophase (m) and the bulk oil phase (b), is facilitated through the formation of the monomer with one silanol group [i.e., $Si(OR)_3OH$]. There is experimental evidence (67) that this partially hydrolyzed alkoxide has amphiphilic properties (see later discussion). Further hydrolysis of the monosilanol species (step 2) generates monomeric species with up to four silanol groups (i.e., silicic acid). All these species [i.e., $Si(OR)_3(OH)$ to $Si(OH)_4$] are designated as ‘‘monomers’’ and are assumed to remain associated with the reverse micellar pseudophase due to their enhanced polar character. These species can participate in particle nucleation and growth. Nucleation (step 3) involves the condensation of monomers, and it can be an intramicellar or intermicellar event. Particle growth (step 4) may occur by addition of monomers to nuclei (an intra- or inter-micellar process), or by aggregation of nuclei (an inter-micellar process).

2.2.3.3. Alkoxide Distribution in the Microemulsion Phase

The view that the monomers are confined to the reverse micellar pseudophase is supported by interfacial tension data (67), which demonstrate that in a two-phase octane/water system, partially hydrolyzed TEOS species partition preferentially into the aqueous phase. The interfacial tension determined at the octane/water interface for samples prepared with precursor ethanolic solutions of different water-to-TEOS molar ratios (h = 0, 0.29, and 0.55) are presented in Figure 2.2.14 (67). As can be seen, for TEOS concentrations below about 4×10^{-3} M, the octane/water interfacial tension is independent of the concentration of TEOS species in the organic phase

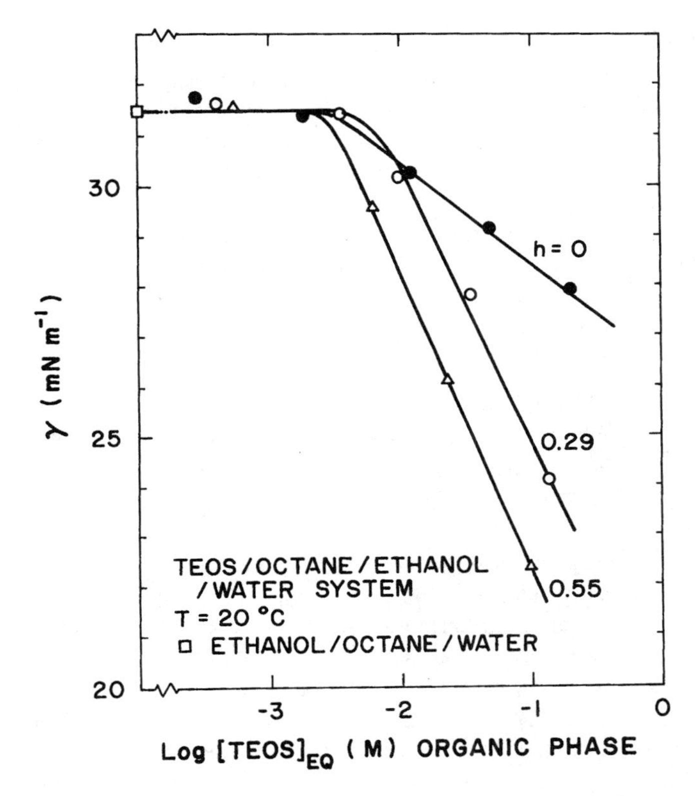

Fig. 2.2.14 Effect of the water-to-TEOS molar ratio (h) in the precursor ethanolic solution on the interfacial tension at the octane/water interface; $[TEOS]_o$ referred to the conjugate organic phase. (From Ref. 67.)

and equal to that of the ethanol/octane/water system. At higher TEOS concentrations, however, the interfacial tension decreases with increase in the TEOS content in the conjugate organic phase; this effect is more significant in samples prepared with higher water-to-TEOS molar ratios (h) in the precursor ethanolic solutions. These results demonstrate that partially hydrolyzed TEOS species are interfacially active.

The fraction of TEOS molecules associated with the reverse micellar pseudophase can be evaluated if the corresponding partition constant, K_p, is known. To determine K_p, the approach taken by D'Aprano et al. (68) to analyze the partition of alcohols in the AOT/heptane reverse micellar system was used (28). Unhydrolyzed TEOS and partially hydrolyzed species are expected to be located in the surfactant layer. The partition constant for TEOS can be expressed as (28):

$$K_p = n_m/n_o\{NP\text{-}5\}_m \quad (4)$$

where n_m and n_o are the moles of solute in the reverse micellar pseudophase and the organic solvent, respectively, and $\{NP\text{-}5]_m$ the molal concentration of NP-5 associated with the reverse micelles. No partition data are available for TEOS or partially hydrolyzed TEOS species in reverse micellar systems. However, by comparing a hydrolyzed TEOS monomer with medium-chain alcohols (C_5–C_7), the partition constant (K_p) of a partially hydrolyzed TEOS molecule (with one silanol group) has been estimated to be in the range of 0.1 to 0.5 kg mol^{-1} in the NP-5 system (28).

If the fraction of solute molecules in the reverse micellar pseudophase, X_m, is expressed as $X_m = n_m/(n_m + n_o)$, then it follows from Eq. (4) that

$$X_m = K_p\{NP\text{-}5\}_m/(1 + K_p\{NP\text{-}5\}_m) \quad (5)$$

where K_p has units of kg mol^{-1}. Thus, given a known total amount of TEOS in the system, and the total number of reverse micellar aggregates (N_m) present at a given R value, the average number of TEOS molecules per reverse micelle (N_t) can be calculated from the fraction X_m associated with the reverse micelles (28). As shown in Table 2.2.3 (28), the N_t values obtained are significantly lower than those calculated by assuming complete association (i.e., an infinite K_p value) between TEOS and the reverse micellar pseudophase. Furthermore, since the number of water molecules per reverse micelle (N_w) is only a function of R (i.e., not affected by TEOS partition), the average water-to-TEOS molar ratio (h) in each reverse micelle now increases with R. For example, for $K_p = 0.6$ kg mol^{-1}, the h ratios at R values of 0.41 and 6.05 are respectively 27 and 298: much higher than the initial constant value ($h = 6.3$) calculated from the overall water and TEOS concentrations. Thus, due to TEOS partitioning, the amount of water at the reaction site is well in excess over that required by stoichiometry for SiO_2 formation.

Table 2.2.3 Effect of TEOS Partition on Reverse Micellar Aggregate Statistics[a]

$[NP\text{-}5]_m$ (M)	R	N[b]	$N_m \times 10^{-18}$	N_w	N_t (K_p[c] $= \infty$)	N_t (K_p[c] $= 0.06$)	N_t (K_p[c] $= 0.1$)
0.40	0.4	19	64.0	8	1.2	0.3	0.07
0.14	1.2	47	9.2	54	8.6	0.9	0.17
0.09	1.8	71	3.9	128	20.3	1.4	0.25
0.07	2.3	89	2.5	203	32.3	1.8	0.31
0.05	3.0	113	1.5	334	53.1	2.3	0.39
0.04	4.2	160	0.7	677	107.4	3.2	0.55
0.03	6.1	227	0.4	1370	217.8	4.6	0.78

[a] Data for samples with concentrated NH_4OH and (initial) $h = 6.3$. *Source:* Ref. 28.
[b] Aggregation numbers from Kitahara data (69).
[c] K_p in kg mol^{-1}.

2.2.3.4 Kinetics of Microemulsion-Mediated Alkoxide Hydrolysis

Recalling Eq. (1), it can be seen that the hydrolysis of silicon alkoxide molecules generates alkanol molecules and consumes water molecules. Chang and Fogler (23,24) investigated the rate of hydrolysis of TEOS by monitoring the alkoxide and ethanol concentrations with Fourier transform infrared spectroscopy (FTIR). The absorption bands for the Si-O-C stretching (795, 967 cm^{-1}) gave the TEOS concentration, those for C-C-O stretching (882, 1050 cm^{-1}) yielded the ethanol concentration, and the band for the H-O-H deformation (1640 cm^{-1}) was used to monitor the H_2O concentration. With increase in reaction time, the intensity of the Si-O-C and H-O-H absorption bands decreased, while that of the C-C-O bands increased. It is interesting to note that the 795 and 967 cm^{-1} bands vanished after relatively long reaction times, signifying the complete consumption of TEOS. Further, it is noteworthy that no significant absorption bands were found for the Si-OH group (910–830 cm^{-1} region). This indicates the absence of large amounts of $RSi(OH)^n$. The implication is that the hydroxy groups resulting from the alkoxide hydrolysis condensed rapidly to give polymeric silica species.

Based on the change in intensity of the Si-O-C band at 967 cm^{-1}, the rate of hydrolysis of TEOS was determined to be first order with respect to the alkoxide concentration:

$$d[\mathrm{TEOS}]/dt = -k'_{\mathrm{H}}\,[\mathrm{TEOS}] \tag{6}$$

The apparent rate constant (k'_{H}) was also found to be linearly dependent on ammonia and hydroxide concentrations. The dependence on H_2O concentration was, however, slight. It is interesting to compare this finding with previous results obtained for homogeneous alcoholic solutions (17), i.e., reaction orders of 1–1.5 for the concentration of water. Chang and Fogler (23) rationalized the lower reaction orders obtained in the microemulsion-based experiments in terms of the higher local concentrations. For example, with a total concentration of 0.58 *M* water in the microemulsion, the local H_2O concentration in the water pools may approach that of bulk water, i.e., 55 *M*. A further consideration is the relatively high water-to-alkoxide molar ratio associated with the alkoxide partitioning, as noted in Section 2.2.3.3 (see Table 2.2.3).

Figure 2.2.15 shows plots of the apparent rate constant (k'_{H}) versus the water-to-surfactant molar ratio (R). It can be seen that when R was varied by keeping [NP-4] constant while varying [H_2O] (lower curve), the apparent rate constant (k'_{H}) increased with R. On the other hand, when R was varied by maintaining [H_2O] constant while varying [NP-4], then k'_{H} decreased as R was increased. The first effect reflects the greater availability of ''free'' water molecules with increase in R. The second trend may be rationalized by viewing each reverse micelle as a ''reactant.'' Then, increase in surfactant concentration increases the number of reverse micelles. Chang and Fogler (23) suggest that the surfactant polar groups (the

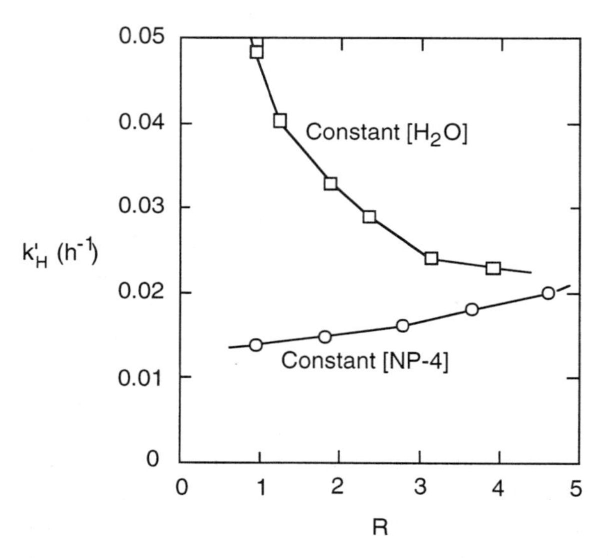

Fig. 2.2.15 Effect of the water-to-surfactant molar ratio (R) on the apparent hydrolysis rate constant (${}_{H}k'$) for TEOS in the system NP-4/heptane/water/ammonia. (From Ref. 23.)

ether -O- and hydroxyl -OH groups) interact through hydrogen bonding with hydroxide ions of the aqueous pseudophase and with the oxygen atoms of TEOS molecules and that this facilitates TEOS/OH^- molecular contact.

2.2.3.5 Hydrolysis-Controlled Growth Kinetics

In principle, silica growth kinetics may be controlled by (1) slow release of monomer via alkoxide hydrolysis in the particle-free reverse micelles, (2) slow surface reaction of monomer addition to the growing particle, and (3) slow transport processes as determined by the dynamics of intermicellar mass transfer. There is strong experimental evidence to support the view that the rate of silica growth in the microemulsion environment is controlled by the rate of hydrolysis of TEOS (23,24,29). Silica growth kinetics can be analyzed in terms of the overall hydrolysis and condensation reactions:

$$Si(OR)_4 + 4H_2O \xrightarrow[k_H]{OH^-} Si(OH)_4 + 4ROH \tag{7}$$

$$Si(OH)_4 \underset{OH^-\ k_D}{\overset{k_C\ OH^-}{\rightleftharpoons}} SiO_2(s) + 2H_2O \tag{8}$$

where k_H, k_C, and k_D are respectively the rate constants for hydrolysis, condensation, and dissolution. In homogeneous alcoholic media, the hydrolysis rate is typically of first order with respect to TEOS, water, and catalyst (OH^-) concentrations (17,70,71), as represented in Eq. (9):

$$r = d[\text{TEOS}]/dt = -k_H [\text{TEOS}] [H_2O] [OH^-] \tag{9}$$

where [TEOS], $[H_2O]$, and $[OH^-]$ are the molar concentrations of the indicated species. Under conditions of excess water and constant catalyst concentration, the hydrolysis rate (r) can be expressed as:

$$r = -k'_H [\text{TEOS}] \tag{10}$$

where $k_H' = k_H[H_2O][OH^-]$ is the pseudo-first-order hydrolysis rate constant.

Assuming that condensation involves only the participation of silicic acid (the monomer), then under conditions of constant water and catalyst concentrations, the resulting rate expressions are (29):

$$\frac{d[\text{TEOS}]}{dt} = -k'_H[\text{TEOS}] \tag{11}$$

$$\frac{d[\text{Si(OH)}_4]}{dt} = k'_H[\text{TEOS}] + -k'_C A([\text{Si(OH)}_4]_{eq} - [\text{Si(OH)}_4]) \tag{12}$$

$$\frac{1}{V_o}\frac{dm_p}{dt} = -k'_C A([\text{Si(OH)}_4]_{eq} - [\text{Si(OH)}_4]) \tag{13}$$

where k'_C is the overall condensation rate constant, A is the total surface area of the solid particles, $[Si(OH)_4]$ and $[Si(OH)_4]_{eq}$ are the molar concentrations of silicic acid at time t and at equilibrium, respectively, m_p represents the total moles of silica in the product particles at time t, and V_o is the total volume of the microemulsion phase. Under conditions where the hydrolysis process is slow as compared with condensation, it can be shown that (29):

$$\frac{1}{V_o}\frac{dm_p}{dt} = -\frac{d[\text{TEOS}]}{dt} = k'_H[\text{TEOS}] = k_g[\text{TEOS}] \tag{14}$$

where k_g is the first-order growth rate constant.

Figure 2.2.16 presents number-average particle diameters (d_n) obtained as a function of reaction time at selected R values (29). In general, the rate of particle growth in the microemulsion phase is slower than that observed for homogeneous media (17). The slow growth rates observed in the microemulsion may be associated with the compartmentalization of reagents in this microheterogeneous system. The particle size versus time data (Figure 2.2.16) were analyzed by assuming first-order kinetics with respect to TEOS concentration. As seen in Figure 2.2.17, particle growth followed first-order kinetics for all R values investigated, up to about 21 ($R = 0.68$) to 30 h ($R = 5.37$) of reaction, in agreement with Eq. (14). Additional support for hydrolysis-controlled growth kinetics [Eq. (14)] comes from the work of Chang and Fogler (23); these investigators found that the apparent hydrolysis rate constant had the same numerical value as the growth rate constant.

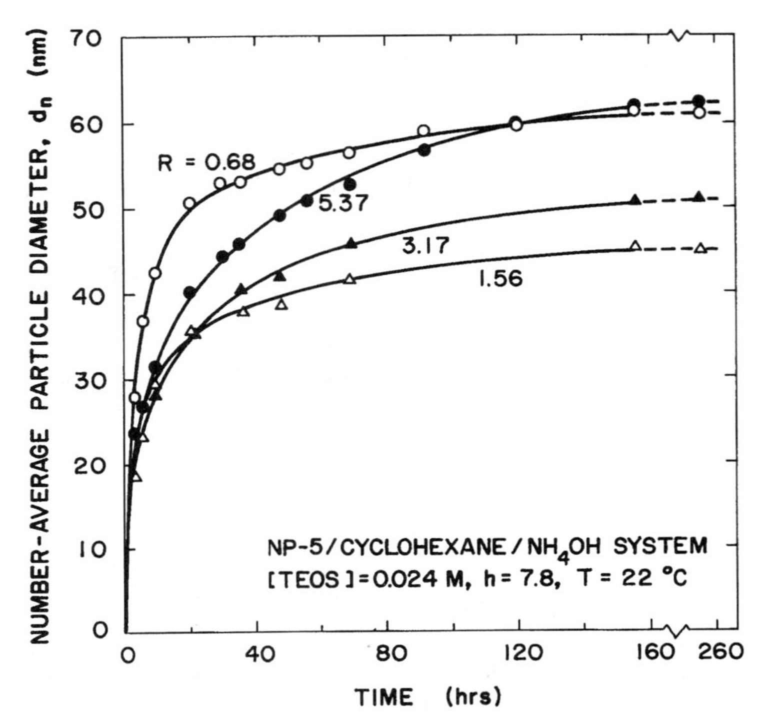

Fig. 2.2.16 Time evolution of the mean diameter of SiO_2 particles produced in the NP-5/cyclohexane/water/ammonia system at different R values. (From Ref. 29.)

2.2.3.6 Microemulsion Pseudophase Growth Model

The first-order growth rate constants (k_g) obtained from the initial slopes in Figure 2.2.17 decrease as R increases (29). In principle, hydrolysis should be favored at large R, since under these conditions water is mostly free and a well-defined hydrophilic domain exists within the reverse micelles (54–57). The observed decrease in the growth rate constant (k_g) as R increases is, therefore, in apparent contradiction to this expectation. This apparent discrepancy can be resolved by considering the microheterogeneity of the microemulsion fluid phase (29). The observed growth kinetics can be interpreted in terms of a pseudophase model (72–74), which assumes that the reaction of interest (hydrolysis in this case) takes place in two separate pseudophase domains. In this model, the microemulsion medium (of total volume V_o) is viewed as a two-phase system consisting of a reverse micellar pseudophase (of volume V_m) and a bulk oil phase (of volume V_b). Furthermore, each reactant is distributed between both phases according to a partition constant P_i.

The observed reaction rate (r) is treated as an averaged value over the whole volume of the system, and it is expressed in terms of the respective reaction rates

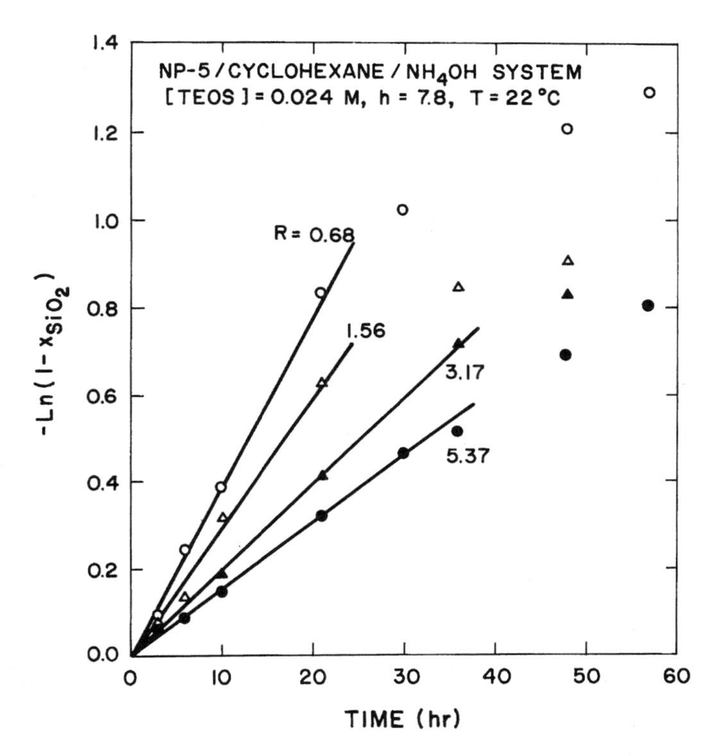

Fig. 2.2.17 First-order growth kinetics of SiO_2 particles in the NP-5/cyclohexane/water/ammonia system at different R values. (From Ref. 29.)

in the reverse micellar pseudophase (r_m) and the bulk oil phase (r_b) and the volume fraction of each pseudophase (ϕ_m and ϕ_b) (29):

$$r = r_m \phi_m + r_b (1 - \phi_m) = k_{obs} [\text{TEOS}]_o[\text{H}_2\text{O}]_o[\text{OH}^-]_o \tag{15}$$

where k_{obs} is the experimentally observed rate constant, and $[\text{TEOS}]_o$, $[\text{H}_2\text{O}]_o$, and $[\text{OH}^-]_o$ refer to the concentrations of the indicated species in the total microemulsion volume (V_o).

The reasonable assumption can be made that hydrolysis in the bulk oil phase is negligible (i.e., $r_b = 0$), since water and OH^- ions reside mainly in the reverse micellar pseudophase. Therefore, Eq. (15) becomes:

$$r = r_m\phi_m = k_m[\text{TEOS}]_m[\text{H}_2\text{O}]_m[\text{OH}^-]_m\phi_m = k_{obs}[\text{TEOS}]_o[\text{H}_2\text{O}]_o [\text{OH}^-]_o \tag{16}$$

where the subscript m indicates quantities referred to the reverse micellar pseudophase. The volume fraction of the reverse micellar pseudophase (ϕ_m) can be expressed as (29):

$$\phi_m = [H_2O]_o(\nu_s/R + \nu_{aq}) \quad (17)$$

where ν_s is the molar volume of the surfactant and ν_{aq} is the effective molar volume of the ammonia solution. The water-to-surfactant molar ratio (R) is given by $R = [H_2O]_o/[NP\text{-}5]_o$, where $[NP\text{-}5]_o$ is the micellized surfactant concentration with respect to the total microemulsion volume (V_o). The concentration of a given species, S, referred to the total microemulsion volume (i.e., $[S]_o$) can be related to its local concentration in the reverse micellar pseudophase (i.e., $[S]_m$) through the corresponding partition constant, P_i. Accordingly, for a microemulsion system of low concentrations of surfactant and aqueous phase, the expression relating k_{obs} with k_m is obtained as (29):

$$k_{obs} = k_m P_T P_W P_{OH} [H_2O]_o (\nu_s/R + \nu_{aq}) \quad (18)$$

From Eq. (9), (14), (15), and (18), the growth rate constant, k_g, can be expressed as (29):

$$k_g = k_m P_T P_W P_{OH} [H_2O]_o^2[OH^-]_o (\nu_s/R + \nu_{aq}) \quad (19)$$

Under conditions of constant water and hydroxyl ion concentrations, Eq. (19) becomes:

$$k_g = K(\nu_S/R + \nu_{aq}) \quad (20)$$

where the constant K is given by

$$K = k_m P_T P_W P_{OH} [H_2O]_o^2[OH^-]_o \quad (21)$$

According to Eq. (20), the growth rate constant k_g should be independent of R ($k_g \propto K\nu_{aq}$) at large R values (i.e., at low surfactant concentrations if the concentration of water is fixed), while k_g should be inversely proportional to R ($k_g \propto K\nu_S/R$) at low R values (i.e., at high surfactant concentrations). The trend predicted by Eq. (20) correlates very well with the experimentally observed effect of R on the rate constant k_g (29); at low R values, k_g decreases with increase in R, but it tends to a plateau at large R values. The pseudophase model can be further examined by plotting Eq. (20) as k_g versus $1/R$. Then the ratio of the molar volumes of the surfactant and aqueous phase (ν_S/ν_{aq}) can be determined from the slope ($K\nu_S$) and intercept ($K\nu_{aq}$). A value of about 5.7 for ν_S/ν_{aq} was obtained by this method (28), which is considered to be in reasonable agreement with the expected value of 14, given the assumptions made in the model.

2.2.4 Summary and Conclusions

It can now be said that the microemulsion-mediated silicon alkoxide sol-gel process has come of age. The ability to form monodisperse spherical silica particles (20–39) and monolithic gels (40–53) by this method has been amply demonstrated. Recipes are available to prepare materials with predetermined characteristics, especially particle size and polydispersity. Potential applications of these microemulsion-derived materials through further processing is already inspiring new research. The preparation of various kinds of organic/silica and inorganic/silica nanocomposite structures has been reported (21,22,33–38,52,53). However, whether any of these developments will eventually lead to practical applications will depend on whether they provide economically attractive reaction environments or materials properties (e.g., particle size and morphology, and encapsulation capability) that are unattainable with other methods.

Microemulsion phase stability remains a challenge. As noted earlier [Eq. (1)], the alkoxide hydrolysis reaction results in the consumption of water and the release of alcohol molecules. These changes tend to destabilize microemulsions (30,41,53), thereby frustrating attempts to prepare large quantities of materials (by use of increased amounts of alkoxide). New microemulsion formulations are needed that are more robust in the alkoxide hydrolysis environment. One approach is to use novel combinations of established surfactants and cosurfactants. For example, Chang et al. (35–37) report that introduction of hexanol stabilizes NP-9 microemulsions against potential destabilization by the ethanol reaction product. A potentially fruitful approach might also be to design new surfactants that are more compatible with the alkoxide hydrolysis reaction. A related issue concerns the contamination of the silica product by the surfactant. Espiard et al. (21,22) have drawn attention to the case of AOT-based surfactants, where the complete removal of residual sodium ions proved to be difficult. To overcome this problem, these researchers investigated the sodium-free system of isopropanol/hexane/water. Unfortunately, serious phase instability problems were encountered.

Also worthy of further attention are investigations into the kinetics and mechanisms of the alkoxide hydrolysis and condensation reactions, and the corresponding particle growth and gel formation processes, particularly as related to micellar dynamics and the roles of the surfactant, cosurfactant, and oil. Particle aggregation has been invoked to rationalize the increase in particle size with R observed at large R (Fig. 2.2.13) (27). No direct evidence has, however, been provided to support this suggestion. Studies along the lines of Robinson et al. (75), involving determination of interdroplet exchange rates and the corresponding particle sizes for a particular microemulsion composition, would be helpful. Also of interest will be investigations into alkoxide–surfactant interactions. The presence of such interactions has been invoked by some researchers to explain the effects of alkoxide on particle evolution (23,52).

It is generally accepted that one of the attractive features of the microemulsion environment for materials synthesis is the stabilization of the produced particles by the microemulsion surfactants. However, in the specific case of alkoxide/microemulsion systems, there have been no investigations into the manner in which this stabilization is effected. For example, when the particle size exceeds the microemulsion droplet size, are the particles expelled from the water pools, or do the particles rather induce the enlargement of the microemulsion water droplets? There have been no investigations into the role of surfactant adsorption in the colloidal stabilization of the silica nanoparticles. To what extent are the product silica particles stabilized by direct adsorption of surfactant molecules? In this connection, systematic comparison of surfactant-type merits some attention. In basic solution, the silica surface surface is expected to be negatively charged (17,66). Is this negative charge relevant when considering particle stabilization by cationic (e.g., CTAB), anionic (e.g., AOT), or nonionic (e.g., NP-5) surfactants?

The as-prepared silica product occurs as nanoparticles dispersed in a microemulsion solution, or as a gel containing entrained microemulsion solution. For further application of this product, it is necessary to separate the solid material from the microemulsion solution. Much of the research on microemulsion-mediated silica synthesis was motivated by the challenge to utilize these solutions as morphological templates. Thus, product recovery was an issue only as it was necessary to obtain samples for characterization, e.g., by TEM. For microscopic observation, it has often proved adequate to place a drop of the silica-bearing microemulsion on a TEM grid, allowing slow solvent evaporation to occur. For practical applications of microemulsion-derived materials, however, much more will be needed. Where the silica product is in the form of nanoparticles, the necessary recovery steps include gross solid–liquid separation (e.g., centrifugation or rotary evaporation), followed by washing (for surfactant removal) and drying. In the case of silica gels, washing for surfactant removal will be important. Systematic investigations into the effects of washing and drying on materials characteristics will undoubtedly be very useful.

REFERENCES TO SECTION 2.2

1. N Ichinose, ed. Introduction to Fine Ceramics. New York: Wiley, 1987.
2. N Ichinose, Y Ozaki, S Kashu. Superfine Particle Technology. New York: Springer-Verlag, 1988.
3. E Matijevic. MRS Bull Dec:18–20, 1989.
4. M Haruta, B Delmon. J Chim Phys 83:859–868, 1986.
5. C Hayashi. Phys Today Dec:44–51, 1989.
6. RP Andres, RS Averback, WL Brown, LE Brus, WA Goddard III, A Kaldor, SG Louie, M Moscovits, PS Peercy, SJ Riley, RW Siegel, F Spaepen, Y Wang. J Mater Res 4: 704–736, 1989.
7. M Boutonnet, J Kizling, P Stenius, G Maire. Colloids Surf 5:209–225, 1982.
8. K Kandori, N Shizuka, M Gobe, K Kon-no, A Kitahara. J Disper Sci 8:477–491, 1987.
9. JH Fendler. Chem Rev 87:877–899, 1987.
10. JB Nagy. Colloids Surf 35:201–220, 1989.
11. AJI Ward, SE Friberg. MRS Bull Dec:41–46, 1989.
12. MJ Hou, DO Shah. In: YA Attia, BM Moudgil, S Chander, eds., Interfacial Phenomena in Biotechnology and Materials Processing. Amsterdam: Elsevier, 1988, pp 443–458.
13. A Khan-Lodhi, BH Robinson, T Towey, C Hermann, W Knoche, U Thesing. In: DM Bloor, E Wyn-Jones, eds., The Structure, Dynamics and Equilibrium Properties of Colloidal Systems. Dordrecht: Kluwer, 1990, pp 373–383.
14. K Osseo-Asare, FJ Arriagada. In: GL Messing, S Hirano, H Hausner, eds. Ceramic Powder Science III. Westerville, OH: American Ceramic Society, 1990, pp 3–16.

15. MP Pileni. J Phys Chem. 97:6961–6973, 1993.
16. K Osseo-Asare. In: KL Mittal, P Kumar, eds. Microemulsions: Fundamental and Applied Aspects. New York: Marcel Dekker, 1999, Chap. 18.
17. CJ Brinker, GW Scherer. Sol-Gel Science. San Diego: Academic Press, 1990.
18. W Stober, A Fink, E Bohn. J Colloid Interface Sci 26:62, 1968.
19. M Yanagi, Y Asano, K Kandori, K Kon-no. Abst 39th Symp Div Colloid Interface Chem, Chem Soc Jpn, 1986, p 396.
20. H Yamauchi, T Ishikawa, S Kondo. Colloids Surf 37:71–80, 1989.
21. P Espiard, JE Mark, A Guyot. Polym Bull 24:173–179, 1990.
22. P Espiard, A Guyot, JE Mark. J Inorg Organometallic Polym. 5:391–407, 1995.
23. CL Chang, HS Fogler. AIChE J 42:3153–3163, 1996.
24. CL Chang, HS Fogler. Langmuir 13:3295–3307, 1997.
25. K Osseo-Asare, FJ Arriagada. Colloids Surf 50:321–339, 1990.
26. FJ Arriagada, K Osseo-Asare. In: HE Bergna, ed., The Colloid Chemistry of Silica. Adv Chem Ser vol 234. Washington, DC: American Chemical Society, 1994, pp 113–128.
27. FJ Arriagada, K Osseo-Asare. J Colloid Interface Sci 211:210–220, 1999.
28. FJ Arriagada, K Osseo-Asare. Colloids Surf, in press.
29. K Osseo-Asare, FJ Arriagada. J Colloid Interface Sci, in press.
30. FJ Arriagada, K Osseo-Asare. Colloids Surf 69:105–115, 1992.
31. FJ Arriagada, K Osseo-Asare. J Colloid Interface Sci 170:8–17, 1995.
32. FJ Arriagada and K Osseo-Asare. J Dispers Sci Technol 15:59–71, 1994.
33. TK Jain, I Roy, TK De, A Maitra. J Am Chem Soc 120:11092–11095, 1998.
34. DJ Chaiko. Separation Sci Technol 27:1389–1405, 1992.
35. SY Chang, L Liu, SA Asher. Mater Res Symp Proc 346:875–880, 1994.
36. SY Chang, L Liu, SA Asher. J Am Chem Soc 116:6739–6744, 1994.
37. SY Chang, L Liu, SA Asher. J Am Chem Soc 116:6745–6747, 1994.
38. JH Adair, T Li, T Kido, K Havey, J Moon, J Mecholsky, A Morrone, DR Talham, MH Ludwig, L Wang. Mater Sci Eng R23:139–242, 1998.
39. SA Johnson, PJ Ollivier, TE Mallouk. Science 283:963–965, 1999.
40. SE Friberg, CC Yang. In: FM Doyle, S Raghavan, P Somasundaran, GW Warren, eds. Innovations in Materials Processing Using Aqueous, Colloid and Surface Chemistry. Warrendale, PA: Minerals, Metals & Materials Society, 1988, pp 181–191.
41. SE Friberg, CC Yang, J Sjöblom. Langmuir 8:372–376, 1992.
42. SE Friberg, AU Ahmed, CC Yang, S Ahuja, SS Bodesha. J Mater Chem 2:257–258, 1992.
43. SE Friberg, SM Jones, CC Yang. J Dispers Sci Technol 13:65–75, 1992.
44. SE Friberg, SM Jones, CC Yang. J Disper Sci Technol 13:45–63, 1992.
45. SE Friberg, SM Jones, A Motyka, G Broze. J Mater Sci 29:1753–1757, 1994.
46. SM Jones, SE Friberg. J Dispers Sci Technol 13:669–696, 1992.
47. SE Friberg, Z Ma. J Non-Cryst Solids 147–148:30–35, 1992.
48. SM Jones, A Amran, SE Friberg. J Dispers Sci Technol 15:513–542, 1994.
49. SE Friberg, SM Jones, J Sjoblom. J Mater Synth Process 2:29–44, 1994.
50. SM Jones, SE Friberg. J Non-Cryst Solids 181:39–48, 1995.
51. JH Burban, M He, EL Cussler. AIChE J 41:159–165, 1995.
52. A Martino, SA Yamanaka, JS Kawola, DA Loy. Chem Mater 9:423–429, 1997.
53. HJ Watzke, C Dieschbourg. Adv Colloid Interface Sci 50:1, 1994.

54. J Eastoe, BH Robinson, DC Steytler, D Thorn-Leeson. Adv Colloid Interface Sci 36: 1–31, 1991.
55. J Sjoblom, R Lindberg, SE Friberg. Adv Colloid Interface Sci 95:125–287, 1996.
56. K Shinoda, B Lindman. Langmuir 3:135–149, 1987.
57. K Kon-no. Surface Colloid Sci 15:125–151, 1993.
58. JG Darab, DM Pfund, JL Fulton, JC Linehan, M Capel, Y Ma. Langmuir 10:135–141, 1994.
59. S Clark, PDI Fletcher, X Ye. Langmuir 6:1301–1309, 1990.
60. AS Bommarius, JF Holzwarth, DIC Wang, TA Hatton. J Phys Chem 94:7232–7239, 1990.
61. R Zana, J Lang. In: SE Friberg, P Bothorel, eds. Microemulsions: Structure and Dynamics. Boca Raton, FL: CRC Press, 1987, p 153.
62. PDI Fletcher, AM Howe, BH Robinson. J Chem Soc Faraday Trans I 83:985–1006, 1987.
63. SS Atik, JK Thomas. J Am Chem Soc 103:3543–3550, 1981.
64. JP Wilcoxon, RL Williamson, R Baughman. J Chem Phys 98:9933–9950, 1993.
65. CF Baes Jr, RE Mesmer. The Hydrolysis of Cations. New York: Wiley, 1976.
66. RK Iler. The Chemistry of Silica. New York: Wiley, 1979.
67. FJ Arriagada, K Osseo-Asare, unpublished research.
68. A D'Aprano, ID Donato, F Pinio, V Turco Liveri. J Sol Chem 18:949, 1989.
69. A Kitahara. J Phys Chem 69:2788, 1965.
70. CH Byers, MT Harris. In: JD Mackenzie, DR Ulrich, eds. Ultrastructure Processing of Advanced Ceramics. New York: Wiley, 1988, p 843.
71. R Aelion, A Loebel, F Eirich. J Am Chem Soc 72:5705, 1950.
72. IV Berezin, K Martinek, AK Yatsimirskii. Russ Chem Rev 42:787, 1973.
73. K Osseo-Asare. Separation Sci Technol 23:1269, 1988.
74. K Osseo-Asare. Colloids Surf 50:373, 1990.
75. TF Towey, A Khan-Lodhi, BH Robinson. J Chem Soc Faraday Trans 86:3757–3762, 1990.

3

Metal Chalcogenides (Sulfides, Selenides, and Tellurides)

TADAO SUGIMOTO, DOROTHEE INGERT,
LAURENCE MOTTE, MARIE-PAULE PILENI,
DAVID J. ELLIOT, KAREN GRIEVE, D. NEIL FURLONG,
and FRANZ GRIESER

3.1 REACTION IN HOMOGENEOUS SOLUTIONS OF METAL IONS

TADAO SUGIMOTO
Institute for Advanced Materials Processing, Tohoku University, Sendai, Japan

This chapter deals with the methods of synthesis, characterization, and growth mechanisms of well-defined uniform particles of metal sufides and selenides formed by direct reaction of metal ions with the chalcogenide ions, released from thioacetamide or selenourea in dilute solutions, or supplied continuously from outside in the form of a high concentration of sulfide ions.

3.1.1. Sulfides

3.1.1.1 Cadmium, Zinc, and Lead Sulfides and Their Mixed Sulfides

Matijević and Wilhelmy (1) prepared uniform spherical polycrystalline particles of cadmium sulfide (CdS) by reaction of Cd^{2+} ions with thioacetamide (TAA) in a dilute acidic media (pH < 2), as shown in the TEM and SEM images of Figure 3.1.1. The reaction finished within 1 h at 26°C. They used seed crystals of CdS to promote the uniformity of the final product, and analyzed the growth kinetics using Nielsen's chronomal. The isoelectric point in terms of pH was determined to be 3.7 by electrokinetic measurement. They also prepared zinc sulfide (ZnS; polycrystalline spheres), whose isoelectric point in pH was 3.0 (2), lead sulfide (PbS; monocrystalline cubic galena) (3), cadmium zinc sulfide (CdS/ZnS; amorphous and crystalline spheres) (3), and cadmium lead sulfide (CdS/PbS; crystalline polyhedra) (3), in a similar manner.

Figure 3.1.2. shows SEM images of the PbS particles. It is of interest that the isometric NaCl-type crystals have rectangular shapes with different orthogonal edge lengths whose ratio is also varied among particles. Since the surface planes are all equivalent {100} faces, the anisotropic growth may not be explained in terms of growth control by adsorption of solute or of impurity. Presumably, this unique growth mechanism may be elucidated in terms of dislocation-induced growth (4), like the rectangular microcrystals of AgBr (5) and AgCl (6).

Figure 3.1.3 shows a TEM image of mixed particles of CdS·xZnS ($x = 21.5$) (upper) and an SEM image of mixes particles of CdS·xPbS ($x = 2.1$) (lower). In the preparation of the ZnS and PbS seed sols used for the preparation of these mixed sulfide particles, only ~4% of the original concentration of Zn^{2+} and ~6% of Pb^{2+} were used up, repectively. For CdS·xZnS prepared in the presence of ZnS seeds,

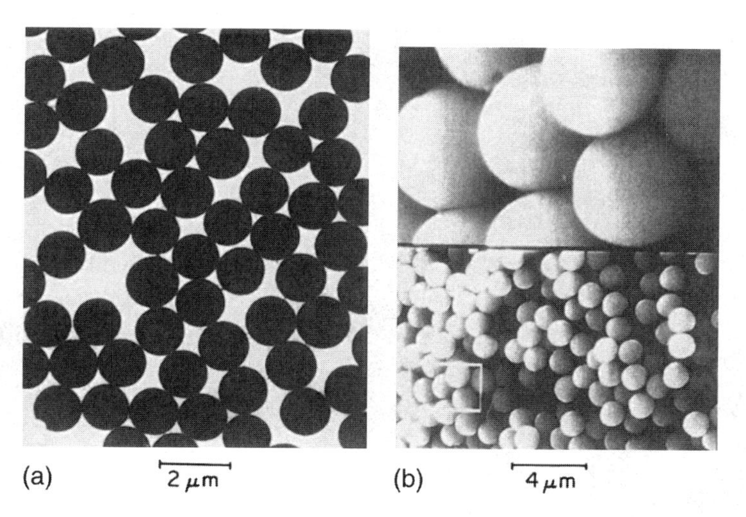

Fig. 3.1.1 CdS particles obtained by addition of (a) 6.25 cm^3 or (b) 12.5 cm^3 of 0.05 mol dm^{-3} TAA to 500 cm^3 of a seed sol at 26°C and subsequent aging for (a) 20 min or (b) 90 min. The seed sol was prepared previously by aging at 26°C for 14.5 h a solution of 1.2×10^{-3} mol dm^{-3} in $Cd(NO_3)_2$, 2.4×10^{-1} mol dm^{-3} in HNO_3, and 5.0×10^{-3} mol dm^{-3} in TAA. (From Ref. 1.)

Fig. 3.1.2 PbS particles obtained by addition of 0.50 cm^3 of 5.0×10^{-2} mol dm^{-3} TAA to 20 cm^3 of a seed sol and aged at 26°C for 20 min. The seed sol was prepared by aging a solution of 5.0×10^{-3} mol dm^{-3} in TAA, 1.2×10^{-3} mol dm^{-3} in $Pb(NO_3)_2$, and 2.4×10^{-1} mol dm^{-3} in HNO_3 at 26°C for 21 h. (From Ref. 3.)

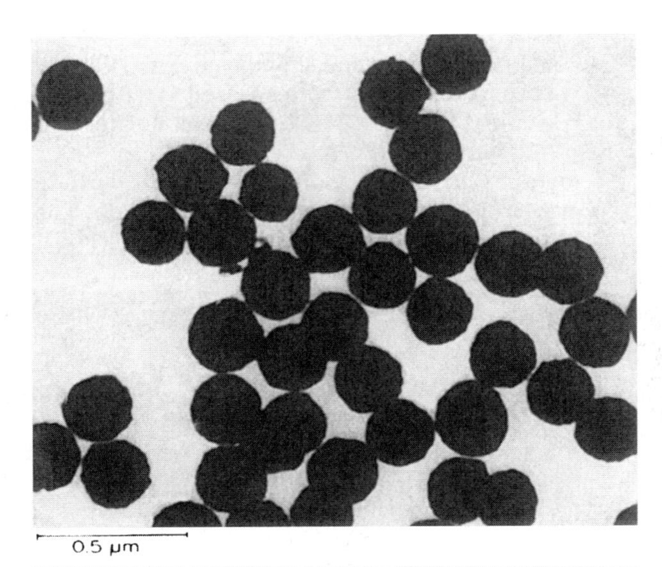

Fig. 3.1.3 Top: CdS·xZnS particles (x = 21.5) obtained by adding ~0.5 cm^3 of a $Cd(NO_3)_2$ solution (5.2×10^{-4} mol dm^{-3} in final concentration) to 20 cm^3 of a ZnS seed sol and aged at 70°C for 100 min, where the ZnS seed sol was prepared by aging a solution of 2.4×10^{-2} mol dm^{-3} in $Zn(NO_3)_2$, 1.1×10^{-1} mol dm^{-3} in TAA, and 6.2×10^{-2} mol dm^{-3} in HNO_3 at 26°C for 5 h. Bottom: CdS·xPbS particles (x = 2.1) obtained by adding 0.50 cm^3 of 4.3×10^{-2} mol dm^{-3} $Cd(NO_3)_2$ and 1.0 cm^3 of 2.1×10^{-2} mol dm^{-3} $Pb(NO_3)_2$ to 20 cm^3 of a PbS seed sol and aging at 80°C for 30 min, where the PbS seed sol was prepared by aging a solution of 1.2×10^{-3} mol dm^{-3} in $Pb(NO_3)_2$, 5.0×10^{-3} mol dm^{-3} in TAA, and 2.4×10^{-1} mol dm^{-3} in HNO_3 at 26°C for 21 h. (From Ref. 3.)

CdS is first deposited onto the seeds, followed by deposition of ZnS, probably due to the solubility of CdS lower than of ZnS. When the precipitation temperature is lower than 60°C, the x-ray diffractometry (XRD) shows no distinct lines but a broad peak, suggesting an amorphous solid. However, higher temperature, such as above 70°C, makes the structure crystalline as a solid solution. These facts may imply that some intraparticle reformation for the solid mixing via the solution phase during the deposition of ZnS, caused by the increasing mixing entropy. In the case of CdS·*x*PbS prepared at 80°C in the presence of PbS seeds, the precipitation of PbS is slightly faster than CdS, and the resulting internal structure is heterogeneous with independent solid phases of CdS (greenockite) and PbS (galena) with limited substitution of metal ions of each other.

The authors assumed that the precipitation of these metal sulfides was controlled by the hydrolysis reaction of TAA promoted by proton in the acidic conditions. However, the hydrolysis of TAA observed in acidic and alkaline ranges is a much slower process than observed in the precipitation of these metal sulfides (7–12), and it may not be accelerated by consumption of S^{2-} ions because of its irreversible nature. In addition, the reaction virtually finished with a great part of the starting metal ions and TAA left unreacted. Also, it has already been verified that the probability of direct reaction of TAA with metal ions is zero or at least negligible from its strong dependence of pH in reactivity (7). Thus, it seems reasonable to consider that the main path is the release of S^{2-} ions from TAA according to the following reaction scheme with the production of acetonitrile (7,13) as found in the reaction of TAA with Cd^{2+} ions in an alkaline media:

$$CH_3CSNH_2 \rightleftarrows CH_3CN + 2H^+ + S^{2-} \tag{1}$$

which halts when TAA reaches the equilibrium with remaining metal ions. Since the release rate of S^{2-} in this process in an acidic medium is low (7), it can be the rate-determining process in the low pH range. In any case, it is of importance in general to clarify the rate-determining step prior to Nielsen's chronomal analysis, since if the precipitation is controlled by the generation of monomer source preceding the diffusion of the monomer toward the surfaces of the particles and its surface reaction on the growing particles, Nielsen's chronomal cannot be applied to such a system. For example, Nielsen's chronomal analysis is inapplicable to the cases in which the total reaction rate is controlled by one of the preceding processes such as hydrolysis of alkoxides, hydrolysis of metal ions, reversible release of metal ions from complexes, reversible release of chalcogenide ions from organic starting materials, decomposition of starting materials, dissolution of precursory solid, etc. If we apply Nielsen's analysis to such a system, the result of the kinetic analysis is that of the preceding rate-determining step, whose treatment should be entirely different, and not of the growth process of the particles. This point should always be kept in mind when we actually analyze the mechanism of a particle growth process or when we read literature.

3.1.1.2 Molybdenum and Cobalt Sulfides

Haruta et al. (14) prepared spherical particles of molybdenum sulfide and cobalt sulfide with a narrow size distribution by reaction of dilute ammonium orthomolybdate or cobalt(II) acetate with sulfide ions liberated from thioacetamide as a reservoir of S^{2-} ions in weakly acidic media. The compositions of these metal sulfides were estimated to be Mo:S:O = 1.0:1.7:3.0 and Co:S:O = 1.0:4.5:6.4 by chemical analysis. Figure 3.1.4 shows an SEM of a thus prepared uniform molybdenum sulfide particles sample. These sulfide particles were of no distinct crystal structure as shown by x-ray diffractometry. The isoelectric points of the Mo sulfide and Co sulfide particles in terms of pH were 1.9 and 3.1, respectively. Both of them are useful as hydrodesulfurization catalysts.

In the systems of Mo and Co sulfides, TAA was assumed to release sulfide ions by hydrolysis accelerated by hydrazine. Since the concentration of S^{2-} in equilibrium with TAA is extremely low despite the exceedingly high release rate constant of S^{2-} in the reversible reaction of Eq. (1), this assumption is reasonable if the concentrations of the free metal ions are too low for the nucleation of these metal sulfides. However, if the role of hydrazine is different than an accelerator of hydrolysis of TAA, and if the deposition rate of the metal sulfide monomers or the release rate of metal ions from the metal ion complexes such as orthomolybdate or cobalt

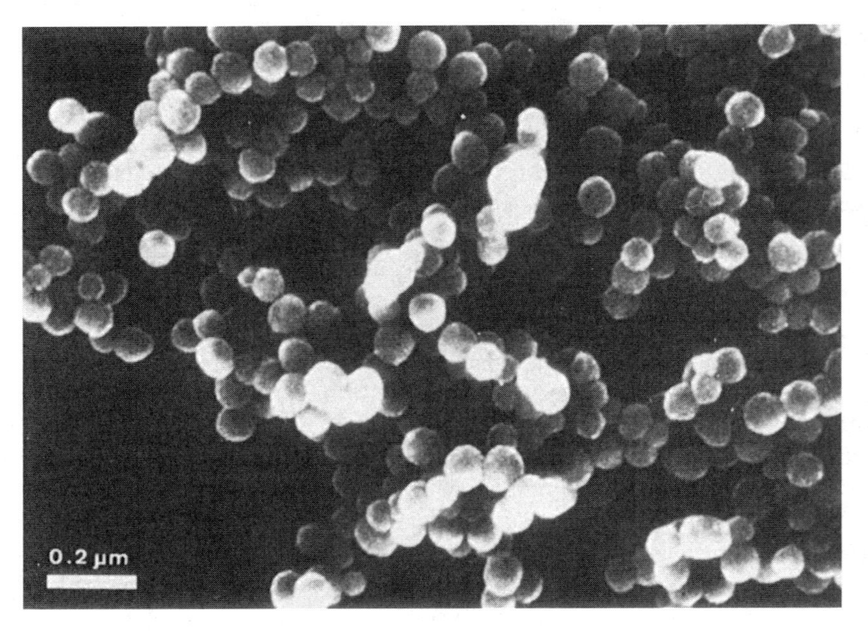

Fig. 3.1.4 SEM image of an example of the uniform Mo sulfide particles. (From Ref. 14.)

acetate is the rate-determining step of the precipitation, there is a possibility of the path of Eq. (1) for the release of S^{2-} ions. One method to distinguish between the reactions of TAA may be the analysis of the by-product of TAA, since CH_3COOH is the main by-product of the hydrolysis, while CH_3CN is produced in the reversible reaction of Eq. (1).

3.1.1.3 Silver-Doped Zinc Sulfide

Submicrometer uniform crystalline spheres of silver-doped zinc sulfide (ZnS:Ag) were prepared by aging 0.04 mol dm^{-3} $Zn(NO_3)_2$ and 2.80×10^{-6} to 1.68×10^{-5} mol dm^{-3} $AgNO_3$ with 0.4 mol dm^{-3} TAA for up to 100 min at initial pH 1.52 and 73°C (15). The authors found that the final particle density decreased with increasing content of silver ions, whereas the total reaction rate was virtually unaffected by the significant difference in the total surface area of the particle. In fact, the final particle diameter increased from 0.3 to 1.1 μm with increase in the content of silver ions from 5.6 to 16.8 μmol dm^{-3}, as shown in Figure 3.1.5. Figure 3.1.6 shows the time evolutions of $[Zn^{2+}]$ and $[Ag^+]$ in the solution phase.

The fact that the total surface area of the particles has virtually no effect within the experimental error range may suggest that the rate-determining step of the overall reaction is not the surface reaction process but a preceding reaction such as the

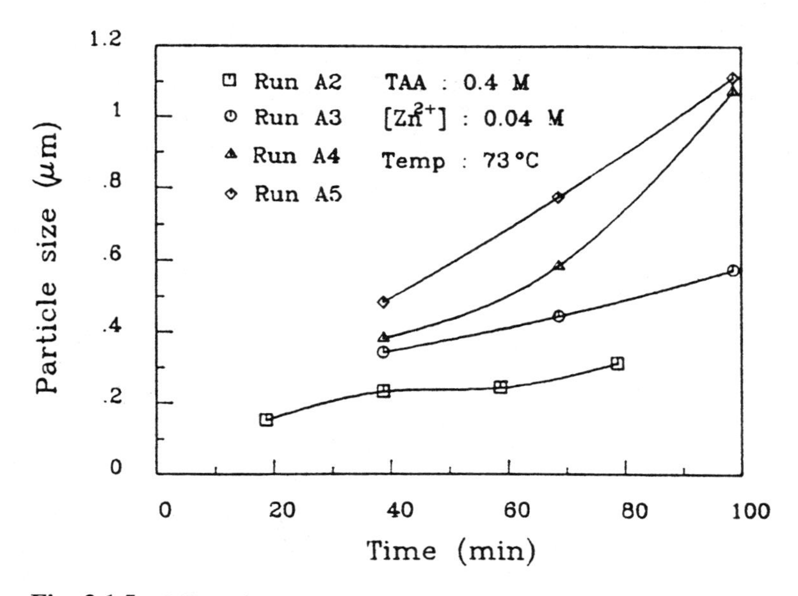

Fig. 3.1.5 Effect of silver content on the particle growth, where $[Ag^+]$ = 5.6, 8.4, 11.2, and 16.8 μmol dm^{-3} for runs A_2, A_3, A_4, and A_5, respectively. (From Ref. 15.)

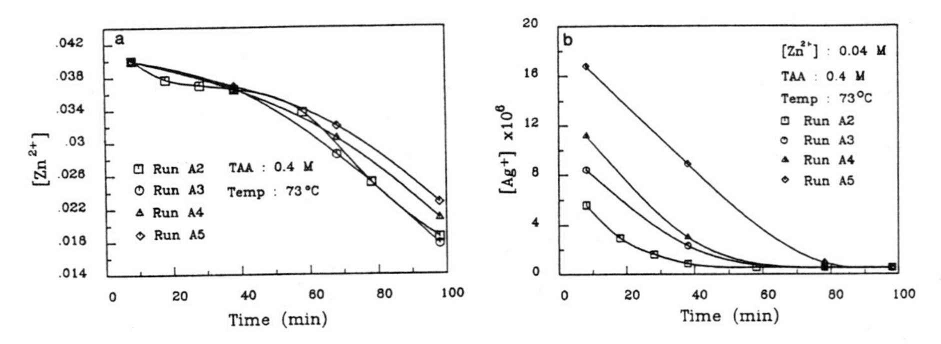

Fig. 3.1.6 Changes of $[Zn^{2+}]$ and $[Ag^{+}]$ with time in the solution phase, where runs A2, A3, A4, and A5 are the same as those in Figure 3.1.5. (From Ref. 15.)

dissociation of TAA in the low pH range. If this is the case, Nielsen's chronomal analysis is of no use in similar systems in general. Also, since $[Ag^{+}]/[Zn^{2+}]^{1/2} >> (K_{sp}^{Ag_2S}/K_{sp}^{ZnS})^{1/2} \simeq (10^{-50}/10^{-25})^{1/2} = 10^{-12.5}$ at 25°C (K_{sp} = solubility product), Ag_2S is deemed to precipitate first even at 73°C, as the authors referred to. In this case, the Ag_2S nuclei may reduce the number of the ZnS particles by heterocoagulation with the ZnS nuclei in the nucleation stage, instead of working as centers for heterogeneous nucleation of ZnS. In addition, the resulting composite particles are expected to have a core/shell structure with a core rich in Ag_2S content.

The polycrystalline particles consist of much smaller crystallites whose size and morphology were found to be strongly affected by the concentration of TAA, probably due to the adsorption of TAA to the surface zinc ions during their growth: i.e., spheres of ~20 nm in average diameter at 0.40 mol dm^{-3} TAA; fibrils of ~50 nm in average length at 0.12 mol dm^{-3} TAA.

3.1.1.4 Palladium Sulfide

Schultz and Matijević (16) prepared nanoparticles of palladium sulfide (PdS) by the continuous double-jet mixing of $PdCl_2$ or $Na_2(PdCl_4)$ and Na_2S. They found that the particle size was 20–30 nm in mean diameter obtained in acidic media (pH = 2–3), but 2–5 nm in alkaline media, probably due to the high equilibrium concentration of sulfide ions S^{2-} by dissociation of H_2S and HS^{-} in the alkaline media (pH = 10–12). A cationic surfactant, cetyl trimethyl ammonium bromide (CTAB), was found to be useful for stabilizing the small particles prepared in alkaline media.

It should be noted here that this system is not completely homogeneous but involves some element of a heterogenous system, since high concentrations of cations and anions are directly reacted with each other at an extremely high supersaturation so that nucleation constantly occurs throughout the whole process. If the dissolution rate of the nascent nuclei is high enough to be instantly dissolved and act as a source of the solute, the system fulfills the requirements for monodisperse particles formation. However, if the dissolution rate is low, concurrent nucleation and growth are inevitable, and thus a highly uniform product cannot be expected. If drastic coagulation is suppressed by a protective agent such as adsorptive surfactant or polymer, we usually obtain very small particles with a somewhat broad size distribution, due to the constant nucleation.

3.1.2 Selenides

3.1.2.1 Cadmium and Lead Selenides

Gobet and Matijević (17) produced monodisperse submicrometer-size particles of cadmium selenide (CdSe) and lead selenide (PbSe) by reversible release of selenide ions from selenourea in solutions of the corresponding metal salts. The equilibrium between selenourea and selenide ions is written as follows:

$$(H_2N)_2CSe \rightleftarrows H_2NCN + 2H^+ + Se^{2-} \qquad (2)$$

where the equilibrium is strongly displaced to the left (18,19). As readily expected, the reaction rate is accelerated with increasing pH due to the increasing equilibrium concentration of selenide ions, and thus the particle size is reduced with increasing pH due to the increasing supply rate of selenide ions. For example, in the system

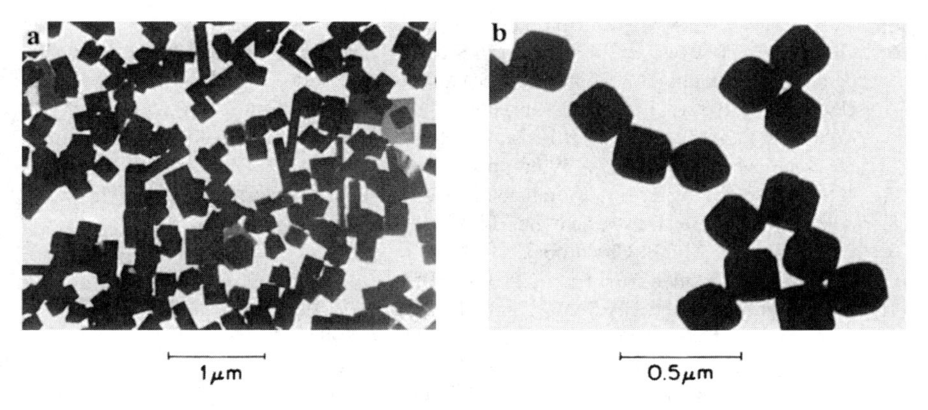

Fig. 3.1.7 TEM images of the PbSe particles prepared in the (a) absence or (b) presence of 82 wt% methanol by use of the reaction between Pb^{2+} ions and Se^{2-} ions released from selenourea. Preparation conditions: (a) [Pb acetate] = 2.0 mmol dm^{-3}, [selenourea] = 1.5 mmol dm^{-3}, [acetic acid] = 1.5 mmol dm^{-3}, temperature = 70°C, time = 1 h, initial pH = 5.0; (b) [Pb nitrate] = 1.8 mmol dm^{-3}, [selenourea] = 1.2 mmol dm^{-3}, temperature = 30°C, time = 2 h, solvent = 82 wt% methanol. (From Ref. 17.)

of CdSe, the onset of nucleation is observed above pH 4.5 at 70°C and the pH dependence of the particle size is rather critical around pH 5 to 6, so that an acetate pH buffer is essential for maintaining the reaction rate and for a high reproducibility in final particle size. Similar to the corresponding metal sulfides, the cadmium selenide particles were spherical polycrystals and the lead selenide particles were single crystals of a variety of rectangular shapes despite the cubic symmetry. However, when a mixed solvent of methanol/water (82/18 by weight) was used instead of water, PbS particles obtained turn out to be monodispersed and nearly cubic ones bound mainly by {100} faces but partly by {110} faces, as shown in Figure 3.1.7. Though the authors gave no special comment on this point, this result may suggest that methanol inhibits the dislocation-induced growth by adsorption to the active sites of the growing surfaces. Although many factors are different in the preparation conditions between the systems with and without methanol in Figure 3.1.7, PbSe particles prepared in aqueous media may maintain the rectangular shape unless the solvent is changed, like PbS particles whose shape is normally rectangular as long as prepared in aqueous media.

REFERENCES TO SECTION 3.1

1. E Matijević, DM Wilhelmy. J Colloid Interface Sci 86:476, 1982.
2. DM Wilhelmy, E Matijević. J Chem Soc Faraday Trans 80:563, 1984.
3. DM Wilhelmy, E Matijević. Colloids Surf 16:1, 1985.
4. T Sugimoto, S Chen, A Muramatsu. Colloids Surf A Physicochem Eng Aspects 135: 207, 1998.
5. A Mignot, E Francois, M Catinot. J Crystal Growth 23:207, 1974; TG Bogg, US patent 4,063,951, 1977; A Mignot, Jpn patent (A) Tokkaisho 58–95,337, 1983.
6. GL House, TB Brust, et al. (Kodak), US patent 5,320,938, 1994; M Saito (Fuji), Jpn patent (A) Tokkaihei 6-308648, 1994; S Yamashita, M Saito, T Ozeki, T Ishizaka (Fuji), US patent 5,498,511, 1996; S Yamashita, T Oyamada, M Saito (Fuji), Jpn patent (A) Tokkaihei 8-211522, 1996; TB Brust, GL House, Proc IS&T's 49th Ann Conf Soc Imag Sci Tech, Minneapolis, MN, 1996, pp 35–37.
7. T Sugimoto, GE Dirige, A Muramatsu. J Colloid Interface Sci 176:442, 1995.
8. AW Hofmann. Ber Deut Chem Ges 11:338, 1878.
9. CV Jorgensen. J Prakt Chem 66:33, 1902.
10. EH Swift, EA Butler Anal Chem 28:146, 1956.
11. EA Butler, DG Peter, EH Swift. Anal Chem 30:1379, 1958.
12. O Rosenthal, TI Taylor. J Am Chem Soc 79:2684, 1957.
13. T Sugimoto, GE Dirige, A Muramatsu. J Colloid Interface Sci 182:444, 1996.
14. M Haruta, J Lamaitre, F Delannay, B Delmon. J Colloid Interface Sci 101:59, 1984.
15. K Chou, T Wu. J Colloid Interface Sci 142:378, 1991.
16. M Schultz, E Matijević. Colloids Surf A Physicochem Eng Aspects 131:173, 1998.
17. J Gobet, E Matijević. J Colloid Interface Sci 100:555, 1984.
18. GA Kitaev, TP Sokolova. Russ J Inorg Chem 15:167, 1970.
19. IK Ostrovskaya, GA Kitaev, AA Velikanov. Russ J Phys Chem 50:956, 1976.

3.2 PREPARATION OF METAL SULFIDES FROM CHELATES

TADAO SUGIMOTO
Institute for Advanced Materials Processing, Tohoku University, Sendai, Japan

Chiu prepared monodisperse crystalline particles of metal sulfides, such as lead sulfide (PbS; cubes; 100 Å) (1), cupric sulfide (CuS; hexagonal bipyramids; 200 Å) (2), and zinc sulfide (ZnS; multifaceted spheres; 0.1–0.4 μm) (3); by introducing hydrogen sulfide gas into dilute acidic solutions of the ethylenediamine tetraacetic acid (EDTA) complexes of the corresponding metal ions (10^{-4}–10^{-3} mol dm^{-3}) for several minutes at room temperature.

EDTA appears to prevent both nucleation and coagulation during the particle growth by shielding the metal ions. Meanwhile, the EDTA complexes liberate metal ions by degree with the progress of particle growth. The use of chelating agents is one of the most promising techniques to produce uniform particles, since it is relatively easy to meet all requirements for monodisperse particles formation.

However, even if we use chelating agents, it is generally impossible to produce uniform particles in highly concentrated solutions of chelates such as those of the order of 10^{-1} mol dm^{-3} or more, due to the tremendous coagulation. Nevertheless, this difficult issue has been cleared by applying a new synthetic method named the ''gel-sol method'' orginally developed for the preparation of monodisperse hematite (α-Fe_2O_3) particles. Gelatin turned out to be a powerful anticoagulant acting as a protective colloid in the synthesis of monodisperse metal sulfide particles in condensed chelate systems, including cadmium sulfide (CdS) (4–6) zinc sulfide (ZnS) (4,6), lead sulfide (PbS) (6), and cupric sulfide (CuS) (6). One of the advantages of gelatin over other anticoagulants is that it can be readily removed by use of a very small amount of proteinase if needed.

In the condensed chelate systems, many kinds of chelating agents were used, such as ethylenediamine-*N,N,N′,N′*-tetraacetic acid (EDTA), nitrilotriacetic acid (NTA), *l*-aspartic acid (AA), trimethylenediamine (TMD), *N,N*-dimethylethylenediamine (DMED), diethylenetriamine (DETA), triethylenetetraamine (TETA), and tris(2-aminoethyl)amine (TAEA) (6). The chelating agents used are listed in Table 3.2.1.

The criteria for the choice of chelating agents were the stability constant of each chelate and the release rate of metal ions. If the stability constant is too low, such as less than 10^{10}, the separation of the nucleation and growth stages are generally difficult due to the too high supersaturation. On the other hand, if the stability constant is too high, such as 10^{18}, the reaction rate is practically too low except for

Table 3.2.1 List of Chelating Agents

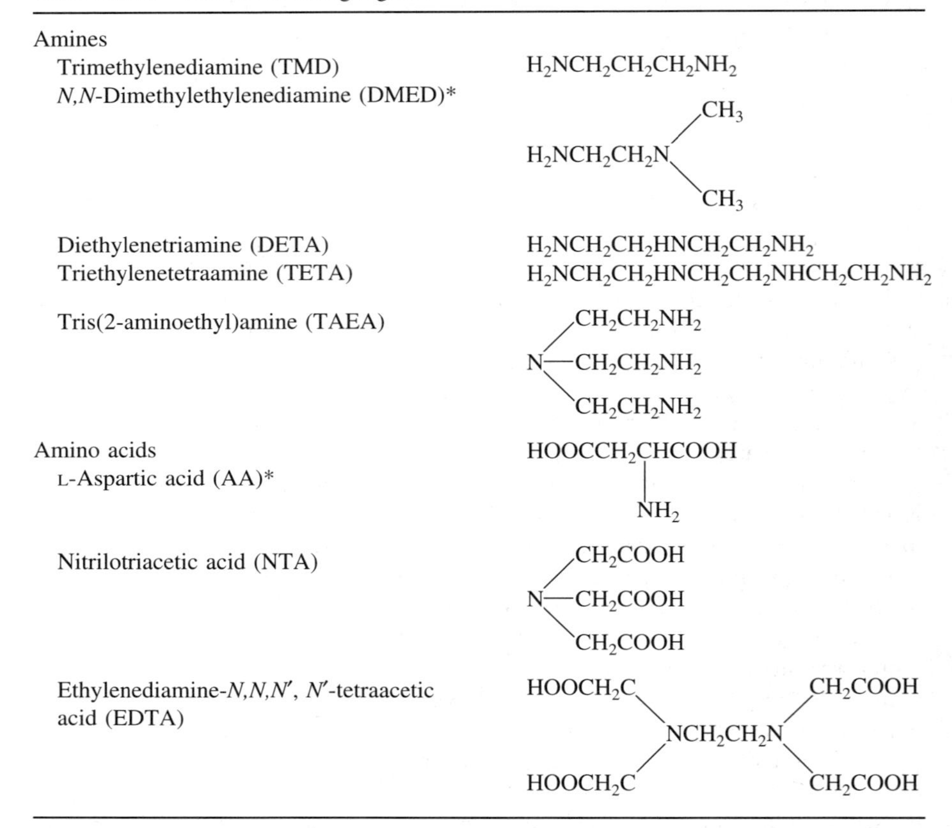

Chelating agent	Formula
Amines	
Trimethylenediamine (TMD)	$H_2NCH_2CH_2CH_2NH_2$
N,N-Dimethylethylenediamine (DMED)*	$H_2NCH_2CH_2N(CH_3)_2$
Diethylenetriamine (DETA)	$H_2NCH_2CH_2HNCH_2CH_2NH_2$
Triethylenetetraamine (TETA)	$H_2NCH_2CH_2HNCH_2CH_2NHCH_2CH_2NH_2$
Tris(2-aminoethyl)amine (TAEA)	$N(CH_2CH_2NH_2)_3$
Amino acids	
L-Aspartic acid (AA)*	$HOOCCH_2CH(NH_2)COOH$
Nitrilotriacetic acid (NTA)	$N(CH_2COOH)_3$
Ethylenediamine-*N,N,N′, N′*-tetraacetic acid (EDTA)	$(HOOCH_2C)_2NCH_2CH_2N(CH_2COOH)_2$

Note: The chelating agents marked with an asterisk (*), DMED and AA, were used only for the formation of CuS.
Source: Ref. 6.

the CuS systems. For the chelating agents giving the intermediate stability constants, the release rate of metal ions should be so low as to be able to achieve a low steady concentration of free metal ions below the critical supersaturation level during the growth of the metal sulfides. Since chelates of the lower stability constants give higher supersaturation in the nucleation stage as well, they normally yield smaller particles. This can be used for size control of the products. Figure 3.2.1 shows the relation between the stability constants of metal chelates and corresponding yields of metal sulfide particles including CdS, ZnS, and PbS under different reaction conditions. The standard conditions for the preparation of the metal sulfides are as follows (standard conditions A): TAA (5 cm^3 of 1.2 mol dm^{-3}) solution containing

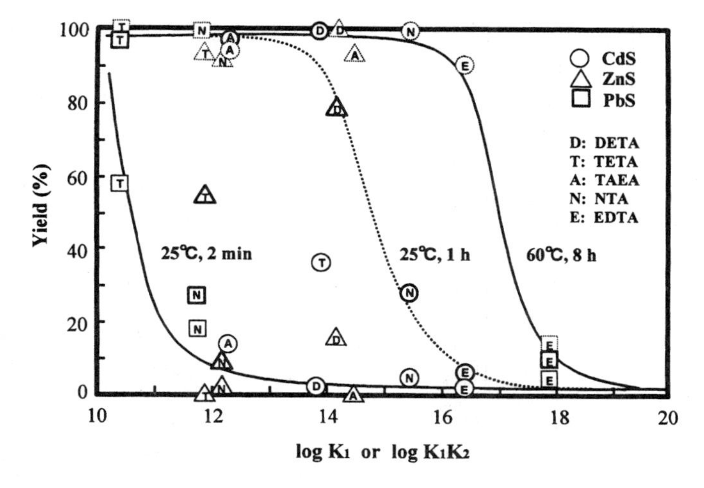

Fig. 3.2.1 Relationship between the stability constants of metal chelates and the corresponding yields of metal sulfide particles including CdS, ZnS, and PbS. The symbols bound by thin, thick, and dotted frames show the yields at different conditions of (a) 25°C, 2 min; (b) 25°C, 1 h; and (c) 60°C, 8 h, respectively. The solid line curves represent the yields at conditions (a) and (c), while the broken line curve is the upper limit of the yields at conditions (b). (From Ref. 6.)

1 wt% gelatin was added quickly to 20 cm^3 of a metal–chelate solution, consisting of 0.30 mol dm^{-3} metal acetates, 2.6 mol dm^{-3} NH_3, 1 wt% gelatin, and a chelating agent whose concentration was 0.33 mol dm^{-3} for those of group I (TAEA, EDTA) or 0.66 mol dm^{-3} for those of group II (TMD, DMED, DETA, TETA, AA, NTA), respectively, where the maximum number of ligands to a metal ion is one for group I and two for group II. The pH of the metal–chelate solutions was adjusted to 9.5 with acetic acid or NaOH prior to the addition of TAA. The mixed solution was aged at 25°C for different durations according to the reaction rate in a thermostated water bath under agitation. In Figure 3.2.1, the temperature and/or aging time was altered under otherwise standard conditions. Also, Figure 3.2.2 shows the reaction rates for the formation of CdS particles in terms of ln([Cd–chelate]/[Cd–chelate]$_0$) with chelating agents TMD, TETA, and NTA, where 0.10 mol dm^{-3} $Cd(CH_3COO)_2$ and 0.22 mol dm^{-3} chelating agents were used instead of the standard concentrations of 0.30 mol dm^{-3} $Cd(CH_3COO)_2$ and 0.66 mol dm^{-3} chelating agents. The final particle size dramatically increased as 2, 10, and 80 nm with TMD, TETA, and NTA, respectively.

If one chooses a chelating agent with a relatively high stability constant, such as EDTA, one may find some ideal separation of nucleation and growth stages as a decisive prerequisite for monodispersed particles formation, since a drastic change of supersaturation can be expected in a system with metal ions around the stoichio-

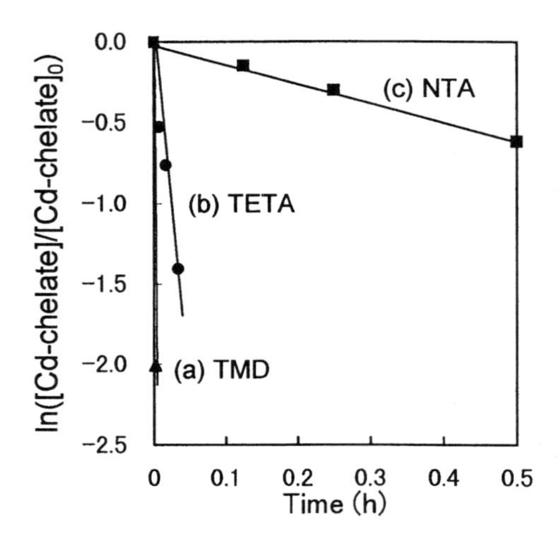

Fig. 3.2.2 Plot of ln([Cd–chelate]/[Cd–chelate]$_0$) against aging time with chelating agents: (a) TMD, (b) TETA, and (c) NTA. An amount of 5 cm^3 of 1.2 mol dm^{-3} TAA was added to each 20 cm^3 of Cd^{2+} chelate solution containing 0.10 mol dm^{-3} $Cd(CH_3COO)_2$ and 0.22 mol dm^{-3} chelating agent. The other conditions were the same as the standard conditions A. (From Ref. 6.)

metric ratio to the coexisting chelating agent, after the initial consumption of the metal ions for nucleation. In this case, if the metal ions are in a slight excess to the chelating agent, the final particle size would be extremely small, because of the precipitation of a great number of nuclei due to the high concentration of free metal ions. The TEM images in Figure 3.2.3 demonstrate the effect of the ratio of EDTA to $Cd(OH)_2$ on the final particle size. In this case, 5 cm^3 of 1.2 mol dm^{-3} TAA in a 1 wt% gelatin solution was added to 20 cm^3 of a Cd–EDTA complex solution, containing 0.300 mol dm^{-3} in $Cd(OH)_2$, 0.303 mol dm^{-3} in EDTA·2Na, 2.0 mol dm^{-3} in CH_3COONH_4, 0.6 mol dm^{-3} in NH_3, and 1 wt% in gelatin, at 60°C with stirring and aged under additional agitation for a duration sufficient for the reaction to be completed (standard conditions B). The $Cd(OH)_2$ in the Cd–EDTA solution is, of course, completely dissolved prior to the addition of TAA.

In all of these systems, thioacetamide (TAA) was commonly used as a source of sulfide ions, which was found to be in the following chemical equilibrium with a very low concentration of sulfide ions, as shown in Chapter 3.1 (7):

$$CH_3CSNH_2 \rightleftarrows CH_3CN + 2H^+ + S^{2-}$$

When the S^{2-} ions were consumed by the reaction with Cd^{2+} ions, TAA was found to release S^{2-} ions so fast that the rate-determining step of the total reaction was

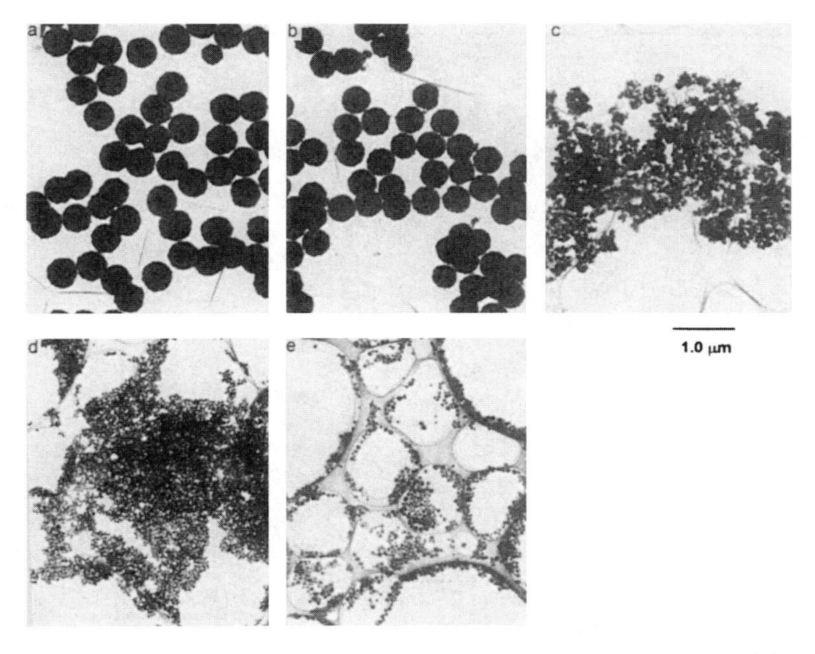

Fig. 3.2.3 TEM images of CdS particles showing the effect of the [EDTA]/[$Cd(OH)_2$] ratio: (a) 1.01, (b) 1.00, (c) 0.99, (d) 0.30, and (e) 0. The aging times are (a) 8 h, (b) 7 h, (c) 6 h, (d) 2 h, and (e) 2 min, respectively. (From Ref. 5.)

the release process of Cd^{2+} ions from the Cd–EDTA chelate (5). Hence, the formation rate of the CdS particles was completely determined by the first-order release rate of Cd^{2+} ions from the Cd–EDTA chelate, as shown in Figure 3.2.4 (5). The first-order dissociation of the Cd–EDTA complex indicates that the reverse association reaction is negligible, consistent with the fact that the release of Cd^{2+} ions is the rate-determining step.

Another important aspect of these systems is the role of highly concentrated ammonia, included in all of these systems as an accelerator of the growth of these metal sulfides. Ammonia has a strong effect of increasing the growth rate of silver bromide by increasing the apparent solubility of AgBr, though the overall rate constant of the surface reaction for the growth of AgBr particles was rather lowered by its adsorption at the growing surface. On this analogy, the significant promotion of the growth of the metal sulfides may be mainly due to the increase of their apparent solubility. Since the stability constants of ammonia complexes are generally much lower than typical chelates of EDTA, NTA, etc., they can promptly release metal ions with the consumption of free metal ions for the particle growth. Hence, ammonia complexes of metal ions play an important role as an accelerator of particle growth,

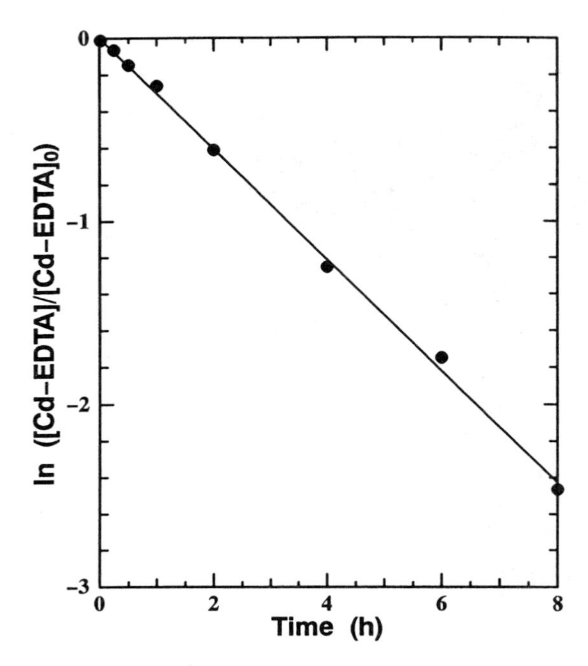

Fig. 3.2.4 Plot of ln([Cd–EDTA]/[Cd–EDTA]$_0$) against aging time. (From Ref. 5.)

as an intermediary of metal ions between a chelate of a high stability constant and the growing particles without increasing the supersaturation ratio. Figure 3.2.5 shows TEM images of CdS particles in the presence or absence of ammonia, where the 2.0 mol dm^{-3} CH_3COONH_4 and 0.6 mol dm^{-3} NH_3 in standard conditions B (a) were replaced by either 2.0 mol dm^{-3} CH_3COONa plus NaOH (b) or NaOH only (c) in systems without ammonia but the initial pH was adjusted with NaOH close to that of the ammonia system. Table 3.2.2 summarizes the effects of ammonia on the yield, particle number, and growth rate of the CdS particles. In the absence of ammonia, the final product was polydispersed particles with a yield of only ~2% after aging at 60°C for 8 h, in contrast to the highly monodispersed product with a yield of more than 90% in the presence of ammonia. The production of the polydisperse particles in the absence of ammonia is due to the high supersaturation ratio of free Cd^{2+} ions above the critical level caused by the exceedingly low growth rate of CdS particles. Figure 3.2.6 schematically shows the role of the ammonia complexes for the formation of monodisperse particles of CdS in a Cd–EDTA system (5).

All of these metal sulfide particles except lead sulfide are polycrystalline spherical particles consisting of much smaller randomly oriented subcrystals, while lead

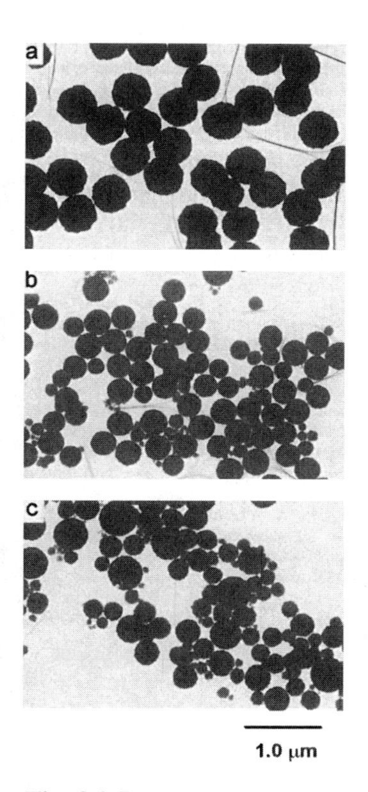

Fig. 3.2.5 TEM images of the resulting CdS particles from various pH buffer systems: (a) CH_3COONH_4-NH_3, (b) CH_3COONa-NaOH, and (c) NaOH only. The other parameters are fixed at the standard conditions B. (From Ref. 5.)

Table 3.2.2 Effects of the Type of the pH Buffer Systems

Type of pH buffer system	Initial pH	Final pH	Yield (mol%)	Particle diameter (μm)	Particle number (dm^{-3})	Growth rate (Å/s)
CH_3COONH_4-NH_3 (standard condition)	8.14	7.84	91.5	0.51	9.55×10^{13}	1.59
CH_3COONa-NaOH	8.62	6.81	2.19	0.32	9.18×10^{12}	0.056
NaOH only	8.34	5.45	1.80	0.31	8.30×10^{12}	0.053

Source: Ref. 5.

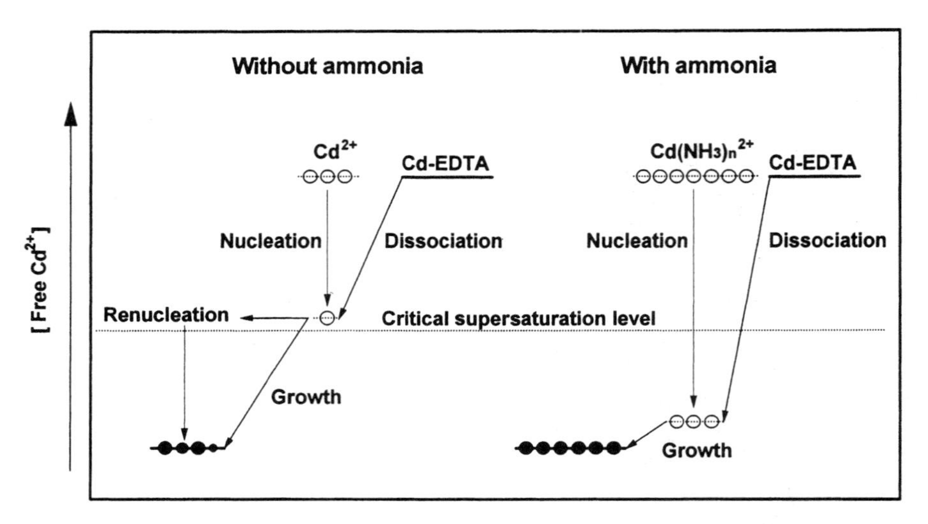

Fig. 3.2.6 Schematic diagram of equilibrium and steady levels of free Cd^{2+} concentration in the absence or presence of NH_3, showing the role of NH_3 on the nucleation and growth of the monodisperse CdS particles, where the open and closed circles represent the solute ions and growing particles, respectively. (From Ref. 5.)

sulfide particles are of a rectangular monocrystalline form. Hence, the polycrystalline particles are grown through a repeated nonepitaxial nucleation of the surface nuclei followed by their limited growth. In this case, the surfaces of the spherical secondary particles serve as only triggers for the surface nucleation like heterogeneous nucleation. For the CdS particles, the size of the subcrystals of each secondary particle increased with the progress of their growth, as revealed by high-resolution electron microscopy on the ultrathin sections of the secondary particles prepared with a microtome (8). This fact suggests that the nucleation rate on the surfaces of the secondary particles goes down more rapidly with the lowering supersaturation than the reduction of the growth rate of the subcrystals from these surface nuclei. In other words, the size of the subcrystals is determined by their relative growth rate against the surface nucleation rate.

In cupric sulfide systems, a high stability constant of a chelate does not necessarily mean a slow release of metal ions, due to the extremely high rates in both association and dissociation of the chelate. For instance, the reactions of chelates, $Cu(TMD)_2{}^{2+}$ and $Cu(DETA)_2{}^{2+}$, with TAA under standard conditions A finish within 2 min at 25°C to yield rather small CuS particles of ~40 to 50 nm, despite the respective high stability constants, $10^{16.9}$ and $10^{21.3}$, comparable to or even much higher than the stability constants of EDTA chelates of the other kinds of metal ions, such as Cd^{2+}, Zn^{2+}, and Pb^{2+}, which are much slower in releasing these metal

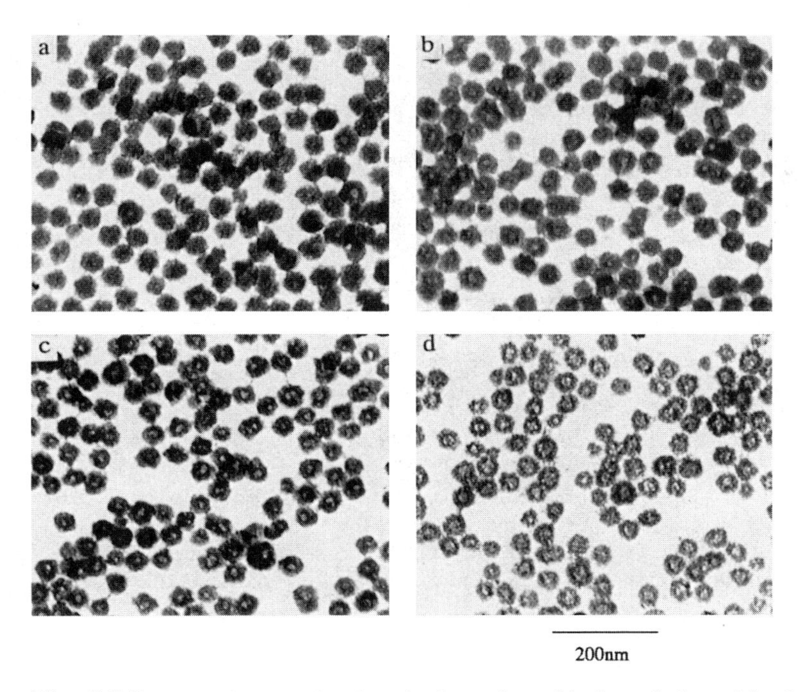

Fig. 3.2.7 TEM images showing the formation of hollow CuS particles by dissolution under the standard conditions A with TMD as a chelating agent. The aging times are (a) 2 min, (b) 1 h, (c) 2 h, and (d) 4.5 h. (From Ref. 6.)

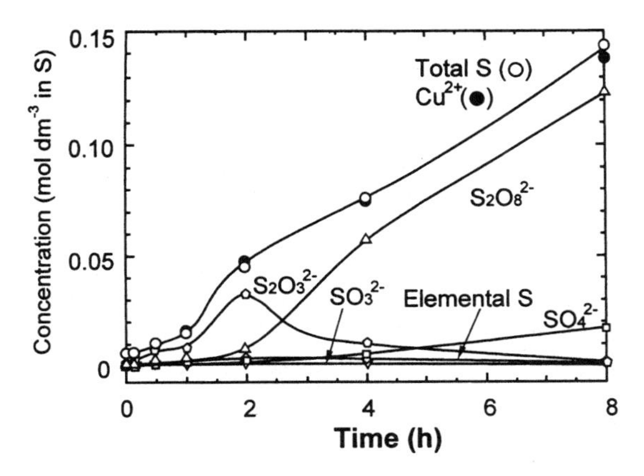

Fig. 3.2.8 Change of the concentrations of the sulfur element of different sulfur species and total concentrations of sulfur and copper in the solution phase with aging time in the absence of gelatin under otherwise standard conditions A with TMD. (From Ref. 6.)

ions. Interestingly, the internal sulfide ions of the CuS particles produced covered with an adsorption layer of gelatin are preferentially oxidized by oxygen with the aid of ammonia and dissolved into the solution phase with Cu^{2+} ions, yielding monodisperse hollow particles of CuS (6). Figures 3.2.7 and 3.2.8 show the time evolution of CuS particles to form the hollow particles with TMD under the standard conditions at 25°C and the concentration changes of sulfur species in the solution phase in a corresponding system but without gelatin.

REFERENCES TO SECTION 3.2

1. G Chiu, EJ Meehan. J Colloid Interface Sci 49:160, 1974.
2. G Chiu. J Colloid Interface Sci 62:193, 1977.
3. G Chiu. J Colloid Interface Sci 83:309, 1981.
4. T Sugimoto, GE Dirige, A Muramatsu. J Colloid Interface Sci 180:305, 1996.
5. T Sugimoto, GE Dirige, A Muramatsu. J Colloid Interface Sci 182:444, 1996.
6. T Sugimoto, S Chen, A Muramatsu. Colloids Surf A Physicochem Eng Aspects 135: 207, 1998.
7. T Sugimoto, GE Dirige, A Muramatsu. J Colloid Interface Sci 176:442, 1995.
8. J Yang, D Shindo, G Dirige, A Muramatsu, T Sugimoto. J Colloid Interface Sci 183: 295, 1996.

3.3 PREPARATION OF MONODISPERSED CdS PARTICLES BY PHASE TRANSFORMATION FROM $Cd(OH)_2$

TADAO SUGIMOTO
Institute for Advanced Materials Processing, Tohoku University, Sendai, Japan

If a solid precursor does not form a gel structure, it is necessary to use subsidiary substances forming a gel-like structure or acting as a protective colloid, such as gelatin, polyethylene glycol, etc. For example, uniform particles of cadmium sulfide (CdS; polycrystalline spheres; 40 nm in mean diameter) consisting of randomly oriented subcrystals of 8.6 nm were prepared from a suspension of 0.5 mol dm^{-3} $Cd(OH)_2$ crystals with 0.55 mol dm^{-3} thioacetamide (TAA) and 1.0 mol dm^{-3} NH_4NO_3 in the presence of 1 wt% gelatin at 20°C and pH 8.5 adjusted with NH_3 (1,2). Figure 3.3.1 shows SEM images of $Cd(OH)_2$ as a starting material and thus-prepared uniform CdS particles as the product (2). It was found that the block layer of gelatin adsorbed to the solid surfaces of $Cd(OH)_2$ inhibited the direct attack of S^{2-} ions on the Cd^{2+} ions in the solid surface layer by controlling the diffusion of S^{2-} ions toward the solid surfaces, since only aggregates of exceedingly small particles of CdS retaining somewhat deformed shape of the original $Cd(OH)_2$ particles were obtained in the absence of gelatin. In addition, the gelatin may work as an inhibitor of coagulation among CdS particles, as obvious from evidence of the action of gelatin as a powerful anticoagulant in the formation of monodispersed CdS particles from chelates of Cd^{2+} in a condensed system (3–5). It was also found that ammonia played a decisive role as a complexing agent of Cd^{2+} ions to promote the growth of the CdS particles and inhibit the direct reaction of S^{2-} ions with the solid surfaces of $Cd(OH)_2$ by promoting the reaction between Cd^{2+} and S^{2-} in the solution phase or on the surfaces of growing CdS particles, since $Cd(NH_3)_n{}^{2+}$ complexes are thought to act as carriers of Cd^{2+} ions, which promptly release Cd^{2+} ions when the concentration of free Cd^{2+} ions is lowered. It was found that sulfide ions preferentially reacted with the Cd^{2+} ions of the $Cd(OH)_2$, particularly when the pH was raised to a high level where the equilibrium concentration of Cd^{2+} ions is low.

The reaction was so fast as to finish within $\sim$1 min, as shown by the transmission electron microscopy (TEM) images in Figure 3.3.2, for which each withdrawn sample on a copper grid was instantly quenched with liquid nitrogen and freeze-dried without melting.

The nucleation stage corresponded to the instantaneous reaction of the Cd^{2+} ions, initially in equillibrium with the $Cd(OH)_2$, with S^{2-} promptly liberated from

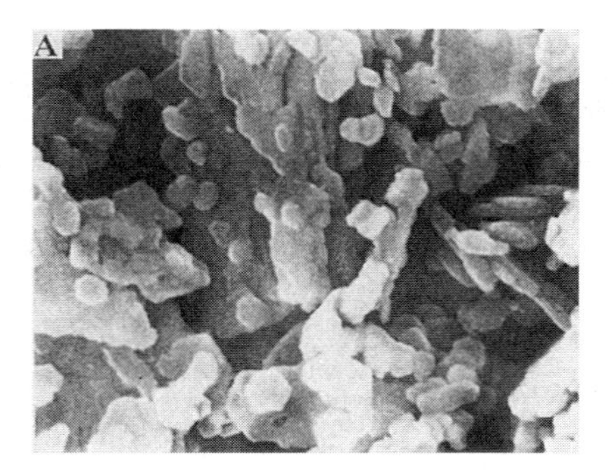

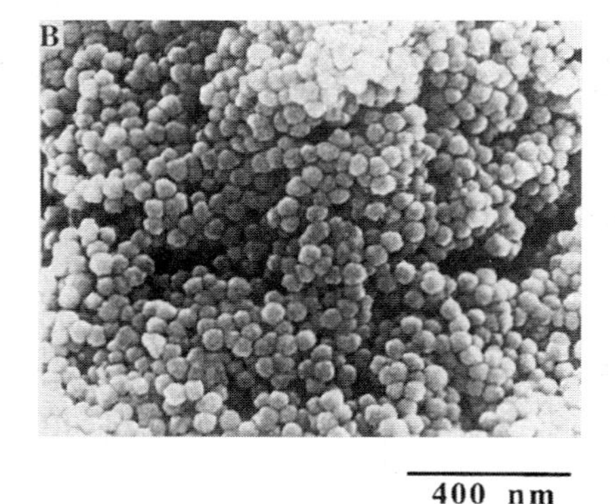

Fig. 3.3.1 Scanning electron micrographs showing the $Cd(OH)_2$ crystals (A) used as a starting material and the uniform CdS particles (B) as the product. (From Ref. 2.)

TAA immediately after the introduction of TAA. The rate-determining step of the growth process was the dissolution of the $Cd(OH)_2$ particles, so that the supersaturation was sufficiently lowered in the growth stage, as has been detected by potentiometry of the activity of Cd^{2+} ions with a Cd/CdS electrode (Fig. 3.3.3) (2). The initial pH of the $Cd(OH)_2$ suspension was 8.50, adjusted with NH_3. The initial Cd potential was 54 ± 2 mV. Both Cd potential and pH dropped rapidly after the addition of TAA. There were two inflection points for the drop of the Cd potential, suggesting three distinctive stages for the reaction of the Cd^{2+} ions. The instantaneous drop of the Cd potential down to approximately -74 mV just after the addition of TAA may correspond to the nucleation stage due to the reaction of TAA with the existing Cd^{2+} ions in the solution phase, and the following plateau until 1 min seems to be due to a steady state of the consumption of the Cd^{2+} ions and the induced dissolution of $Cd(OH)_2$, corresponding to the growth stage of the CdS particles. The inflection

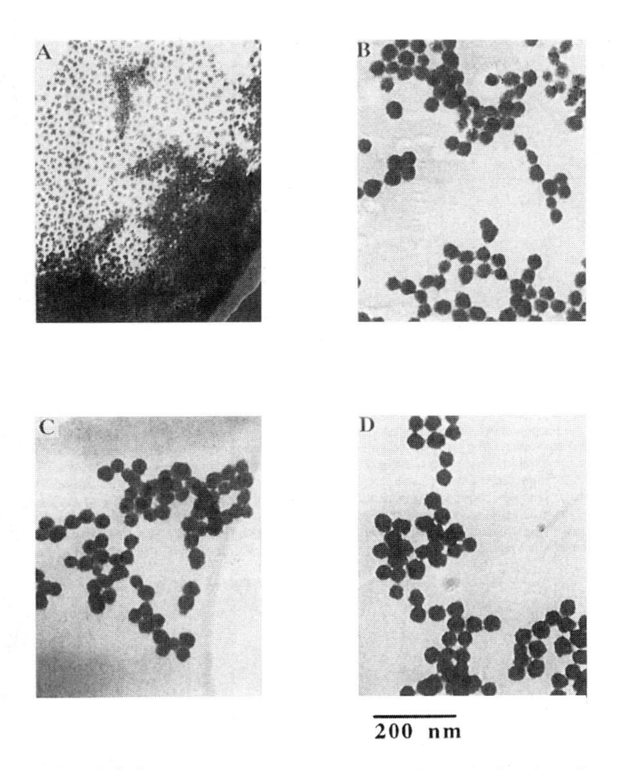

Fig. 3.3.2 Transmission electron micrographs showing the rapid growth process of the uniform CdS particles with aging time under the standard conditions: (A) 15 s, (B) 30 s, (C) 1 min, and (D) 1 h. (From Ref. 2.)

point at around 1 min may correspond to the complete dissolution of the $Cd(OH)_2$, and the following descent of the Cd potential may be due to the reaction of the Cd^{2+} ions remaining in the solution in the absence of $Cd(OH)_2$. This observation is consistent with the findings shown in Figure 3.3.2, wherein small polydispersed CdS particles were formed after as little as 15 s of aging and rapidly grew into larger spherical particles within 1 min. The sufficiently low steady concentration of Cd^{2+} ions during the growth stage is important for the formation of the uniform CdS particles, because it prevents nucleation during the growth stage. The drop of pH is due to the reaction of the Cd^{2+} ions, originally present in equilibrium with $Cd(OH)_2$, with hydrogen sulfide released from TAA ($Cd^{2+} + H_2S \rightarrow CdS + 2H^+$). However, the decrease in pH was rather limited partly because of the action of the NH_3/NH_4^+ system and gelatin as pH buffers, and partly because of the release of OH^- ions with the dissolution of $Cd(OH)_2$.

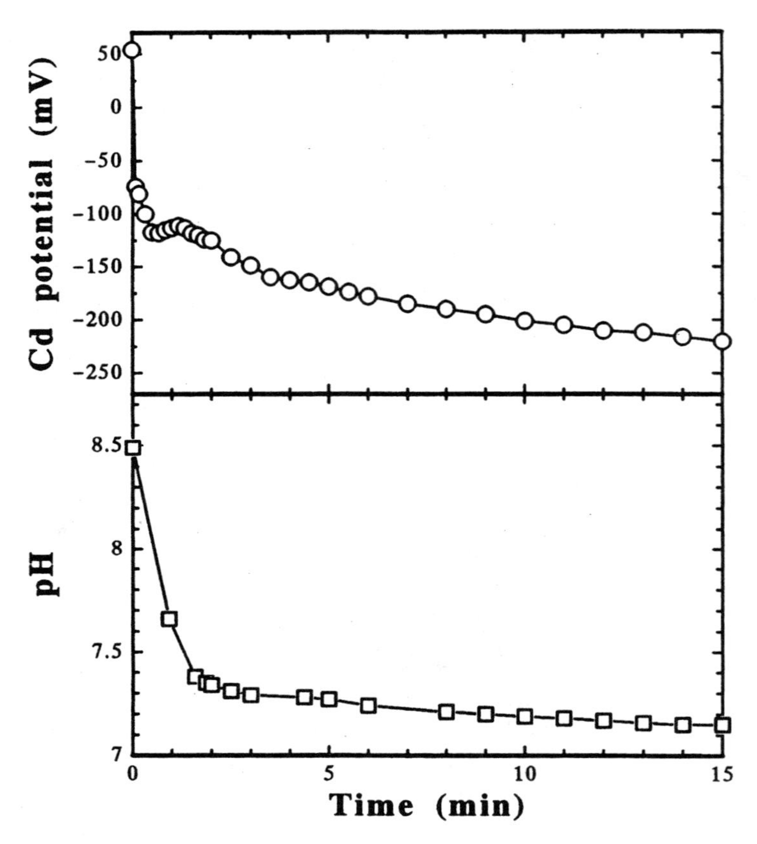

Fig. 3.3.3 Change of the Cd potential and pH of the $Cd(OH)_2$ suspension with the formation of CdS particles under the standard conditions. (From Ref. 2.)

Variation of the initial content of the $Cd(OH)_2$ over two orders of magnitude had virtually no influence on the final particle number, suggesting that the nuclei of the CdS particles were formed only from Cd^{2+} ions initially dissolved in equilibrium with $Cd(OH)_2$ and not on the solid surfaces of $Cd(OH)_2$ (2). It is also suggested from this fact that the nucleation has been finished when the dissolution of the $Cd(OH)_2$ starts.

The presence of a plateau in the change of Cd potential strongly supports that the rate-determining step of the total reaction in the growth stage is the dissolution process of $Cd(OH)_2$ and not the release process of S^{2-} ions from thioacetamide. Since it is now obvious that the release of sulfide ions is so fast as to be almost instantaneous, the reaction for the release of sulfide ions from TAA must be some unknown reaction completely different from the slow irreversible hydrolysis of TAA that has long been believed. After the analysis of the product of TAA, the reaction turned out to be the following reversible reaction, in which the equilibrium is strongly displaced to the left, but S^{2-} ions are promptly released to recover the equilibrium when consumed (2,4):

$$CH_3CSNH_2 \rightleftarrows CH_3CN + H_2S$$

Figure 3.3.4 shows Fourier transform infrared (FTIR) spectra of acetonitrile standard (A) and the organic product from the rection of TAA with Cd^{2+} ions (2). The stability constant defined by

$$K \equiv \frac{[CH_3CSNH_2]}{[CH_3CN]\ [H^+]^2[S^{2-}]}$$

was found to be $10^{31.82}$ at 25°C and $10^{25.85}$ at 60°C (4).

Figure 3.3.5 shows a high-resolution transmission electron micrograph showing a close-up view of a part of a CdS particle. One may find that the particle consists of randomly oriented crystallites, as is obvious from the clear lattice image of each

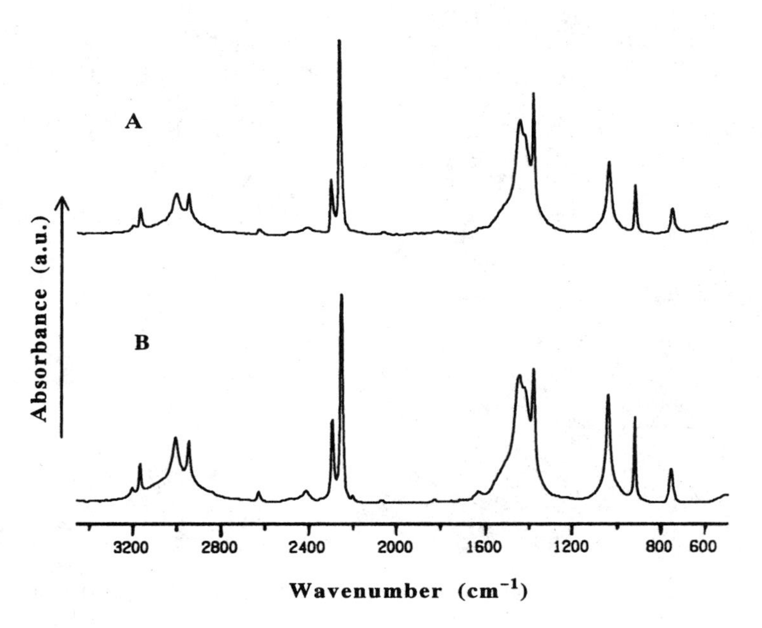

Fig. 3.3.4 FTIR spectra of acetonitrile standard (A) and the organic product from the reaction of TAA with Cd^{2+} ions (B). (From Ref. 2.)

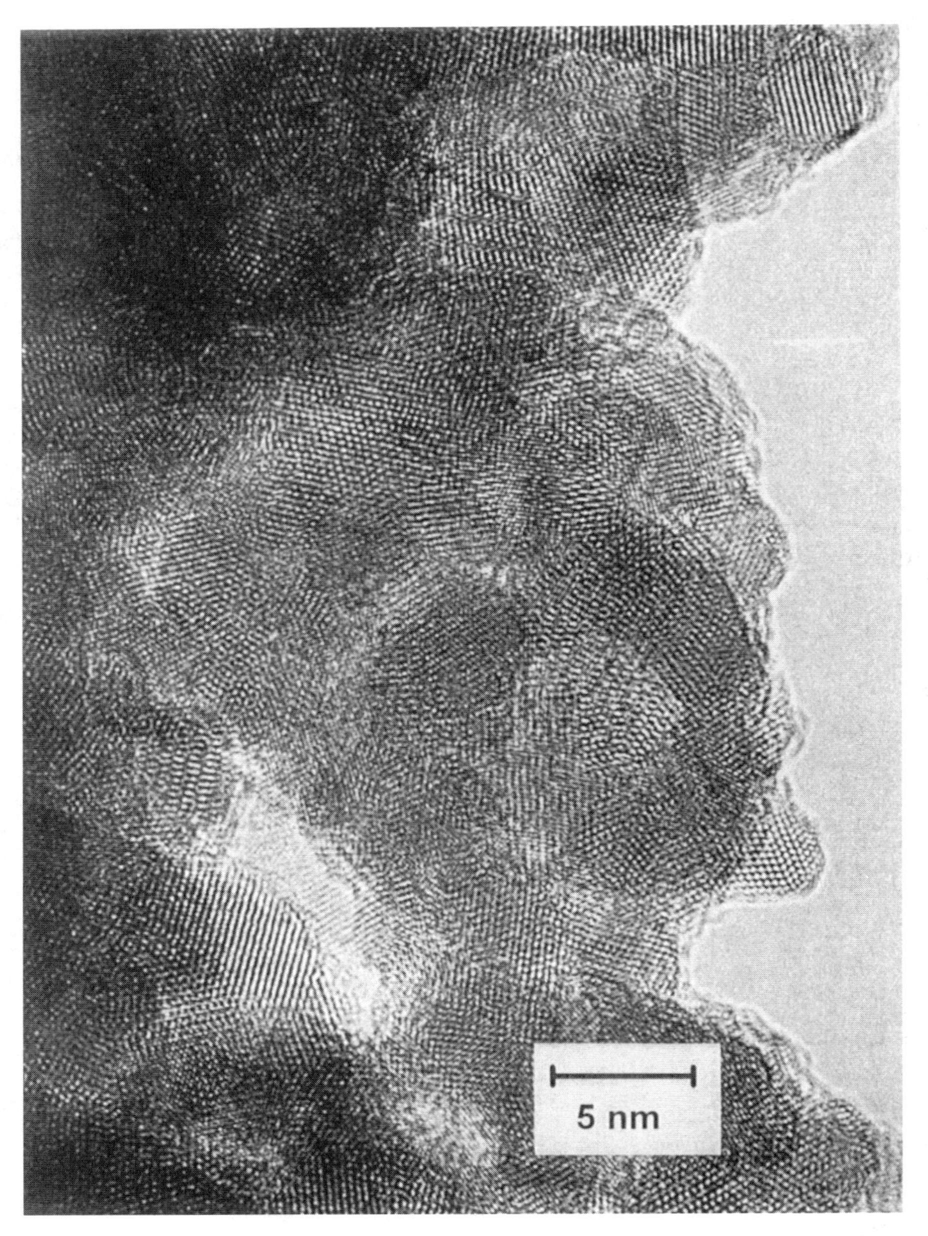

Fig. 3.3.5 High-resolution transmission electron micrograph showing a close-up view of a part of a CdS particle. (From Ref. 2.)

crystallite. From this picture, one may be apt to conclude that the spherical CdS particles are formed by aggregation of preformed primary particles. However, for concluding the growth mechanism of colloidal particles, scrupurous growth analyses are generally needed, especially when the particles are polycrystals composed of randomly oriented subcrystals. The seeding method for the distinction between the normal growth mode by the deposition of the monomeric solute and that by aggregation of preformed primary particles cannot be applied to this system, because the rate-determining step is the dissolution of the $Cd(OH)_2$ so that the total reaction is not accelerated by seeding even though the reaction is based on the deposition of the monomeric solute. Nevertheless, the growth of the CdS particles was found to proceed through the direct deposition of the solute from the effect of the addition mode of TAA on the size distribution of the product. When TAA was introduced by a two-step addition of a half of the total amount for each step at a time interval of 15 min, the final size distribution became bimodal. When it was introduced continuously for 20 min at a constant rate, the resulting size distribution of the product was found very broad. These facts clearly demonstrated that the new nuclei generated during the growth stage of the preformed particles simply acted as centers for the growth of new particles without aggregation to the preformed particles. The Cd potentials monitored with the two-step addition of TAA and with the continuous addition are shown in Figure 3.3.6 (2). The steady Cd potential during the continuous addition of TAA is kept significantly lower than the equilibrium level, consistent with the conclusion that the dissolution process of $Cd(OH)_2$ solid is the rate-determining step of the total reaction.

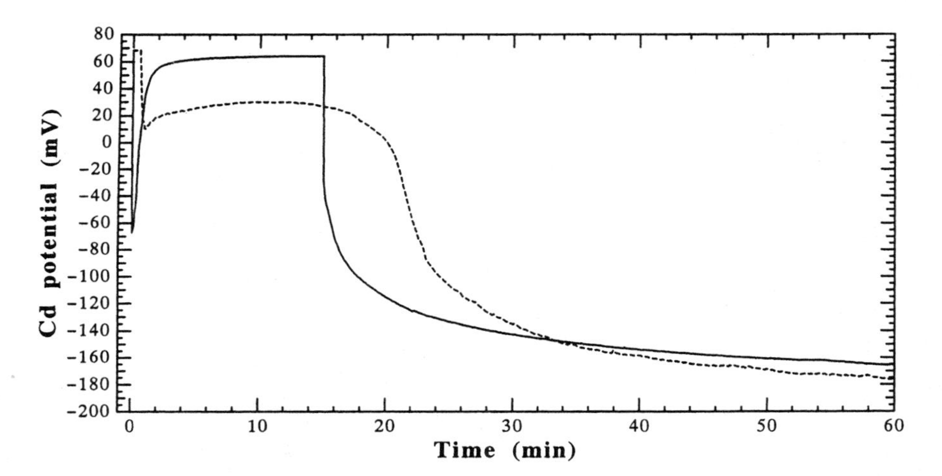

Fig. 3.3.6 Change of Cd potentials during the two-step addition of TAA (solid line) and continuous addition (broken line). (From Ref. 2.)

REFERENCES TO SECTION 3.3

1. T Sugimoto, GE Dirige, A Muramatsu. J Colloid Interface Sci 173:257, 1995.
2. T Sugimoto, GE Dirige, A Muramatsu. J Colloid Interface Sci 176:442, 1995.
3. T Sugimoto, GE Dirige, A Muramatsu. J Colloid Interface Sci 180:305, 1996.
4. T Sugimoto, GE Dirige, A Muramatsu. J Colloid Interface Sci 182:444, 1996.
5. T Sugimoto, S Chen, A Muramatsu. Colloids Surf A Physicochem Eng Aspects 135: 207, 1998.

3.4 CdS AND CdTe NANOPARTICLES MADE IN REVERSE MICELLES: PREPARATION MODES AND OPTICAL PROPERTIES

DOROTHEE INGERT, LAURENCE MOTTE, and MARIE-PAULE PILENI
Université Pierre et Marie Curie, Paris, France

3.4.1 Introduction

Nature has taken advantage of the rich liquid-crystalline behavior of amphiphilic liquids to create ordered fluid biomembrane structures and to modulate dynamic processes in cells (1,2). A key step in the control of mineralization employed by almost all organisms is the initial isolation of a space. Then, under controlled conditions, minerals formation is induced within this space. It is usually delineated by cellular membranes, vesicles, or predeposited macromolecular matrix frameworks. Filling up these spaces with amorphous minerals would appear to require a quite different strategy from that in filling spaces with crystalline material. The simplest way to fill a space with crystals is to create as high a local supersaturation as possible, and then induce nucleation or let the system spontaneously reach a state of lower energy by crystallization, while at the same time removing the excess solvent. This situation is observed in spherulites of calcium carbonate that spontaneously form in metastable supersaturated solutions. ''Droplets'' of calcium carbonate and water separate from solution and with time, or upon drying crystallize, leading to spherulites.

In terms of particle growth, some analogies between surfactant self assemblies and natural media can be proposed. In both cases, this growth needs a supersaturated medium where the nucleation can take place.

Increasingly chemists are contributing to the synthesis of advanced materials with enhanced or novel properties by using colloidal assemblies as templates. Colloid chemistry is particularly well suited to this objective since nanoparticles, by definition, are colloidal and since processing of advanced materials involve reactions at solid–solid, solid–liquid or solid–gas interfaces (3–5).

In solution, surfactant molecules self assemble to form aggregates. At low concentration the aggregates are generally globular micelles (6), but these micelles can grow on increasing surfactant concentration and/ or upon addition of salt, alcohols etc. In this case, micelles have been shown to grow to elongated more or less flexible rod-like micelles (7–10), in agreement with theoretical predictions for micellization (11,12). In addition to their academic interest, the ordering transitions might well be at the origin of mechanical instabilities in dispersion of mesoscopic aggregates subjected to high constant shear. Owing to the remarkable polymorphism of surfactant aggregates, quite a number of these transitions have been observed in amphiphilic systems. As illustrative examples, we can mention the sponge (isotropic) to lamellar (smectic) transition or the onion (liposome) to lamellar (smectic) transition (13). The amphiphilic molecules spontaneously self-assemble to form highly flexible locally cylindrical aggregates, with the average size reaching several micrometers (14,15). The details of the morphology (curvature of the film) of the mixture at a local scale fluctuate widely. The contribution of the entropy of the folded film is predominant in the free energy of the solution, while the morphology has little influence. The interfacial curvature is toward the water, and by convention we describe this as negative mean curvature. Phases with negative mean interfacial curvature are known as type II or inverse (16).

The fabrication of assemblies of perfect nanometer-scale crystallites (quantum crystal) identically replicated in unlimited quantities in such a state that they can be manipulated and understood as pure macromolecular substances is an ultimate challenge in modern materials research with outstanding fundamental and potential technological consequences.

The preparation and characterization of these colloids have thus motivated a vast amount of work (17). Various colloidal methods are used to control the size and/or the polydispersity of the particles, using reverse (3) and normal (18,19) micelles, Langmuir–Blodgett films (4,5), zeolites (20), two-phase liquid–liquid system (21), or organometallic techniques (22). The achievement of accurate control of the particle size, their stability, and a precisely controllable reactivity of the small particles are required to allow attachment of the particles to the surface of a substrate or to other particles without leading to coalescence and hence losing their size-induced electronic properties. It must be noted that, manipulating nearly monodispersed nanometer size crystallites with an arbitrary diameter presents a number of difficulties.

Reverse micelles are well known to be spherical water in oil droplets stabilized by a monolayer of surfactant. The phase diagram of the surfactant sodium bis(2-ethylhexyl) sulfosuccinate, called Na(AOT), with water and isooctane shows a very large domain of water in oil droplets and often forms reverse micelles (3,23). The water pool diameter is related to the water content, $w = [H_2O]/[AOT]$, of the droplet by (23) $D(\text{nm}) = 0.3w$. From the existing domain of water in oil droplets in the phase diagram, the droplet diameters vary from 0.5 nm to 18 nm. Reverse micelles are dynamic (24–27) and attractive interactions between droplets take place.

The intermicellar exchange process, governed by the attractive interactions between droplets, can be modified by changing the bulk solvent used to form reverse micellar solution (26). This is due to the discrete nature of solvent molecules and is attributed to the appearance of depletion forces between two micelles (the solvent is driven off between the two droplets) (26). When the droplets are in contact forming a dimer, they exchange their water contents. This exchange process is associated with the interface rigidity, which corresponds to the bending elastic modulus of the interface (27). Hence, in collisions the droplets exchange their water contents and again form two independent droplets.

The intermicellar potential, deduced from Baxter model, decreases with the number of carbon atoms (25). This has been explained in term of solvent penetration: bulk solvent molecules having a small number of carbon atoms penetrate easily in the surfactant alkyl chains, which are then well solvated. This induces a decrease in the intermicellar interactions. The increase in the number of carbon atoms of the bulk solvent induces a decrease in the solvation of the surfactant and then an increase in the intermicellar attractive potential.

This process has been used to make nanosized material by either chemical reduction of metallic ions or coprecipitation reactions. These various factors (water content, intermicellar potentials) control the size of the particles.

Surfactants having a positive curvature, above a given concentration usually called the critical micellar concentration, cmc, self-assemble to form oil-in-water aggregates called normal micelles. The surfactant most often used is sodium dodecyl sulfate, Na(DS) or SDS. To make particles, the counterion of the surfactant is replaced by ions which participate in the chemical reaction. These are called functionalized surfactants.

3.4.2 CdS Nanocrystals

Cadmium sulfide suspensions are characterized by an absorption spectrum in the visible range. In the case of small particles, a quantum size effect (28–37) is observed due to the perturbation of the electronic structure of the semiconductor with the change in the particle size. For the CdS semiconductor, as the diameter of the particles approaches the excitonic diameter, its electronic properties start to change (28,33,34). This gives a widening of the forbidden band and therefore a blue shift in the absorption threshold as the size decreases. This phenomenon occurs as the cristallite size is comparable or below the excitonic diameter of 50–60 Å (34). In a first approximation, a simple ''electron hole in a box'' model can quantify this blue shift with the size variation (28,34,37). Thus the absorption threshold is directly related to the average size of the particles in solution.

In aqueous solution, the mixture of solutions containing cadmium and sulfide ions induces a precipitation of CdS semiconductor. When adding a protecting polymer such as sodium hexametaphosphate (HMP) in the solution, no precipitation is observed and a yellow solution remains optically clear, indicating the formation of CdS clusters. In reverse micelles, similar behavior of the latter is observed, as shown later.

Various materials have been synthetized in reverse micelles (3). Cadmium sulfide and cadmium selenide semiconductors (38–47) were the first materials prepared by this method. This has been extended to semiconductor alloys such as

$Cd_{1-y}Zn_yS$ (48–50) or dilute magnetic semiconductors like $Cd_{1-y}Mn_yS$ (51,52). Recently, as described later, CdTe has been prepared via a soft chemistry, for the first time.

3.4.2.1 Influence of the Cadmium Derivative Used as Reactant

AOT/Cd(NO$_3$)$_2$/Isooctane/Water Reverse Micelles in the Presence and the Absence of HMP. In the presence of an excess of cadmium ions, $[Cd^{2+}]/[S^{2-}] = 2$, the absorption spectra obtained at various water content, *w*, in Na(AOT) reverse micelles in the presence and in the absence of HMP show a red shift with increasing the water content, *w*. For a given *w* value, a blue shift in the presence compared to the absence of HMP is observed. As described in the litterature (28,34,37), the average size of the particles can be deduced from the absorption onset. The size of the semiconductor is always less than that obtained in aqueous solution. The presence in reverse micelles of HMP as a protecting agent allows a reduction in the size of the particle.

In the presence of an excess of sulfide ion, $[Cd^{2+}]/[S^{2-}] = \frac{1}{2}$, and in the absence of any protecting agent such as HMP, the absorption threshold of the cadmium sulfide particles reaches that of the particles obtained in aqueous solution (41) (490 nm) and the presence of HMP makes it possible to obtain smaller particle sizes (absorption threshold of the order of 470 nm).

These data confirm the protecting effect of HMP and Na(AOT), which plays a role in limiting of the growth of the particles (43).

Using Mixed Reverse Micelles [Cd(AOT)$_2$/Na(AOT)/Isooctane/Water]. In the presence of an excess of cadmium ions, $[Cd^{2+}]/[S^{2-}] = 2$, Figure 3.4.1 shows behavior similar to that observed in Na(AOT) reverse micelles in the presence of HMP. An increase in the average size with the water content is observed. Below the absorption onset several bumps are observed (Fig. 3.4.1) and can be clearly recognized in the second derivative. These weak absorption bands correspond to the excitonic transitions. This clearly shows a narrow size distribution. At low water content, the first excitonic peak is well resolved and is followed by a bump. The second derivative shows a very high intensity of this bump (insert, Fig. 3.4.1A). With small crystallites, according to the data previously published (53,54), several bumps due to several excitonic peaks are expected. The insert of Figure 3.4.1A shows only one bump. This is due to the very small size of the particles, so the others bumps are blue shifted and are not observable under our experimental conditions. By increasing the water content, that is, by increasing the size of the particles, several bumps are observed (insert Fig. 3.4.1B). This confirms the fact that, at $w = 5$, the bumps are blue shifted, and indicates a very narrow distribution in the size of the particles. The intensity of these bumps decreases with increasing the water content *w* (insert Fig. 3.4.1), indicating a decrease in the number of excitonic transitions when

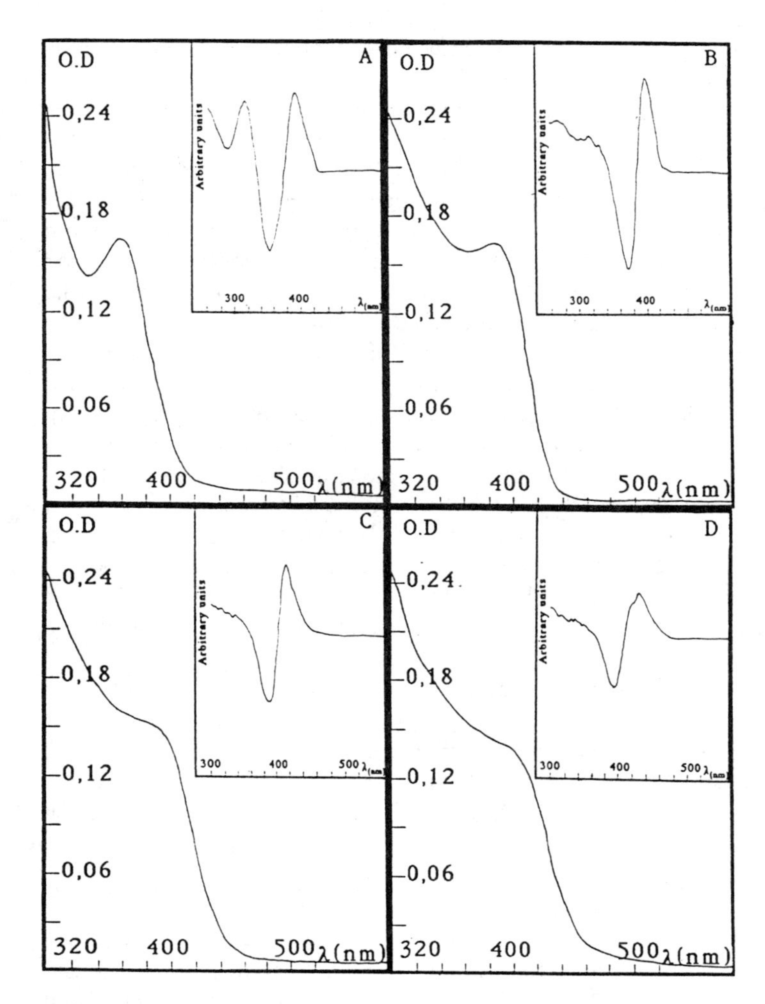

Fig. 3.4.1 Variation of the absorption spectrum of CdS in reverse micelles with the water content. [AOT] = 0.1 *M*, $[(AOT)_2Cd] = 2 \times 10^{-4}$ *M*, $[Cd^{2+}]/[S^{2-}] = 2$; (A) $w = 5$; (B) $w = 10$; (C) $w = 20$; (D) $w = 40$; inserts, second derivatives of the photoabsorption—the minimum indicates the excitonic peak.

the size of the particle increases. This is in agreement to the theoretical calculations previously reported for the Q particles (53,54).

Electron microscopy has been performed using a sample synthesised at $w = 10$, $[Cd^{2+}]/[S^{2-}] = 2$, and characterized by 430-nm absorption onset, which corresponds to a CdS diameter equal to 25 Å. The microanalysis study shows the characteristic lines of sulfide and cadmium ions, indicating that the observed particles are CdS semiconductor crystallites. The electron diffractogram shows concentric circles, which are compared to a simulated diffractogram of bulk CdS. A good agreement between the two spectra is obtained, indicating the particles keep zinc blend crystalline structure (fcc) with a lattice constant equal to 5.83 Å.

In the presence of an excess of sulfide ions, $[Cd^{2+}]/[S^{2-}] = \frac{1}{2}$, a strong change in the absorption spectra at low water content is observed compared to that obtained for a ratio of $[Cd^{2+}]/[S^{2-}]$ equal to 2. By increasing the water content, the sharp peak disappears and a similar behavior as in the case of excess of cadmium is observed, i.e., a red shift in the absorption spectrum. The sharp peak observed at low water content increases with the relative amount of sulfide ions (45). This peak is attributed to sulfide clusters (55) formed on the CdS particles because of the high local concentration of sulfide ions. The disappearance of this peak when increasing the water content could be explained by the fact that sulfide clusters, with negative charges, are repelled to the center of the droplets and redissolve themselves inside the water pool.

The average radius is deduced. Figure 3.4.2 shows a change in the size of the particle with the molar ratio of cadmium and sulfide ions concentrations. The largest size particles are obtained for $[Cd^{2+}]/[S^{2-}] = 1$ and the smallest for $[Cd^{2+}]/[S^{2-}] = 2$. It can be noticed that the size of CdS is always smaller when one of the two reactants are in excess ($[Cd^{2+}]/[S^{2-}] = \frac{1}{4}, \frac{1}{2}, 2$). This confirms that the crystallisation process is faster when one of the reacting species is in excess (56).

3.4.2.2 Influence of the Preparation Mode on the Size Distribution

The fluorescence of cadmium sulfide is strongly dependent on the particle size and the presence or absence of sulfide vacancies (57,58). The fluorescence of cadmium sulfide particles observed in solution in water (37,59,60) and in acetonitrile solution (59) is dependent on the particle method of preparation of CdS particles:

1. Two types of fluorescence spectra are observed in aqueous solution and in the presence of an excess of cadmium ions. In the absence of any stabilizer the fluorescence is characterized by two very weak bands (41), one centered at 450 nm and attributed to the direct recombination of charge carriers (61) from shallow traps, and the other very broad at about 650 nm, which is not clearly attributed. The presence of a stabilizing agent, such as HMP (59,60), makes it possible to increase the sulfide vacancies at the surface resulting in a more intense fluorescence band, centered in the region of 550 nm. This band is attributed to the recombination of

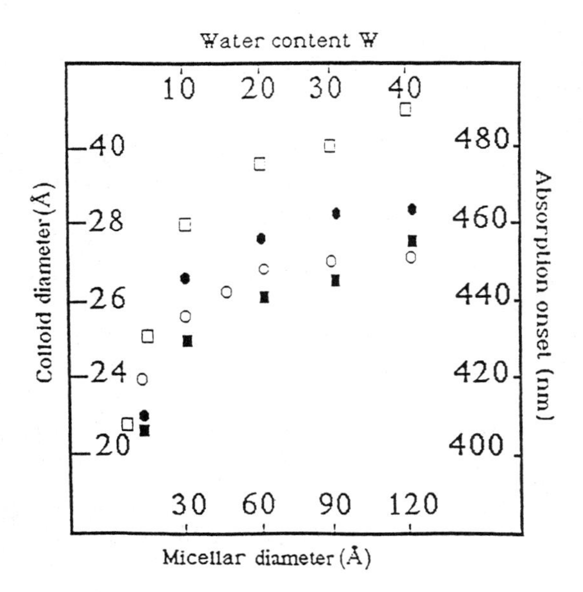

Fig. 3.4.2 Variation of the size of the CdS particles and the absorption onset with the water content and with the size of the droplet. [Na(AOT)] = 0.1 *M*, for $[Cd^{2+}]/[S^{2-}]$ ratio equal to $\frac{1}{4}$, $\frac{1}{2}$, and 1, $[Cd(AOT)_2]$ = 1 × 10^{-4} *M*, and for $[Cd^{2+}]/[S^{2-}]$ = 2, $[(AOT)_2Cd]$ = 2 × 10^{-4} *M*. $[Cd^{2+}]/[S^{2-}] = \frac{1}{4}$, (○); $[Cd^{2+}]/[S^{2-}] = \frac{1}{2}$, (●); $[Cd^{2+}]/[S^{2-}] = 1$, (□); $[Cd^{2+}]/[S^{2-}] = 2$, (■).

charges previously trapped at the surface in sulfide vacancies (62). As the energy of these deeper traps is related to the particle size (58), the shift with excitation wavelengths of the emission maxima to longer wavelengths is explained by the increase in the size of the colloids, which brings about a decrease in the energy of these traps (58).

2. In the presence of an excess of sulfide ions two fluorescence emissions are observed. The first is centered at 450 nm and is attributed to the direct recombination of charge carriers. The second emission band, observed at around 650 nm, depends on the particle size. This second emission band is very weak and is very often quenched by the presence of species absorbed at the interface. By analogy it could be attributed to cadmium ion vacancies.

As the fluorescence is dependent on the particle sizes (58), a shift of the excitation threshold with the emission wavelength indicates the presence of particles of different sizes in the solution which shows size distribution of the aggregates (41,43). In the case where this excitation spectrum is analogous to the absorption spectrum, the threshold is a measure of the average size of the emitting particles.

The change of the fluorescence maximum with the excitation wavelength gives a qualitative measure of the polydispersity (33,34).

$Na(AOT)$–$Cd(NO_3)_2$–Isooctane–H_2O Reverse Micelles in the Presence and in the Absence of HMP. In the presence of an excess of cadmium, $[Cd^{2+}]/[S^{2-}] = 2$. The fluorescence spectra normalized at the maximum emission and obtained at various water content and various excitation wavelength are centered at 500 nm, at low water content. The emission maxima are not changed with the excitation wavelengths ($\Delta\lambda = 10$ nm). By increasing the water content, a red shift of the fluorescence spectra, due to the emission of biggest particles, is observed, and the emission maxima strongly depend on the excitation wavelengths ($\Delta\lambda = 40$ nm). This indicates polydispersity among the particles. This is confirmed from the excitation spectra observed at various emission wavelengths: At low water content ($w = 10$), the fluorescence excitation threshold is not drastically changed ($\lambda = 405 \pm 2$ nm), whereas at high water content ($w = 30$), a shift in the fluorescence excitation threshold is obtained by changing the emission wavelength ($\lambda = 415 \pm 10$ nm).

Using Mixed Reverse Micelles [$Cd(AOT)_2$–$Na(AOT)$–Isooctane–Water]. In the presence of excess of cadmium ions, $[Cd^{2+}]/[S^{2-}] = 2$, Figures 3.4.3A and 3.4.3B show the fluorescence spectra at various water content. The unchanged spectra with the excitation wavelengths indicate a lower size distribution of the particles formed in mixed reverse micelles than that obtained in the case of the $Cd(NO_3)_2/\pm$HMP/AOT/isooctane/water reverse micellar system. The shift of the emission maximum with the water content, *w,* can be related to the increase of the average size of the particles. For a sample synthesized for a given experimental

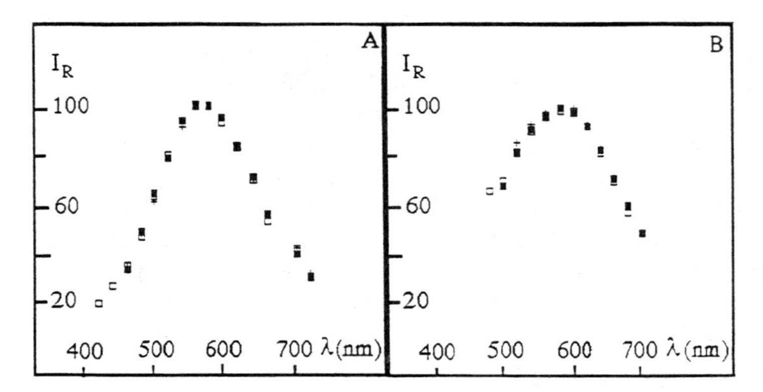

Fig. 3.4.3 Variation of the relative fluorescence spectra (normalized at the emission maximum) with the water content at various excitation wavelength: (A) $[(AOT)_2Cd] = 2 \times 10^{-4}$ *M*, $w = 10$; λexc: □ = 380 nm; ■ = 400 nm; + = 420 nm. (B) $[(AOT)_2Cd] = 2 \times 10^{-4}$ *M*, $w = 30$; λexc: □ = 380 nm; ■ = 410 nm; + = 430 nm.

condition (fixed w and ratio $[Cd^{2+}]/[S^{2-}]$ the fluorescence excitation spectra are unchanged with the emission wavelengths. This confirms the monodispersity in the size of the CdS particles.

3.4.2.3 Change of Particle Sizes with Various Parameters

Influence of Intermicellar Interactions. By replacing isooctane by cyclohexane as the bulk solvent, the intermicellar potential decrease inducing a decrease in the exchange micellar rate by a factor of 10 (24,63) whereas with decane the exchange micellar rate increases by a factor of 2 (27). The size of the droplets remains the same by replacing the bulk solvent.

Synthesis of CdS has been performed by using cyclohexane, isooctane or decane as the bulk solvent. The data are compared to those obtained with isooctane. Figure 3.4. shows the absorption onset and the CdS diameter obtained at various water contents, using $Cd(NO_3)$ reactant (Fig. 3.4.4A and B) or functionalized surfactant $Cd(AOT)_2$ (Fig. 3.4.4C and D). At low water content, $w < 10$, similar sizes are obtained, whatever bulk solvent is used. These indicate that for low water content, size control is not governed by exchange micellar rate but only by preparation mode and by the nature of ions in excess (Cd^{2+} or S^{2-}).

At high water content, $w \geq 10$, by using $Cd(NO_3)_2$ as a reactant, the size of the particles increases with the change of the bulk solvent as follows: cyclohexane, isooctane, decane. Hence, at a given droplet size (w = constant), the increase in the intermicellar potential induces an increase in the average diameter of the particles.

Influence of the Additives. Complexing agents like Kryptofix 222 and Kronenether are known to effect a dramatic change in the interactions of cations with their counter ions. The addition of these macrocycle compounds to a micellar Na(AOT) solution is supposed to gather alkaline cations (Na^+) within macrocycle cavities near the interface.

Addition of Kryptofix 222 and Kronenether to reverse micellar system induces no changes in the droplet size and an increase in the droplet–droplet interactions. The complexation of cations Na^+ of AOT led to a decrease in counterion binding, and consequently repulsive interactions between polar head groups of AOT surfactant are increasing. This could induce a more flexible interface of reverse micelles.

The addition of macrocycles on CdS synthesis in reverse micelles induces a strong change in absorption spectra. For a given water content, a red shift of absorption onset is observed in presence of macrocycles. This effect is more pronounced in presence of Kryptofix 222 and when the CdS nanocrystallite synthesis is realized in presence of an excess of sulfide S^{2-} ions ($x = \frac{1}{2}$). This red shift is characteristic of an increase in the average nanocrystallite size. It can be noticed that absorption of CdS particles synthetized in reverse micelles in presence of an excess of cadmium Cd^{2+} ions ($x = 2$) is reduced in presence of macrocycles. This indicates a decrease in the yield of CdS particles and is attributed to complexation of functionalized

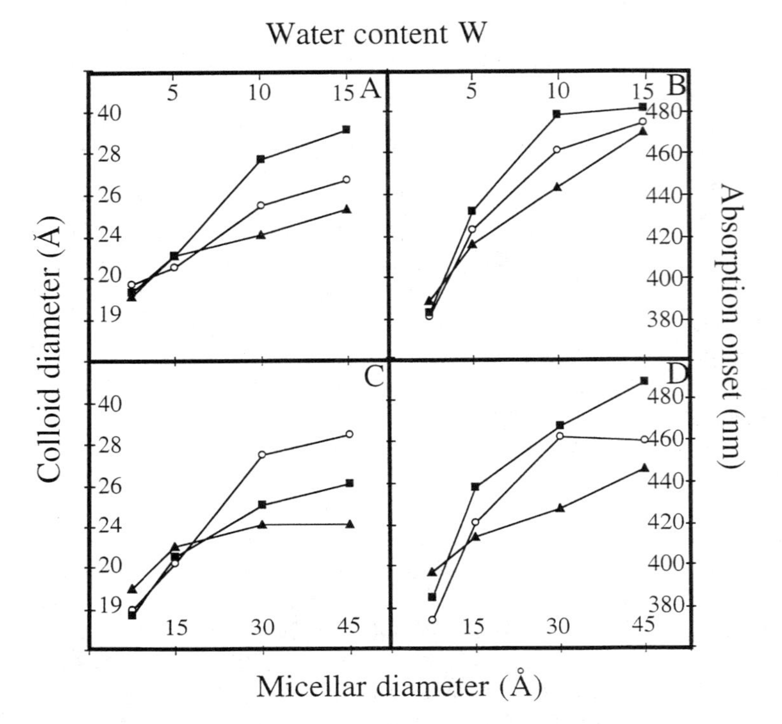

Fig. 3.4.4 Variation of the size of the CdS particles and the absorption onset with the water content and with the size of the droplet, [AOT] = 0.1 *M*: (A) $[Cd^{2+}]/[S^{2-}] = 2$, $[CdNO_3] = 2 \times 10^{-4}$ *M*. (B) $[Cd^{2+}]/[S^{2-}] = \frac{1}{2}$, $[CdNO_3] = 10^{-4}$ *M*. (C) $[Cd^{2+}]/[S^{2-}] =$ 2, $[(AOT)_2Cd] = 2 \times 10^{-4}$ *M*. (D) $[Cd^{2+}]/[S^{2-}] = \frac{1}{2}$, $[(AOT)_2Cd] = 10^{-4}$ *M*. Bulk solvent used: isooctane (○), cyclohexane (▲), decane (■).

surfactant counterion $[Cd(AOT)_2]$ by macrocycles. This complexation induces nucleation centers less accessing for CdS synthesis, and this favors bigger particles and size polydispersity.

The influence of the addition of cetyl trimethyl ammonium chloride, CTAC, to the reverse micellar solution affects the droplet size and micellar interactions, as demonstrated by the DQLS experiment (64). Addition of CTAC to micellar system at a given water content leaves the droplet size unchanged, whereas a decrease in the intermicellar attraction has been observed. This decrease is more important for high CTAC concentrations. This has been interpreted to steric repulsion induced by the long hydrocarbon tail of CTAC (C_{16}). Thus, the addition of this compound to CdS synthesis could modify the nucleation and/or growth process. The experiments were performed by solulization of CTAC in the micellar solution containing either sodium sulfide or $Cd(AOT)_2$.

When CTAC is solubilized in micellar solution with sulfide S^{2-} ions, at low water contents ($w < 10$), the presence of CTAC induces a strong decrease in CdS nanocrystallite size. For a given water content, the absorption spectra are blue shifted when the syntheses are performed in the presence of CTAC compared to that obtained in its absence. The temporal evolution of absorption at 250 nm is approximated to nucleation rate of CdS. It slows down in the presence of CTAC. This blue shift is more pronounced at low water content and high CTAC concentration. Hence it is observed a decrease in the particle size by increasing CTAC concentration. This can be related to the decrease in the intermicellar potential in the presence of CTAC (64).

At high water content, $w = 10$, a drastic decrease in the nucleation rate with the increase in CTAC concentration is observed. Furthermore, a red shift in the absorption spectra occurs, indicating an increase in the average particle size. The change in the nucleation rate with increasing CTAC concentration could be explained by a change in the average location of the reactants inside the micellar solution. At high CTAC concentrations ($>2 \times 10^{-3}$ M), strong interactions between cation CTA^+ and anion S^{2-} could occur, inducing the formation of CTA_2S. In such conditions, the two reactants Cd^{2+} and S^{2-} are localized at the micelle interface [functionalized surfactant $Cd(AOT)_2$ and CTA_2S]. In previous studies it has been demonstrated (65,66) that the change in the average location induces changes in the chemical reaction. Hence, the location of the two reactants at the micellar interface [$Cd(AOT)_2$ and CTA_2S] slows down the nucleation process. Notice that at low water content the nucleation process is not drastically changed. This is attributed to the fact that the amount of water in the droplets is low enough to hydrate CTAC and hence prevent the formation of CTA_2S.

When CTAC is solubilized in micellar solution with cadmium Cd^{2+} ions, a better resolution in the excitonic peak with increasing CTAC concentration is observed. The sharp peak is more intense for low water content and for high CTAC concentration. This clearly shows a narrow size distribution.

At low water content, $w = 3$ and $w = 5$, a decrease in nanocrystallite size (blue shift of absorption onset) is observed in presence of low CTAC concentrations ($<2 \times 10^{-3}$ M). This is directly related to the decrease in interdroplet attractive interactions induced by CTAC addition.

At high water content, $w = 10$, a similar size is obtained by comparison with the CdS synthesis in absence of CTAC.

The nucleation rate is slowed down with increasing CTAC concentration, notably at a water content w equal to 3. However, this phenomenon is less important compared to what is obtained previously by solubilizing CTAC in micellar solution with sulfide S^{2-} ions.

3.4.3 CdTe Nanocrystals

Synthesis by soft chemistry made the first disodium telluride, Na_2Te. This product is highly poisonous and oxidable. To make Na_2Te we used the two procedures described in Ref. 69. The difference between these two is based mainly on the reaction time and on the amount of naphthalene used as catalyst. One procedure gives disodium telluride containing a large amount of impurities. The only procedure valuable for making pure Na_2Te is that described next. The reaction is carried out under nitrogen atmosphere by using a standard Schlenk technique. A Schlenk container is charged with tellurium powder (48 mmol), sodium (96 mmol), naphthalene (0.8 mmol) and 20 mL tetrahydrofurane, THF, which has been distilled and degassed by freeze–thaw process under nitrogen previously. The reaction mixture is stirred for 4 days at room temperature and filtered. The product is washed with 20 mL of freshly degassed THF and dried under vacuum. Na_2Te is then in powder form, which is very oxidable. So all the procedure described next to make CdTe nanoparticles is done in a glass box.

Syntheses of CdTe are made in reverse micelles. The coprecipitation takes place by mixing two micellar solutions having the same water content, w = $[H_2O]/[AOT]$: 0.1 M Na(AOT) containing Na_2Te and a mixed micellar solution containing $Cd(AOT)_2$ (69) and Na(AOT). An excess of cadmium ions is used in the synthesis ($x = [Cd^{2+}]/[Te^{2-}] = 2$). The overall concentrations of Na_2Te and $Cd(AOT)_2$ in micellar solution are 1.5×10^{-4} M and 3×10^{-4} M, respectively. A few seconds after mixing these two degassed micellar solutions, CdTe nanocrystals are formed. The syntheses are performed at various droplet sizes. On dodecanthiol addition to the micellar solution, a selective surface reaction between the thio derivative and cations takes place. The coated particles are then extracted from micellar solution and the surfactant used to form the colloidal dispersion is removed by degassed ethanol addition. The coated particles precipitate. After centrifugation the substrate is removed. The coated particles are in powder form and are stable under air. The coated particles are then redispersed in a mixture of two solvents (isopentane/methylcyclohexane, 3 v/v), forming an optically clear glass at low temperature. The coated particles dispersed in solution are stable for a few days (7 days). By keeping them longer time, the stability of the particles markedly changes. At the opposite, in the powder form the particles are stable during several months.

Two procedures are used:

Procedure I. Immediately after dodecanethiol addition, the particles are extracted and treated as already described.

Procedure II. After mixing the two micellar solutions, the particles are kept in micellar solution under nitrogen for 48 h. Dodecanethiol is then add and extracted as before.

Remark: When syntheses are made in the presence of excess of tellurium ($x = [Cd^{2+}]/[Te^{2-}] = \frac{1}{2}$), CdTe nanoparticles are formed. After dodecanethiol addition and during the washing process, a black precipitate appears due to the formation of telluride oxide. It is then impossible to separate the CdTe nanoparticles from TeO_2 aggregates.

The particle size, determined from the transmission electron microscopy pattern, is controlled by the size of the water droplets in which syntheses are made. Similar behavior has been observed for various nanoparticles, as described previously in the case of CdS particles. As expected, the aging of particles (procedure II) induces an increase in the particle size compared to what observed by procedure I. Table 3.4.1 gives the average size of the particles and the size distribution for the two procedures used. Hence procedure I permits a change in the particle size from 2.6 nm to 3.4 nm, whereas it is from 3.4 to 4.1 nm by procedure II. Electron diffraction studies confirmed that the CdTe nanocrystallites exhibited the bulk-like, zinc-blende crystal structure (69). The coated particles are analyzed by energy dispersion spectroscopy (EDS); they are composed of 51% Cd and 49% Te. The electron diffraction, energy dispersion spectroscopy, and average size distribution (13%) (Table 3.4.1) remain unchanged with the two procedures used (I and II) and for various particle sizes.

As observed for direct semiconductors, the CdTe absorption spectrum is size dependent. A red shift in the absorption spectrum with increasing the particle size

Table 3.4.1 Average Diameter D (nm) of CdTe Nanoparticles Obtained at Various Water-in-Oil Droplet Sizes d(nm) and Through the Two Procedures

	d (nm)		
	1.5	3	12
Procedure I:			
D (nm)	2.6	3.1	3.4
σ (%)	17.8	13.2	11.6
E_g (1) (eV)	1.99	1.96	1.92
E_g (2) (eV)	2.17	2.07	2.02
Procedure II:			
D (nm)	3.4	3.8	4.1
σ (%)	12.7	11.9	11.1
E_g (1) (eV)	1.92	1.89	1.81
E_g (2) (eV)	2.02	1.97	1.87
ω_{PL} (300 K) (eV)	2.04	1.99	—
E_g (77 K) (eV)	2.07	2.01	1.89
ω_{PL} (77 K) (eV)	2.09	2.03	1.91
Φ_{PL}	1	0.98	0.11

Note: σ (%) is the size distribution. E_g (1) and E_g (2) are the energy band gap determined either from [$\sigma h\nu = (h\nu - E_g)^{1/2}$] or from the first derivation of the absorption spectrum at 300 K. E_g(77 K) is the energy band gap determined at 77 K. ω_{PL}(300 K) and ω_{PL}(77 K) are the PL peak position at 300 K and 77 K. The relative fluorescence quantum yield, determined at 77 K, is Φ_{PL}.

is observed. This is attributed to a quantum size effect (28–37,41). The energy band gap E_g is deduced either from the following equation (34):

$$\sigma h\nu = (h\nu - E_g)^{1/2}$$

or from the first derivative of the absorption spectrum. Table 3.4.1 gives the values of the energy band gap determined by these two methods and for the various procedures. It increases with decreasing CdTe particle size. When comparison is possible, Table 3.4.1 shows an unchanged value of the band gap with the fabrication mode. However, the excitonic peak is better defined for particles produced by procedure II (Figs. 3.4.5A and B). Noticed that the absorption spectra of particles differing by their sizes and preparation modes present a shoulder at 750 nm, which cannot be attributed to oxide formation. Preliminary EXAFS data confirm the absence of oxide. This is valid in the range of error of the experimental data. Hence no obvious explanations can be given to explain this.

Photoluminescence (PL) spectra markedly change with the procedure used. Figure 3.4.5C and D show the PL spectra recorded, at room temperature, for particles having same average size (3.4 nm) and synthetized by the two procedures. Figure

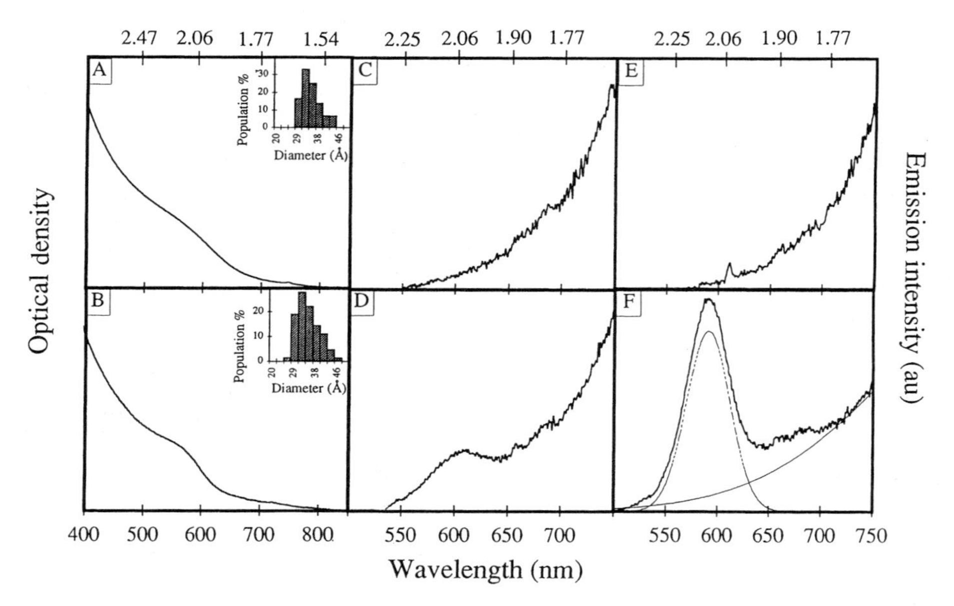

Fig. 3.4.5 Absorption spectra of CdTe nanoparticles having an average size of 3.4 nm and prepared by procedure I (A) and procedure II (B). PL were spectra recorded from particles prepared by procedure I and procedure II, recorded at room temperature (C and D) and at 77 K (E and F), respectively.

3.4.5C shows a long-tailed PL spectrum. This is due to the recombination of charge carriers in the surface traps (29–34,70,71). At the opposite, Figure 3.4.5D shows a shoulder at 610 nm and the trap emission. Assuming that this spectrum is due to the sum of two PL spectra that are described by two Gaussians, the photoluminescence maximum at high energy, $\omega_{PL}(300\ K)$, is deduced. The energy band gap and the maximum of photoluminescence are very close (Table 3.4.1). This permits one to attribute the photoluminescence at higher energy to the direct recombination of free electron and hole. Notice that that for the first time, the direct transition is observed by using reverse micelles as synthesis media. The previous data obtained with CdS (41,43,45) and CdZnS (49,50) nanoparticles shows either at room or low temperature trap emission. This has been explained by the fact that reverse micelles produce particles not sufficiently controlled with a low stabilization (34). Similar behavior is obtained with CdS, CdMnS (51,52), CdZnS (49,50), and CdTe (Fig. 3.4.5C) when the particles are extracted immediately after dodecanethiol addition from reverse micelles. Figure 3.4.5D shows that the appearance of the direct transition is observed when the particles are kept inside the micellar solution during 48 h (procedure II). This direct transition can be explained by formation of a passivated layer due to Oswald ripening. This permits a decrease of the surface defects and then the traps emission. These data have to be related to those obtained with CdSe previously: The CdSe nanoparticles are produced through inorganic syntheses at 280°C (22). A direct transition is observed, at room temperature, when the particles are coated by ZnS (72,73). This has been explained by a decrease in the number of CdSe defects with shell formation. Similarly, direct transition has been observed by adsorbate addition at the particle interface (74). The procedure described here permits one to obtain the direct transition without addition of external adsorbate.

As expected, on decreasing the temperature to 77 K, the PL spectrum of CdTe made from procedure I remains similar to that obtained at room temperature and is attributed to recombination of charge carriers in the surface traps. On the other hand, the PL spectrum of CdTe made from procedure II shows a well-defined peak centered at 590 nm. Similar behavior is observed for various particle sizes (from 3.4 nm to 4.1 nm). Table 3.4.1 shows a red shift in the maximum of the PL spectrum due to the direct transition by increasing the particle size. This is due to a quantum size effect. The PL peak position and the band edge measured from absorption are very close (Table 3.4.1). This confirms that the PL peak observed at high energy arises from the recombination of a free electron and hole. The relative fluorescence quantum yield of the direct transition, Φ_{PL}, is obtained by assuming two Gaussian curves (one due to the direct transition and second due to trap emissions). Table 3.4.1 shows a decrease in the quantum yield, Φ_{PL}, with increasing the particle size.

On increasing temperature, a red shift in the PL spectrum is observed. Similar behavior has been observed in bulk materials and is attributed to change in the relative position of the valence and conduction bands due to the dilatation of the lattice. The maximum of the PL spectrum, ω_{PL}, follows similar variation as the energy band gap and is expressed as (75):

$$\omega_{PL}(T) = \omega_{PL}(0) - AT^2/(B + T)$$

where $\omega_{PL}(T)$ and $\omega_{PL}(0)$ are maximum PL at T and 0 K, respectively, and A and B are constants related to the material.

The best fit is obtained with $B = 100$ K and $A = 2.4 \times 10^{-4}$ eV K^{-1}. These values are closed to those obtained from the bulk phase ($B = 180$ K and A $= 4 \times 10^{-4}$ eV K^{-1}, respectively) (76). On decreasing temperature, the PL intensity increases and its maximum is shifted to higher energy. To determine the peak intensities and positions, the PL spectrum is simulated as before by assuming two Gaussians. The temperature dependence of the photoluminescence due to direct transition obeys an Arrhenius law given by (34):

$$I_{PL}(77\ K)/I_{PL}(T) = AT^{-0.5} \exp(-E_a/kT)$$

where $I_{PL}(77$ K), I_{PL} (T) and E_a are the relative PL intensity at 77 K and T (K) and the activation energy respectively. Thermal quenching of PL intensity describes nonradiative processes and is well described by Arrhenius behavior with an activation energy of 70 meV.

Hence, when particles are extracted immediately after coating (procedure I), the PL spectra of CdTe differing by their sizes are characterized by shallow traps due to surface defects. If the particles are kept in micellar solution during 48 h before extraction (procedure II), excitonic PL spectra are observed. The shift of the photoluminescence with increasing the particle size is due to a quantum size effect. The appearance of the direct luminescence at high energy cannot be attributed to a quantum size effect. In fact, particles having same average size show different luminescence (Fig. 3.4.5). This can be attributed to formation of a passivation layer, which decreases the surface defects and favors the direct free electron–hole transition.

REFERENCES TO SECTION 3.4

1. L Addadi, S Weiner. Angew Chem Int Ed Engl 31:153, 1992.
2. S Mann. In: DW Bruce, D O'Hare, eds. Inorganic Materials. New York: John Wiley & Sons, 1992.
3. MP Pileni. J Phys Chem 97:6961, 1993.
4. JH Fendler, F Meldrum. Adv Mater 7:607, 1995.
5. JH Fendler. Chem Mater 8:1616, 1996.
6. C Tanford. The Hydrophobic Effect. New York: Wiley, 1973.
7. SJ Chen, DF Evans, BW Ninham, DJ Mitchell, FD Blum, SJ Pickup. Phys Chem 90: 842, 1986.
8. DF Evans, DJ Mitchell, BW Ninham. J Phys Chem 90:2817, 1986.
9. IS Barnes, ST Hyde, BW Ninham, PJ Derian, M Drifford, TN Zemb. J Phys Chem 92: 2286, 1988.
10. G Porte, J Appell, Y Poggi. J Phys Chem 84:3105, 1980.
11. DJ Mitchell, BW Ninham. J Chem Soc Faraday Trans 2:77, 601, 1981.

12. SA Safran, LA Turkevich, PA Pincus. J Phys Lett 45:L69, 1984.
13. O Diat, D Roux, J Phys II:3, 9, 1993.
14. JF Berret, DC Roux, G Porte, P Lindner. Europhy Lett 25:521, 1994.
15. JN Israelachvili, DJ Mitchell, BW Ninham. J Chem Soc Faraday Trans 2:72, 1525, 1976.
16. MP Pileni, ed. Reactivity in Reverse Micelles Amsterdam: Elsevier, 1989.
17. G Schmid, ed. Clusters and Colloids. Weinhem: VCH, 1994.
18. I Lisiecki, F Billoudet, MP Pileni. J Phys Chem 100:4160, 1996.
19. N Moumen, P Veillet, MP Pileni. J Mag Mag Mat 149:42, 1995.
20. N Herron, Y Wang, M Eddy, GD Stucky, DE Cox, K Moller, T Bein. J Amer Chem Soc 111:530, 1989.
21. M Brust, D Walker, D Bethell, DJ Schiffrin, R Whyman. J Chem Soc Chem Commun 801, 1994.
22. CB Murray, DJ Norris, MG Bawendi. J Am Chem Soc 115:8706, 1993.
23. SA Safran, LA Turkevich, PA Pincus. J Phys Lett 45:L69 1984.
24. TF Towey, A Khan-Lodl, BH Robinson. J Chem Soc Faraday Trans. 2 86:3757–3762, 1990.
25. C Robertus, JGH Joosten, YK Levine. J Chem Phys 93(10):7293–7300, 1990.
26. G Cassin, JP Badiali, MP Pileni. J Phys Chem 99:12941–12946, 1995.
27. TK Jain, G Cassin, JP Badiali, MP Pileni. Langmuir 12:2408–2411, 1996.
28. LE Brus. J Chem Phys 79:5566, 1983.
29. R Rossetti, JL Ellison, JM Bigson, LE Brus. J Chem Phys 80:4464, 1984.
30. AJ Nozik, F Williams, MT Nenadocic, T Rajh, OI Micic. J Phys Chem 89:397–399, 1985.
31. MG Bawendi, ML Steigerwald, LE Brus. Annu Rev Phys Chem 41:477, 1990.
32. A Henglein. Chem Rev 89:1861, 1989.
33. Y Wang, N Herron. Phys Rev B 41:6079, 1990.
34. Y Wang, N Herron. J Phys Chem 95:525, 1991.
35. Y Kayanuma. Phys Rev B 38:9797, 1988.
36. PE Lippens, M Lannoo. Phys Rev B 39:10935, 1989.
37. A Henglein. J Chim Phys 84:441, 1987.
38. M Meyer, C Walberg, K Kurchara, JH Fendler. J Chem Soc Commun 90:90, 1984.
39. P Lianos, K Thomas. Chem Phys Lett 125:299, 1986.
40. P Lianos, and K Thomas. J Colloid Interface Sci 117:505, 1987.
41. C Petit, MP Pileni. J Phys Chem 92:2282, 1988.
42. ML Stigerwald, AP Alivisatos, JM Gibson, TD Harris, R Kortan, AJ Muller. J Am Chem Soc 110:3046, 1988.
43. C Petit, P Lixon, MP Pileni. J Phys Chem 94:1598, 1990.
44. BH Robinson, AN Khan-Lodhi, TF Towey. J Chem Soc Faraday Trans 86:3757, 1990.
45. L Motte, C Petit, P Lixon, L Boulanger, MP Pileni. Langmuir 8:1049, 1992.
46. C Karayigitoglu, M Tata, T John-Vivay, GL McPherson. Colloids Surf A, 82:151, 1994.
47. K Suriki, M Harada, A Shioi. J Chem Eng Jpn 109:245, 1996.
48. T Hirai, H Sato, I Komasawa. Ind Eng Chem Res 33:3262, 1994.
49. J Cizeron, MP Pileni. J Phys Chem 99:17410, 1995.
50. J Cizeron, MP Pileni. J Phys Chem 101:8887, 1997.
51. L Levy, JF Hochepied, MP Pileni. J Phys Chem 100:18322, 1996.

52. L Levy, N Feltin, D Ingert, MP Pileni. J Phys Chem 1997.
53. L Katsikas, A Eychmüller, M Giersig, H Weller. Chem Phys Lett 172:201, 1990.
54. HM Schmidt, H Weller. Chem Phys Lett 129:615, 1986.
55. D Barnes, AS Kenyon, EM Zaiser, VK Lamer. J Colloïd Sci 2:349, 1947.
56. CH Fisher, H Weller, C Lume-Pereira, E Janata, A Heinglein. Ber Bunsenges Phys Chem 90:46, 1986.
57. T Dannhauser, M O'Neil, K Johansson, D Whitten, G McLendon. J Phys Chem 90: 6074, 1986.
58. N Chestnoy, TD Harri, R Hull, LE Brus. J Phys Chem 90:3393, 1986.
59. JJ Ramsden, M Graetzel. J Chem Soc Farraday Trans. I 90:919, 1984.
60. R Rossetti, LE Brus. J Phys Chem 86:4470, 1982.
61. L Spanhel, H Weller, AJ Fojtik, A Henglein. Ber Bunsenges Phys Chem 91:88, 1987.
62. PV Kamat, NM Dimitrijevic, RW Fessenden. J Phys Chem 91:396, 1987.
63. PD Fletcher, BH Robinson, J Tabony. J Chem Soc Faraday Trans. I 82:2311, 1986.
64. I Lisiecki, M Borjling, L Motte, B Ninham, MP Pileni. Langmuir, 11:2385, 1995.
65. MP Pileni, P Brochette, B Hickel, BJ Lerebours. Colloïd Int Sciences 98:549, 1984.
66. F Michel, MP Pileni. Langmuir 10:390. 1994.
67. KT Higa, DC Harris. Organometallics 8:1674, 1989.
68. C Petit, P Lixon, MP Pileni. Langmuir 7:2620, 1991.
69. K Zanio. Semiconductors and Semimetals. New York: Academic Press, 1978, vol 13.
70. CRM Oliveira, AM Paula, FO Plentz Filho, JA Medeiros Neto, LC Barbosa, OL Alves, EA Menezes, JMM Rios, HL Fragnito, CH Brito Cruz, CL Cesar. Appl Phys Lett 66(4): 439, 1995.
71. AM Paula, LC Barbosa, CHB Cruz, OL Alves, JA Sanjurjo, CL Cesar. Appl Phys Lett 69(3):357, 1996.
72. BO Dabbousi, J Rodriguez-Viejo, FV Mikulec, JR Heine, H Mattoussi, R Ober, KF Jensen, MG Bawendi. J Phys Chem B 101:9463, 1997.
73. AR Kortan, R Hull, RL Opila, MG Bawendi, ML Steigerwald, PJ Carroll, LE Brus. J Am Chem Soc 112:1327, 1990.
74. H Weller. Angew Chem Int Ed Engl 32:41, 1993.
75. YP Varshni. Physica 34:149, 1967.
76. CC Kim, M Daraselia, JW Garland, S Sinavananthan. Phys Rev B 56(8):4786, 1997.

3.5 PREPARATION OF METAL CHALCOGENIDES IN LB FILMS

DAVID J. ELLIOT
CSIRO, Molecular Science, South Clayton, Victoria, Australia

D. NEIL FURLONG
Royal Melbourne Institute of Technology, Melbourne, Australia

KAREN GRIEVE and FRANZ GRIESER
University of Melbourne, Parkville, Victoria, Australia

3.5.1 Introduction

In recent years there has been massive interest in the properties of metal chalcogenide (MC) semiconductors, owing, in part, to their potential and realized applications in optical, optoelectronic, electronic, photoelectronic, and photocatalytic devices.* Some of the devices include photovoltaic cells, infrared detectors, optical switches, and frequency doublers. This interest has intensified with the realization that the properties of semiconductors are tunable via the size-quantized effect (1–4), which can be summarized as follows. When a semiconductor is irradiated at the appropriate wavelength, an electron/hole pair, otherwise known as the Wannier exciton, is created. The exciton is delocalized over a dimension known as the Bohr radius. When the dimensions of the semiconductor approach that of the Bohr radius, the energy of the exciton increases due to its spatial confinement. This is manifested in a blue shift in the optical absorbance spectra of the semiconductor. Numerous methods for quantifying optical absorbance characteristics of MCs for the purpose of identifying Q-state effects have been used in the literature. The onset of absorbance and the band gap (E_g) are two such parameters that require a linear extrapolation of the absorbance curve to the energy (wavelength) axis of the plot. A less arbitrary method involves determining the minima in either the first or second derivative of the absorbance/energy plots in the region of the onset of absorbance of the semiconductor. In Figure 3.5.1 (5) the sizes of CdS particles, observed by electron microscopy (EM) or x-ray diffractometry (XRD) studies, have been plotted as a function of the minima of the first derivatives. The energy of absorbance increases above that of the bulk material (2.4 eV for CdS) (the wavelength decreases) as the size of the particle decreases below the Bohr radius (~6.0 nm for CdS). A curve has been fitted to the data using a theoretical relationship between particle diameter and wavelength

* For the purposes of this review, the term *metal chalcogenide* refers to combinations of divalent metal ions with dianionic S, Se, and Te members of the chalcogenide family.

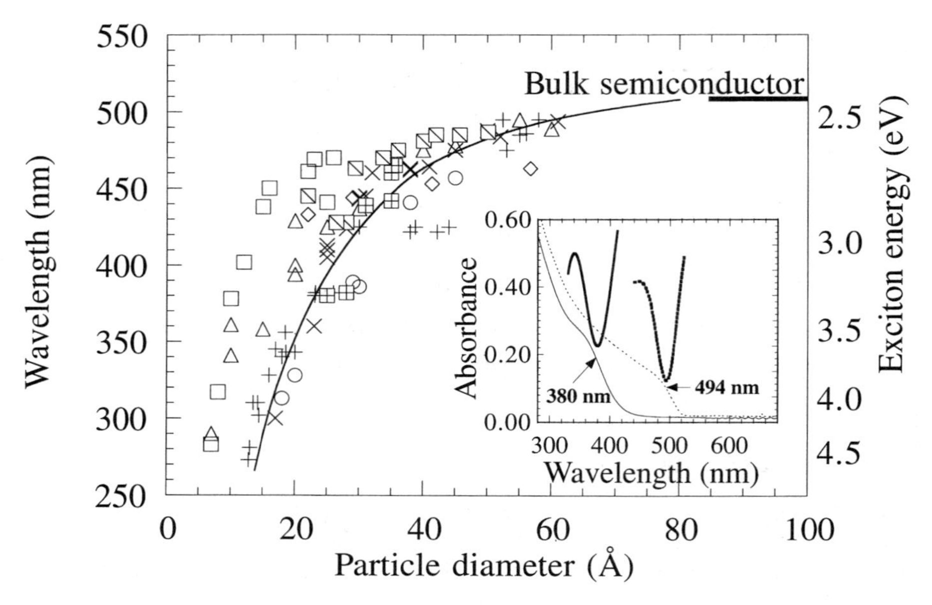

Fig. 3.5.1 The minimum of the first derivative of UV/visible absorbance spectra of CdS particles as a function of the particle diameter. The data points have been collected from literature sources where particle sizes were determined by EM or XRD. If specific data about the minimum of the first derivative were not expressly provided, they were estimated from spectra supplied. The estimated error in this technique is less than ±5 nm. For clarity, only data from groups with the greatest number of data points have been used here. A curve has been fitted to the data using a theoretical relationship between particle diameter and wavelength using the effective mass model (6). Inset: Absorbance spectra of colloidal CdS produced by exposure of a CdAr film, or a Cd^{2+}/HMP solution, to H_2S. The minima of the first derivative (380 nm in film, 494 nm in solution) correspond to particle sizes of approximately 2.5 nm and 6.0 nm, respectively. (From Ref. 5.)

derived by Sphanel et al. (6) using the effective mass model and assuming spherical symmetry of CdS. In the inset of Figure 3.5.1 the minima of the first derivative of absorbance spectra of colloidal CdS in a Langmuir–Blodgett (LB) film and in solution are 380 nm and 494 nm, respectively. The particles in solution are just on the edge below which Q-state effects are observed (510 nm), whereas the particles in the films are of a size well within the Q-state regime.

The basic recipe for producing Q-state metal chalcogenide semiconductors involves the exposure of metal ions to a source of the chalcogenide in the presence of some stabilizing medium (7 and ref. therein). This is depicted in Eq. (1) for the specific case of divalent metal ions reacting with the sulfide ion:

$$M^{2+} + S^{2-} + \text{stabilizing medium} \rightarrow \text{Q-state MS} \quad (1)$$

Stabilizing media have included surfactants, polymers, polyelectrolytes, chemical capping agents such as thiols or amines, glasses, zeolites, and microporous metal oxide films. Surfactants as a stabilizing system have been in the form of LB films, self-assembled films, Langmuir monolayers, inverse micelles, and vesicles. With all these media, size confinement is thought to occur through one or both of two mechanisms. With matrices like microporous metal oxide films or glasses the semiconductor particles are physically confined or restricted in their movement and growth due to the physical dimensions the media provide for particle growth. With surfactants it is primarily the chemical interaction between the head group of the surfactant and the surface of the semiconductor particle that is thought to restrict growth. This ''chemical capping'' is quite effective, for example, when thiols are used to arrest the growth of species such as CdS. LB films containing metal ions provide an environment whereby conceivably both mechanisms operate to restrict particle growth. In addition to the interaction of the head group with the semiconductor surface, the planes defined by the head groups in an LB bilayer could physically restrict the movement of growing semiconductor particles or the diffusion of molecules to growing particles. In addition to semiconductor materials, metals have also been produced in LB films (8–12), and the same mechanisms for restriction of growth are likely to apply.

LB films offer many advantages over other media for the investigation of Q-state particle formation and for potential device application. The chemistry of metal-ion incorporation into LB films is well characterized. Also, film thickness can be regulated to a molecular level, facilitating quantitative analysis of the films and the materials produced in the films. Also, optically clear films can be made on a variety of substrates, allowing for a variety of analytical techniques and device application.

3.5.2 Film and MC Preparation

The techniques for construction of LB films containing Q-state metal chalcogenide particles can be divided into two broad categories, distinguishable by at which stage in the film construction the metal chalcogenide is produced.

3.5.2.1 MC Produced Before Film Construction

Three versions of this concept have been used to prepare LB film containing MCs. In the first, pioneered by Fendler (13–16), a compressed Langmuir monolayer is used as a template for the preparation of Q-state particles. For instance, diffusion of H_2S through a monolayer of dodecylbenzenesulfonic acid (17) or a bis amine type surfactant (14) into a Cd^{2+} solution resulted in the growth of Q-state CdS under the monolayer. The vertical dipping technique (Fig. 3.5.2) was then used to transfer the monolayer, with associated semiconductor, to a substrate to give multilayer LB films.

In the second approach, aqueous solutions of Q-state CdS are produced by exposure of Cd^{2+} ions to H_2S in the presence of a stabilizing agent such as sodium

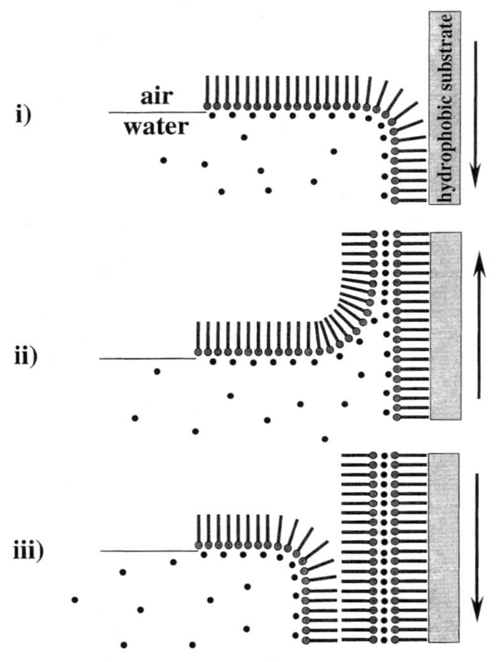

Fig. 3.5.2 Schematic representation of the vertical dipping method of preparing LB films from compressed monolayers of fatty acids on aqueous subphases of divalent metal ions.

hexametaphosphate. This preparation is then used as a subphase under a compressed monolayer of dioctadecyldimethylammonium chloride or bromide (DDAC and DDAB) for the preparation of LB films via the vertical dipping technique. Electrostatic forces between the negatively charged, polyphosphate-stabilized CdS and the positively charged ammonium headgroup of the DDAC are responsible for the incorporation of the CdS into the films.

Finally, a colloidal MC can be complexed with an amphiphile and the whole complex can then be used for a spreading solution to create a Langmuir monolayer. For example, LB films have been prepared from $CHCl_3$ dispersions of trioctylphosphine oxide-stabilized CdSe (18).

3.5.2.2 MC Produced After Film Construction

The general case involves spreading of an amphiphile onto a subphase containing a metal ion. The metal ion is complexed to the surfactant at the air/water interface and the monolayer is then transferred to a substrate to build up a multilayer film using the vertical dipping technique or some variation. Subsequent exposure of the film to a dihydrogen chalcogenide (H_2X; X = S, Se, Te) results in the formation of the metal chalcogenide in the film. In a variation, LB films composed of the amphiphile without any metal ions are first constructed using the vertical dipping technique with a purely aqueous subphase. This film is then immersed into a solution containing the desired metal ion. Diffusion of metal ions into the resulting film and subsequent ion exchange then provide metal-ion-containing LB films. This process has been referred to as *intercalation.* Examples include fatty acid films exposed to Cu^{2+} (19,20), Cd^{2+} (21), and Hg^{2+} (21–23) solutions.

The use of a fatty acid monolayer on a subphase of divalent metal ions compromises the archetypal preparation of LB films containing metal ions. Complexation of subphase metal ions to the fatty acid monolayer occurs through fatty acid proton/metal ion exchange and is pH dependent. For Cd^{2+} ions, at 2×10^{-4} *M,* for example, the onset of the exchange is at approximately pH 5 and stoichiometric exchange occurs at pH 6. This stoichiometric reaction is depicted for the general case for divalent metal ions and fatty acids in Eq. (2):

$$MX_2 + HOOC(CH_2)_nCH_3 \rightarrow M^{2+}\{^-OOC(CH_2)_nCH_3\}_2 + 2HX \quad (2)$$

A range of divalent metal ions including Pb from the main block elements, Cd, Co, Cu, Hg, Mn, Ni, Ru, Ni, Pd, Pt, and Zn from the transition metal series, and Ca and Ba from the group IIa metals have been incorporated into LB films via this sort of exchange (7 and ref. therein). From this point on, such fatty acid/M^{2+} type films will be designated as M^{2+}-FA films (metal–fatty acid films) for the general case or by the metal ion followed by an abbreviation for the deprotonated fatty acid used. For example, PbSt, CdAr, and ZnBe represent Pb stearate [lead octadecanoate or $Pb^{2+}\{^-OOC(CH_2)_{16}CH_3\}_2$], Cd arachidate [cadmium eicosanoate or $Cd^{2+}\{^-OOC(CH_2)_{18}CH_3\}_2$], and Zn behenate [zinc docosanoate, or $Zn^{2+}\{^-OOC(CH_2)_{20}CH_3\}_2$], respectively.

The technique of vertical dipping for construction of multilayer LB films containing metal ions is depicted in Figure 3.5.2. Centrosymmetric bilayers, with metal ions sandwiched between the polar head groups in the film, are deposited onto vertically mounted substrates passed down and then up through the metal/surfactant monolayer at the air/water interface. The structure of the film is represented by the bis chelate coordination of the carboxylate function of the fatty acid around a divalent metal ion. This type of structure has been determined for a number of metal ions including Cd, Co, Mn, Pb and Zn (24 and ref. within). The structure of CoSt (25), inferred from infrared (IR) and XRD studies, is depicted in Figure 3.5.3a (25). Centrosymmetric films, where the area of the film transferred to the substrate

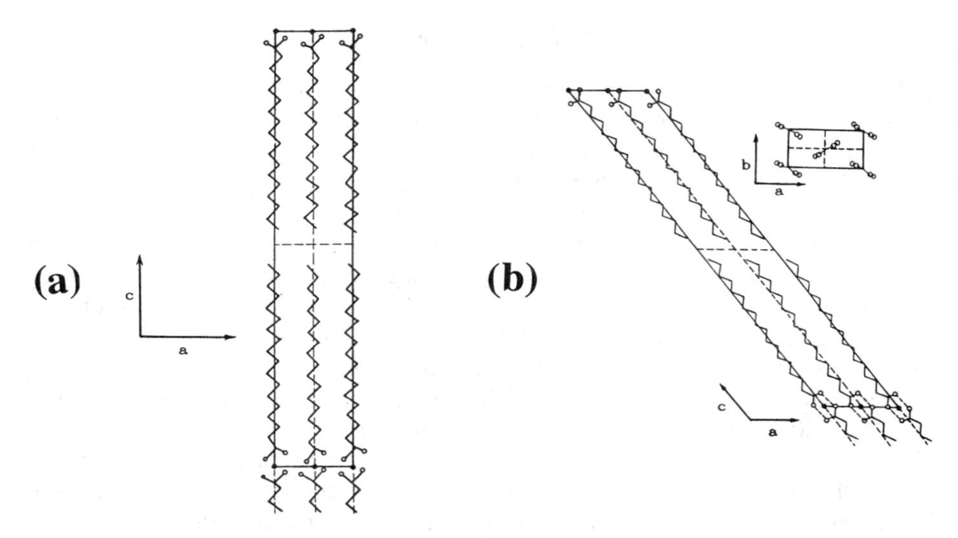

Fig. 3.5.3 Local microstructure of (a) CoSt LB film and (b) CoS-StH film (CoSt exposed to H_2S). Closed circles and open circles designate cobalt and oxygen atoms, respectively. (From Ref. 25.)

matches the substrate area for both up and down portions of the dip cycle, are referred to as *Y type* films.

Variations on the vertical dipping technique have been utilized to construct films containing divalent metal ions. For example, the quartz crystal microbalance (QCM) has been used to evaluate the horizontal lifting method of CdSt LB film construction (26). In this method, the QCM quartz plate was touched to monolayers compressed on a subphase and lifted horizontally. Y-type transfer (transfer ratio of 1) was demonstrated with two centrosymmetric monolayers deposited for each cycle. A combination of the vertical and horizontal dipping techniques has been utilized to prepare multilayer films from an amphiphilic porphyrin compound (27).

Owing to the large variety of surfactants, metal ions, and complex metal ions that have been incorporated into LB films, variations of the stoichiometry given in Eq. (2) are plentiful. Some of these are outlined in the following section. With fatty acid films, metal ions and complex metal ions containing something other than a divalent charge include Ag^+, Fe^{3+}, Ti(IV) from the transition metal series, U(III) from the actinide series, and M^{3+} from the lanthanide series (7).

3.5.2.3 Surfactants Used to Construct LB Films Containing MCs

Surfactants used to make films containing MCs can be categorized as neutral, cationic, and anionic.

Anionic. The anionic category can generally be described by an acid function attached to a hydrophobic tail. Variations of the straight-chain fatty acid type surfactant include diynoic moieties in the hydrocarbon tail (28–33), fluorination of the hydrocarbon chain (34), and introduction of an azobenzene group into the hydrocarbon chain (34). Surfactants containing other acid functions in the head group include dodecylbenzenesulfonic acid (17), dihexadecylhydrogen phosphate (34), and amphiphiles containing the *o, m,* and *p* phthalate moiety (28,32). Amphiphiles made from esterification of poly(maleic acid) with octadecanol (PMAO) have been used to prepare LB films containing Pb^{2+}, and subsequent exposure to H_2S resulted in formation of Q-state PbS (35,36). Other amphiphilic polymers containing a carboxylate function on a side chain have been used for the preparation of Cd^{2+}-containing LB films and subsequent conversion to Q-state CdS particles (37).

Cationic. Positively charged amphiphiles used for LB film construction contain a secondary, tertiary, or quaternary nitrogen in the head group. The fabrication of metal-containing LB films with the cationic surfactants is generally restricted to complex metal anions. Dimethydioctadecylammonium bromide (DDAB) has been used, for example, to complex a number of chlorometallates of platinum and palladium (10). The stoichiometries for complete exchange of metal ions or complex metal ions for surfactant-associated ions for DDAB are given in Eq. (3):

$$2(DDA^+Br^-) + (MCl_n)^{2-} \rightarrow DDA_2(MCl_n) + 2\ Br^- \tag{3}$$

As DDAB contains a quaternary N in the head group, the cationic charge is independent of pH. The stoichiometry in Eq. (3) for the $[MCl_n]^{2-}$ ($n = 4$, M = Pt, Pd; $n = 6$, M = Pt) complex ions has been established by XPS and optical absorbance measurements. The $(PtCl_6)^{2-}$ anion has also been incorporated into LB films using a tertiary amine surfactant (38). In this case the charge of the head group, and therefore the complexation to the anion, is pH dependent.

Neutral. A bis(ethylenediamine) structure has been incorporated into the surfactant molecule n-$C_{16}H_{33}C(H)[CON(H)(CH_2)_2NH_2]_2$ in order to incorporate metal ions in an LB film structure via coordination instead of ionic complexation as occurs for anionic/cationic amphiphiles (14). Also, films of *n*-octadecylacetoacetate containing Cu^{2+} have been prepared, and exposure to H_2S has resulted in the formation of a copper sulfide (39). Ditetradecyl-*N*-[4-[[6-(*N,N′,N′*-trimethyl-ethylenediamino)-hexyl]oxy]benzoyl]-L-glutamate (DTG), which also contains the ethylenediamine unit, was used to make self-assembled films containing Cd^{2+} (40).

3.5.2.4 Self-Assembled Films

Although not strictly LB films, there are other types of self-assembled films containing Q-state MCs that resemble LB films. One example involves the self-assembly of the amphiphile DTG into an organized film by slow evaporation of solvent from a dispersion of the amphiphile (40). The structure of the ''cast'' film has the head-to-head centrosymmetric structure associated with Y type LB films. Immersion into a Cd^{2+} solution resulted in the intercalation of Cd^{2+} ions between the head groups of the amphiphile in the bilayer film, and subsequent exposure to H_2S yielded Q-state CdS particles.

The technique of alternating polyelectrolyte film construction has also been adapted to incorporate semiconductors into layered films. For example, multilayer films have been constructed by alternately dipping a quartz substrate into a solution of poly(diallylmethylammonium chloride) and then a solution of a stabilized CdS or PbS colloid (41). The layer-by-layer self-assembly of alternating polymer and metal sulfide is at least partially driven by the electrostatic attraction of the cationic polymer and the negative charge of the stabilized MC colloid particles.

3.5.2.5 Formation of Metal Chalcogenide in LB Films

As mentioned earlier, there have been two strategies for producing Q-state semiconductors in LB films. The preparation of Q-state particles, which can then be incorporated into LB films, has been reviewed elsewhere (1,4) and is not discussed here in any detail. The reactions of M^{2+} containing LB films with dihydrogen chalcogenides H_2X ($X = S, Se, Te$) is generally considered the primary method of MC fabrication in LB films and the reaction has been studied extensively for a large range of metal ions. The reaction is frequently represented as in Eq. (4):

$$M^{2+}\{^{-}OOC(CH_2)_nCH_3\}_2 + H_2X(g) \rightarrow MX + 2HOOC(CH_2)_nCH_3 \qquad (X = S, Se, Te) \tag{4}$$

Along with the MC, the fatty acid is regenerated upon exposure to the dihydrogen-chalcogenide. Much quantitative and qualitative experimental data have been reported that supports the stoichiometry depicted in Eq. (4). As discussed in a later section, however, there is also considerable evidence that the reaction does not proceed to completion as depicted in Eq. (4).

The reactions with H_2S of complex metal anions incorporated into LB films made from DDAB have also been studied (10). The stoichiometry of the reaction can be represented as given in Eq. (5):

$$DDA_2(MCl_n) + [(n - 2)/2]H_2S \rightarrow MS_{(n-2)/2} + 2(DDA^+Cl^-) + (n - 2)\,HCl \tag{5}$$

(where $M = Pt$, $n = 4$ or 6; $M = Pd$, $n = 4$). While reaction to form some sort of metal sulfide is evident from XPS and absorbance data, the stoichiometry is difficult to establish using XPS due to the incorporation of excess sulfur in the film.

It is well established, mainly from UV/visible absorbance spectra, that the semiconductors formed in LB films from the reactions in Eq. (4) and (5) are in the Q-state regime. The optical properties of the generated MC and mechanistic and kinetic aspects of the reactions are discussed in a later section.

3.5.2.6 Further Intercalation/Sulfidation

Immersion of fatty acid films into solutions of metal ions, as already indicated, results in an intercalation of the metal ions into the planes formed by the carboxylate head groups of the fatty acids. This can be accomplished in M-FA films where the FA has been regenerated by exposure to H_2S [Eq. (4)] or in FA films deposited without any metal ions. Subsequent exposure of the films to H_2S has been shown to result in the formation of the metal sulfide. This intercalation/sulfidation (i/s) cycle can be repeated several times to increase the concentration of the metal sulfide in the film. This process has been investigated for CdS (34,39,42,43), PbS (39,43,44), ZnS (39,43), and HgS (45) produced in M-FA films, using Fourier-transform infrared (FTIR) and UV/visible spectroscopies, QCM gravimetry, and atomic force microscopy (AFM).

From UV/visible spectroscopy it is evident that there is a clear trend of redshifting the absorbance spectra, and hence an increase in particle size, of the metal sulfides produced from i/s cycling. While there is consensus of a progressive increase in the absorbance of an MS with repeated i/s cycling, the absorbance increase is not always reported to be linear with the number of cycles. In Figure 3.5.4 (5) for CdS/ArH films over five i/s cycles, both the trends of increasing and redshifting of the absorbance are demonstrated. The intercalation/sulfidation of Pb^{2+} ions into a CdS/StH film (5 Cd^{2+}/H_2S i/s cycles) was followed by UV/visible spectroscopy (39). The growth of PbS apparently proceeded without changing the optical properties of the CdS preexisting in the film.

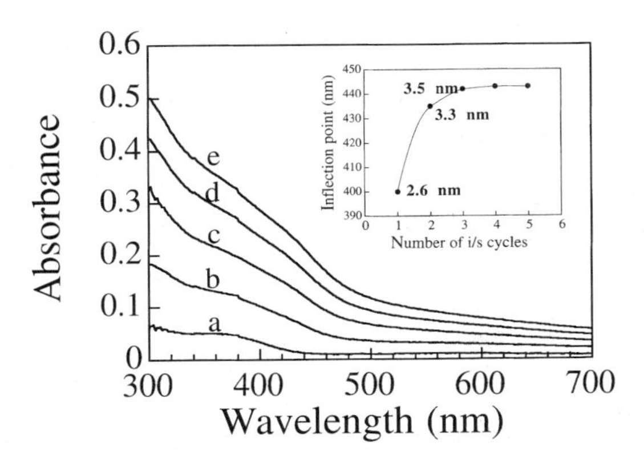

Fig. 3.5.4 The UV/visible absorption spectrum of CdS formed in a 19-layer ArH LB film after 1–5 (a–e) intercalation cycles. The increase in the total absorption, the red shift in the inflection point (minimum in the first derivative), and the increase in scatter can be seen. Inset: Inflection point wavelength as a function of i/s cycles. The average size of the particles in nm, as estimated from UV/visible absorption spectra, are shown. (From Ref. 5.)

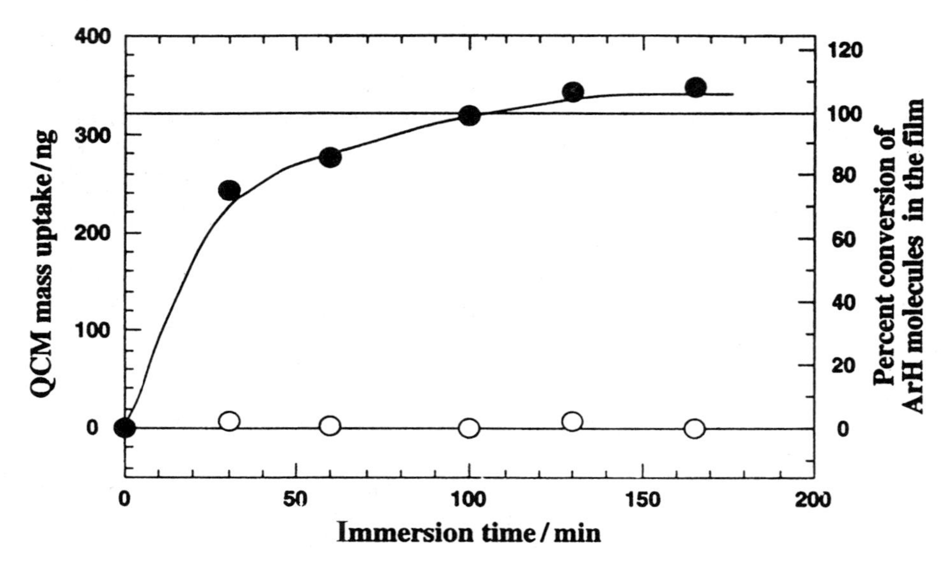

Fig. 3.5.5 QCM mass uptake versus immersion time of 20-layer CdAr films in a stirred solution of 4×10^{-4} *M* $KHCO_3$. Closed circles represent a film that has been exposed to H_2S with the excess H_2S removed by vacuum. The open circles are for a CdAr film not exposed to H_2S. The right-hand ordinate axis shows the percentage conversion of ArH molecules to CdAr for the extensively gassed film calculated using Eq. (2). It has been assumed that the HCl produced in Eq. (2) has not been retained in the film. (From Ref. 46.)

The kinetics of intercalation have been followed by FTIR and QCM gravimetry. For example, the frequency change of a QCM with a 20-layer CdAr film (exposed to H_2S) was followed as a function of immersion time in a $CdCl_2$ solution (Fig. 3.5.5) (46). The percent conversion [calculated from mass of film on QCM and stoichiometry in Eq. (3)] reaches 100% within about 100 min immersion. In contrast, time-resolved FTIR spectra of CdBe/H_2S and HgBe/H_2S films immersed into Cd^{2+} and Hg^{2+} solutions, respectively, indicated that the intercalation of metal ions into the LB film of BeH takes 24–48 h to complete (21). The kinetics of the sulfidation reaction [Eq. (4)] are discussed in a later section.

FTIR has also been used to study changes in H_2S-exposed LB films during further intercalation/sulfidation cycles (39,43,47,48). Diffusion of M^{2+} ions into FA films results in the complete disappearance of the $\nu(C{=}O)$ peak (associated with the -COOH group of the fatty acid) with concurrent appearance of the $\nu_s(CO_2^-)$ and $\nu_a(CO_2^-)$ stretches.

3.5.2.7 Metal Chalcogenides Formed in LB Films

It is evident from the chemical literature that the most thoroughly studied system has been CdS in fatty acid films (21,23,29–31,33,34,42,43,46–67). This reflects the well-established optical and photoelectrochemical properties of the material for reference, the high quality of LB films available using Cd^{2+} ions and fatty acids, and the useful properties of CdS for applications in optical and photoelectronic devices and photocatalysis. Other metal sulfides studied include CoS (25), Cu_xS_y (5,9,20,39), HgS (21–23,45,64), PbS (36,39,44,54,56,68–73), PdS (10,74), PtS_n (10,74), and ZnS (39,53). The use of other chalcogenides has been studied in combination only with Cd^{2+}. Examples include CdSe (18,30,31,60,75), CdTe (30,31), and the mixed chalcogenides, CdS_xSe_{1-x} (30,31,60), CdS_xTe_{1-x} (30), and $CdSe_xTe_{1-x}$ (30).

3.5.3 Techniques for Investigating MCs in LB Films

3.5.3.1 UV/Visible Spectroscopy

UV/visible spectroscopy is undoubtedly the most important and widely used technique for identifying and quantifying Q-state effects for MCs in LB films. Correlations between particle size and features in the absorption spectrum have been used frequently. Examples of such a curves used for CdS are shown in Figure 3.5.1. UV/visible spectroscopy has indicated Q-state effects for other MCs in LB films, including CdSe (18,30,31,75), CdTe (30,31), CuS_x (9), HgS (23,45), PbS (35,36,39,44,69–71), MS (M = Pt, Pd) and PtS_2 (10), and ZnS (39,53). The UV/visible spectra of CdX (X = S, Se, Te) made by exposure of Cd^{2+}/diynoic FA films to H_2X are given in Figure 3.5.6 (31). The absorption onsets of the Cd

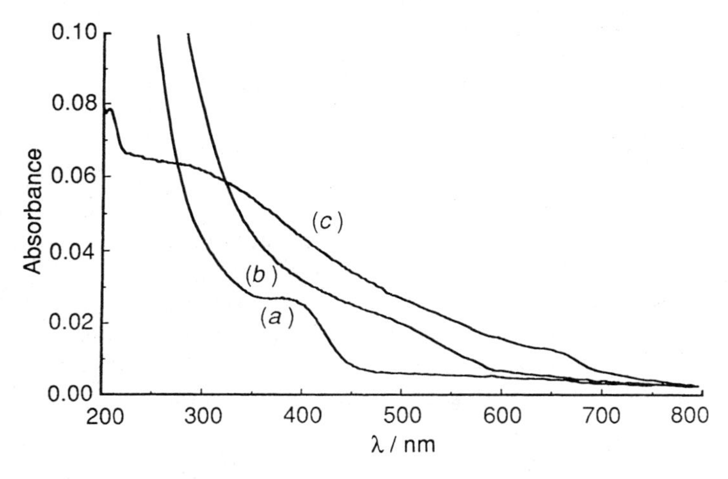

Fig. 3.5.6 Absorbance spectra of 190-layer cadmium-10, 12-diynoate LB films after exposure to (a) hydrogen sulfide, (b) hydrogen selenide, or (c) hydrogen telluride. The spectra have been corrected for background (quartz and film). (From Ref. 31.)

chalcogenides in Figure 3.5.6 are blueshifted relative to that of the bulk materials (CdS/450 nm, CdSe/600 nm, CdTe/740 nm), clearly demonstrating the Q-state effect. In addition to CdS, a size/optical absorbance correlation has been established for CdSe, PbS, and ZnS, and particle sizes have been assigned from absorbance spectra. The bulk band gap of CdSe is 1.67 eV or 743 nm, and the spectra in Figure 3.5.6 obviously indicate a size-quantization effect. In fact, from CdSe particle size/exciton energy correlations similar to that in Figure 3.5.1 for CdS, the CdSe particles were determined to be between 1.1 and 1.3 nm in diameter (75). For the films containing M(II) (M = Pt, Pd, or Cu) and Pt(IV), exposure to H_2S resulted in an orange-brown color in the film. As all these bulk metal-sulfide species are narrow-band-gap semiconductors and are generally black, this orange-brown color constitutes evidence for a blueshifting of the optical properties characteristic of Q-state particle formation. However, systematic studies of the correlation between their optical properties and particle size do not exist and particle size cannot be assigned from optical spectra. UV/visible spectroscopy has provided evidence for the formation of mixed chalcogenide species by exposure of CdX (X = S, Se) species to dihydrogen chalcogenides H_2X (X = S, Se, Te) (30,60). Thus the species $CdS_xSe_{(1-x)}$, $CdS_xTe_{(1-x)}$, and $CdSe_xTe_{(1-x)}$ were formed by exchange of one chalcogenide for another. In the case of $CdS_xSe_{(1-x)}$ a CdS core/CdSe shell structure was rationalized based on a surface passivation reaching saturation at approximately one or two monolayers of CdSe on the surface (60). It was also concluded that the optical properties of the mixed chalcogenide were determined by the CdSe shell. The redshift in the absorbance of CdS, concurrent with H_2Se exposure, was expected given the lower E_g of CdSe relative to CdS.

UV/visible spectroscopy has been used to quantify characteristics of LB films such as metal ion complexation, and intercalation, and MC formation. For example, the absorbance spectra of films made from the surfactant DDAB on subphases of complex metal ions $(MCl_n)^{2-}$ (M = Pt, n = 4 or 6; M = Pd, n = 4) contained the two ligand-to-metal charge transfer (LMCT) bands characteristic of the $(MCl_n)^{2-}$ ions (10). The stoichiometry in Eq. (3) was also confirmed through comparison of the extinction coefficients of the $(MCl_n)^{2-}$ ions in the films and in solution. For LB films prepared by the vertical dipping method, linear plots of MS absorbance, at a fixed wavelength, as a function of LB layer number have been obtained for CdS (17,43,63), CdSe (30), and HgS (45), confirming a uniform transfer of bilayers over repeated dip cycles. This has also been seen for CdS in self-assembled films made by alternately immersing a substrate into a solution of a cationically charged polymer and a CdS solution (76). Increases in absorbances of CdS (34,39,42,43), PbS (39,43,44), ZnS (39,43), and HgS (45) (Fig. 3.5.7) have also been obtained for M-FA films as a function of the number of i/s cycles. Progressive redshifting of the spectra of the metal sulfides was also observed with increased i/s cycles, indicative of particle growth (Fig. 3.5.4). However, the assignment of particle size from optical spectra of films subjected to repeated i/s cycling is probably tenuous, as an increase in absorbance due to scatter is apparent in the spectra (Fig. 3.5.4).

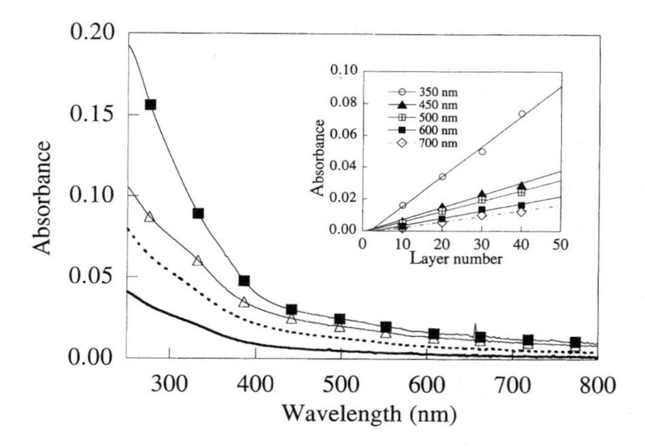

Fig. 3.5.7 Absorbance (corrected for ArH and the quartz substrate) of 10-layer (solid line), 20-layer (dashed line), 30-layer (triangles) and 40-layer (squares) HgAr films exposed to H_2S for 15 min. Inset: Absorbance of HgS in ArH (corrected for ArH and quartz) as a function of layer number at 350, 400, 500, 600, and 700 nm. (From Ref. 45.)

Cd^{2+}-FA films have been prepared with a surfactant containing a polymerizable diacetylenic function in the hydrocarbon tail (30,31,33). Films, polymerized using UV light, were blue in color. Exposure of these films to H_2X resulted in the formation of Q-state MC particles in the film concurrent with a change in film color to red. The color change has been attributed to some conformational change in the polymer structure brought about by the formation of the metal sulfide. UV/visible spectroscopy has also been used to investigate the kinetics of the reactions of M^{2+}-FA films with H_2X and to probe potential interactions between CdS and either HgS (64) or PbS (39) made in fatty acid films.

3.5.3.2 FTIR

FTIR has been mainly used to obtain structural details of films and to monitor intercalation of metal ions into the film structure and the subsequent reactions of the films with dihydrogen chalcogenides. Both transmission (FTIR-T) and reflection-absorbance (FTIR-RA) modes have been utilized. For the most part these studies have involved films of fatty acids with divalent metal ions. The key features of the FTIR spectra of these films include the asymmetric and symmetric stretching modes of the carboxylate group: $\nu_s(CO_2^-)$ and $\nu_a(CO_2^-)$, associated with the M^{2+}/carboxylate complex, and the carbonyl stretching mode $\nu(C{=}O)$ of the protonated fatty acid. The disappearance of the $\nu(CO_2^-)$ (1500–1600 cm^{-1}) and appearance of the $\nu(C{=}O)$ bands ($\sim$1700 cm^{-1}), concurrent with the formation of the metal chalcogenide and regeneration of the fatty acid, have been used to evaluate

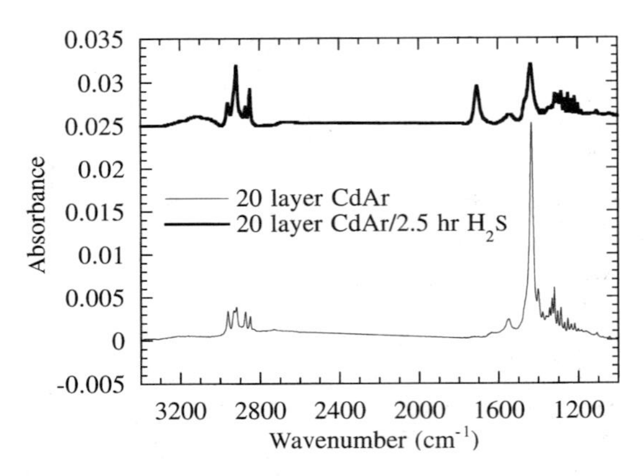

Fig. 3.5.8 Grazing-angle FTIR spectra of a 20-layer CdAr film and an analogously prepared film exposed to H_2S for 2.5 h. (From Ref. 48.)

the extent of the reaction as depicted in Eq. (4) (Fig. 3.5.8) (48). In addition, the exposure of M-FA films to H_2S resulted in the appearance of a broad signal at approximately 3100 cm^{-1}, which has been attributed to the hydroxyl stretching mode $\{\nu(OH)\}$ of the protonated fatty acid.

Structural information of LB films has also been obtained from FTIR studies. In the carboxylate form of the fatty acid, the relative intensities of the $\nu_s(CO_2^-)$ and $\nu_a(CO_2^-)$ signals are dependent on the orientation of the chain axis. The dipole moments of the $\nu_s(CO_2^-)$ and $\nu_a(CO_2^-)$ stretches are parallel to and perpendicular to the chain axis, respectively. In transmission mode the electric vector of the IR radiation interacts strongly with dipole moments parallel to the substrate. This means that in transmission mode the $\nu_s(CO_2^-)$ will be most intense, and the $\nu_a(CO_2^-)$ the weakest, for films with the chain axis perpendicular to the substrate. The opposite is true for the FTIR-RA mode. There is general consensus that in M-FA films the chain axis is approximately perpendicular to the substrate while the protonated form of the acid after exposure to H_2S has a tilt relative to the substrate. Further discussion of FTIR as an investigative tool into the reaction of M^{2+}-FA films with dihydrogen chalcogenides is given in later sections.

Information regarding crystal packing has also been obtained from FTIR spectra. For example, for the CoSt/H_2S system, changes in CH_2 scissoring vibrational band with H_2S exposure suggested a change from hexagonal to orthorhombic packing (25).

FTIR has also been used to confirm hydration of $(PdCl_4)^{2-}$ ions incorporated into LB films made from the cationic surfactant DDAB (10) and the coordination modes of amine ligands of complex metal ions of M(II) (M = Pt, Pd) in LB films of ArH (74). Films of amphiphilic oligomers, made from poly(maleic)acid and an octadecanol ester, with intercalated Pb^{2+} ions have been investigated by FTIR (35).

3.5.3.3 XRD

X-ray studies have been used mainly to investigate film structure of M-FA films, before and after exposure to the dihydrogen chalcogenide, and also for films subjected to repeated i/s cycles. In some cases the morphology of the MC formed in the film is argued on the basis of x-ray evidence, as discussed in a later section.

The terms *basal spacing, d spacing, long spacing of the until cell,* and *bilayer spacing* are commonly cited parameters, obtained from XRD, used to describe the repeating unit in the crystalline structure of the LB film. The distinction between these terms is unclear, however, and it is difficult to have a critical understanding of the literature. For example, in a report of CdAr exposed to H_2S, two types of layered structures with ''bilayer spacings'' of 5.52 nm and 4.4 nm were observed (52). Before exposure to H_2S a bilayer spacing of 5.52 nm was found. Similarly, for CdSt and PbSt sequentially exposed to H_2S the ''*d* spacing'' of 4.90 nm is progressively replaced with a system of reflections giving $d = 3.85$ nm (54). The ''*d* spacing'' of CoSt (5.0 nm) changes to 4.0 nm after H_2S (25). It is known that for M^{2+}-FA films the surfactant chains are essentially perpendicular to the substrate and after reaction with H_2S the surfactants tilt relative to the substrate consistent with generation of the fatty acid. This is depicted in Fig. 3.5.3 (25) for the structure of CoSt, determined by an XRD study, before and after exposure to H_2S. Thus in the preceding examples the parameter cited is measured normal to the substrate and the decrease in this parameter reflects the tilting of the surfactant, concurrent with metal sulfide and fatty acid formation. There are several other studies, however, that cite a significant increase (35,36,39,43,44,58) or at least retention (57,69) of the ''*d* spacing'' upon exposure of M^{2+}-FA films to H_2S. These studies apparently measure the *d* spacing parallel to the surfactant axis.

XRD studies of CdSt before and after H_2S have been conducted for films prepared from StH monolayers compressed at different surface pressures before transfer. The XRD patterns for films deposited at 20 mN m^{-1}, 30 mN m^{-1}, and 37.5 mN m^{-1}, before and after H_2S exposure, are given in Fig. 3.5.9 (66). For the films prepared at the two lowest surface pressures a new phase appears after H_2S exposure and two long spacings, 4.0 nm and 5.0 nm, were determined. For the film deposited at 37.5 mN m^{-1} no new phase was observed after exposure to H_2S. These results were interpreted in terms of retention of the uniform plane of electron-dense Cd atoms from CdS formed in films deposited at the highest pressure, whereas CdS aggregation is implicated in the films deposited at lower pressures. XRD was also used to compare film quality of PbSt films deposited by the vertical dipping technique as a function of dipping speed (70). Pure Y type films were observed for films deposited at the fastest dipping rate (10 cm min^{-1}), while a deterioration of the transfer was observed for lower dipping rates.

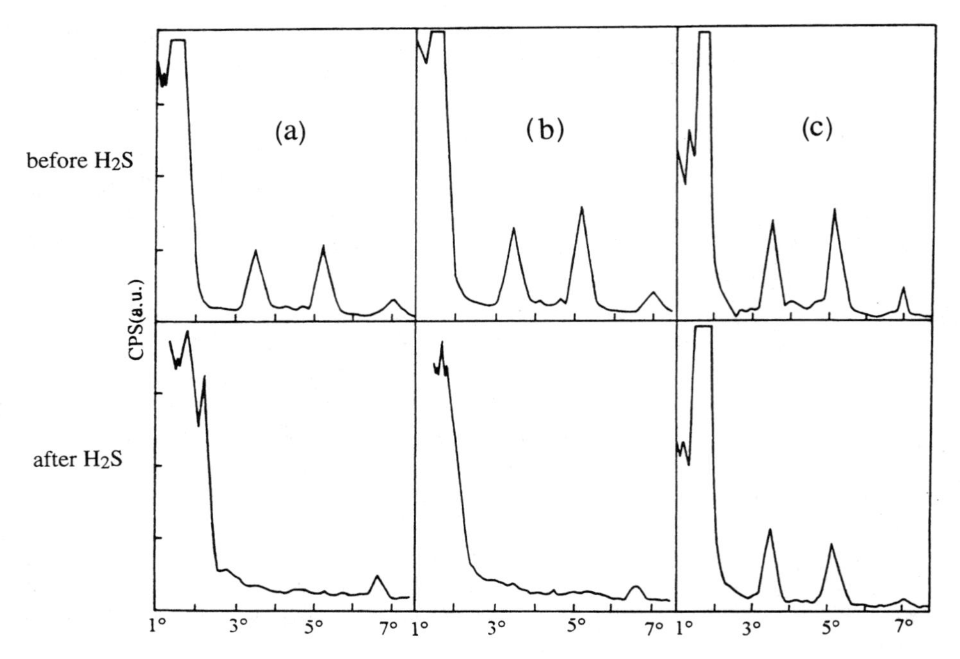

Fig. 3.5.9 X-ray diffraction pattern of 9-layer CdSt LB films deposited on quartz under (a) 25 mN m^{-1}, (b) 30 mN m^{-1}, and (c) 37.5 mN m^{-1}, before and after exposure to H_2S. (From Ref. 66.)

3.5.3.4 Quartz Crystal Microbalance

The quartz crystal microbalance (QCM) is a piezoelectric device consisting of a thin (e.g.) quartz wafer sandwiched between two electrodes. A potential applied across the electrodes results in an oscillation of the quartz. The frequency of the oscillation, which can be measured accurately, is sensitive to mass loading. The relationship between frequency and mass loading is described by the Sauerbray equation:

$$\Delta F = -2F_o^2 \Delta m/A(\rho_q\mu_q)^{0.5} \qquad (6)$$

where ΔF is the frequency change, F_o is the resonant frequency of the crystal, Δm is the mass loading, A is the electrode area, μ_q is the shear modulus of the quartz (2.95×10^{11} dyn cm^{-3}), and ρ_q is the density of quartz (2.65 g cm^{-3}). For a 9-MHz QCM (with an area of 2.05×10^{-5} m^2), a 1-Hz decrease in frequency corresponds to 1.12×10^{-9} g loading. The QCM has been mainly used for investigating film transfer (26,45,58,77,78), intercalation of metal ions into fatty acid films (Fig. 3.5.5)

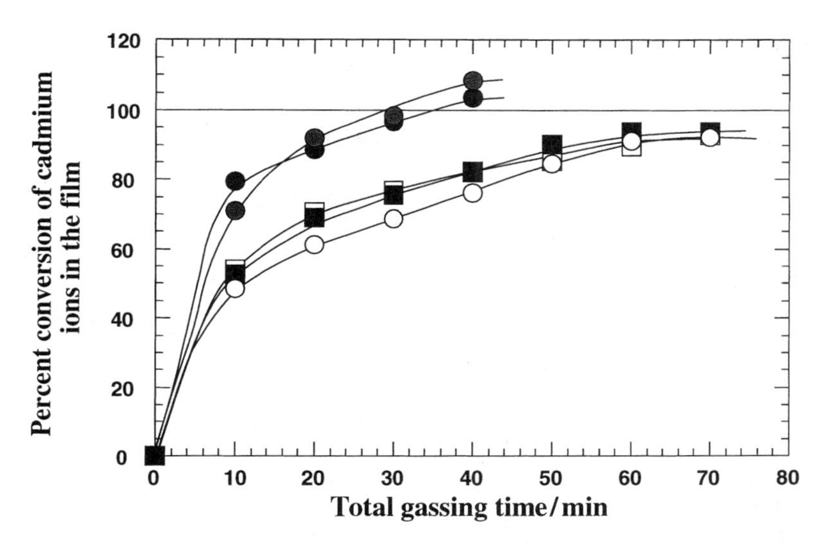

Fig. 3.5.10 Percent conversion of Cd^{2+} ions in the film, assuming stoichiometry in Eq. (4), versus total H_2S exposure time for different thickness of CdAr films. Filled squares, 10-layer CdAr; circles, 20-layer CdAr films; open squares, 40-layer CdAr film. QCMs with LB films were exposed to H_2S in a sealed cell and then removed from the cell and left in air for a period prior to frequency measurements. (From Ref. 46.)

(42,46), and the reactions of the M-FA films with H_2X (Fig. 3.5.10) (42,45,46,58,59,75).

In addition, the QCM has been used measure the binding of Pb^{2+} ions to the carboxylate head group of a Langmuir monolayer of ArH (24), and the thermal stability of CdAr films, before and after H_2S exposure (29).

3.5.3.5 Transmission Electron Microscopy

Transmission electron microscopy (TEM) has been an underutilized yet valuable tool in particle size characterization of MC particles in LB films. Monolayer films of trioctylphosphine oxide-capped CdSe (18), spread as a monolayer on an aqueous subphase, were transferred to a TEM grid. A close-packed hexagonal arrangement of 5.3-nm (σ ~4%) crystallites was found. TEM images were also obtained for HMP-stabilized CdS incorporated in BeH/octadecylamine films (79) and for CdS formed under an amine-based surfactant monolayer and transferred to a TEM grid (14). In one study, direct viewing of CdS and CdSe particles made from Cd^{2+}-FA films on TEM grids was not possible due to poor phase contrast between the particles and the film (30). Diffraction patterns were observed, however, that were consistent with crystalline β-CdS or CdSe. Approximately spherical particles of CdSe could be viewed directly when films were washed with $CHCl_3$ to remove the surfactant. These images are not direct proof of particle shape because solvent washing may affect MC morphology. However, direct observation of spheroidal Au (8) and AgI (11) particles formed in LB films has been reported by TEM. Particles of CdS made from CdAr film on a TEM grid exposed to H_2S have also been directly observed when the film was removed by heating to 70°C under vacuum (10 kPa) for 24 h (46).

3.5.3.6 Atomic Force Microscope

The atomic force microscope (AFM) has been used to investigate LB film quality and other properties and to obtain sizes and distributions of MC produced within LB films. For example, in an image of a three-layer CuAr film on mica, large pits were evident (9). In a cross-section analysis a stepped drop of ~7.5 nm to the substrate was consistent with three monolayers of an M-Ar film (~2.5 nm per layer). The thickness of LB films can also be obtained in good-quality films by excavating down to the substrate by the AFM tip in contact mode (45,80).

The normal procedure for obtaining AFM images of MCs formed in LB films involves removal of the surfactant matrix with a solvent such as $CHCl_3$. That this procedure removes the surfactant while leaving behind the MC has been demonstrated by absorbance spectra of H_2S-exposed CdAr films before and after $CHCl_3$ washing (58,64). An image is presented in Figure 3.5.11 (42). The z dimensions of the particles were determined, by section analysis, to be 2.3 ± 0.7 nm, which is in good agreement with the dimension indicated by exciton energy/particle diameter curve (Fig. 3.5.1) and the optical absorbance spectrum of $CHCl_3$-washed films. An electron micrograph of an H_2S-exposed CdAr film on a TEM grid that was heated to 70°C under vacuum also indicated that the majority of particles were between 2 and 4 nm in diameter (46). Similar particle dimensions were found for PdS (74), CuS_x (9), and HgS (45) on mica using AFM imaging of solvent-washed films. Histograms of CdS particle sizes for CdAr films subjected to repeated i/s cycles indicate a progressive increase in the size of the particles (42), which, at least qualitatively, agrees with the UV/visible spectra of CdAr films subjected to the same treatment (34,42,43).

AFM images were obtained for films constructed, on freshly cleaved mica, from compressed monolayers of DDAB on a subphase of HMP-stabilized CdS (81). Particles, with dimensions of 8 ± 3 nm, were seen to be evenly distributed. The determined area of 58 nm^2/particle coincided well with the area per molecule determined for DDAB from its spreading isotherm, implying 1:1 particle/surfactant stoichiometry. This result is puzzling given that freshly cleaved mica is hydrophilic and therefore any particles would be buried under a layer of the hydrophobic tails of the DDAB and unaccessable to the AFM tip.

AFM images of the CuS_x (9) and PdS (74) particles on mica gave mean particle sizes of 4.1 nm and 4.8 nm, respectively. The images were obtained after washing the H_2S-exposed CuAr films with $CHCl_3$.

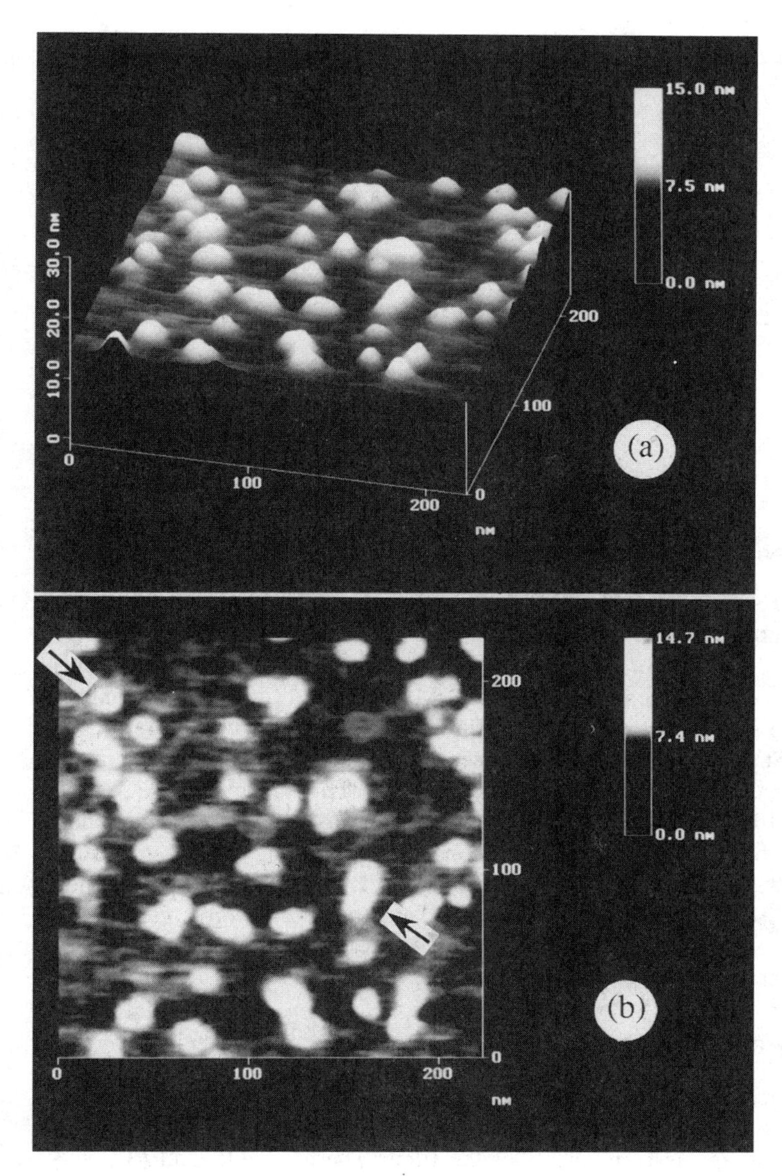

Fig. 3.5.11 AFM contact mode image of CdS particles made in CdAr bilayer film deposited on mica, exposed to H_2S and subsequently immersed in $CHCl_3$ to remove excess surfactant. (a) Surface plot. (b) AFM top view showing spatial distribution and X-Y morphology. Arrows indicate cross-section taken through image. (c) Cross-sectional scan of particles in (a) and (b) showing z height. (From Ref. 42.)

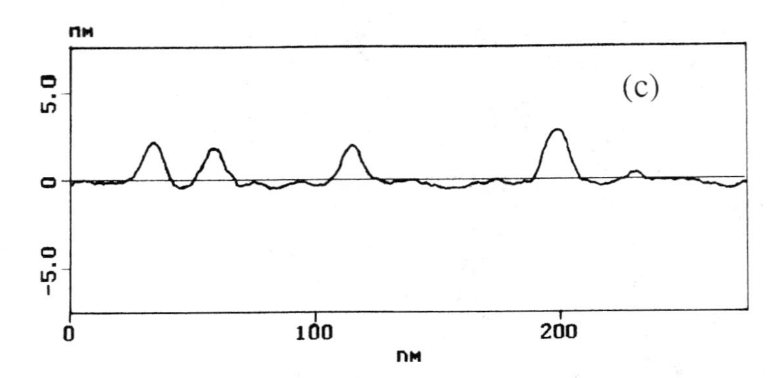

Fig. 3.5.11 Continued

3.5.3.7 Scanning Tunneling Microscope

Scanning tunneling microscope (STM) investigations of MC particles have been performed both for MC produced in LB film from M-FA/H_2X reactions (55,56,58,71,73,81) and for films prepared from stabilized MC particles introduced into LB films (81). An even distribution of HMP-stabilized CdS particles (approximately 8 ± 3 nm diameter) in DDAB films was found by both AFM and STM images (81). String-like attachments between particles, observed in the STM images, were interpreted as weak interparticle interaction. Individual particles of CdS were also identified in CdAr bilayers (55,58) and multilayers (58) on graphite exposed to H_2S, and in one instance atomic resolution was obtained and the hexagonal structure of CdS was observed (58). Curren/voltage (*I/V*) characteristics of individual MS particles, isolated between the scanning tip and the pyrolytic graphite substrate, have been measured by STM (55,56,81). Further discussion of these studies is provided in a later section.

3.5.3.8 X-ray Photoelectron Spectroscopy

X-ray photoelectron spectroscopy (XPS) has been used to obtain stoichiometric information about MC/LB films and also to identify possible oxidation states of, primarily, the inorganic components of the films. These include CdS (37,64,79), CdSe (75), CuS_x (9,19,39), HgS (22,45,64), PbS (39,68,70), and MS_n (M = Pt, n = 2 or 4; M = Pd, n = 2) (10,74). Sulfur:metal and carbon:metal ratios comprise the main stoichiometric information obtained from XPS data. While there are claims to the contrary (39), there is little support from XPS data for a stoichiometric reaction between divalent metal ions in LB films with H_2X as depicted in Eq. (4) (i.e., chalcogenide:metal = 1). In most instances the chalcogenide:metal ratio is found to be less than 1. In some cases the chalcogenide:metal ratio was found to be greater than 1 and oxidation of chalcogenide was postulated.

Carbon:metal ratios have been used to confirm the stoichiometry of metal ion complexation in LB films as depicted in Eq. (2) and (3). Marshbanks et al. (82), however, have concluded from FTIR monitoring of changes in LB films of fatty acids during XPS analysis that the x-ray beam removed surfactant and preferentially removed carboxylate groups from the fatty acid. At least one other study has commented on lower than expected carbon:metal ratios (45). Qualitatively at least, XPS has provided useful information from carbon:metal ratios. For example, the pH dependence of Cd^{2+}/H^{+} exchange is reflected by higher carbon:Cd^{2+} ratios for Cd^{2+}/calixerene films made from subphases at lower pH than that expected for complete exchange (37). XPS was used to obtain Cd:Hg ratios for films prepared from a BeH monolayer on a mixed Cd^{2+}/Hg^{2+} subphase (64).

Regarding the chemical nature of species in LB films, XPS has been used to identify mixed sulfide/polysulfides in nonstoichiometric sulfides of copper (9,19), elemental Se in CdAr films exposed to H_2Se (75), and the nature of nitrogen-containing species in LB films made from ArH on subphases of amine-complexed metal ions of M^{2+} [M = Cu (9), Pt, Pd (74)]. The decomposition of PbS in StH acid films has also been investigated by XPS (68,70). Of two XPS studies of HgS formation in FA films (22,45), one was able to demonstrate that the β-HgS form (*meta*-cinnabar) was present.

XPS has been of use in revealing the oxidation of MC species in LB films. For example with films containing CdS, constructed from a mixed BeH/octadecyl amine monolayer on a subphase of stabilized CdS particles, only one oxygen state was found (79). By contrast, three oxygen states were found for the colloid in solution, indicating that the CdS in the film is protected from oxidation. This result correlated well with an observation that copper sulfide made in an LB film is resistant to oxidation when compared to copper sulfide prepared analogously in solution (9). The decomposition, via oxidation, of PbS made in films of StH has been followed by XPS (68,70). One study found that the decomposition rate of PbS was much slower for films deposited at faster rate (70). It was suggested that films deposited at the faster deposition rate were more ordered and this presented a barrier to PbS decomposition via oxidation.

3.5.3.9 Ellipsometry and Surface Plasmon Resonance

These techniques have been used to determine changes in the thickness of M-FA films concurrent with MC formation. For three different studies of Cd-FA films ellipsometry has given an average thickness change per layer of 0.28 nm (33), 0.3 nm (63), and 0.2 nm (53). A smaller thickness change of 0.13 nm (59) per layer was determined by surface plasmon resonance (SPR) for CdAr films exposed to H_2S. For CuBe and ZnBe films exposed to H_2S, a thickness change per layer of only 0.09 nm was determined by ellipsometry (53). Both techniques assume that the film is a uniform thin slab with a constant refractive index. In the processing of the data, no account is taken of the fact that with particle formation the film is likely to become distorted. With such film changes it is not appropriate to treat the film as a uniform refractive index medium, and it is questionable that the calculated thickness changes have any true physical meaning.

In addition to the film thickness measurements, the kinetics of the reactions of MBe (M = Cd, Cu, Zn) films with H_2S were monitored in real time with ellipsometry (53).

3.5.3.10 Miscellaneous

In addition to the techniques already discussed, aspects of MC formation in LB films have been investigated by Rutherford backscattering (RBS) (65) and fluorimetry (17,49,50,76,79,81,83,84). The results from the RBS analysis of H_2S-exposed CdAr films provided evidence for the formation of spheroidal CdS particles. Investigations of the fluorescence properties of MC particles in LB films are discussed later.

3.5.3.11 Monolayer and Film Transfer Characteristics

Π/A isotherms of amphiphile monolayers on aqueous subphase have been used to determine occupied molecular areas (A_{mol}) and other (implied) structural properties. The area of the monolayer can be monitored during film transfer to obtain information on film structure. An example is seen in Figure 3.5.12 (10) for a DDAB monolayer, sustained at a surface pressure of 20 mN m^{-1} on a K_2PtCl_4 subphase. A uniform transfer is demonstrated over 10 dip cycles (average transfer of 1.03), and a Y-type film is evident from the same transfer for both up and down portions of

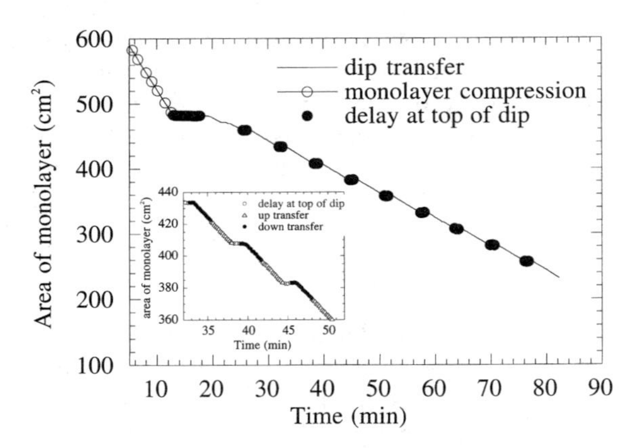

Fig. 3.5.12 Area of DDAB air–water monolayer on K_2PtCl_4 subphase as a function of time through compression of monolayer and 10 dip cycles of a quartz plate. Inset: Area of monolayer as a function of time through dip cycles 2–4. (From Ref. 10.)

the dip cycle. Brewster angle microscopy has been used to obtain images of amphiphile capped CdS monolayers at various stages in the monolayer compression (81).

3.5.4 Mechanistic Investigations

The formation of metal sulfide particles by exposing LB films containing metal ions to H_2S can be broken down into two basic processes as depicted in Eqs. (7) and (8):

$$M^{2+} + H_2S \rightarrow MS + 2H^+ \tag{7}$$

$$MS \rightarrow [MS]_n \tag{8}$$

The initial step of the reaction, Eq. (7), provides the individual metal sulfide molecules via reaction of the M^{2+} ions with H_2S. In the case of films derived from fatty acids, the two carboxylate functions, associated with the M^{2+} ion, are the sink for the two protons released from the reaction. The diffusion and coalescence of the individual MS molecules to give MS particles are depicted in Eq. (8). Despite an abundance of literature concerning Q-state particle formation in LB films, there has been little discussion relating to mechanistic aspects of how the nature of the LB support matrix effects the processes depicted in Eq. (7) and (8). The remainder of this section outlines the mechanistic and kinetic insights gained into these processes over the course of study of metal chalcogenide formation in LB films.

3.5.4.1 Particle Morphology

Size quantization of an MC can occur in three dimensions. If dimensions of the semiconductor are restricted in one dimension, only a ''quantum well'' (''sheets'') is the result. Size restriction in two and three dimensions gives ''quantum wires'' and ''quantum dots'' or ''Q-state particles,'' respectively. Undoubtedly, quantum size effects are seen in the UV/visible absorbance spectra of MCs produced in LB films; however, such spectra do not allow definition of morphology. A spheroidal morphology of the semiconductor particles, such as CdS, is often assumed without discussion or evidence, presumably for the convenience of ascribing particle size from UV/visible spectra (Fig. 3.5.1). There are, however, a number of publications that address the question of particle morphology directly. Of these there are a significant number of publications claiming quantum well formation as the result of confinement of the MC to the plane defined by the head-to-head structure of a Y-type film. Claims of the MC forming a continuous sheet one molecule thick cannot be taken seriously, as the molecular area (A_{mol}) occupied by a single CdS, for example, is far less than the A_{mol} of fatty acid in a crystalline LB film (~21 nm^2). An even more improbable representation is given in Figure 3.5.3b (25), where monomolecular CoS molecules are depicted between the FA head groups. A more likely representation of CdS forming in the planes defined by the head groups of the fatty acid would be disks or domains of CdS quantum wells with vacant spaces between. Evidence for quantum well formation generally relates to XRD and FTIR measurements of M-FA films before and after exposure to H_2S (25,43,47,61,66,69), which indicate retention of the crystalline nature of the film after formation of the metal sulfide. The most convincing evidence consists of x-ray diffraction data for CdSt films deposited from different surface pressures (61,66). Retention of diffraction peaks, before and after H_2S, was observed only for the film deposited at the highest pressure (37 mN m^{-1}), while at the lower pressures (20, 25, and 30 mN m^{-1}) a completely different diffraction pattern was observed (Fig. 3.5.9) (66). Most of the intensity of the reflections is due to the planes of electron-dense Cd atoms, and it is argued that retention of the peaks after H_2S exposure indicates retention of these Cd atom planes in the resulting CdS. The observed decrease in the intensity of the peaks was not discussed at length, although in other XRD studies a noted decrease in peak intensities was discussed in terms of some disordering, possibly resulting from CdS particle aggregation.

Moriguchi et al. (43) noted a correlation between the change in the basal plane spacing with H_2S exposure of CdSt, and the ionic radius of sulfide, and used this as evidence for CdS monolayer formation. Measurements of high lateral conductivity in metal ion fatty acid films exposed to H_2S (20,23) and of photoelectric properties (21) have also been used to invoke the concept of continuous sheets of metal sulfide forming.

There is also, however, considerable evidence that MC species formed in LB films aggregate to form spheroidal particles. X-ray studies of metal ion/fatty acid films before and after exposure to H_2S can be interpreted in terms of MS nanoparticles distributed throughout the film without completely destroying the original layered structure of the LB film (20,51,52,54,57). In one study of CdAr-fullerene/H_2S films, retention of diffraction peaks was observed before and after H_2S exposure; however, the decrease in intensities of the reflections was interpreted in terms of CdS aggregation (57). An interesting point has been raised regarding FTIR spectra of these films. A plane of CdS would prevent the "facing dimers" structure known for FA films, and this would shift the $\nu(C{=}O)$ vibrational mode from 1707 cm^{-1} to approximately 1720 cm^{-1} (57). This is not observed in any of the FTIR studies discussed earlier. The results from Rutherford backscattering (RBS) experiments of CdAr films exposed to H_2S have provided evidence for the disordering of the LB film as the result of spheroidal CdS formation (65). Perhaps the most compelling argument for particles comes from TEM, STM, and AFM imaging. TEM images of particles have been obtained of AgI and Ag(0) (11), Au(0) (8), and CdS (63) produced in LB films containing Ag(I), Au(III), and Cd(II) ions exposed to KI, $NaBH_4$, CO, and H_2S, respectively. The mechanism by which AgI and Au form and grow in LB films is likely to be the same as for MC species. AFM images of spheroidal particles have been obtained for $M^{2+}-FA$ [M = Cd (42), Cu (9), Hg (45), Pd (74)] films exposed to H_2S and subsequently washed with $CHCl_3$. Particle size distribution from the AFM images of CdS have correlated with optical absorbance spectra of H_2S-exposed CdAr films prior to $CHCl_3$ washing, although this does not constitute direct evidence for particle formation. In the case of the Au film reactions, AFM images of the CO exposed films demonstrated many raised bumps in the film with the same dimensions and distribution as the Au particles observed by TEM (8). Individual particles have also been identified by STM for M^{2+}-FA films exposed to H_2S (55,56,71).

In conclusion, the most reliable and consistent evidence supports MC particle formation from the reactions of M^{2+}-FA LB films with dihydrogen chalcogenides. There is some indirect evidence for quantum well formation in LB films prepared under certain conditions.

3.5.4.2 Kinetics

A number of publications have addressed the kinetics of the reaction of metal ions in M-FA films with H_2S using FTIR (22,23,39,43,47,48,53,61,66,70) and UV/visible (10,43,45,46,74,85) spectroscopy, XPS (19), ellipsometry (53), QCM gravimetry (42,45,46,58,59,64), and XRD (54). As with MC morphology studies, there is contradictory kinetic evidence and an obvious void of recognition between independent groups of work. For example, there are sets of data that support very long reaction times (hours and sometimes several days) (22,23,46,54,58,62,66,70) and very short reaction times (minutes) (39,42,43,45,46,48,53,59,64). Peng et al. have provided the most systematic and comprehensive investigation of the effects of various parameters on reaction rates (61,66,70) using FTIR spectroscopy and XRD, and most discrepancies can be resolved within the context of these parameters. For example, they discovered that faster dipping speeds for film transfer, higher surface pressures of Langmuir monolayers for film transfer, and lower H_2S pressures all support longer reaction times. The chemical nature of the species in the film can also have an effect on reaction rates, as explained in detail in a later section. Other discrepancies regard the extent of the reaction and the dependence of reaction rate on film thickness.

The use of FTIR to study reaction kinetics and chemical equilibria is based on the disappearance of a ν(COO—) vibrational mode of the deprotonated fatty acid in concurrence with the appearance of the ν(C=O) mode of the protonated FA. This type of data has been used to support reaction times of a few minutes to several hours or even days. Also, the extent of the reaction has been reported to be stoichiometric (22,23,39,43,47,53), based on the complete disappearance of the ν(COO—) vibrational mode, whereas other studies indicate less than 100% conversion (Fig. 3.5.8) of the carboxylate to the protonated form (48,61,66,70). The effect of film structure on reaction rate has been discussed (61,66,70); however, discrepancies of the extent of the reaction of M-FA with H_2S have not been previously addressed. FTIR spectroscopy has also been used to determine the effect of film thickness on reaction rate. For example, for 11-, 21-, and 31-layer CdSt films deposited at 37.5 mN m^{-1}, reactions with H_2S were found to stop at 42, 78, and 140 h, respectively (66).

The use of the QCM to measure the mass uptake by a M-FA film due to reaction of the metal ions in LB films with H_2S [Eq. (4)] is powerful for investigating the kinetics and stoichiometry of the reaction. The frequency can be converted to a mass change [Eq. (6)] and expressed as a percent conversion based on the mass of film deposited on the QCM and the reaction stoichiometry in Eq. (2) and (4). In some studies the profiles for mass uptake of CdAr (42,46,59) and HgAr (45) films as a function of H_2S exposure time were not measured in real time. The films were exposed to H_2S for a period of time, removed from the H_2S, and left to sit in air for several minutes before each QCM frequency measurement was made. In these studies it was found that the reaction equilibrium reached a level corresponding to 100% conversion of metal ions to metal sulfide within minutes. It is obvious, however, that these times are artificially small, as the reaction of H_2S with the metal will continue in the film after removal from the H_2S and this will contribute to the mass on the QCM. Changes in UV/visible spectra for analogous films prepared on quartz, and exposed to H_2S in the same manner, corroborated the QCM results. The experimental procedure for these UV/visible studies also was not done in real time. While the experimental procedures used for these QCM and UV/visible studies preclude any real quantitative assessment of kinetics, they at least qualitatively suggest a relatively quick reaction time of M-FA films with H_2S. In Figure 3.5.10 it is apparent that the rate of reaction is independent of film thickness (46).

Real-time mass increases of MBe (M = Cd, Hg) films on a QCM have been measured as a function of H_2S exposure time (58,64). The results from one of the studies, presented in Figure 3.5.13 (58), show the mass increase reached at equilibrium was proportional to film thickness and corresponded to 83–88% conversion of Cd^{2+} ions to CdS. The rate at which equilibrium was reached was strongly dependent on film thickness, in contradiction to the results presented in Figure 3.5.10 (46).

In the other real-time study, extended H_2S exposure (>17 min for HgBe, >125 min for CdBe) gave a mass change suggestive of greater than 100% conversion, and there was continual downward drift in the QCM frequency for the entire H_2S exposure. Part of the total frequency drop was recovered by subsequently placing the QCM under vacuum. This can probably be attributed to the removal of adsorbed H_2S from the film. The permanent frequency decrease can then be attributed to metal sulfide formation and another, slower process likely to be sulfur deposition via decomposition of the H_2S. Mass increases not associated with metal sulfide formation were also demonstrated for QCMs with films of CaBe or BeH exposed to H_2S. XPS evidence also suggests the accumulation of sulfur in LB films with exposure to H_2S for several h (10). The possibility of additional processes contributing to frequency changes of the QCM were not considered for the data presented in Figure 3.5.13. There is an obvious discrepancy between the time required for the 20-layer CdAr film in Figure 3.5.13 to reach "saturation" (84 min) and the 20-layer CdBe in the other real-time study (<10 min). A possible cause of this discrepancy may rest with the difference in the surface pressures used to compress the monolayers before film deposition (27–30 mN m^{-1} for Figure 3.5.13 vs. 20 mN m^{-1}).

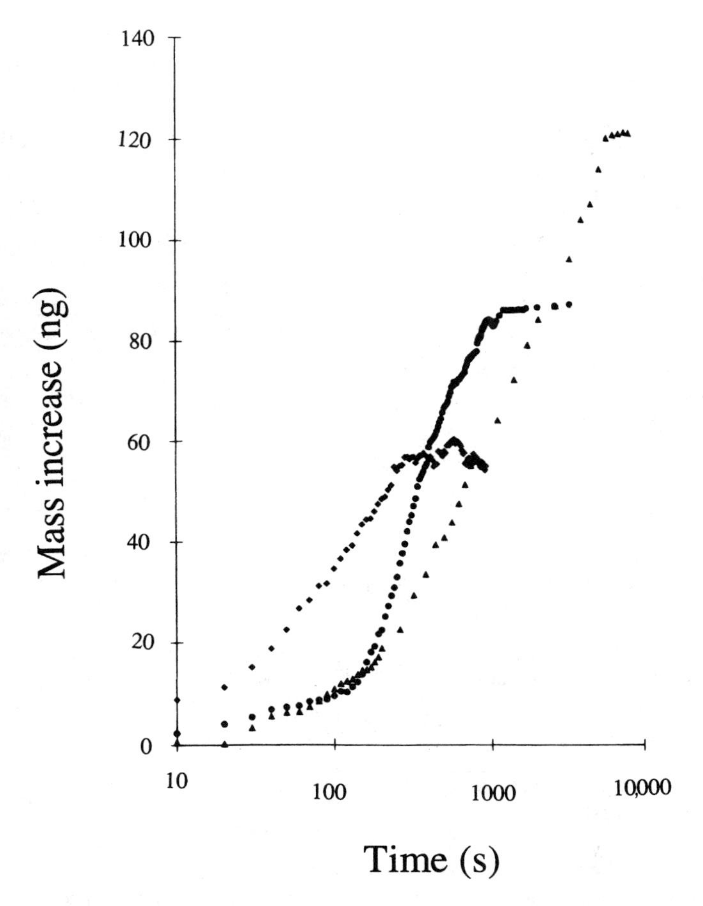

Fig. 3.5.13 Real-time QCM monitoring of the reaction between H_2S and CdAr in three different samples (20 bilayers, triangles; 15 bilayers, circles; 10 bilayers, diamonds). Films were deposited onto both sides of the resonators. (From Ref. 58.)

UV/visible absorbance measurements of H_2S-exposed M-FA films (M = Cd or Hg) have provided information about reaction rates. Delineation of the two processes depicted in Eq. (7) and (8) is difficult, however, as they can occur simultaneously. The reactions of LB films (CdAr) and cast self-assembled films of the cadmium salt of ditetradecyl-*N*-[4-([6-*N*, *N*′,*N*′-trimethylethylenediamino)-hexyl]oxy) benzoyl]-L-glutamate ($Cd(DTG)_2$) with H_2S have been followed by time-resolved UV/visible absorbance (85). The most profound observation from this investigation is with respect to the effect that water has on the rate of growth of the CdS particles

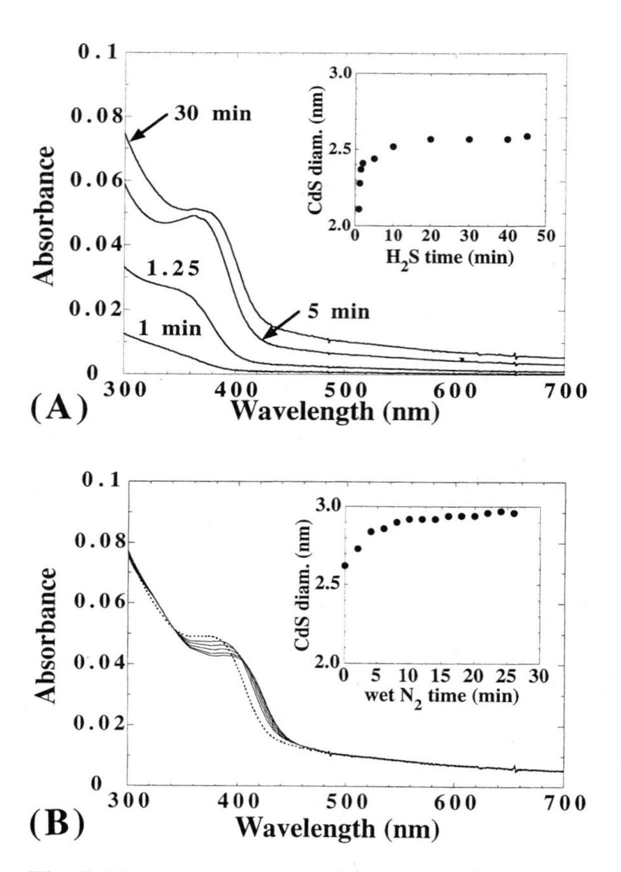

Fig. 3.5.14 UV-visible absorbance spectra recorded while a 20-layer CdAr film was sequentially exposed to (A) dry H_2S gas and then (B) wet N_2. In (B) the upper to lower spectra in the region between 345 and 405 nm (and the lower to upper spectra in the region above 405 nm) correspond to exposure times of 2, 4, 8, 16, and 24 min, respectively. The absorbance contributions due to the quartz and the LB film have been subtracted from the spectra. (From Ref. 85.) The insets in (A) and (B) show the CdS particle diameter (from absorbance spectra) as a function of exposure time to dry H_2S and wet N_2, respectively.

[Eq. (8)]. This is illustrated in Figure 3.5.14 (85), where the real-time spectra of CdAr are given at time intervals during dry H_2S exposure and subsequent exposure of the same film to wet N_2. As expected, for the H_2S exposure, there is a rapid increase in the absorbance intensity associated with CdS formation. After the first 2-min exposure to H_2S the changes in the spectra (broadening and redshifting) can be interpreted in terms of particle growth and increasing polydispersity. For a $Cd(DTG)_2$ film exposed to dry H_2S there was an increase in the absorbance up to the 19.5 h, as well as a red shift in the exciton energy (330 nm at 2 h to 370 at 19.5 h). The difference in the rates of the primary reaction [Eq. (7)] between CdAr and $Cd(DTG)_2$ may be related to film structure and the chemical environment surrounding the Cd^{2+} ion. For both CdAr and $Cd(DTG)_2$ films exposed to dry H_2S, subsequent introduction of wet nitrogen resulted in a red shift and broadening of the CdS peak indicative of particle growth and increasing polydispersity. This effect was more profound for the $Cd(DTG)_2$ film than for the CdAr film. A possible mechanism has been proposed for water-facilitated growth of CdS, which is analogous to Ostwald ripening in colloidal solutions. The formation of ions [e.g., Cd^{2+} or $Cd(OH)^+$, and S^{2-} or HS^-] was postulated from the interaction of H_2O with the surface of a particle. Complexation of the ions on the larger CdS particle results in growth of these particles and dissolution of smaller ones. If, in general, CdS growth in LB films involves ionic species, it is likely to be restricted to the hydrophilic planes defined by the carboxylate head groups of the fatty acid in a Y-type film. It is not possible, however, with the current state of knowledge, to exclude interlayer diffusion of neutral species through the hydrophobic portions of the LB film. The growth of CdS in CdSt films was also observed as a function of H_2S exposure time in a separate study (43).

Another real-time study of the reaction of M-FA films with H_2S utilized ellipsometry to monitor changes in film thickness concurrent with metal sulfide formation (53). The reactions appeared to reach equilibrium within the same period of time (within 2 h), with a change per monolayer of 0.2 nm for CdBe and 0.9 nm for both CuBe and ZnBe. Their ellipsometry results, in agreement with Peng et al. (66), also show a dependence of the reaction rate on the H_2S pressure and the surface pressure at which the films were deposited.

As with the evidence for MC morphology in LB films, the discrepancies in studies of reaction kinetics can probably be attributed to variations in film structure incurred by film construction parameters.

3.5.4.3 Role of Coordination Sphere of Metal Ion on Its Reaction Rate with Hydrogen Sulfide

It has been determined that the coordination sphere of the metal ion in the film has an effect on the rate of its reaction with H_2S (10,74). For example, CdAr (42,46,59) and HgAr (45) UV/visible absorbance and QCM measurements indicate the reaction depicted in Eq. (4) reaches its endpoint within minutes of exposure to H_2S. While these were not real-time studies, they at least showed qualitatively that the reaction is relatively quick. A similar set of absorbance measurements for DDA_2MCl_n films demonstrates that the same reaction requires several days to reach completion (10). A possible cause for the differing reaction rates was given with the $DDAB/(PdCl_4)^{2-}$ system. As opposed to $(PtCl_6)^{2-}$ and $(PtCl_4)^{2-}$ ions, the rate of hydrolysis of the $(PdCl_4)^{2-}$ ion is such that during the course of the dipping experiment the hydrolyzed species $[PdCl_{4-n}(H_2O)_n]^{n-2}$ [or mixed aqua/hydroxy species $PdCl_x(OH)_y(H_2O)_z$]

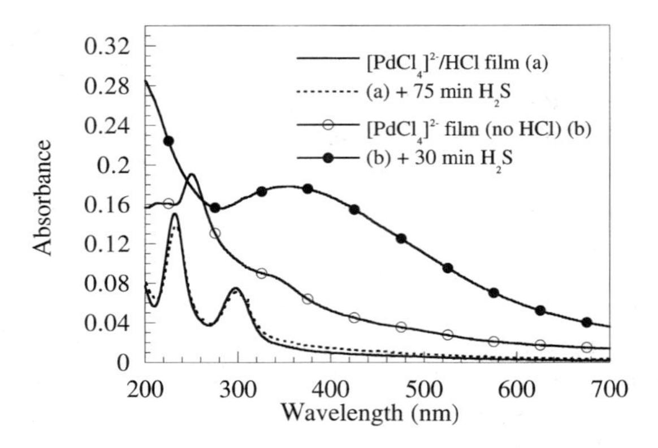

Fig. 3.5.15 Absorbance spectra of 18-layer films from an arachidic acid monolayer on a 2.66×10^{-4} *M* K_2PdCl_4/0.1 *M* HCl subphase and a 3.15×10^{-4} *M* K_2PdCl_4 (no HCl) subphase and after exposure of these films to H_2S for 75 and 30 min, respectively. (From Ref. 10.)

are predominent in the subphase. This is evident when comparing the absorbance spectra of films prepared from $(PdCl_4)^{2-}$ and $(PdCl_4)^{2-}$/HCl subphases (Fig. 3.5.15) (10). The addition of HCl suppresses the hydrolyses and the two LMCT bands of the intact $(PdCl_4)^{2-}$ ion are clearly evident in the absorbance spectrum. The FTIR spectra of the films deposited from the DDAB/$(PdCl_4)^{2-}$ and DDAB/$(PdCl_4)^{2-}$/HCl also demonstrate this difference. Broad 3000–3600 cm^{-1}, 500–800 cm^{-1}, and 800–1100 cm^{-1} bands unique to the film prepared in the absence of HCl can be attributed to the symmetric/asymmetric O-H stretching modes of a hydroxy or aqua ligand, rocking and M-O stretching modes of a coordinated aqua ligand, and M-O-H bending mode, respectively. There is a marked difference in the way the respective absorbance spectra respond to H_2S exposure. With the film containing the hydrolyzed $(PdCl_4)^{2-}$ ion there is a rapid change in the absorbance spectrum and the film appears orange-brown within minutes, concurrent with the formation of PdS (Fig. 3.5.15). In contrast, within the same time frame of H_2S exposure, the film containing the intact $(PdCl_4)^{2-}$ shows very little change. Films were also prepared from arachidic acid and subphases containing a series of Pt(II) and Pd(II) amine complexes (74). In all cases the reactions with H_2S proceed within minutes. Thus in the series of films studied, where there is no base in the coordination sphere of the metal ion in the film [i.e., unhydrolyzed $(MCl_n)^{2-}$ ions in DDAB film], the reaction of the metal ion with H_2S is much slower than with ions containing a base in the coordination sphere of the metal [i.e., hydrolyzed $(PdCl_4)^{2-}$ ion or metal ion in contact with carboxylate function of fatty acid]. This implies that the rate-determining step of the reaction is the deprotonation of the H_2S by a base in the coordination sphere of the metal. Peng et al. (60,61,66) have pointed out the importance of film structure on the kinetics of H_2S reactions with metal ions in LB films, and it is possible that the rate effects just discussed are structurally rather than chemically determined.

3.5.4.4 Role of Surfactant Film in Restricting Particle Growth

As mentioned in the introduction, the general method for the preparation of Q-state metal sulfides particles involves exposure of metal ions to a source of sulfide in the presence of some stabilizing medium. The mechanism by which the growth of the semiconductors is restricted is thought to be through ''chemical capping'' and/or physical confinement. Evidence for the chemical capping of metal sulfide particles in LB films and the implication of its role in the stabilization of Q-state particles has been discussed recently (29,48). In one study, films of ArH, CdAr, and H_2S-exposed CdAr on QCMs were heated at 10 kPa pressure. The rate of the frequency change of the QCM for each of these films was measured as a function of temperature as shown in Figure 3.5.16 (29). The rate of frequency increase can be related to the rate of mass loss due to the removal of the film with heating. The CdAr complex has a higher melting point than ArH, and therefore the higher rate of frequency change for the ArH film is not suprising. The CdAr film exposed to H_2S shows behavior similar to the ArH film, which supports the chemistry described in Eq. (4), where ArH is regenerated from CdAr after exposure to H_2S. The mass of CdS produced in a film can be calculated from the mass of CdAr on the QCM and from the stoichiometry from Eq. (2) and (4). Even after extensive outgassing at high temperatures (70–80°C), a residual mass was found over and above what is expected for the CdS. The plot of differential mass (residual mass on QCM minus calculated

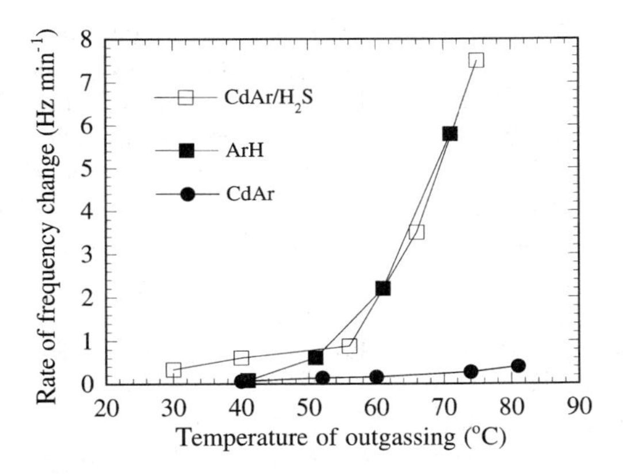

Fig. 3.5.16 Rate of removal of LB films from QCM (an increasing QCM frequency indicates a decreasing mass of film) as a function of temperature for CdAr, H_2S-exposed CdAr, and ArH. (From Ref. 29.)

mass CdS) versus the calculated mass of CdS indicates a linear relationship. It is likely that this differential mass is ionized arachidic acid associated with the surface of the particle.

Further evidence for a residual amount of deprotonated ArH in H_2S-exposed films of metal arachidate films has been obtained from grazing-angle FTIR spectra of metal arachidate films before and after exposure to H_2S (48). In the FTIR spectrum of a 20-layer CdAr film (Fig. 3.5.8) the asymmetric and symmetric stretching modes of the carboxylate group, ν_s (CO_2) and $\nu_a(CO_2)$, are evident at 1433 cm^{-1} and 1550 cm^{-1}, respectively. After exposure to H_2S, new bands at 1708 cm^{-1} and 3031 cm^{-1} have appeared, corresponding to the carbonyl stretch $\nu(C{=}O)$ and hydroxyl stretch $\nu(OH)$ of the protonated carboxylic acid. While reduced in intensity, the $\nu_s(CO_2)$ and $\nu_a(CO_2)$ stretches are still evident after extensive exposure to H_2S, implying that not all the CdAr molecules in the film are converted to ArH. Similar results have been obtained for ArH films containing Pd(II) and Pt(II) (74), exposed to H_2S. Other FTIR measurements also indicate that less than 100% of carboxylate functions are reprotonated at the endpoint of the reaction of CdAr films exposed to H_2S (52,61,66,70).

FTIR measurements have also been made on H_2S-exposed CdAr films that have been heated at 80°C under vacuum. The results are congruent with CdAr-coated QCMs exposed to H_2S and subsequently heated under vacuum, and in addition provide evidence for the association of the ionized arachidic acid with the CdS particles. After heating, the $\nu(CO)$ and $\nu(OH)$ bands are almost entirely absent, indicating that ArH molecules are preferentially sublimed from the film. Since the intensities of the $\nu_s(CO_2)$ and $\nu_a(CO_2)$ bands are not markedly different after heating, it is clear that arachidic acid molecules present as the cadmium salt are removed at much slower rates than those in the protonated form. A change in the relative intensities of the $\nu_s(CO_2)$ and $\nu_a(CO_2)$ bands would indicate orientation changes of the arachidate moiety relative to the substrate. Naselli et al. (86) have shown that there are virtually no changes in the FTIR spectrum of CdAr after heating to 80°C. It is therefore unlikely that unreacted layers of CdAr exist in the H_2S exposed film. More likely is the association of the ionized Ar^- with the surface of the CdS particle. This association is likely to involve the interaction between the anionic capping surfactant and cadmium ions on the surface of the particles. This sort of interaction is thought to be responsible for the stabilization of aqueous Q-state CdS with reagents such hexametaphosphate (6) and thiophenol (87).

XPS also provides evidence that, at its endpoint, the reaction of metal ions in LB films with H_2S is not stoichiometric as depicted in Eq. (4). For example, XPS analysis for a number of MBe films (M = Cd or Hg) to H_2S for 1 h gave an average S:M ratio of 0.76 instead of 1 as predicted by Eq. (4). Both QCM and UV/visible absorbance measurements indicate that 1 h of H_2S exposure is more than enough for the reaction to reach its endpoint. In another XPS investigation of films of calixerenes containing Cd^{2+} ions, S:Cd ratios of 0.84 ± 0.1, on average, were obtained (37).

The studies just cited are in conflict with other FTIR (22,23,39,43,47,53) and XPS (39,43,45) studies, which suggests that the reaction depicted in Eq. (4) is stoichiometric. Kinetic/structural relationships have been established, and it is possible that film construction parameters that affect kinetics also affect the chemical equilibrium in the reaction depicted in Eq. (4).

3.5.4.5 Tuning Optical Properties

It has been established that manipulation of MC optical properties can be effected in LB films through control of particle size or of the chemical nature of species in the film.

Size Control. For LB films where the MC is prepared prior to film construction, the MC size can be controlled by a number of well-known mechanisms. For example, the size of HMP-stabilized CdS is dependent on the pH of the solution before the H_2S is introduced. Similarly, for CdS grown under a monolayer of an amine type surfactant, particle size is pH dependent (14).

For MCs produced in LB films, a number of size/film preparation correlations have been observed. The most systematic and far-ranging method involves subjecting M-FA films to H_2S and then subsequent i/s cycles. In one study the average particle diameter of CdS particles in ArH films was found to progressively increase from 2.3 nm from the first H_2S exposure to 9.8 nm after 4 additional i/s cycles (42). The particle sizes were determined using the AFM for CdAr films deposited on mica that have been exposed to H_2S and subsequently washed with $CHCl_3$. It is not known how washing affects the particle size distribution. The absorbance spectra for analogously treated CdAr films on quartz corroborated the trend of increasing particle size with repeated $CdCl_2/H_2S$ cycling (Fig. 3.5.4, inset) (5). In this case, however, a particle diameter of only 3.5 nm was found after 5 i/s cycles. Absorbance spectra of CdS (5,39,43), ZnS (39), and PbS (39) produced in StH films through repeated i/s cycles also showed a progressive redshift with an increasing number of i/s cycles (39).

For CdS made in LB films, Moriguchi et al. (34) have elegantly demonstrated a correlation between the occupied molecular area (A_{mol}) of the amphiphile and band gap (E_g) of the CdS particles. This effect is shown in Figure 3.5.17 (34) for films of 4 amphiphiles with A_{mol} ranging from approx. 0.2 nm^2 to approx 0.4 nm^2, for 6 i/s cycles. The molecular area is another way of expressing the concentration of Cd^{2+} per unit area, and thus a correlation between $[Cd^{2+}]$ and CdS size is expressed. This observation agrees with another study of PbS formation in LB films made from an amphiphilic poly(maleic acid)/octadecanol ester (35,36). It was found that with increasing carboxylic acid:hydrocarbon chain ratio, and hence increasing $[Pb^{2+}]$, an increase in PbS particle size was observed. However there are reports that the $[Cd^{2+}]$ in the film does not effect the size of the resulting CdS. For example, the same CdS size was found in films prepared with diynoic fatty acid surfactants (A_{mol} = 0.2 nm^2) and a surfactant with a phthalic acid head group (A_{mol} = 0.4

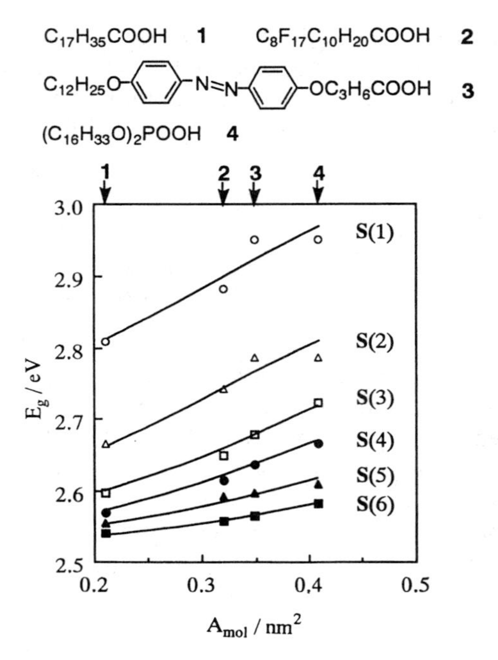

Fig. 3.5.17 Dependence of the energy gap (E_g) of CdS formed in LB films of amphiphiles 1–4 on the occupied molecular area (A_{mol}). (From Ref. 34.)

nm^2) (30). Furthermore, when the Cd^{2+} concentration was reduced by introducing Ca^{2+} ions, no reduction in the CdS particle size was observed. A reduction in CdS particle size was observed when dihexadecylphosphate (DHP) and Ca^{2+} were introduced into the Langmuir monolayer and the subphase, respectively. It has also been observed the optical properties of CdS made in CdAr do not vary for films deposited from Cd^{2+} subphases in the pH range 5–6. In this pH range the H^+/Cd^{2+} exchange extends from 0 to 100%. At pH 7, CdS particles are larger (2.5 nm) than CdS produced from ArH monolayers on Cd^{2+} subphases at pH 5–6 (2 nm) (88). The transfer of the films deposited at pH 7 degrades with increasing layer number, whereas in the pH 5–6 region Y-type transfer is maintained, and therefore the size of CdS particles is more likely to be an effect of film structure rather than Cd^{2+} concentration in the film.

Heating of an H_2S-exposed film of CdAr (deposited from pH 7 subphase) above 70°C was accompanied by an increase in particle size from 2.5 to 3.4 nm (48). Washing of an H_2S-exposed CdAr (deposited at pH 6) film with $CHCl_3$ caused a red shift in the absorbance spectrum of the CdS corresponding to an increase in particle size (58,64). LB films containing M^{2+} ions have been prepared with polymeric amphiphiles containing a fatty acid group (35–37). UV/visible spectra of these H_2S-exposed films indicate the formation of the corresponding metal sulfide with smaller dimensions than the metal sulfides formed in the corresponding metal ion/straight-chain fatty acid film. The rigidity of the polymeric films is expected to be greater than for the FA films, and this may impose greater restrictions on particle growth relative to FA films.

There are contradictory reports of the effect of film thickness on CdS particle for Cd-FA films exposed to H_2S. For example, Moriguchi et al. (34) reported larger particles (redshifted UV/visible spectrum) for bilayer Cd dihexadecyl phosphate films exposed to H_2S relative to multilayer films. This implies that stacking in the film imposes some barrier to particle growth. XRD also indicated more disordering in the bilayer films exposed to H_2S. For CdAr films, however, the same group reported that the absorption onset of CdS made in 19-, 29-, and 39-layer films is independent of film thickness (43). At least three studies (30,37,45,63) report that the size of CdS grown in LB films is independent of film thickness; however, these studies did not extend their study to include bilayer films, and thus their results do not necessarily contradict those found by Moriguchi et al.

Chemical Control. The formation of mixed chalcogenide/mixed metal semiconductors is another way of fine tuning optical spectra. For example, exposure of CdS in ArH films to H_2Se resulted in a redshifting in the absorbance spectrum of CdS attributed to the formation of CdS_nSe_{1-n} species (60). XPS results revealed that after extensive exposure to H_2Se a stable average composition of $CdS_{0.4}Se_{0.6}$ resulted, and a CdS core/CdSe shell structure was hypothesized. In a similar manner, mixed chalcogenides of cadmium $CdS_nTe_{(1-n)}$ and $CdSe_nTe_{1-n)}$ were prepared in nonacosa-10,12-diyonate multilayer films (30,31). The absorbance spectrum of CdS has also been altered through doping with Hg^{2+} ions. Mercury sulfide has a much lower solubility than CdS, and immersion of CdS particles into an Hg^{2+} solution should result in exchange of Cd^{2+} ions at the surface of the CdS particles, leading to a passifying thin layer of HgS resistant to further exchange. This ''surface exchange saturation'' has been observed (89), for example, in the absorbance spectra of CdS colloids prepared from solution, as a function of exposure time to Hg^{2+} ions. This type of ion exchange was investigated with Q-state CdS prepared in LB films in a recent study (64). Two 20-layer CdBe films from a subphase at pH 6.0 were exposed to H_2S. One of these films was gently stirred in chloroform for 30 min, a treatment shown (42,58,64) to remove the organic material while leaving behind the CdS. The intact CdBe film (H_2S exposed) was subjected to a vacuum for several hours to remove any excess H_2S in the film. Both substrates were then immersed in a 2.5 $\times 10^{-4}$ *M* $HgCl_2$ solution with gentle stirring. The resulting changes in the absorbance spectra for the $CHCl_3$-treated substrate after Hg^{2+} immersion are shown in Figure 3.5.18 (64). There is a progressive increase in absorbance over the entire wavelength range, and the characteristic absorption edge of CdS disappears. That ''surface exchange saturation'' has occurred is evident by superimposition of the

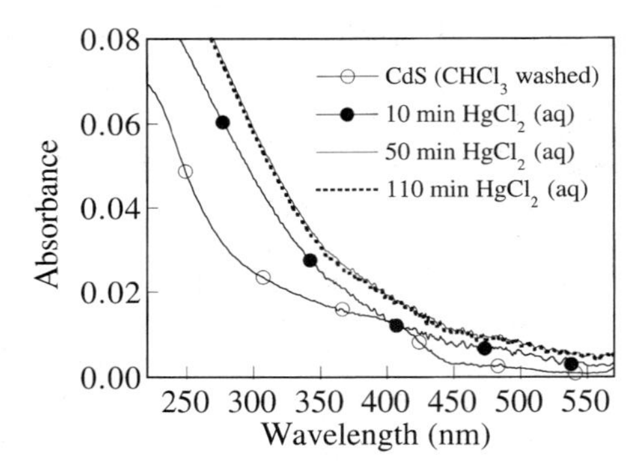

Fig. 3.5.18 Absorbance curves (corrected for the absorbance of the quartz) for H_2S-exposed CdBe, immersed in $CHCl_3$ and subsequently immersed in 2.5×10^{-4} *M* $HgCl_2$ for 0, 10, 50, and 110 min (total immersion times). (From Ref. 64.)

absorbance spectra between 50 min and 110 min immersion. For the intact film, immersion into Hg^{2+} resulted in an eventual loss of the CdS absorption edge and a very large increase in the overall absorbance. The films were visibly cloudy, and the increase in absorbance can probably be attributed to scatter.

There has only been one attempt to produce a mixed metal chalcogenide via a mixed metal ion-FA film precursor (64). LB films of mixed cadmium–mercury–behenate (CdHgBe) were fabricated on hydrophobed quartz plates from monolayers of BeH on subphases containing a mixture of Cd^{2+} and Hg^{2+} ions. The pH of the subphase was adjusted to between 6.0 and 6.2 to ensure complete exchange of fatty acid protons for subphase metal ions. The films were exposed to H_2S for 60 min and then analyzed by XPS to obtain the Cd:Hg and S:M ratios. There was a good correlation between the Cd:Hg ratio in the film and that in the subphase. UV/visible absorbance spectra of the films were also measured before and after exposure to H_2S. It has been demonstrated with various aqueous CdS/HgS colloids that where there is contact or mixing of the two colloids, the resulting UV/visible spectra are not the simple arithmetic addition of their respective spectra (64 and ref. therein). It is on this basis that the spectra from H_2S-exposed CdHgBe films were analyzed to determine the degree of interaction of CdS and HgS particles formed in LB films. The spectra observed closely matched simulated spectra for discrete CdS and HgS constructed by adding together CdS and HgS spectra in fractions suggested by the XPS data, and it was thus concluded that individual CdS and HgS particles form, rather than some composite. The individual CdS and HgS formation, as opposed to some mixed or interacting species, was rationalized on the basis of the relative rates of reaction of the Cd^{2+} and Hg^{2+} with H_2S. The real-time mass uptake of films of HgBe and CdBe on QCMs with H_2S exposure (64) suggested that the reaction of Hg^{2+} ions in LB films is substantially quicker than with Cd^{2+} ions. Therefore, in the mixed $Cd_xHg_{1-x}Be_2$ films it is possible that all the Hg^{2+} is converted to Q-state HgS particles before a significant portion of the Cd^{2+} has reacted with H_2S. As indicated earlier, there is considerable evidence (29,48) that the Q-state particles formed in LB films are stabilized by a surface coating of fatty acid, and this could prevent coalescence of the stabilized HgS particles with CdS particles that are still mobile and growing in the film.

In another study, films of StH/CdS were subjected to repeated Pb^{2+} intercalation and sulfidation cycles. UV/visible spectra reveal that the spectrum of the CdS is not affected by the formation and growth of the PbS (39).

3.5.5 Properties of Semiconductor Particles in Films

3.5.5.1 Fluorescence

The fluorescence of CdS particles in LB films has been measured (17,49,50,76,79,81,83,84) with an emphasis on surface passivation of the CdS to decrease the density of surface trapping sites. The major consequence is an increase in the quantum efficiency of the ''exciton fluorescence'' or band edge emission, or at least enhancement of the emission from ''shallow traps'' relative to ''deep traps.'' Two approaches have been used to achieve surface passivation. One approach utilized chemical reagents that remove defects or trapping sites on an HMP-stabilized CdS colloid. As an example of this approach, a colloidal CdS solution was used as subphase under a BeH/octadecylamine (1:3) monolayer, and the monolayer with associated CdS was transferred to a quartz substrate (79). A 220-nm blue shift (from 710 to 488 nm) in the emission spectrum was observed compared to the CdS colloid. This was attributed to binding of low-energy surface trapping sites by the octadecylamine. ''Fluorescence activation'' of HMP-stabilized CdS has also been effected before incorporation into LB films (76,81) by coating of the CdS particles with a layer of $Cd(OH)_2$ (6). Immersion of the films into solutions of triethylamine, Sr^+, or Ag^+ ions also enhanced the excitonic fluorescence of the CdS.

The second successful approach toward passivating CdS in LB films, and thereby enhancing the high-energy fluorescence, has been to expose the films to a 1-MeV H^+ beam (49,50,83). The CdS was produced via exposure of a CdAr film to H_2S.

3.5.5.2 Electronic and Photoelectronic Properties

Studies include wet-cell photoelectrochemical measurements (42,60,72,90), STM measurements on single MS particles in thin films (55,56,81), and conductivity measurements of metal chalcogenides in LB films (20,21,23). Many such studies are driven by the search for cheaper methods and materials for the fabrication of semiconductors suitable for photoelectrochemical devices. Moreover, the ability to tune optical properties via the Q-state effect and the versatility of LB fabrication make the LB films an attractive medium for semiconductor production. The photoelectrochemical properties of LB films containing Q-state CdS particles in a wet cell have been investigated (42,60). The photoelectrochemical cell (PEC) consisted of a Pyrex vessel with a platinum counterelectrode, a standard calomel electrode (SCE) as the reference electrode, and an ITO/glass/CdS-containing LB film as the photoelectrode. Experiments were conducted at 21°C, at pH 7.25, using 1.0 *M* Na_2SO_3 as electrolyte. The open circuit voltage upon illumination of the PEC is generated by excitation of CdS or CdS_xSe_{1-x} to give an electron/hole pair [Eq. (9)]:

$$CdS + h\nu \rightarrow CdS(e^-/h^+) \tag{9}$$

Current flows when the electron/hole pair is prevented from recombining directly via percolation of the photogenerated electron to the ITO electrode and by scavenging of the hole by the solution electrolyte. The circuit is completed by reduction of the oxidized electrolyte species (SO_4^{2-} or $S_2O_6^{2-}$) at the platinum electrode.

In order to investigate the effect of changing the CdS optical properties on the photoelectrochemical response of the cell, the CdS particle size was progressively increased by repeated $CdCl_2$(aq)/H_2S exposure. Also CdS particles with a monolayer shell of CdSe, $CdS_xSe_{(1-x)}$, were produced by exposure of CdS particles to H_2Se as described earlier. Theoretically, any trend in the measured open circuit voltage (V_{oc}) of the cell should follow the trend of band gap or exciton energies of a semiconductor. The V_{oc} as a function of particle size for CdS formed in 9-layer CdAr LB films is given in Figure 3.5.19 (5). The size of the particles was estimated from UV/visible absorption spectra. The magnitude of V_{oc} in the PEC increased from approximately 0.23 mV for the 2.54-nm CdS particles to 0.87 for the 4.2-nm parti-

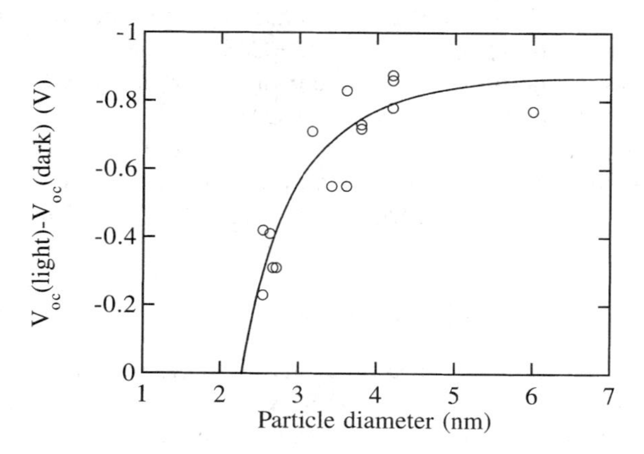

Fig. 3.5.19 Open-circuit voltage (V_{oc}) for CdS formed in 9-layer CdAr films as a function of particle size (estimated from UV/visible absorption spectra). The electrolyte was 1.0 *M* Na_2SO_3 at pH 7.25, and platinum and standard calomel electrodes were used as the counter and reference electrodes, respectively. (From Ref. 5.)

cles. This apparent contradiction was rationalized on the basis of energy loss in the electron as the photogenerated electron diffuses to the ITO surface through the high-resistance hydrocarbon film. In the film containing larger particles, the concentration of CdS in the film is greater, which may result in lower resistance conduits via connection between CdS particles. Exposure of the CdS particles to H_2Se to give a CdS_xSe_{1-x} species results in a decrease in V_{oc} of approximately 100 mV, congruent with the red shift in the optical spectra. The short-circuit current (I_{sh}) was significantly larger for CdS_xSe_{1-x} films relative to CdS, probably due to the increase in absorbance in the visible region of the spectrum upon exposure of CdS to H_2Se. For the PECs, examined efficiencies of between 0.2 to 1% were observed. The efficiency was calculated as the product of the open-circuit voltage, short-circuit current, and fill factor divided by the photon flux of the lamp integrated over the wavelength range of absorption. Photoelectric effects have also been noted for several metal sulfides fabricated in bilayer lipid membranes (72,90).

The conductivity and photoconductivity of copper sulfide (20) and HgS (21,23) in LB films made from fatty acids have been measured. These measurements confirmed the semiconducting nature of species in the film resulting from exposure of the M^{2+}-containing films to H_2S. Conductivities observed were greater than expected based on bulk materials properties because of high surface/bulk ratios and high doping levels. The conductivity measurements were also used to suggest the formation of a continuous monomolecular film of MS as opposed to discrete segregated MS particles. However, in one case the authors admit the results are not very reproducible (20).

An STM probe has been used to isolate individual MS (M = Cd, Pb) particles and to measure electronic phenomena (55,56,81). The MS films were prepared either by exposure of metal ion/fatty acid films to H_2S (55,56) or by transfer of a compressed DDAB-complexed CdS monolayer (81). All the films were transferred onto highly oriented pyrolytic graphite (HOPG) for the STM measurements. A junction was created at an individual CdS particle with the STM tip as one electrode and the graphite as the other, and the current/voltage characteristics of the particles were measured. For the particle prepared in the fatty acid films the *I*/*V* curves exhibit step-like features characteristic of monoelectron phenomena. In the case of the DDAB-coated CdS particles the *I*/*V* measurements demonstrated *n*-type semiconductor behavior. The absence of steps in this system is probably a reflection of the larger size of the particles in the DDAB films (8 nm by AFM) compared to the 2-nm particle size typically found for MS particles formed in fatty acid films.

3.5.5.3 Nonlinear Optical Properties

In addition to the blueshifting of the optical absorbance of Q-state semiconductor particles, a linear optical effect, there are nonlinear optical effects demonstrated by Q-state semiconductors. Two types have been observed by MCs prepared in organized films. One is the third harmonic generation or ‘‘frequency tripling’’ (44). A third-order nonlinear susceptibility of 10^{-12} esu was measured for PbS prepared in an StH film.

Transient absorbance bleaching and recovery is another nonlinear optical property observed from a Q-state MC in an organized film (40,91). A potential application of this phenomenon is optical switching. For reliable photobleaching experiments with CdS, the requirement of an absorbance of at least 2 at 400 nm is required. This precluded the use of LB films prepared by the vertical dipping method utilizing the stoichiometry depicted in Eq. (2) and subsequent exposure to H_2S. In Figure 3.5.1 (insert), the absorbance of the CdS in a 40-layer ArH film, at 400 nm, is 0.0152. This means that a 5000-or-more-layer film would be required to meet the absorbance requirements just described. Not only would this be impractical in terms of film preparation time, but the optical clarity of the films is generally compromised with

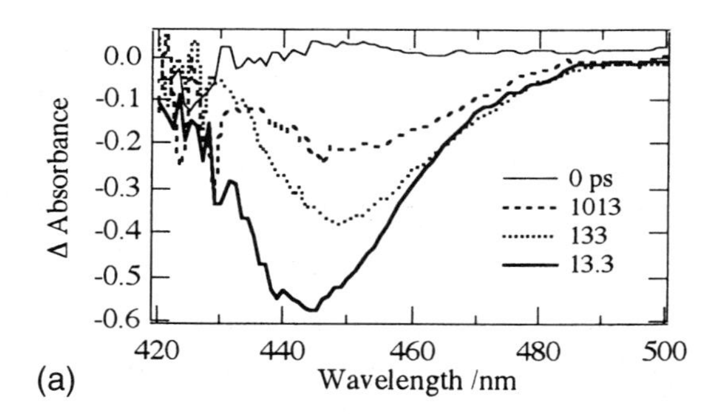

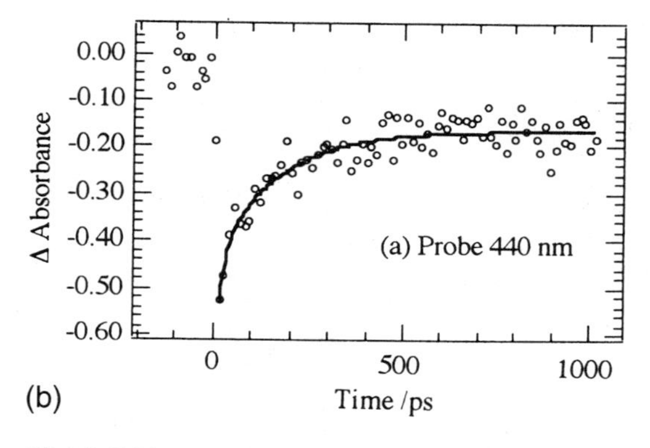

Fig. 3.5.20 (a) Time-resolved bleaching spectra of CdS nanoparticles in DTG film obtained at different times. Pump energy = 40 μJ/pulse. (b) Time profile of transient bleaching for CdS in DTG film at 440 nm. (From Ref. 40.)

films greater than 100 layers thickness. These problems have been surmounted by producing CdS in films in self-assembled ''cast'' films of ditetradecyl-*N*-[4-([6-(*N,N',N'*-trimethyl-ethylenediamino)hexyl]oxy)benzoyl]-L-glutamate (DTG), which resembles an LB film in structure. The bleaching recovery of CdS in the DTG film, shown in Figure 3.5.20 (40), displayed multiexponential kinetics. One of the studies was an examination of the bleaching recovery of CdS in Nafion and in aqueous solutions as well as in a DTG film (40). It was concluded that the bleaching recovery times were decreased with decreasing concentration of traps on the surface of the particles. This conclusion was derived from the fact that solutions of ''fluorescence-activated'' CdS particles have much faster recovery times than ''unactivated'' CdS colloidal solutions, and the CdS in Nafion and self-assembled films, which demonstrate very weak fluorescence.

3.5.6 Summary

Two strategies have been utilized for the preparation of LB films containing MCs distinguished by the stage in the film construction at which the semiconductor is prepared. Most studies have focused on the exposure of M^{2+}-FA LB films to a dihydrogen chalcogenide. Variations have included initial preparation of the semiconductor prior to film construction and the use of surfactants containing neutral or positively charged head groups.

The primary techniques for characterizing MCs and the LB films include various spectroscopic (UV/visible, FTIR, and XPS) and microscopic (STM, AFM, TEM) techniques, as well as XRD, and QCM gravimetry.

Investigations of the reaction kinetics and equilibrium of metal ions in M^{2+}-FA LB films with H_2S and the morphology of the resulting metal sulfide have resulted in what appears to be on first glance an abundance of contradictions. For example, FTIR and UV/visible spectroscopic, XRD, ellipsometry, and QCM gravimetric evidence has been used to suggest that the reaction depicted in Eq. (4) can take anywhere from a couple of minutes to several days to reach completion. This can be reconciled by considering the effects of various parameters, including surface pressure at which the film is deposited, rate of film transfer, pressure of H_2S, and chemical nature of the surfactant and metal coordination sphere, on reaction kinetics. FTIR and XPS evidence has been used to claim that reaction depicted in Eq. (4) is either stoichiometric or nonstoichiometric. Evidence of nonstoichiometry was interpreted, in some studies, as being evidence for the role of the surfactant in the restricting the growth of metal sulfide particles in LB films. Evidence has been presented for both two-dimensional sheets or domains and spheroidal particles of metal sulfides forming in M^{2+}-FA films. Again the effect of various film and reaction parameters has been implicated in the morphology of the resulting metal sulfide particles, and it is likely that under most conditions used to prepare LB films the metal sulfides formed in these films are spheroidal in geometry.

LB films have been shown to be an appropriate medium for growth of a large range of MC materials exhibiting Q-state effects. The tuning of the optical properties of the Q-state particles has been accomplished either by manipulation of the particle size or by changing the chemical nature of the MC. MC particle size/film structure correlations have been established, but the most systematic and widely used method of manipulating particle size has been through repeated sulfidation/intercalation cycling of M^{2+}-FA films. The fluorescence, electronic, photoelectronic, and nonlinear optical properties of MCs in LB films have been investigated to some extent, but there is still scope for further detailed work.

REFERENCES TO SECTION 3.5

1. H Weller. Adv Mater 5:88, 1993.
2. JH Fendler, FC Meldrum. Adv Mater 7:607, 1995.
3. F Grieser, DN Furlong, RS Urquhart, DJ Elliot. NATO ASI Ser 3(12):733, 1996.
4. A Henglein. Chem Rev 89:1861, 1989.
5. K Grieve. PhD thesis, University of Melbourne, Parkville, Victoria, Australia, in preparation.
6. L Spanhel, M Haase, H Weller, A Henglein. J Am Chem Soc 106:5649, 1987.
7. DJ Elliot, K Grieve, DN Furlong, F Grieser. Adv Coll Inter Sci, in press.
8. DJ Elliot, DN Furlong, F Grieser, P Mulvaney, M Giersig. Colloids Surf 129–130:141, 1997.
9. DJ Elliot, DN Furlong, F Grieser. Colloids Surf 141:9, 1998.
10. DJ Elliot, DN Furlong, TR Gengenbach, F Grieser, RS Urquhart, CL Hoffman, JF Rabolt. Colloids Surf 103:207, 1995.
11. C Fan, T Lu, B Li, L Jiang. Thin Solid Films 286:37, 1996.
12. J Leloup, P Maire, A Barraud, A Ruadel-Teixier. J Chim Phys 82:695, 1985.
13. J Yang, FC Meldrum, JH Fendler. J Phys Chem 99:5500, 1995.
14. KC Yi, JH Fendler. Langmuir 6:1519, 1990.
15. XK Zhao, Y Yuan, JH Fendler. J Chem Soc Chem Commun 18:1248, 1990.
16. XK Zhao, L McCormick, JH Fendler. Chem Mater 3:922, 1991.
17. NA Kotov, FC Meldrum, C Wu, JH Fendler. J Phys Chem 98:2735, 1994.
18. BO Dabbousi, CB Murray, MF Rubner, MG Bawendi. Chem Mater 6:216, 1994.
19. H Chen, X Chai, Q Wei, Y Jiang, T Li. Thin Solid Films 178:535, 1989.
20. J Leloup, A Ruaudel-Texier, A Barraud. Thin Solid Films 210–211:407, 1992.
21. C Zyberajch, A Ruadel-Teixier, A Barraud. Synth Metals 27:B609, 1988.
22. C Zyberajch-Antoine, A Barraud, H Roulet, G Dufour. Appl Surf Sci 52:323, 1991.
23. C Zyberajch, A Ruadel-Teixier, A Barraud. Thin Solid Films 179:9, 1989.
24. N Matsuura, DJ Elliot, DN Furlong, F Grieser. Colloids Surf 126:189, 1997.
25. X Luo, Z Zhang, Y Liang. Langmuir 10:3213, 1994.
26. Y Okahata, K Ariga, K Tanaka. Thin Solid Films 210/211:702, 1992.
27. SV Batty, T Richardson, P Pocock, L Rahman. Thin Solid Films 266:96, 1995.
28. DN Furlong, F Grieser. Chem Aust 59:617, 1992.
29. DN Furlong, R Urquhart, F Grieser, K Tanaka, Y Okahata. J Chem Soc Faraday Trans 89:2031, 1993.

30. F Grieser, DN Furlong, D Scoberg, I Ichinose, N Kimizuka, T Kunitake. J Chem Soc Faraday Soc 88:2207, 1992.
31. DJ Scoberg, F Grieser, DN Furlong. J Chem Soc Chem Commun 7:515, 1991.
32. DJ Scoberg, DN Furlong, CJ Drummond, F Grieser, J Davy, RH Prager. Colloids Surf 58:409, 1991.
33. R Zhu, Y Wei, C Yuan, S Xiao, Z Lu. Solid State Commun 84:449, 1992.
34. I Moriguchi, F Shibata, Y Teraoka, S Kagawa. Chem Lett 9:761, 1995.
35. LS Li, L Qu, L Wang, R Lu, X Peng, Y Zhao, TJ Li. Langmuir 13:6183, 1997.
36. X Peng, R Lu, Y Zhao, L Qu, H Chen, T Li. J Phys Chem 98:7052, 1994.
37. AV Nabok, T Richardson, F Davis, CJM Stirling. Langmuir 13:3198, 1997.
38. P Ganguly, DV Paranjape, M Sastry. J Am Chem Soc 115:793, 1993.
39. I Moriguchi, H Nii, K Hanai, H Nagaoka, Y Teraoka, S Kagawa. Colloids Surf 103: 173, 1995.
40. H Inoue, R Urquhart, T Nagamura, F Grieser, H Sakaguchi, DN Furlong. J Photopol Sci Technol 11:73, 1998.
41. NA Kotov, I Dekany, JH Fendler. J Phys Chem 99:13065, 1995.
42. HS Mansur, F Grieser, MS Marychurch, S Biggs, RS Urquhart, DN Furlong. J Chem Soc Faraday Trans 91:665, 1995.
43. I Moriguchi, K Hosoi, H Nagaoka, I Tanaka, Y Teraoka, S Kagawa. J Chem Soc Faraday Trans 90:349, 1994.
44. I Moriguchi, K Hanai, Y Teraoka, S Kagawa, S Yamada, T Matsu. Jpn J Appl Phys 34:L323, 1995.
45. DJ Elliot, DN Furlong, TR Gengenbach, F Grieser. Colloids Surf 102:45, 1995.
46. RS Urquhart, DN Furlong, H Mansur, F Grieser, K Tanaka, Y Okahata. Langmuir 10: 899, 1994.
47. I Moriguchi, I Tanaka, Y Teraoka, S Kagawa. J Chem Soc Chem Commun 19:1401, 1991.
48. RS Urquhart, CL Hoffman, DN Furlong, NJ Geddes, JF Rabolt, F Grieser. J Phys Chem 99:15987, 1995.
49. K Asai, T Yamaki, S Seki, K Ishigure, H Shibata. Thin Solid Films 284–285:541, 1996.
50. K Asai, T Yamaki, K Ishigure, H Shibata. Thin Solid Films 277:169, 1996.
51. JK Basu, MK Sanyal. Phys Rev Lett 79:4617, 1997.
52. A Dhanalbalan, H Kudrolli, SS Major, SS Talwar. Solid State Commun 99:859, 1996.
53. FN Dulstev, LL Sveshnikova. Thin Solid Films 288:103, 1996.
54. V Erokhin, L Feigin, G Ivakin, V Klechkovskaya, Y Lvov, N Stiopina. Makromol Chem Macromol Symp 46:359, 1991.
55. V Erokhin, P Facci, S Carrara, C Nicolini. J Phys D Appl Phys 28:2534, 1995.
56. V Erokhin, P Facci, S Carrara, C Nicolini. Thin Solid Films 284–285:891, 1996.
57. CT Ewins, B Stewart. Thin Solid Films 284–285:49, 1996.
58. P Facci, V Erokhin, A Tronin, C Nicolini. J Phys Chem 98:13323, 1994.
59. NJ Geddes, RS Urquhart, DN Furlong, CR Lawrence, K Tanaka, Y Okahata. J Phys Chem 97:13767, 1993.
60. HS Mansur, F Grieser, RS Urquhart, DN Furlong. J Chem Soc Faraday Soc 91:3399, 1995.
61. Z Pan, J Liu, X Peng, T Li, Z Wu, M Zhu. Langmuir 12:851, 1996.
62. Z Pan, X Liu, S Zhang, G Shen, L Zhang, Z Lu, J Liu. J Phys Chem 101:9703, 1997.

63. ES Smotkin, C Lee, AJ Bard, A Campion, MA Fox, TE Mallouk, SE Webber, JM White. Chem Phys Lett 152:265, 1988.
64. DJ Elliot, DN Furlong, F Grieser. Colloids Surf 155:101, 1999.
65. T Yamaki, K Asai, K Ichigure. Chem Phys Lett 273:376, 1997.
66. Z Pan, X Peng, T Li, J Liu. Appl Surf Sci 108:439, 1997.
67. C Liu, AJ Bard. J Phys Chem 93:3232, 1989.
68. X Peng, Q Wei, Y Jiang, X Chai, T Li, J Shen. Thin Solid Films 210–211:401, 1992.
69. X Peng, S Guan, X Chai, Y Jiang, T Li. J Phys Chem 96:3170, 1992.
70. X Peng, H Chen, S Kan, Y Bai, T Li. Thin Solid Films 242:118, 1994.
71. R Zhu, G Min, Y Wei, HJ Schmitt. J Phys Chem 96:8210, 1992.
72. XK Zhao, S Baral, R Rolandi, JH Fendler. J Am Chem Soc 110:1012, 1988.
73. GW Min, XM Yang, ZH Lu, W Yu. J Vac Sci Technol B 12:1984, 1994.
74. DJ Elliot, DN Furlong, TR Gengenbach, F Grieser, RS Urquhart, CL Hoffman, JF Rabolt. Langmuir 11:4773, 1995.
75. RS Urquhart, DN Furlong, TR Gengenbach, NJ Geddes, F Grieser. Langmuir 11:1127, 1995.
76. Y Tian, C Wu, JH Fendler. J Phys Chem 98:4913, 1994.
77. K Ariga, Y Okahata. Langmuir 10:3255, 1994.
78. DN Furlong. In: BP Binks, ed. Modern Characterization Methods of Surfactant Systems. Surfactant Science Series. Vol. 83. New York: Marcel Dekker, 1999, p. 471.
79. Z Du, Z Zhang, W Zhao, Z Zhu, J Zhang, Z Jin, T Li. Thin Solid Films 210–211:404, 1992.
80. HG Hansma, SAC Gould, PK Hansma, HE Gaub, ML Longo, JAN Zasadzinski. Langmuir 7:1051, 1991.
81. Y Tian, JH Fendler. Chem Mater 8:969, 1996.
82. TL Marshbanks, HK Jugduth, WN Delgass, EI Franses. Thin Solid Films 232:126, 1993.
83. K Asai, K Ishigure, H Shibata. J Lumin 63:215, 1995.
84. T Yamaki, K Asai, K Ichigure, H Shibata. Radiat Phys Chem 50:199, 1997.
85. RS Urquhart, N Matsuura, F Grieser, DN Furlong. Langmuir, in press.
86. C Naselli, JF Rabolt, JD Swalen. J Phys Chem 82:2136, 1985.
87. N Herron, Y Wang, H Eckbert. J Am Chem Soc 112:1322, 1990.
88. Unpublished results.
89. A Mews, A Eychmuller, M Giersig, D Schoos, H Weller. J Chem Phys 98:934, 1994.
90. S Baral, JH Fendler. J Am Chem Soc 111:1604, 1989.
91. H Inoue, RS Urquhart, T Nagamura, F Grieser, H Sakaguchi, DN Furlong. Colloids Surf 126:197, 1997.

4

Silver Halides

TADAO SUGIMOTO, KIJIRO KON-NO, and KEISAKU KIMURA

4.1 CONTROLLED DOUBLE-JET PROCESS

TADAO SUGIMOTO
Institute for Advanced Materials Processing, Tohoku University, Sendai, Japan

Berry et al. (1–3), Moisar and Klein (4), and Claes and Berendsen (5) developed a unique method for preparation of monodisperse particles of silver bromide (AgBr) and silver chloride (AgCl), called controlled double-jet precipitation, as characterized by simultaneous introduction of silver nitrate and corresponding halide solutions into a gelatin solution at a precisely regulated rate. In the double-jet process, very fine primary nuclei or embryos are generated in a domain of the gelatin solution where these reactant solutions are injected. In the meantime, the primary nuclei are dispersed into the bulk solution region where relatively large stable nuclei grow at the expense of the smaller unstable nuclei by Ostwald ripening (6). Thus, in this open system, a nucleation zone and a bulk zone for particle growth coexist in the same solution throughout the precipitation process, as shown in Fig. 4.1.1, where the small circles and the larger cubes represent the primary nuclei of AgBr and the stable nuclei, respectively.

In the early stage, the number of stable nuclei increases with the growing supersaturation by the dissolution of the unstable nuclei. When a sufficient number of somewhat grown stable nuclei have been built up in the bulk phase, they become able to absorb the whole solute provided by the constant dissolution of the stationary primary nuclei. From this moment, the growing particles cease to increase in number, whereas the primary nuclei generated in the nucleation zone begin to act simply as a monomer source. In other words, two distinct stages of the nucleation (i.e., accumulation of stable nuclei) and growth are observed in this system, like usual homogeneous monodisperse systems. This is the reason for the formation of monodisperse particles in this system. In this system, gelatin plays a definite role as a protective colloid to prevent coagulation among the primary nuclei as well as the growing particles at a high ionic strength. The dynamic behavior of the embryos, unstable nuclei, and stable nuclei in the nucleation process is illustrated in Fig. 4.1.2 (6).

Figure 4.1.3 shows a more concrete scheme of a typical controlled double-jet (CDJ) system, in which the flow rate of the $AgNO_3$ solution is previously programed, while the flow rate of the KBr solution is precisely controlled to keep the excess concentration of Br^- ions constant in response to the silver potential monitored with a silver electrode (7).

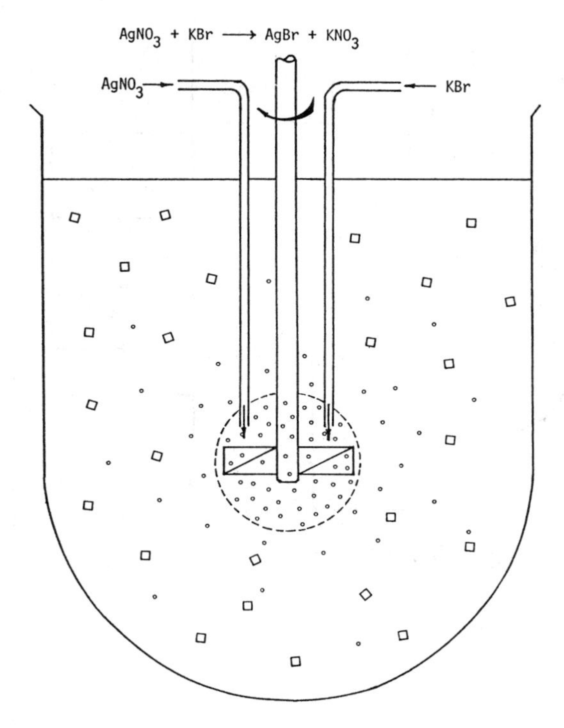

Fig. 4.1.1 Scheme of a double-jet precipitation system of AgBr particles.

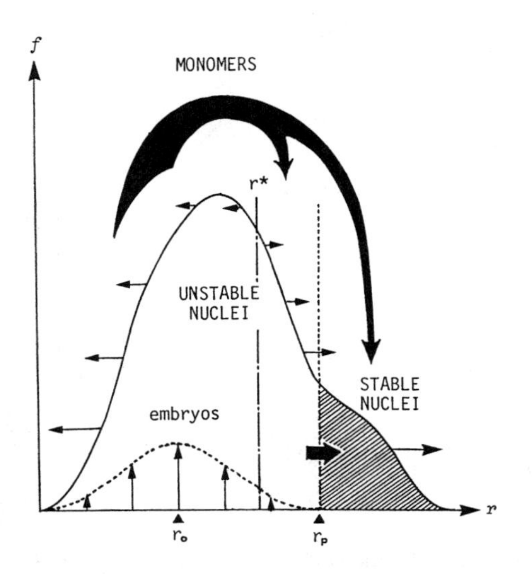

Fig. 4.1.2 Model of the nucleation process in an open system. (From Ref. 6.)

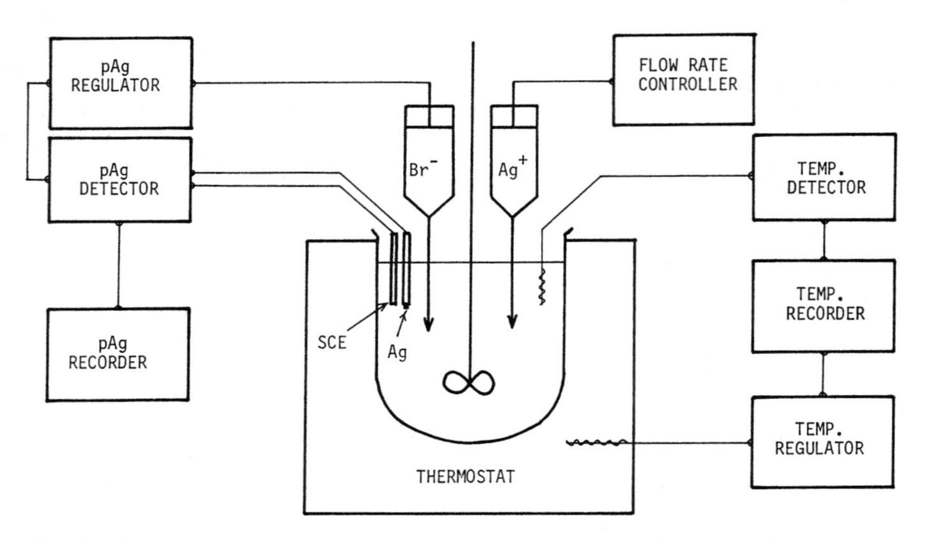

Fig. 4.1.3 Scheme of controlled double-jet (CDJ) apparatus. (From Ref. 7.)

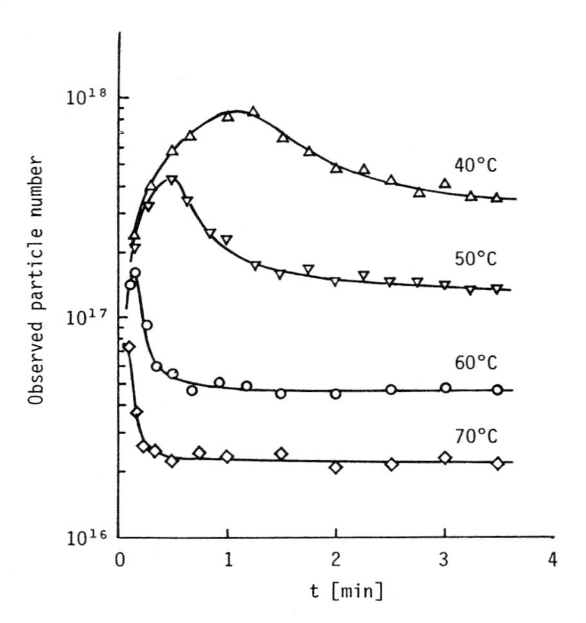

Fig. 4.1.4 Changes in observed particle number with time at different temperatures ($Q = 10^{-3}$ mol s^{-1}, pBr 3.0). (From Ref. 6.)

As readily expected from the behavior of unstable and stable nuclei in Fig. 4.1.2, the particle number initially increases up to a peak and then turns to decrease as actually observed in a AgBr system. Figure 4.1.4 shows typical changes in particle number with time at different temperatures in a double jet system, in which 1.000 mol dm^{-3} $AgNO_3$ and 1.002 mol dm^{-3} KBr were introduced simultaneously into a well-stirred 1000 cm^3 of 2 wt% gelatin solution at a constant rate of 10^{-3} mol s^{-1} for $AgNO_3$ ($Q = 10^{-3}$ mol s^{-1}; pBr 3.0) (6). In this double-jet system, it was also found that the particle radius changing with time was independent of the flow rate of the $AgNO_3$ solution with addition of a KBr solution at almost the same rate, as shown in Fig. 4.1.5. This fact means that the number of the generated stable nuclei is proportional to the feed rate of the reactants, as actually verified by Fig. 4.1.6 (6). These experimental results for the AgBr system were in excellent agreement with the following theoretical prediction for diffusion-controlled growth:

$$n_{+}^{\infty} = \frac{1.567\ QRT}{8\pi D\sigma V_{\mathrm{m}} C_{\infty}}$$

where n_{+}^{∞} is the final particle number, Q the feed rate of the reactants, R the gas

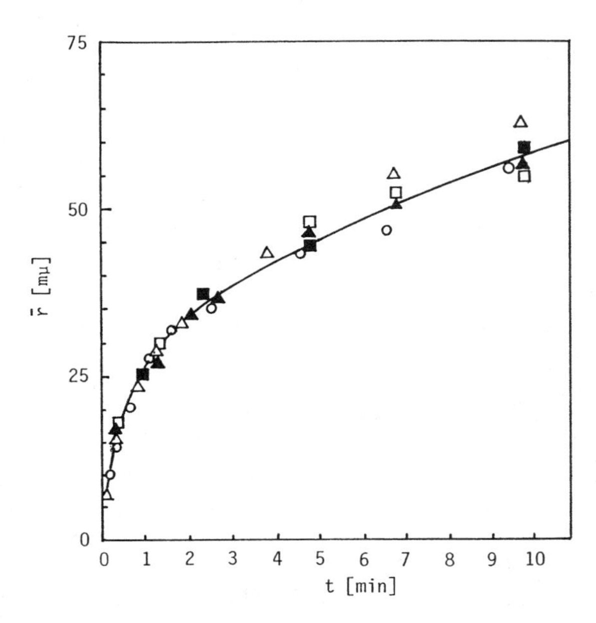

Fig. 4.1.5 Change of the mean particle radius of the observed particles with time at different flow rates of 1 *N* $AgNO_3$ at 70°C and pBr 3.0: (○) 10, (△) 20, (▲) 30, (□) 40, (■) 60 cm^3 min^{-1}. (From Ref. 6.)

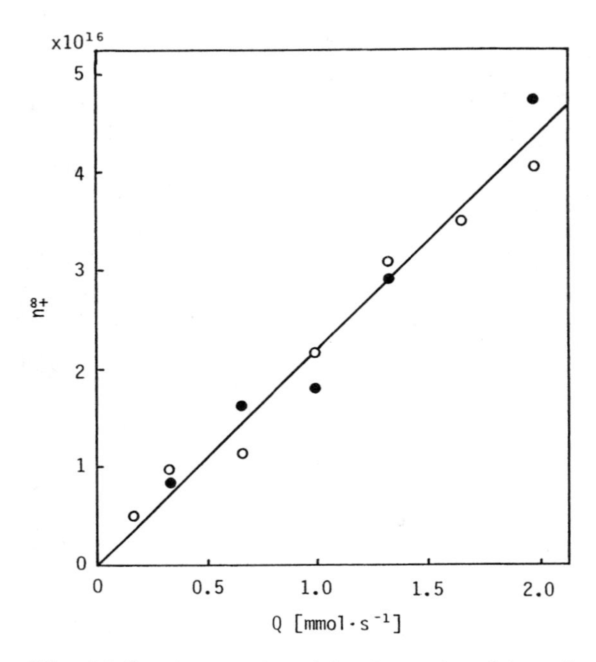

Fig. 4.1.6 Final number of the observed particles, n_{+}^{∞}, as a function of Q at 70°C and pBr 3.0. The closed circles are repeats at the same Q values. (From Ref. 6.)

constant, T the absolute temperature, D the diffusion coefficient of solute, V_m the molar volume of the particles, and C_∞ the solubility of the particles (6).

Figure 4.1.7 shows electron micrographs of AgBr particles prepared with this system at the respective pBr values (pBr $\equiv -\log[Br^-]$) (7). One may find a salient effect of pBr on the crystal habit of the AgBr particles in Fig. 4.1.7; i.e., the shape changes from octahedron bound by {111} faces to cube bound by {100} faces via cuboctahedron bound by both {111} and {100} faces with increasing pBr. This is because the growth rates of the {100} and {111} faces strongly depend on pBr. Figures 4.1.8 (8) and 4.1.9 (9) show the pBr dependence of the growth rates of the {100} and {111} faces in the absence and presence of ammonia, respectively. As a rule, the growth rate of the {111} face is lower than that of the {100} face in the low pBr range, but higher than the latter in the high pBr range, so that octahedral particles are formed in the low pBr range and cubic particles in the high pBr range. However, in the absence of ammonia, both {100} and {111} faces grow in the diffusion-controlled growth mode within pBr 2.6–3.5 and thus the growth rates of both faces are apparently equal, but the particle shape continuously changes even in this pBr range (8). The particle form in this pBr range is deemed to be explained

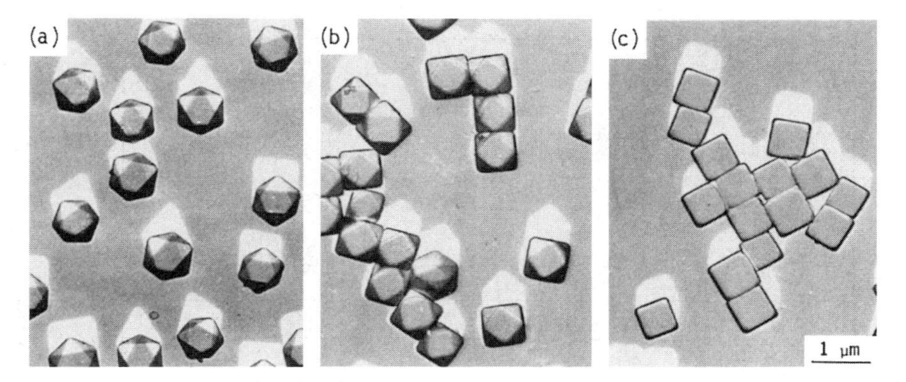

Fig. 4.1.7 AgBr particles prepared at pBr (a) 2.0, (b) 2.8, and (c) 4.0 by the CDJ technique. (From Ref. 7.)

in terms of the equilibrium form, in contrast to the reaction-controlled growth mode in which the particle form belongs to the growth form determined by the relative growth rate of each crystal face. Hence it seems that the cuboctahedral particles in Fig. 4.1.7, prepared at pBr 2.8, may be classified into the equilibrium form in shape. Since the linear growth rate is inversely proportional to the particle radius in a diffusion-controlled growth mode, the absolute standard deviation of the size distribution is expected to be reduced in this pBr range. This self-sharpening of size distribution is also expected of AgCl in almost all pCl range because of its definite diffusion-controlled growth in a wide range of pCl (10), though the particle shape is always cubic because of the thermodynamically stable {100} face of AgCl. In the presence of a high concentration of ammonia, there is no common diffusion-controlled range for the {100} and {111} faces, as is obvious from Fig. 4.1.9. In this case, the growth mode is the reaction-controlled one for all pBr, and thus the particle shape is determined only by the relative growth rates of these two kinds of faces.

For the precise control of the particle shape of AgBr, the controlled double-jet technique with accurate pBr regulation is indispensable, particularly in the high pBr range, in which the excess concentration of bromide ions is extremely small.

Although the flow rate of a silver nitrate solution in the CDJ system is usually set constant, the flow rate is changed, if necessary, according to the predetermined program. For instance, if one wants to conduct the precipitation most efficiently without renucleation during the growth stage, one should control the flow rate of the $AgNO_3$ solution to keep the supersaturation for the growth of silver halide particles at a level slightly below the critical supersaturation for nucleation. In this case, the linear growth rate, dr/dt, for diffusion-controlled growth is given by

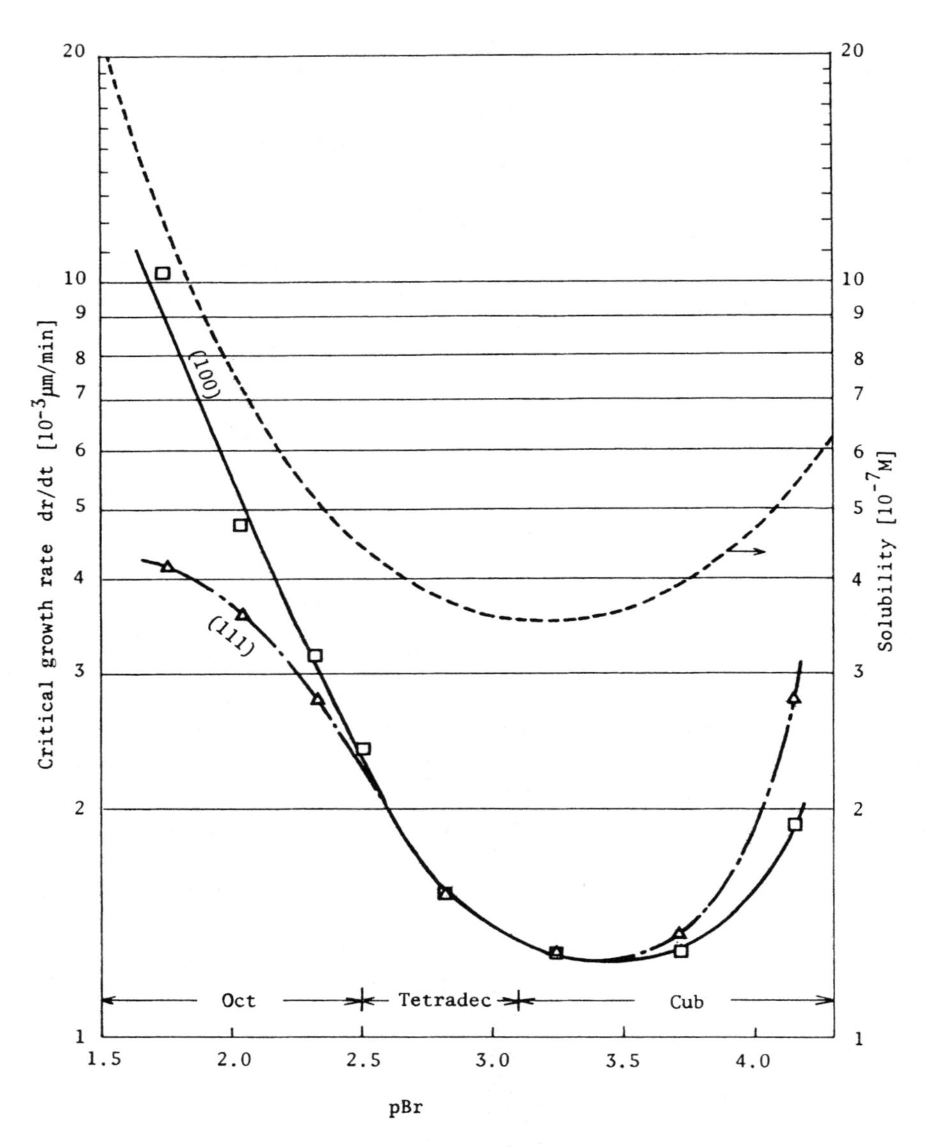

Fig. 4.1.8 Growth rates of the {100} and {111} faces and the solubility of AgBr versus pBr at 60°C. (From Ref. 8.)

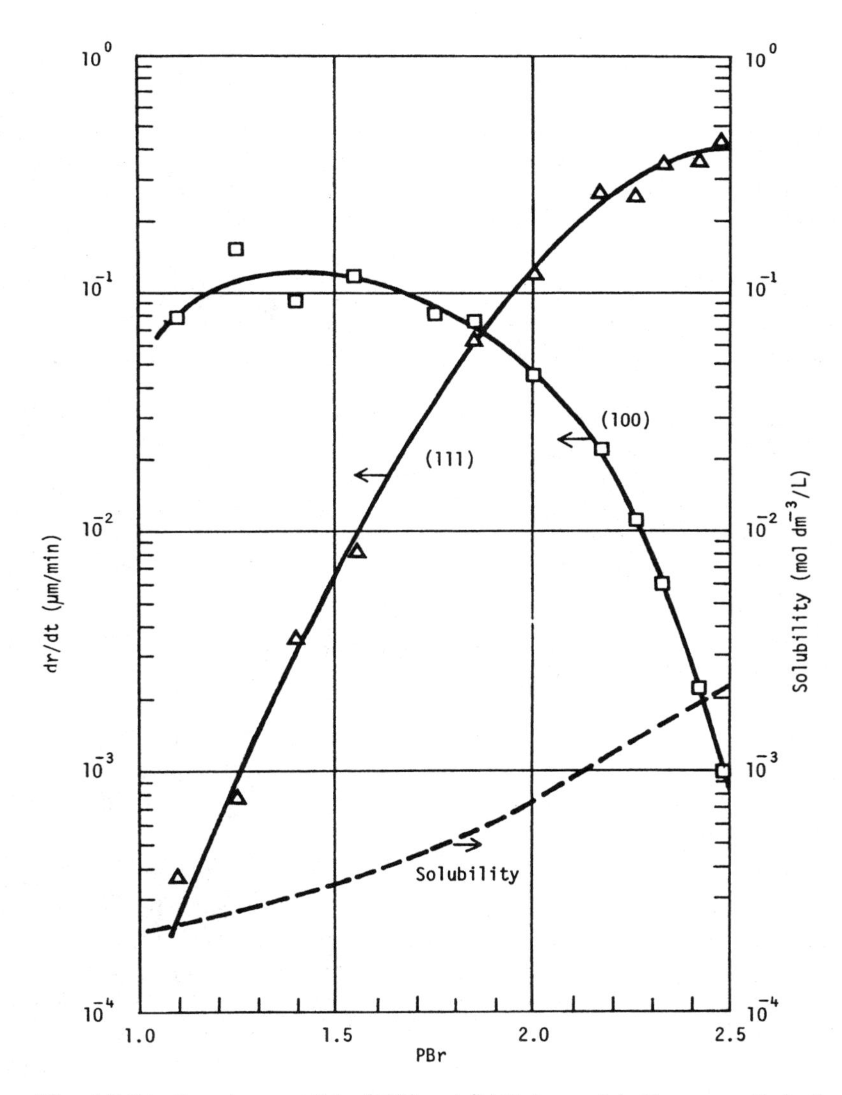

Fig. 4.1.9 Growth rates of the {100} and {111} faces of AgBr versus pBr in the presence of 0.5 *M* NH_3 and 1.0 *M* NH_4NO_3 at 50°C and pH 8.48. (From Ref. 9.)

$$\frac{dr}{dt} = \frac{DV_m}{r}\Delta C^*$$

where r is the particle radius, D the diffusion coefficient of solute, V_m the molar volume of the particle, and ΔC^* the maximum supersaturation or difference between the supersaturation slightly below the critical level and the solubility of the particles. If ΔC^* is assumed to be constant, one obtains

$$r = \sqrt{2DV_m \, \Delta C^* t}$$

If the number concentration of the particles is denoted by n, the maximum consumption rate of solute, $-dC/dt$, is given by

$$-\frac{dC}{dt} = \frac{4\pi r^2 n}{V_m}\frac{dr}{dt}$$
$$= 4\pi n \sqrt{2(D \, \Delta C^*)^3 V_m t}$$

In the case of reaction-controlled growth, dr/dt is given by

$$\frac{dr}{dt} = kV_m \, \Delta C^*$$

where k is the reaction rate constant. Hence,

$$r = kV_m \, \Delta C^* t$$

and

$$-\frac{dC}{dt} = 4\pi n k^3 V_m^2 \, (\Delta C^*)^3 t^2$$

Therefore, the most efficient preparation of monodispersed particles can be achieved by regulating the feed rate of reactants ($= -dC/dt$) as proportional to $t^{1/2}$ in the case of diffusion-controlled growth and to t^2 in the case of reaction-controlled growth, as illustrated in Fig. 4.1.10.

Strictly speaking, ΔC^* depends on the particle size, since the solubility of a particle increases with decreasing particle size by the Gibbs–Thomson effect. If the supersaturation is low during the particle growth by diffusion control, the self-sharpening of the size distribution is lessened by the Gibbs–Thomson effect. Hence, if the bulk concentration of solute is kept high throughout the precipitation by raising the addition rate of the reactants with the particle growth by diffusion control, the size distribution broadening by the Gibbs–Thomson effect is expected to be minimized (11). Thus, the benefit of the increasing addition rate of the reactants according to the growth mode is not limited only to the efficient synthesis of the monodispersed particles, but is also in favor of the promotion of the self-sharpening of the size distribution. In fact, gradual increase of the addition rate of the reactants with time

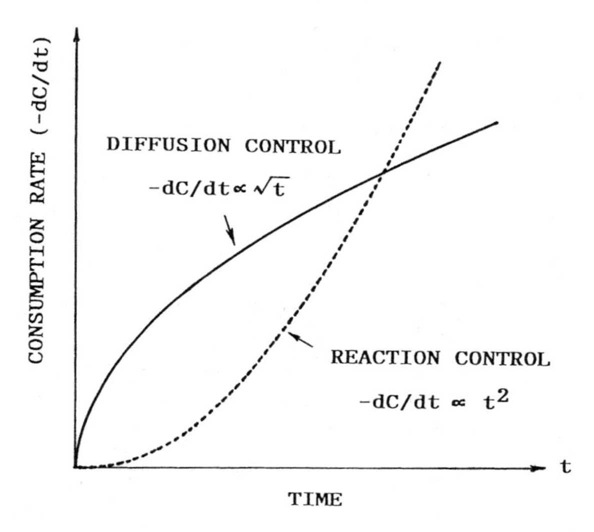

Fig. 4.1.10 Maximum consumption rates of solute as a function of time for diffusion-controlled growth and reaction-controlled growth.

has been proposed in patents in expectation of promoting the self-shapening effect in the manufacture of silver halide emulsions in photographic industry (12,13).

REFERENCES TO SECTION 4.1

1. CR Berry. J Opt Soc Am 52:888, 1961.
2. CR Berry, SJ Marino, CF Oster Jr. Photogr Sci Eng 5:332, 1961.
3. CR Berry, DC Skillman. Photogr Sci Eng 6:159, 1962.
4. E Moisar, E Klein. Ber Bunsenges Phys Chem 67:949, 1963.
5. F Claes, R Berendsen. Photogr Korresp 101:37, 1965.
6. T Sugimoto. J Colloid Interface Sci 150:208, 1992.
7. T Sugimoto. M R S Bull 15(12):23, 1989.
8. T Sugimoto. J Colloid Interface Sci 93:461, 1983.
9. T Sugimoto. J Colloid Interface Sci 91:51, 1983.
10. RW Strong, JS Wey. Photogr Sci Eng 23:344, 1979.
11. T Sugimoto. Adv Colloid Interface Sci 28:65, 1987.
12. JD Lewis. US Patent 4,067,739, 1978.
13. M Saito. US Patent 4,242,445, 1980.

4.2 OSTWALD RIPENING PROCESS

TADAO SUGIMOTO
Institute for Advanced Materials Processing,
Tohoku University, Sendai, Japan

It is possible to grow uniform silver halide particles by Ostwald ripening in a closed system in which two kind of particles definitely different in mean size coexist and the larger particles are grown by dissolution of the smaller ones based on the Gibbs–Thomson effect (1). If the smaller particles and larger ones are denoted by A and B, the rates of change in particle radius of the respective particles A and B are given by

$$-\frac{dr_A}{dt} = K_A V_m (C_A - C_b) \tag{1}$$

$$\frac{dr_B}{dt} = K_B V_m (C_b - C_B) \tag{2}$$

where K_A and K_B are the reaction rate constants of particles A and B; C_A and C_B are the solubilities of particles A and B; V_m is the molar volume of the solid; and C_b is the bulk concentration of the solute. At the steady state of the dissolution of particles A and the growth of particles B, it holds that

$$S_A \frac{dr_A}{dt} + S_B \frac{dr_B}{dt} = 0 \tag{3}$$

where S_A and S_B are the total surface areas of particles A and B, respectively. Hence one obtains

$$\frac{dr_B}{dt} = \frac{K_A K_B S_A \overline{V}_m (C_A - C_B)}{K_A S_A + K_B S_B} \tag{4}$$

If the growth rate of particles B is governed by the dissolution rate of particles A (dissolution-controlled growth), it must holds that $K_A S_A << K_B S_B$. In this case, Eq. (4) reduces to

$$\frac{dr_B}{dt} = \frac{S_A}{S_B} K_A \overline{V}_m (C_A - C_B) \tag{5}$$

and

$$C_b \simeq C_B \tag{6}$$

If the growth rate of particles B is determined by the deposition rate of the solute onto particle B (deposition-controlled growth), then $K_A S_A >> K_B S_B$ so that the Eq. (4) is approximated by

$$\frac{dr_B}{dt} = K_B \overline{V}_m (C_A - C_B) \tag{7}$$

and

$$C_b \simeq C_A \tag{8}$$

Consequently, dr_A/dt is proportional to S_A/S_B for the dissolution-controlled growth, whereas it is independent of S_A/S_B for the deposition-controlled growth. Obviously, information on the growth mechanism is obtained only in the case of deposition-controlled growth.

If one uses tabular double-twin particles of AgBr as particles B together with much smaller particles of normal AgBr crystal as particles A, one may be able to study the growth mechanism of the tabular particles, characterized by the exclusively rapid lateral growth with negligibly small virtical growth to the basal planes at a sufficiently low pBr (pBr $\equiv -\log[Br^-]$) and the increasing absolute standard deviation of the size distribution with the growth. In order to explain the characteristics of the growth of the double-twin tabular grains, the spherical diffusion model was proposed for the growth of the tabular grains, as illustrated in Fig. 4.2.1, where r

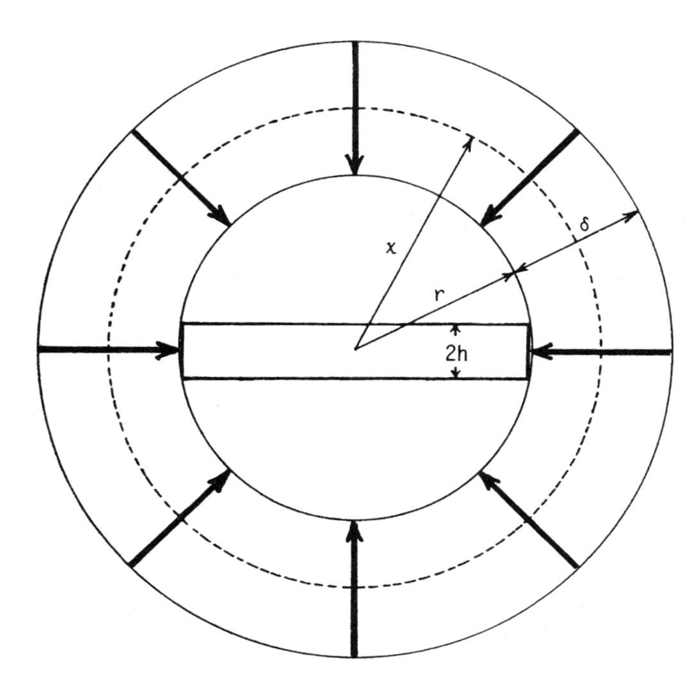

Fig. 4.2.1 Spherical diffusion model for the growth of a tabular grain. (From Ref. 6.)

is the disk radius of the tabular grain, x is the radius of a spherical surface hypothesized within the diffusion layer of solute, and δ is the thickness of the diffusion layer of solute (2). In this model, the tabular grains are assumed to be grown only in each lateral direction by diffusion control with convergent diffusion of solute. The total flux of solute across a hypothetical sphere with a radius of x is written as

$$J = 4\pi x^2 D \frac{dC}{dx} \tag{9}$$

Since J is independent of x in the steady state, integration of C with respect to x from $r + \delta$ to r gives

$$J = \frac{4\pi D r(r + \delta)}{\delta}(C_b - C_B) \tag{10}$$

On the other hand, if a half of the thickness of the tabular grain is denoted by h,

$$J = \frac{4\pi r h}{V_m} \frac{dr}{dt} \tag{11}$$

Combination of Eqs. (10) and (11) yields

$$\frac{dr}{dt} = k\left(1 + \frac{r}{\delta}\right) \tag{12}$$

where

$$k \equiv \frac{DV_m}{h}(C_b - C_B) \tag{13}$$

If the size dependence of C_B by the Gibbs–Thomson effect can be ignored, k is a constant for all particles in a system, and dr/dt increases linearly with r. From Eq. (12),

$$\ln \frac{r + \delta}{r_0 + \delta} = \frac{k}{\delta} t \tag{14}$$

where r_0 is the initial value of r. On the other hand, the relationship between the standard deviation of the size distribution, Δr, and the mean radius $\bar{r}$ of all tabular grains in the system is also derived from Eq. (12) as

$$\frac{\Delta r}{\bar{r} + \delta} = \frac{\Delta r_0}{\bar{r}_0 + \delta} \tag{15}$$

where $\bar{r}_0$ is the initial mean radius. Therefore, $\Delta r/(\bar{r} + \delta)$ is kept constant during the growth of the tabular grains, though the absolute standard deviation of the size distribution is enlarged. The thickness of the diffusion layer δ is essentially determined by the hydrodynamics of the Brownian motion of colloidal particles (3).

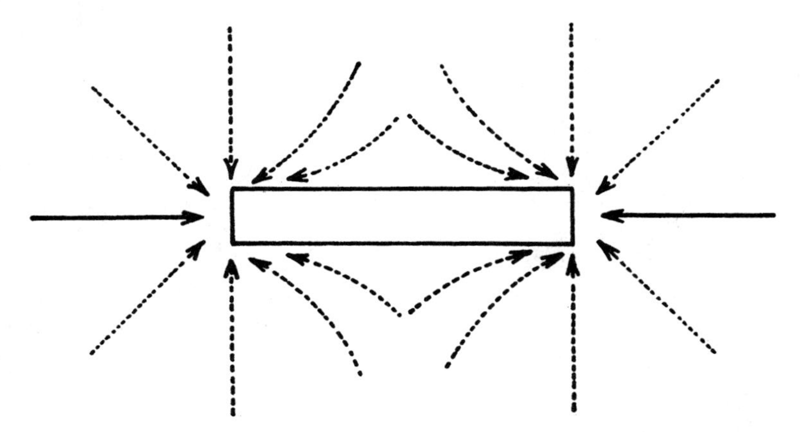

Fig. 4.2.2 A more realistic profile of the diffusive solute flux about a tabular grain. (From Ref. 6.)

However, if the particles are so densely populated in a system that half the mean interparticle separation (the surface-to-surface distance between two particles) is smaller than the dynamically determined thickness of the diffusion layer, δ may be approximated by half the interparticle separation (4,5). If the number concentration of particles A is sufficiently high and much greater than that of the tabular grains, δ is given by a half of the mean interparticle separation of the particles A. This is a simplified model of the actual feature of the diffusion flux of solute around a tabular grain shown in Fig. 4.2.2 (6). In contrast, Karpinski and Wey (7) proposed a cylindrical diffusion model in which the solute is assumed to diffuse only laterally from the circumferential side of a tabular grain. This model, however, assumes only the diffusion flux normal to the side faces indicated by the solid-line arrows in Fig. 4.2.2 and ignores the rather greater contribution of the tangential diffusion flux of the dotted arrows, leading to a conclusion that the absolute standard deviation of the size distribution is kept constant during the growth, which may be different from the actual observation.

In order to test the spherical diffusion model, fairly uniform tabular double-twin particles of AgBr were actually grown by dissolution of a great amount of very fine particles of AgBr normal crystal through Ostwald ripening (2,6). Figure 4.2.3 shows a transmission electron microscopy (TEM) image of replicas of tabular particles ($\bar{r}_0 = 0.410$ μm) mixed with much smaller particles ($\bar{r}_0 = 0.0425$ μm) of AgBr in the initial stage (molar ratio $= 1:4$) (a) and after aging for 30 min at pBr 1.52 (pBr $= -\log[\mathrm{Br}^-]$) and 70°C (b) (2). This method is based on the high activity of the troughs of the side planes of the double-twinned tabular particles as a center for the surface nucleation and the high solubility of the coexisting fine particles.

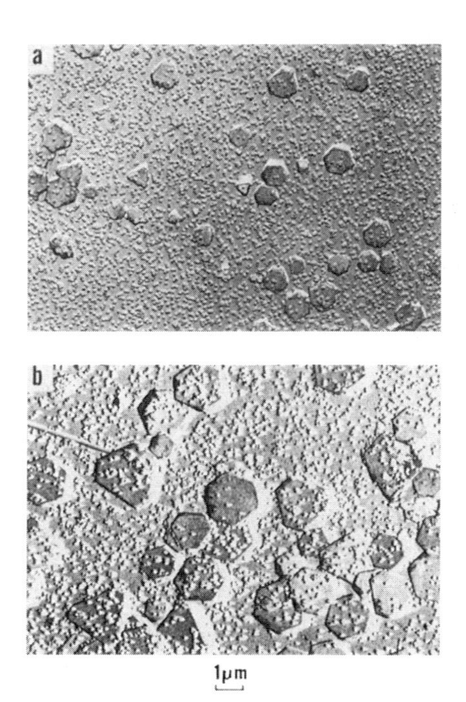

Fig. 4.2.3 A mixture of tabular AgBr grains ($\bar{r}_0$ = 0.410 μm; $\bar{h}_0$ = 0.117 μm) and fine spherical ones ($\bar{r}_0$ = 0.0425 μm) at the molar ratio of 1:4. Electron micrographs: (a) at the start of aging; (b) after aging for 30 min at pBr 1.52 and 70°C ([particles A] = 0.08 *M*, [gelatin] = 5%, [KNO_3] = 1 *M*). (From Ref. 2.)

Figure 4.2.4 is for the lateral growth rate of the tabular grains (particles B) as a function of the molar ratio of [particle A]/[particles B] where [particles A] was fixed, showing the characteristic of the deposition-controlled growth with the increasing molar ratio. Hence this system is suitable for the analysis of the growth mechanism of the tabular grains. It was also found that the lateral growth rate of particles B strongly depends on [particles A] when [particles B] is fixed, corresponding to the change of δ, and that the activation energy of the growth rate constant was about 4.5 kcal mol^{-1} in the low pBr range below 3. Hence it is obvious that the tabular grains are grown by diffusion control in the low pBr range. The activity of the side troughs for the lateral growth of the tabular particles was found to increase dramatically with increasing concentration of the excess bromide ions or decreasing pBr, as clearly shown in Fig. 4.2.5. As a result, a linear relationship was found between aging time and $\ln[(\bar{r} + \delta)/(\bar{r}_0 + \delta)]$ in a low pBr range less than 3, as shown in Fig. 4.2.6. Also, the coefficient of variation or the relative standard deviation of the

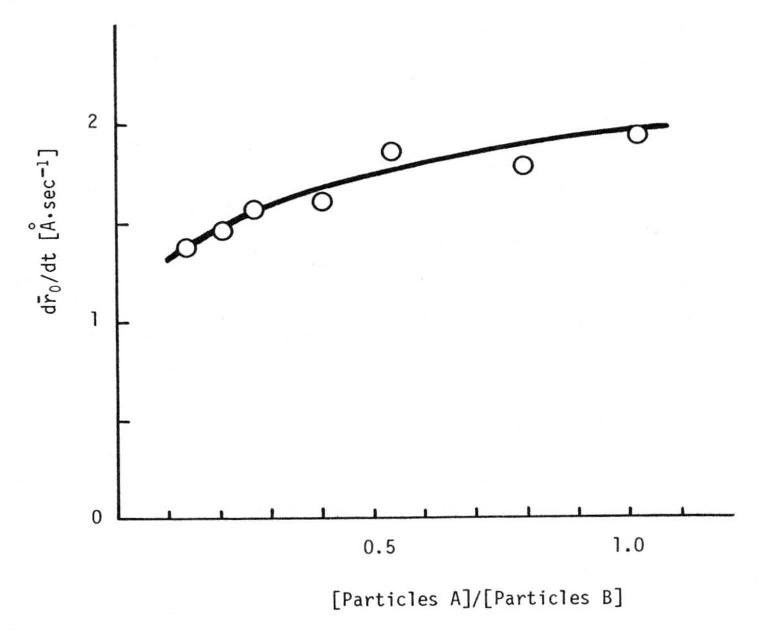

Fig. 4.2.4 Lateral growth rate of the tabular grains (particles B) as a function of [particles A]/[particles B] at pBr 1.52 and 70°C, where [particles B] was changed while [particles A] is fixed. (From Ref. 2.)

size distribution was kept almost constant during their growth in a low pBr range less than 3 as shown in Fig. 4.2.7, in accord with the spherical diffusion model, while it was reduced at high pBr due to the diminishing activity of the side troughs. Since the spontaneous nucleation does not occur in this system, the broadening of the size distribution due to nucleation does not take place.

This method can be applied to the growth of normal monodispersed single-crystal particles of silver halides without degrading the uniformity of the original particles. In this case, ammonia is useful as an accelerator of the Ostwald ripening (8). For example, Fig. 4.1.9 in the preceding section for the pBr dependence of the growth rates of the {100} and {111} faces of AgBr was obtained by the Ostwald ripening procedure using small cubic particles of $\bar{r}_{100} = 0.115$ μm as particles A with larger cubic particles of $\bar{r}_{100} = 0.462$ μm or octahedral particles of $\bar{r}_{111} = 0.376$ μm as particles B, in the presence of 0.5 *M* NH_3 and 1.0 *M* NH_4NO_3 at 50°C, where $\bar{r}_{100}$ and $\bar{r}_{111}$ are the mean distances from the center of a particle to the {100} and {111} faces, respectively (8).

As is obvious from Fig. 4.1.9 in the preceding section, the growth rate of the {100} face of AgBr decreases with increasing pBr in the presence of ammonia, while

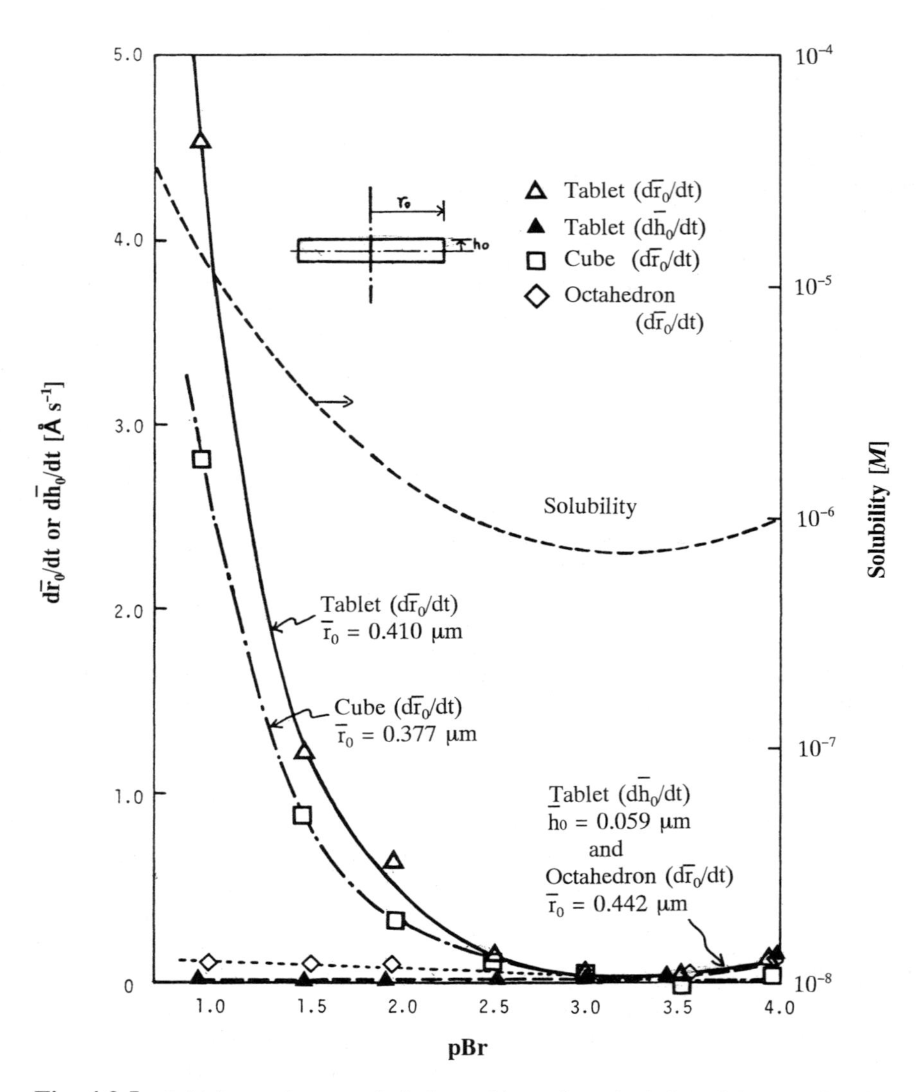

Fig. 4.2.5 Initial growth rates of tabular, cubic, and octahedral particles as a function of pBr, where [particles A]/[particles B] = 4 and temperature = 70°C. (From Ref. 2.)

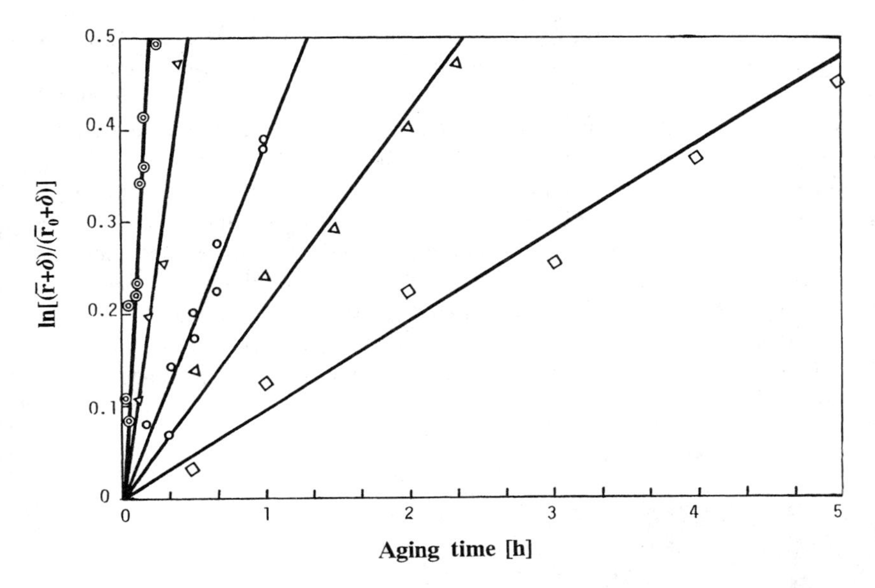

Fig. 4.2.6 Relationship between aging time and $\ln[\bar{r} + \delta)/(\bar{r}_0 + \delta)]$ for pBr 1.00 (◎), 1.52 (▽), 2.00 (○), 2.30 (△), and 2.52 (◇). (From Ref. 2.)

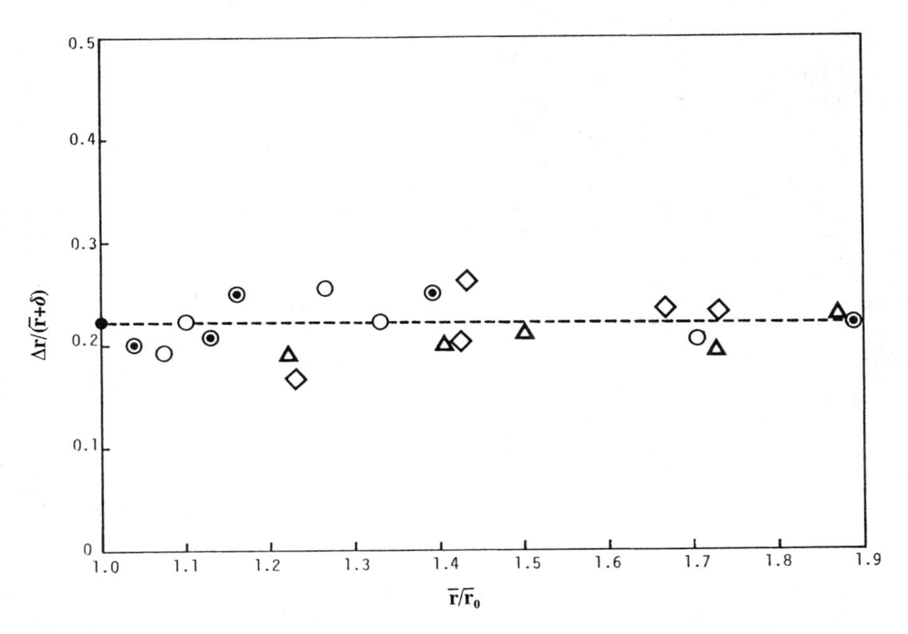

Fig. 4.2.7 Relationship between $\bar{r}/\bar{r}_0$ and $\Delta r/(\bar{r} + \delta)$ for pBr 1.52 (◉), 2.00 (○), 2.30 (△), and 2.52 (◇). (From Ref. 2.)

the growth rate of the {111} face increases. Hence it is expected that if we mix small cubic particles of AgBr with larger octahedral particles at a low pBr and allow the mixture to be aged, the {100} faces of the small cubic particles may grow by dissolution of the corners of the octahedral ones with a high solubility until they reach a quasi-equilibrium in solubility between the {100} faces of the small and large particles. Similarly, if we employ a combination of small octahedra with larger cubes in this ripening at a high pBr, we may find the growth of the {111} faces of the small octahedra by dissolution of the corners of the larger cubes. Since smaller particles are expected to grow at the expense of the larger particles in these cases, the phenomenon was named ''reversed Ostwald ripening.'' This reversed Ostwald ripening, associated with narrowing in size distribution, was actually observed as shown in Figs. 4.2.8 and 4.2.9, where the size distributions were obtained with a Coulter counter (9).

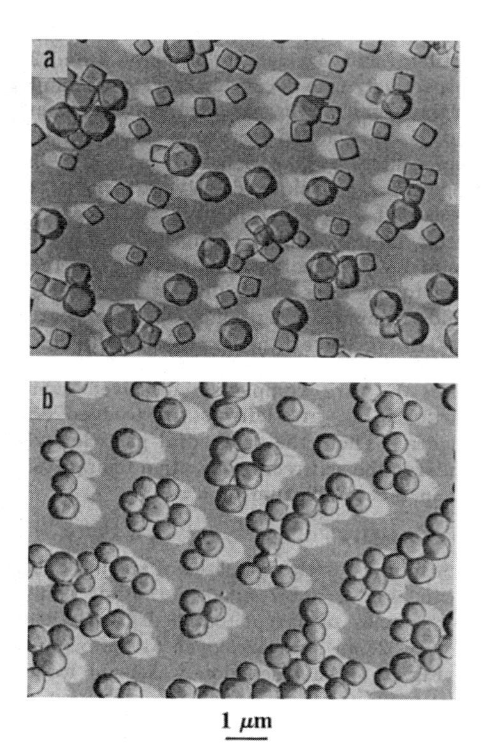

Fig. 4.2.8 Electron micrographs of a mixture of large octahedral particles ($\bar{r}_{111} = 0.327$ μm) and small cubic ones ($\bar{r}_{100} = 0.247$ μm) (a) before and (b) after 60-min aging at 50°C and pBr 1.40, respectively ([AgBr] = 1 *M*, [NH_3] = 0.5 *M*, [NH_4NO_3] = 0.1 *M*, [gelatin] = 1 wt%, pH = 8.48.). (From Ref. 9.)

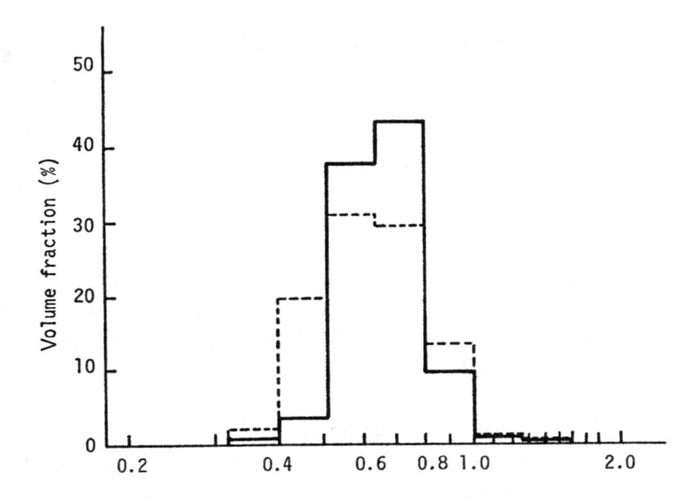

Fig. 4.2.9 Histograms of the size distributions of the particles shown in Fig. 4.2.8. Original and final size distributions are shown by broken and solid lines, respectively. The diameter of an equivalent sphere having the same volume as a nonspherical particle was obtained with a Coulter counter. (From Ref. 9.)

REFERENCES TO SECTION 4.2

1. T Sugimoto. Adv Colloid Interface Sci 28:65, 1987.
2. T Sugimoto. Photogr Sci Eng 28:137, 1984.
3. T Sugimoto. AIChE J 24:1125, 1978.
4. R Jagannathan, JS Wey. J Crystal Growth 51:601, 1981.
5. R Jagannathan, JS Wey. Photogr Sci Eng 26:61, 1982.
6. T Sugimoto. J Imag Sci 33:203, 1989.
7. PH Karpinski, JS Wey. J Imaging Sci 32:34, 1988.
8. T Sugimoto. J Colloid Interface Sci 91:51, 1983.
9. T Sugimoto. J Soc Photogr Sci Technol Jpn 46:307, 1983.

4.3 REACTION IN MICROEMULSIONS

KIJIRO KON-NO
Science University of Tokyo, Tokyo, Japan

Silver halide particles have been synthesized by mixing two reversed micelles, one containing aqueous silver nitrate and other containing aqueous sodium or potassium chlorides. The NaCl formation reaction in Aerosol OT–alkane solutions was estimated to complete in a very short period of 60 ms by the stopped-flow measurements (1–3). Since this time is on the same order of magnitude as the time scale for the intermicellar exchange of materials, one can assume that the reaction of precipitation is faster in bulk water, and hence, the rate of nucleation and the particle growth are controlled by the collision, fusion, and split of the droplets. In order to examine the particle growth in micellar droplets, the number of nuclei per water droplet (Nn/Nm), mean number of Ag^+ per water droplet (Ag^+/Nm), and the sum of probabilities containing i ions or more per water droplet ($\sum p_k$), where the probability of containing more than 100 ions per water droplet is negligible, are calculated as a function of $AgNO_3$ concentration at a given Wo in Aerosol OT–n-heptane system by Monnoyer and Nagy (4). The values obtained are presented in Table 4.3.1. A constant value of F for all the $AgNO_3$ concentrations in Table 4.3.1 corresponds to the optimum value of i, which is the number of AgBr units required to form a "viable" nucleus, and then if it is assumed that $i > 3$, the F value indicated to be too high for the lowest initial $AgNO_3$ concentrations. In this system however, the value of $i = 1$ is approximately reasonable for $F \approx 4 \times 10^{-4}$–$10^{-3}$ at Wo (= $[H_2O]/[\text{surfactant}]$) = 8 and 10 microemulsions, and for 7×10^{-3} at Wo = 12.4 and 20 microemulsions, respectively. The low value of F in Table 4.3.1 shows that only approximately 1 out of every 1000 water droplets leads to the formation of a nucleus for Wo = 8 and 10 microemulsions, and approximately 1 out of every 100 water droplets for Wo = 12.4 microemulsion. On the other hand, Shah and coworkers reported that the mass of Ag^+ and Cl^- or Br^- in each droplet before reaction is only 1/200 and 1/300 of the amount required to grow each AgCl and AgBr particle of 5 nm in diameter, respectively (2,3). These small values emphasize the fact that nucleation occurs at very beginning of the mixing time of two microemulsions, and once the nuclei are formed they grow to reach a certain size owing to the fast intermicellar exchange of nuclei. Such growth process may be also suggested from the fact that the final particle size is generally larger than the size of initiated water droplet without particles. Recently, Sato and coworkers have examined the particle growth from the kinetics on the basis of the intermicellar exchange coagulation process (5). As coagulation proceeds via second-order kinetics, the second-order rate constant k_c for particle coagulation is calculated from the rate of decrease in

Table 4.3.1 Parameters for the Formation of AgBr Particle in Aerosol OT–*n*-Heptane System

$[AgNO_3]$ (*M*)	Particle size (nm)	N_n/N_m	N_{Ag_+}/N_m	$\sum_i^{100} P_k$	*F*
Wo = 8					
0.063	7.38 ± 1.92	2×10^{-4}	1.17	0.4756	4.2×10^{-4}
0.125	9.61 ± 2.94	3×10^{-4}	2.25	0.8003	3.7×10^{-4}
0.250	10.18 ± 3.70	4×10^{-4}	3.57	0.9445	4.2×10^{-4}
0.500	10.43 ± 3.82	7×10^{-4}	1.52	0.6104	1.1×10^{-3}
Wo = 10					
0.063	6.25 ± 1.12	7×10^{-4}	1.52	0.6104	1.1×10^{-3}
0.125	7.37 ± 1.96	1×10^{-3}	4.23	0.9711	1.0×10^{-3}
0.250	10.33 ± 3.18	8×10^{-4}	5.89	0.9945	8.0×10^{-4}
0.500	11.91 ± 3.23	1×10^{-3}	10.06	0.9999	1.0×10^{-3}
Wo = 12.4					
0.063	6.04 ± 1.36	1×10^{-2}	2.88	0.8909	1.1×10^{-2}
0.125	5.61 ± 0.86	8×10^{-3}	5.96	0.9948	8.0×10^{-3}
0.250	8.08 ± 1.92	2×10^{-3}	10.09	0.9999	2.0×10^{-3}
0.500	8.17 ± 2.73	6×10^{-3}	23.85	1.0000	6.0×10^{-3}
Wo = 20					
0.063	6.53 ± 1.47	5×10^{-3}	11.28	1.0000	5.0×10^{-3}
0.125	6.16 ± 1.12	1×10^{-2}	17.53	1.0000	1.0×10^{-2}

Note: N_n/N_m, Number of nuclei per water droplet.
N_{Ag_+}/N_m, Mean number of nuclei per water droplet.
$\sum_i^{100} P_k$, Square sum of probabilities containing one Ag^+ or more per water droplet. Here *i* is the number of AgBr units necessary to form a nucleus.
F, Optimum value of *i* ion to form a nucleus.

the particle concentration. Here, the particle concentration was calculated, assumed that all the reactants are converted to sherical and monodisperse particles and that the density of particles was independent of the particle size. When the values of k_c obtained were plotted against particle size, which is estimated from band gap energy of particles, the values were almost independent of small AgI particles below 4.5 nm, and close to an estimated value for the rate constant of intermicellar exchange reaction of reversed micelles, $k_{ex} = 55.6 \times 10^7 \ M^{-1} \ s^{-1}$ (6). As a result, it was found that small AgI particles grow by the rapid intermicellar coagulation process. For AgI particles greater than 4.5 nm and AgBr particles over the whole range of size, however, the values of k_c were smaller than the k_{ex} value, indicating that the coagulation processes for these particles are not controlled by the exchange rate. Then Sato et al. further proposed a model such that the particle coagulation occurs in micellar droplets containing two or more particles and thus the coagulation rate is proportional to the concentration of micelles. The first-order rate constants, k_{mc}, estimated from the micelle concentration for AgBr particles and larger AgI particles were independent of the reactant concentration, indicating that the coagulation kinetics is controlled by the concentration of droplets containing two or more particles rather than the intermicellar coagulation. When the k_{mc} values were compared for AgBr and AgI at a given particle size, the values for AgBr decreased with decreasing Wo, but the values of AgI particles behaved the opposite way. This difference is due to the formation of larger AgI particles than AgBr particles. However, such analysis was not performed for AgCl particles because AgCl particles dissolve by impurities in Aerosol OT, and further the particle size could not estimated from the absorption spectrum. On the other hand, the growth rate of AgCl and AgBr particles qualitatively performed from the change of transmittance of particle solution with reaction time increased with increasing chain length of alkane, such as hexane, heptane, and decane, used as bulk solvent (2,3), whereas it was opposite for the chain length of alcohol used (2). However, the growth rate of particles was enhanced with increasing the amount of pentanol in *n*-octane system, but the addition of the long-chain nonionic surfactant Arlacel-20 into the *n*-decane system slowed it down. Such difference in growth rate should be reflected on the particle size, as mentioned later.

In the system Wo = 8 of alkane solutions of Aerosol OT, spherical and uniform AgCl particles of 5 to 10 nm in diameter were precipitated from the reaction between $AgNO_3$ and NaCl (2). On the other hand, nearly spherical and uniform AgBr particles of 5–7 nm were also produced in an Aerosol OT system of Wo = 4.63 in *n*-hexane, and they increased with increasing the chain length of alkane used as bulk solvent: namely, 5–8 nm in *n*-heptane and 5–10 nm in *n*-octane, respectively (3). The size of AgBr particles produced in *n*-hexane also increased with increasing Wo. The size of AgBr particles produced in heptane solution of Aerosol OT was 11 nm at Wo = 10 and 25 nm at Wo = 20 (7). In the Aerosol OT–isooctane system, the particle diameter estimated from band gap energy was 2.4–2.7 nm for AgBr in the range Wo = 3–10 and 2.8–6.0 nm for AgI in the range Wo = 3–30. The size of these particles increased with increasing the concentration of reactants, but the size of AgI particles was significantly larger than that of AgBr particles under same preparation conditions (5). In order to clarify the relationship between the particle size and the extent of Wo, Tanabe and coworkers synthesized AgCl particles as a function of Wo in the Aerosol OT–cyclohexane system at 30°C (8). The size of AgCl particles was <2 nm in the reversed micelle region of Wo = 3, at which the water droplets are rigidly bound to the ionic polar groups of surfactant, but in the swollen micelle regions of Wo = 6–8 at which the droplets are held together with the hydrated ionic polar groups the size increased to 2.0–2.4 nm. On the other hand, the size of particles formed at Wo = 20–25, a region of W/O

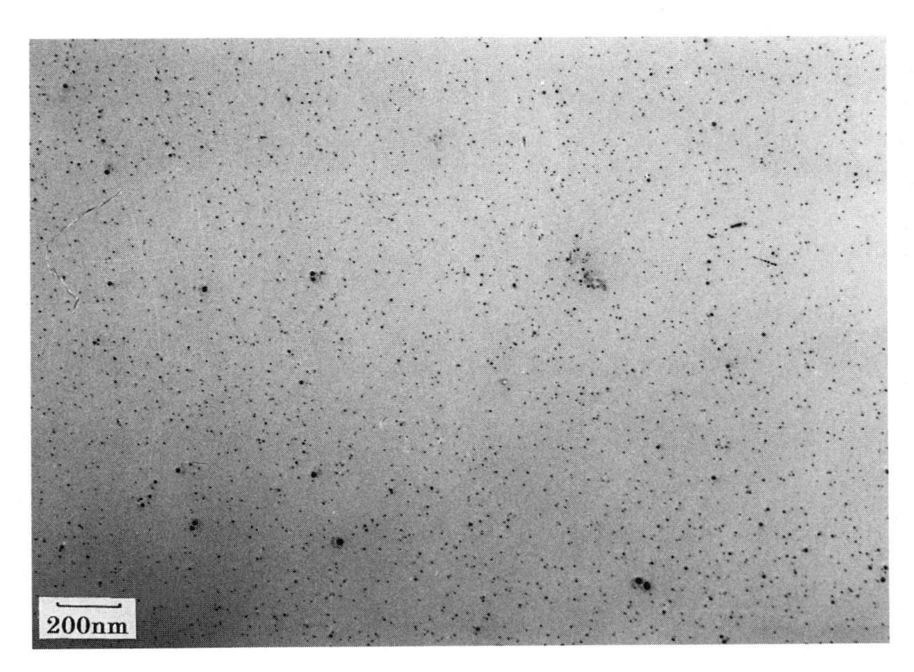

Fig. 4.3.1 Transmission electron micrograph of AgCl particles precipitated in 0.10 mol kg^{-1} cyclohexane system. Wo = 25; [Ag^+] and [Cl^-]/[Aerosol OT] = 0.01.

microemulsions containing bulk-like water, was 4.1–4.9 nm. Such stepwise increases in the size may indicate that the particle size is not only controlled by the size of micelle water droplets, but may also be restricted by the states of water droplets. A transmission electron micrograph of representative AgCl particles is shown in Fig. 4.3.1. When AgCl or AgBr particles are synthesized in polyoxyethylene (6) nonylphenyl ether–and hexaethylene glycol dodecyl ether–cyclohexane systems, $AgCl_2^-$–or $AgBr_2^-$–surfactant complexes were found to be precipitated together with AgCl or AgBr particles (9). After the complexes were centrifuged, the size of the particles obtained increased with increasing Wo, but a stepwise increase in particle size with Wo, as seen in Aerosol OT systems (8), was not observed. Compared to the particle size precipitated in Aerosol OT micelles, that of polyoxyethylene (6) nonylphenyl ether was significantly larger under same preparation conditions. Such differences in the particle size may due to a different atmosphere of water droplets, which are microreactors. The sizes of particles obtained in each reversed micelle system are summarized in Table 4.3.2.

In particle formation in ionic and nonionic reversed micelles as mentioned earlier, simple electrolytes such as $AgNO_3$ and NaCl were used as reactants. If the

Table 4.3.2 Size of Silver Halide Particles in Various Reversed Micelle Systems

Particle	Reaction system[a]	Wo	Micellar state[b]	Diameter (nm)	Reference
AgCl	AOT–*n*-alkane	8		5–10	2
	AOT–cyclohexane	3	RMs	<2.0	8
		6	SMs	2.0 ± 0.4	8
		8	SMs	2.4 ± 0.7	8
		20	MEs	4.1 ± 1.0	8
		25	MEs	4.9 ± 1.2	8
	NP-6–cyclohexane	3	RMs	7.5 ± 0.4	9
		6	SMs	9.2 ± 0.9	9
		8	SMs	17.1 ± 3.8	9
		29	MEs	23.1 ± 14.3	9
	DP-6–cyclohexane	3	RMs	6.0 ± 0.7	9
		6	SMs	9.1 ± 1.5	9
AgBr	AOT–*n*-hexane	4.63		5–7	3
		7.94		20	3
	AOT–*n*-heptane	4.63		5–8	3
	AOT–*n*-heptane	10		11 ± 0.33	7
		20		25	7
	AOT–*n*-octane	4.63		5–10	3
	AOT–isooctane	3–10		2.4–2.7[c]	5
AgI	AOT–isooctane	3–30		2.8–6.0[c]	5

[a] AOT, Aerosol OT; NP-6, polyoxyethylene(6) nonylphenyl ether; DP-6, hexaethylene glycol dodecyl ether.

[b] RMs, reversed micelles; SMs, swollen micelles; Mes, W/O microemulsions.

[c] Estimated from bandgap energy.

halogen species of particles are supplied from the counterions of cationic surfactants, one may expect to form a different shape or size of particle because the formation of particles occurs at interfaces of the water droplets. Synthesis of AgCl particles was attempted by mixing two reversed micelle solutions, one polyoxyethylene (6) nonylphenyl ether containing aqueous $AgNO_3$ and other dioleyldimethylammonium chloride (10). When AgCl particles were prepared as a function of molar ratio of dioleyldimethylammonium chloride to $AgNO_3$, spherical particles were precipitated in solutions below $[Cl^-]/[Ag^+] = 19$, whereas in solution above that ratio, cubic and uniform AgCl particles containing metal silver dots, as showed in Fig. 4.3.2, were produced. The particle size, expressed by edge length of particle, increased rapidly above Wo = 5.5–6 as seen in Fig. 4.3.3. From a comparison of the solubilization diagram of water, it is found that the rapid increase is attributable to the transition from the water droplets rigidly bound to the ionic polar groups of surfactant to the water droplets interacting with the hydrated ionic polar groups by the hydrogen

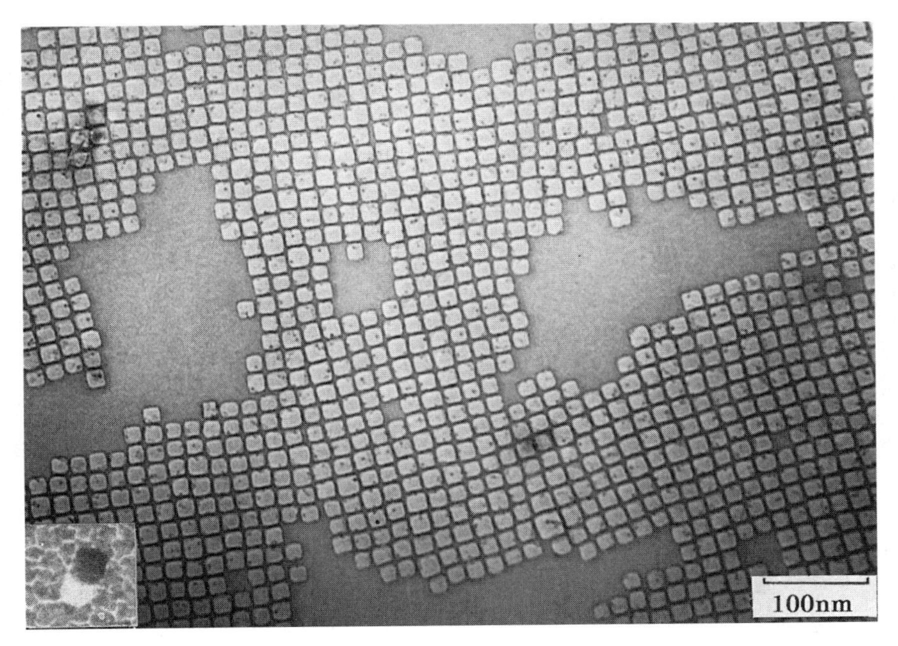

Fig. 4.3.2 AgCl particles containing metal silver dots precipitated in 0.10 mol kg^{-1} mixtures of dioleyldimethylammonium chloride and polyoxyethylene(6) nonylphenyl ether–cyclohexane system. Wo = 3; [dioleyldimethylammonium chloride]/[$AgNO_3$] = 33.3.

bond. Such morphology change in particles, unfortunately, is found to be not due to the kind of the source of Cl species or the difference in the location for particle formation, because similar cubic particles were also precipitated in cyclohexane solutions of polyoxyethylene(6) nonylphenyl ether systems above 30 molar ratio of KCl to $AgNO_3$. Therefore the formation of cubic particles may be controlled by the extent of molar ratio of Cl^- to Ag^+. The formation of Ag dots in AgCl particles is also found to be not due to the reduction of Ag^+ coordinated with the double bond in the dioleyldimethylammonium chloride molecule, because Ag dots were also precipitated even using didodecyldimethylammonium chloride without a double bond. When dioleyldimethylammonium chloride was replaced by butyldodecyldimethylammonium bromide or iodide, respectively, the resulting particle shapes depended also on the extent of molar ratio of Br^- or I^- to Ag^+ (11). In the case of butyldodecyldimethylammonium bromide, spherical AgBr particles were formed in the system below $[Br^-]/[Ag^+] = 5$, but in the range of $[Br^-]/[Ag^+] = 6–20$ it became platelike particles. Similar spherical AgI particles were also produced in systems of $[I^-]/[Ag^+] = 1–4$, and followed by rod- and platelike particles at $[I^-]/[Ag^+] = 5–10$ and $[I^-]/[Ag^+] = 11–20$, respectively. When these rod- and

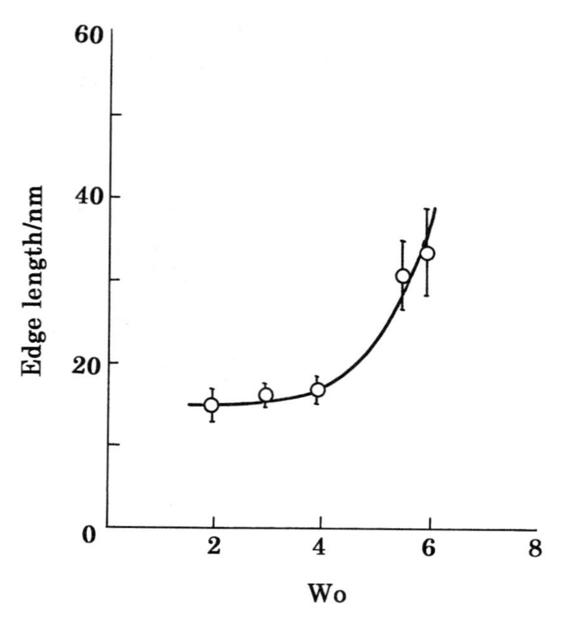

Fig. 4.3.3 Change of particle size with Wo in 0.10 mol kg^{-1} mixtures of dioleyldimethylammonium chloride and polyoxyethylene(6) nonylphenyl ether–cyclohexane system. [Dioleyldimethylammonium chloride]/[$AgNO_3$] = 33.3.

platelike particles, identified as $AgNO_3$–surfactant complexes by Fourier transform infrared (FTIR) and x-ray diffractometry (XRD) measurements, were washed with methanol, the platelike bromide salt complexes generated AgBr particles, whereas the rod- and platelike iodide salt complexes yielded mixtures of β-type AgI with hexagonal structure and γ-type AgI with face-centered cubic structure, respectively. Washing of spherical AgI particles precipitated in systems of $[I^-]/[Ag^+]$ = 1–4 led to mixtures of β- and γ-type AgI.

Calorimetric study on the precipitation of AgCl, AgI, Ag_2S, and silver tetraphenyl borate (AgBF) by the reaction of $AgNO_3$ with NaCl, KI, ammonium disulfide, and sodium tetraphenyl boride was carried out as a function of Wo at various concentration of Aerosol OT and reagent salts in *n*-heptane at 25°C (12). The molar enthalpies (ΔH) of precipitation of AgCl particles were independent of Aerosol OT concentration, but the values for all the particles became more negative with increasing Wo and leveled off at higher Wo values. At the highest Wo values, the observed molar enthalpies of precipitation of AgCl, AgI, and AgBF are less than the corresponding values in bulk water. The small ΔH values for the particles at the lower Wo indicate the formation of smaller particles. Such size limitations are also evidenced from the shift of ΔH–salt concentration curves: The more exothermic values of ΔH as salt concentration increases indicate an increase in the particle size.

The stability of of silver halide sols in the Aerosol OT system was examined from the variation of transmittance of the dispersion of particles after the reaction was completed (2,3). The stability of AgCl and AgBr particles increased with decreasing alkyl chain length of alkane used as bulk solvent. In *n*-dodecane systems, however, AgCl particles were aggregated after 5 days, but they were easily dispersed again by sonification (2). The stable dispersion of particles in the solvents of shorter chain alkane, which is a good solvent for Aerosol OT, is due to the steric stabilization by the surfactant film adsorbed on the particle surface. Such a stabilizing effect was also found by adding the longer chain alcohol and nonionic surfactant Arlacel-20 into Aerosol OT systems (2). However, the dispersion of AgI particles precipitated in the Aerosol OT system of Wo = 6 in isooctane was unstable, but that of Wo = 30 was very stable (5).

REFERENCES TO SECTION 4.3

1. M Dvolaizky, R Ober, C Taupin, R Anthore, X Auvray, C Petipas, C Wiliams. J Dis Sci Technol 4:29, 1983.
2. MJ Hou, DO Shah. In: YA Attia, BM Mouggil, S Chander, eds. Interfacial Phenomena in Biotechnology and Materials Processing. Amsterdam: Elsevier, 1988, p 443.
3. CH Chew, LM Gan, DO Shah. J Dis Sci Technol 11:593, 1990.
4. P Monnoyer, A Fonseca, JB Nagy. Colloid Surf A 100:233, 1995.
5. H Sato, T Hirai, I Komasawa. J Chem Eng Jpn 29:501, 1996.
6. T Hirai, H Sato, I Komasawa. Ind Eng Chem Res 33:3262, 1994.
7. K Johansson, AP Marchetti, GL McLendon. J Phys Chem 96:2873, 1992.
8. J Tanabe, F Tang, K Kawai, K Kon-no. 67th Annual Meeting of Chemical Society of Japan, Tokyo, 1994, p 408.
9. J Tanabe, E Kawai, K Kon-no. 46th Symposium of Colloid and Interface Science of Japan, Tokyo, 1993, p 316.
10. E Ishimizu, K Kawai, K Kon-no. 69th Annual Meeting of Chemical Society of Japan, Tokyo, 1996. p 140.
11. E Ishimizu, K Kawai, K Kon-no. Yukagaku, in press.
12. A D'Aprano, F Pinio, V Turco Liveri. J Soln Chem 20:301, 1991.

4.4 REACTION WITH THIOL

KEISAKU KIMURA
Himeji Institute of Technology, Akou-gun, Hyogo, Japan

Compounds of the I–VII group in the periodic table are known to exhibit good ionic conductivity and have attracted much attention as possible candidates for solid electrolytes. A typical family of compounds is LiI, CuCl, CuBr, and AgI. Historically, polycrystalline solid electrolytes were noticed to show significantly higher ionic conductivity than bulk crystals, since a half century ago. Furthermore, a large increase in conductivity was reported for the system of the mixture of a solid electrolyte such as CuCl (1) and AgI (2) with submicrometer particles of several sorts of insulating materials. In this case, the size of the metal halide itself was on the order of a micrometer or larger. It was also reported that the enhanced conductivity was approximately proportional to the inverse of the size of the electrolyte substances (2). Hence it is natural to make an effort to obtain fine particles of metal halides in order to get better conductivity.

Silver halides are usually known as a photosensitive material, and the development of their preparation technique has been carried out based on the photographic viewpoint; that is, many techniques so far reported are in the solution phase. In view of the solid electrolyte, the first nanosized AgI crystals were recently reported by our group using a gas evaporation technique(physical method) (3,4) in the size range from micrometers to several tens of nanometers. In Fig. 4.4.1 is shown size histograms of the small AgI crystals produced by this method. The peak position of the histogram of the smallest sample is around 70 nm (designated 140 nm in Fig. 4.4.1). The ionic conductivity of this sample is much higher by three orders of magnitude than in the bulk sample. Triggered by this result, much effort has been devoted toward getting a large amount of nanometer-sized AgI. In this context, this section describes the preparation of nanometer-sized particles of silver halide.

4.4.1 Overview of Surface-Modified Nanocolloids

There are several ways to control the size of AgI in the regime of nanometers. Typical examples are using gelatin (5), stabilizing polymers (6–8), and thiol (9,10). Among these, the effect of the thiol ligand is marvellous. The strong interaction of thiol with metal elements was first observed in the modification of the surface of the CdS electrode (11). Immediately after this report, it was applied to a systematic synthesis of nanosized materials of CdS (12–15), metals (16–18), and other compounds (19–21). The basic idea is that the surface of the nucleus is modified with the thiolate anions to prevent further crystal growth. Thus the ratio of the concentration of thiolate anions to a target materials determines the final size of the substances. This

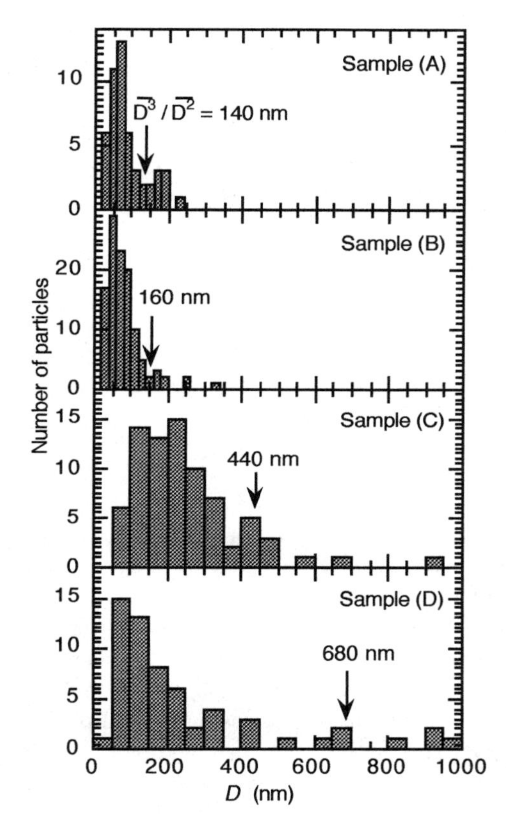

Fig. 4.4.1 Histograms of diameters (D) of small AgI primary crystals in four samples A–D. Vertical arrows indicate the averaged position, $<D^3>/<D^2>$, which is an indicator of the specific surface area. (From Ref. 4.)

strategy is possible only when thiol reacts selectively with the surface metal atoms in a self-assembled way. Table 4.4.1 shows some examples of nanocolloids produced by using thiols as a size controller. Until now, only a limited number of compounds have been prepared by this method. The smallest size reported was on the order of 1 nm or less for thiol-modified metallic particles.

4.4.2 Methods

The size of microcrystals is controlled by the $[I^-]/[RSH]$ ratio, in which RSH denotes 1-thioglycerol (CH_2 $OHCHOHCH_2SH$). A typical preparation with a ratio of 9.0 was as follows: 20 ml of a 0.2 M $AgNO_3$ aqueous solution was rapidly added into

Table 4.4.1 Examples of Thiol-Derivatized Nanoparticles

Group		Reference
I–VII	AgI	9,10
II–VI	CdS	12–15,20,22–24
	CdSe	20,21
	CdTe	19,20
	ZnS	25,26
Metal	Au	16–18,27–29
	Ag	30–32
	Pt	32
Others	PbS	33,34

a solution containing 18 ml 0.2 *M* KI and 2 ml 0.2 *M* iodide–RSH mixture under vigorous stirring. After 20 min of reaction, the resulting turbid yellow suspension of 40 mL was sealed into cellulose tubing (Visking tube, tube size 24/32) allowing dialysis against flowing fresh water for 24 h in the dark. The pH of the suspension changed from 2.13 to 7.16 during this process. After preparation, the suspensions of these samples were turbid and displayed different colors depending on the $[I^-]/[RSH]$ ratio. The color of sample I was white-grey, whereas that of sample VI was yellow. These suspensions were very stable in the dark, since no color change was perceived after more than 6 months. After freeze drying of the precipitate, 0.92 g yellow powder, which corresponds to a yield of 98%, was obtained. The preparation conditions for various samples were shown in Table 4.4.2.

4.4.3 Growth Mechanism and Reaction Kinetics

AgI crystals grow in aqueous solution via four steps: (1) formation of molecules and complexes, (2) formation and growth of nuclei, (3) aggregation, and (4) ripening. These reaction scheme can be written as

$$Ag^+ + I^- \rightarrow AgI \tag{1}$$

$$h\text{AgI} \rightarrow \{\text{AgI}\}_h \quad (h = i,j) \tag{2}$$

$$\{\text{AgI}\}_i + \{\text{AgI}\}_j \rightarrow \{\text{AgI}\}_{i+j} \tag{3}$$

Aggregation and ripening correspond to Eq. (3). The overall reaction is written as

$$m\text{Ag}^+ + m\text{I}^- \rightarrow \{\text{AgI}\}_m \tag{4}$$

The formation of thiol-modified AgI particles is considered to be a consecutive reaction; the first step is the reaction of Eq. (4) and the second one is

Table 4.4.2 Preparation Conditions, Particle Size, and Composition

	Preparation condition				Particle composition[a]	
Sample	$[I^-]/[Ag^+]$	$[RSH]/[Ag^+]$	$[I^-]/[RSH]$	Diameter[b] (nm)	β-Type (%)	γ-Type (%)
	Experiments with $[Ag^+] = [I^-] + [RSH]$					
I	0.4	0.6	0.67	7	74	26
II	0.5	0.5	1.00	10	67	33
III	0.6	0.4	1.50	11	60 (54)[c]	40 (46)
IV	0.7	0.3	2.33	12	56 (44)	44 (56)
V	0.8	0.2	4.00	14	49 (30)	51 (70)
VI	0.9	0.1	9.00	15	40 (37)	60 (63)
VII	1.0			~280	32	68
	Experiments with $[Ag^+] = [I^-] + \frac{1}{2}[RSH]$					
VIII	0.6	0.8	0.75	7	77	23
IX	0.7	0.6	1.17	10	65	35
X	0.8	0.4	2.00	11	45	55
XI	0.9	0.2	4.50	13	41	59

Source: Ref. 10.

[a] Derived from *R* values according to Ref. 5. See text for the definition of *R*.

[b] Determined by Scherrer's equation.

[c] Compostion values in parentheses (%) were from XRD pattern simulation with a Rietveld method.

$$\{AgI\}_m + nAg^+ + tRSH \rightarrow (RS)_t(Ag^+)_n\{AgI\}_m + tH^+ \quad (5)$$

where RSH acts as a terminating agent for the crystal growth and is adsorbed on the particle surface. We should note that there is a side reaction that gives a molecular complex, RSAg, because RSH reacts with Ag^+ to form RSAg simultaneously. This is a competitive reaction with Eq. (4).

$$Ag^+ + RSH \rightarrow RSAg + H^+ \quad (6)$$

This parallel reaction is unavoidable in the present reaction mode. However, it is possible to eliminate this reaction in such a way to enhance reactions (4) and (5) while decreasing (6). This can be achieved by increasing the $[I^-]/[RSH]$ ratio. We can also apply the fractional precipitation method utilizing the difference of solubility to a mixed solvent.

4.4.4 Characterization

It is not always easy to characterize the structure and the composition of nanometer-sized materials, especially in terms of their surface structure. Hence we had engaged in several physicochemical techniques to prove size of particles and a possible structure. These are high-resolution electron microscopy (HRTEM), electron diffraction (ED), x-ray diffraction (XRD), energy-dispersive x-ray analysis (EDX), thermogravimetric analysis (TGA), elemental analysis, Fourier transform infrared spectroscopy (FTIR), and optical measurement. Six AgI samples in Table 4.4.2 (I–VI) were studied systematically, while five other samples (VII–XI) were tested occasionally.

TEM. Transmission electron microscopy is a direct method to determine the size and structure of nanoparticles. Figure 4.4.2 shows an electron micrograph of sample VI sampling from a 200 times dilution of initial suspension. The particles are spherical with a peak diameter of 16 nm and with a standard deviation of 9 nm, which is consistent with those determined by XRD (15 nm). The TEM image shows that the morphology is different from that of AgI prepared by just mixing the $AgNO_3$ and KI in diluted solutions (6). The latter method gives the shape of regular triangles with rounded corners or irregular polygons whose formation can be explained by the coalescence and recrystallization of smaller particles through ''Ostwald ripening'' processes. This difference in crystal habit indicates that RSH strongly adsorbs on the AgI particle surface, which inhibited the anisotropic growth of the special crystal facet of the particles. Irrespective of the wide size distribution (FWHM ~9 nm), the mean particle size was still controlled by the $[I^-]/[RSH]$ ratio even when the total concentration of Ag^+ or I^- reached 0.1 *M*.

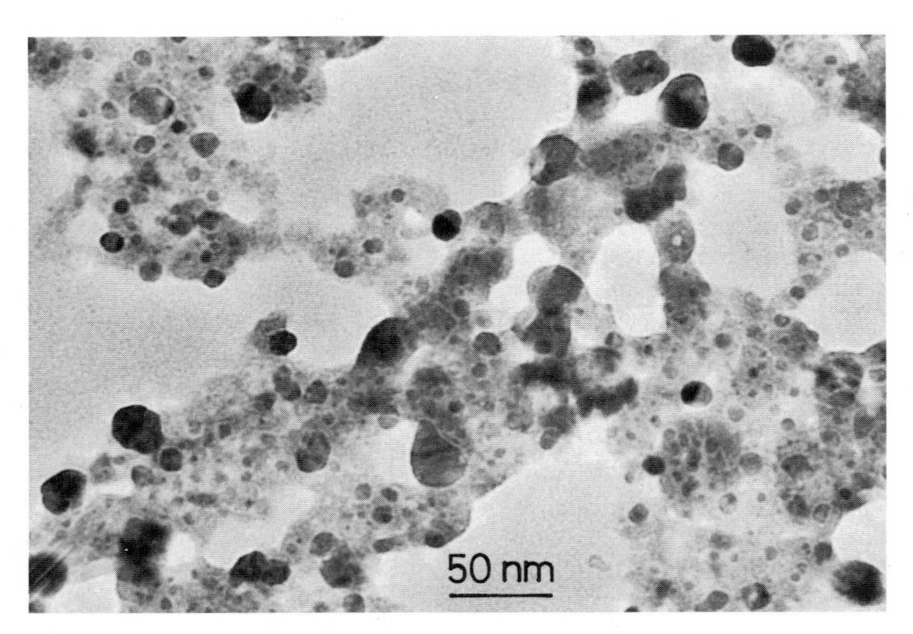

Fig. 4.4.2 Transmission electron micrograph of AgI ultrafine particles prepared with an initial $[I^-]/[RSH]$ ratio of 9.0. (From Ref. 9.)

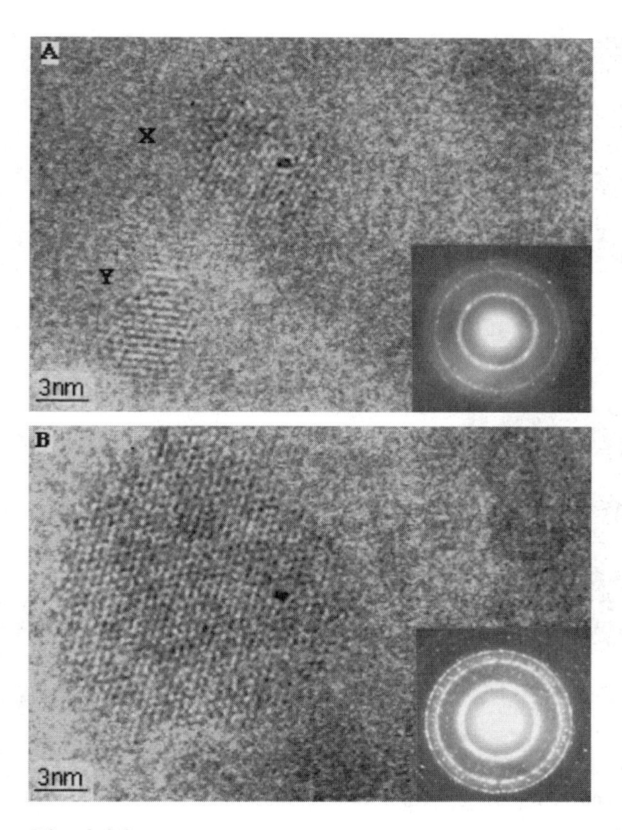

Fig. 4.4.3 High-resolution electron micrograph and electron diffraction pattern. (A) Sample II; the mean diameter and lattice fringe of particles X and Y were 7.7 nm, 0.374 nm, and 7.1 nm, 0.397 nm, respectively. (B) Sample V; the 15.4-nm particle was viewed along the [001] zone axes with the {100} lattice spacing of 0.397 nm. (From Ref. 10.)

From high-resolution (HRTEM) images (see Fig. 4.4.3), it is shown that the most frequently observed lattice fringes in samples I–VI were 0.375 ($\pm$5) nm, which is indexed to the lattice spacing of the β-(002) or γ-(111) plane. Lattice spacings at 0.231 and 0.397 nm were detected, the former belonging to the β-(110) or γ-(220) plane of AgI and the latter to the β-(100) plane of AgI. Figure 4.4.3A shows two particles with lattice spacings of 0.374 and 0.397 nm, respectively from sample II. Figure 4.4.3B shows another example from sample V with the {100} spacing of 0.397 nm of hexagonal AgI. The diffraction rings from both particles are indexed to the known β- and γ-AgI. The images of atomic resolution of the crystallite frequently revealed the standard hexagonal pattern of the β-(001) plane, especially for samples V and VI. However, some deviations of the interplanar angles and lattice spacings

Fig. 4.4.4 High-resolution electron micrographs and power spectra of sample VI: (A) Hexagonal structure in the [001] orientation with distortions d_{100}, d_{010}, d_{1-10} = 0.396, 0.377, 0.403 nm; (B) cubic structure close to the [110] orientation with distortions d_{111} = 0.407 nm, d_{-111} = 0.394 nm, and d_{200} = 0.332 nm. (From Ref. 10.)

from the bulk values have appeared. Figure 4.4.4A shows an HRTEM image of sample VI as an example. The particle viewed along the [001] axes had lattice spacings of d_{100} = 0.396 nm, d_{010} = 0.377 nm, and d_{1-10} = 0.403 nm. The angles between the identified planes were 60°, 65°, and 55°. A similar slight deviation was found in other samples, indicating that these distortions belong to an intrinsic structural distortion of AgI nanocrystal. Figure 4.4.4B shows an example of the face-centered cubic structure in the [110] orientation. The corresponding lattice spacings were d_{111} = 0.407 nm, d_{-111} = 0.394 nm and d_{200} = 0.332 nm. The angles between these planes were 73°, 54°, and 53°. The lattice constants calculated from these lattice spacings were 0.664, 0.643, and 0.665 nm, respectively. These constants were close to the standard value, 0.649 nm, of cubic AgI, but still slightly distorted. This deviation suggest that there is a strong interaction between Ag^+ and thiolate. The same kind of distortion was also reported for thiol-stabilized CdS nanocrystals (35).

XRD. Although TEM can directly determine the size and shape of the particles, XRD analysis was further applied to determine systematically the size of the particles by Scherrer's method. Figure 4.4.5 shows the x-ray diffraction patterns for samples I–VI. All observed peaks were broadened compared with those of bulk AgI. Three major diffraction peaks at $2\theta = 23.7°$, $39.3°$, and $46.4°$ can be ascribed to either β-AgI (Fig. 4.4.5, position g) or γ-AgI (Fig. 4.4.5, position h). The calculated particle sizes from Scherrer's equation are listed in Table 4.4.2. It is apparent that the strongest peak at $2\theta = 23.7°$ comprises two peaks from β- or γ-AgI phases. Hence the relative composition of both phases was obtained by studying the relative intensities of the three peaks at $2\theta = 22.4°$, $23.7°$, and $25.5°$. The results are also listed in Table 4.4.2 in the column for particle composition. Some samples were also analyzed by the simulation method using a Rietveld method to estimate the composition and are given in the parentheses in Table 4.4.2. It is known that the γ phase is metastable at room temperature. These results confirm that the thiolates adsorb strongly to the surface of AgI and form stable hexagonal structured nanoparticles.

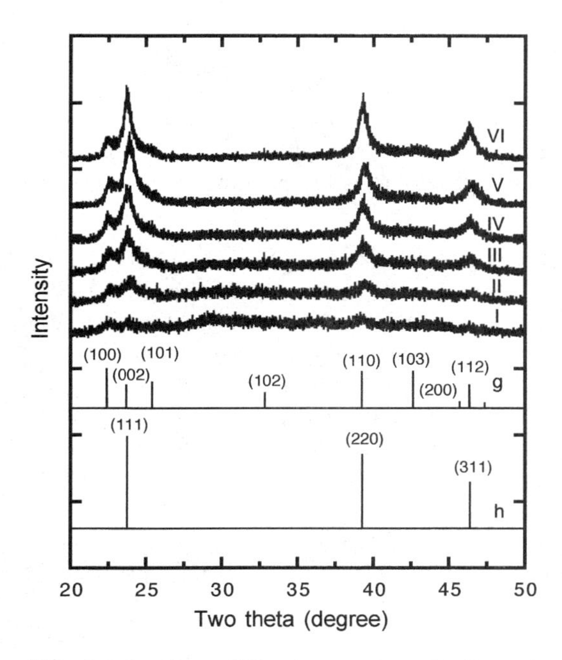

Fig. 4.4.5 X-ray diffraction patterns of AgI powder of samples I–VI in Table 4.4.2. Locations g and h on the abscissa show the position and relative intensity of the standard XRD pattern of β- and γ-AgI, respectively. (From Ref. 10.)

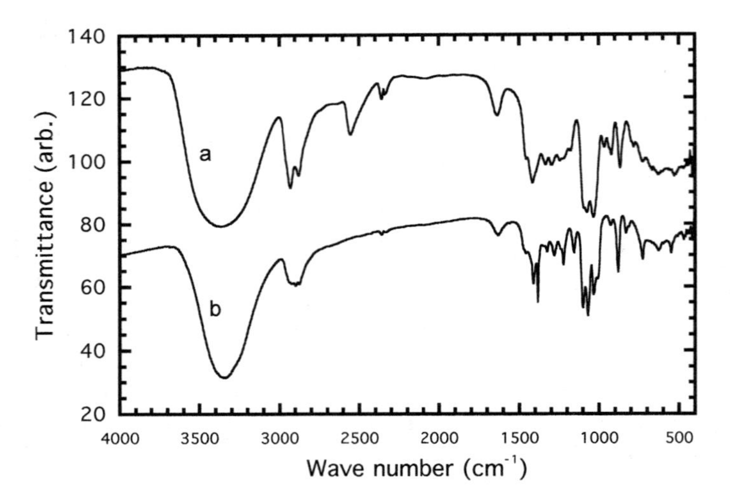

Fig. 4.4.6 FTIR spectra of (a) pure RSH and (b) powdered sample I.

FTIR. The FTIR spectrum of sample I as a representative is shown in Fig. 4.4.6. The spectral features of FTIR of all the samples resembled that of RSH (spectrum a) except that the S–H stretching vibration mode at 2558 cm^{-1} was absent in the spectra. This fact strongly suggests that thiol binds to AgI surface with its sulfur atom. The broad O–H vibration band at 3370 cm^{-1} was still observed. This finding also supports that the mercapto group is the bonding group and not the O-H group.

Elemental Analysis. The chemical composition analysis gives much information concerning the surface structure of the particles, and the results are given in Table 4.4.3. The summation of all detected elements (Ag, I, C, H, O, S) for samples I–VI was almost 100%, indicating that the other elements did not exist in a significant amount. The mass ratio in Table 4.4.3 was converted into a molar ratio, giving an empirical formula that was approximately equal to the apparent formula calculated from the initial composition of $AgNO_3$, KI, and RSH within experimental error as were also given in the table. The general formula of the particle was found to be represented by $AgI_x(SC_3H_7O_2)_{1-x}$ (X = 0.4–0.9 for samples I–VI). It should be noted that the formula of the organic phase is $SC_3H_7O_2$, different from $SC_3H_8O_2$ (= RSH); that is, one hydrogen atom is extracted from an RSH molecule during reaction. This fact is in accord with the FTIR observations.

The content of organics in the particle from elemental analysis seems too high for assuming a monolayer covering of organics over particle surface and the sample comprised solely of surface-modified particles. Taking into account that all samples I–VI were synthesized with a condition of excess Ag^+, we present two different

Table 4.4.3 Composition and Empirical Formula of AgI Powder Together with Theoretical Formula for Proposed Structure A or B in Fig. 4.4.7

Sample	Elemental analysis mass (%), Ag/I/C/H/O/S	$\sum$	Empirical formula
I	45.0/23.0/9.4/1.8/9.2/8.9	97.3	$AgI_{0.43}(S_{1.0}C_{3.0}H_{7.0}O_{2.2})_{0.62}$
II	45.0/27.9/7.7/1.5/7.8/5.9	95.8	$AgI_{0.52}(S_{0.9}C_{3.0}H_{7.1}O_{2.3})_{0.51}$
III	45.6/33.2/6.3/1.3/6.2/5.8	98.4	$AgI_{0.62}(S_{1.0}C_{3.0}H_{7.3}O_{2.2})_{0.42}$
IV	45.6/38.9/4.5/0.9/4.5/3.4	97.8	$AgI_{0.72}(S_{0.8}C_{3.0}H_{7.1}O_{2.2})_{0.30}$
V	46.0/43.6/3.1/0.5/3.1/2.5	98.8	$AgI_{0.81}(S_{1.0}C_{3.0}H_{6.2}O_{2.2})_{0.20}$
VI	46.7/48.5/1.5/0.3/1.5/0.9	99.4	$AgI_{0.88}(S_{0.7}C_{3.0}H_{7.2}O_{2.2})_{0.10}$

	Structure A		Structure B	
Sample	Theoretical formula[b]	ΔAg:ΔRS[a]	Theoretical formula[b]	ΔAg:ΔRS[a]
I	$Ag_{0.50}I_{0.43}(RS)_{0.07}$	50:55	$Ag_{0.49}I_{0.43}(RS)_{0.13}$	51:49
II	$Ag_{0.58}I_{0.52}(RS)_{0.07}$	42:45	$Ag_{0.58}I_{0.52}(RS)_{0.11}$	42:40
III	$Ag_{0.68}I_{0.62}(RS)_{0.07}$	32:36	$Ag_{0.68}I_{0.62}(RS)_{0.12}$	32:30
IV	$Ag_{0.78}I_{0.72}(RS)_{0.07}$	22:24	$Ag_{0.78}I_{0.72}(RS)_{0.12}$	22:18
V	$Ag_{0.87}I_{0.81}(RS)_{0.07}$	13:14	$Ag_{0.87}I_{0.81}(RS)_{0.12}$	13:8
VI	$Ag_{0.94}I_{0.88}(RS)_{0.07}$	6:4	$Ag_{0.95}I_{0.88}(RS)_{0.12}$	5: −2

Source: Ref. 10.

[a] ΔAg and ΔRS represent the residual amoung (%) of Ag and RS after subtraction of the theoretical formula from the empirical one, respectively.

[b] Iodide content was kept the same in the empirical and theoretical formulas.

models for the surface structures of thiol-modified AgI particles, shown in Fig. 4.4.7. Structure A supposes that all the Ag^+ sites on the AgI particle surface are occupied by adsorbed thiolates and all the surface I^- ions are bound by Ag^+ ions to compensate the excess negative charge caused by adsorbed RS^- ions. Structure B represents the highest occupation of thiolates on the AgI surface, such that all the Ag^+ sites in structure A are bound to RS^- ions.

The theoretical composition ratio of Ag^+, I^-, and RS^- ions can be calculated for structure A and B by assuming that the Ag^+ and I^- ions are evenly spread on the AgI surface. Using the known ionic radius for Ag^+ (1.15 Å) and I^- (2.20 Å) and the density of bulk AgI to be 5.68 g/cm^3, the number ratios of $Ag^+/I^-/RS^-$ for structures A and B are calculated to be $(0.61R + 6.5)/0.61R/6.5$ and $(0.61R + 6.5)/0.61R/13$, respectively, where R is the radius in angstroms of the AgI particle. In the lower part of Table 4.4.3, the calculated composition ratios based on models A and B are listed, all of which are much lower than those of the empirical formula.

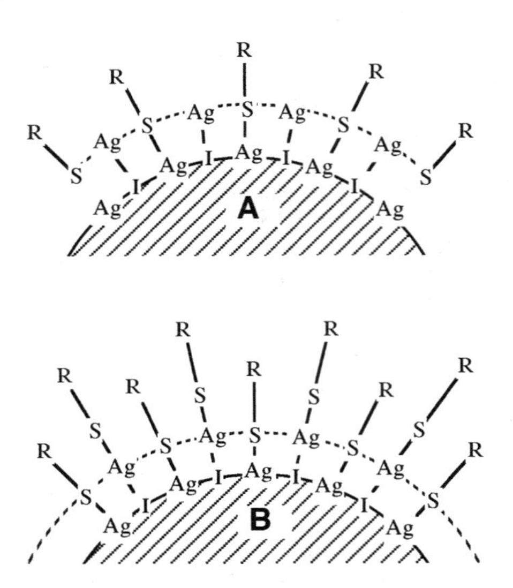

Fig. 4.4.7 Models for the surface structures of thiol-modified AgI particles. The shadow represents the inner part of particles. The solid and dashed lines show the surface and adsorbed ions, respectively. Structure A stands for the half-filled model and B for highest coverage model. (From Ref. 10.)

The molar ratio of the excess amount of Ag^+ (defined by ΔAg) to that of RS^- ions(defined by ΔRS) are close to unity in all the samples, indicating that the side reaction [Eq. (6)] and then its product RSAg should be taken into account to interpret the data of elemental analysis. This RSAg molecular complex was found to be amorphous by XRD observation. It was also found that a particle with a size of 15 nm was almost free from RSAg by comparing the content of empirical formula with that of the theoretical one in Table 4.4.3. The RSAg complex cannot be removed in the present procedure for smaller particles whose size is less than 10 nm due to its small K_{sp} value, but one can eliminate it by a careful control of the pH of suspensions.

TGA. Thermogravimetric analysis (TGA) and differential thermal analysis (DTA) are other means to confirm the above structural models. Figure 4.4.8 shows the thermal analysis data for sample I. Curve (a) shows a TG datum of a mass loss about 22% after heating over 350°C. The derivative curve (b) of mass loss curve (a) clearly shows that there are at least four steps during the decomposition of the sample. This finding was further confirmed by the DTA data curve (c) shown in the same figure. It is clearly seen that there are four endothermic peaks. The DTA and TGA curves were similar for all samples. Note that the relative ratios of mass

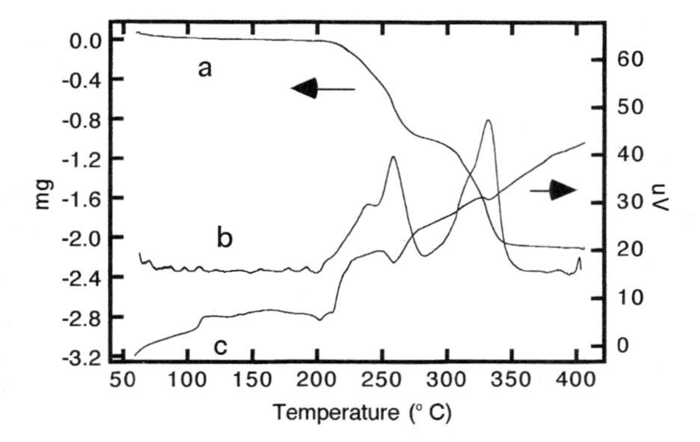

Fig. 4.4.8 TGA and DTA result of sample I. (a) Mass loss curve; (b) Derivative curve of (a); (c) Differential thermal analysis curve. (From Ref. 10.)

loss in these four steps were almost identical for different samples, suggesting that the four steps result from the bond break in the organics.

After heating, the sample turned bulklike with black color, and EDX data showed that the residues contained S in addition to a large amount of I and Ag. X-ray diffraction patterns of the residues gave the main diffraction peaks of AgI, indicating that the residues were mainly AgI crystallite. Total mass losses of samples I–VI obtained from TGA lie between the total organics content and total mass amount of C + H + O in the samples from the elemental analysis, for which the data are tabulated in Table 4.4.4. The difference may be caused by the residual S in the particle surface. This result supports that heating mainly resulted in the decomposition of organics in the powders.

Optical Spectra. All six samples tested (I–VI) showed typical absorption features of AgI crystals with a peak or shoulder around 422 nm, and this is shown in Fig.

Table 4.4.4 Comparison of Mass Loss from TG of AgI Samples with Total Organics Content and That of the Sum of C + H + O from the Elemental Analysis

	I	II	III	IV	V	VI
Mass loss (%)	21.78	18.41	14.63	11.33	7.30	4.23
Total organics (%)	29.30	22.90	19.60	13.30	9.20	4.20
C + H + O (%)	20.40	17.00	13.80	9.90	6.70	3.30

Source: Ref. 10.

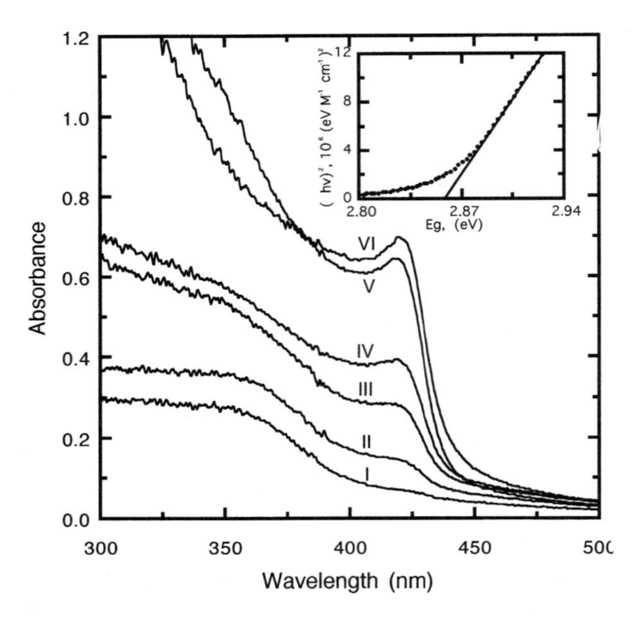

Fig. 4.4.9 Absorption spectra of samples I–VI in Table 4.4.2. Inset shows that the band gap of the sample V is 2.86 eV. (From Ref. 10.)

4.4.9. The exact absorption threshold of the absorption spectra was determined from the intercept of $(\sigma h\nu)^2$ versus $h\nu$ plots (where σ is the absorption coefficient) for the direct band gap and was found to be 2.82–2.86 eV regardless of size. From a relationship between E_g and particle size of AgI after Meisel et al. (36), we found that the size of these particles was 6–7 nm. For the smallest particles, sample I, there is no peak in the spectra. Other than this finding, there is no apparent ''quantum confinement effect'' such as peak shift in the sample whose sizes are larger than 5 nm.

Instead, optical measurement was applied to clarify the formation mechanism of AgI nanoparticles in diluted suspensions. Figure 4.4.10A shows the absorption spectra of the 1-day aged suspension containing 3.33×10^{-4} *M* Ag^+ and 6.67×10^{-4} *M* I^- as a function of RSH concentration. When the content of RSH increased from 0 to 6.67×10^{-3} *M* (curves a–d), the absorption spectra of AgI particles blue-shifted, suggesting the quantum size effect. The relationship between the particle diameter, *d* (Å), and the concentration of RSH, *c* (*M*), was plotted in the inset, which shows the double-logarithmic linear line of *d* versus *c*. The aggregation number is found to be proportional to the cube of the size. The same relationship was reported on the formation of CdS nanoparticles (37). These correlations indicate that AgI and CdS have the same ionic nature and have the same reaction modules of thiols.

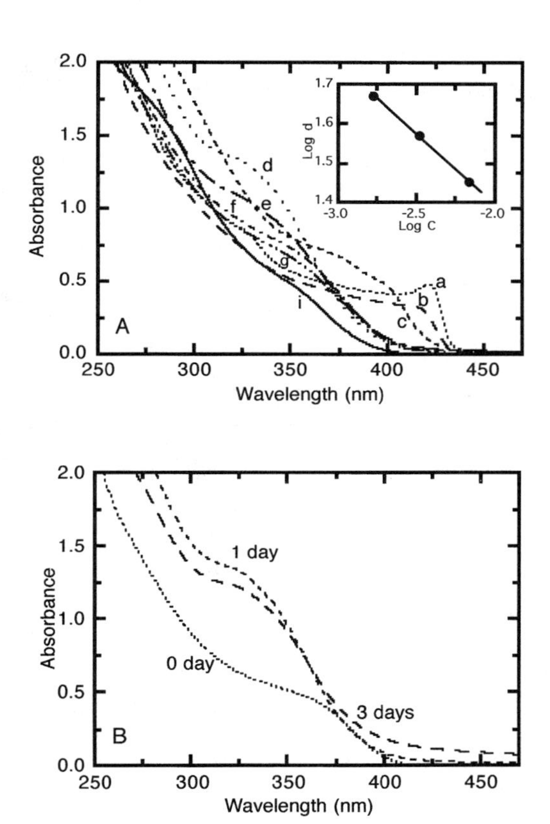

Fig. 4.4.10 Top: Absorption spectra of AgI nanoparticles prepared after 1 day of aging with 3.33×10^{-4} *M* Ag^+ and 6.67×10^{-4} *M* I^- in the presence of RSH. The concentration of RSH were: (a) 0 *M,* (b) 1.67×10^{-3} *M,* (c) 3.33×10^{-3} *M,* (d) 6.67×10^{-3} *M,* (e) 1.00×10^{-2} *M,* (f) 1.33×10^{-2} *M,* and (g) 1.67×10^{-2} *M.* For comparison, curve i represents the absorption spectrum of a freshly prepared solution containing 3.33×10^{-4} *M* Ag^+ and 1.67×10^{-2} *M* RSH. The inset shows a double-logarithmic plot of mean diameter of AgI particles versus concentration of RSH for curves (b), (c), and (d). Bottom: The time course of the absorption spectra of condition (d). (From Ref. 10.)

An extension of this finding assumes that thiolates may be used for the synthesis of other I–VII compound nanoparticles.

The absorption band at 320–330 nm seen in Fig. 4.4.10B was reported in a AgI particle formed by pulse radiolysis technique. The unstable species thus formed was attributed to initial aggregates of AgI molecules whose lifetime was on the order of milliseconds. Therefore in the present experiment, these transient species were successfully stabilized with RSH and can survive for several days before further aggregation. Their size was estimated to be in the range of 2–3 nm as judged from the absorption onset at 405 nm. The isolation of this smallest species remains a future project.

REFERENCES TO SECTION 4.4

1. T Jow, JB Wagner Jr. J Electrochem Soc 126:1963, 1979.
2. K Shahi, JB Wagner Jr. J Electrochem Soc 128:6, 1981.
3. T Ida, H Saeki, H Hamada, K Kimura. Surf Rev Lett 3:41, 1996.
4. T Ida, K Kimura. Solid State Ionics 107:313, 1998.
5. CR Berry. Phys Rev 161:848, 1967.
6. T Tanaka, H Saijo, T Matsubara. J Photogr Sci 27:60, 1979.
7. A Henglein, M Gutierrez, H Weller, A Fojtik, J Jirkovsky. Ber Bunsen-Ges Phys Chem 93:593, 1989.
8. KR Gopidas, M Bohorquez, PV Kamat. J Phys Chem 94:6435, 1990.
9. S Chen, T Ida, K Kimura. Chem Commun 2301, 1997.
10. S Chen, T Ida, K Kimura. J Phys Chem B 102:6169, 1998.
11. MJ Natan, JW Thackeray, MS Wrighton. J Phys Chem 90:4089, 1986.
12. Y Nosaka, K Yamaguchi, H Miyama, H Hayashi. Chem Lett 605, 1988.
13. PV Kamat, NM Dimitrijevic. J Phys Chem 93:4259, 1989.
14. N Herron, Y Wang, H Eckert. J Am Chem Soc 112:1322, 1990.
15. VL Colvin, AN Goldstein, AP Alivisatos. J Am Chem Soc 114:5221, 1992.
16. M Brust, M Walker, D Bethell, DJ Schiffrin, R Whyman. J Chem Soc Chem Commun 801, 1994.
17. RL Whetten, JT Khoury, MM Alvarez, S Murthy, I Vezmar, ZL Wang, PW Stephens, CL Cleveland, WD Luedtke, U Landman. Adv Mater 8:428, 1996.
18. T Yonezawa, M Sutoh, T Kunitake. Chem Lett 619, 1997.
19. T Rajh, OI Micic, AJ Nozik. J Phys Chem 97:11999, 1993.
20. CB Murray, DJ Norris, MG Bawendi. J Am Chem Soc 115:8706, 1993.
21. CB Murray, CR Kagan, MG Bawendi. Science 270:1335, 1995.
22. M Kundu, AA Khosravi, SK Kulkarni, P Singh. J Mater Sci 32:245, 1997.
23. T Ogata, H Hosokawa, T Oshiro, Y Wada, T Sakata, H Mori, S Yanagida. Chem Lett 1665, 1992.
24. H Hosokawa, T Ogata, Y Wada, K Murakoshi, T Sakata, H Mori, S Yanagida. J Chem Soc Faraday Trans 92:4575, 1996.
25. S Mahamuni, AA Khosravi, M Kundu, A Kshirsagar, A Bedekar, DB Avasare, P Singh, SK Kulkarni. J Appl Phys 73:5237, 1993.
26. AA Khosravi, M Kundu, BA Kuruvilla, GS Shekhawat, RP Gupta, AK Sharma, PD Vyas, SK Kulkarni. Appl Phys Lett 67:2506, 1995.
27. MM Alvarez, JT Khoury, TG Schaaff, MN Shafigullin, I Vezmar, RL Whetten. J Phys Chem 101:3706, 1997.
28. TG Schaaff, MN Shafigullin, JT Khoury, I Vezmar, RL Whetten, WG Cullen, PN First. J Phys Chem B 101:7885, 1997.
29. A Badia, W Gao, S Singh, L Demers, L Cuccia, L Reven. Langmuir 12:1262, 1996.
30. SA Harfenist, ZL Wang, MM Alvarez, I Vezmar, RL Whetten. J Phys Chem 100:13904, 1996.

31. S Connolly, S Fullam, B Korgel, D Fitzmaurice. J Am Chem Soc 120:2969, 1998.
32. KV Sarathy, G Raina, RT Yadav, GU Kulkarni, CNR Rao. J Phys Chem B 101:9876, 1997.
33. T Torimoto, H Uchida, T Sakata, H Mori, H Yoneyama. J Am Chem Soc 115:1874, 1993.
34. T Torimoto, T Sakata, H Mori, H Yoneyama. J Phys Chem 98:3036, 1994.
35. W Vogel, J Urban, M Kunda, SK Kulkarni. Langmuir 13:827, 1997.
36. KH Schmidt, R Patel, D Meisel. J Am Chem Soc 110:4882, 1988.
37. CH Fischer, A Henglein. J Phys Chem 93:5578, 1989.

5

Metal Sulfates

TADAO SUGIMOTO

5.1 REACTION IN HOMOGENEOUS SOLUTIONS

TADAO SUGIMOTO
Institute for Advanced Materials Processing, Tohoku University, Sendai, Japan

5.1.1 Barium Sulfate

Andreasen et al. (1) prepared monodispersed spherical particles of barium sulfate ($BaSO_4$: rhombic crystal system) of 2–3 μm in mean diameter by mixing $BaCl_2$ solution containing HCl with an equal volume of H_2SO_4 solution containing HCl through quickly pouring the two solutions into the third beaker at the same time at 20°C. Typically, the solutions were 30 cm^3 of a $BaCl_2$ solution (1/18 mol dm^{-3} in $BaCl_2$ and 10/3 mol dm^{-3} in HCl) and 30 cm^3 of a H_2SO_4 solution (1/18 mol dm^{-3} in H_2SO_4 and 10/3 mol dm^{-3} in HCl) (1). The unique way of mixing served to a great extent in producing the uniform particles, as shown in a micrograph in Fig. 5.1.1. The hydrochloric acid was added for increasing the solubility of barium sulfate

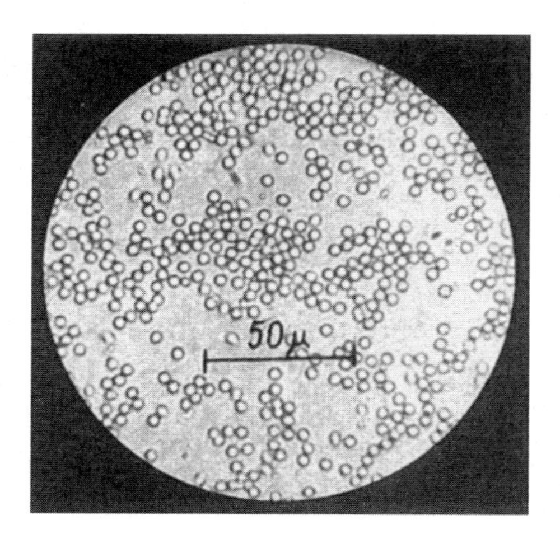

Fig. 5.1.1 Monodispersed $BaSO_4$ spheres prepared by pouring equimolar solutions of $BaCl_2$ and H_2SO_4, with a high concentration of HCl contained in both solutions, into a beaker at the same time at 20°C, where 30 cm^3 of a $BaCl_2$ solution (1/18 mol dm^{-3} in $BaCl_2$ and 10/3 mol dm^{-3} in HCl) and 30 cm^3 of a H_2SO_4 solution (1/18 mol dm^{-3} in H_2SO_4 and 10/3 mol dm^{-3} in HCl) were used. (From Ref. 1.)

in order to increase the final particle size. This is in accord with the rule of size control of monodispersed particles, since the increase in solubility leads to the increase of the volumic growth rate $\dot{v}$ at the end of the nucleation in the mass balance of $n = QV_m/\dot{v}$, where n is the final particle number, Q is the supply rate of the reactants in mole per unit time, and V_m is the molar volume of the solid. They also found that the final size was dramatically decreased by replacing the water increasingly with propylalcohol (i.e., from 3 μm to ~0.3 μm with increasing proportion of alcohol from 0 to 60 vol%). This is due to the decrease in $\dot{v}$ by the decrease in solubility of barium sulfate.

Moreover, they prepared monodispersed ellipsoidal particles of barium sulfate of 3–4 μm in mean diameter by oxidation of sodium thiosulfate ($Na_2S_2O_3$) with hydrogen peroxide (H_2O_2) to gradually release SO_4^{2-} ions ($2S_2O_3^{2-} + 4H_2O_2 \rightarrow SO_4^{2-} + S_3O_6^{2-} + 4H_2O$), in the presence of Ba^{2+} ions (1). Typically, a mixed solution (20°C), consisting of 100 cm^3 of $BaCl_2 \cdot 2H_2O$ solution (10 g $dm^{-3} \simeq 0.041$ mol dm^{-3}) and 5 cm^3 of H_2O_2 solution, and another solution (20°C), consisting of 100 cm^3 of $Na_2S_2O_3 \cdot 5H_2O$ solution (20 g $dm^{-3} \simeq 0.040$ mol dm^{-3} SO_4^{2-}) and distilled water and/or a solution of some additive such as 0.5 mol dm^{-3} citrate salt, are mixed together in a half-liter conical beaker for about 6 s and strongly shaken for about 20 s, followed by constant reciprocating motion to keep the solution uniform. In this procedure, the volume of water added to the thiosulfate solution was varied to adjust the total volume to 250 cm^3 all the time. When citrate salt was added, the particle size was reduced with increasing content of citrate probably due to the strong adsorption of citrate ions. Figure 5.1.2 shows a transmission electron

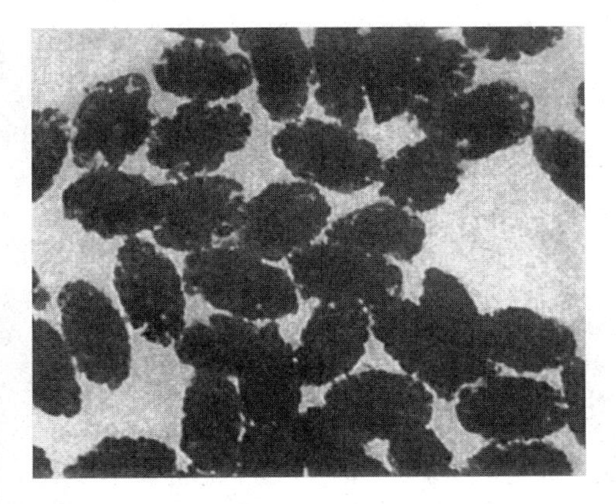

Fig. 5.1.2 Monodispersed $BaSO_4$ ellipsoids of 0.14 μm mean size, prepared by homogeneous precipitation with the slow release of SO_4^{2-} ions through decomposition of thiosulfate ions by hydrogen peroxide in the presence of Ba^{2+} ions at 20°C. To reduce the mean size, the reaction was carried out with sodium citrate (0.1 mol dm^{-3}). (From Ref. 1.)

micrograph (TEM) of monodispersed ellipsoidal particles prepared this way with ~0.14 μm mean size in the presence of 0.1 mol dm^{-3} sodium citrate.

Petres et al. (2) studied the effect of adsorptives, such nonionic surfactant Triton X-100 (alkyl aryl polyether alcohol), anionic surfactant Aerosol MA (dihexyl sodium sulfosuccinate), and complexing agents (citrate and ethylenediamine tetracetic acid (EDTA), on the morphology of $BaSO_4$ particles prepared on the basis of the gradual decomposition of thiosulfate with hydrogen peroxide after Andreasen et al. (1) and found that both Triton X-100 and EDTA yielded uniform ellipsidal particles with sizes decreasing with increasing content of these additives, while Aerosol MA gave rectangular particles with jagged surfaces, and citrate produced under suitable conditions spherical particles. For the stabilization of $BaSO_4$ particles, literature on ionic surfactants, such as dodecylpyridinium halides, sodium dodecylsulfate, and potassium laurate, is also available (3–5).

Takiyama (6) prepared highly monodispersed spindle-type particles of barium sulfate ($BaSO_4$) by decomposing the Ba–EDTA complex with hydrogen peroxide. In this reaction, the initial concentration of the Ba–EDTA complex is a decisive factor in separation of nucleation and growth stages.

Petres et al. (7) found nonionic surfactants such as Triton X-100 (alkyl aryl polyether alcohol) and Triton X-305 were useful for stabilization of $BaSO_4$ particles prepared by the method of Takiyama. They also observed the internal structure of thus prepared $BaSO_4$ particles by electron microscopy on their ultrathin sections 30–50 nm thick sliced with an ultramicrotome, and found the internal structure porous with a mean pore size ~3 nm (8). One may find the same porous structure in the direct TEM of the small ellipsoidal particles of $BaSO_4$ in Fig. 5.1.2 as well by Andreasen et al. (1). Though Petres et al. (6,8) elucidated the growth mechanism in terms of aggregation of preformed primary particles from the internal structure, more detailed growth analysis will probably be needed for concluding this problem.

5.1.2 Basic Iron Sulfate

For the formation of monodispersed basic iron sulfate ($Fe_3(SO_4)_2(OH)_5{\cdot}2H_2O$) particles, ferric complexes such as $Fe(OH)^{2+}$, $Fe_2(OH)_2{}^{4+}$, and $FeSO^+{}_4$ were supposed to be responsible for the particle formation (9,10). Since these complexes initiate nucleation in the form of monomers or dimers, they appear to have a high degree of freedom and activity for the formation of a crystal structure. In fact, they are known to be of a haxagonal crystal symmetry, and sulfate ions are built in as a component of the crystal lattice. Table 5.1.1 shows typical preparation conditions of the basic iron sulfate particles (9). The monoclinic basic iron sulfate ($Fe_4(SO_4)(OH)_{10}$) particles found in this table are more or less coprecipitated with the hexagonal crystals when $Fe_2(SO_4)_3$ is used as a starting material at relatively high temperatures. Figure 5.1.3a and Fig. 5.1.3b show TEM images of monodispersed particles of $Fe_3(SO_4)_2(OH)_5{\cdot}2H_2O$ with a small amount of $Fe_4(SO_4)(OH)_{10}$ and their carbon replicas, prepared by aging a 0.088 mol dm^{-3} solution of $Fe_2(SO_4)_3$ at 98°C for 3 h. As found in the TEM of replica in Fig. 5.1.3b, the apparently

Table 5.1.1 Typical Conditions for the Formation of Monodispersed Basic Iron Sulfate Particles

System number	Solution aged	Temperature and method of aging	Time of aging (h)	Major chemical composition	Crystal symmetry
1	0.09 *M* $Fe_2(SO_4)_3$	98 ± 5°C oven	4.0	$Fe_3(SO_4)_2(OH)_5 \cdot 2H_2O$ plus lesser amount $Fe_4(SO_4)(OH)_{10}$	Hexagonal Monoclinic
2	0.09 M $Fe_2(SO_4)_3$ + 0.013 *M* NaOH	98 ± 5°C oven	21.5	$Fe_4(SO_4)(OH)_{10}$ plus small amount $Fe_3(SO_4)_2(OH)_5 \cdot 2H_2O$	Monoclinic Hexagonal
3	0.09 *M* $Fe_2(SO_4)_3$	98 ± 0.1°C oil bath; heating rate = 1.5°C/min	1.2 (at 98°C)	$Fe_3(SO_4)_2(OH)_5 \cdot 2H_2O$ plus small amount $Fe_4(SO_4)(OH)_{10}$	Hexagonal Monoclinic
4	0.09 M $Fe_2(SO_4)_3$ + 0.167 *M* urea (NH_2CONH_2)	98 ± 5°C oven	3.0		
5	0.18 *M* $FeNH_4(SO_4)_2$		2.4 (at 80°C)		
6	0.18 *M* $Fe(NO_3)_3$ + 0.27 *M* Na_2SO_4		1.5 (at 80°C)		
7	0.18 *M* $Fe(NO_3)_3$ + 0.53 *M* Na_2SO_4	80 ± 0.1°C oil bath; heating	1.5 (at 80°C)	$Fe_3(SO_4)_2(OH)_5 \cdot 2H_2O$	Hexagonal
8	0.18 *M* $Fe(NO_3)_3$ + 0.53 *M* Na_2SO_4	rate = 1.5°C/min	5.0 (at 80°C)		
9	0.18 *M* $Fe(NO_3)_3$ + 0.53 *M* K_2SO_4		0.6 (at 80°C)		
10	0.18 *M* $Fe(NO_3)_3$ + 0.53 *M* $(NH_4)_2SO_4$		1.0 (at 80°C)		

Source: Ref. 9.

pseudocubic particles and hexagonal ones in Fig. 5.1.3a are the same particles bound by both square faces with a chipped corner and triangular faces, where the former repose on the chipped square faces and the latter on triagular faces. If we apply the result of OPML–XRD analysis on basic aluminum sulfate ($Al_3(SO_4)_2(OH)_5 \cdot 2H_2O$) particles of the same crystallographic symmetry as $Fe_3(SO_4)_2(OH)_5 \cdot 2H_2O$, the chipped square faces and the triangular faces may be assigned to the {012} and {001} faces of the hexagonal crystal, respectively (11). Figure 5.1.3c and Fig. 5.1.3d show TEM images of $Fe_3(SO_4)_2(OH)_5 \cdot 2H_2O$ particles and their carbon replica, prepared by aging a mixed solution of 0.18 mol dm^{-3} in $Fe(NO_3)_3$ and 0.53 mol dm^{-3} in Na_2SO_4 in an oil bath heated from room temperature to 80°C at a constant rate of 1.5°C min^{-1} for 1.5 and 2 h, respectively. The apparently spherical particles in Fig. 5.1.3c are actually oblate spheroids with two flat triagular faces, as is obvious from the TEM of the carbon replicas in Fig. 5.1.3d. The round corners are thought to be due to the adsorption of SO_4^{2-} on the anology of the basic aluminum sulfate

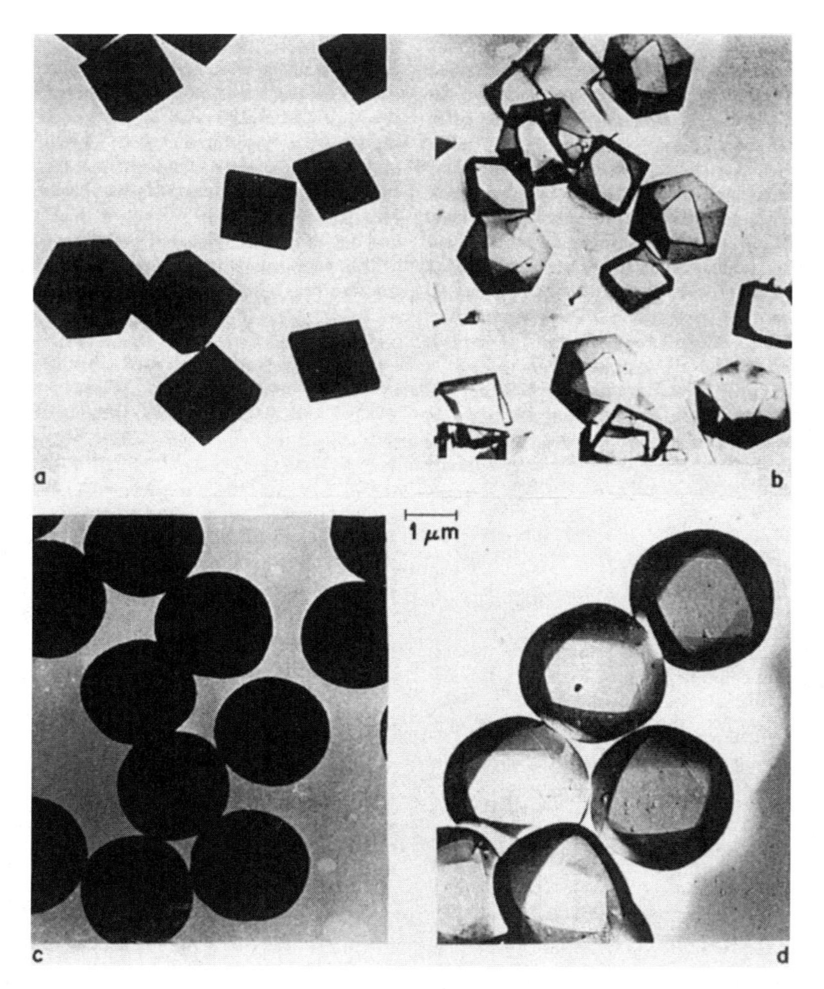

Fig. 5.1.3 TEM images (a) basic iron sulfate particles prepared by aging a 0.088 mol dm^{-3} $Fe_2(SO_4)_3$ at 98°C in an oven for 3 h; (b) carbon replica of the same particles as described in (a); (c) basic iron sulfate particles prepared by aging a solution of 0.18 mol dm^{-3} in $Fe(NO_3)_3$ and 0.53 mol dm^{-3} in Na_2SO_4 in oil bath heated from room temperature to 80°C at a constant rate of 1.5°C min^{-1} and aged at 80°C for 1.5 h; (d) carbon replica of the same particles as in (c), but aged for 2 h at 80°C. (From Ref. 9.)

particles (11). However, the development of the {001} faces of the oblate spheroids may not attributed to the effect of SO_4^{2-}, but to the adsorption of OH^- to the {001} faces, if the analogy of basic aluminum sulfate can be applied to the basic iron sulfate, since arrested growth of the {001} faces of basic aluminum sulfate particles of the same crystal structure becomes significant with increasing pH (11).

In the formation of the basic iron sulfate particles, coexisting cations play a dominant role. For example, the particle size decreases in the order K_2SO_4 > $(NH_4)_2SO_4$ > Na_2SO_4 after the same period of aging, and the lattice parameters are found to change systematically, suggesting that these cations are incorporated into the lattice structure by replacing an H_2O with a hydroxide of one of these cations: c_0/a_0 = 2.31, 2.27, 2.36, and 2.36 for $Fe_3(SO_4)_2(OH)_5 \cdot 2H_2O$, $NaFe_3(SO_4)_2(OH)_6$, $KFe_3(SO_4)_2(OH)_6$, and $NH_4Fe_3(SO_4)_2(OH)_6$, respectively (9). In extreme cases with some other sulfate salts, such as $Li_2(SO_4)_2$, Cs_2SO_4, $MgSO_4$, $NiSO_4$, and $CuSO_4$, precipitation of sol is completely inhibited (9).

5.1.3 Basic Cerium Sulfate

Hsu et al. (12) prepared rodlike basic cerium(IV) sulfate particles of 1–3 μm in length from dilute solutions of $Ce(SO_4)_2$ in acidic media at 90°C (see Fig. 5.1.4). When the concentration of SO_4^{2-} is lowered, monodispersed polycrystalline spherical particles of CeO_2 are obtained under similar conditions. Figure 5.1.5 shows the precipitation domains of the respective species at 90°C for 12 h. Even for the forma-

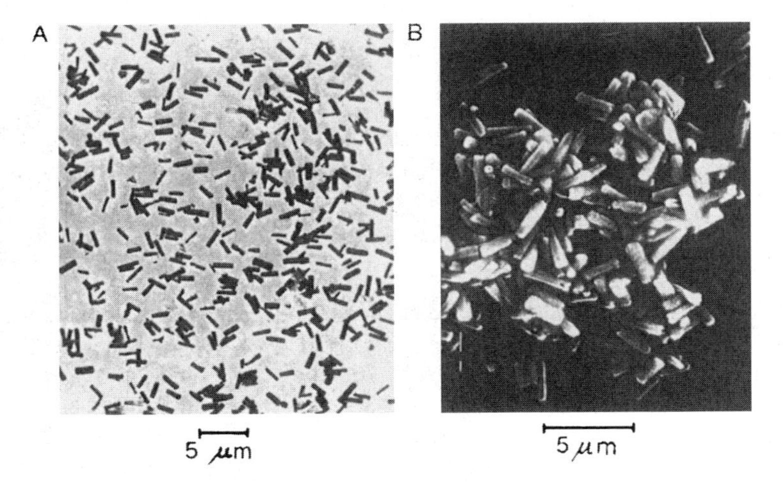

Fig. 5.1.4 (A) TEM and (B) SEM of rodlike particles of basic cerium sulfate prepared by aging 2.5×10^{-3} mol dm^{-3} $Ce(SO_4)_2$, 1.0×10^{-2} mol dm^{-3} H_2SO_4, and 0.4 mol dm^{-3} Na_2SO_4 at 90°C for 12 h. (From Ref. 12.)

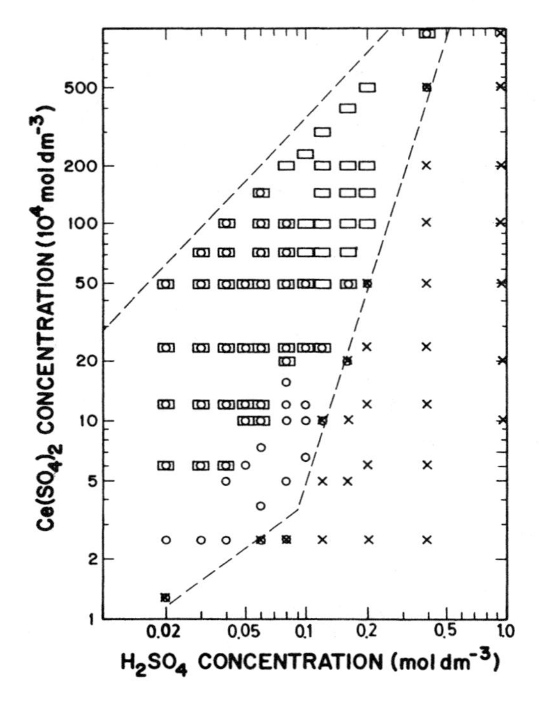

Fig. 5.1.5 Domain diagram of precipitates from solutions containing $Ce(SO_4)_2$ and H_2SO_4 aged for 12 h at 90°C: spheres (○); rods (▭); rods mixed with spheres (▣); a very small amount of spheres (⊗); no particles (x). (From Ref. 12.)

tion of the uniform CeO_2 particles in which sulfate ions is insufficient for forming basic cerium sulfate, sulfate ions may play an important role, like the formation of monodispersed spherical particles of amorphous $Cr(OH)_3$ (13) and $Al(OH)_3$ (14).

5.1.4 Basic Zirconium Sulfate

Uniform submicrometer-size spheres of basic zirconium sulfate ($Zr_2(OH)_6SO_4 \cdot 2H_2O$) and oxybasic zirconium carbonate ($Zr_2O_2(OH)_2CO_3 \cdot 2H_2O$) were prepared by aging a mixed solution consisting of 5×10^{-3} mol dm^{-3} $Zr(SO_4)_2$, 5×10^{-2} mol dm^{-3} HNO_3, 1.8 mol dm^{-3} urea, and 3 wt% PVP for 5 h at 50°C and by aging the same solution but at 80°C, respectively (15). Here PVP was particularly useful as an anticoagulant. On calcination, both precursor particles of amorphous structure were crystallized to tetragonal zirconia (ZrO_2) at ~600–700°C with their original spherical shape retained. Figure 5.1.6 shows DTA curves (10°C min^{-1}) of samples prepared at the indicated temperatures (a) and their cooling curves (b). The

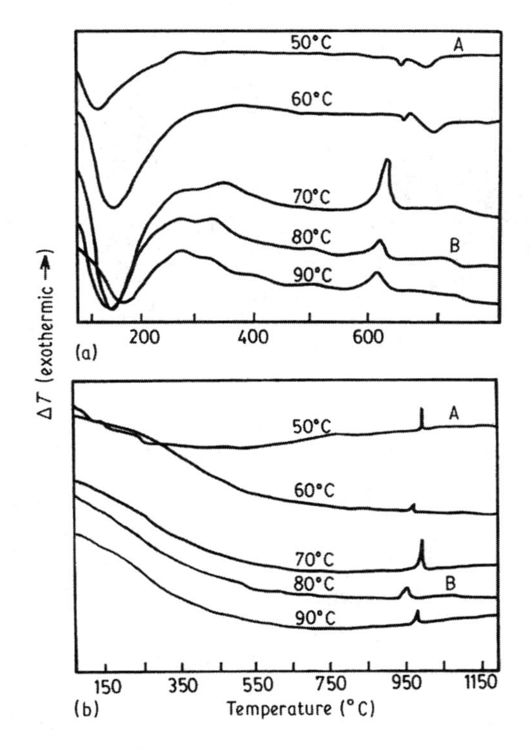

Fig. 5.1.6 DTA curves (10°C min^{-1}) of samples prepared by aging a mixed solution, consisting of 5×10^{-3} mol dm^{-3} $Zr(SO_4)_2$, 5×10^{-2} mol dm^{-3} HNO_3, 1.8 mol dm^{-3} urea, and 3 wt% PVP, for 5 h at the indicated temperatures (a) and their cooling curves (b). $Zr_2(OH)_6SO_4{\cdot}2H_2O$ (50, 60°C); $Zr_2O_2(OH)_2CO_3{\cdot}2H_2O$ (70, 80, 90°C). (From Ref. 15.)

upper two curves for basic zirconium sulfate (50, 60°C) in Fig. 5.1.6a are characterized by the endothermic peaks in 80–270°C due to the release of the hydrated water and the exothermic peak at 650°C due to the transition to crystalline tetragonal ZrO_2 overlapped with the broad endothermic band due to the gradual decomposition of the sulfate from 640 to 695°C. On the other hand, the endothermic peaks in the range of 80 to 270°C for the lower three curves (70, 80, 90°C) of the oxybasic zirconium carbonate particles in Fig. 5.1.6a are due to a combination of dehydration and decomposition of carbonate. The exthothermic peak at 325°C suggests the formation of crystalline $Zr_2O_3(OH)_2$, and the sharp exthothermic peak at ~610°C corresponds to the transition to tetragonal ZrO_2. The exthothermic peaks around 1000°C in the cooling curves for all samples in Fig. 5.1.6b may correspond to the transition from tetragonal to monoclinic structure. As the decomposition of urea below 70°C is extremely slow, the pH in the system for preparation of the basic zirconium sulfate

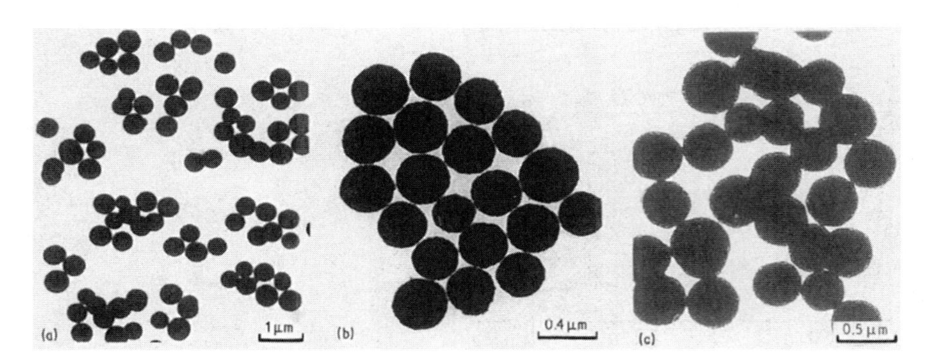

Fig. 5.1.7 TEM images of as-prepared spherical particles of basic zirconium sulfate (a), those calcined at 600°C for 3 h (b), and basic zirconium sulfate particles coated with $Y(OH)CO_3$ (c). The coated particles were exposed to a high-energy electron beam. (From Ref. 15.)

at 50°C was found to remain constant near 2 throughout the total reaction for 5 h. Under this condition, urea is believed to work as an acceptor of protons to promote indirectly the formation of hydrated zirconium complex such as $Zr(OH)(H_2O)^{3+}$ as a precursor to the basic zirconium sulfate precipitate (15–17). At a higher temperature such as 80°C in the acidic media, the decomposition of urea is accelerated to yield NH_3 and CO_2 (18, 19), causing an increase in the pH—e.g., up to 7 after aging for 5 h at 80°C in the present system. In this case, the $SO_4{}^{2-}$ ions coordinated to Zr^{4+} ions are likely to be replaced by OH^- at pH > 4 and by a high concentration of $CO_3{}^{2-}$, yielding oxybasic zirconium carbonate (17, 20, 21).

While composite particles of $ZrY_{0.8}O_{3.2}$ of a mixed cubic–tetragonal structure were obtained by calcination (800°C) of precursor particles of $ZrY_{0.8}(OH)_{3.8}(CO_3)_{1.3}$ prepared by aging a mixed solution of 5.0×10^{-3} mol dm^{-3} $Zr(SO_4)_2$, 4.0×10^{-3} mol dm^{-3} $Y(NO_3)_3$, 5.0×10^{-2} mol dm^{-3} HNO_3, 3.0 wt% PVP, and 1.8 mol dm^{-3} urea at 80°C for 5 h, a similar attempt to prepare precursor particles of a mixed composition containing sulfate in place of carbonate by aging at 50°C was unsuccessful (15). Instead, it was possible to coat basic zirconium sulfate particles with $Y(OH)CO_3$ by aging a solution of 1.0×10^{-3} to 4.0×10^{-3} mol dm^{-3} $Y(NO_3)_3$ and 0.6 mol dm^{-3} urea in the presence of 0.15 to 0.60 g dm^{-3} $Zr_2(OH)_6$-$SO_4 \cdot 2H_2O$ powder at 90°C for 2 h.

Figure 5.1.7 shows TEM images of as-prepared spherical particles of basic zirconium sulfate (a), those calcined at 600°C for 3 h (b), and basic zirconium sulfate particles coated with $Y(OH)CO_3$ (c).

5.1.5 Basic Copper Sulfate

Fairly large tabular particles of $Cu_4(OH)_6SO_4 \cdot H_2O$ and acicular particles of Cu_4(-$OH)_6SO_4$ are obtained by aging a mixed solution of 1.0×10^{-2} mol dm^{-3} $Cu(NO_3)_2$,

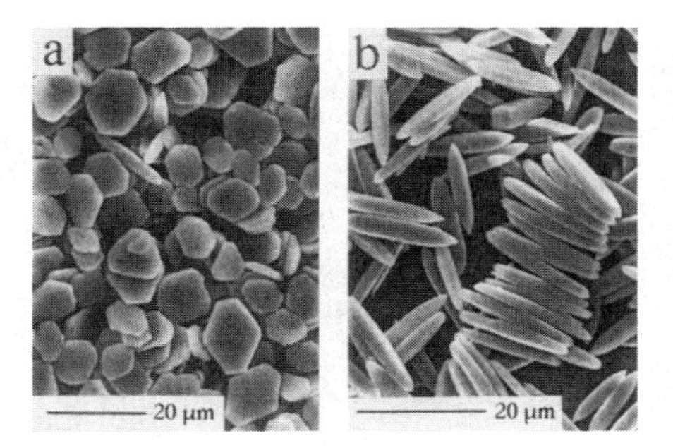

Fig. 5.1.8 SEM images of the tabular particles of (a) $Cu_4(OH)_6SO_4{\cdot}H_2O$ and (b) acicular particles of $Cu_4(OH)_6SO_4$, obtained by aging a mixed solution of 1.0×10^{-2} mol dm^{-3} $Cu(NO_3)_2$, 1.2 mol dm^{-3} ethylenediamine, and 1.0×10^{-1} mol dm^{-3} Na_2SO_4 at 100°C for 15 min and 30 min, respectively. (From Ref. 22.)

1.2 mol dm^{-3} ethylenediamine, and 1.0×10^{-1} mol dm^{-3} Na_2SO_4 at 100°C for 15 min and 30 min, respectively, where the pH values after aging for 0, 15, and 30 min are 6.3, 6.0, and 5.9, respectively (22). Their SEM images and XRD profiles are shown in Figs. 5.1.8 and 5.1.9, respectively. Though both are monoclinic, they are considerably different in lattice parameters as revealed from the XRD data, even

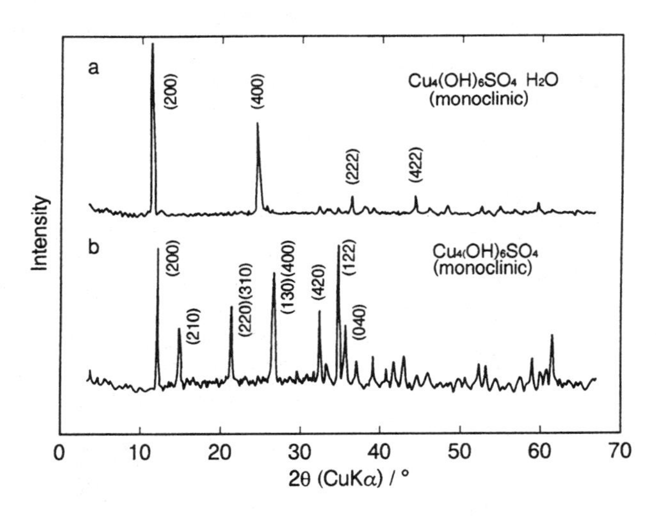

Fig. 5.1.9 XRD profiles of the tabular and acicular particles shown in Fig. 5.1.8. (From Ref. 22.)

if the orientation of the anisometric particles on a XRD sample holder is taken into account. Since the tabular particles are precipitated first and the acicular particles are formed by recrystallization of the former with the progress of time in the same solution, the tabular crystals are metastable ones.

In view of the high uniformity of the acicular particles, it is likely that their nucleation has finished by the time when the precipitation of the tabular particles is completed, followed by their growth under a low supersaturation, in equilibrium with the tabular particles, without renucleation.

REFERENCES TO SECTION 5.1

1. AHM Andreasen, K Skeel-Christensen, B Kjaer. Kolloid-Z 104:181, 1943.
2. JJ Petres, GJ Dežlić, B Težak. Croat Chem Acta 40:213, 1968.
3. NA Held, KN Samochwalov. Kolloid-Z 72:13, 1935.
4. BD Cuming, JH Schulman. Australian J Chem 12:413, 1959.
5. M Miura, H Fujita, Y Watari. J Sci Hiroshima Univ Ser A-II 28:41, 1964.
6. K Takiyama. Bull Chem Soc Jpn 31:950, 1958.
7. JJ Petres, GJ Dežlić, B Težak. Croat Chem Acta 38:277, 1966.
8. JJ Petres, GJ Dežlić, B Težak. Croat Chem Acta 41:183, 1969.
9. E Matijević, RS Sapieszko, JB Melville. J Colloid Interface Sci 50:567, 1975.
10. RS Sapieszko, RC Patel, E Matijević. J Phys Chem 81:1061, 1977.
11. T Sugimoto, H Itoh, H Miyake. J Colloid Interface Sci 188:101, 1997.
12. WP Hsu, L Rönnquist, E Matijević. Langmuir 4:31, 1988.

13a. R Demchak, E Matijević. J Colloid Interface Sci 31:257, 1969.
13b. E Matijević, AD Lindsay, S Kratohvil, ME Jones, RI Larson, NW Cayey. J Colloid Interface Sci 36:273, 1971.
13c. A Bell, E Matijević. J Inorg Nucl Chem 37:907, 1975.
14a. R Brace, E Matijević. J Inorg Nucl Chem 35:3691, 1973.
14b. DL Catone, E Matijević. J Colloid Interface Sci 48:291, 1974.
14c. WB Scott, E Matijević. J Colloid Interface Sci 66:447, 1978.

15. B Aiken, WP Hsu, E Matijević. J Mater Sci 25:1886, 1990.
16. PK Das Gupta, SP Moulik. J Phys Chem 91:705, 1987.
17. CF Baea, RE Mesmer. The Hydrolysis of Cations. New York: John Wiley, 1976.
18. RC Warner. J Biol Chem 142:705, 1942.
19. WHR Shaw, JJ Bordeaux. J Am Chem Soc 77:4729, 1955.
20. HH Willard, NK Tang. J Am Chem Soc 59:1190, 1937.
21. HH Willard, HC Fogg. J Am Chem Soc 59:1197, 1937.
22. S Hamada, Y Kudo, I Ishiyama. J Jpn Soc Colour Mater 69:658, 1996.

5.2 FORMATION OF BASIC ALUMINUM SULFATE BY PHASE TRANSFORMATION FROM CONDENSED HYDROXIDE GEL

TADAO SUGIMOTO
Institute for Advanced Materials Processing, Tohoku University, Sendai, Japan

It was rather surprising that monodispersed basic aluminum sulfate (BAS) microcrystals ($Al_3(SO_4)_2(OH)_5{\cdot}2H_2O$; hexagonal crystal system) were formed in a gel–sol system (1), since Matijević et al. (2,3) had prepared monodispersed amorphous spheres of aluminum hydoxide particles in a similar system but in a dilute homogeneous solution of aluminum salt in the presence of sulfate ions. As explained in Ref. 1, the precipitation of the completely different species in the gel–sol system seems to be due to the much higher concentrations of sulfate and aluminum ions and the presence of the condensed aluminum sulfatohydroxide gel as an enormous capacity of a reservoir of the solutes, leading to the onset of nucleation and subsequent constant growth of basic aluminum sulfate (BAS).

The standard gel–sol procedure is based on the aging of a highly condensed gel at 100°C for 72 h, whose overall composition is $[Al^{3+}] = 1$ mol dm^{-3}, $[SO_4^{2-}]/[Al^{3+}] = 2/3$, and $[NaOH]/[Al^{3+}] = 1.8$ (initial pH $= 3.4$ at room temperature). The transmission electron micrographs in Fig. 5.2.1 clearly display the growth of the spheroidal cubic particles of basic aluminum sulfate by dissolution of the amorphous solid of sulfate-containing aluminum hydroxide with the progress of time up to 3 days. On the other hand, Fig. 5.2.2 illustrates the compositional change in the solution phase of the gel. The initial descent and subsequent ascent in $[Al^{3+}]$ and $[SO_4^{2-}]$ correspond to the development of gel network with increasing temperature and its subsequent dissolution with the increasing solubility of the gel as a function of temperature. The succeeding rapid drop of $[Al^{3+}]$ and $[SO_4^{2-}]$ after ~2 h and the last slow decline thereof correspond to the nucleation of BAS and the steady growth of BAS particles with the dissolution of the hydroxide gel, respectively. The significantly low $[SO_4^{2-}]/[Al^{3+}]$ ratio about 1/4 to 1/5, much less than 2/3, in the solution phase in the last steady state suggests the strong adsorption of SO_4^{2-} to the surfaces of the growing BAS particles. The sulfate ions must be adsorbed in the form of a polymeric complex, since the occupation area per sulfate ion is calculated to be only ~0.1 Å^2 from the adsorbed amount and the total surface area of the BAS at the end of the reaction.

In this gel–sol system, pH and concentration of sulfate ions are the decisive factors for the reaction rate and monodispersity of the product. The reaction rate for

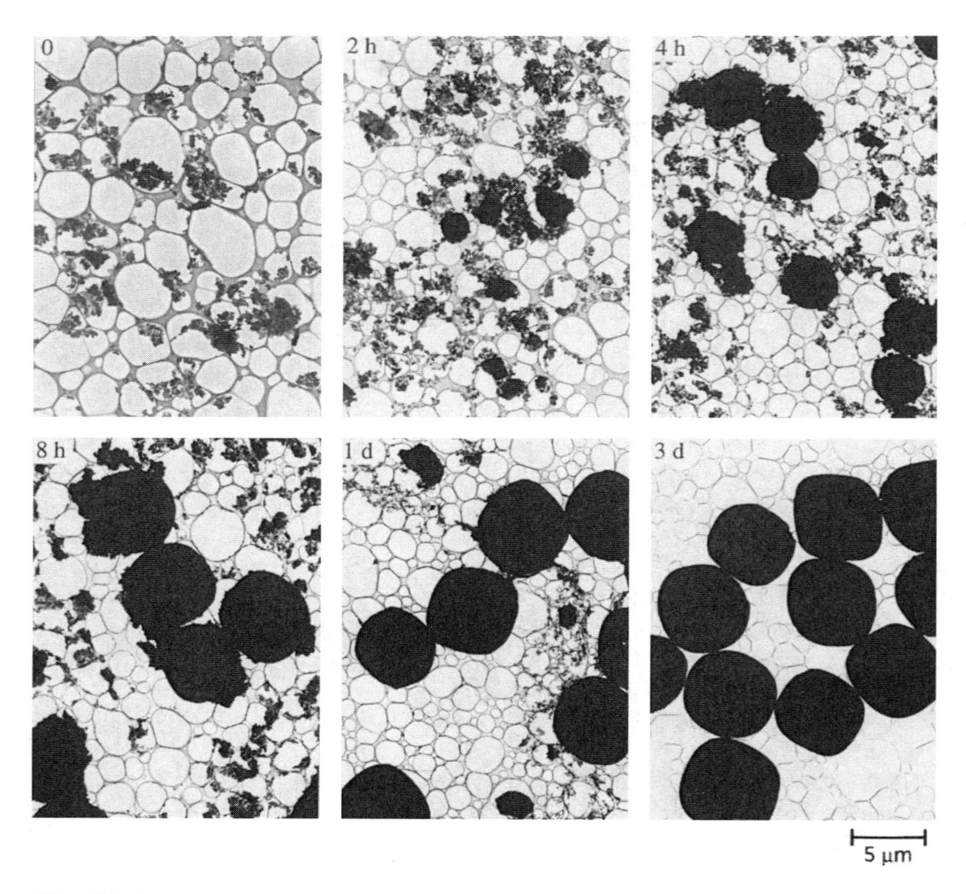

Fig. 5.2.1 Transformation of the solid phase from amorphous aluminum hydroxide to spheroidal cubic particles of basic aluminum sulfate with aging time under the standard conditions: $[Al^{3+}] = 1.0\ mol\ dm^{-3}$, $[SO_4^{2-}]/[Al^{3+}] = 2/3$, $[NaOH]/[Al^{3+} = 1.8$, temperature $= 100°C$. (From Ref. 1.)

the formation of BAS is significantly lowered with increasing pH, so that if the initial pH is raised up to ~6 with $[NaOH]/[Al^{3+}] = 2.4$, the reaction is not completed even after aging for 8 days at 100°C. If initial pH > 7.5, no precipitation of BAS is observed within 8 days. If the initial pH is set below the standard one ($= 3.4$), the reaction is accelerated with decreasing pH down to ~3, probably due to the increase in the concentration of the precursor complex to BAS. As a result, the mean size is lowered with broadening size distribution, due to the concurrent nucleation with growth. Yet, as the initial pH becomes still lower than 3, the reaction appears

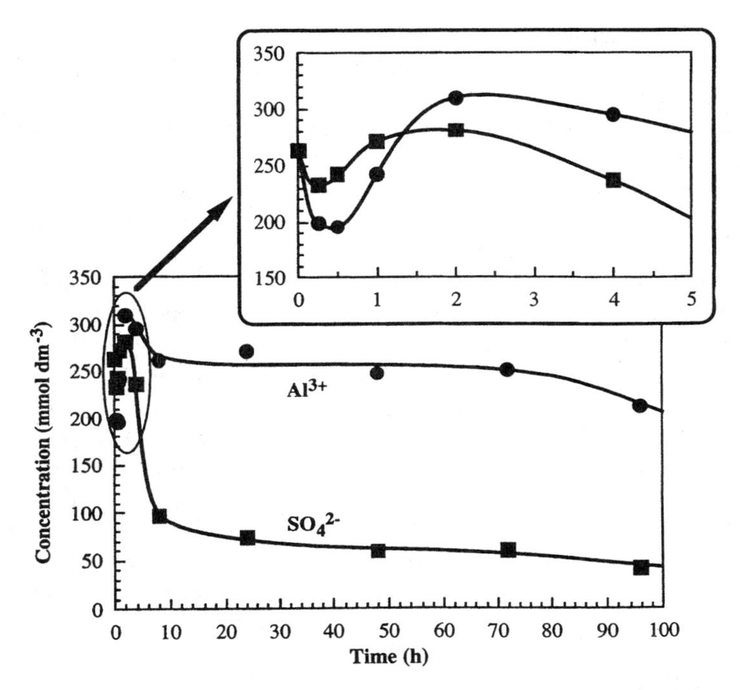

Fig. 5.2.2 Concentrations of Al^{3+} and SO_4^{2-} ions in the solution phase as a function of aging time in the standard system for the formation of the monodispersed basic aluminum sulfate particles. (From Ref. 1.)

to be rather decelerated and the yield of BAS tends to be lowered. These facts suggest that the concentration of the precursor complex to BAS becomes maximum around pH 3 when $[SO_4^{2-}]/[Al^{3+}] = 2/3$. However, a slightly higher pH, 3.4, may be the optimum for the formation of monodispersed particles. On the other hand, as $[SO_4^{2-}]$ increases, the growth rate of BAS is lowered by adsorption of the sulfate ions. Moreover, as the sign of the size distribution broadening by renucleation during the growth is not observed despite the retarded growth by the adsorption of sulfate ions, the concentration of the precursor complex to BAS also seems to be kept lower than the critical level for the spontaneous nucleation, because of the excessively high content of sulfate. For example, if the $[SO_4^{2-}]/[Al^{3+}]$ ratio is increased from 2/3 to 1, the reaction is not completed within 3 days, with a great deal of hydroxide gel left unreacted. However, when the $[SO_4^{2-}]/[Al^{3+}]$ ratio becomes lower than 2/3, the yield of BAS is, of course, lowered due to the deficiency in sulfate ions for forming the composition of $Al_3(SO_4)_2(OH)_5 \cdot 2H_2O$, and the size distribution becomes much broader, probably due to the excessively high concentration of the precursor complex leading to the renucleation during the growth stage. Hence, at least, for the monodispersity and yield, the standard ratio of $[SO_4^{2-}]/[Al^{3+}] = 2/3$ is the most suitable. For the concrete data and further analysis on the effects of pH and the content of sulfate ions upon the growth rate and monodispersity of the BAS particles, see Ref. 1.

Another aspect of pH and $[SO_4^{2-}]$ is their remarkable effect on the crystal habit of the BAS particles. The scanning electron micrographs (SEM) in Fig. 5.2.3 are those of BAS particles of different shapes prepared with variation of the initial pH and total concentration of SO_4^{2-}. The samples A-1 and A-3 (B-3) differ only in the $[NaOH]/[Al^{3+}]$ ratio: i.e., 2.4 (initial pH = 6.1) for A-1, while 1.8 (initial pH = 3.4) for A-3 (B-3). Obviously, the particles of A-1 are oblate, while those of A-3 (B-3) are spheroidal cubes, suggesting a retarded growth of a specific crystal face at a relatively high pH. On the other hand, sample A-3 (B-3) differs from B-5 only in $[SO_4^{2-}]/[Al^{3+}]$ ratio: i.e., 2/3 for A-3 (B-3), but 1/4 for B-5. One may conclude from this comparison that adsorption of sulfate ions rounds off the corners and edges of the cubic particles by retarding the growth of these sites. Sample B-5[3] was obtained under the same conditions as those for sample B-5, but the pH of the former was reduced from 3.2 to 0.4 with HNO_3 120 min after the start of aging, and the aging was continued for a total of 1 week. The added amount of nitric acid was 6 cm^3 of 6.4 mol dm^{-3} HNO_3 to 24 cm^3 of the system for sample B-5. It is evident that the triangular faces on the corners of the cubic particles disappear by the reduction of pH. According to the corresponding x-ray diffraction (OPML–XRD) profiles in Fig. 5.2.4, consisting of only the selective diffraction peaks of the lattice planes of the oblate or polyhedral particles completely oriented parallel to the sample holder plate in a monolayer (4), the flat faces of the oblate

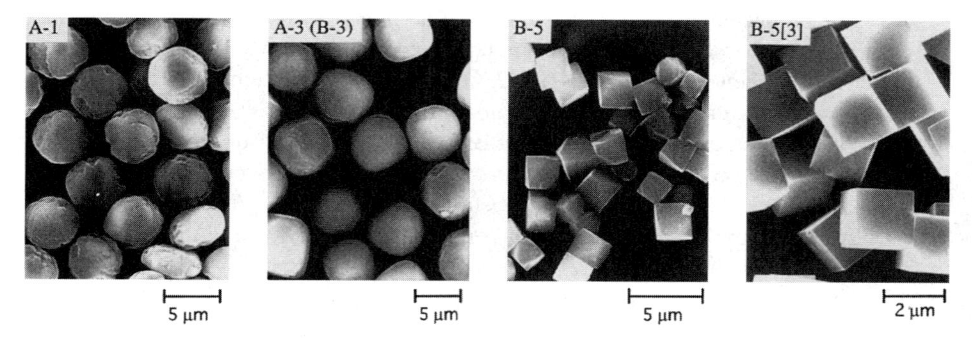

Fig. 5.2.3 SEM images of the basic aluminum sulfate particles of A-1, A-3 (B-3), B-5, and B-5[3], characterized by ($[NaOH]/[Al^{3+}]$; $[SO_4^{2-}]/[Al^{3+}]$) as (2.4; 2/3), (1.8; 2/3), (1.8; 1/4), and (1.8 + HNO_3; 1/4). The particles of sample A-1 were separated from the coexisting aluminum hydroxide gel. (From Ref. 1.)

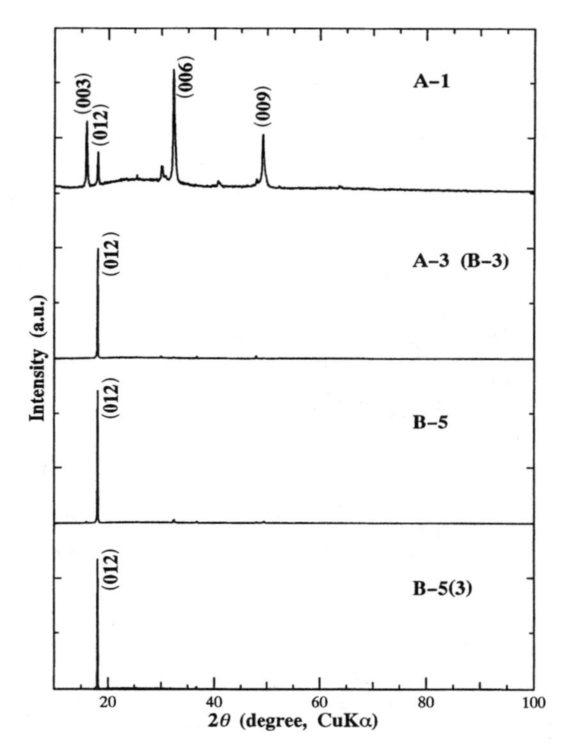

Fig. 5.2.4 OPML–XRD profiles of the same samples in Fig. 5.2.3. (From Ref. 1.)

particles of sample A-1 and the triangular faces of the truncated cuboidal particles in B-5 are assigned to the {001} faces of the hexagonal crystal system, while the flat faces of the spheroidal cubes in A-3 (B-3), the truncated square faces in B-5 and the square faces of the nearly perfect cubes in B-5[3] are the {012} faces. The diffraction peak of the {012} faces in the OPML–XRD of sample A-1 seems to be due to the oblate particles stacked together in the upright position on the substrate, as usually observed in TEM images of platelet-type particles. Therefore, OH^- ions are deemed to be strongly adsorbed to the {001} faces and inhibit the growth of the {001} faces, whose growth rate is much higher than that of the {012} in a strongly acidic medium as in the case of B-5[3]. The effect of pH on the crystal habit of BAS is confirmed in Fig. 5.2.5, in which the particles were formed at a fixed $[SO_4^{2-}]/[Al^{3+}]$ ratio = 1/4 with varying $[NaOH]/[Al^{3+}]$ = 1.5, 1.8, 2.4, and 2.7 for C-1, C-2, C-3, and C-4, respectively. The corresponding initial pH values are 3.0, 3.2, 3.7, and 6.9. The scanning electron micrographs clearly demonstrate the development of the {001} faces with increasing pH. This trend is observed in the

Fig. 5.2.5 SEM images of the basic aluminum sulfate particles of samples C-1, C-2, C-3, and C-4 systematically controlled in crystal habit at $[SO_4^{2-}]/[Al^{3+}] = 1/4$ by variation of the overall $[NaOH]/[Al^{3+}]$ ratio as 1.5, 1.8, 2.4, and 2.7, respectively. The respective initial pH values are 3.0, 3.2, 3.7, and 6.9. (From Ref. 1.)

typical morphological change of hematite particles (α-Fe_2O_3) as well, in the same hexagonal crystal system, from pseudocubes bound by the {012} faces in acidic media to platelets bound by the basal planes of the {001} faces and the side planes of the {012} faces in alkaline media (5,6).

If 1.67 mol dm^{-3} Cl^- is present under otherwise the same conditions as the standard sample A-3 (B-3) in Fig. 5.2.3, prolate particles are obtained as shown in Fig. 5.2.6 with the revolution axis collinear with the c axis of the hexagonal system. The arrested growth in the direction normal to the c axis by Cl^- is not observed in the growth of hematite, in which Cl^- ions arrest the growth of the {012} faces of hematite to yield the pseudocubic particles. The surfaces of the prolate particles are rather rough as compared to the other BAS particles, but they are thought to be single crystals, or at least very close to single crystals, as their XRD profiles are as sharp as those of the particles of samples B-5 and A-3. Hence, extraneous anions may not be incorporated in the BAS, in contrast to the cases of hematite.

As described earlier, sharp-edged cubic particles are obtained under the conditions of a low content of sulfate such as $[SO_4^{2-}] = 0.25$ mol dm^{-3} ($[SO_4^{2-}]/[Al^{3+}] = 1/4$) with a low pH such as 0.4, corresponding to the conditions for the preparation

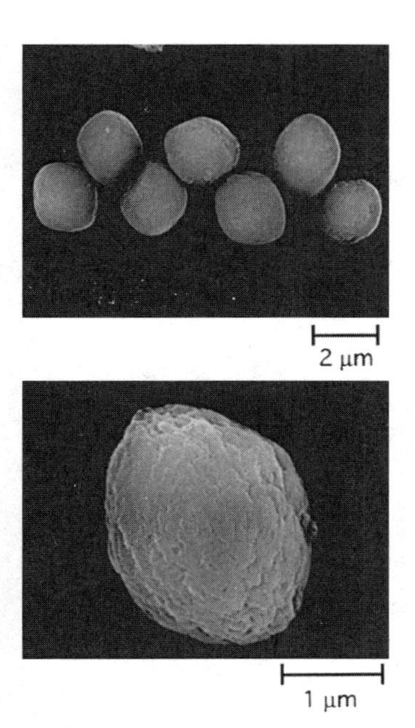

Fig. 5.2.6 SEM images of the ellipsoidal particles of basic aluminum sulfate prepared in the presence of 1.67 mol dm^{-3} chloride ions under the otherwise standard conditions. (From Ref. 1.)

of sample B-5[3] in Fig. 5.2.3. The preparation of the monodispersed cubic particles of sample B-5[3] was based on a proposal of the ''supersaturation quenching'' (7): i.e., a technique for the preparation of monodispersed particles through deliberate reduction of supersaturation by changing pH, temperature, solute concentration, etc. in the early growth stage to minimize the renucleation during the growth stage. For sample B-5[3], the supersaturation quenching was performed by reducing pH with nitric acid 120 min after the start of aging in the same system as that for the preparation of sample B-5, but with no addition of nitric acid in the latter. For the supersaturation quenching the timing, as well as the reduction level of the supersaturation, is a key factor. Figure 5.2.7 shows the effects of the timing of supersaturation quenching, in which HNO_3 was added after 80, 100, and 120 min for B-5[1], B-5[2], and B-5[3], respectively. Obviously, the final particle size decreases with increasing interval between the start of aging and the addition of nitric acid, due to the increase in the number of nuclei generated during the interval. In addition, the relative standard deviation of the size distribution is reduced as the quenching is delayed, showing a reduced probability of nucleation after quenching due to the promoted consumption of the solute for particle growth with the increase in total surface area. As different from ordinary oxide or hydroxide particles, the concentration of the precursor complex to BAS is a function of both pH and $[SO_4^{2-}]$ at a given content of aluminum ions, and thus the pH dependence of the precursor complex is not simple. For example, the concentration of the precursor complex appears to increase with decreasing pH, at least to pH 3, in contrast to ordinary oxide or hydroxide particles, whose precursor complexes are decreased with decreasing pH. Thus, for the supersaturation quenching of the BAS system, it was necessary to reduce the pH to a very low value, much lower than 3. The exceedingly low pH has a disadvantage of a low yield of the final product, since a high concentration of free Al^{3+} ions is required at a low pH to satisfy the solubility product; e.g., the yields of samples B-5[1], B-5[2], and B-5[3] are in the range of 0.76 to 0.78%, much lower than 22% for B-5, and 79% for A-3 (B-3).

Like the gel–sol system of monodispersed hematite (α-Fe_2O_3) particles (8), systematic control of the particles size by seeding is possible for the BAS system as well. Figure 5.2.8 exhibits an example of the size control of the BAS particles by adding BAS seeds of 12 nm in mean diameter to the standard gel prior to the aging. The seeds were prepared by grinding 4 g of a BAS powder, obtained by the standard procedure and previously moistened with 0.5 cm^3 of distilled water, in the agate pot of a planetary mill with agate balls for 30 min. The ground powder was dispersed in 30 cm^3 of distilled water by ultrasonication for 30 min, and was then centrifuged at 18,000 rpm for 30 min; the supernatant liquid was used as a stock seed suspension. By this seeding technique, it is possible to control the mean size of the BAS in a range from 6 to 0.4 μm.

Another aspect of the seeding experiment is that it provides us with important information on the particle growth mechanism, as mentioned in Section 1.3 in Chapter 1. The total reaction for the formation of the BAS particles is greatly accelerated

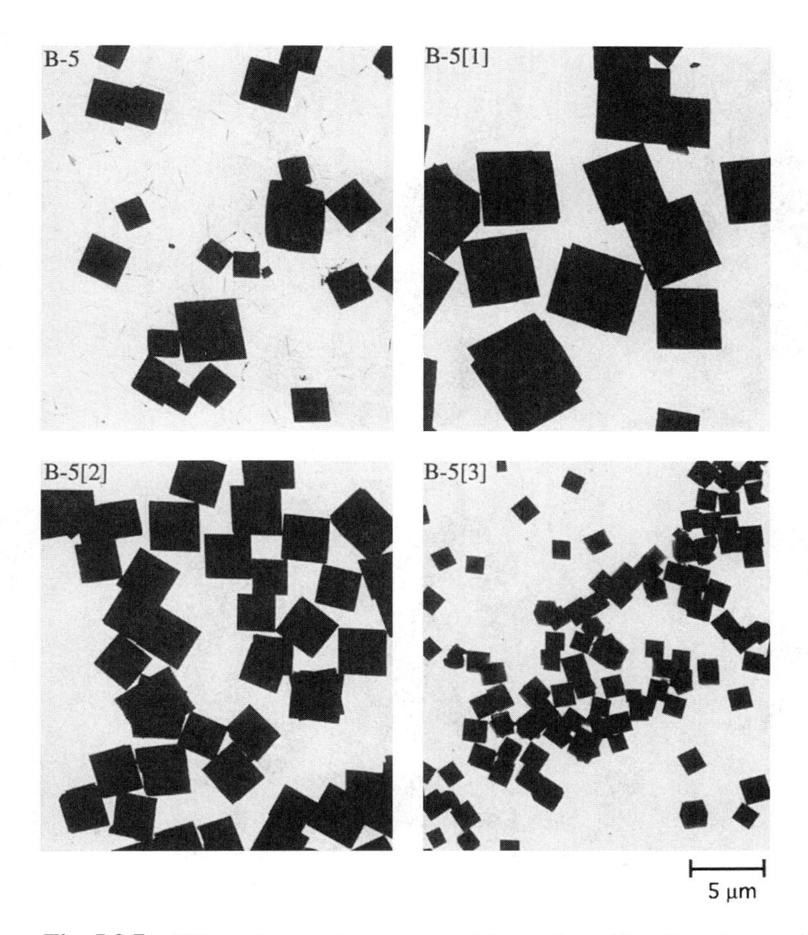

Fig. 5.2.7 Effects of supersaturation quenching on the uniformity and mean size of the final products in the B-5 system. Samples B-5, B-5[1], B-5[2], and B-5[3] are those unquenched and quenched with a fixed amount of nitric acid after 80, 100, and 120 min of aging, respectively. The added amount of nitric acid was 6 cm^3 of 6.4 mol dm^{-3} HNO_3 to 24 cm^3 of each sample B-5. (From Ref. 1.)

by the presence of the seeds; e.g., the reaction was completed within 1 day for all seeded samples in Fig. 5.2.8. This fact leads to two important conclusions: that the growth of the BAS particles proceeds through the dissolution–recrystallization mechanism, and that the rate-determining step in the growth of the BAS particles is the deposition process of the solute to the BAS particles and not the dissolution process of the hydroxide gel. These conclusions are common to the gel–sol system for the formation of hematite particles (5). In this case the concentration of the solute

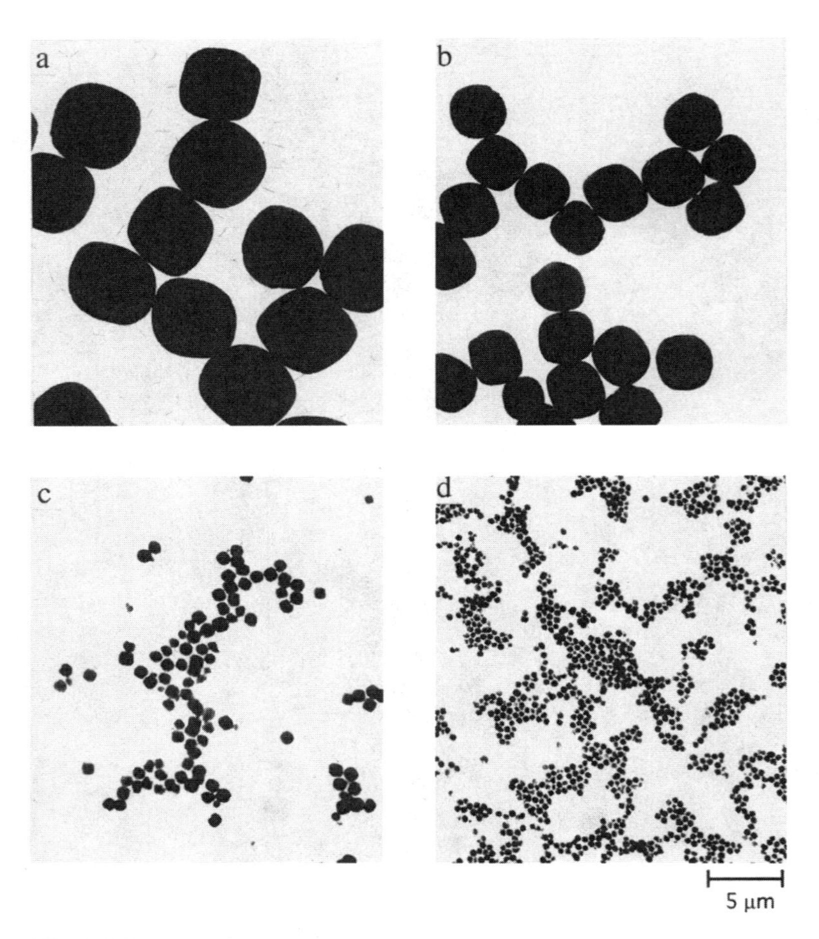

Fig. 5.2.8 Size control of the basic aluminum sulfate particles by seeding in the standard system A-3 (B-3). The added amounts of the seeds in terms of the equivalent volume of the stock seed suspension in 24 cm^3 of the standard hydroxide gel were: (a) 0, (b) 1.2×10^{-2}, (c) 2.4×10^{-1}, and (d) 2.4 cm^3. (From Ref. 1.)

is virtually in equilibrium with the hydroxide gel, and the discrepancy between the solubilities of the hydroxide gel and the growing particles is the driving force for the particle growth. Though the thermodynamic supersaturation for particle growth, in terms of the difference in solubility product between the hydroxide gel and the growing particles, must be kept constant with the change of pH in the presence of the hydroxide gel, a strong pH dependence of the growth rate suggests the decisive role of the precursor complex in the kinetic acceleration of the growth process.

REFERENCES TO SECTION 5.2

1. T Sugimoto, H Itoh, H Miyake. J Colloid Interface Sci 188:101, 1997.
2. R Brace, E Matijević. J Inorg Nucl Chem 35:3691, 1973.
3. DL Catone, E Matijević. J Colloid Interface Sci 48:291, 1974.
4. T Sugimoto, A Muramatsu, K Sakata, D Shindo. J Colloid Interface Sci 158:420, 1993.
5. T Sugimoto, K Sakata, A Muramatsu. J Colloid Interface Sci 159:372, 1993.
6. T Sugimoto, S Waki, H Itoh, A Muramatsu. Colloids Surfaces A Physicochem Eng Aspects 109:155, 1996.
7. T Sugimoto Adv Colloid Interface Sci 28:65, 1987.
8. T Sugimoto, Y Wang, A Muramatsu. Colloids Surfaces A Physicochem Eng Aspects 134:265, 1998.

6

Metal Phosphates and Apatites

TATSUO ISHIKAWA

6.1 REACTION IN HOMOGENEOUS SOLUTIONS FOR THE SYNTHESIS OF METAL PHOSPHATE PARTICLES

TATSUO ISHIKAWA
Osaka University of Education, Osaka, Japan

6.1.1 Introduction

Metal phosphates are valuable materials as pigments, catalysts, adsorbents, ion exchangers, sensors, and bioceramics. Aluminum phosphates are noble porous materials composed of uniform-sized mesopores that are anticipated to be applicable for catalysts with a high selectivity. Zirconium phosphates can be synthesized to have a layered structure in which various inorganic molecules can be intercalated, and the resultant inorganic organic complexes give rise to noble natures. Phosphates of cobalt and nickel are popular violet and green pigments, respectively. Since fine phosphate particles are used in these applications, their functions depend on the morphology and surface structure. In this chapter, the preparation, structure, and properties of various metal phosphate particles are described.

6.1.2 Synthesis of Metal Phosphates

Metal phosphates can be prepared by wet and dry methods. The homogeneous precipitation in aqueous systems is often employed to obtain uniformly sized and well-crystallized particles. The solid reactions can be used for preparation of some metal phosphates; however, the particles with controlled morphology are difficult to synthesize. Moreover, the solid reaction consumes more energy than the reaction in aqueous system, except the hydrothermal reaction at elevated temperature. Recently the preparation from aqueous solution at low temperature received attention in view of saving energy and as an application for a wide variety of substances. Therefore the preparation of metal phosphate particles by the precipitation method is described next. Although phosphates include ortho-, pyro-, and polyphosphates, only orthophosphates are dealt with here.

6.1.3 Precipitation of Metal Phosphates

The precipitation reactions of metal phosphates in aqueous systems are shown generally by the scheme in Fig. 6.1.1 (1). As seen in the scheme, the reaction of H_3PO_4 with hydroxides like NaOH yields soluble monobasic (NaH_2PO_4), dibasic

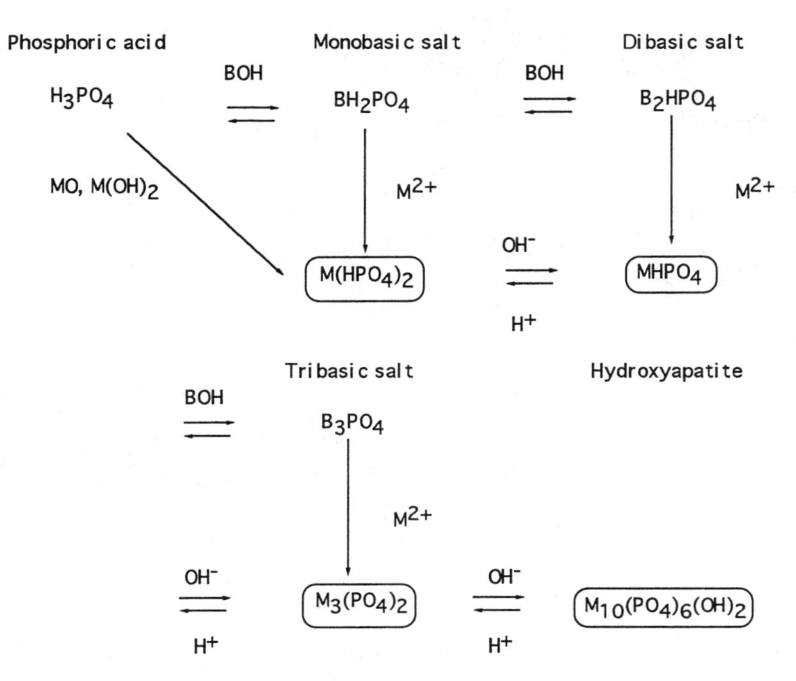

Fig. 6.1.1 Scheme for precipitation reactions of various metal phosphates. B is Na^+, K^+, NH_4^+ and CH_3- or C_2H_5-. M^{2+} is a divalent metal ion. (From Ref. 1.)

(Na_2HPO_4), and tribasic salts (Na_3PO_4). The addition of various metal ions to solutions dissolving these salts precipitates monobasic, dibasic, and tribasic metal phosphates as shown in the boxes. As described next, these precipitation reactions are widely used for preparation of different kinds of metal phosphate particles.

6.1.4 Phosphates of Divalent Metals

Uniform fine particles of phosphates of various divalent metal ions have been prepared by precipitation from aqueous solutions. For instance, spherical cobalt phosphate particles are prepared by aging solutions containing $CoSO_4$ and NaH_2PO_4 in the presence of urea and sodium dodecyl sulfate (SDS) (2). Figure 6.1.2 shows transmission electron microscopy (TEM) and scanning electron microscopy (SEM) images of the spherical cobalt phosphate particles of uniform size. No precipitation occurred in the absence of urea, which is a homogeneous precipitation agent. Urea elevates the pH of heated solutions by decomposing to NH_4^+ and CO_2 in acidic and to NH_3 and CO_3^{2-} in neutral and basic media. When pH rises sufficiently, the precipitation reaction of Co^{2+} with phosphate ions takes place. The spherical particles

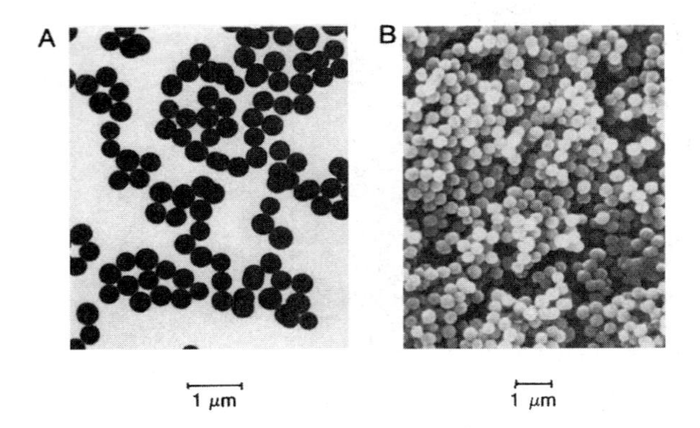

Fig. 6.1.2 (A) TEM and (B) SEM pictures of cobalt phosphate particles formed by aging a solution containing 5.0×10^{-3} mol dm^{-3} $CoSO_4$, 5.0×10^{-3} mol dm^{-3} NaH_2PO_4, 1.0 mol dm^{-3} urea, and 1.0×10^{-2} mol dm^{-3} SDS at 80°C for 3h. (From Ref. 2.)

within a narrow size distribution result in solutions of a molar ionic ratio $[Co^{2+}]/[PO_4^{3-}] \approx 1.5$, which is the same stoichiometric composition as of the formed solids, $Co_3(PO_4)_2 \cdot xH_2O$ (x = 3.4–3.7). A similar result was reported for trivalent metal phosphate, such as aluminum phosphate, where spherical particles in a narrow size distribution were generated in solutions of $[Al^{3+}]/[PO_4^{3-}] \approx 1$ (3). On the other hand, spherical phosphate particles of Cd^{2+}, Ni^{2+}, and Mn^{2+} in a narrow size distribution were formed by aging solutions of a molar ratio of $[M^{2+}]/[PO_4^{3-}] \approx$ 1 that differs from the molar ratios of the formed particles (4). SDS is an effective additive in the precipitation of spherical metal phosphates of narrow size distribution. It is possible that this surfactant affects the nucleation process and prevents particle aggregation. The addition of a nonionic (polyoxyethylene octyl ether) or of polyvinyipyrolidone gave aggregated solids. Therefore, the sulfate groups of SDS seem to play an important role in the precipitation process. To make this clear, the effects of anionic, cationic, and nonionic surfactants on the formation of spherical cobalt phosphate particles were examined (5). Figure 6.1.3 compares the TEM pictures of the particles formed with SDS, cetyltrimetylammonium chloride (CTAC), and polyoxyethylene(20) nonylphenyl ether (NP-20). It can be clearly seen that the particles without surfactant (A) and with NP-20 (D) are aggregates of spherical particles while those with SDS (B) and CTAC (C) are separated spherical particles; that is, the latter two surfactants are effective in the production of spherical cobalt phosphates with controlled particle size. It is noteworthy that all the spherical particles of phosphates of Ni^{2+}, Co^{2+}, Cd^{2+}, and Mn^{2+} are amorphous and contain OH^- and/or hydrated H_2O.

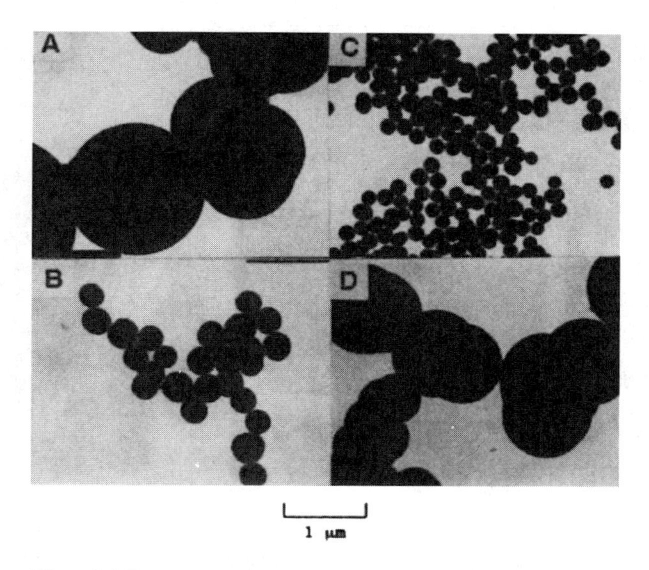

Fig. 6.1.3 TEM micrographs of cobalt phosphate particles precipitated without (A) and with 1.0×10^{-3} mol dm^{-3} of SDS (B), CTAC (C), and NP20 (D). [$CoSO_4$] = 5.0×10^{-3} mol dm^{-3}, [NaH_2PO_4] = 5.0×10^{-3} mol dm^{-3}, and [urea] = 1.0 mol dm^{-3}. (From Ref. 5.)

It is of interest to know the structural change of the particles by heating. Figure 6.1.4 depicts a differential thermal analysis (DTA) curve of the spherical cobalt phosphate particles that clearly shows an endothermic peak at 180°C due to the release of H_2O and an endothermic peak at 580°C indicating phase transformation. The x-ray diffraction (XRD) pattern of the crystalline material obtained by heating at 600°C is characteristic of $Co_3(PO_4)_2$ (JCPDS 13-503). A similar thermal change is observed in the cases of phosphates of Ni^{2+} and Cd^{2+}, though the temperature of phase transformation is ~800°C higher than that of cobalt phosphate. The morphology change of the spherical metal phosphate particles with heating was examined by TEM. Figure 6.1.5 displays the TEM images of the cobalt phosphate particles treated at different temperatures. The morphology of the particles is preserved up to 500°C when the effect of crystallization becomes apparent. While still essentially spherical, there is an obvious indication of a change in the internal structure (Fig. 6.1.5D). The specific surface areas are essentially the same up to 500°C, but it is suddenly changed by treating at 600°C, in agreement with the DTA and XRD results. A similar morphology change can be recognized in nickel phosphate particles (6).

Information on the inner structure of spherical metal phosphate particles is required to understand the formation mechanism of these particles. The estimation of porosity of the particles is an available mean for this subject. Kandori et al. (6)

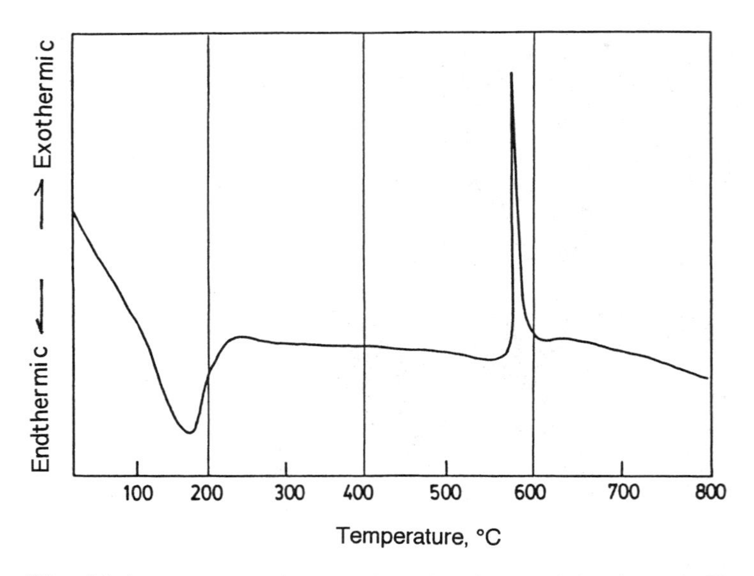

Fig. 6.1.4 DTA curve for the cobalt phosphate particles shown in Figure 6.1.2.

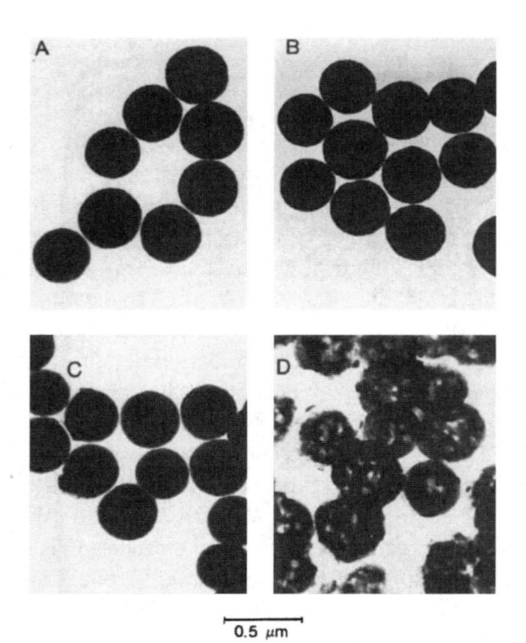

Fig. 6.1.5 TEM pictures of the cobalt phosphate particles calcined in air at different temperatures. (A) Original particles prepared under the conditions given in Figure 6.1.2; (B) 300°C; (C) 500°C; and (D) 600°C.

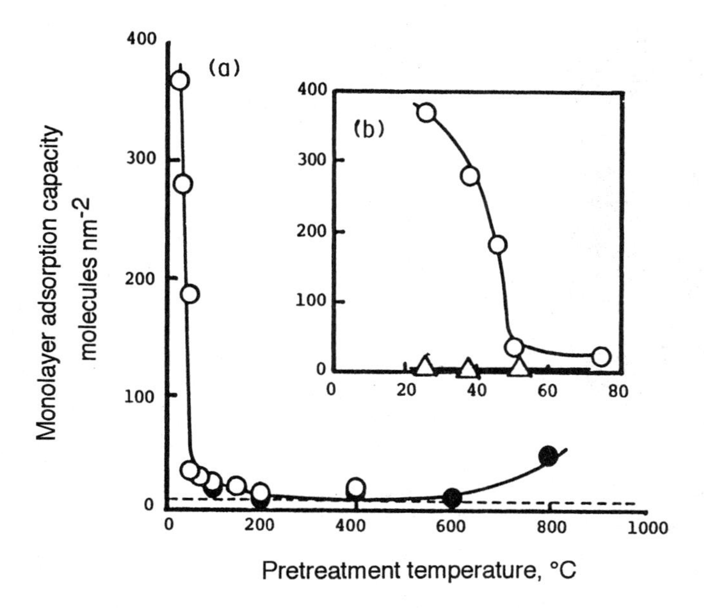

Fig. 6.1.6 (a) Plots of monolayer adsorption capacity of H_2O against pretreatment temperature. Open and filled circles represent the particles treated in vacuo and in air for 2 h, respectively. (b) Comparison of monolayer adsorption capacity of H_2O (○) and CCl_4 (△) at the low treatment temperature region.

studied the pore structure of uniform spherical cobalt phosphate particles by adsorption of molecules with different sizes. Figure 6.1.6 shows a plot of the monolayer adsorption capacity of H_2O and CCl_4 per unit surface area determined by N_2—that is, a measure of porosity—against outgassing temperatures. The particles outgassed below 45°C, especially at 25°C, show high capacity for H_2O and extremely low capacity for CCl_4, that is, a high selectivity for H_2O adsorption. This result indicates the existence of ultramicropores accessible to H_2O but not to N_2 and CCl_4, because the diameters of H_2O, N_2, and CCl_4 molecules are 0.276, 0.353, and 0.464 nm, respectively. It is of interest that the particles treated at 25°C show an XRD pattern characteristic of a layered structure that turns amorphous on treating above 50°C, showing that the layered structure is markedly unstable. The high adsorption selectivity for H_2O would be ascribed to slit-shaped ultramicropores in the layers. In the case of nickel phosphate, such a layered structure is not observed, though the ultramicropores accessible only to small H_2O molecules are formed and the spherical particles consist of primary fine particles as seen from the TEM picture of nickel phosphate particles in Fig. 6.1.7. For the spherical cobalt phosphate particles, such fine primary particles were not detected by TEM. Therefore the spherical nickel phos-

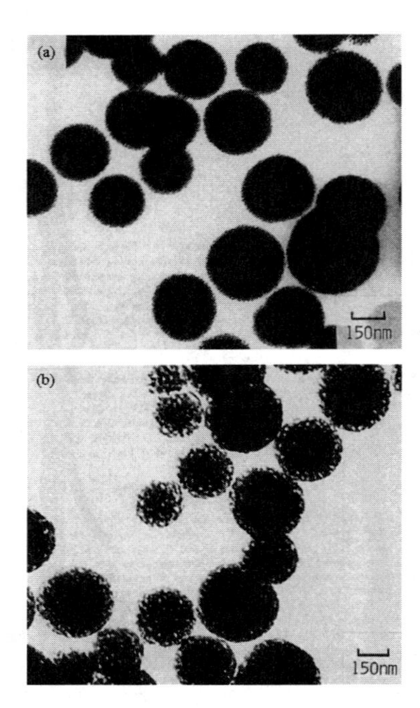

Fig. 6.1.7 TEM pictures of nickel phosphate particles, (a) before and (b) after being calcined at 600°C for 2 h.

phate particles possess inner structure distinct from that of the cobalt phosphates ones. To elucidate this structural difference, the pore sizes of the spherical particles of nickel and cobalt phosphates were determined by N_2 adsorption. The cobalt phosphate particles have only ultramicropores inaccessible for N_2 but not for H_2O. The ultramicropores are collapsed by treatment at elevated temperature, and the pore size increases to be accessible to larger molecules than H_2O. On the other hand, as seen from Fig. 6.1.8, the nickel phosphate particles outgassed at 25–400°C possess micropores with a mean radius of 2 nm, while the materials calcined in air at 100–800°C show a wider pore size distribution than the outgassed ones. These results indicate that the spherical nickel phosphate particles are formed by aggregation of the primary particles and that spherical nickel phosphate particles with mesopores can be prepared by treatment at elevated temperatures.

6.1.5 Aluminum Phosphate

Aluminophosphate molecular sieves with uniformly sized pores are a new porous material composed of $AlPO_4$ and called ALPO. Recently much attention has been

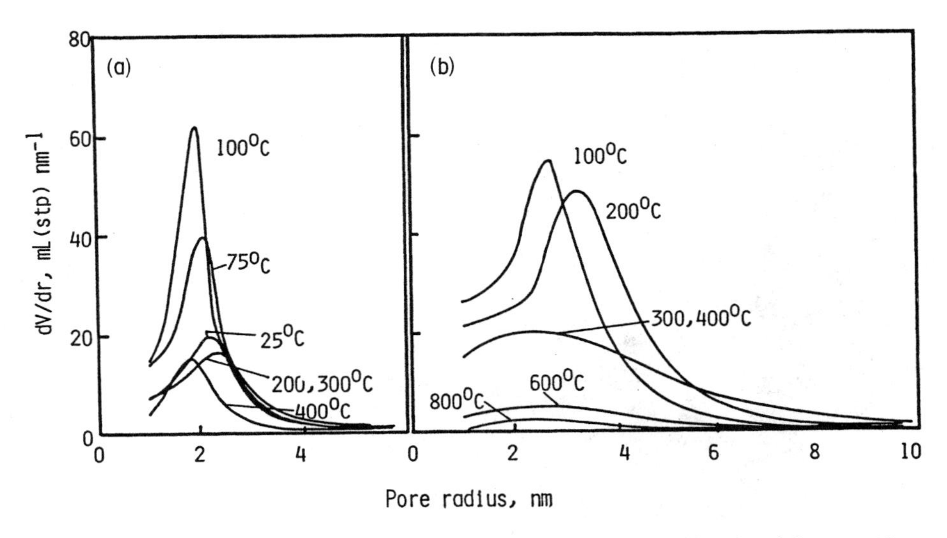

Fig. 6.1.8 Pore size distribution of uniform spherical nickel phosphate particles treated at various temperatures (a) in vacuo and (b) in air.

given to ALPO as an adsorbent and catalyst. Some aluminum phosphate particles are known to exhibit high catalytic activity and stronger surface acidity than silica alumina cracking catalysts (7). On the other hand, Wilson et al. have reported that aluminum phosphate can be produced as a porous crystalline material, like zeolite (8). A series of aluminum phosphate molecular sieves (named $ALPO_4$-*n*) can be endowed with Bronsted acidity by isomorphous substitution of hetero atoms into their frameworks (9). Recently, $ALPO_4$-18 substituted with silicon and divalent metal (designated as SAPO-18 and MAPO-18, respectively) was synthesized, and it was claimed that these molecular sieves are excellent solid acid catalysts for methanol conversion to light alkanes (10). However, all these molecular sieve particles are not uniform in either shape or size. If we could prepare the particles with uniform shape and size, these characteristic properties would be more emphasized by the uniformity of the surface and inner structures of these particles, which would permit us to design many highly efficient products.

Kandori et al. prepared spherical aluminum phosphate particles by aging a solution of 3.98×10^{-2} mol/dm^3 $Al(NO_3)_3$ and 3.98×10^{-2} mol/dm^3 Na_2HPO_4 at 100°C (11). The particle size increased with progressing aging and became constant after aging for 19 h in this condition. A TEM photograph of the particles formed after aging for 19 h is displayed in Fig. 6.1.9A. Figure 6.1.9B shows a TEM picture of the particles produced after aging for 2 h. It is obviously recognized that the particles grew by the agglomeration of small primary particles, similar to the

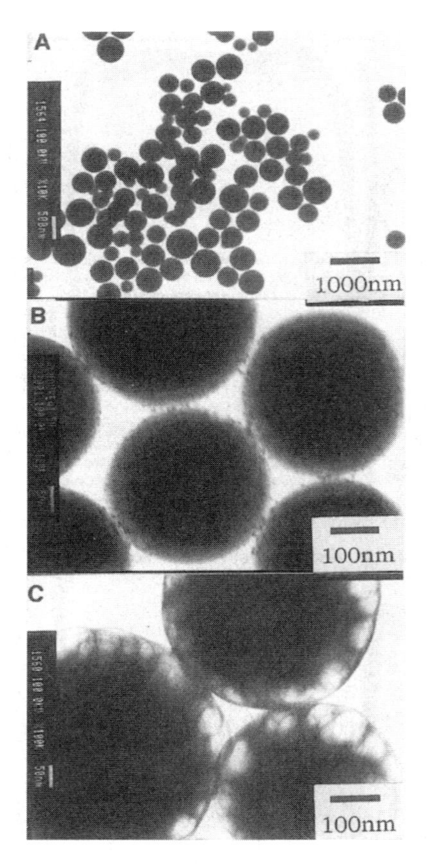

Fig. 6.1.9 TEM pictures of aluminum phosphate particles produced by aging a solution of 3.98×10^{-2} mol dm^{-3} $Al(NO_3)_3$, 3.98×10^{-2} mol dm^{-3} Na_2HPO_4, and 0.1 mol dm^{-3} HNO_3 at 100°C for (A) 19 h and (B) 2h. (C) Sample A exposed to a high electron beam. (From Ref. 11.)

spherical cadmium, nickel, and manganese phosphate particles mentioned earlier (4). The large void spaces are visualized on a TEM picture of the samples (Fig. 6.1.9C) exposed to high electron beam irradiation during TEM observation. The XRD measurement revealed that the spherical aluminum phosphate particles dried in vacuo at room temperature are amorphous. On the other hand, Matijevic et al. reported that the spherical particles, prepared by aging a solution dissolving 5×10^{-2} mol/dm^3 $Al(NO_3)_3$, 5×10^{-2} mol/dm^3 Na_2HPO_4, and 3.5×10^{-2} HNO_3 at 98°C for 138 h, were crystalline and showed the XRD pattern of variscite (3). The porosity of the spherical aluminum phosphate particles was examined by the adsorption of N_2 and H_2O. The specific surface area of the spherical particles determined by the BET N_2 method nearly equaled that estimated from their mean particle radius, revealing that the particles are nonporous to N_2 molecules, whereas the spherical particles outgassed at room temperature irreversibly adsorbed large amount of H_2O. This implies the presence of pores accessible not to N_2 molecules but to H_2O ones. These ultramicropores are thought to be openings among the primary particles agglomerated. Small H_2O molecules with a large dipole moment are able to enter the ultramicropores. Therefore, the adsorption results support the TEM observation that the spherical aluminum phosphate particles are formed by agglomeration of the primary particles, as are the spherical particles of other metal phosphates except for cobalt phosphate.

To control the size of the spherical aluminum phosphate particles, urea can be used as a homogeneous precipitation agent. The decomposition of urea at elevated temperatures plays an essential role in controlling the particles size.

Figure 6.1.10 shows the mean diameter of the spherical aluminum phosphate particles prepared by aging a solution containing 3.98×10^{-2} mol/dm^3 $Al(NO_3)_3$, 3.98×10^{-2} mol/dm^3 Na_2HPO_4, 0.1 mol/dm^3 HNO_3, and varied amounts of urea at 100°C for 19 h (12). The mean particle size steeply decreases from 560 to 100 nm at urea concentration below 0.2 mol/dm^3 and then decreases monotonously to 36 nm as the urea concentration is increased to 1 mol/dm^3. Morales et al. (13) proposed a mechanism of aluminum phosphate precipitation by the reaction of aluminum salts with H_3PO_4, which follows steps (a)–(d):

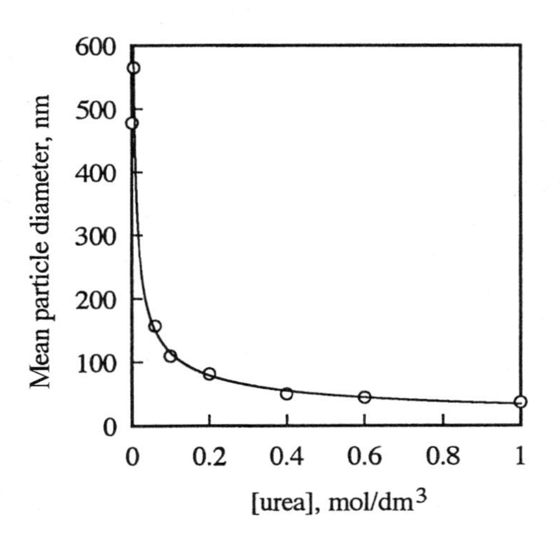

Fig. 6.1.10 Plots of the mean diameter of spherical aluminum phosphate particles against urea concentration. (From Ref. 12.)

(a) Formation of monomers:

$$Al^{3+} + H_2O = Al(OH)^{2+} + H^+ \quad (1)$$

$$H_3PO_4 = H_2PO_4^- + H^+ \quad (2)$$

$$H_2PO_4^- + Al(OH)^{2+} + H_2O = AlPO_4{\cdot}2H_2O\ (aq) + H^+ \quad (3)$$

(b) Nucleation:

$$n(AlPO_4{\cdot}2H_2O)\ (aq) = (AlPO_4 \cdot 2H_2O)_n(s)$$

(c) Crystal growth.
(d) Proton adsorption.

Since Na_2HPO_4 is used instead of H_3PO_4 in the preparation of the spherical particles described earlier, reactions (2) and (3) should be reactions (2′) and (3′), respectively:

$$Na_2HPO_4 = HPO_4^{2-} + 2Na^+ \quad (2')$$

$$HPO_4^{2-} + Al(OH)^{2+} + H_2O = AlPO_4{\cdot}2H_2O\ (aq) \quad (3')$$

Moreover, they mentioned that the precipitation of aluminum phosphate is regulated by competing effects between nucleation and proton saturation of crystal surfaces. Even if the supersaturation is high, the latter causes blockages on the crystal surfaces and arrests crystal growth, reducing the particle size. On the contrary, the adsorption of protons onto the crystal surface causes deprotonation of phosphoric acid and hydrolysis of Al^{3+} ions to $Al(OH)^{2+}$ by a continuous rise in pH, increasing the concentration of growth units to give a large number of small particles. The monomer formation in step (a) consists of deprotonation reactions. Therefore, the drop in the particle size upon addition of urea is caused by accelerating the reaction with NH_4^+ ions produced by decomposition of urea at elevated temperatures in acidic solution.

6.1.6 Summary

Spherical particles of various metal phosphate particles can be prepared by precipitation using urea as a homogeneous precipitation agent. Surface-active agents, such as SDS and CTAC, are effective in preparation of uniform-size spherical particles. The formed spherical particles are amorphous and contain OH^- and H_2O, except cobalt phosphate particles with layered structure. These particles are agglomerates of primary particles, and have pores of different sizes ranging from ultramicropore to mesopore.

REFERENCES TO SECTION 6.1

1. HW Mooney, MA Aia. Chem Rev 61:433, 1961.
2. T Ishikawa, E Matijevic. J Colloid Interface Sci 123:122, 1988.
3. EP Katsanis, E Matijevic. Collods Surfaces 5:43, 1982.

4. II Sprinsteen, E Matijevic. Colloid Polym Sci 267:1007, 1987.
5. K Kandori, E Matsuda, A Yasukawa, T Ishikawa. J Jpn Soc Colour Mater 68:75, 1995.
6. K Kandori, H Nakashima, T Ishikawa. J Colloid Interface Sci 160:499, 1993.
7. KK Kearby. US Patent 3,342,750, 1967.
8. ST Wilson, BM Lok, CA Messina, TR Cannan, EM Flanigen. J Am Chem Soc 104: 1146, 1982.
9. DB Akolekar. J Chem Soc Faraday Trans 90:1041, 1994.
10. J Chen, JM Thomas. J Chem Soc Chem Commun 603, 1994.
11. K Kandori, T Imazato, A Yasukawa, T Ishikawa. Colloid Polym Sci 274:290, 1996.
12. K Kandori, N Ikeguchi, A Yasukawa, T Ishikawa. J Colloid Interface Sci 182:425, 1996.
13. G Morales, RR Clemente, E Matijevic. J Colloid Interface Sci 151:555, 1992.

6.2 DIFFERENT REACTIONS FOR THE SYNTHESIS OF APATITE PARTICLES

TATSUO ISHIKAWA
Osaka University of Education, Osaka, Japan

6.2.1 Introduction

Apatites are metal basic phosphates for which the chemical formula is $M_{10}(PO_4)_6(OH)_2$ [M = divalent metal]. The most typical apatite is calcium hydroxyapatite, $Ca_{10}(PO_4)_6(OH)_2$, designated CaHAP. This material receives attention concerning biomineralization because it is a main component of vertebrate animals' bones and teeth. There have been voluminous investigations in dental and medical sciences. The synthesized CaHAP is of interest in bioceramics, as an adsorbent for biomaterials, and as an ion exchanger, chemical sensor, and catalyst. The characteristics of the CaHAP particles are thought to be influenced by the particle morphology, since the surface structure of the CaHAP particles is changed by their particle shapes, as speculated from the crystal structure of CaHAP described in detail later. Therefore, morphology control of CaHAP particles is important in each application of this material.

6.2.2 Preparation of CaHAP Particles

6.2.2.1 Precipitation Reaction

Precipitation is mostly employed for the preparation of CaHAP particles (1). The precipitation reactions used for this method were shown in Figure 6.1.1. Aqueous solutions of calcium salt and phosphoric acid or basic phosphate are mixed at pH 8–10, and the resulting precipitates are aged under various conditions. The Ca/P ratio, H_2O content, impurity, and morphology of the produced CaHAP particles are strongly influenced by the mixing manner of the reactants and by the aging conditions, including temperature, time, and pH. In industrial production, 25–80% H_3PO_4 solution is added to a suspension of 2–10% $Ca(OH)_2$ up to Ca/P = 1.67 with stirring. The resulting precipitation is left at room temperature for more than 1 day and then is filtrated and dried. The mean size and specific surface area of the prepared particles are ~0.5 μm and ~100 m^2/g. The preparation of CaHAP at laboratory scale is performed as follows. A calcium salt solution is mixed with a phosphate solution and the resulted suspension is aged at pH 4–10. Figure 6.2.1 displays TEM pictures of CaHAP particles formed at different pH and 100°C (2). The particle length is decreased by elevating the pH. CaHAP with a smaller Ca/P ratio forms at a lower

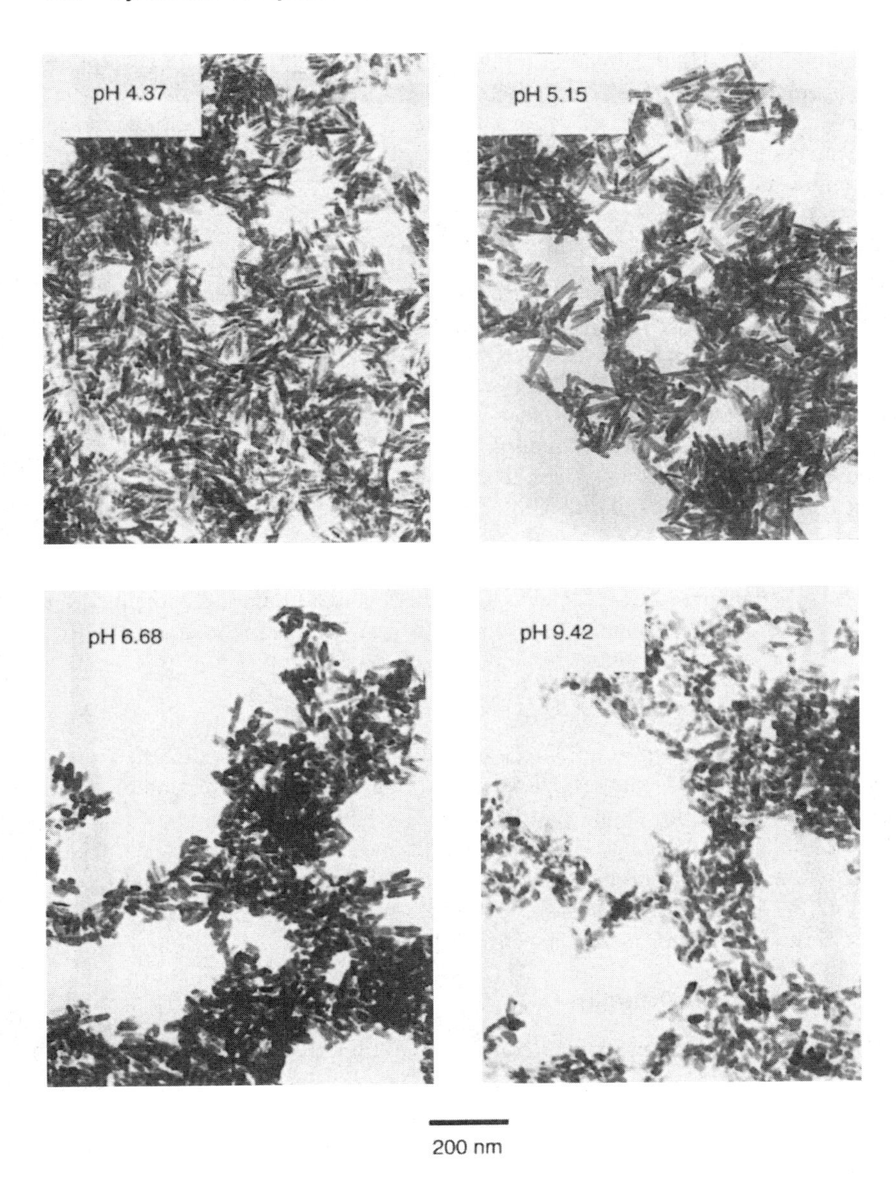

Fig. 6.2.1 TEM pictures of CaHAP particles formed at different pH. (From Ref. 2.)

pH of aging. The CaHAP formed at a high pH contains CO_3^{2-} due to absorption of atmospheric CO_2 during the preparation. The PO_4^{3-} in CaHAP cyrstals is replaced by CO_3^{2-}, resulted in a high Ca/P ratio. The particles produced by precipitation can be further crystallized and grown by further hydrothermal treatment at elevated temperatures.

6.2.2.2 Homogeneous Precipitation

For synthesis of pure particles with controlled morphology, homogeneous precipitation is widely used. Urea is employed as a homogeneous precipitation agent to synthesize CaHAP (3,4). The urea added to solutions containing Ca^{2+} and PO_4^{3-} decomposes to yield CO_2 and NH_4^+ on heating and raises the solution pH, leading to precipitation of octacalcium phosphate, $Ca_8H_2(PO_4)_6{\cdot}5H_2O$ (OCP), or dicalcium phosphate, $CaHPO_4$ (DCP), which finally transforms into CaHAP. However, CaHAP prepared with urea contains an appreciable amount of CO_3^{2-} because CO_2 evolved from urea is dissolved as CO_3^{2-} in the solutions and is easily incorporated into CaHAP. In order to synthesize CaHAP free from CO_3^{2-}, another precipitation agent generating no CO_2 must be employed. Amides are available for this purpose. Yasukawa et al. (5) synthesized CaHAP using formamide and acetamide as precipitation agents instead of urea. Formamide is hydrolyzed at an elevated temperature as follows:

$$HCONH_2 + H_2O \rightarrow HCOOH + NH_3$$

The resulting NH_3 raises the solution pH as well as urea. In Figure 6.2.2 the crystal phases of the materials are shown as a function of aging time and solution pH after aging. It is clear from this figure that the condition giving CaHAP, shown by the solid circles, relates to the solution pH, aging time, and kinds of amides. The CaHAP obtained by this method is highly crystallized needlelike particles. Acetamide gives more crystallized larger particles than formamide, though the formation pH with the former is somewhat higher than that with the latter.

6.2.2.3 Hydrolysis Reaction

CaHAP particles can be prepared by conversion of slightly soluble calcium phosphates, such as $CaHPO_4{\cdot}2H_2O$ (6) and α-$Ca_3(PO_4)_2$ (7), in aqueous media below 100°C. The characteristic of this method is that the obtained particles retain the shape of the precursor particles. $CaHPO_4{\cdot}2H_2O$ is transformed to CaHAP by hydrolysis at pH 8 and 40°C for 3 h. The formed nonstoichiometric CaHAP (Ca/P $=$ 1.50) is converted to a stoichiometric one (Ca/P $=$ 1.67) after further treatment at 40°C and pH $>$ 9 for 3 h. On the other hand, the hydrolysis of α-$Ca_3(PO_4)_2$ at pH 5.5–10 and 80°C for 2 h gives CaHAP with Ca/P ranging from 1.50 to 1.67, depending on the solution pH. The specific surface areas of the produced particles are 10–60 m^2/g. Needlelike CaHAP particles are prepared by hydrolysis of $CaHPO_4{\cdot}2H_2O$ with refluxing for 1 month in distilled and deionized water. The supernatant must be

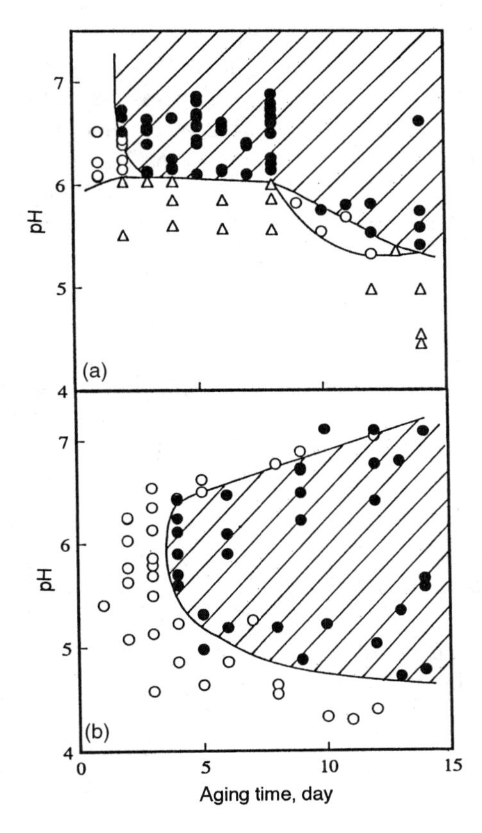

Fig. 6.2.2 Crystal phases of the materials formed by aging with formamide (a) and acetamide (b) with different pH of the supernatants after aging. ● CaHAP, ○ CaHAP-DCP mixture, △ DCP. (From Ref. 5.)

changed four to six times during the refluxing to reduce the amount of acid generated (8).

6.2.2.4 Hydrothermal Reaction

Highly crystallized CaHAP particles are synthesized by hydrothermal reactions at 300–700°C under 8.6–200 MPa for several days (1,9). The CaHAP particles produced are stoichiometric single crystals and rod- or needlelike particles elongated along the *c* axis. The reaction systems are CaHAP–H_2O, $Ca_2P_2O_7$–CaO–H_2O, $Ca(NO_3)$–NaOH–KH_2PO_4–H_2O, H_3PO_4–$Ca(NO_3)_2$–H_2O, etc. Hydrothermal treatment is frequently used for crystallization of low crystalline CaHAP particles and for preparation of uniformly shaped particles. The precursor CaHAP gels formed by a precipitation method are dispersed in aqueous solution adjusted at pH 10 and treated with stirring at 200°C for 10 h. By this treatment, the CaHAP particles turn into uniformly sized hexagonal prisms of 25 $\times$ 90 nm (10). On the other hand, the hydrothermal treatment of the precursor CaHAP particles at 140°C for 2 h yields needlelike CaHAP particles of 100–200 nm length and Ca/P < 1.67 (11). Treatment in the presence of citric acid at 200°C for 3 h gives large whisker particles of 10–30 μm length and 0.5 μm diameter (12).

6.2.2.5 Spray Pyrolysis Method

Spherical CaHAP particles are synthesized by spray pyrolysis. The spherical particles are suitable for high-performance liquid chromatography (HPLC) adsorbents. The atomized solutions dissolving $Ca(NO_3)_2$ and $(NH_4)_2HPO_4$ are evaporated into the hot part of a furnace, to be decomposed at 600°C (13). The CaHAP particles formed are hollow agglomerates of fine particles.

6.2.2.6 Solid-State Synthesis

Since stoichiometric CaHAP is stable above ~1000°C, CaHAP particles can be synthesized by solid reactions at high temperatures. An advantage of this method is that the CaHAP particles produced are stoichiometric and keep the shape of the precursor phosphate particles. One of the solid reactions used for CaHAP preparation is as follows:

$$3Ca_3(PO_4)_2 + CaCO_3 + H_2O \text{ (vapor)} \rightarrow Ca_{10}(PO_4)_6(OH)_2 + CO_2$$

This reaction is completed at 1200°C for 2 h in the presence of excess $CaCO_3$ (14). The residual CaO after the reaction is removed by washing with an NH_4Cl aqueous solution. The produced CaHAP particles are highly crystallized, and the specific surface area is ~2 m^2/g. The other reaction systems are $Ca_2P_2O_7$–$CaCO_3$–H_2O (15) and $CaHPO_4$–$CaCO_3$–H_2O (16).

6.2.2.7 Other Methods

Besides the already mentioned methods, many techniques of CaHAP preparation have been developed; agar gel method (17), electrodeposition (18), chelate decomposition (19), sol–gel process (20), microemersion (21), and so on. Since these methods yield particles with different morphologies, stoichiometries and crystallinities, the desired particles can be prepared by selecting an appropriate method.

6.2.3 Crystal Structure of CaHAP

Apatite is the mineral name for a series of substances with the general formula $M_{10}(RO_4)_6X_2$ as shown in Table 6.2.1. Thus, apatites possess a variety of compositions. Among these apatites, CaHAP is most popular. The crystal structure of CaHAP

Table 6.2.1 Chemical Compositions of Various Apatites

General formula: $M_{10}(RO_4)_6X_2$
M^{2+}: Ca, Sr, Pb, Mn, Zn, Mg, Fe, Cd, etc.
R: P, As, V, S, Si, etc.
X^-: OH, F, Cl, Br, etc.

is illustrated in Fig. 6.2.3 (22). OH^- ions line up along the c axis. There are two kinds of Ca^{2+} ions: Ca_I aligning along the c axis as pointed by the arrows and Ca_{II} surrounding OH^-, which are named respectively as column Ca and screw one. Ca_I is more easily dissolved in aqueous solutions than Ca_{II}, so that Ca deficiency occurs in the Ca_I sites. The Ca^{2+}, $PO_4{}^{3-}$, and OH^- of CaHAP can be replaced by other

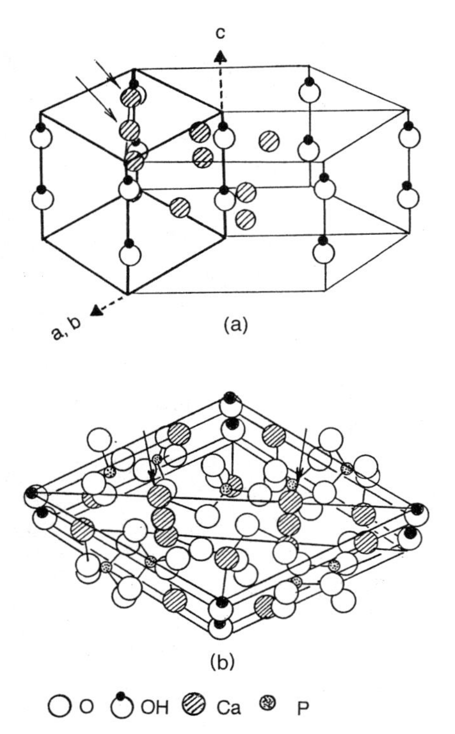

Fig. 6.2.3 Crystal structure of CaHAP. (a) The hexagonal model showing a part of OH^- and Ca^{2+} ions. (b) The unit cell ($a = b = 0.942$ nm, $c = 0.688$ nm). The arrows point to Ca_I. (From Ref. 22.)

ions as speculated from Table 6.2.1. The Ca_I in CaHAP crystals is exchanged with various cations given in Table 6.2.1 by immersing CaHAP particles in solutions of the cations. Thus, CaHAP is available to remove harmful heavy metal ions from the solutions, such as Pb^{2+}, Cd^{2+} and Hg^{2+} (23–25). The OH^- ions of CaHAP are replaced by anions. It is well known that the exchange of OH^- with F^- reduces the solubility of CaHAP in acid, which relates to prevention of dental caries by treatment with F^-. Because of the ion exchangeability, the synthesis of stoichiometric CaHAP with Ca/P = 1.67 is rather hard and requires particular attention. The nonstoichiometry is frequently caused by Ca deficiency in the lattice, as already mentioned. The composition of Ca-deficient CaHAP is written as $Ca_{10-x}(HPO_4)_x(PO_4)_{6-x}(OH)_{2-x} \cdot nH_2O$ $(0 < x \leq 1)$. The charge imbalance from the Ca deficiency is compensated by various processes, such as OH^- deficiency, protonation of $PO_4{}^{3-}$ to $HPO_4{}^{2-}$ or $H_2PO_4{}^-$, replacement of $PO_4{}^{3-}$ by $CO_3{}^{2-}$ or $HCO_3{}^-$, occupation of Ca^{2+} defects by other metal ions, and so forth. The Ca^{2+} and OH^- sites opened are occupied by H_2O molecules. For this reason, incorporation of various ions into the CaHAP lattice leads to formation of nonstoichiometric CaHAP particles. Actually, most CaHAP crystals in hard tissues and synthesized by conventional methods are nonstoichiometric. It has been established that various natures of CaHAP particles, such as catalytic activity, thermal stability, solubility in acid, etc., strongly depend on their nonstoichiometry.

6.2.4 Solid Solution Particles of Apatites

As described earlier, Ca^{2+} ions of CaHAP can be substituted with other metal ions. Ion exchange of CaHAP is interested in preparation of catalysts, removal of harmful metal ions in water, and incorporation of heavy metal ions such as Pb^{2+} and Cd^{2+} and radioactive Sr^{2+} ions into bones. Substitution with metal ions can be usually done by coprecipitation and ion-exchange methods. The former is available for preparing solid solution particles of different kinds of apatites. The formation of the solid solutions is restricted by charge and size of metal ions. Divalent ions with ionic radii close to or larger than 0.099 nm of Ca^{2+}, such as Sr^{2+} (0.112), Ba^{2+} (0.134), Pb^{2+} (0.120), and Cd^{2+} (0.097), easily replace the Ca^{2+} of CaHAP, so that solid solutions of apatites of these ions with CaHAP can be synthesized (26–28). On the other hand, metal ions smaller than Ca^{2+}—for instance, Mg^{2+} (0.066), Fe^{3+} (0.064), and Ni^{2+} (0.069)—are incorporated into CaHAP crystals to a limited extent. Figure 6.2.4 displays TEM images of the particles of CaSrHAP solid solutions synthesized by a coprecipitation method (29). CaHAP (A) and SrHAP (D) particles are respectively short rods and needles, and SrHAP particles are much larger and more crystallized than CaHAP ones. The solid solution particles (B and C) are agglomerates of small primary particles showing high specific surface area and low crystallinity, which is attributed to crystal distortion due to the size difference between Ca^{2+} and Sr^{2+}. A similar relation is found for SrBaHAP particles (30).

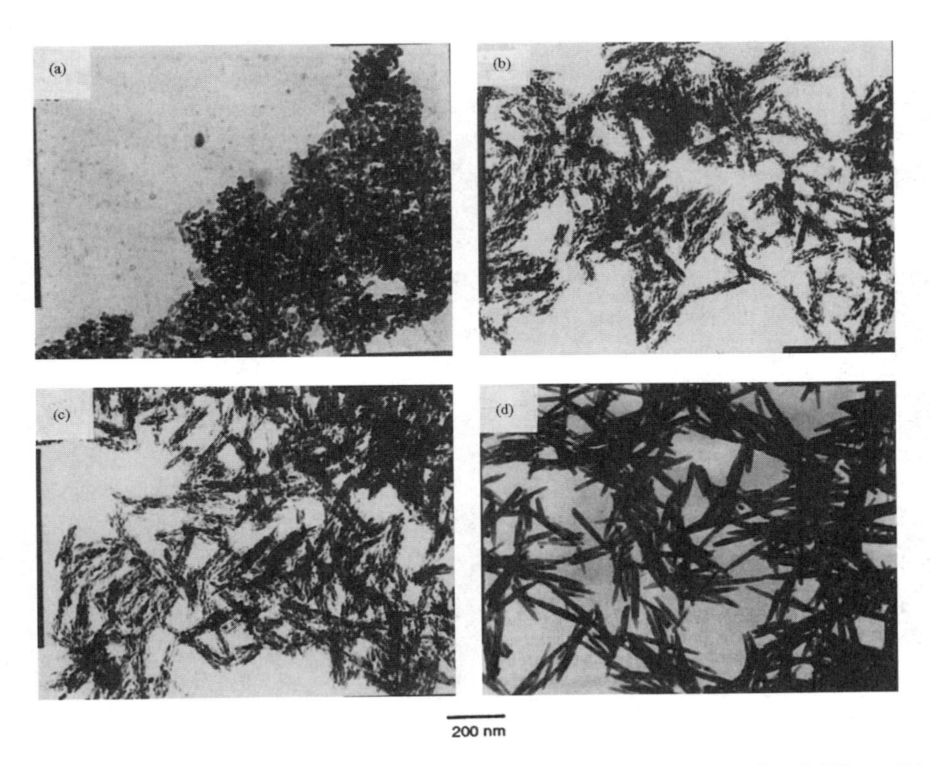

Fig. 6.2.4 TEM pictures of SrCaHAP particles with various atomic ratios Sr/(Ca + Sr): (a) 0, (b) 0.26, (c) 0.56, (d) 1.00. (From Ref. 29.)

6.2.5 Influence of Metal Ions on CaHAP Formation

The metal ions smaller than Ca^{2+}, such as Mg^{2+}, are contained in the CaHAP lattice only to a limited degree, different from Sr^{2+} and Ba^{2+}, but affect the formation of CaHAP particles. Since Mg^{2+} is slightly contained in teeth to change the properties of enamels, such as crystallinity and solubility, the influence of this ion on the formation and structure of CaHAP was investigated (31–34). Figure 6.2.5 displays TEM images of the CaHAP particles substituted with Mg^{2+} at different atomic ratios Mg/(Ca + Mg) = X_{Mg} (35). The particle size increases with an increase of X_{Mg}, while the particles are agglomerates of the small particles. The crystallinity of the particles is lowered by Mg substitution to be amorphous at $X_{Mg} > 0.3$, which clearly implies that Mg^{2+} prevents the crystallization of CaHAP particles. A maximum specific surface area is observed at $X_{Mg} = 0.16$. On the other hand, trivalent Cr^{3+} markedly promotes the crystal growth of CaHAP distinct from Mg^{2+}, as seen in Fig.

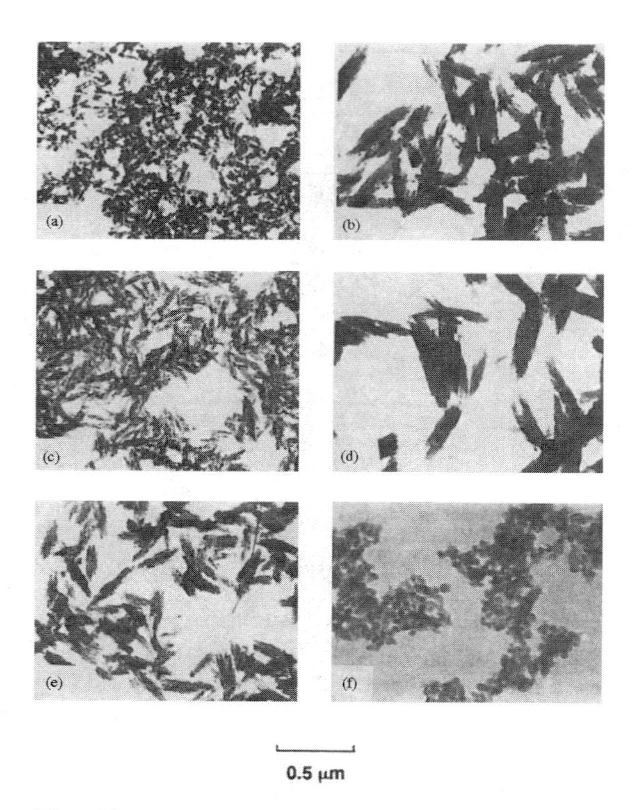

Fig. 6.2.5 TEM pictures of MgCaHAP particles with different atomic ratios Mg/(Ca + Mg): (a) 0, (b) 0.09, (c) 0.16, (d) 0.23, (e) 0.31, (f) 0.38. (From Ref. 35.)

6.2.6, showing TEM pictures of the particles formed with different quantities of Cr^{3+} (36). Other trivalent metal ions, such as Fe^{3+}, Al^{3+}, and La^{3+}, also promote the crystal growth of CaHAP but less than Cr^{3+} (37).

6.2.6 Surface Structure of CaHAP

Knowledge of the surface structure of CaHAP particles is fundamentally needed not only in medical and dental sciences but also in application of synthetic CaHAP particles to bioceramics and adsorbents for biomaterials, because the affinity of CaHAP surface to biomaterials is an important factor in all the cases. The surface structure of CaHAP was investigated by various means including infrared (IR) (38,39), NMR (40), TPD (41), and XPS (42). Among these methods, IR spectroscopy is most appropriate for the surface characterization of CaHAP particles.

Fig. 6.2.6 TEM pictures of CaHAP particles substituted with Cr^{3+} at different atomic ratios Cr/(Ca + Cr): (a) 0, (b) 0.041, (c) 0.13, (d) 0.23, (e) 0.50, (f) 1.0. (From Ref. 36.)

Figure 6.2.7(a) shows the transmission IR spectra in vacuo of CaHAP particles outgassed at varied temperatures (38). The strong sharp peak at 3570 cm^{-1} of all the spectra is assignable to the stretching mode of OH^- in crystals. The spectrum of the sample treated at a low temperature of 150°C shows a broad peak around 3300 cm^{-1} that disappears upon outgassing at 500°C and is assigned to the OH stretching mode of H_2O strongly adsorbed and involved in crystals. A weak band is seen at 3670 cm^{-1}. The enlarged spectra of this part shown in Figure 6.2.7(b) demonstrate that the 3670-cm^{-1} band consists of three bands at 3682, 3670, and

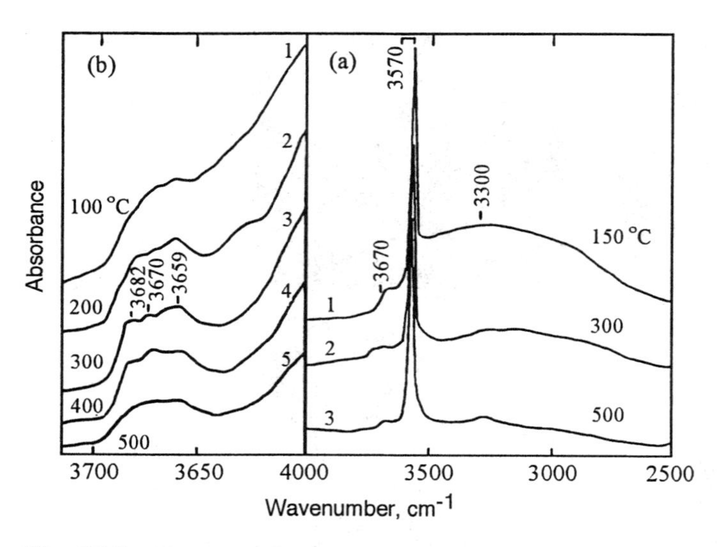

Fig. 6.2.7 IR spectra of CaHAP (Ca/P = 1.64) outgassed at different temperatures. (From Ref. 39.)

3659 cm^{-1}. These bands can be assigned to the surface P-OH. The P-OH originates from the surface HPO_4^{2-} and $H_2PO_4^-$ of CaHAP particles. The spectra shown in Figure 6.2.7 were taken on the rod-shaped CaHAP particles elongating along the *c* axis. The predominant crystal faces of the particle surface, which are more than 90% of the total particle surface, are *ac* and *bc* planes, of which the structures are the same as seen in the crystal structure of CaHAP (Figure 6.2.3). Figure 6.2.8 depicts the structure of *ac* and *bc* planes, showing that PO_4^{3-}, Ca^{2+}, and OH^- are exposed on the surface. However, the particles prepared in an aqueous system or stocked in air adsorb H_2O, and these surface ions are hydrated, so that the removal of the hydrating H_2O requires outgassing at elevated temperature. The surface PO_4^{3-} is thought to turn into HPO_4^{2-} and $H_2PO_4^-$ by protonation to balance the surface charge. The spectra in Fig. 6.2.7(b) indicate that there are different kinds of surface P-OH. However, the detailed assignments of these P-OH groups remain unclear at the present. On the other hand, no band due to the surface OH^- appears in the spectra in Figure 6.2.7. Although F^-–OH^- exchange was examined to elucidate the reason for this, only the 3570-cm^{-1} band of the OH^- in crystals was reduced by this treatment and the other bands did not vary. Consequently, the surface OH^- band would be included in the 3570-cm^{-1} band of OH^- in crystals, which suggests that the surface OH^- hydrogen bonds to O of the neighboring HPO_4^{2-} or $H_2PO_4^-$ as well as the bulk OH^-. The surface Ca/P molar ratio of CaHAP determined by XPS is usually less than the 1.67 of the stoichiometric ratio, indicating that Ca

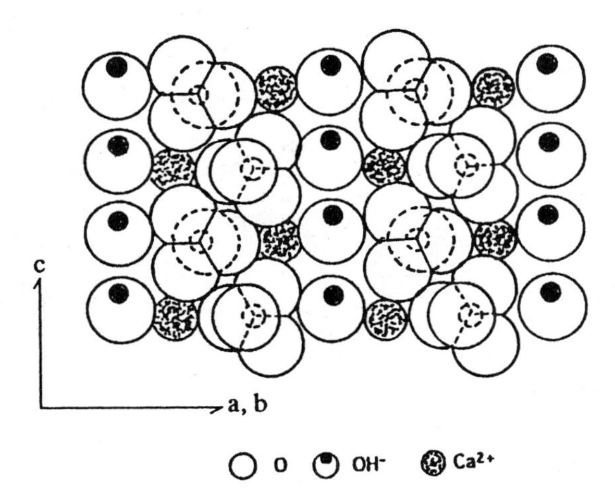

Fig. 6.2.8 Structure of *ac* and *bc* planes of CaHAP crystal.

defects are formed in the surface phase and cause the protonation of surface PO_4^{3-}. Furthermore, there is a possibility that a part of the surface OH^- ions are replaced by H_2O molecules to compensate the Ca deficiency. Table 6.2.2 summarizes the assignment of the bands of the spectra in Figure 6.2.7, which is confirmed by various means including H–D exchange, ion exchange, molecular adsorption, etc.

The surface structure of CaHAP can be changed by substituting Ca^{2+} with other metal ions. Figure 6.2.9 shows the IR spectra of MgCaHAP solid solution particles (35). It is clearly seen that the surface P-OH bands vary as Ca^{2+} is replaced by Mg^{2+}; the state of surface P-OH of MgCaHAP particles is affected by the kind of cations. Similar results were obtained on the other solid solution particles, such

Table 6.2.2 Assignment of the IR Bands of CaHAP with Ca/P = 1.64

cm^{-1}	Assignment
3682	Surface P-OH
3670 (vw)	Surface P-OH
3659	Surface P-OH
3570 (s)	Bulk OH^-
3300 (b)	Adsorbed and bound H_2O

Note: (b) broad, (vw) very weak, (s) strong.

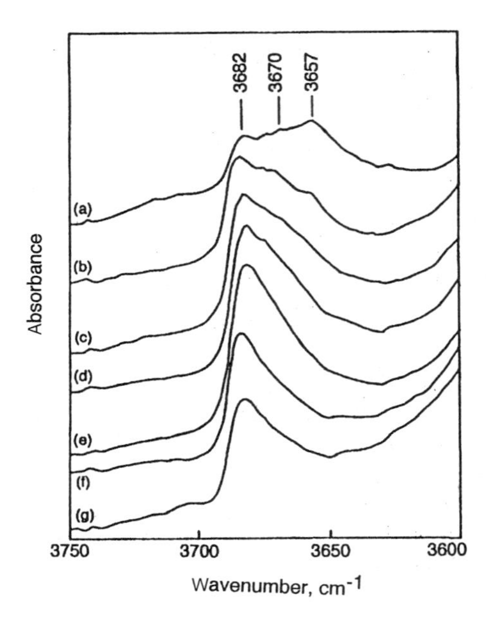

Fig. 6.2.9 IR spectra of the MgCaHAP particles with different atomic ratios Mg/(Ca + Mg): (a) 0, (b) 0.05, (c) 0.09, (d) 0.14, (e) 0.16, (f) 0.23, (g) 0.31. (From Ref. 35.)

as SrCaHAP (29) and BaSrHAP (30). Consequently, substitution with metal ions modifies the surface of CaHAP particles.

6.2.7 Surface Modification

Modification of the CaHAP surface is expected to give a novel function to CaHAP particles. The surface of the synthetic CaHAP particles is hydrophilic because PO_4^{3-} and OH^- are exposed on the surface as already mentioned. The modification with molecules having long alkyl groups, such as alkyl phosphates, is able to make the CaHAP surface hydrophobic and controls the affinity of CaHAP particles to biomaterials, such as proteins and lipids, resulting in a better compatibility of CaHAP with animal organisms. Moreover, the CaHAP particles modified are dispersed in polymers as a filler cement used in medical and dental therapies. Two methods for modification of CaHAP by alkyl phosphates are employed: coprecipitation and anion-exchange methods (43,44). The CaHAP particles were modified by treating with monohexyl, monooctyl, and monodecyl phosphate solutions, for which the solvents were a mixture of acetone and water (45–47). XRD patterns of the modified materials showed that the modified CaHAP particles have a surface-layered phase consisting of multilayers of alternating octacalcium phosphate (OCP)-like phase and

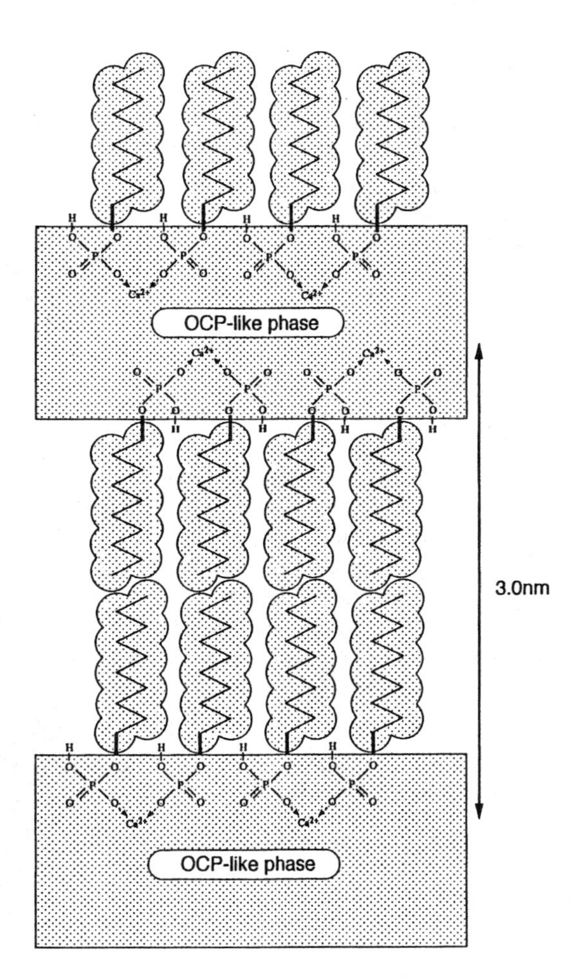

Fig. 6.2.10 Schematic structure of the layered phase on CaHAP particles modified with decyl phosphate. (From Ref. 46.)

bimolecular layer of alkyl groups as shown in Figure 6.2.10. Since a similar layered structure was found on calcium akyl phosphate formed by reaction of Ca^{2+} with alkyl phosphate in aqueous system (48), the surface layers on CaHAP particles are inferred to be formed by dissolution and recrystallization. The particle width was increased by the modification, while the length was not changed. Therefore, these layers grow perpendicularly to the *c* axis. The hydrophobicity of the modified particles was confirmed by H_2O adsorption and preferential dispersion between water and hexane. Recently, Tanaka et al. modified CaHAP particles with hexamethyldisilazane, $[(CH_3)_3Si]_2NH$, and found the formation of the surface Si-OH (49). The surface modification of CaHAP particles was carried out by ion exchange. The substitution with Cr^{3+} and Fe^{3+} generates surface Cr-OH (50) and Fe-OH (51), and the particles substituted with silicate are covered by surface Si-OH (52).

6.2.8 Molecular Adsorption on CaHAP

It is well known that CaHAP exhibits characteristic behaviors as acid and base catalysts in various reactions, such as dehydration (53) and decomposition of alcohols (54), isomerization of 1-butene (55), and methane oxidation (56). Information on adsorption of CaHAP is indispensable in applications for catalysts and adsorbents. The adsorption of several molecules including H_2O, CH_3OH, CH_3I, and CO_2, on CaHAP was investigated (57,58). The adsorption of H_2O and CH_3OH is of interest in the catalytic reactions using CaHAP. The interaction of CO_2 to CaHAP receives attention concerning CO_2 reserve by CaHAP (59), denaturation of bones by CO_2 (60), and formation of long life carbonate radicals by x-ray irradiation (61). Furthermore, the adsorption of these molecules is a useful method to characterize CaHAP particles. The adsorption of proteins on CaHAP in aqueous media was studied concerning applications for HLPC adsorbents for biomaterials and the affinity of artificial bones and tooth in vivo (62–66). From these adsorption studies, only the adsorption of H_2O and CO_2 on CaHAP are interpreted in detail next.

6.2.8.1 H_2O Adsorption

A typical isotherm of H_2O on CaHAP particles outgassed at 300°C is shown in Fig. 6.2.11 (39). This isotherm belongs to type II in the IUPAC classification. Figure

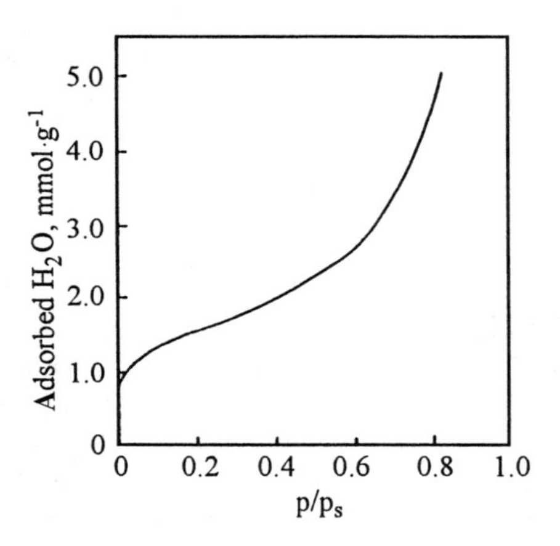

Fig. 6.2.11 Adsorption isotherm of H_2O on CaHAP (Ca/P = 1.59) at 25°C. (From Ref. 39.)

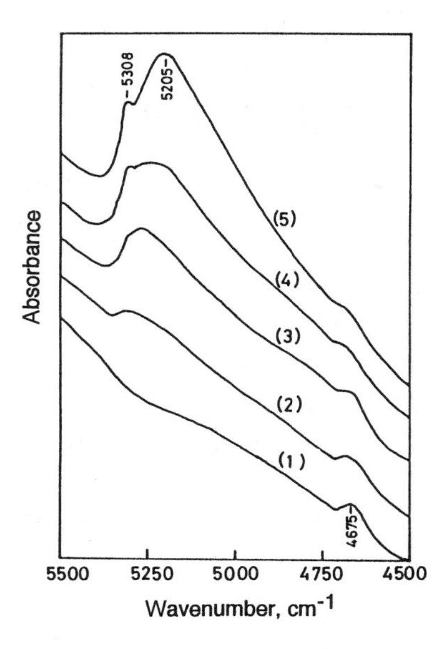

Fig. 6.2.12 IR spectra of CaHAP adsorbing various amounts of H_2O: (1) 0, (2) 0.34, (3) 0.81, (4) 1.1, and (5) 2.3 × 10 mmol g^{-1}. The temperature of pretreatment was 300°C. (From Ref. 39.)

6.2.12 shows the change of near infrared (NIR) spectra with quantity of adsorbed H_2O (39). The 4675-cm^{-1} band appearing at low adsorbed amount is assigned to a combination band of the surface P-OH groups. The intensity of this band is plotted against the adsorbed amount in Figure 6.2.13. The intensity of the band decreases to about one-third of the total absorbance when the adsorbed amount is increased from 0.43 to 1.1 mmol/g, meaning that two-thirds of the total surface P-OH groups interact with the adsorbed H_2O. Above 1.1 mmol/g, the intensity of the band is essentially unchanged. From this result there seem to be at least three steps of H_2O adsorption: (1) decomposed, hydrated, or strongly adsorbed H_2O (<0.43 mmol/g); (2) H_2O adsorbed by hydrogen bonding with the surface P-OH groups (0.43–1.1 mmol/g); and (3) H_2O adsorbed by hydrogen bonding and an adsorption interaction such as dipolar and dispersive interactions with the CaHAP surface and/or the adsorbed H_2O. As seen in Figure 6.2.12, a broad band appears around 5250 cm^{-1}. This band grows stronger proportional to the amount of adsorbed H_2O from 0.43 mmol/g and splits into 5205 and 5308 cm^{-1}. The former band is assignable to the hydrogen-bonded OH groups of the adsorbed H_2O and the latter one to the free OH groups of the adsorbed H_2O. These results allow us to postulate a model of the configuration of the surface P-OH groups and adsorbed H_2O as shown in Figure 6.2.14.

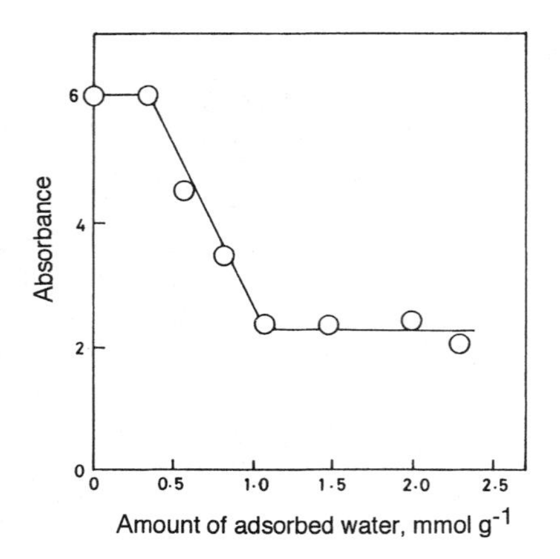

Fig. 6.2.13 Plots of the intensity of the 4675-cm^{-1} band against the amount of adsorbed H_2O.

6.2.8.2 CO_2 Adsorption

Figure 6.2.15 shows the adsorption isotherms of CO_2 on the CaHAP particles with different Ca/P ratios (67). Before the measurement of the first isotherms, shown by the open symbols, the samples were outgassed at 300°C for 2 h. The second set of isotherms, shown by the solid symbols, were measured on the samples outgassed at 25°C for 2 h after taking the first isotherms. The first and second isotherm sets are parallel, and the adsorbed amount of the second isotherm is less than that of the first one, which signifies that a part of the CO_2 adsorbed is irreversibly adsorbed. The amount of the irreversibly adsorbed CO_2 (noted as n_i) was evaluated by subtracting the adsorbed amount at 500 torr in the second isotherm from that in the first one. The n_i values are plotted against Ca/P ratios by the open symbols in Figure 6.2.16. Similar results were reported on SrCaHAP (29), SrHAP (68), and MgCaHAP

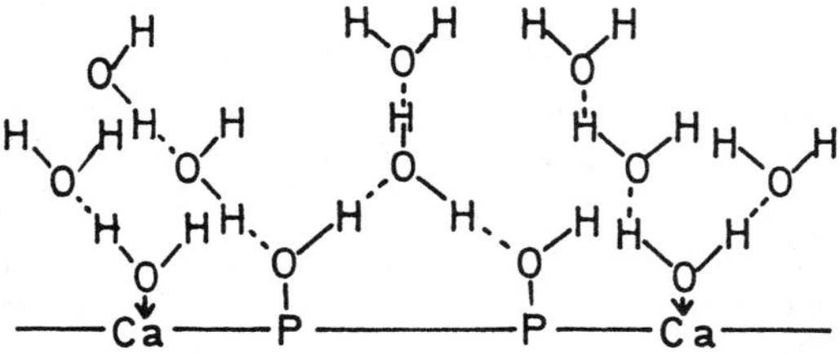

Fig. 6.2.14 Model of the adsorption of H_2O on CaHAP.

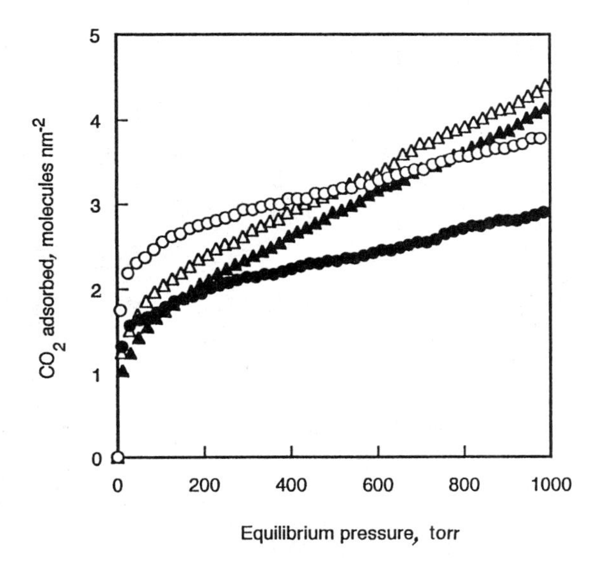

Fig. 6.2.15 Adsorption isotherms of CO_2 on CaHAP particles with Ca/P = 1.67 (○,●) and 1.60 (△,▲). The open and solid symbols represent the first and second isotherms, respectively. (From Ref. 67.)

(35). It should be noted that n_i is minimized at Ca/P = 1.6, where the intensity of the surface P-OH bands is maximum. This reveals that the surface P-OH interferes with the irreversible adsorption of CO_2, rather than acting as irreversible adsorption sites for CO_2. The adsorption mechanism of CO_2 on CaHAP was investigated by in situ IR spectroscopy. It was found that the surface CO_3^{2-} and H_2O resulted from the physisorbed CO_2. This fact suggests that CO_2 is irreversibly adsorbed by the reaction with the surface OH^-:

$$2OH^- + CO_2 \rightarrow CO_3^{2-} + H_2O \qquad (1)$$

Deitz et al. proposed the reaction:

$$OH^- + CO_2 \rightarrow HCO_3^{2-} \qquad (2)$$

from the isosteric heat of adsorption (69). However, since reaction (2) yields no H_2O, this reaction would be followed by reaction (3), generating CO_3^{2-} and H_2O:

$$HCO_3^- + OH^- \rightarrow CO_3^{2-} + H_2O \qquad (3)$$

HCO_3^- is an adsorption intermediate as confirmed by IR spectroscopy.

From the IR results, a mechanism of CO_2 adsorption on the CaHAP particles shown in Figure 6.2.17 can be proposed. As described already, the surface of CaHAP

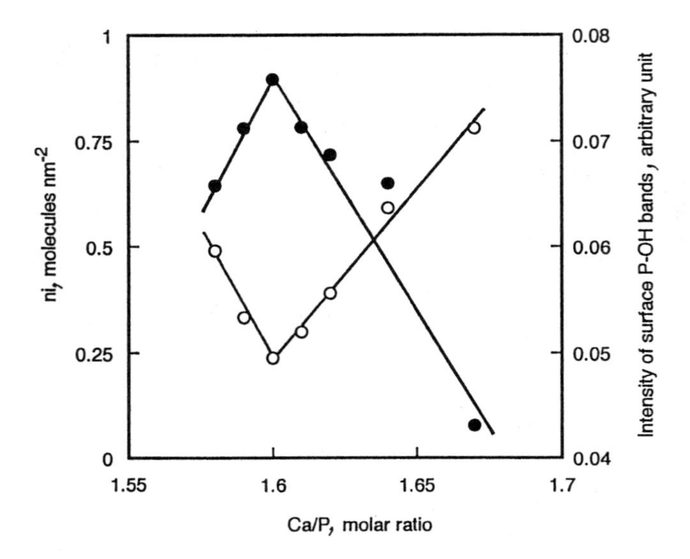

Fig. 6.2.16 Plots of the intensity of the surface P-OH bands (●) and the amount of irreversibly adsorbed CO_2 (n_i, ○) against Ca/P ratio. (From Ref. 67.)

particles consists of OH^-, HPO_4^{2-}, and $H_2PO_4^-$. The surface OH^- ions align along the *c* axis, as do the bulk ones, as shown in Figure 6.2.8. The physisorbed CO_2 reacts with the surface OH^- to produce the surface HCO_3^- species (B) that reacts with the neighboring OH^- to yield CO_3^{2-} and H_2O (C and D). The overall surface reaction is represented by Eq. (1).

6.2.9 Summary

CaHAP particles can be prepared by various wet and dry methods. The particles with desired morphology can be obtained by selecting a suitable method. Because

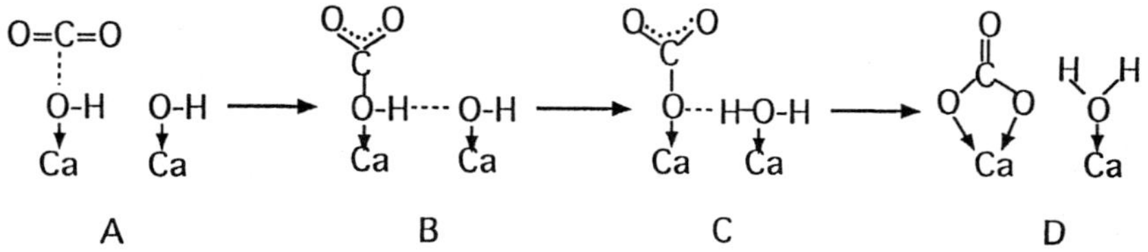

A B C D

Fig. 6.2.17 Mechanism of irreversible adsorption of CO_2 on CaHAP particles.

of high ion exchangeability of CaHAP, stoichometric material is rather hard to synthesize. The surface of CaHAP particles possesses P-OH groups acting as adsorption sites for molecules. The hydrophilic CaHAP particles turn hydrohobic by modification with alkyl phosphates.

Acknowledgments

The author is thankful for the collaboration of Dr. K. Kandori and Dr. A. Yasukawa of the Osaka University of Education.

REFERENCES TO SECTION 6.2

1. JC Elliott. Structure and Chemistry of the Apatites and Other Calcium Orthophosphates. Amsterdam: Elsevier, 1994.
2. ZH Cheng, A Yasukawa, K Kandori, T Ishikawa. J Chem Soc Faraday Trans 94:1501, 1998.
3. A Mortier, J Lemaitre, L Rodrique, PG Rouxhet. J Solid State Chem 78:215, 1989.
4. M Kinoshita, K Itatani, S Nakamura, A Kishioka. Gypsum & Lime 227:19, 1991.
5. A Yasukawa, H Takase, K Kandori, T Ishikawa. Polyhedron 13:3071, 1994.
6. A Atokinson, PA Bradford, IP Selmes. J Appl Chem Biotechnol 23:517, 1973.
7. H Monma, S Ueno, M Tsutsumi, T Kanazawa. Yogyo-Kyokai-Shi 66:590, 1978.
8. RA Young, DW Holcomb. Calcif Tissue Int 34:S17, 1982.
9. M Yoshimura, H Suda. In: PW Brown, B Constantz, eds. Hydroxyapatite and Related Materials. London: CRC Press, 1994, p 45.
10. K Ioku, M Yoshimura, S Somiya. Nippon Kagaku Kaishi 1565, 1988.
11. L Yubao, K De Groot. J Mater Sci Mater Med 5:326, 1994.
12. N Asaoka, H Suda, M Yoshimura. Nippon Kagaku Kaishi 25, 1995.
13. K Itatani, O Takahashi, A Kishioka, M Kinoshita. Gypsum & Lime 213:77, 1988.
14. H Monma, T Kanazawa. Nippon Kagaku Kaishi 339, 1972.
15. BO Fowler. Inorg Chem 13:207, 1974.
16. A Schleede, B Meppen, B Jorgensen. Angew Chem 52:316, 1939.
17. M Tanahashi, K Kamiya, T Suzuki, H Nasu. J Mater Sci Mater Med 3:48, 1992.
18. H Monma, Y Kitami, M Tsutsumi. Trans Mater Res Soc Jpn 14A:781, 1994.
19. K Kandori, N Horigami, A Yasukawa, T Ishikawa. J Am Ceram Soc 80:1157, 1997.
20. A Deptuta, W Lada, T Olczak, A Borello, C Alvani, A di Bartolomeo. J Non-Cryst Solids 147:537, 1992.
21. D Walsh, JD Hopwood, S Mann. Science 264:1576, 1994.
22. K Sudarsanan, RA Young. Acta Cryst B25:1534, 1969.
23. T Suzuki, T Hatsushika, Y Hayakawa. J Chem Soc Faraday Trans I 77:1059, 1981.
24. T Suzuki, T Hatsushika, M Miyake. J Chem Soc Faraday Trans I 78:3605, 1982.
25. T Suzuki, K Ishigaki, M Miyake. J Chem Soc Faraday Trans I 80:3157, 1984.
26. C Lagergren, D Carlstrom. Acta Chem Scand 11:545, 1957.
27. R Klement, H Haselbeck. Anorg Allg Chem 336:113, 1965.
28. RMH Verbeeck, CJ Lassuyt, HJM Heijligers, FCM Driessens, JWGA Vrolijk. Calcif Tissue Int 33:243, 1981.

29. T Ishikawa, H Saito, A Yasukawa, K Kandori. J Chem Soc Faraday Trans 89:3821, 1993.
30. A Yasukawa, M Kidokoro, K Kandori, T Ishikawa. J Colloid Interface Sci 191:407, 1997.
31. S Thiradilok, F Feagin. Ala J Med Sci 15:144, 1978.
32. C Robinson, JA Weatherell, AS Hallworth. Caries Res 15:70, 1981.
33. JA Weatherell, C Robinson. Proc Finn Dent Soc 78:81, 1982.
34. RA Terpstra, FCM Driessens. Calcif Tissue Int 39:348, 1986.
35. A Yasukawa, S Ouchi, K Kandori, T Ishikawa. J Mater Chem 6:1401, 1996.
36. M Wakamura, K Kandori, T Ishikawa. Polyhedron 16:2047, 1997.
37. M Wakamura, K Kandori, T Ishikawa, Colloids Surfaces A 142:107, 1998.
38. SJ Joris, CH Amberg. J Phys Chem 20:3172, 1971.
39. T Ishikawa, M Wakamura, S Kondo. Langmuir 5:140, 1989.
40. JP Yesinowski, RA Wolfgang, M Mobley. In: DN Misra, ed. Adsorption on and Surface Chemistry of Hydroxyapatite. New York: Plenum, 1984, p 151.
41. P Somasundaran, GE Agar. J Colloid Interface Sci 24:433, 1967.
42. K Konishi, M Kambara, MH Noshi, M Uemura. J Osaka Dent Univ 21:1, 1987.
43. A Lebugle, M Subirade, V Delpech. In: PW Brown, B Constantz, eds. Hydroxyapatite and Related Materials. London: CRC Press, 1994, p 231.
44. K Kandori, A Fujiwara, A Yasukawa, T Ishikawa. Colloids Surfaces A 150:161, 1999.
45. T Ishikawa, H Tanaka, A Yasukawa, K Kandori. J Mater Chem 5:1963, 1995.
46. H Tanaka, A Yasukawa, K Kandori, T Ishikawa. Langmuir 13:821, 1997.
47. H Tanaka, A Yasukawa, K Kandori, T Ishikawa. Colloids Surfaces A 125:53, 1997.
48. H Tanaka, T Watanabe, M Chikazawa, K Kandori, T Ishikawa. Colloids Surfaces A 139:53, 1998.
49. H Tanaka, T Watanabe, M Chikazawa, K Kandori, T Ishikawa. J Colloid Interface Sci 206:205, 1998.
50. M Wakamura, K Kandori, T Ishikawa. Colloids Surfaces A 142:107, 1998.
51. T Ishikawa, H Saito, A Yasukawa, K Kandori. Bull Chem Soc Jpn 69:899, 1996.
52. T Ishikawa, M Wakamura, T Kawase, S Kondo. Langmuir 7:596, 1991.
53. SJ Joris, CH Amberg. J Phys Chem 20:3167, 1971.
54. H Monma. J Catal 75:200, 1982.
55. Y Imizu, M Kadoya, H Abe, H Itoh, A Tada. Chem Lett 415, 1982.
56. S Sugiyama, T Minami, T Moriga, H Hayashi, K Koto, M Tanaka, J B Moffat. J Mater Chem 6:459, 1996.
57. ME Dry, RA Beebe. J Phys Chem 20:3172, 1971.
58. T Ishikawa. Stud Surf Sci Catal 99:301, 1996.
59. CF Poyart, Bursaux, A Freminet. Respir Physiol 25:89, 1975.
60. RM Blitz, ED Pellegrino. Clin Orthop Rel Res 25:279, 1977.
61. G Bacquet, VQ Truong, M Vignoles, JC Trombe, G Bonel. Calcif Tissue Int 33:105, 1981.
62. K Kandori, S Sawai, Y Yamamoto, H Saito, T Ishikawa. Colloids Surfaces 68:283, 1992.
63. K Kandori, M Saito, H Saito, A Yasukawa, T Ishikawa. Colloids Surfaces A 94:225, 1995.
64. K Kandori, M Saito, T Takebe, A Yasukawa, T Ishikawa. J Colloid Interface Sci 174: 124, 1995.

65. K Kandori, T Shimizu, A Yasukawa, T Ishikawa. Colloids Surfaces B 5:81, 1995.
66. K Kandori, M Saito, T Takebe, A Yasukawa, T Ishikawa. J Colloid Interface Sci 191: 498, 1997.
67. ZH Cheng, A Yasukawa, K Kandori, T Ishikawa. Langmuir 14:6681, 1998.
68. T Ishikawa, H Saito, K Kandori. J Chem Soc Faraday Trans 88:2937, 1992.
69. VR Deitz, FG Carpenter, RG Arnold. Carbon 1:245, 1964.

7

Metal Carbonates

EGON MATIJEVIĆ, RONALD S. SAPIESZKO,
and KIJIRO KON-NO

7.1 FORMATION OF MONODISPERSED METAL (BASIC) CARBONATES IN THE PRESENCE OF UREA

EGON MATIJEVIĆ
Clarkson University, Potsdam, New York

RONALD S. SAPIESZKO
Aveka, Inc., Woodbury, Minnesota

7.1.1 Introduction

One efficient and convenient way to produce uniform particles of metal (basic) carbonates has been by aging metal salt solutions in the presence of urea at moderately elevated temperatures. Under these conditions, urea decomposes by releasing carbonate ions (CO_3^{2-}, HCO_3^-) with simultaneous increase in the pH.

The hydrolysis of urea under varying conditions of temperature, concentration, acidity, etc. has been described by numerous authors (1–17). It was found that the reaction rate increases with rising temperature and urea concentration (3,4,6), as well as in the presence of an acid (5,6) or a base (4,9,14).

Investigations of the decomposition of urea in aqueous solutions over the temperature range of 25 to 45°C show very little degradation taking place during 3 days of aging (15). However, the decomposition takes place 25 times faster at 90°C than at 60°C (11). In both cases the reaction is of the first order, resulting in the formation of ammonium cyanate, according to:

$$CO(NH_2)_2 \rightarrow NH_3 + HNCO \rightarrow NH_4^+ + NCO^- \quad (1)$$

(H—N=C=O is the more stable isomer of cyanic acid.) This reaction represents the slow step of the urea hydrolysis.

In acidic solutions the cyanate ion rapidly hydrolyzes according to:

$$NCO^- + 2H^+ + H_2O \rightarrow NH_4^+ + CO_2 \quad (2)$$

while in neutral and basic solutions the fast step of the hydrolysis is

$$NCO^- + OH^- + H_2O \rightarrow NH_3 + CO_3^{2-} \quad (3)$$

Since the rate of these processes can be carefully controlled, it is possible to establish conditions that would result in uniform particles by precipitation of sparingly soluble compounds, the composition of which is affected by cations and anions

Table 7.1.1 Summary of Monodispersed Metal (Basic) Carbonate Particles Obtained by Aging Metal Salt Solutions in the Presence of Urea

Compound	Anion present	Particle shape	Solution pH	Reference
$Cu_2(OH)_2CO_3$	NO_3^-	Spherical	6–8	18, 19
$La(OH)CO_3$	NO_3^-/Cl^-	Spherical	6–7	20
$Sm(OH)CO_3$	NO_3^-/Cl^-	Spherical	6–8	20
$Eu(OH)CO_3$	NO_3^-/Cl^-	Spherical	6–8	20
$Gd(OH)CO_3$	NO_3^-/Cl^-	Spherical	7–9	20
$Tb(OH)CO_3$	NO_3^-/Cl^-	Spherical	6–8	20
$Y(OH)CO_3$	Cl^-	Spherical		21, 22, 23, 24
$Y_xCe_y(OH)CO_3 \cdot H_2O$	Cl^-	Spherical		21
$Y_2(CO_3)_3NH_3 \cdot 3H_2O$	Cl^-	Rodlike		21
$Ce_2O(CO_3)2$	NO_3^-	Ellipsoids Platelets	6–8	20
$Zn_5(OH)_6(CO_3)_2$	NO_3^-, Cl^-	Rodlike		25
	SO_4^{2-}	Spherical		25
$NiCO_3 \cdot Ni(OH)_2 \cdot H_2O$	NO_3^-, SO_4^{2-}	Spherical		26
$MnCO_3$	SO_4^{2-}	Cubic		27
$CdCO_3$	NO_3^-/Cl^-	Cubic		28
$Y(OH)CO_3 + Cu_3(OH)_2(CO_3)_2$	NO_3^-	Spherical		30
$Y_2(CO_3)_3 \cdot 3H_2O + Cu_4(OH)_6CO_3$	NO_3^-	Spherical		30
$Y(OH)CO_3 \cdot H_2O + Al(OH)_3$	NO_3^-, SO_4^{2-}	Spherical		31
$La_x(OH)_y(CO_3)_z + Cu_p(OH)_r(CO_3)_s$	NO_3^-	Spherical		30
$ZrY_{0.8}(OH)_{3.8}(CO_3)_{1.3} \cdot H_2O$	NO_3^-, SO_4^{2-}	Spherical		32

present in the solution (Sec. 1.1). Thus, depending on the properties of metal ions of interest, the resulting solids may consist of only hydroxides or carbonates, or they may have complex natures, including basic carbonates, metal amino carbonates, etc. If certain anions are coordinated in preference to the carbonate ion, different kinds of composite solids, such as basic sulfates or phosphates, appear as end products. In this section, only studies that resulted in particles containing carbonate ion as one of the constituents are reviewed.

Table 7.1.1 summarizes a number of systems described in the literature.

The particles formed are in most cases spherical, although rods, ellipsoids, platelets, and hexagonal structures have also been produced. Solids composed of spherical particles are as a rule amorphous, while those of other morphologies are crystalline. In general, aging times of 2 h at approximately 100°C were sufficient to produce the desired results; however, in some instances much longer times were necessary to complete the precipitation process.

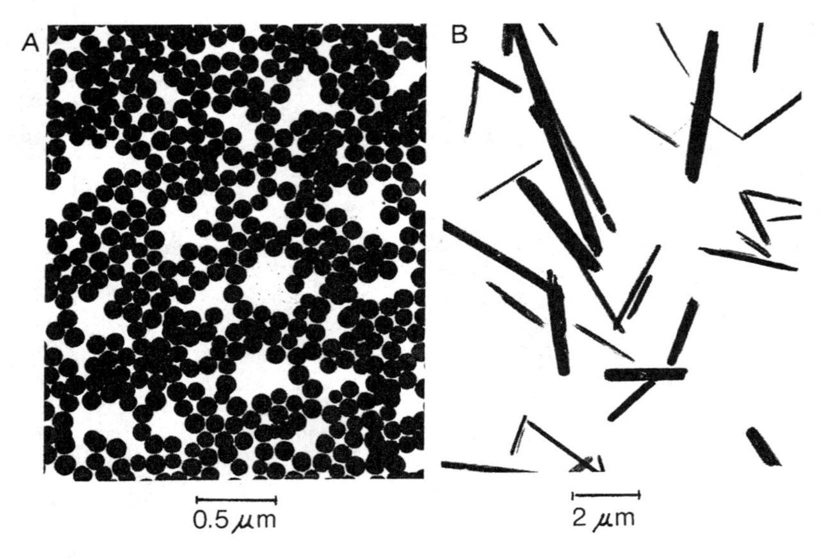

Fig. 7.1.1 Transmission electron micrographs (TEM) of (A) $Y(OH)CO_3 \cdot H_2O$ particles obtained by aging at 90°C for 2.5 h in a solution of 1.5×10^{-2} mol dm^{-3} YCl_3 and 0.5 mol dm^{-3} urea; (B) $Y_2(NH_3)(CO_3)_3 \cdot 3H_2O$ particles obtained by aging at 115°C for 18 h a solution of 3.0×10^{-2} mol dm^{-3} YCl_3 and 3.3 mol dm^{-3} urea. (From Ref. 21.)

7.1.2 Simple (Basic) Carbonates

Figure 7.1.1 illustrates two dispersions precipitated by aging solutions of YCl_3 in the presence of urea. While the reactants are the same, varying their concentrations yielded particles different in shape, chemical composition, and crystallinity (21).

By carefully heating the powders of metal basic carbonates, the particles can be converted to oxides while preserving their morphology and dispersibility. Figure 7.1.2 is a scanning electron micrograph of Gd_2O_3 obtained by calcining at 600°C for 24 h spherical particles of $Gd(OH)CO_3$ (20).

Aging $MnSO_4$/urea solutions yielded directly crystalline cubic $MnCO_3$ (rhodochrosite) particles (Fig. 7.1.3) (27). Interestingly, Hamada et al. obtained the same kind of dispersions by reacting solutions of $MnSO_4$ and NH_4HCO_3 (29).

On calcination at 700°C in air, the manganese carbonate particles are converted to Mn_2O_3 (bixbyite) and in nitrogen to MnO (manganosite) (27).

7.1.3 Internally Composite Particles

Essentially the same process can be used for the preparation of monodispersed particles of internally mixed composition. To do so, solutions containing two or more different metal salts are aged at elevated temperatures in the presence of urea.

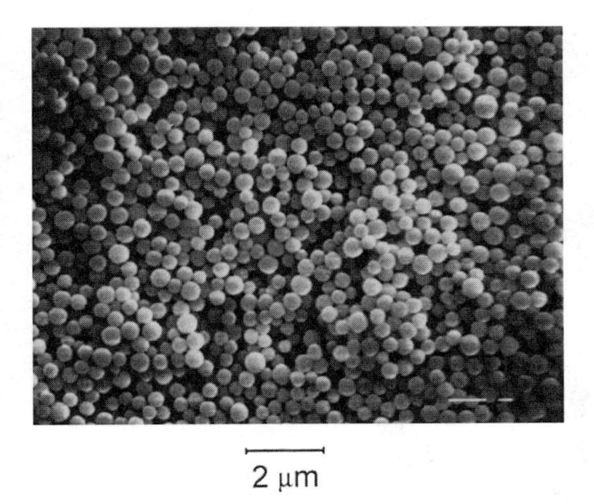

Fig. 7.1.2 Scanning electron micrograph (SEM) of Gd_2O_3 particles, obtained by the calcination at 600°C for 24 h of spherical $Gd(OH)CO_3$ particles prepared by aging at 85°C for 1.5 h a solution containing 5.6×10^{-3} mol dm^{-3} $GdCl_3$ and 0.5 mol dm^{-3} urea. (From Ref. 20.)

Fig. 7.1.3 SEM of $MnCO_3$ particles, obtained by aging at 85°C for 1.5 h a solution containing 0.16 mol dm^{-3} $MnSO_4$ and 0.4 mol dm^{-3} urea. (From Ref. 27.)

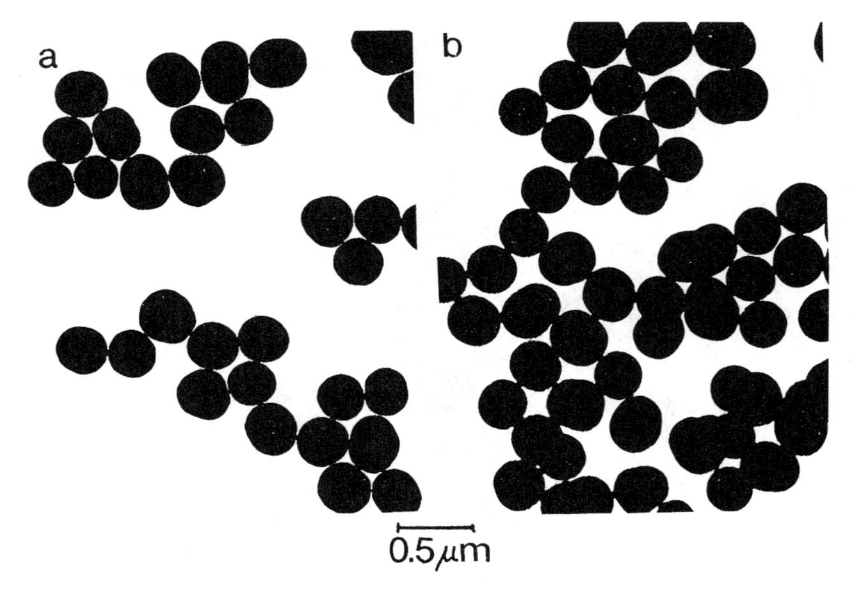

Fig. 7.1.4 TEM of particles obtained by aging at 90°C for 3.5 h a solution containing 3.84 $\times$ 10^{-3} mol dm^{-3} $Al(NO_3)_3$, 0.5 mol dm^{-3} urea, 5.0 $\times$ 10^{-3} mol dm^{-3} HNO_3, 2 wt% polyvinylpyrrolidone (PVP), and (a) 2.16 $\times$ 10^{-3} mol dm^{-3} $Al(NH_4)(SO_4)_2$ and 6.0 $\times$ 10^{-3} mol dm^{-3} $Y(NO_3)_3$; or (b): 2.88 $\times$ 10^{-3} mol dm^{-3} $Al(NH_4)(SO_4)_2$ and 2.0 $\times$ 10^{-3} mol dm^{-3} $Y(NO_3)_3$. (From Ref. 31.)

One such system consisting of uniform spheres precipitated in solutions of $Y(NO_3)_3$, $Al(NO_3)_3/Al(SO_4)_3$, and urea is shown in Figure 7.1.4 (31). The chemical analysis of the resulting solids showed that their overall composition corresponded to a mixture of $Y(OH)CO_3{\cdot}H_2O$ and $Al(OH)_3{\cdot}0.5H_2O$ with a small amount of occluded sulfate ions. However, it is important to note that these particles are internally inhomogeneous, despite their morphological and size consistency; their composition changes with growth, as exemplified in Figure 7.1.5. These data show that the molar ratio [Al]/[Y] decreases exponentially with time from a huge to a small excess of aluminum in the solids (31); consequently, the bulk of the final particles differs from that at the periphery. This observation is common when the precipitation in solutions of mixed compositions is carried out slowly. The reason for this chemical inhomogeneity can be easily understood in view of the differences in the concentrations of reactants in solutions and in the solubilities of individual constituents in the mixed systems.

On calcination of the original powders at elevated temperatures, particles of mixed oxides are obtained, which retain their sphericity and dispersibility, and in some instances they become internally chemically homogeneous. In the described

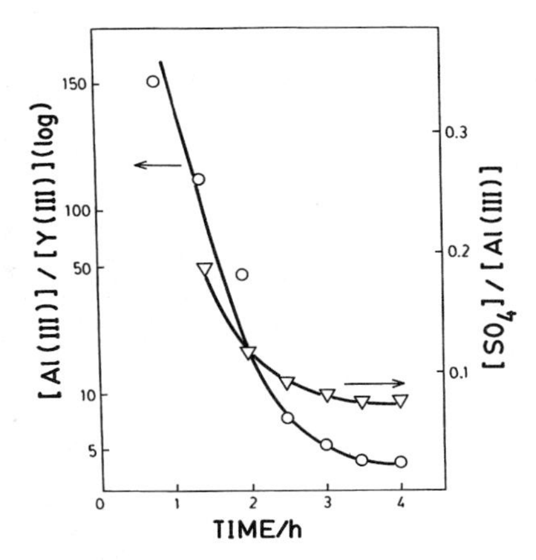

Fig. 7.1.5 Changes of molar ratios [Al(III)/[Y(III)] and [SO_4]/[Al(III)] in solids obtained by aging at 90°C for increasing periods of time solutions having the same composition as in Figure 7.1.4(b). (From Ref. 31.)

system, fast heating can yield the garnet composition and structure ($Y_3Al_5O_{12}$), if the precursor particles of the required molar ratio [Y]/[Al] = 3/5 are used in the precipitation process (31).

Similarly, solutions containing $Zr(SO_4)_2$ and $Y(NO_3)_3$ aged in the presence of urea at 80°C yielded spherical particles of the overall composition $ZrY_{0.8}(OH)_{3.8}(CO_3)_{1.3}{\cdot}H_2O$, with molar ratios of metals similar to their concentrations in the reacting solutions (32). It is noteworthy that at 50°C the precipitate contained only zirconium, while no presence of yttrium could be detected. It was also established that at 80°C the particle size was sensitive to the zirconium salt concentration, but not to Y^{3+} ion concentration, suggesting that the Zr^{4+} ions initiate the nucleation stage, whereas the Y^{3+} ions are incorporated in the particles during the subsequent growth.

On calcination of this prepared powder, particles having the composition $ZrY_{0.8}O_{3.2}$ were obtained. The electrokinetic measurements with aqueous dispersions of the latter showed an isoelectric point at pH 6.8, characteristic of Y_2O_3. This example further substantiates the inhomogeneity within the particles, but also indicates that heating, as carried out in this case, did not produce internal uniformity.

A systematic study of Cu(II)–Y(III) and Cu(II)–La(III) systems has shown that uniform spheres can be produced by aging the corresponding metal nitrate solutions in the presence of urea (30). In both cases the internal composition of the resulting particles varied with their size in the course of their growth (Fig. 7.1.6). Once the entire precipitation process was completed, the molar ratios of both metals in the solids corresponded to the same ratio initially introduced into the reacting solutions.

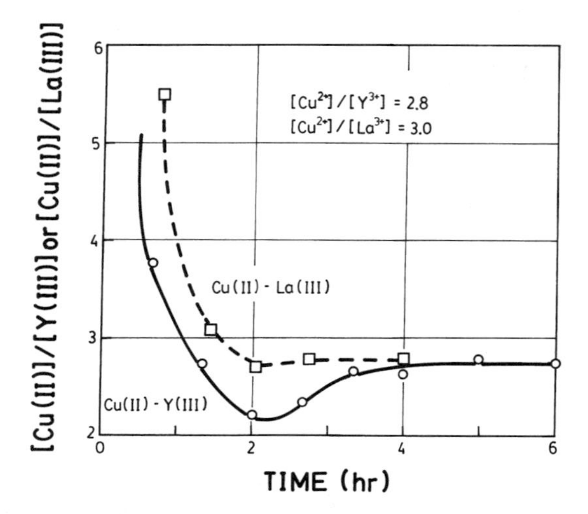

Fig. 7.1.6 Changes of molar ratios [Cu(II)]/[Y(III)] or [Cu(II)]/[La(III)] in particles as a function of the time of aging at 90°C the following solutions: (———) 8.0×10^{-3} mol dm^{-3} $Y(NO_3)_3$, 0.4 mol dm^{-3} urea, and $[Cu^{2+}]/[Y^{3+}] = 2.8$; (- - -) 8.0×10^{-3} mol dm^{-3} $La(NO_3)_3$, 0.5 mol dm^{-3} urea, and $[Cu^{2+}]/[La^{3+}] = 3.0$ (30).

Recently, spherical yttrium iron garnet (YIG) particles over a broad range of sizes were obtained by hydrolysis at 90°C of acidic urea solutions containing the respective metal salts (33).

7.1.4 Coated Particles

It is frequently desirable to cover a given particle with a layer of different chemical composition. In doing so, one can alter the surface reaction sites, as well as optical, magnetic, conductive, and other properties of the dispersed matter. Finally, if a material of a given particle shape is needed, but cannot be obtained directly, it is possible to use cores of a different composition but of the desired morphology and coat them with a shell of the chemical compound of interest. This approach combines the needed overall morphology with the surface reactive sites of choice.

Again, decomposition of urea in an acidic metal salt solution in the presence of preformed particles can yield homogeneous layers of variable thicknesses of metal basic carbonates on the core materials of different chemical composition. In order to obtain uniformly coated particles (rather than a mixture of the latter and of the independently precipitated coating material), a balance between the amount of the initial dispersed matter and the concentration of the reacting solutes must be optimized.

One important aspect of such coating processes is the generality of the procedure. It would appear that specific surface characteristics of the preformed particles are not necessarily essential for the successful deposition of the new layer. For example, yttrium basic carbonate coatings were produced on zirconium basic sulfate

(32), silica (34), polystyrene latex (35), hematite (36,37), and silicon nitride (38) cores. Figure 7.1.7 illustrates the hematite system prepared under conditions given in the legend. The uniformity of the coating is quite apparent; it is also clearly seen that no extraneous particles are present in this dispersion.

Using the same procedure, spherical nanosize particles of hematite were coated with yttrium basic carbonate and showed that various surface thermodynamic properties of these systems were essentially those of yttria (37).

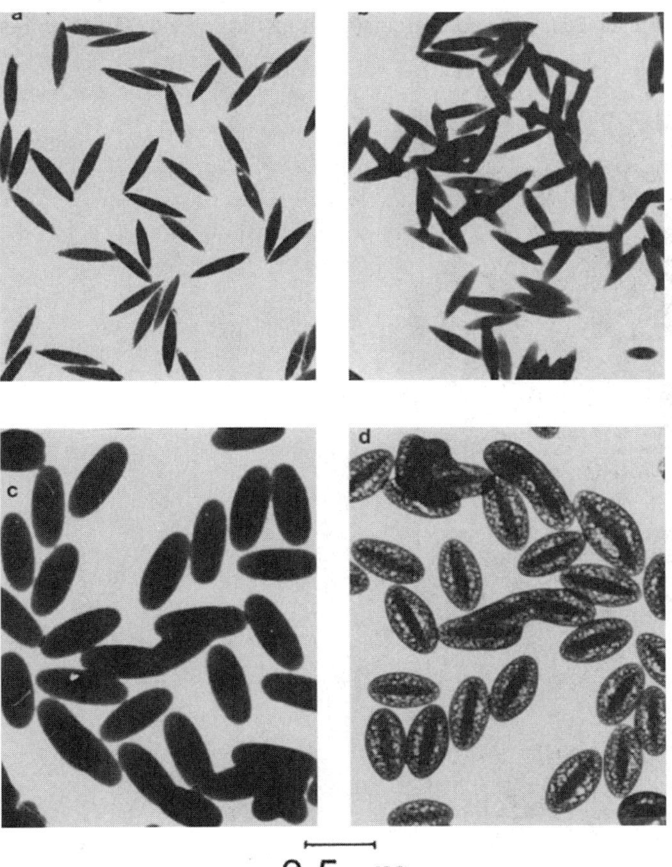

Fig. 7.1.7 TEM of coated hematite particles obtained by aging at 90°C for 2 h aqueous dispersions containing 1.8 mol dm^{-3} urea and (a) 2.05×10^2 mg dm^{-3} ellipsoidal α-Fe_2O_3 and 1.0×10^{-3} mol dm^{-3} $Y(NO_3)_3$; (b) 1.03×10^2 mg dm^{-3} α-Fe_2O_3 and 1.0×10^{-3} mol dm^{-3} $Y(NO_3)_3$; (c) 1.03×10^2 mg dm^{-3} α-Fe_2O_3 and 5.0×10^{-3} mol dm^{-3} $Y(NO_3)_3$; (d) the same particles as in (c) exposed to a high-intensity electron beam. (From Ref. 36.)

In most cases the coating alone may be changed on subsequent treatments (e.g., by calcination at elevated temperatures), such as by converting basic carbonates into oxides. However, in some instances the shell and the core may interact to yield a different compound. For example, silica particles coated with $Y(OH)CO_3$ on heating to 1000°C reacted to yield yttrium silicate, $Y_2Si_2O_7$, which was restricted to the shell, as long as the amount of silica in the core was in molar excess (32).

Under appropriate conditions, manganese carbonate cubes dispersed in $NiSO_4$ solutions containing urea were covered with a layer identified as $NiCO_3 \cdot Ni(OH)_2 \cdot H_2O$ (Fig. 7.1.8A) (27), which on calcination transformed into Mn_2O_3 cores

Fig. 7.1.8 SEM of $MnCO_3$ particles covered with a layer identified as $NiCO_3 \cdot Ni(OH)_2 \cdot H_2O$, obtained by aging at 85°C for 2 h an aqueous dispersion containing 0.4 g dm^{-3} $MnCO_3$ particles illustrated in Fig. 7.1.3, 0.26 mol dm^{-3} $NiSO_4$, and 0.4 mol dm^{-3} urea. (b) The same particles as in (a) calcined at 350°C for 6 h in hydrogen flow. (From Ref. 27.)

coated with NiO. When the latter powder was heated at 350°C for 6 h in a stream of hydrogen, both the core and the coating were reduced to the corresponding metals, with overall retention of the particle morphology (Fig. 7.1.8B) (27).

REFERENCES TO SECTION 7.1

1. J Walker, FJ Hambley. J Chem Soc 67:746, 1895.
2. CE Fawsitt. Z Physik Chem 41:601, 1902.
3. GJ Burrows, CE Fawsitt. J Chem Soc 105:609, 1914.
4. EA Werner. J Chem Soc 113:84, 1918.
5. TW Price. J Chem Soc 115:1354, 1919.
6. EA Werner. J Chem Soc 117:1078, 1920.
7. HH Willard, NK Tang. J Am Chem Soc 59:1190, 1937.
8. HH Willard, HC Fogg. J Am Chem Soc 59:1197, 1937.
9. RC Warner. J Biol Chem 142:705, 1942.
10. KJ Laidler, JP Hoare. J Am Chem Soc 72:2489, 1950.
11. WHR Shaw, JJ Bordeaux. J Am Chem Soc 77:4729, 1955.
12. WHR Shaw, DG Walker. J Am Chem Soc 78:5769, 1956.
13. AR Amell. J Am Chem Soc 78:6234, 1956.
14. KR Lynn. J Phys Chem 69:687, 1965.
15. HL Welles, AR Giaquinto, RE Lindstrom. J Pharm Sci 60:1212, 1971.
16. JC Macdonald, J Serphillips, JJ Guerrera. J Phys Chem 77:370, 1973.
17. PK DasGupta, SP Moulik. J Phys Chem 91:5826, 1987.
18. S Kratohvil, E Matijević. J Mater Res 6:766, 1991.
19. I Haq, E Matijević. Colloids Surf 81:153, 1993.
20. E Matijević, WP Hsu. J Colloid Interface Sci 118:506, 1987.
21. B Aiken, WP Hsu, E Matijević. J Am Ceram Soc 71:845, 1988.
22. DJ Sordelet, M Akinc. J Colloid Interface Sci 122:47, 1988.
23. YS Her, E Matijević, WR Wilcox. Powder Technol 61:173, 1990.
24. YS Her, E Matijević, WR Wilcox. J Mater Res 7:2269, 1992.
25. M Castellano, E Matijević. Chem Mater 1:78, 1989.
26. L Durand-Keklikian, I Haq, E Matijević. E Colloids Surf A 92:7267, 1994.
27. I Haq, E Matijević, K Akhtar. Chem Mater 9:2659, 1997.
28. A Janeković, E Matijević. J Colloid Interface Sci 103:436, 1985.
29. S Hamada, Y Kudo, J Okada, H Kano. J Colloid Interface Sci 118:356, 1987.
30. F Ribot, S Kratohvil, E Matijević. J Mater Res 4:1123, 1989.
31. WP Hsu, G Wang, E Matijević. Colloids Surf 61:255, 1991.
32. B Aiken, WP Hsu, E Matijević. J Mater Sci 25:1886, 1990.
33. RHM Godoi, M Jafelicci Jr, RF Marques, LC Varanda. Mater Res Soc Symp Proc 517: 583, 1998.
34. H Giesche, E Matijević. J Mater Res 9:436, 1994.
35. N Kawahashi, E Matijević. J Colloid Interface Sci 138:534, 1990.
36. B Aiken, E Matijević. J Colloid Interface Sci 126:645, 1988.
37. RC Plaza, L Zurita, JDG Durán, F Gonzales-Caballero, AV Delgado. Langmuir 14: 6850, 1998.
38. AK Garg, LC De Jonghe. J Mater Res 5:136, 1990.

7.2 REACTION IN MICROEMULSIONS

KIJIRO KON-NO
Science University of Tokyo, Tokyo, Japan

Metal carbonate particles such as $CaCO_3$, $BaCO_3$, and $SrCO_3$ have been synthesized by bubbling CO_2 through ionic and nonionic reversed-micelle solutions containing the corresponding aqueous metal hydroxides. In polyoxyethylene(6) nonylphenyl ether–cyclohexane system, the growth rate of $BaCO_3$ particles was faster by a factor of four in the W/O microemulsion system of Wo(= $[H_2O]/[\text{surfactant}]$) = 72 than in bulk water, and hence the reaction was completed in 30 (1). The formation of $BaCO_3$ particles during the bubbling CO_2 was observed at Wo = 72 by transmission electron microscopy. Consequently, when a small amount of CO_2 was bubbled, small ellipsoidal particles were precipitated on the edge of needlelike materials, which are considered to be $Ba(OH)_2$–surfactant complexes, and with an increasing amount of CO_2 bubbled they grew slowly to final ellipsoidal particles so the needle-like materials disappeared. The mechanism of formation of $CaCO_3$ or $BaCO_3$ particles in ionic and nonionic surfactants systems was investigated qualitatively by calculating the number of micelles (N_m) solubilizing aqueous $Ca(OH)_2$ and $CaCO_3$ particles (N_p) existing in unit volumes of surfactant solution from initial micellar weight (Mw) and mean diameter of final particles, assuming that aqueous metal hydroxides are solubilized equally in each micelle (2–5). The parameters of reversed micelle and $CaCO_3$ particles in a cyclohexane solution of the Ca salt of Aerosol OT are presented in Table 7.2.1 as a representative example (2). The values of N_p are much smaller than N_m at any Wo, suggesting the formation of particles by fusion among nuclei or microcrystallines of particles formed in each micelle. The value of N_m/N_p, which is the number of micelles required to form one particle, was on the order of 10^8 in any systems. This mechanism may be suggested from the fact that the sizes of particles are always much larger than those of initial micelle droplets at any Wo. Such a formation process for $CaCO_3$ particles, in which the final size of the particles is determined by the fusion between nucleated micelles, was modeled by a deterministic population balance framework after specifying the minimum amount of Ca^{2+} required to form a nucleus, assuming that interaction between calcium-containing micelles occurs due to Brownian collision and fusion (6). The experimental result on the number of micelles required to form a $CaCO_3$ particle the size of the $CaCO_3$ particles reported by Kandori and coworkers (2) was predicted successfully by the model. The values of N_m/N_p predicted by the model are also presented in Table 7.2.1. A similar growth mechanism of particles operated for $CaCO_3$ and $BaCO_3$ particles in the W/O microemulsions of polyoxyethylene(6) nonylphenyl ether and the Ba salt of Aerosol OT in cyclohexane, respectively (3,4).

Table 7.2.1 Numbers of Micelles and $CaCO_3$ Particles in Cyclohexane Solutions of the Ca Salt of Aerosol OT

	Micelle			Particle			
Wo	State	Mw	N_m/mL	Diameter (nm)	N_p/mL	N_m/N_p	Model N_m/N_p
5	SMs	9.4×10^3	4.88×10^{18}	48.0 ± 11.4	7.65×10^{10}	0.63×10^8	—
7.5	SMs	13.2×10^3	3.60×10^{18}	62.8 ± 17.0	3.41×10^{10}	10.6×10^8	—
10	MEs	23.5×10^3	2.11×10^{18}	115.6 ± 39.0	0.55×10^{10}	3.86×10^8	5.05×10^8
15	MEs	46.3×10^3	1.17×10^{18}	129.2 ± 33.6	0.40×10^{10}	2.93×10^8	1.79×10^8
20	MEs	97.4×10^3	0.65×10^{18}	126.8 ± 35.4	0.41×10^{10}	1.46×10^8	1.02×10^8

Note: SMs, swollen micelles; MEs, W/O microemulsions.

For $CaCO_3$ particles in W/O microemulsions of hexaethylene glycol dodecyl ether, however, the values of N_m/N_p lie between 0.044 and 0.28, indicating that the particles form by the destruction of micelle-solubilizing aqueous $Ca(OH)_2$ rather than by the intermicellar exchange process (5). In the case of the formation of $CaCO_3$ particles in the Ca salt of Aerosol OT–cyclohexane system, as seen in Table 7.2.2, the particles

Table 7.2.2 Change of Average Diameter of Reversed Micelles and Particles in Ca and Ba Salts of Aerosol OT in Cyclohexane

	Micelle			Micelle		
Wo	State	Diameter (nm)	$CaCO_3$ Diameter (nm)	State	Diameter (nm)	$BaCO_3$ Diameter (nm)
1	RMs	3.8	54.8 ± 17.5			
2				RMs	4.0	63.0 ± 13.4
4				SMs	4.0	75.0 ± 23.3
5	SMs	5.5	48.0 ± 11.4			
6				SMs	4.2	88.0 ± 25.0
7				MEs	5.0	104.9 ± 25.0
7.5	SMs	5.8	62.8 ± 17.0			
8				MEs	5.8	164.8 ± 28.3
10	MEs	7.5	115.6 ± 39.0			
15	MEs	8.6	129.2 ± 33.6			
20	MEs	11.6	126.8 ± 35.4			
25	MEs	13.9	110.4 ± 20.0			
30	MEs	17.0	120.4 ± 24.4			

Note: RMs, reversed micelles; SMs, swollen micelles; MEs, W/O microemulsions.

did not form in the reversed-micelle region below Wo = 4, but spherical and relatively uniform $CaCO_3$ particles were precipitated in both the regions of swollen micelles and W/O microemulsion (2). In the swollen micelle region, the particle size increased from 48 ± 11 to 63 ± 17 nm as Wo increased from 5.0 to 7.5, and in the W/O microemulsion region above Wo = 10 they almost converged to 110 ± 20 to 130 ± 33 nm. The reason that the particles are not precipitated below Wo = 4, irrespective of the presence of reversed micelles, may be due to the undissociation of $Ca(OH)_2$, because water inside the micelle was tightly bound with ionic polar groups of surfactant, which formed reversed micelles. On the other hand, a stepwise increase in size in the vicinity of Wo = 10 is due to the transformation from swollen micelles containing water interacting with the hydrated polar groups, to W/O microemulsions including bulklike water. These results may indicate that the particle size is not only affected by Wo, that is, the size of the micellar water droplet, but also is controlled by the states of water droplets inside the micelles. Such control of particle size by micellar droplet was also observed for $CaCO_3$ particles precipitated in the cyclohexane solutions of polyoxyethylene(6) nonylphenyl ether (4). Single Gaussian-type histograms were found in the size of the particles precipitated in colorless reversed-micelle solutions formed by adding 0.70–2.10 mL of aqueous $Ca(OH)_2$ into the surfactant solution of 100 g, but in the blue translucent W/O microemulsion above 2.10 mL of aqueous $Ca(OH)_2$ they became clearly bimodal with a different ratio (3). From dynamic light-scattering measurements, it was found that the precipitation of bimodal particles is induced by the formation of bimodal micelles, including aqueous $Ca(OH)_2$. However, the size of $CaCO_3$ particles precipitated in cyclohexane solution of hexaethylene glycol dodecyl ether was not always controlled by the micellar droplet size, since more uniform and small particles of 9 or 10 ± 2 nm in diameter were produced in colorless solutions, irrespective of the absence of micelles, and further, particles of 6 or 7 ± 2 nm are produced in blue translucent W/O microemulsions (5). The dispersion system of $CaCO_3$ particles was very stable, and this stability can be ascribed to the adsorbed layer of surfactant molecules on the particles, giving a higher entropic repulsion energy (8). In contrast to the $CaCO_3$ particles, $BaCO_3$ particles precipitated in the polyoxyethylene(6) nonylphenyl ether–cyclohexane system were rodlike with an axis of ~23–39 nm in the colorless reversed-micelle region of Wo = 2.25–9 and were relatively uniform and ellipsoidal in the blue translucent W/O microemulsion region of Wo = 18–72 (1). These particle sizes were almost independent of the extent of Wo in both the regions. In the region transforming from the reversed micelles to the microemulsions were produced mixtures of rodlike and ellipsoidal particles. In the reversed-micelle region, the sizes of rodlike particles depended on the concentrations of surfactant and $Ba(OH)_2$, the oxyethylene chain length of surfactant, and temperature; $BaCO_3$ particle size showed a minumum at 0.125 mol kg^{-1} polyoxyethylene(6) nonylphenyl ether (7), whereas at 0.10 mol kg^{-1} $Ba(OH)_2$ the particle size showed a maximum at 55°C (1). The $BaCO_3$ particle size decreased with increasing oxyethylene chain length of polyoxyethylene nonylphenyl ethers in cyclohexane (1). However, spherical and uniform $BaCO_3$ particles were precipitated in Ba salt of Aerosol OT systems of any Wo in cyclohexane (4). The size of the particles increased slowly with increasing Wo in both the regions of reversed micelles and swollen micelles, whereas at Wo = 7, where W/O microemulsions form, the particle size increased rapidly, as seen for $CaCO_3$ particles produced from the Ca salt of Aerosol OT (2). The particle size of $CaCO_3$ and $BaCO_3$ particles produced with Ca and Ba salts of Aerosol OT–cyclohexane are summarized in Table 7.2.2. $SrCO_3$ particles were also synthesized as a function of Wo and surfactant concentrations in the same nonionic surfactant systems used in the preparation of $CaCO_3$ particles (9). In the polyoxyethylene(6) nonylphenyl ether–cyclohexane system, spindlelike $SrCO_3$ particles having a line of apsides of 347 ± 45 to 307 ± 61 nm and minor axis of 55 ± 18 to 67 ± 11 nm were produced in colorless reversed-micelle solutions, whereas in blue translucent W/O microemulsions they became dendrites. In hexaethylene glycol dodecyl ether systems, however, rodlike and spindlelike particles were produced in reversed-micelle and microemulsion solutions, respectively. The size of rodlike particles was independent of the concentration of surfactant, but that of spindlelike particles increased with increasing surfactant concentration.

X-ray diffraction spectra of $CaCO_3$ particles produced in nonionic surfactant–cyclohexane solutions (2,5) showed that the particles are mixtures of calcite, vaterite, and aragonite. In polyoxyethylene(6) nonylphenyl ether systems (2), calcite and vatelite were formed in colorless reversed-micelle solutions, but in blue translucent W/O microemulsions aragonite, instead of vaterite, was formed together with calcite. However, calcite and vaterite were produced in both the colorless and blue translucent solutions in hexaethylene glycol dodecyl ether systems (5). The fractions of calcite and vaterite were almost the same in colorless solutions of both the surfactants, but in blue translucent solutions of polyoxyethylene(6) nonylphenyl ether that of calcite and aragonite depended on the amount of aqueous $Ca(OH)_2$ added, as seen in Table 7.2.3. The metastable forms such as vaterite and aragonite were transformed exponentially to calcite with time at 30°C. The rate constants of the transformation (k), estimated from the slopes of the straight lines of plots of $\log(P/P_0)$ versus time, are presented in Table 7.2.4, where P_0 and P are the fractions of metastable forms at $t = 0$ and t. The range of k values obtained was on the order of $10^{-6}\ s^{-1}$ in any system, whereas larger k values were found for the transformation of vaterite than for that of aragonite in polyoxyethylene(6) nonylphenyl ether systems. This discrepancy in k values agreed with the magnitude of the activation energy of the transformation of each crystalline form in bulk water, namely, $452 \pm 19\ kJ\ mol^{-1}$ for aragonite to calcite and $208 \pm 8\ kJ\ mol^{-1}$ for vaterite to calcite. These results indicate that the micro water droplets reaction field in the interior of reversed micelles not only restrains the size of particle produced therein, is but also controls the crystalline forms of the particles and the transformation velocity of the crystalline forms.

Table 7.2.3 Change of Fraction of Crystalline Forms with Amount of Aqueous $Ca(OH)_2$ Added

Amount of aqueous $Ca(OH)_2$ added[a]	Solubilization region	Crystalline forms (%)		
		Calcite	Vaterite	Aragonite
NP-6[b] system				
0.70	Colorless	66	34	0
1.40	Colorless	59	41	0
2.10	Colorless	61	39	0
2.80	Blue translucent	100	0	0
4.20	Blue translucent	88	0	12
5.60	Blue translucent	89	0	11
7.0	Blue translucent	60	0	40
9.80	Blue translucent	41	0	59
DP-6[c] system				
0.70	Colorless	66	34	0
0.98	Colorless	61	39	0
1.47	Blue translucent	26	74	0
1.68	Blue translucent	20	80	0
1.89	Blue translucent	30	70	0

[a] As mL/100 g surfactant; $[Ca(OH)_2] = 0.02\ mol\ kg^{-1}$.
[b] NP-6, polyoxyethylene nonylphenyl ether.
[c] DP-6, hexaethylene glycol dodecyl ether.

Table 7.2.4 Rate Constants of Transformation of Crystalline Forms

Amount of aqueous $Ca(OH)_2$ added[a]	Solubilization region	Transformation to calcite	$k\ (\times 10^{-6}\ s^{-1})$
NP-6[b] system			
1.40	Colorless	Vaterite	6.6
9.80	Blue translucent	Aragonite	2.8
DP-6[c] system			
0.98	Colorless	Vaterite	4.7
1.68	Blue translucent	Vaterite	1.1

[a] As mL/100 g surfactant; $[Ca(OH)_2] = 0.02\ mol\ kg^{-1}$.
[b] NP-6, polyoxyethylene nonylphenyl ether.
[c] DP-6, hexaethylene glycol dodecyl ether.

REFERENCES TO SECTION 7.2

1. K Kon-no, I Koide, A Kitahara. Nippon Kagaku Kaishi 815, 1984.
2. K Kandori, K Kon-no, A Kitahara. J Colloid Interface Sci 122:78, 1988.
3. K Kandori, N Shizuka, K Kon-no, A Kitahara. J Dis Sci Technol 8:477, 1987.
4. K Kandori, K Kon-no, A Kitahara. J Dis Sci Technol 9:61, 1988.
5. K Kon-no, N Shizuka. Shikizai 63:65, 1990.
6. R Bandyopadhyara, R Kumar, KS Gandhi, D Ramkrishna. Langmuir 13:3610, 1997.
7. I Koide, K Kon-no, A Kitahara. Shikizai 63:132, 1990.
8. K Kandori, K Kon-no, A Kitahara. J Colloid Interface Sci 115:579, 1987.
9. K Kon-no, N Shizuka, A Kitahara. Shikizai 63:132, 1990.

8

Nitrides

SABURO IWAMA

8.1 REACTION IN GAS PHASES

SABURO IWAMA
Daido Institute of Technology, Nagoya, Japan

8.1.1 General Properties

Ultrafine nitride particles of metals and semiconductors are receiving increasing interest as promising materials in many industrial fields. Here a general view of properties is given for typical and feasible nitrides.

8.1.1.1 Aluminum Nitride

Aluminum nitride (AlN) has a hexagonal structure of wurtzite type, and combines the following physical properties: high thermal conductivity of 320 W m^{-1} K^{-1} (theoretical value) (1) comparable to that of metals, high electrical resistivity of greater than 10^{14} Ω cm, and thermal expansion coefficient of 4.1×10^{-6} K^{-1}, matching silicon. These are excellent properties as a substrate material for silicon in very large scale integrated (VLSI) circuits. Both thermal conductivity and electrical resistivity, however, are known to be affected severely by oxygen and metallic impurities included in sintered AlN (2,3). The oxygen inclusion originates mainly in a sintering additive such as CaO, which is necessary for sintering the commercial AlN powder at submicrometer size, and in the residual alumina used as the starting material in carbothermal reduction and nitridation of α-Al_2O_3 (4).

8.1.1.2 Titanium Nitride

Titanium nitride (Ti-N) has two phases: δ phase (TiN) with a cubic structure, and ϵ phase(Ti_2N) with a tetragonal one. TiN has a broad composition range from about $TiN_{0.6}$ to $TiN_{1.16}$, which reflects the variation of lattice parameter from 0.4217 nm at both border compositions to 0.4240 nm at the stoichiometric composition (5). TiN is a hard refractory material with the melting point of about 3220 K and microhardness of 2450 kg mm^{-2}, acid and corrosion proof, and with low electrical resistivity, 25.0 $\mu\Omega$ cm. Furthermore, the high thermal conductivity of about 29 W m^{-1} K^{-1} is retained up to about 1900 K (6). These features are widely applied to the metallurgical coating of cutting tools. Ultrafine particles (UFPs) of TiN are often used as reinforcements in metal, ceramic, and polymer matrix composites (7). Hydrogen sorption–desorption characteristics of metal–TiN nanocomposite particles are being studied for use as a new catalyst (8).

8.1.1.3 Iron Nitride

Iron nitride (Fe-N) includes nitrogen atoms in the interstitial sites of iron lattice and has the following six phases: γ phase with a face-centered cubic (fcc) structure, α' phase with a body centered tetragonal (bct) structure, α'' phase ($Fe_{16}N_2$) with a bct structure of metastable state, γ' phase (Fe_4N) with fcc structure having a nitrogen atom in the body-centered site of the γ-phase unit cell, ϵ phase (Fe_xN, $3 \geqq x > 2$) with a hexagonal structure, and ζ phase (Fe_2N) with an orthorhombic structure. The formation of these phases depends on the nitrogen content, temperature, thermal treatment and preparation condition, particularly in the case of α'' phase. The γ-Fe, which is stable above 860 K, transforms to α' phase by quenching, but to γ'-Fe_4N + α-Fe by annealing. The α'' phase appears together with γ'-Fe_4N and α-Fe in the annealing process at 393 K for several week (9). The most attractive property of iron nitride is its magnetism. The γ-Fe ultrafine particles (UFPs) show a paramagnetic property even at the low temperature of 1.3 K (10). Other phases of the Fe-N system have a ferromagnetic property. Among them, the iron atom in the α'' phase has a magnetic moment of $2.9 \pm 0.2\ \mu_B$ (11), which is larger than the 2.221 μ_B of α-Fe. The γ'-Fe_4N also has a large magnetic moment compared to that of α-Fe. Hardness and better stability than iron are important features of iron nitride for its application to recording media.

8.1.1.4 Silicon Nitride

Silicon nitride (Si_3N_4) has two phases with hexagonal structure: α- and β-Si_3N_4, with the latter being a high-temperature phase. Strong, mostly covalent bonding between silicon and nitrogen is responsible for the high strength and high toughness of silicon nitride, which has the following physical and mechanical properties: a decomposition temperature of about 2170 K, thermal conductivity of about 17 W $m^{-1}\ K^{-1}$ at 1273 K, electrical resistivity of greater than $10^{14}\ \Omega$ cm, thermal expansion coefficient of $2\text{–}3 \times 10^{-6}\ K^{-1}$, Vickers hardness of 1700–2700 kg mm^{-2}, bending strength of $5\text{–}10 \times 10^3$ kg cm^{-2}, and the compressive strength of $5\text{–}8 \times 10^3$ kg cm^{-2}. The mechanical strength of Si_3N_4 is retained at a relatively high temperature, and this makes silicon nitride a most feasible material for high-temperature structural ceramics. Because of the covalent bonding property mentioned earlier, fabrication of dense Si_3N_4 requires some added sintering aids such as MgO or Y_2O_3, unless a simultaneous application of high temperature and high pressure is adopted using hot isostatic pressing (HIPing) or hot pressing.

8.1.1.5 Gallium Nitride

Gallium nitride (GaN) has a hexagonal structure of wurtzite type with a wide direct bandgap of 3.4 eV at 300 K. GaN-based III–V nitrides with wide band gaps are the potential candidates for device applications in the blue and ultraviolet wavelengths

(12). Recently, nanometer-sized semiconductor crystals have been of great interest for the formation of quantum confinement.

The physical and mechanical properties of all the final products made of nitride UFPs depend intimately on the smallness and uniformity of its particle size and also on the purity of nitride particles used. For three decades the major efforts have been related to the development of powder synthesis technology.

8.1.2 Synthesis Techniques of Metal and Semiconductor Nitrides

Ultrafine particles (UFPs) of metal and semiconductor nitrides have been synthesized by two major techniques; one is the reactive gas condensation method, and the other is the chemical vapor condensation method. The former is modified from the so-called gas condensation method (or gas-evaporation method) (13), and a surrounding gas such as N_2 or NH_3 is used in the evaporation chamber instead of inert gases. Plasma generation has been widely adopted in order to enhance the nitridation in the particle formation process. The latter is based on the decomposition and the subsequent chemical reaction of metal chloride, carbonate, hydride, and organics used as raw materials in an appropriate reactive gas under an energetic environment formed mainly by thermal heating, radiofrequency (RF) plasma, and laser beam. Synthesis techniques are listed for every heat source for the reactive gas condensation method and for the chemical vapor condensation method in Tables 8.1.1 and 8.1.2, respectively.

8.1.3 Detailed Procedure and Characterization of the Products

8.1.3.1 Reactive Gas Condensation Method

DC-Arc Plasma Furnace. A schematic illustration of this method for producing UFPs by means of DC-arc plasma is shown in Figure 8.1.1. UFPs of AlN + Al and TiN were synthesized by using N_2 gas at 0.1 MPa under an arc current and voltage of 140–250 A and 25–30 V, respectively (14,15). The feature of an arc plasma is a high-temperature plasma, where the electron temperature (T_e) is nearly equal to the ion temperature (T_i). Nitrogen atoms dissolved into the molten metal beneath the arc have an important role in particle formation. By using an N_2 + NH_3 atmosphere, the AlN ratio to AlN + Al for the UFPs increased from 0.3 to 0.95 with increasing NH_3 partial pressure (16). The Fe-N system (17) and GaN (12,18) UFPs can be synthesized by the similar method. A schematic model (17) of the nitriding zone is illustrated in Figure 8.1.2. The formation mechanism of nitride UFPs in this method is explained as follows:

1. Formation of metal vapor by ''an enhanced evaporation'' promoted by nitrogen circulation between atomic nitrogen in the arc and supersaturated nitrogen in molten metal beyond the arc.

Table 8.1.1 Synthesis Techniques of Ultrafine Nitride Particles by Reactive Gas Condensation

Synthesis technique	Raw materials	Surrounding gas	Synthetic products	References
1. DC-arc plasma furnace	Pure metals	N_2, NH_3-Ar	Al-AlN	14, 16, 19, 20, 21, 22, 77
			Ti-TiN	14, 15
			Co-TiN, Ni-TiN	23, 24
			γ-Fe, γ'-Fe_4N, Fe_3N	17
			GaN	12, 18
2. RF-plasma torch reactor	Al (powder)	N_2-Ar	AlN	27
	Ti (powder)		TiN	25
	NiTi alloy (powder)		Ni-TiN	28
3. Electron beam evaporation	Pure metals	N_2, NH_3-He	AlN, TiN	30, 32
			ZrN, HfN, VN, NbN, Ta_2N, CrN, Mo_2N, W_2N	31, 32
4. Laser ablation	Pure metals	N_2	AlN	35
			Ti-TiN	33, 34, 36
5. Resistive heating combined with reaction zone	Pure metals	NH_3	γ-Fe(N), γ'-Fe_4N	37
		NH_3–N_2	Mg_3N_2	41
		N_2	AlN, γ-Fe(N),	38
			γ'-Fe_4N, ϵ-Fe_xN, α,β-Si_3N_4	42
			InN, GaN	
6. Reactive sputtering	Pure metals	N_2-Ar	CoN	44
			Ni_3N-AlN	43
			Fe-AlN	45, 46, 47

2. Reaction of metal vapor with nitrogen in the atmosphere to form nitride UFPs, depending on the nitrogen affinity of metals.

A twin torch plasma furnace, where DC anode and cathode arcs were coupled together above an aluminum melt, was developed for synthesis of AlN UFPs in order to control the aluminum evaporation rate and the concentration of nitrogen atoms in the plasma column independently (19). A two-stage transferred-arc plasma reactor was built for AlN synthesis, where aluminum is evaporated in a transferred-arc plasma chamber and then reacted in a separate tubular reactor, allowing a better control of the reaction conditions (20). Arc plasma technique has been modified and

Table 8.1.2 Synthesis Techniques of Ultrafine Nitride Particles by Chemical Vapor Condensation

Synthesis technique	Raw materials	Synthetic products	Refs.
1. Thermally activated CVC	$FeCl_2$-NH_3	α-Fe, γ-Fe(N), γ′-Fe_4N, ε-Fe_3N	48
	$Fe(CO)_5$-NH_3	ε-Fe_3N	49
	(HMDS)-NH_3	a-Si-C-N-O	55
	$Ga_2(NMe_2)_6$-NH_3	GaN	56
	$Al(OH)(C_4H_4O_4)\cdot\frac{1}{4}H_2O$	AlN	52
	$AlCl_3$-N_2-NH_3	AlN	54
	AlH(NR)-NH_3	AlN	50
	BAC, BAL	AlN	51
2. Plasma-enhanced CVC	$Al(CH_3)_3$-NH_3	AlN	60
	$SiCl_4$-NH_3-Ar	α,β-Si_3N_4	58
	$Si(CH_3)_4$-$SiCl_3CH_3$-NH_3-H_2 $SiCl_4$-C_2H_4-NH_3-H_2	Si-C-N	78
	B_2H_6-H_2-NH_3	h-BN	61
	$SiCl_4$-NH_3-H_2-Ar	Si_3N_4	57, 59
	TTIP-NH_3-N_2	TiN, Ti_2N	62
3. Laser-induced CVC (cw CO_2 laser)	SiH_4-NH_3	a-Si_3N_4	63
	$(CH_3)_3SiNHSi(CH_3)_3$	Si-C-N	68, 69
	SiH_4-C_2H_4-NH_3	SiC-Si_3N_4 composite powder	80
	$Fe(CO)_5$-NH_3	γ′-Fe_4N, γ-Fe	73
	$Fe(CO)_5$-NH_3-C_2H_4	γ′-Fe_4N, ε-Fe_3N γ′-Fe_4(N,C), ε-Fe_3(N,C)	74
	$Mo(CO)_6$-NH_3	Mo_2N	79
	SiH_2Cl_2-NH_3	a-Si_3N_4	66
	SiH_4-$(CH_3)_2NH$-NH_3	Si-C-N	72
	C_2H_2-N_2O-NH_3	C-N	76
	BCl_3-NH_3	BN	75

developed to synthesize metal-nitride composite UFPs such as Al-AlN (21,22), Co-TiN, and Ni-TiN (23,24).

RF-Plasma Torch Reactor. A diagram of a radiofrequency (RF) plasma torch reactor (25) is shown in Figure 8.1.3. The apparatus consists of an RF-plasma reactor with frequency of 4.0 MHz, a powder feeder, a gas supply system, and an exhaust system. The torch is composed of a work coil and three concentric quartz tubes for three independent gas flows: outer, inner, and center carrier gas flows. For TiN UFPs synthesis, N_2 is added to the outer flow of Ar, and titanium powder sieved in advance to a size less than 25 μm is fed into the plasma by the carrier gas of Ar

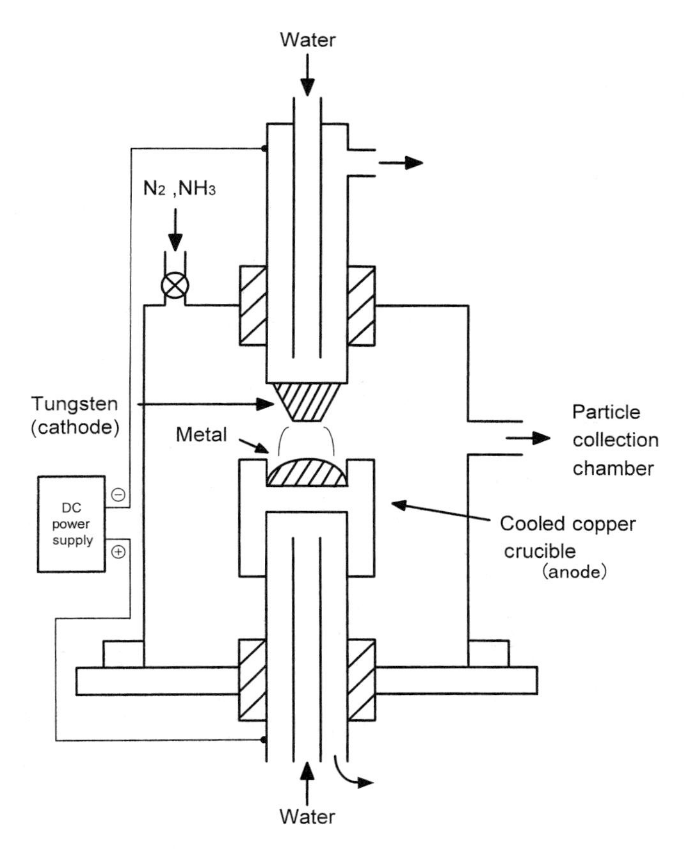

Fig. 8.1.1 Schematic illustration of the DC-arc plasma apparatus.

through the center tube. Typical operating conditions are shown in Table 8.1.3. The chemical analysis revealed a high oxygen content in the product, which was identified with δ-TiN by x-ray diffraction. It was concluded that high oxygen content in the product is not attributable to the large specific surface area of TiN UFPs but to the purity of the flowing gases. For minimizing the oxygen content, the following operating conditions are preferable: a higher feeding rate of powder, a higher power of RF generator, and a higher flow rate of N_2 required to form the stoichiometric TiN. Although the radial powder injection method (26) is found to be more effective for evaporating refractory metal powder than the axial injection, it is important to find the optimum conditions to generate a stable, high-density, high-temperature plasma ($T_i \sim T_e$, 7×10^3–1×10^4 K) in the individual apparatus.

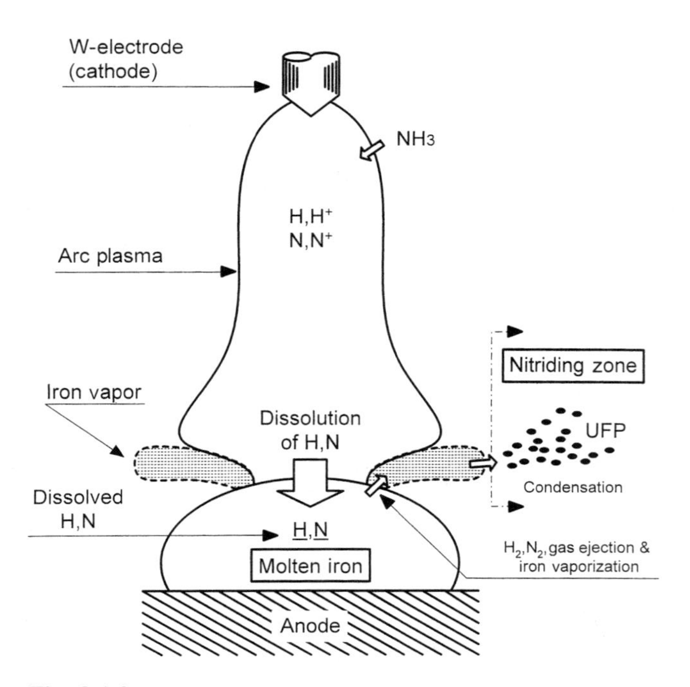

Fig. 8.1.2 Schematic model of nitriding zone of iron vapor and/or condensed iron particles in ''reactive plasma–metal'' reaction. (From Ref. 17, p. 412, with kind permission from The Japan Institute of Metals.)

AlN UFPs were synthesized (27) by injecting Al powder with NH_3 gas into the tail of RF-plasma generated at 13.56 MHz. Composite UFPs of Ni-TiN with dumbbell-like or dicelike morphology were produced by using Ni-Ti alloy powder as the raw material (28).

Electron Beam Evaporation. An electron beam heating was adopted (29) in the gas condensation method. A schematic illustration of the powered electron beam evaporation apparatus (30) is shown in Figure 8.1.4. The N_2 or NH_3 gas pressure in the evaporation chamber is kept at about 130 Pa, while the gas pressure in the gun chamber is maintained below 10^{-2} Pa by the differential evacuation system. In most cases, the evaporation was carried out by heating the head of a bundle of source metal wires by a focused electron beam. The accelerating voltage of the electron beam was 80 kV and the beam current was set at 5 to 20 mA, depending on the source material. The bundle consisted of 50 wires, with the diameter of each wire being 0.2 to 0.3 mm. This crucibleless evaporation method thus has the advantage of being free from impurities coming from the crucible at an elevated temperature.

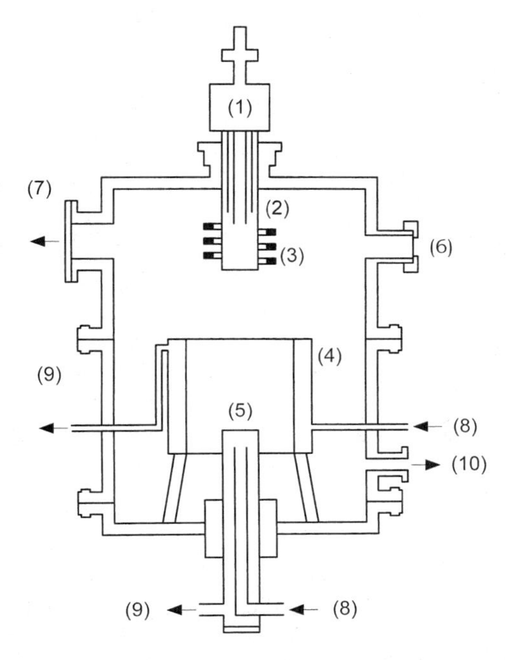

Fig. 8.1.3 Reactor chamber designed for powder processing studies: (1) torch head, (2) three concentric quartz tubes, (3) work coil, (4) water-cooled Pyrex cylinder, (5) water-cooled copper quenching plate, (6) window, (7) to generator, (8) water in, (9) water out, (10) to exhaust system. (Reprinted from J Mater Sci, 65, The synthesis of ultrafine titanium nitride in an r.f. plasma. Copyright 1979, with kind permission from Kluwer Academic Publishers.)

Table 8.1.3 Operating Conditions

1. Gas flow rate
Inner Ar = L/min
Outer Ar = 30 L/min
Outer N_2[a] = 0.5–2.0 L/min
Powder carrier Ar = 2.0 L/min
2. Plate powder output = 7.5 [kV] × 4.1 [A]
3. Powder feed rate[a] = 0.022–0.31 g/min

[a] Variable.

Source: Ref. 25. Kluwer Academic Publishers, J Mater Sci 14:1626, 1979, with kind permission from Kluwer Academic Publishers.

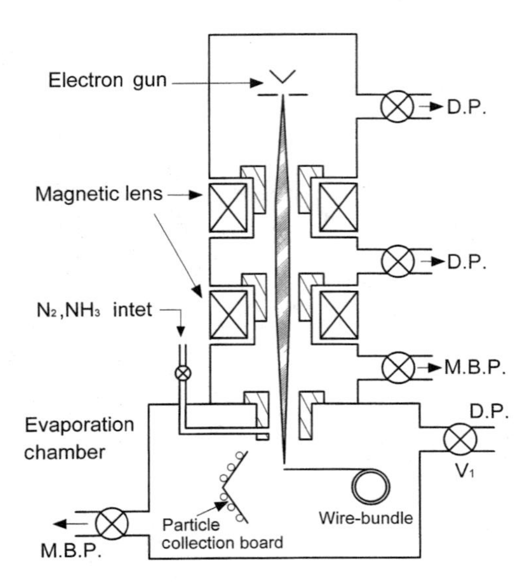

Fig. 8.1.4 Schematic illustration of the evaporation apparatus with electron beam heating, DP and MBP mean diffusion pump and mechanical booster pump, respectively. (Reprinted from J Cryst Growth, 56 S Iwama et al., Ultrafine powders of TiN and AIN produced by a reactive gas evaporation technique with electron beam heating, pp 265–269. Copyright 1982, with permission from Elsevier Science.)

Results from XRD or electron diffraction of UFPs of 11 elements are listed in Table 8.1.4. Ten kinds of nitride UFPs except for indium nitride can be synthesized by this method (30–32). Most nitride UFPs do not show a clear crystal habit, which is due to a fairly small particle size of 5 nm at most. Only TiN UFPs formed in N_2 gas show a cubic crystal habit, as shown in Figure 8.1.5.

From the comparison of phases of synthesized UFPs with those of residual materials after vaporizing in N_2 and NH_3 gas, the following process can be suggested for the formation of nitride UFPs by the electron beam evaporation:

1. Nitridation of source metals occurs in the range of about 1 cm from the head of wire bundle, except for tungsten. Electron beam plasma generated along the beam path may greatly contribute to the nitridation process at an elevated temperature.
2. Vaporization proceeds from the head of wire bundle. However, the vaporization mechanism of the nitride, i.e., whether the vaporization takes place through the decomposition of the nitride into metal vapor and nitrogen atoms or not, is not clear.

Table 8.1.4 Characteristics of Ultrafine Particles Formed by Evaporating Source Metals in an Atmosphere of N_2 or NH_3 at About 1 Torr

Source metal	Atmospheric gas	Obtained material	Crystal structure	Lattice parameter (nm)
Al	NH_3	AlN	Wurtzite type	$a_0 = 0.311 \pm 0.001$[a] $c_0 = 0.498 \pm 0.001$[a]
In	N_2, NH_3	In	Tetragonal	
Ti	N_2, NH_3	TiN	NaCl type	$a_0 = 0.4232 \pm 0.0002$[a]
Zr	N_2	ZrN	NaCl type	$a_0 = 0.460 \pm 0.005$[a]
Hf	N_2	HfN	NaCl type	$a_0 = 0.457 \pm 0.005$[a]
V	N_2, NH_3	VN	NaCl type	$a_0 = 0.415 \pm 0.005$[b]
Nb	N_2, NH_3	NbN	NaCl type	$a_0 = 0.426 \pm 0.005$[b]
Ta	N_2, NH_3	?	Amorphous	
Cr	N_2, NH_3	CrN	NaCl type	$a_0 = 0.417 \pm 0.005$[b]
Mo	N_2, NH_3	Mo_2N }	Partially disordered	$a_0 = 0.419 \pm 0.005$[b]
W	NH_3	W_2N }	NaCl type	$a_0 = 0.420 \pm 0.005$[b]

[a] Obtained from x-ray diffraction.
[b] Obtained from electron diffraction.
Source: Refs. 31 and 32. Reprinted from J Cryst Growth, 66, S Iwama et al., Growth of ultrafine particles of transition metal nitrides, pp 189–194. Copyright 1984, with permission from Elsevier Science.

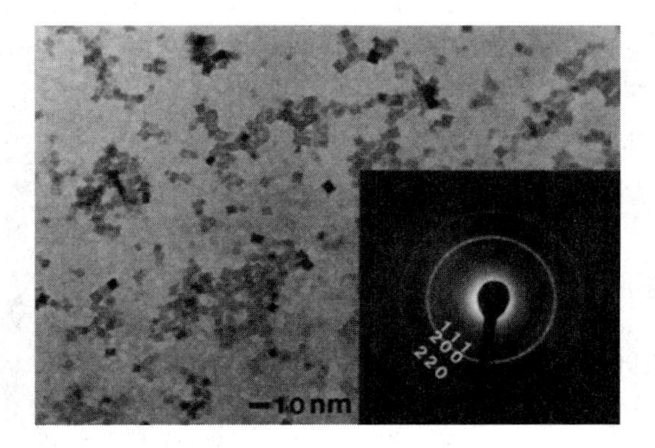

Fig. 8.1.5 Electron micrograph of TiN UFPs synthesized in N_2 gas at 130 Pa and the corresponding electron diffraction pattern with cubic structure.

The reason indium nitride UFPs cannot grow by this method is probably the low decomposition temperature (620°C) of InN. The similar consideration can be done on tungsten nitride and molybdenum nitride. The phases of W_2N and Mo_2N are recognized in UFPs, but no WN and MoN phases are detected; their decomposition temperature is about 800°C.

Laser Ablation. A mixture of UFPs composed of TiN and Ti was synthesized by laser ablation of titanium in N_2 gas at 0.1 MPa using an Nd YAG laser (wavelength: 1.06 μm) with peak power of 10 kW for 200-ns pulse (33). This method, however, was not successful in obtaining nitride UFPs of Mo, Al, Fe, and Si. Pure TiN UFPs were produced using a high-power YAG laser generator: maximum power of 80 J/pulse, repetition rate of 4–10 pulses/s (34). The particle size of TiN UFPs obtained was about 20 nm at an ambient N_2 gas pressure of 13.3 kPa, and decreased with decreasing N_2 gas pressure. AlN UFPs can be synthesized by an excimer laser (12 J/cm^2) ablation of aluminum under N_2 gas pressure of 10 kPa and the subsequent calcination at 900°C for 2 hr under N_2 flow (35).

One of the advantages of laser ablation is to provide a clean and a powerful energy. Furthermore, the energy density can be controlled easily by beam defocusing on a target material.

The $(TiN)_n^+$ clusters produced by a laser-induced plasma reactor source were investigated by means of time-of-flight mass spectrometry. The mass spectral abundance indicates that the clusters have cubic structures resembling subunits of the fcc lattice of TiN. The primary stoichiometries observed are $(TiN)_n^+$ ($n = 1–126$), except for Ti_nN_{n-1} ($n = 14, 63$) (36).

Resistive Heating Combined with Reaction Zone. The simplest way to form nitride UFPs on a laboratory scale is to adopt the reactive gas condensation method with a resistive heating evaporator. UFPs of the Fe-N system were synthesized by evaporating iron from a tungsten basket heated to 1600–2300°C in NH_3 gas pressure of 27 kPa (37). Crystalline phases of α-Fe, γ-Fe including about 2.5wt% of nitrogen, and γ'-Fe_4N depend on the NH_3 gas pressure and the evaporation temperature.

In order to realize a separate control of vaporization of metals and nitridation of UFPs, a reactive gas condensation apparatus with a microwave plasma reactor was developed (38). The apparatus consists of an evaporation chamber with a resistive heating unit and an N_2 gas inlet, a plasma reactor supplying a 2.45 GHz microwave by a cavity resonator, and a particle collection chamber connected to an exhausting vacuum pump as shown in Figure 8.1.6. A microwave plasma is a low-temperature plasma characterized with a high electron temperature ($T_e > 10^4$ K) but a low ion temperature ($T_i < 10^3$ K). This feature contributes to yielding active radicals effective in nitridation. Phase diagrams of synthesized AlN and Fe-N systems at a typical N_2 gas velocity of 3 m/s in the plasma region are shown in Figure 8.1.7, (a) and (b), respectively. Pure AlN UFPs were synthesized in the restricted parameter condition hatched in the diagram of microwave power versus N_2 gas pressure. As for the Fe-N system, two-mixed phase regions, one composed of α-

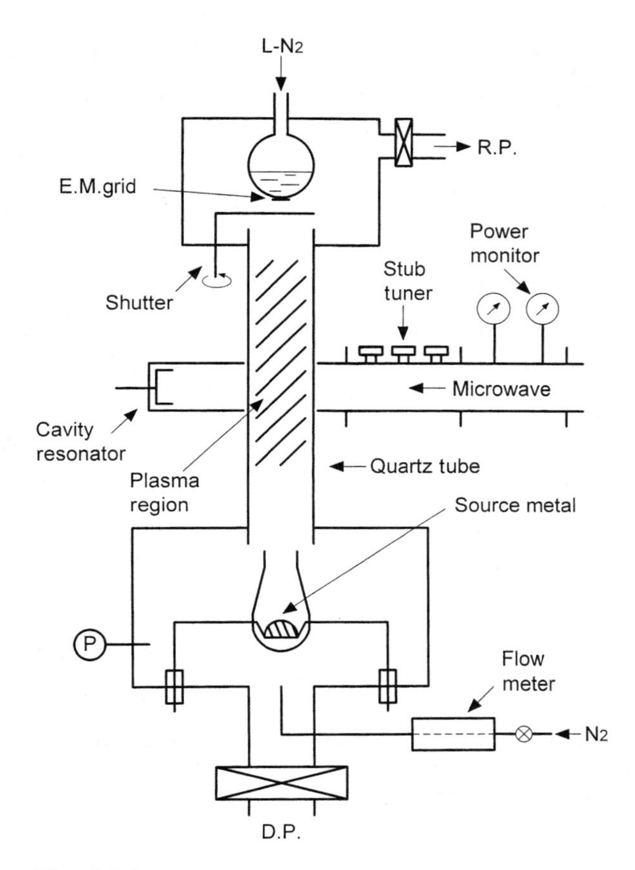

Fig. 8.1.6 Schematic illustration of the evaporation apparatus with a microwave plasma reaction zone. (From Ref. 38. Reprinted with permission of the Society of Materials Science Japan.)

Fe, γ-Fe, and γ'-Fe_4N, and the other of γ'-Fe_4N, ϵ-Fe_xN ($2 < x \leqq 3$), and ζ-Fe_2N, were formed depending on the parameter conditions shown in Figure 8.1.7(b). Crystalline silicon nitride UFPs were synthesized under the conditions at 0.4 kPa N_2 gas pressure, 1 m/s the gas velocity, and 1.2 kW microwave power. Electron diffraction indicates that the UFPs are composed of a mixed phase of α-Si_3N_4 and β-Si_3N_4 as shown in Figure 8.1.8. An unreacted Si phase was detected in addition to α- and β-Si_3N_4 when the gas velocity increased to 3 m/s. The α to β ratio in the mixed phase decreased with increasing N_2 gas pressure in general. Particles tended to grow in the plasma region. The particle size of nitride UFPs obtained by this method was distributed from 20 to 300 nm, which was found to be larger by a factor of 5 to 10

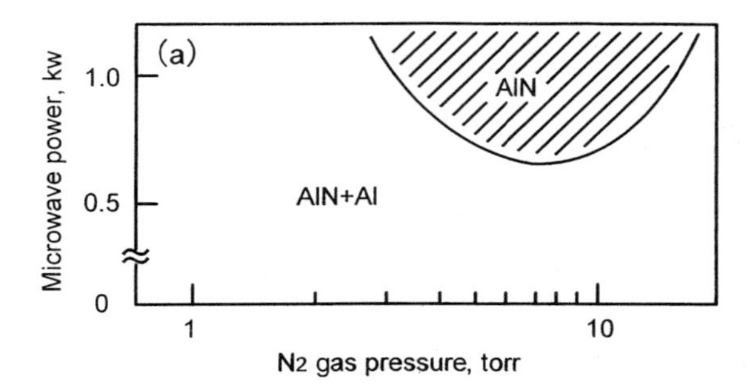

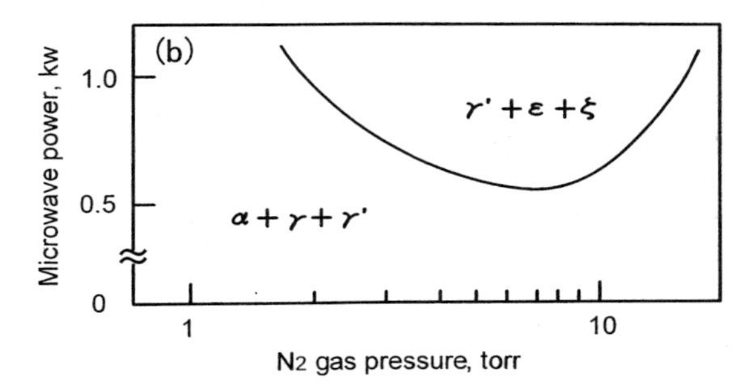

Fig. 8.1.7 Phase diagrams of UFPs of (a) Al–N system and (b) Fe-N system relating to the microwave power and the pressure of N_2 gas at velocity 3 m/s. (From Ref. 38. Reprinted with permission of the Society of Materials Science Japan.)

than that of particles collected at the front of the plasma region. It should be noticed that the silicon nitride UFPs as prepared by microwave plasma processing showed crystalline Si_3N_4, in contrast to an amorphous state of most silicon nitride UFPs formed by other synthesis techniques. This microwave plasma processing has been developed into a new type with a high-velocity flowing gas plasma to provide an unique process for phase transformation and/or surface modification of UFPs (39,40).

Magnesium nitride (Mg_3N_2) particles on a submicrometer scale were synthesized by evaporating magnesium metal in a pressure of 1 kPa of mixed $NH_3 + N_2$ gas flow kept at 800°C by a cylindrical furnace (41). Ge and In UFPs, which were size selected in advance with a differential mobility analyzer, were reacted with NH_3 gas in a tube furnace at ~1000°C to form size-selected GaN and InN UFPs (42).

Reactive Sputtering. Nanocomposite films of Ni_3N/AlN, CoN/BN, and CoN/Si_3N_4 were synthesized by reactive sputtering of a nickel aluminide, a cobalt boride,

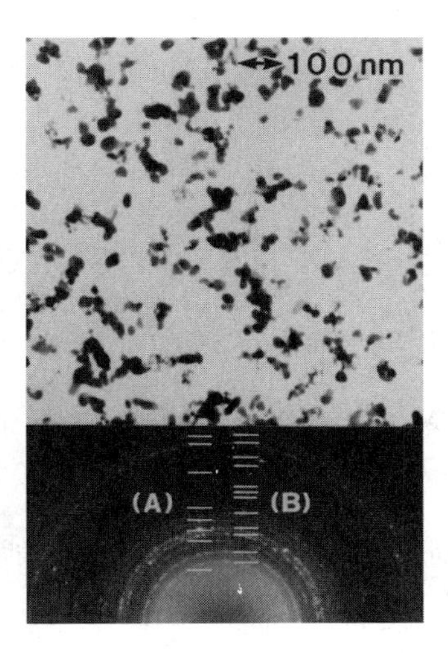

Fig. 8.1.8 Electron micrograph of Si_3N_4 UFPs and the electron diffraction pattern, where two series of Debye rings, (A) and (B), correspond to α- and β-Si_3N_4, respectively. (From Ref. 38. Reprinted with permission of the Society of Materials Science, Japan.)

and a cobalt silicide, respectively, in N_2 gas plasma generated in a parallel-plate DC glow discharge (43,44). Thermal decomposition profile measured by mass analyzer and weight loss indicated that CoN releases half its nitrogen in the conversion to Co_2N at 340°C and decomposes to cobalt metal at 400°C. Magnetic response shows paramagnetic behavior even at 30 K.

Multilayered composite films of Fe/AlN were synthesized by using multi-target-type RF sputter deposition equipment (45). Fe and AlN were alternatively deposited on a room-temperature substrate in 1 Pa of sputter gas of Ar and N_2, respectively. Four kinds of specimen denoted as $[Fe(495.7)nm/AlN(434.3)nm]_1$, $[Fe(178.2)nm/AlN(169.4)nm]_3$, $[Fe(64.8)nm/AlN(61.6)nm]_8$, and $[Fe(6.5)nm/AlN(6.2)nm]_{35}$, where numerals in parentheses represent single-layer thickness and subscripts show number of stacking, were examined by XRD, He ion yield extended x-ray absorption fine structure (EXAFS), and Mössbauer spectroscopy. Expansion of the lattice parameter of α-Fe accompanied by the decrease of grain size occurred significantly with the decrease of single-layer thickness. Saturation magnetization of 3- and 8-stacking specimens was as large as 1.2 to 1.1 times that of bulk α-Fe from the magnetic measurement by vibrating sample magnetometer (VSM) apparatus (46). Furthermore, a component with a slightly larger hyperfine field than that for α-Fe was explicitly observed in Mössbauer spectra of 3- and 8-stacking specimens. These observations brought about a conclusion that iron nitride formed at the interface of multilayered films may play an important role in magnetic properties of the composite films (45,47).

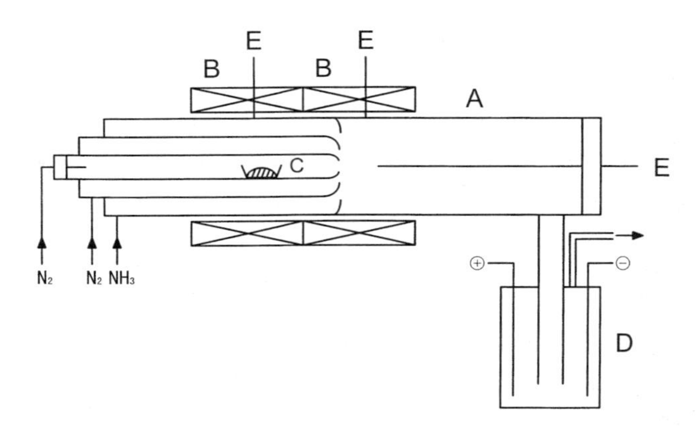

Fig. 8.1.9 Schematic illustration of the apparatus for thermally activated chemical vapor condensation: A, reactor; B, furnace; C, $FeCl_2$; D, electrostatic collector; E, thermocouple. (From Ref. 48. Reprinted with permission of the Chemical Society of Japan.)

8.1.3.2 Chemical Vapor Condensation Method

Thermally Activated Chemical Vapor Condensation. A typical apparatus for thermally activated chemical vapor condensation is illustrated in Figure 8.1.9. UFPs of the Fe-N system were synthesized using $FeCl_2{\cdot}nH_2O$ as a source material by this method (48). A precursor $FeCl_2$, which was prepared in advance by baking $FeCl_2{\cdot}nH_2O$ at 250°C for 2 h in N_2 gas flow, was vaporized from an alumina boat at 800°C, and then reacted with NH_3 gas at the reaction zone kept at a constant temperature of 780, 880, or 980°C. The main products at 780 and 880°C were ε-Fe_3N and α-Fe, while at 980°C they were γ'-Fe_4N, γ-Fe, and α-Fe.

Iron nitride UFPs for magnetic fluids were synthesized by a vapor–liquid chemical reaction between iron carbonyl [$Fe(CO)_5$] and NH_3. Figure 8.1.10 illustrates the experimental apparatus (49). The reactor contains a kerosene solution of $Fe(CO)_5$ with added surfactant of amine. Two-stage heat processing at different temperatures was performed, each stage being supposed to proceed according to the following processes:

$$Fe(CO)_5 + NH_3 \rightarrow Fe_2(CO)_6(NH_2)_2 + Fe_3(CO)_9(NH)_2 + CO + H_2 \quad (1)$$

under preheating at 90°C for 1 h and

$$Fe_2(CO)_6(NH_2)_2 + Fe_3(CO)_9(NH)_2 \rightarrow \varepsilon\text{-}Fe_3N + CO + NH_3 + H_2 \quad (2)$$

under heating at 185°C. The ε-Fe_3N UFPs obtained were highly uniform in size, smaller than 10 nm, and well dispersed without agglomeration. Magnetization curves of iron nitride magnetic fluids showed high saturation magnetic flux densities ranged up to 2330 G in 4 πM_s and no hysteresis.

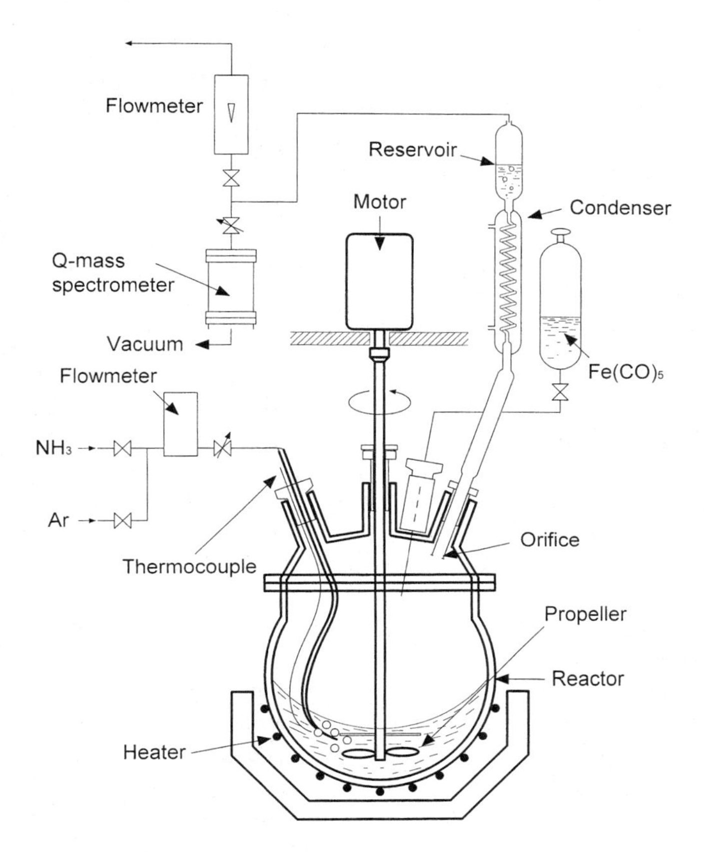

Fig. 8.1.10 Schematic illustration of the apparatus for vapor–liquid reaction. (Reprinted from J Mag Mag. Mater, 122, Iron-nitride magnetic fluids prepared by vapor-liquid reaction and their magnetic properties, I Nakatani et al., pp 10–14. Copyright 1993, with permission from Elsevier Science.)

Aluminum nitride UFPs have been synthesized by thermal decomposition from many kinds of precursor such as polyminoalane$(1/n)\cdot[\mathrm{A1H(NR)}]_n$ (50), aluminum polynuclear complexes of basic aluminum chloride (BAC) or basic aluminum lactate (BAL) (51), and (hydroxo)(succinato) aluminum(III) complex, $Al(OH)(C_4H_4O_4)\cdot\frac{1}{4}H_2O$ (52). These precursors were calcined under N_2 or NH_3 gas flow. The calcination temperatures, which depend on the individual precursor, can be lower by 600–200°C than the 1700°C in the conventional carbothermal reduction method. The XRD measurements at intermediate stages of the calcination process showed the phase change from an amorphous state to a trace of γ-alumina with very fine grains and finally to wurtzite-type AlN (51,52). Lowering the calcination temperature was attributable to a consequence of the intimate mixing of γ-alumina with carbon at the molecular level during the calcination process (52). The reactivity of γ-alumina was found to be much higher than those of α- and θ-alumina and $Al(OH)_3$ (53).

The highest heat conductivity of AlN is guaranteed by the lowest oxygen content in it (2). For minimizing the oxygen content in AlN UFPs, aluminum chloride ($AlCl_3$) was used as the precursor (54).

Amorphous UFPs of n-$SiC_xN_yO_z$ and crystalline GaN UFPs can be synthesized by thermal decomposition in NH_3 gas from the precursor hexamethyldisilazane (55), and from an amide precursor, [$Ga_2(NMe_2)_6$,Me=CH_3] (56), respectively.

Plasma-Enhanced Chemical Vapor Condensation. The DC-arc plasma torch and the RF-plasma torch reactors characterized by a high-temperature plasma, described earlier, have also been used as the thermal source in plasma-enhanced chemical vapor condensation (CVC). In this case, gas-phase reactants were injected into the plasma region, instead of the source powder feeding. Silicon nitride UFPs were synthesized from the reactants of silicon tetrachloride: $SiCl_4$, NH_3, H_2, and Ar, in a DC-arc plasma torch reactor (57). The as-synthesized powder, with the composition 37–39% N(-NH_4Cl), 53–56% Si, 3–5% O, 2.5–4.0% NH_4Cl, was pure white and amorphous, with particle size 30 to 60 nm. Crystallization occurred around 1500°C under nitrogen atmosphere to produce mainly α-Si_3N_4 and some β-Si_3N_4 and NH_4Cl.

Figure 8.1.11 illustrates the diagram of the apparatus with an inductively coupled RF-plasma torch reactor for the synthesis of silicon nitride UFPs from $SiCl_4$ and NH_3 as reactants, and Table 8.1.5 represents the experimental conditions (58). The quasi-amorphous Si_3N_4 UFPs thus prepared transformed to crystallite α-Si_3N_4, accompanied by grain growth and reduction in specific surface area, by postannealing above 1450°C.

A hybrid plasma reactor was developed for the synthesis of fine ceramic powders (59). The reactant $SiCl_4$ was injected into a DC-arc plasma jet and decomposed completely in a hybrid plasma: an RF-plasma superimposed on the DC-arc plasma. The reaction with the second reactant NH_3 and/or CH_4 gas, which was injected into the tail flame of the plasma, formed Si_3N_4 UFPs and/or Si_3N_4 + SiC mixed UFPs, with structures that were amorphous.

Besides the already mentioned techniques, a low-temperature plasma has been adopted to enhance the reaction in CVC. Through the synthesis of AlN UFPs by an RF-plasma-enhanced CVC using trimethylaluminum [$Al(CH_3)_3$] and NH_3 as reactants, the effect of experimental parameters on the rate of powder formation, particle size, and structure was examined (60). A high RF current was primarily connected to a high electron density, which activated the gas-phase reaction to promote the powder formation rate. The increase of both susceptor temperature and $Al(CH_3)_3$ concentration also increased the powder formation rate and enhanced the grain growth, where both mechanisms—coalescence by particle collision and vapor deposition on to particle surfaces—were believed to occur.

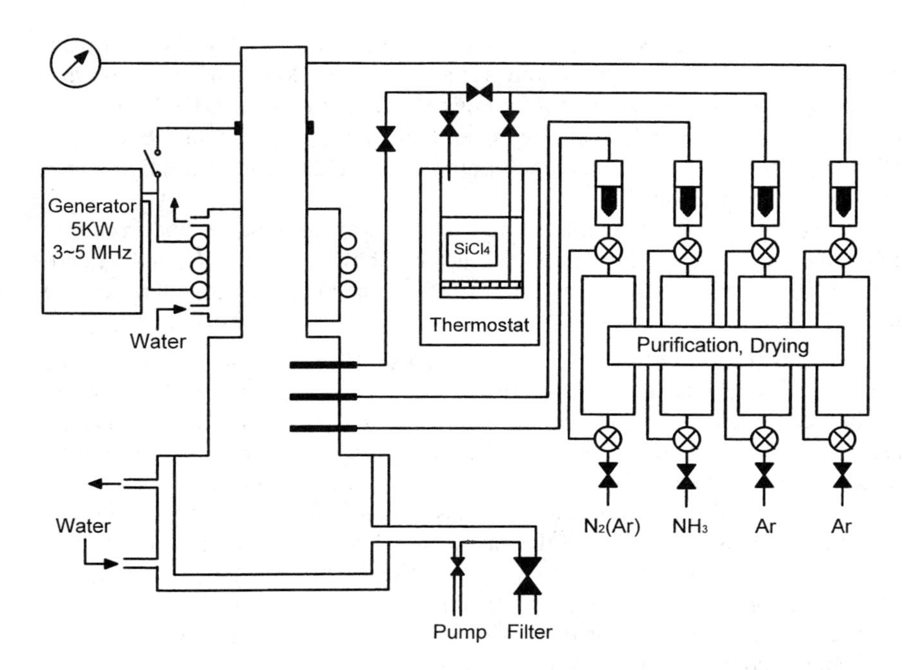

Fig. 8.1.11 Schematic illustration of the apparatus for plasma-enhanced chemical vapor condensation. (From Ref. 58. Reprinted with permission of the Ceramic Forum International/ Ber. DKG.)

Table 8.1.5 Processing Conditions for the Synthesis of Silicon Nitride UFPs by an Inductively Coupled RF-Plasma Torch Reactor

Frequency	3–5 MHz
HF-capacity	3–4 kW
Plasma gas Ar	400–1200 L/h
NH_3	100–1000 L/h
$SiCl_4$	40–650 g/h
Carrier gas Ar for $SiCl_4$	100–200 L/h
$NH_3/SiCl_4$	12–38

Source: Ref. 58. Reprinted with permission of the Ceramic Forum International/Ber. DKG.

Boron nitride UFPs with hexagonal structure (h-BN) were synthesized from a diborane (B_2H_6)–ammonia gas mixture as the reactant in a low-temperature plasma generated by an RF-glow discharge (61). Fourier transform infrared (FTIR) analysis revealed that the BN UFPs contain hydrogen bonded in NH_2, NH, and BH forms, with NH_2 being the most dominant configuration, in addition to absorption bands characteristic of hexagonal BN.

Titanium nitride UFPs synthesized by a microwave plasma method from titanium tetraisopropoxide and NH_3 or N_2 as reactants showed cubic (TiN) and tetragonal (Ti_2N) structures, depending on the operational conditions, particularly on the relative nitrogen gas flow rates (62).

Laser-Induced Chemical Vapor Condensation. The kinds of input energy for chemical vapor condensation are pure thermal heating in thermally activated CVC (described earlier), and a combination of thermal and plasma heating in plasma-enhanced CVC, respectively. A laser beam, however, supplies its energy to a reaction system through a nonthermal process; the absorption of the beam energy, $h\nu$, by the reactant gases. Then reactants with a strong absorption coefficient for the laser beam have to be chosen as at least one of the gas species. The reactions are accomplished in an effectively ''wall-less environment,'' because the hot zone is confined to the interaction region of the reactant gas stream and the laser beam. This offers an advantage of being free from the contamination that comes from hot walls and that is inevitable to some extent in other pyrolyses.

Figure 8.1.12 shows a schematic illustration of laser-induced CVC (63). For the synthesis of silicon nitride, the reaction between silane (SiH_4) and ammonia

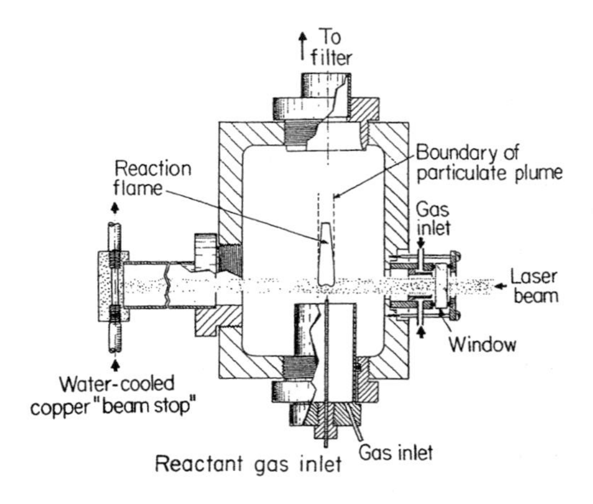

Fig. 8.1.12 Schematic illustration of the apparatus for laser-induced chemical vapor condensation. (From Ref. 63. Reprinted with permission of the American Ceramic Society, PO Box 6136, Westerville, OH 43086-6136. Copyright 1982 by the American Ceramic Society. All rights reserved.)

Table 8.1.6 Processing Conditions for the Synthesis of Silicon Nitride UFPs by a Laser-Induced CVC

Laser intensity (W/cm^2)	760
Pressure, Pa (atm)	2×10^4 (0.2)
SiH_4 (cm^3/min)	11
NH_3 (cm^3/min)	110
Argon (cm^3/min)	1000
Gas velocity at nozzle max (cm/s)	1140
Gas velocity at a laser beam (cm/s)	949
Reaction zone temp. (°C) pyrometer	867

Source: Ref. 63. Reprinted with permission of the American Ceramic Society, PO Box 6136, Westerville, OH 43086-6136.

proceeds by absorbing a continuous wave CO_2 laser light ($\lambda = 10.6\ \mu m$), and forms amorphous Si_3N_4 UFPs according to the following gas-phase reaction:

$$3SiH_4(g) + 4NH_3(g) \xrightarrow{h\nu} Si_3N_4(s) + 12H_2(g) \tag{3}$$

The typical processing conditions and powder characteristics are shown in Table 8.1.6 and Table 8.1.7, respectively. The compaction property of this powder was

Table 8.1.7 Characteristics of Synthesized Si_3N_4 UFPs

Surface area (m^2/g), BET	117
Particle size (nm), BET equivalent	17.6
Mean particle size (nm) ± std. dev., TEM	16.8 ± 3.9
Size range (nm), TEM	10–25
Color	Light brown
Si (wt %)[a]	72[c]
N (wt %)[a]	26.00
O (wt %)[b]	0.05
Free Si (wt %)	35
Crystallinity	Amorphous

[a] Wet chemical analysis, Luvak, Inc., Boylston, MA.
[b] Neutron activation analysis, Union Carbide, Tuxedo, NY.
[c] 60 wt% Si for stoichiometric Si_3N_4.
Source: Ref. 63. Reprinted with permission of the American Ceramic Society, PO Box 6136, Westerville, OH 43086-6136.

investigated at high pressures and various temperatures followed by pressureless sintering (64). Visually transparent compacts were fabricated by pressing the starting powder without outgassing in liquid nitrogen under 5 GPa. After pressureless sintering for 1 h at 1400°C in N_2 gas flow, the hardness increased to over 2000 kg/mm^2, keeping the visual transparency. A study on microstructure and grain-boundary characterization of hot isostatic pressed high-purity Si_3N_4 was carried out by means of high-resolution transmission electron microscopy (65). The exposed Si_3N_4 powder sample, hot isostatic pressed at 193 MPa for 1 h at 1950°C, reached a density of 92% ρ_{Th}, and showed the grain-boundary phase of glassy SiO_2, which might contribute to densification via a solution–reprecipitation mechanism. In contrast, no grain-boundary phase was observed in samples hot isostatic pressed from unexposed powders, resulting in a higher hot isostatic pressing temperature to 2050°C to achieve a density of 70% ρ_{Th}.

The SiH_4 in reaction (3) can be replaced by an organosilicon compound. From chlorinated silane (SiH_2Cl_2) and NH_3, silicon nitride UFPs were synthesized according to the following three routes, depending on both the temperature and the NH_3 injection site—below, into, and above the laser beam (66):

$$SiH_2Cl_2 + 4NH_3 \rightleftarrows Si(NH)_2(s) + 2NH_4Cl + 2H_2 \qquad (4a)$$

at low temperature before the irradiation by laser,

$$3Si(NH)_2\ (s) \overset{h\nu}{\rightleftarrows} Si_3N_4\ (s) + 2NH_3 \qquad (4b)$$

at 1100–1500 K,

$$3SiH_2Cl_2 + 4NH_3 \overset{h\nu}{\rightleftarrows} Si_3N_4\ (s) + 6HCl + 6H_2 \qquad (5)$$

at 1100–2000 K,

$$SiH_2Cl_2 \overset{h\nu}{\rightleftarrows} Si(s,l) + 2HCl \qquad (6a)$$

above 2170 K or before NH_3 is mixed with SiH_2Cl_2, and

$$3Si(s,l) + 4NH_3 \overset{h\nu}{\rightleftarrows} Si_3N_4\ (s) + 6H_2 \qquad (6b)$$

below 2170 K.

A nanometer-sized silicon nitride, compacted into bulk sample, showed a strong piezoelectric constant: 2613×10^{-12} [C/N]. It is interpreted by the charge accumulation in the interfaces and the surfaces of microvoids (67).

Nanocomposite Si–C–N ternary UFPs have been synthesized for aiming at improving the properties of the final powders, increasing safety, and lowering the production cost. In the laser-induced synthesis, many gaseous mixtures are used as reactants, such as hexamethyl-disilazane ($(CH_3)_3SiNHSi(CH_3)_3$) (68,69), methylamine (CH_3NH_2), and dimethylamine ($(CH_3)_2NH$) (70–72).

UFPs of the Fe-N system can be synthesized from iron pentacarbonyl [$Fe(CO)_5$] and NH_3 as reactants by a 1000-W continuous wave CO_2 laser irradiation. The NH_3 gas is the absorbent of the laser beam in this case. At the lower synthesis temperature, below 650°C, UFPs of γ'-Fe_4N with particle size of 10–25 nm grew dominantly. Above 1150°C, however, the growth of γ-Fe UFPs with larger particle size of 30–100 nm was predominant (73). Iron carbonitride (ICN) UFPs were also synthesized from the ternary reactants of $Fe(CO)_5$, NH_3, and C_2H_4. The structure of ICN UFPs was hexagonal with ϵ-$Fe_3(N,C)$ phase. A large saturation magnetization up to 142 emu/g was obtained and was ascribed to the carbon layer on ICN UFPs (74).

Laser-induced CVC has been applied to the synthesis of UFPs of novel materials such as BN from the BCl_3–NH_3 system (75) and C_xN_y from the C_2H_2–N_2O–NH_3 system (76).

REFERENCES TO SECTION 8.1

1. MP Borom, GA Slack, JW Syzmaszek. Am Ceram Soc Bull 51:852, 1972.
2. K Shinozaki, N Iwase. Tech Pap Assoc Finish Processes SME(fer)FC 16–22, 1986.
3. N Kuramoto, H Taniguchi, I Aso. Proc 36th Electronic Conf, New York, IEEE, 424, 1986.
4. T Sakai, M Iwata. Yogyo-Kyokai-Shi 82:181, 1974 (in Jpn).
5. P Ehrlich, Z Anorg Allg Chem 259:1, 1949.
6. JF Lynch, CG Ruderer, WH Duckworth. Engineering Properties of Selected Ceramic Materials. The American Ceramic Society, Westerville, OH 1966.
7. PA Janeway. Adv Ceram 28:28, 1990.
8. Y Sakka, H Okuyama, S Ohno. Nanostruct Mater 8:465, 1997.
9. KH Jack. Proc R Soc London A208:200, 1951.
10. T Yamaoka, M Mekata, H Takaki. J Phys Soc Jpn 35:63, 1973.
11. TK Kim, M Takahashi. Appl Phys Lett 20:492, 1972.
12. HD Li, HB Yang, GT Zou, S Yu, JS Lu, SC Qu, Y Wu. J Cryst Growth 171:307, 1997.
13. (For example) R Uyeda. Progr Mater Sci 35:1, 1991.
14. M Uda, S Ohno. Chem Soc Jpn 1984:862, 1984 (in Jpn).
15. Y Sakka, S Ohno, M Uda. J Am Ceram Soc 75:244, 1992.
16. M Uda, S Ohno, H Okuyama. Yogyo-Kyokai-Shi 95:76, 1987 (in Jpn).
17. S Ohno, H Okuyama, K Honma, K Takagi, T Honjyo, M Ozawa. J Jpn Inst Metals 59: 408, 1995 (in Jpn).
18. HD Li, HB Yang, S Yu, GT Zou, YD Li, SL Liu, SR Yang. Appl Phys Lett 69:1285, 1996.
19. H Ageorges, S Megy, K Chang, J-M Baronnet, JK Williams, C Chapman. Plasma Chem Plasma Process 13:613, 1993.
20. FJ Moura, RJ Munz. J Am Ceram Soc 80:2425, 1997.
21. A Inoue, BG Kim, K Nosaki, T Yamaguchi, T Masumoto. J Appl Phys 71:4025, 1992.
22. HD Li, HB Yang, GT Zou, S Yu. Adv Mater 9:156, 1997.
23. Y Sakka, S Ohno. Appl Surf Sci 100/101:232, 1996.
24. Y Sakka, S Ohno. Nanostruct Mater 7:341, 1996.

25. T Yoshida, A Kawasaki, K Nakagawa, K Akashi. J Mater Sci 14:1624–1630, 1979.
26. T Yoshida, K Nakagawa, T Harada, K Akashi. Plasma Chem Plasma Process 1:113, 1981.
27. K Baba, N Shohata, M Yanazawa. Appl Phys Lett 54:2309, 1989.
28. Y Sakka, H Okuyama, T Uchikoshi, S Ohno. Nanostruct Mater 8:465, 1997.
29. S Iwama, E Shichi, T Sahashi. Jpn J Appl Phys 12:1531, 1973.
30. S Iwama, K Hayakawa, T Arizumi. J Cryst Growth 56:265, 1982.
31. S Iwama, K Hayakawa, T Arizumi. J Cryst Growth 66:189, 1984.
32. S Iwama, K Hayakawa, T Arizumi. J Soc Mater Sci Jpn 32:943, 1983 (in Jpn).
33. H Xu, S Tan, L Sun. Mater Lett 12:138, 1991.
34. R Okada, M Haneda, S Hioki, T Araya, A Matsunawa, S Katayama. J Jpn Soc Powder Powder Metal 42:1184, 1995 (in Jpn).
35. GP Johnston, RE Muenchausen, DM Smith, SR Foltyn. J Am Ceram Soc 75:3465, 1992.
36. ZY Chen, AW Castleman Jr. J Chem Phys 98:231, 1993.
37. K Yamauchi, S Yatsuya, K Mihama. J Cryst Growth 46:615, 1979.
38. S Iwama. J Soc Mater Sci Jpn 36:1162, 1987 (in Jpn).
39. K Hayakawa, S Iwama. J Cryst Growth 99:188, 1990.
40. S Iwama, T Fukaya, K Tanaka, K Ohshita, Y Sakai. Nanostruct Mater 12:241, 1999.
41. T Murata, K Itatani, FS Howell, A Kishioka, M Kinosita. J Am Ceram Soc 76:2909, 1993.
42. MH Magnusson, K Deppert, JO Malm, C Svensson, L Samuelson. J Aerosol Sci 28(Suppl 1):S471, 1997.
43. L Maya, T Thundat, JR Thompson, RJ Stevenson. Appl Phys Lett 67:3034, 1995.
44. L Maya, M Paranthaman, JR Thompson, T Thundat, RJ Stevenson. J Appl Phys 79: 7905, 1996.
45. S Kikkawa, M Fujiki, M Takahashi, F Kanamaru, H Yoshioka, T Hinomura, S Nasu, I Watanabe. Appl Phys Lett 68:2756, 1996.
46. M Fujiki, S Kikkawa, M Takahashi, F Kanamaru, T Hinomura, S Nasu, H Yshioka. J Jpn Soc Powder Powder Metal 43:1420, 1996 (in Jpn).
47. S Kikkawa, M Fujiki, M Takahashi, F Kanamaru. Ceramic Microstructure: Control at the Atomic Level, eds. AP Tomsia, A Glaeser. New York: Plenum Press, 1998, p 605.
48. T Tanaka, K Tagawa, A Tasaki. Chem Soc Jpn 1984:930, 1984 (in Jpn).
49. I Nakatani, M Hijikata, K Ozawa. J Mag Mag Mater 122:10, 1993.
50. M Seibold, C Russel. J Am Ceram Soc 72:1503, 1989.
51. N Hashimoto, Y Sawada, T Bando, H Yoden, S Deki. J Am Ceram Soc 74:1282, 1991.
52. WS Jung, SK Ahn. J Mater Chem 4:949, 1994.
53. A Tsuge, H Inoue, M Kasori, K Shinozaki. J Mater Sci 25:2359, 1990.
54. C Li, L Hu, W Yuan, M Chen. Mater Chem Phys 47:273, 1997.
55. W Chang, G Skandan, SC Danforth, M Rose, AG Balogh, H Hahn, and B Kear. Nanostruct Mater 6:321, 1995.
56. KE Gonsalves, SP Rangarajan, G Carlson, J Kumar, K Yang, M Benaissa, MJ Yacaman. Appl Phys Lett 71:2175, 1997.
57. F Allaire, S Dallaire. J Mater Sci 26:6736, 1991.
58. I Dörfel, HD Klotz, R Mach, E Schierhorn. Ceramic Forum Int/Ber.DKG 70:227, 1993.
59. HJ Lee, K Eguchi, T Yoshida. J Am Ceram Soc 73:3356, 1990.

60. KH Kim, CH Ho, H Doerr, C Deshpandey, RF Bunshah. J Mater Sci 27:2580, 1992.
61. J Costa, E Bertran, JL Andujar. Diamond Related Mater 5:544, 1996.
62. P Mehta, AK Singh, AI Kingon. Mater Res Soc Symp Proc 249:153, 1992.
63. WR Cannon, SC Danforth, JH Flint, JS Haggerty, RA Marra. J Am Ceram Soc 65:324, 1982.
64. A Pechenik, GJ Piermarini, SC Danforth. J Am Ceram Soc 75:3283, 1992.
65. P Lu, SC Danforth, WT Symons. J Mater Sci 28:4217, 1993.
66. RA Bauer, JGM Becht, FE Kruis, B Scarlett, J Schoonman. J Am Ceram Soc 74:2759, 1991.
67. WX Wang, DH Li, ZC Liu, SH Liu. Appl Phys Lett 62:321, 1993.
68. RW Rice. J Am Ceram Soc 69:C-183, 1986.
69. RW Rice, RL Woodin. J Mater Res 4:1538, 1989.
70. M Cauchetier, O Croix, M Lance, C Robert, M Luce. In: G de With, RA Terpstra, R Metselaar, eds. Euro-Ceramics, Vol. 1. London: Elsevier Applied Science, 1989, p 130.
71. R Alexandrescu, I Morjan, E Borsella, S Botti, R Fantini, TD Makris, R Giorgi, S Enzo. J Mater Res 6:2442, 1991.
72. E Borsella, S Botti, R Fantini, R Alexandrescu, I Morjan, C Popescu, TD Makris, R Giorgi, S Enzo. J Mater Res 7:2257, 1992.
73. XQ Zhao, F Zheng, Y Liang, ZQ Hu, YB Xu, GB Zhang. Mater Lett 23:305, 1995.
74. XQ Zhao, Y Liang, ZQ Hu, BX Liu. J Appl Phys 79:7911, 1996.
75. M Luce, O Croix, YH Zhou, M Cauchetier, M Sapin, L Boulanger. Euro-Ceramics 2(1): 233, 1991.
76. R Alexandrescu, F Huisken, G Pugna, A Crunteanu, S Petcu, S Cojocaru, R Cireasa, I Morjan. Appl Phys A 65:207, 1997.
77. Y Sakka, H Okuyama, T Uchikoshi, S Ohno. Nanostruct Mater 5:577, 1995.
78. K Szulzewsky, C Olschewski, I Kosche, HD Klotz, R Mach. Nanostruct Mater 6:325, 1995.
79. R Ochoa, WT Lee, PC Eklund. Prepr Papers Am Chem Soc Div Fuel Chem 40:360, 1995.
80. M Suzuki, Y Maniette, Y Nakata, T Okutani. J Am Ceram Soc 76:1195, 1993.

9

Metals

NAOKI TOSHIMA, FERNAND FIÉVET, MARIE-PAULE PILENI, and KEISAKU KIMURA

9.1 REACTIONS IN HOMOGENEOUS SOLUTIONS

NAOKI TOSHIMA
Science University of Tokyo at Yamaguchi, Onoda, Japan

9.1.1 Introduction

Fine metal particles have received much attention in recent years from the viewpoints of chemical, physical, and biological interests (1–4). They are one of the most promising advanced materials. Compared with metal oxide or metal salts, metals have the highest electric and thermal conductivity, considerably higher weight and melting point, and usually excellent catalytic properties. These properties of metals cannot be replaced by other materials. Thus, even after the rapid growth of plastic, bulk metals keep their important position as one of the most common raw materials.

When we consider the metals of nanoscopic size, fine metal particles from micrometer to nanometer size can be synthesized by both physical and chemical methods. The former method provides the fine metal particles by decreasing the size by addition of energy to the bulk metal, while in the latter methods, fine particles can be produced by increasing the size from metal atoms obtained by reduction of metal ions in solution. Since chemical reactions usually take place in homogeneous solution in any case, this chapter includes most of the cases of synthesis and growth of fine metal particles. However, the polyol process, reaction in microemulsions, and formation in the gas phase are omitted, since they are described in later chapters by specialists in those fields.

9.1.2 Synthesis of Fine Metal Particles in Homogeneous Solution

Colloidal dispersions of fine metal particles can usually be prepared by reduction of metal ions. The first scientific report to synthesize colloidal dispersion of metals was presented by Faraday, who prepared metal colloids without stabilizers (5). In his case counteranions may have played the role of the stabilizer. In most recent cases, however, stabilizers are usually added to the system to stabilize the colloidal dispersions.

In this section, the general concept and practical procedures to synthesize the colloidal dispersions of fine metal particles in homogeneous solution are described.

9.1.2.1 Synthesis Without Stabilizers

Faraday prepared a colloidal dispersion of gold fine particles by reduction of gold(III) ions with white phosphorus. Recently, Turkevich prepared the gold sol by reduction with citrate (6). His group prepared Pt and Pt/Au colloids by the same method as well, discussing the growing process of the metal nucleus. Other reductants involve formaldehyde, hydrazine, hydrogen, hydrogen peroxide, carbon monoxide, etc. To prepare colloidal dispersions of fine metal particles by this method, very careful experiments are required. For example, the vessel should by completely cleaned by using steam after washing with concentrated nitric acid, and at least twice-distilled water should be used at any time as well. In this method the stabilizers are not used during the preparation. Probably the counteranions used for the synthesis may work to give a negative charge to the fine particles, which can prevent the aggregation of particles forming precipitates, by repulsive interaction among the negatively charged particles. Thus, the colloidal dispersions prepared by this method are often not stable enough. To stabilize these unstable colloids, the ions remaining in the system are required to be removed from the system by using osmotic membranes and/or ion exchange resins. The latter are often used in recent years for their easiness. The other method to stabilize the unstable sols is addition of stabilizers after the synthesis of the sol. Faraday used gelatin for this purpose. Synthetic water-soluble polymers like poly(vinylalcohol) and poly(*N*-vinyl-2-pyrrolidone) are more recently used as well.

9.1.2.2 Synthesis with Stabilizers

Alcohol Reduction. Colloidal dispersions of metal nanoparticles can be prepared by an alcohol reduction method in the presence of stabilizers. A typical procedure is as follows: Solutions of precious metal salts or complexes, such as H_2PtCl_6, $HAuCl_4$, $PdCl_2$, $RhCl_3$, and $RuCl_3$, in ethanol/water (1/1, v/v) are treated by heat until refluxing for a few hours in the presence of a water-soluble polymer like poly(*N*-vinyl-2-pyrrolidone) (PVP) (Scheme 9.1.1). Ethanol works as a reductant, and can be replaced with other primary or secondary alcohol. The alcohol content is changeable. The concentration of noble metal ions is changeable as well, but is often about 0.66 mmol dm^{-3}. Degree of polymerization of the polymer is also changeable. In the case of PVP, the K-30 grade, whose degree of polymerization is

$$MX_m + m/2\ C_2H_5OH \longrightarrow M^0 + m\ HX + m/2\ CH_3CHO$$

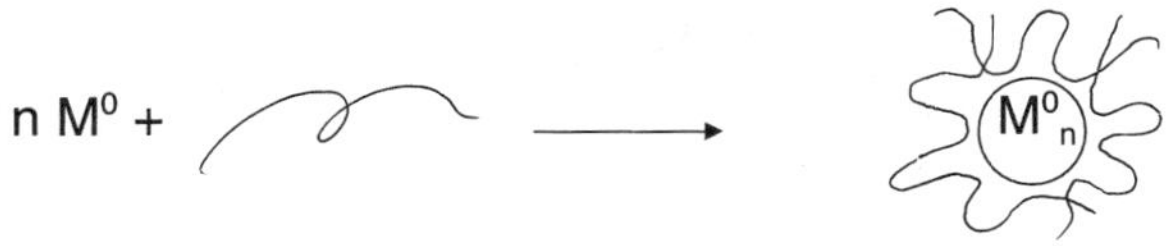

$$n\ M^0 + \text{(polymer)} \longrightarrow (M^0_n)$$

Scheme 9.1.1

~40,000, is often used. The molar ratio (R) of PVP in monomer units to metal is changeable as well, but should be above 1 and is usually 40. If R is less than 1, it is difficult to obtain a stable dispersion (7). The polyol process, which is described in a separate chapter, is a kind of alcohol reduction method. Use of alcohols with a high boiling point can increase the reaction temperature, which can promote a less active reaction. For example, light transition metal ions can be reduced by alcohol only at high temperature.

Photoreduction. The same solution of noble metal salts or complexes in ethanol/water as that used for the alcohol reduction can be used for photoreduction after a degassing procedure of three freeze–thaw cycles or nitrogen bubbling. Irradiation of the solution in a Pyrex Schlenck tube with visible light for a few hours usually provides a colloidal dispersion similar to that of alcohol reduction. Any kind of light source is applicable. A 500-W super-high-pressure mercury lamp is often used for convenience. The photoreduction method has an advantage for preparing colloidal metal, even in pure water. The photoreduction in water is initiated by the absorption of light by noble metal salts, resulting in the cleavage of metal–halogen bonds, for example (Scheme 9.1.2). Sometimes ultraviolet light is required for this cleavage (8). A quartz vessel must be used in this case instead of a Pyrex one.

In the presence of alcohols, photoreduction occurs more easily. Ketones like acetone can act as a sensitizer in this case (9). Surfactants can be used as protective colloids instead of polymers (10). Irradiation with ^{60}Co γ-rays is more effective for the reduction of metal ions (11).

Hydroborate Reduction. Lithium or sodium tetrahydroborate and diborane can be used for reduction of metal ions, especially light transition metal ions, to produce colloidal metals. For example, colloidal copper protected by polymer was prepared by reduction of copper(II) sulfate by a large excess of sodium tetrahydroborate in the presence of PVP or other polymers (12). A similar procedure for nickel(III) chloride produced nickel boride, not zero-valence nickel metal particles.

Ligand-stabilized metal nanoparticles such as triphenylphosphine-stabilized gold nanoparticles were usually synthesized by reduction of the corresponding metal ions or complexes with excess diborane. Tetraalkylammonium salts of hydro-

$$H_2PtCl_6 \xrightarrow{h\nu} Pt + 2\,HCl + 2\,Cl_2$$

$$2\,Cl_2 + 2\,H_2O \longrightarrow 2\,HCl + 2\,HClO$$

$$H_2PtCl_6 + 2\,H_2O \xrightarrow{h\nu} Pt + 4\,HCl + 2\,HClO$$

Scheme 9.1.2

$$PdCl_2 + 2\ N(C_8H_{17})_4BEt_3H \xrightarrow{THF} Pd(nanoparticle) + 2\ N(C_8H_{17})_4Cl + 2\ BEt_3 + H_2$$

$$PtBr_4 + 4\ N(C_8H_{17})_4\ BEt_3H \xrightarrow{THF} Pt(nanoparticle) + 4\ N(C_8H_{17})_4Br + 4\ BEt_3 + 2H_2$$

$$RhCl_3 + 3\ N(C_8H_{17})_4BEt_3H \xrightarrow{THF} Rh(nanoparticle) + 3\ N(C_8H_{17})_4Cl + 3\ BEt_3 + 1.5\ H_2$$

Scheme 9.1.3

trialkylborane have been used for preparation of metal nanoparticles stabilized by tetraalkylammonium halide in organic solvents (Scheme 9.1.3) (13). Miyake et al. proposed poly(acrylonitrile) (PAN) or copolymer containing a thiol group as a coordinating stabilizer to prepare a fine gold particles (14). The coordination of the nitrile group of PAN to gold(III) ion prevents the reduction of the gold ion, $(AuCl_4)^-$, by alcohol, but not by borohydride, since borohydride is a stronger reductant than alcohol. The gold nanoparticles obtained were much smaller than other polymer-stabilized gold nanoparticles such as PVP-stabilized ones, probably because PVP cannot coordinate to gold while PAN can. The thiol copolymer also works well as a stabilizer for preparing fine gold particles.

Hydrogen Reduction. Nord, a pioneer in using the colloidal noble metal particles as catalysts for organic reactions, prepared colloidal dispersions of noble metal particles by reduction of the corresponding metal hydroxide by molecular hydrogen (15). In the case of platinum sol, for example, refluxing for minutes after addition of NaOH to an aqueous solution of K_2PtCl_4 and polyvinylalcohol precedes the hydrogen reduction (Scheme 9.1.4). Kiwi and Grätzel improved this procedure (16). They used H_2PtCl_6 instead of K_2PtCl_4 and removed the large particles by applying a centrifuge technique to obtain small platinum particles with an average radius of 11 nm.

$$K_2PtCl_4 + 2\ NaOH \longrightarrow Pt(OH)_2 + 2\ NaCl + 2\ KCl$$

$$Pt(OH)_2 + H_2 \xrightarrow{PVA} Pt(nanoparticle) + 2H_2O$$

Scheme 9.1.4

On the other hand, zero-valence organometallic compounds, instead of metal salts, can also be used as a good precursor for metal nanoparticle synthesis. From zero-valence organometallic compounds, particles can be generated at relatively low temperature without using reducing reagents. Hydrogenation of olefin-type ligands in zero-valence organometallic compounds is a smart method for the synthesis of organosols of metal nanoparticles. Hydrogenation of cyclooctadiene (COD) and cyclooctatriene (COT) of Ru(COD)(COT) complexes gives ruthenium nanoparticles in the presence of PVP or other polymers (17). Copper nanoparticles could also be obtained by treatment of (cyclopentadienyl)(*tert*-BuNC)Cu with CO in the presence of PVP or other polymers (18).

Electrochemical Reaction. The electrochemical process is a unique preparative method of metal nanoparticles and has been recently developed by Reetz et al. (19). In this method, metal ions are produced from a bulk metal electrode in an electrochemical cell.

For example, alkyl ammonium-stabilized metal nanoparticles were generated by electrochemical process. A target bulk metal sheet is settled as an anode in an electrochemical cell as shown in Figure 9.1.1. Metal cations are generated at the anode and move to the cathode. Metal ions are reduced there by electrons generated from the cathode to form zero-valence metal atoms. In many cases, the zero-valence metal atoms are deposited onto the cathode metal sheet (usually platinum) or precipi-

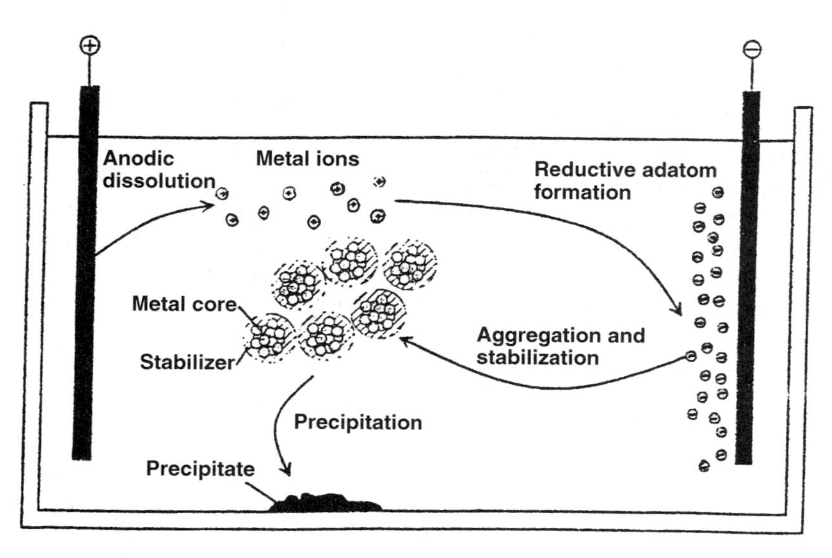

Fig. 9.1.1 Formation of tetraalkylammonium-stabilized metal nanoparticles by an electrochemical method. (From Ref. 19.)

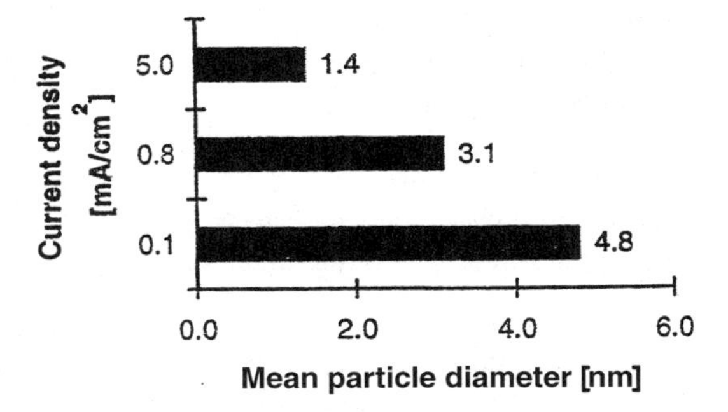

Fig. 9.1.2 Dependence of average diameter of Pd nanoparticles upon the current density in an electrochemical method. (From Ref. 19.)

tated. Under the present conditions, however, the metal atoms are concentrated at the cathode, aggregating to form metal nanoparticles. Since the particles are stabilized by alkyl ammonium, they are dispersed into the electrolytes. In this process the particle size can be easily controlled by controlling the current density. Figure 9.1.2 illustrates the dependence of average diameter of Pd nanoparticles on the current densities during the preparation. The current density directly influences the reduction potential at the cathode.

Other Methods. Other reductants like hydrazine, sodium metal, etc. can be used for the reduction of metal ions. Decomposition of metal salts or complexes by heat treatment is sometimes used for synthesis of fine particles as well. In this case the valence of metals in the fine particles should be carefully examined. Recently, irradiation of ultrasonic wave was applied to the synthesis of colloidal dispersions of metal fine particles.

Stabilizers are usually used during the reduction of metal ions to stabilize the colloidal dispersions of fine metal particles. The coordination interaction is the main factor to stabilize the metal particles. Thus, polymers with coordinating groups are good stabilizers. The choice of coordinating groups should depend on the kind of metal.

9.1.2.3 Synthesis of Bimetallic Particles

Metal catalysts composed of more than two different metal elements are of interest from both technological and scientific viewpoints for improving the catalyst quality or properties (20). In fact, bimetallic (or multimetallic) catalysts have long been valuable for in-depth investigations of the relationship between catalytic activity and catalyst particle structure (21). Sinfelt et al. have made a series of studies on bimetallic nanoparticle catalysts supported on inorganic supports, for example,

Ru/Cu (22) and Pt/Ir (23) on silica. They analyzed the detailed structure of these samples by an extended x-ray absorption fine structure (EXAFS) technique, showing an alloy structure for the nanoparticles with a diameter of 1–3 nm, associated with their special properties.

Here, the detailed synthetic methods and procedures of colloidal dispersions of bimetallic nanoparticles in a homogenous solution are described because the bimetallic system is one of the recent greatest topics of the fine metal particles (2).

Coreduction of Mixed Ions. Coreduction of mixed ions is the simplest method to synthesize bimetallic nanoparticles. However, this method cannot be always successful. Au/Pt bimetallic nanoparticles were prepared by citrate reduction by Miner et al. from the corresponding two metal salts, such as tetrachloroauric(III) acid and hexachloroplatinic(IV) acid (24). Reduction of the metal ions is completed within 4 h after the addition of citrate. Miner et al. studied the formation of colloidal dispersion by ultraviolet–visible (UV-Vis) spectrum, which is not a simple sum of those of the two monometallic nanoparticles, indicating that the bimetallic nanoparticles have an alloy structure. The average diameter of the bimetallic nanoparticles depends on the metal composition. By a similar method, citrate-stabilized Pd/Pt bimetallic nanoparticles can also be prepared.

An alcohol reduction method has been applied to the synthesis of polymer-stabilized bimetallic nanoparticles. They have been prepared by simultaneous reduction of the two corresponding metal ions with refluxing alcohol. For example, colloidal dispersions of Pd/Pt bimetallic nanoparticles can be prepared by refluxing the alcohol–water (1 : 1 v/v) mixed solution of palladium(II) chloride and hexachloroplatinic(IV) acid in the presence of poly(*N*-vinyl-2-pyrrolidone) (PVP) at about 90–95°C for 1 h (Scheme 9.1.5) (25). The resulting brownish colloidal dispersions are stable and neither precipitate nor flocculate over a period of several years. Pd/Pt bimetallic nanoparticles thus obtained have a so-called core/shell structure, which is proved by an EXAFS technique (described in Section 9.1.3.3).

Pt/Ru bimetallic nanoparticles were prepared by coreduction of the corresponding metal salts in the presence of glucose (26). EXAFS data indicated the existence of a Pt–Ru bond in this case, proving the formation of bimetallic nanoparticles. Pd/Rh bimetallic nanoparticles, stabilized by PVP, were also prepared by the alcohol reduction method (27).

$$H_2PtCl_6 + 2\ C_2H_5OH \longrightarrow Pt + 6\ HCl + 2\ CH_3CHO$$

$$PdCl_2 + C_2H_5OH \longrightarrow Pd + 2\ HCl + CH_3CHO$$

Scheme 9.1.5

$$Pd(OCOCH_3)_3 + CuSO_4 + PVP \longrightarrow PVP\text{-}Cu/Pd(OH)_n$$

$$PVP\text{-}Cu/Pd(OH)_n \xrightarrow[\text{glycol}]{} PVP\text{-}Cu/Pd \text{ (nanoparticle)}$$

Scheme 9.1.6

Nanoparticles of light transition metals are more difficult to prepare because of lower redox potential of the metal ions than those of the precious metal ions. In other words, light transition metal ions are difficult to reduce to zero valence, and zero-valence light transition metals are easily oxidized to ions.

Polymer-stabilized bimetallic nanoparticles containing both a light transition metal element and a precious metal element can also be prepared by a modified alcohol reduction method. For example, Cu/Pd bimetallic nanoparticles were successfully prepared with various Cu:Pd ratios by refluxing a glycol solution of the hydroxides of Cu and Pd in the presence of PVP or by thermal decomposition of metal acetates.

In the case of the modified alcohol reduction procedure (28), metal hydroxides are prepared first by pH control of the ionic solutions (Scheme 9.1.6). In practice the procedure is as follows: Copper sulfate, palladium acetate, and PVP are dissolved in glycol, followed by adjustment of the pH value to 9.5–10.5 by dropwise addition of NaOH to obtain metal hydroxide. The color of the solution changed from yellow to green, which indicated formation of metal hydroxide. Refluxing the solution at 198°C with a nitrogen flow to remove water from the system for 3 h gave a stable homogeneous dark-brown solution of PVP-stabilized Cu/Pd bimetallic nanoparticles. Auger and x-ray photoelectron (XPS) spectroscopies indicated that these nanoparticles contain Cu^0 species. PVP-stabilized Ni/Pd and Ni/Pt bimetallic nanoparticles could also be prepared by the similar method (29).

Thermal decomposition of metal acetates in the presence of PVP was proposed by Bradley et al. (30), where the preparative procedure of Esumi et al. (31) was modified. Thus, heating of palladium and copper acetates in a solvent with a high boiling point (ethoxyethanol) provides PVP-stabilized Pd/Cu bimetallic nanoparticles. In this method, not only thermal decomposition but also reduction by ethoxyethanol could be involved. However, the Bradley method can provide Cu/Pd bimetallic nanoparticles that contain less than 50 mol% of Cu, while our method mentioned earlier can provide fine particles with 80 mol% of Cu. In Esumi's original procedure, methyl *iso*-butyl ketone (MIBK) was used as a solvent without a stabilizer. In his method, Cu^{II} was not completely reduced to Cu^0, but Cu_2O was involved in the bimetallic nanoparticles. Probably, thanks to Cu^I species in the surface of the particles, no stabilizer is necessary for the stable dispersion.

Irradiation with γ-rays was also used to synthesize bimetallic nanoparticles. Remita et al. synthesized poly(vinyl alcohol) (PVA)-stabilized Ag/Pt bimetallic nanoparticles by radiolysis of an aqueous mixture of Ag_2SO_4 and K_2PtCl_4 at a concentration of 10^{-4} mol dm^{-3}. A typical Ag plasmon absorption band is observed at ~410 nm with only low intensity at the mole ratio of Ag:Pt = 60:40, indicating the formation of Ag/Pt bimetallic nanoparticles. Poly(acrylic acid) was also used as the stabilizer, although the resulting UV-Vis spectra were quite different.

Henglein also proposed to use γ radiolysis to produce bimetallic nanoparticles of precious metals (32). Silver particles having a gold layer were prepared and the UV-Vis absorption spectra of these bimetallic nanoparticles were intensively investigated. Several precious metal ions are deposited onto Ag particles to produce bimetallic nanoparticles (33). Hg ions can be reduced in the presence of Ag sol, resulting in the formation of Hg shell-type Hg/Ag bimetallic nanoparticles (34). Harriman reported the coreduction of $HAuCl_4$ and H_2PtCl_6 in the presence of a nonionic surfactant (Carbowax) by γ-radiolysis (35). However, photochemical coreduction of $HAuCl_4$ and H_2PtCl_6 in the presence of poly(ethylene glycol) monolaurate generates only a mixture of monometallic nanoparticles (36). Probably, the nonmiscibility of Au and Pt, the very weak coordination of metal ions and/or atoms to the stabilizer, and the slow rate of photoreduction result in the formation of monometallic nanoparticles in this case.

In the coreduction of the corresponding two metal salts by tetraoctylammonium triethylhydroborate in THF, Pt/Ru (37) and Pt/Rh (13,38) bimetallic nanoparticles, stabilized by tetraoctylammonium bromide, were obtained. No metal borate was found in the product.

Successive Reduction of Metal Ions. Successive reduction of two metal salts can be considered as one of the most suitable methods to prepare core/shell structured bimetallic particles (Fig. 9.1.3). In 1970, Turkevich and Kim tried to grow gold on Pd nanoparticles to obtain gold-layered Pd nanoparticles (39). The deposition of one

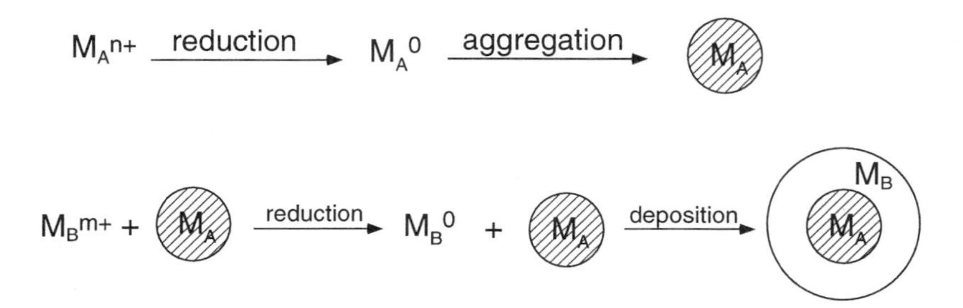

Fig. 9.1.3 Schematic illustration of formation of M_A-core/M_B-shell structured bimetallic particles by successive reduction.

metal element on preformed monometallic nanoparticles of another metal seems to be very effective. For this purpose, however, the second element must be deposited on the surface of preformed particles. If the second element cannot be deposited on the preformed particles, the mixtures of this kinds of particles will be produced.

An attempt to prepare PVP-stabilized Au/Pd bimetallic nanoparticles was made by the successive reduction procedure (40). When preparation of Au nanoparticles precedes the reduction of Pd ion, only mixtures of Pd and Au monometallic nanoparticles were produced. On the other hand, when the preparation of Pd nanoparticles precedes the reduction of Au ions, some bimetallic nanoparticles were found. However, the Au/Pd bimetallic nanoparticles thus obtained did not have a core/shell structure, although the bimetallic nanoparticles obtained by the coreduction have such a structure (Fig. 9.1.4).

Ligand-stabilized Au/Pd (41) and Au/Pt (42) bimetallic nanoparticles were prepared by Schmid et al. by successive reduction. Au colloids, with a diameter of ~18 nm, could be covered by Pt or Pd shells when the chloride precursors, H_2PtCl_6 or $PdCl_2$, were reduced. The color change from red to brown-black may suggest the covering of Au core with Pt and Pd atoms. Addition of the water-soluble ligand p-$H_2NC_6H_4SO_3Na$ stabilized the obtained bimetallic nanoparticles. In the case of Au/Pt bimetallic nanoparticles, uniform heterogeneous agglomerates with a large spherical Au core surrounded by small (~5 nm diameter) spherical Pt particles were obtained, which was proved by EDX measuerments (42). In the Au/Pd bimetallic case, on the contrary, a rather continuous structure was formed. TEM investigations of Au/Pd bimetallic nanoparticles show somewhat irregular structures, but no agglomeration of small particles was observed. The EDX spectrum of such a single

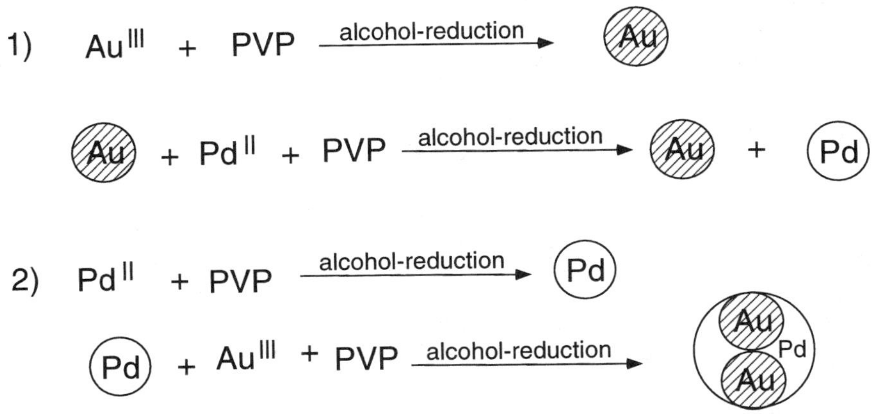

1) Au^{III} + PVP —alcohol-reduction→ (Au)

(Au) + Pd^{II} + PVP —alcohol-reduction→ (Au) + (Pd)

2) Pd^{II} + PVP —alcohol-reduction→ (Pd)

(Pd) + Au^{III} + PVP —alcohol-reduction→ (Au Au Pd)

Fig. 9.1.4 Successive reduction of Au^{IV} and Pd^{II} ions by an alcohol reduction method in the presence of PVP.

particle showed only Pd on the surface. The reverse successive reduction also worked and Pd-core/Au-shell structured bimetallic nanoparticles were produced.

In an earlier study, Turkevich and Kim proposed gold-layered palladium nanoparticles (39). Three types of Au/Pd bimetallic nanoparticles, such as Au-core/Pd-shell, Pd-core/Au-shell, and random alloyed particles, are prepared by the application of successive reduction. Two kinds of layered Pd/Pt bimetallic nanoparticles were also reported by successive reduction (43). However, detailed analyses of the structure of these bimetallic nanoparticles were not carried out at that time. Only the difference of UV-Vis spectra between the bimetallic nanoparticles and the physical mixtures of the corresponding monometallic nanoparticles was discussed.

Sacrificial Hydrogen Reduction. In the previous session, successive reduction of metal ions is sometimes successful to synthesize core/shell structured bimetallic fine particles. However, sometimes it is not successful. For example, Schmid and coworkers could synthesize Au/Pd bimetallic fine particles, while we could not. Compared to these, the size of the Pd core particles is much larger in Schmid's case than in our case. During the reduction of $AuCl_4^-$, the oxidation of Pd^0 atoms may also occur in the system of Schmid and coworkers. However, because of their large size, before the decomposition of Pd core particles, generated Au atoms may precipitate on them, thus also protecting Pd from further oxidation.

Recently, our group has invented a modified successive alcohol reduction process to prepare ''inverted core/shell'' structured bimetallic nanoparticles (44). In practice, PVP-stabilized Pd-core/Pt-shell bimetallic nanoparticles with a size of $\sim$2 nm were obtained. This method is based on the fact that precious metals like Pd have ability to adsorb hydrogen and split it to form metal-H bonds on the surface. Hydrogen atoms adsorbed on precious metal atoms have a very strong reducing ability, implying a quite low redox potential. As illustrated in Figure 9.1.5, Pt ions added into the dispersion of Pd nanoparticles are reduced by hydrogen atoms adsorbed on the preformed Pd nanoparticles without oxidizing Pd atoms to Pd ions, and are deposited on the Pd nanoparticles to form Pd-core/Pt-shell bimetallic nanoparticles.

Reduction of Double Complexes. Except for Ag/Au bimetallic nanoparticles, bimetallic nanoparticle dispersions containing Ag have not been studied extensively. One of the possible reasons is that an Ag^I ion readily reacts with a halide ion to produce water-insoluble silver halide. Also, many other water-soluble metal salts other than halides are not as suitable as precursors for the production of metallic particles by mild reduction.

Torigoe and Esumi proposed silver(I) bis(oxalato)palladate(II) as a precursor of Ag/Pd bimetallic nanoparticles stabilized by PVP (45). Photoreduction of the aqueous precursor in a quartz vessel gave Ag/Pd bimetallic nanoparticles at various concentrations. The particles deviate from spherical ones but are uniform. Each particle contains both metal elements, as confirmed by EDX measurement. The size can be changed with concentration of the precursor. The average composition of

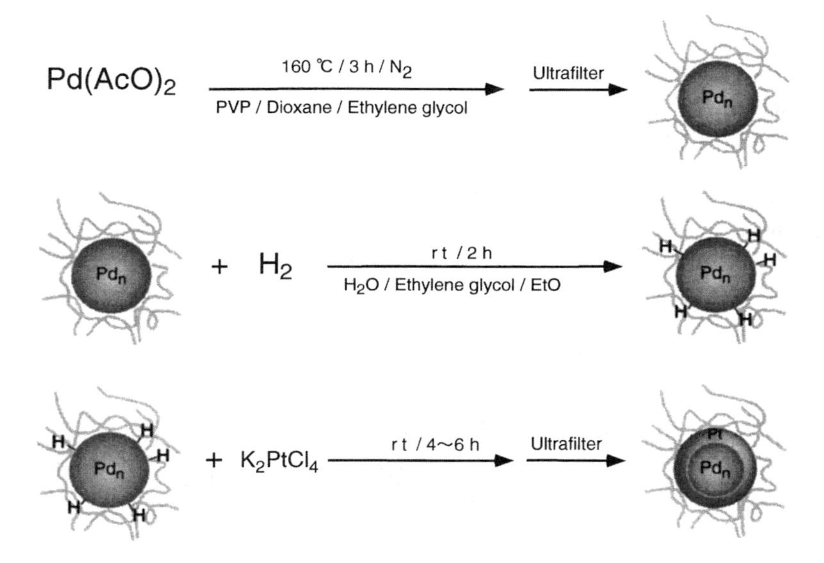

Fig. 9.1.5 Preparation process of Pd-core/Pt-shell (inverted core/shell) structured bimetallic clusters by a sacrificial hydrogen reduction. (From Ref. 69.)

bimetallic nanoparticles became Ag-rich with increasing concentration of the precursor, which could be confirmed by EDX and UV-Vis spectral investigations.

Preparation of PVP-stabilized Ag/Pt bimetallic nanoparticles by borohydride reduction from silver(I) bis(oxalato)platinate(I) was also proposed (46).

Pd/Sn bimetallic nanoparticles were synthesized from $SnCl_2$ and $Pd(NO_3)_2$ (47). In this case, the authors presented the pretreatment of the mixed solution. This pretreatment may produce the double complexes of Sn and Pd ions. The procedure is as follows: Separate boiled clear solutions of $SnCl_2$ and $Pd(NO_3)_2$ with a small portion of concentrated HCl were mixed and evaporated to dryness. The dry solids were digested with concentrated HNO_3 and HCl, forming an orange-red solution, which was again evaporated. The dry solid was then extracted with water and reduced by borohydride in the presence of surfactant. In the Pd–Sn bimetallic nanoparticles thus prepared, tin exists as Sn^{II} or Sn^{IV} ions, as confirmed by XPS.

Electrochemical Synthesis of Bimetallic Particles. Most chemical methods for the preparation of metal nanoparticles are based at first on the reduction of the corresponding metal ions with chemical reagents to form metal atoms and then on the controlled aggregation of the obtained metal atoms. Instead of chemical reduction, an electrochemical process can be used to create metal atoms from bulk metal. Reetz and Helbig proposed an electrochemical method including both oxidation of bulk metal and reduction of the metal ions for size-selective preparation of tetraalkylammonium salt-stabilized metal nanoparticles (see Fig. 9.1.1) (19). The particle size could be controlled by the current density. The Pd/Pt, Ni/Pd, Fe/Co, and Fe/Ni bimetallic nanoparticles could be also obtained by this method (48). Bulk plates of two metal elements were immersed as anodes into electrolyte containing a stabilizer (tetraalkylammonium salt), and a Pt plate was used as a cathode. From the anodes the corresponding metal ions were generated, and the generated metal ions were reduced by electrons generated from the Pt cathode to produce tetraalkylammonium salt-stabilized bimetallic nanoparticles. The advantages of this method include low cost, high yield, easy isolation, and simple control of the metal content of the bimetallic nanoparticles.

9.1.3 Characterization of Metal Particles

9.1.3.1 Size and Shape of the Particles

The size of fine particles, especially nanoparticles, of metal can be measured only by transmission electron microscopy (TEM). Usually a microscope with the magnitude larger than 100,000 is used for the measurement. The sizes of more then 200 individual particles should be measured. The results are shown as histograms, from which the average diameter and standard deviation can be calculated. If the relative standard deviation is less than 10%, the particles can be said to be well monodispersed.

Plasmon absorption of dispersions of coin metal fine particles like gold was often used to estimate the size of particles. The absorption peak can be calculated on the basis of Mie theory. However, this is not always true. The peak position can move, depending on not only the size of the particles but also the environment of the particles and the extent of aggregation of the particles. Thus, UV-Vis absorption is only used just for understanding the rough image of particle dispersions.

Light scattering is also used to estimate the size, size distribution, and aggregation feature (49). However, it is useful only for the dispersions of naked metal particles with enough largeness. Recently developed nanoparticles are usually surrounded by stabilizers like water-soluble polymers, surfactant molecules, etc. Then the light scattering method reflects the whole size of metal nanoparticles with surrounded envelopes (Fig. 9.1.6), and cannot distinguish such whole size from the size of the envelopes without metal nanoparticles.

In order to calculate the hydrodynamic size (Stokes size) of metal nanoparticles with the surrounding envelope, the Taylor dispersion method has been proposed (50). Since metal nanoparticles can be detected by UV-Vis absorption in this method, only the size of the particles involving metal nanoparticles can be determined (Fig. 9.1.7).

Scanning tunneling microscopy (STM) has been used to determine the dimensions of metal nanoparticles stabilized by alkylammonium salt in combination with high-resolution TEM (51). The difference between the diameter determined by STM (d in Fig. 9.1.6) and that determined by TEM (d_M in Fig. 9.1.6) allows estimation

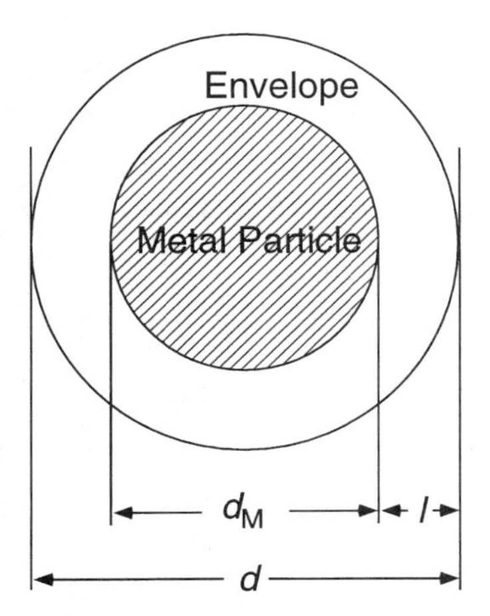

Fig. 9.1.6 Image of a metal nanoparticle surrounded by envelope such as polymer, surfactant, ligand, etc.

of the thickness of the protective layer (l in Fig. 9.1.6). However, the samples for STM were dried, which may change the structures. Thus, the STM image may not present the real structure of protected particles in a solvent.

The greatest advantage of the Taylor dispersion method compared to the STM method for analyzing the entire nanoparticle size involving the protective layer is that the entire size can be directly measured in the solution, when the surrounding molecules on the surface of naked metal nanoparticles rapidly exchange with those ''free'' in the solution. In addition, although the envelope molecules like surfactants can form ''free'' micelles without metal nanoparticles, only the envelope molecules with metal nanoparticles can be measured by the Taylor dispersion method because the diffusion was detected by the UV-Vis absorption of the metal nanoparticles (see Fig. 9.1.7).

9.1.3.2 Composition

In the case of monometallic particles, the metal element is determined by the charged metal ions. The only question is the oxidation state of the element. This can be determined by x-ray photoelectron spectroscopy (XPS). The oxidation state as well as the ratio of each state can be determined by comparing the position and area of the peaks of XPS spectra.

However, if the metal particles are surrounded by stabilizers, the sensitivity of the XPS spectra decreases, which makes it difficult to determine the correct oxidation state and the ratio of each state. In this case the surrounding stabilizers must be removed before the analysis. Sometimes, only washing with solvents is

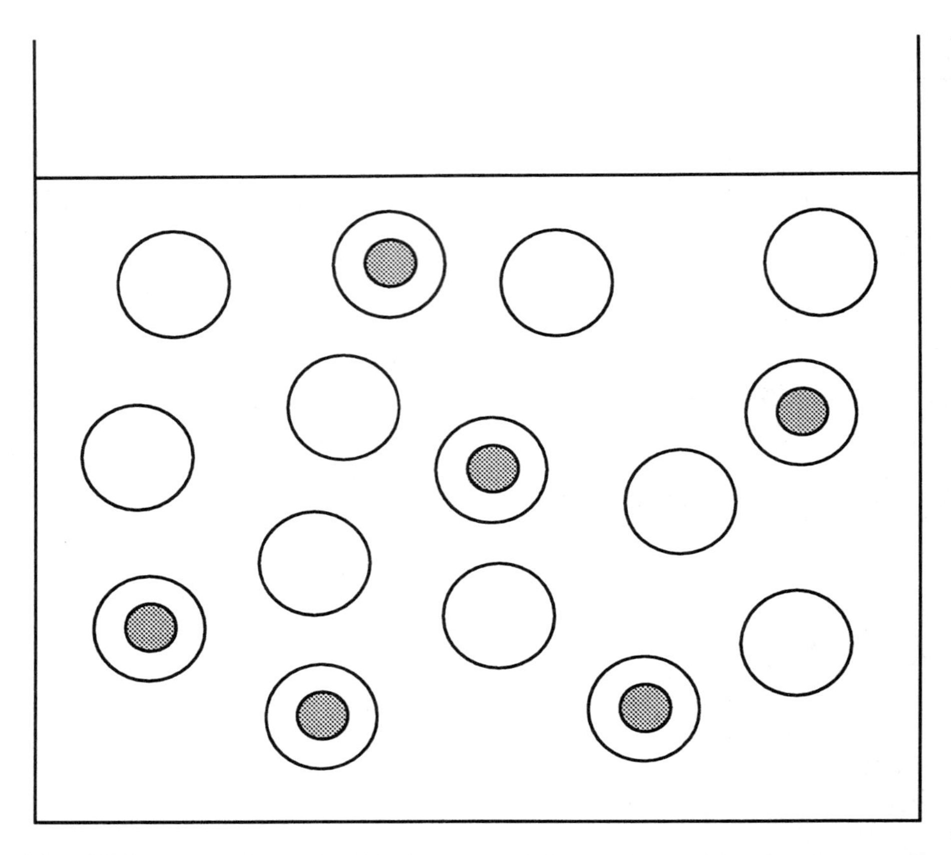

Fig. 9.1.7 Image of the colloidal dispersions of the envelopes with and without metal nanoparticles. Light scattering can measure the average size of both envelopes, and the Taylor dispersion method can do only the size of the envelopes with metal nanoparticles.

enough for this purpose, but sometimes more effective methods are required. For example, etching with an ion beam could be useful. A coordination capture method (52) can be applied to remove most parts of polymer from polymer-protected metal particles. (cf. Fig. 9.1.8) (53). Thus, a silica gel bearing mercaptan ligands (SiO_2-SH) was suspended in polymer-protected metal nanoparticle dispersions. The nanoparticles captured by silica gels were separated by a centrifuge technique, washed with ethanol several times, and dried under vacuum. By this procedure most parts of polymer surrounding the metal particles were removed, and the XPS spectra could be measured with enough accuracy. This coordination capture process may cause some structural change of metal particles, but the results of these samples corresponded well with the other analyses.

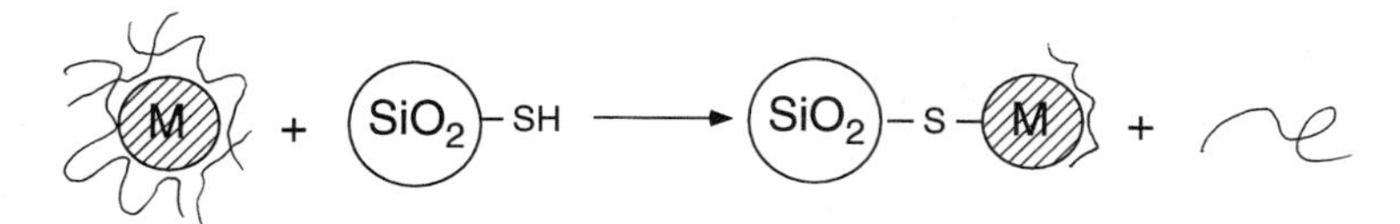

Fig. 9.1.8 Image of the coordination capture method to remove polymers from the metal nanoparticles surrounded by polymers.

The composition of the surface-bound species must be considered; they contribute to the stability of the dispersions of metal nanoparticles. In the case of electrostatically stabilized dispersions, the techniques to measure the interfacial electronic phenomena, including electrophoresis, electroosmosis, etc., are useful (54). In order to understand the composition (as well as structures) of the chemical species bound in the surface of metal particles, spectroscopic measurements used for common organic substances are used as well as the elemental analysis.

Infrared (IR) spectroscopy is a useful technique to understand the structure of ligands of metal complexes. The coordination of PVP on Pd nanoparticles was confirmed by the shift of C═O stretching vibration of PVP by the presence of Pd nanoparticles, as shown in Figure 9.1.9 (55). Surface-enhanced Raman spectroscopy (SERS) can be used for understanding the structure of chemical species adsorbed on coin metal particles. For example, the Raman intensity of molecules adsorbed on gold or silver particles or surfaces is enhanced by factors of up to 10^5. SERS of polymer-protected Ag was first reported by Lee and Meisel (56). In their study, Ag_2SO_4 aqueous solution was reduced by H_2 or $NaBH_4$ in the presence of polyvinylalcohol (PVA) and boiled for 1–3 h. The colloid prepared by this method showed an absorption maximum at 400 nm. Strong Raman signals could be easily observed from a dye molecule (a carbocyanine derivative) in the presence of PVA-Ag colloid, although attempts to observe SERS from several other dyes yielded negative results. Siiman et al. investigated SERS on PVP-protected Ag colloids prepared by the ethanol reduction procedure, which was in fact developed in our laboratory (57). In their study, dabsyl aspartate, a derivative of methyl orange, was used to probe the combined surface Raman and resonance Raman signals on account of its facile binding to PVP. They found that Ag colloids protected by PVP during the reduction process could show an SERS effect when excess base was present. The requirement of base seemed to be linked to the promotion of growth and aggregation of the silver particles, as evidenced by adsorption spectra.

9.1.3.3 Structure

Structural information of monometallic and bimetallic nanoparticles can be mainly obtained by x-ray methods, since each metal element has its own x-ray absorption

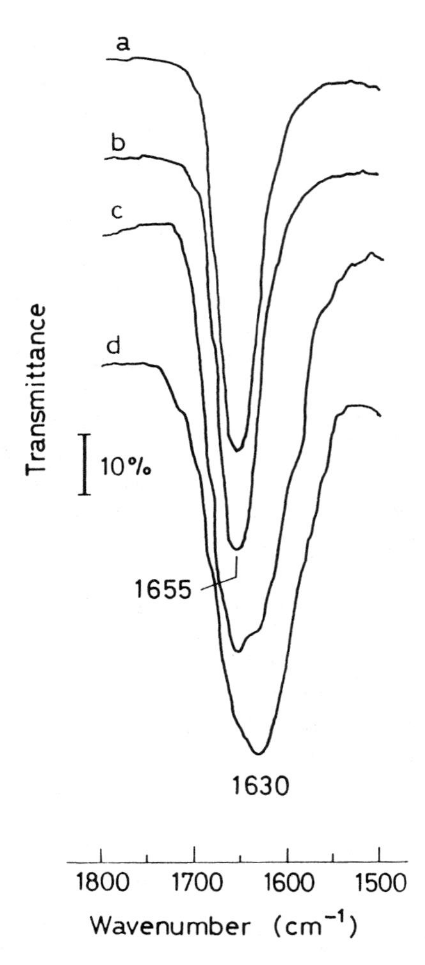

Fig. 9.1.9 Infrared spectra of a poly(*N*-vinyl-2-pyrrolidone)(PVP) film (a) and those involving Pd nanoparticles at the mole ratios of PVP/Pd (*R*) of 41 (b), 5 (c), and 2.25 (d). (From Ref. 55.)

spectrum. Crystallinity of nanoparticles can be considered with x-ray diffraction data. Other common spectroscopic methods are used as well.

X-Ray Diffraction. X-ray diffraction (XRD) is one of the most common methods to analyze the crystal structure of inorganic compounds. Thus, XRD is used for analysis of metal nanoparticles.

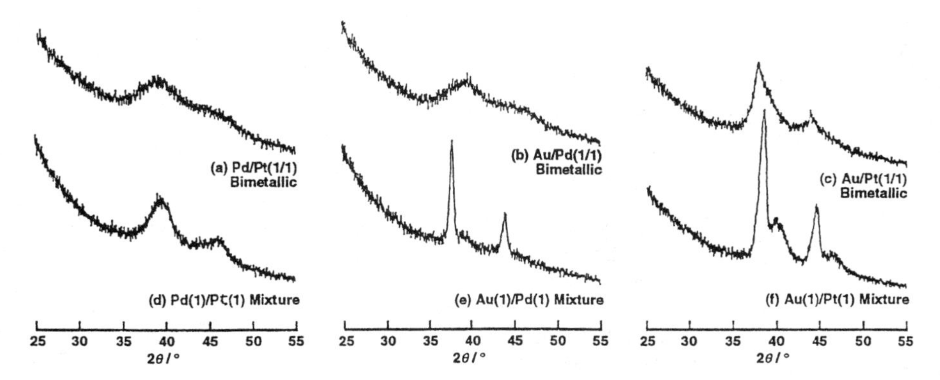

Fig. 9.1.10 X-ray diffractograms of PVP-stabilized (a) Pd/Pt (1/1), (b) Au/Pd (1/1), and (c) Au/Pt (1/1) bimetallic nanoparticles, and physical mixtures of PVP-stabilized monometallic nanoparticles of (d) Pd(1)/Pt(1), (e) Au(1)/Pd(1), and (f) Au(1)/Pt(1). (From Ref. 53.)

The examples are shown in Figure 9.1.10, which gives x-ray diffractograms of three types of physical mixtures of PVP-stabilized Pd, Pt, and Au monometallic nanoparticles, and the corresponding PVP-stabilized bimetallic nanoparticles (53). The diffraction patterns of the physical mixtures are consistent with the sum of two individual patterns, and are clearly different from those of the bimetallic nanoparticles, which have two broader peaks, indicating that several interatomic lengths exist in a single particle. By XRD one can easily understand if the obtained ''multimetallic'' nanoparticles have an alloy structure or are simple physical mixtures of monometallic particles.

X-Ray Photoelectron Spectrometry. X-ray photoelectron spectrometry (XPS) was applied to analyses of the surface composition of polymer-stabilized metal nanoparticles, which was mentioned in the previous section. This is true in the case of bimetallic nanoparticles as well. In addition, the XPS data can support the structural analyses proposed by EXAFS, which often have considerably wide errors. Quantitative XPS data analyses can be carried out by using an intensity factor of each element. Since the photoelectron emitted by x-ray irradiation is measured in XPS, elements located near the surface can preferentially be detected. The quantitative analysis data of PVP-stabilized bimetallic nanoparticles at a 1/1 (mol/mol) ratio are collected in Table 9.1.1. For example, the composition of Pd and Pt near the surface of PVP-stabilized Pd/Pt bimetallic nanoparticles is calculated to be Pd/Pt = 2.06/1 (mol/mol) by XPS as shown in Table 9.1.1, while the metal composition charged for the preparation is 1/1. Thus, Pd is preferentially detected, suggesting the Pd-shell structure. This result supports the ''Pt-core/Pd-shell'' structure. The similar consideration results in the ''Au-core/Pd-shell'' and ''Au-core/Pt-shell'' structure for PVP-stabilized Au/Pd and Au/Pt bimetallic nanoparticles, respectively (53).

Table 9.1.1 Relative Intensity in Quantitative XPS Analyses of the Component Metal Elements of PVP-Stabilized Core/Shell Structured 1/1 (mol/mol) Bimetallic Nanoparticles

Nanoparticle	Relative intensity
Au-core/Pd-shell	Pd/Au = 1.84
Au-core/Pt-shell	Pt/Au = 1.57
Pt-core/Pd-shell	Pd/Pt = 2.06

EXAFS (Extended X-Ray Absorption Fine Structure). Characterization of the surface of metal nanoparticles had been limited to chemical methods, e.g., chemisorption of hydrogen and carbon monoxide. In 1970s, the situation was surprisingly changed due to the advances in x-ray absorption spectroscopy, especially extended x-ray absorption fine structure (EXAFS). Advances in this method have been achieved with the use of synchrotron radiation, which runs effectively at Tsukuba (Japan), Grenoble (France), etc. Now it is one of the most valuable tools to get structural information on bimetallic nanoparticles.

The EXAFS technique was used to establish the structural model of several kinds of PVP-stabilized bimetallic nanoparticles (58). PVP-stabilized Pd/Pt nanoparticles were chosen at the first stage because they are relatively small and uniform in size; especially, the average diameter of the particles with the composition of Pd/Pt = 4/1 (mol/mol) was as small as 1.5 nm (each particle should have only 55 atoms in it) (59). Furthermore, the absorption edge of Pd and that of Pt are far enough apart. EXAFS measurements were thus performed on the Pd K-edge and Pt L_3-edge, providing the coordination number around Pd and Pt atoms, respectively (25,59). At the first step, some monometallic and alloy bulk references were examined by EXAFS technique to understand the difference of values of the contributions of different bonds, i.e., Pd–Pd, Pd–Pt, Pt–Pd, and Pt–Pt. At the same time, the corresponding monometallic Pd and Pt nanoparticles were also analyzed. From these analyses, some fitting parameters, such as ΔE (the difference between the theoretical and experimental threshold energies) and σ (the Debye–Waller factor of the coordination shell), which will be used for two-shell curve fitting of Pd/Pt bimetallic nanoparticles with higher accuracy, were obtained (25). Fourier transforms of Pd K-edge EXAFS of PVP-stabilized Pd/Pt bimetallic nanoparticles at various Pd/Pt ratios are shown in Figure 9.1.11. When the Pd/Pt ratio decreases, the main peak, assigned as a Pd–Pd bond, splits into two peaks, mainly attributed to Pd–Pt and Pd–Pd bonds. Coordination numbers of PVP-stabilized Pd/Pt (4/1) bimetallic nanoparticles determined by the EXAFS method are shown in Table 9.1.2. Each Pd/Pt (4/1) bimetallic nanoparticle is estimated to contains 55 atoms in it and has a three-

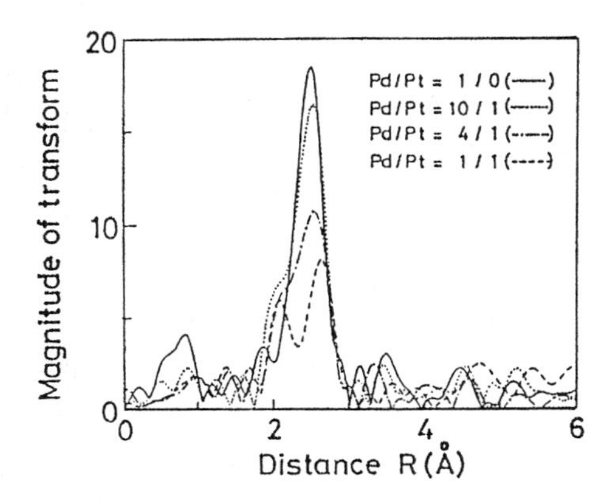

Fig. 9.1.11 Fourier-transformed EXAFS spectra at Pd K-edge of PVP-stabilized Pd/Pt bimetallic nanoparticles at Pd/Pt ratio = 1/0, 10/1, 4/1, and 1/1. (From Ref. 25a.)

layered (two shell type) fcc structure on the basis of the average diameter of 1.5 nm. Two structural models were constructed for this particle, as shown in Figure 9.1.12 (25,59). In the random medel, Pd and Pt atoms are located at random in a particle, and the coordination numbers can be determined only by the composition of the two elements. In the other model, the Pt-core/Pd-shell model, Pt atoms are located in the central core and Pd atoms are near the surface. The number of atoms in the two-layered core is 13 (24%), and 42 atoms (76%) should be located in the surface layer. In this model, the calculated coordination numbers are mostly consistent with the experimentally observed values. In a similar way, the structural model of Pd/Pt (1/1) bimetallic nanoparticles can be constructed (25). These considerations are the first report of complete determination of the structure of metal nanoparticles by using an EXAFS technique.

Table 9.1.2 Coordination Numbers of PVP-Stabilized Pd/Pt (4/1) Bimetallic Nanoparticles and Those Calculated for the Models

			Coordination number N		
Sample	Edge	Bond	Observed	Pt core	Random
Pd/Pt (4/1)	Pd-K	Pd-Pd	4.4 ± 1.0	4.6	6.0
		Pd-Pt	2.3 ± 1.3	2.0	1.9
	Pt-L_3	Pt-Pt	5.5 ± 1.7	5.5	1.9
		Pt-Pd	3.5 ± 1.5	6.5	6.0

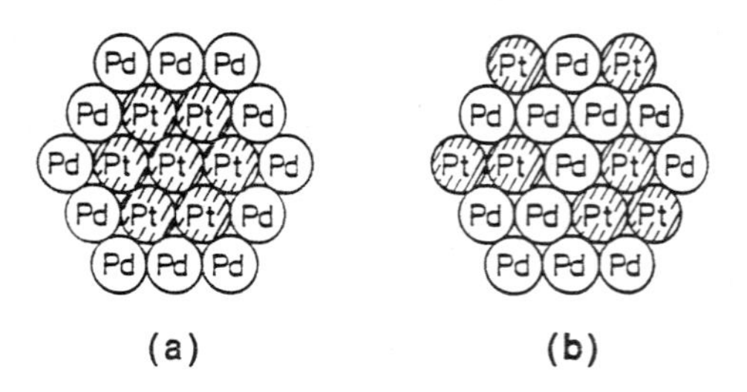

Fig. 9.1.12 Cross section of the model of Pd/Pt(4/1) bimetallic nanoparticle: (a) Pt-core/Pd-shell model and (b) random model. (From Ref. 25a.)

Au/Pd (60) and Pd/Rh bimetallic nanoparticles (61) and the oxidized state of Pd/Pt bimetallic nanoparticles (62) were also investigated by the similar way.

Electron Probe Microanalysis, Energy-Dispersive X-Ray Microanalysis. Electron probe microanalysis (EPMA), or energy-dispersive x-ray microanalysis (EDX, EDAX) is an effective x-ray spectroscopy, especially for microanalysis. Characteristic x-rays are emitted from the sample, after being exposed to an x-ray or electron beam, and the emission is detected to characterize metal particles. Li-drifted Si semiconductor is usually used as an effective detector, converting x-ray photons to current pulses, which are proportional to x-ray energy. Such a system is called EDX or EDAX. This detection system provides higher energy resolution than the detectors used in other wave dispersive x-ray spectroscopies. Thanks to this detector, microarea analysis can be carried out. This system is usually attached to transmission electron microscopy (TEM) or scanning electron microscopy (SEM) for local analyses, because x-rays are one of the by-products in an electron microscope.

The composition of PVP-stabilized Ag/Pd bimetallic nanoparticles, prepared from silver(I) bis(oxalate)palladate(II), was determined by this method (45). The spatial resolution was ~5 nm. The EDX profile showed that each particle has both elements in it, suggesting the formation of bimetallic nanoparticles.

By controlling the size and area of the spot where electron beam is irradiated, the local composition of the metal elements of bimetallic nanoparticles can also be determined (63).

Small-Angle X-Ray Scattering. Small-angle x-ray scattering (SAXS) has been used as one of the most powerful techniques in structural analyses of heterogeneous materials. This technique has been applied to various fields in complex fluids like colloids, surfactants, emulsions, polymers, liquid crystals, membranes, etc., as well as those in the solid state. Heterogeneities occurring in the length scale ranging from 1.0 nm to 100 nm have usually been studied by this technique. Thus, the SAXS technique can be applied to yield further insight into the aggregated structure of polymer-stabilized metal nanoparticles.

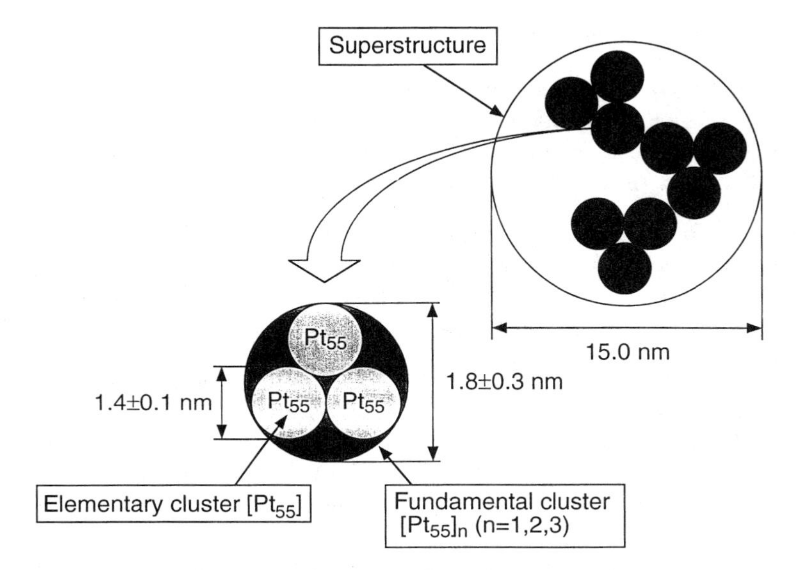

Fig. 9.1.13 Schematic model for the hierarchy in organization of the dispersion of PVP-stabilized Pt nanoparticles. (From Ref. 64.)

For example, the aggregated structures of the solutions containing polymer–metal complexes and the colloidal dispersions of metal nanoparticles stabilized by polymers have been analyzed quantitatively (64). SAXS analyses of colloidal dispersions of Pt, Rh, and Pt/Rh (1/1) nanoparticles stabilized by PVP have indicated that spatial distributions of metal nanoparticles in colloidal dispersions are different from each other. The superstructure (greater than 10.0 nm in diameter), with average size highly dependent on the metal element employed, is proposed. These superstructures are composed of several fundamental clusters with a diameter of ~2.0–4.0 nm, as shown in Figure 9.1.13 for PVP-stabilized Pt nanoparticles.

Infrared Spectroscopy. Infrared (IR) spectroscopy is also used for understanding the structure of the bimetallic nanoparticles. Carbon monoxide can be adsorbed on the surface of metals, and the IR spectra of the adsorbed CO depend on the kind of metal. These properties are used for analyzing the surface structure of metal nanoparticles. The inverted core/shell structure, constructed by sacrificial hydrogen reduction, was probed by this technique (44).

9.1.4 Mechanism of Formation of Metal Nanoparticles

Here the mechanism to form metal nanoparticles from metal ions is discussed. The formation mechanism of core/shell structured bimetallic nanoparticles is described as well (2,65).

9.1.4.1 General Consideration

This section is concerned with the mechanism of formation and growth of metal nanoparticles in homogeneous solutions. As mentioned in the section on synthesis of metal nanoparticles, there are many kinds of synthetic methods. Each method has its own process and mechanism. However, there are four main processes:

1. Reduction of metal ions to metal atoms.
2. Aggregation of metal atoms to form metal nuclei.
3. Growth of metal nuclei to metal nanoparticles.
4. Stabilization of metal nanoparticles by stabilizers.

In some cases, reduction, aggregation, and growth occur at once. Stabilizers such as surfactants, polymers, etc. can be added to the solution before or after the growth of metal nanoparticles.

The particular case in which water-soluble polymers are added to aqueous solutions of metal ions at an initial stage is discussed next.

9.1.4.2 Mechanism of Formation of Monometallic Nanoparticles

Reduction of Metal Ions to Metal Atoms. Usually the synthesis of metal nanoparticles starts from reduction of metal ions to metal atoms. The reduction potential of metal ions depends on the kind of metal ions and on the environment surrounding metal ions, such as ligands and solvents. The reduction can occur in the order of the reduction potential. For example, if a weakly coordinating stabilizer like PVP is used, precious metal ions can be easily reduced by refluxing alcohol. If a strongly coordinating ligand like triphenylphosphine is used, however, precious metal ions cannot be reduced by refluxing alcohol. A stronger reductant, like $NaBH_4$, is required for the reduction. In other words, the first step is not the reduction of naked metal ions, but the reduction of stabilizer-coordinating metal ions.

Photoirradiation or heat treatment is also applied for the reduction of metal ions. In these cases, enough energy to cleave the metal–ligand bond is required for the reduction. Theoretically, the absorption band attributed to the charge transfer from ligand to metal must be excited by the photoirradiation. The coexistence of alcohol can assist the photoreduction. In this case, the corrdination of the alcohol to metal ions may decrease the reduction potential, or the ketyl radical produced may reduce the metal ions as shown in Scheme 9.1.7.

Aggregation of Metal Atoms to Form Metal Nuclei. After the reduction of metal ions, the solution colored by metal ions becomes colorless. At this stage, metal nanoparticles are not produced because the color of the dispersions of metal nanoparticles does not appear at all. If coin metal is used, this color change is much more clear than for the other precious metals.

$$CH_3COCH_3 \xrightarrow{h\nu} [CH_3COCH_3]^*$$

$$[CH_3COCH_3]^* + CH_3CH_2OH \longrightarrow CH_3\dot{C}(OH)CH_3 + CH_3\dot{C}HOH$$

$$M^{n+} + n\ CH_3\dot{C}(OH){\cdot}CH_3 \longrightarrow M^0 + n\ CH_3CO{\cdot}CH_3 + n\ H^+$$

$$M^{n+} + n\ CH_3\dot{C}HOH \longrightarrow M^0 + n\ CH_3CHO + n\ H^+$$

Scheme 9.1.7

Thus, the color change from pale to colorless means the reduction of metal ions to metal atoms or microclusters that have no color at all. There is no evidence of whether the solution consists of single atoms or microclusters. However, it is clear at least that further treatment of this solution with heat or photons can produce a colloidal dispersion of metal nanoparticles, and that, in contrast, the treatment of this solution with oxygen at room temperature provides the colored solution of metal ions. These observations again support the presence of atoms or microclusters in the solution at this stage.

Aggregation of the atoms or microclusters may give metal nuclei. The microcluster itself may work as the nucleus. Although the size of microcluster or nucleus is not clear, the nucleus may consist of 13 atoms, which is the smallest magic number. This idea may be supported by the structural analysis of PVP-stabilized Pt nanoparticles (64) and other systems. In fact, a particle of 13 atoms is considered an elemental cluster. In the case of preparation of PVP-stabilized Rh nanoparticle dispersions by alcohol reduction, formation of very tiny particles, the average diameter of which is estimated to be 0.8 nm, was observed (66). These tiny particles in the metastable state may consist of 13 atoms each and easily increase in size to the rather nanoparticles with average diameter of 1.4 nm, i.e., the particles composed of 55 atoms. This observation again supports the idea that the elemental cluster of 13 atoms is the nucleus.

Growth of Nuclei to Metal Nanoparticles. If the elemental cluster of 13 atoms is the nucleus, the growth of nuclei to metal nanoparticles could proceed by deposition of atoms or microclusters on the surface of nuclei. This process is understandable based on the consideration of the formation of monodispersed nanoparticles. However, structural analysis has often proposed the aggregation of elemental clusters to form fundamental clusters (64). A similar idea is discussed for the structural analysis of bimetallic nanoparticles with cluster-in-cluster structure (40,61).

Thus, we have to consider two processes for the formation of nanoparticles from nuclei or elemental clusters:

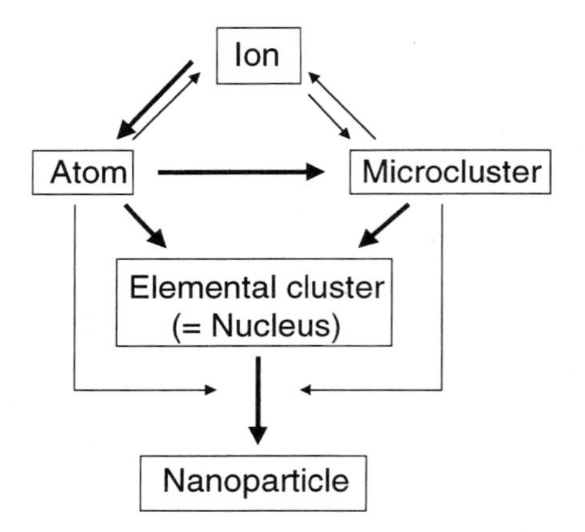

Fig. 9.1.14 Formation process of metal nanoparticle from metal ion in homogeneous solution.

1. Deposition of atoms or microclusters on the nuclei.
2. Aggregation of nuclei or elemental clusters.

Nobody knows which process is more reasonable. The process may depend on the reaction conditions. In fact the successive growing is successful only in the special case (67).

Based on the preceding considerations, the possible formation process of metal nanoparticles from metal ions can be summarized as illustrated in Figure 9.1.14.

9.1.4.3 Mechanism of Formation of Bimetallic Nanoparticles by Coreduction

In the previous section a formation process of monometallic nanoparticles was proposed. Now formation process of bimetallic nanoparticles is considered. The bimetallic nanoparticles systhesized by coreduction have so-called core/shell structure. Thus, the formation mechanism of this core/shell structure is of great interest.

Coreduction of Au and Pt ions by refluxing alcohol in the presence of PVP gives the colloidal dispersions of Au-core/Pt-shell structured bimetallic nanoparticles, as mentioned before. The formation of this bimetallic nanoparticles was traced by in situ UV-Vis spectra (68). The spectral change is shown in Figure 9.1.15, in which the peaks ascertained to be the metal ions disappear at first, and then the broad tailing peaks due to the colloidal dispersions appear. More precisely speaking, the tetrachloroauric(III) acid (at ~320 nm) is reduced first, followed by reduction of hexachloroplatinic(IV) acid (at ~265 nm). This order of reduction is consistent with the standard redox potential of the two metal ions. After the reduction of two

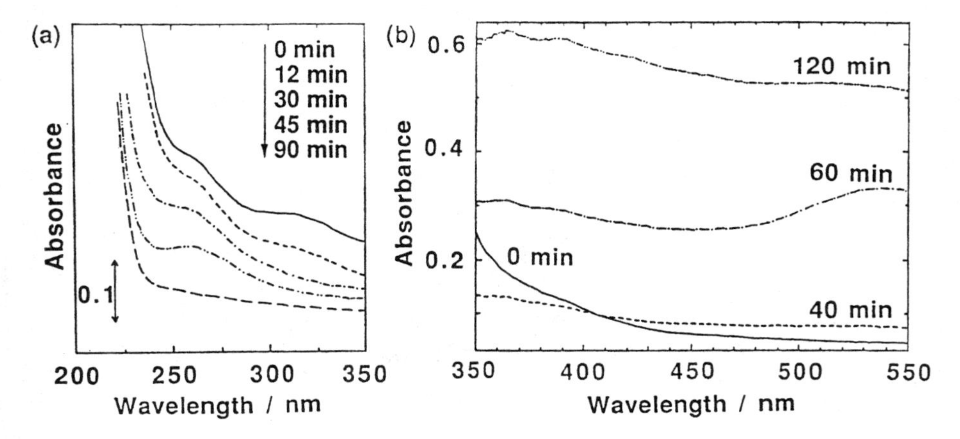

Fig. 9.1.15 UV-visible spectra during the formation of Au/Pt (1/) bimetallic nanoparticles in water/ethanol in the presence of PVP: (a) $\lambda < 350$ nm, (b) $\lambda > 350$ nm.

metal ions, the gold atoms or microclusters aggregate at first, preceding the plasmon absorption peak due to the gold particles, and then the deposition of platinum atoms or microclusters on the gold nuclei follows, resulting in the tailing absorption in the whole area due to the platinum nanoparticles with decreasing plasmon absorption due to the gold particles. In this way the Au-core/Pt-shell bimetallic nanoparticles have been produced (2,68). This process can be illustrated in Figure 9.1.16 (68).

In conclusion, the order of reduction of metal ions is controlled by their redox potential. This is also true in other pairs of precious metals such as Pd/Pt, Au/Pd, etc. (53). In addition, poly(*N*-vinyl-2-pyrrolidone) (PVP) plays an important role for the formation of the core/shell structure. In the case of the Au/Pt system, the aggregation starts from Au but not Pt. This is probably due to the coordinating ability of metals to PVP. The Pt atoms or microclusters coordinating to PVP are more stable than the Au atoms or microclusters, since Au cannot coordinate to PVP. Thus, Au atoms or microcluster aggregate at first after the reduction, and then Pt atoms or microclusters deposit on the Au nuclei. In summary, the core/shell structure is controlled by (1) the redox potential of metal ions, and (2) the coordination ability of metals to PVP, stabilizing polymer.

9.1.4.4 Stabilization of Metal Nanoparticles

Stabilizers play a very important role in the formation of metal nanoparticles. Metal nanoparticles are often synthesized in the presence of stabilizers. This means that the stabilizers play a role not only at the last stage but also during the formation of nanoparticles. In other words, the stabilizers have an interaction not only with metal nanoparticles but also with metal ions, atoms, microclusters, and elemental clusters.

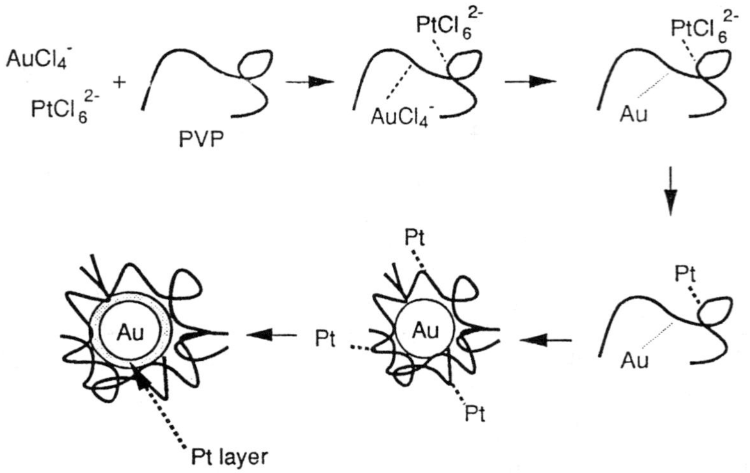

Fig. 9.1.16 The formation process of the PVP-stabilized Au-core/Pt-shell structured bimetallic nanoparticles.

The kinds of interaction depend on the stabilizer. They involve hydrophobic/hydrophilic interaction, electrostatic or ionic interaction, coordination, and hydrogen bonding. PVP is often used as a synthetic stabilizer. In this case, hydrophobic/hydrophilic interaction was thought to be the main interaction. However, this may not true (55). Even though it is weak, the coordination ability of PVP plays an important role not only in stabilizing the metal nanoparticles but also in controlling the core/shell structure in bimetallic nanoparticles (53). Each interaction is weak, but the interaction occurs simultaneously at may points. This is the advantage of PVP. Because of this property, PVP-stabilized metal nanoparticles can work as an excellent catalyst for many organic reactions in solution (65).

9.1.5 Concluding Remarks

Colloidal dispersions of fine metal particles have a long history. Metal nanoparticles are now in the spotlight because of recent developments in nanometer-scale science and technology. Especially the precise structure of the monodispersed bimetallic nanoparticles has become clear quite recently, thanks to the development of EXAFS technology. The mechanism of formation, growth, and structure control is not completely clear yet. In some parts, especially in Section 9.1.4, the discussion may be speculative but is based on the experience of the present author for over 20 years.

Metal nanoparticles have special properties bridging between bulk metal and metal atoms or ions (65). They are the basis of the promising advanced materials.

They will play an important role in miniaturization of electronic devices soon. This information is expected to be helpful for investigators in these fields as well.

REFERENCES TO SECTION 9.1

1. G Schmid. Clusters and Colloids. Weinheim: VCH, 1994.
2. N Toshima, T Yonezawa. New J Chem 22:1179, 1998.
3a. HH Huang, FQ Yan, YM Kek, CH Chew, GQ Xu, W Ji, PS Oh, SH Tang. Langmuir, 13:172, 1997.
3b. SR Emory, WE Haskins, S Nie. J Am Chem Soc 120:8009, 1998.
3c. RS Ingram, MJ Hosteller, RW Murray, T-G Schaaff, JT Khoury, RL Whetten, TP Bigioui, DK Guthrie, PN First. J Am Chem Soc 119:9279, 1997.
4. A Yamagishi, Y Fukushima, eds. Nanostructured Materials in Biological and Artificial Systems. New York: Elsevier, 1998.
5. M Faraday. Philos Trans R Soc Lond 147:145, 1857.
6. J Turkevich, PC Stevenson, J Hiller. Discuss Faraday Soc 11:55, 1951.
7. H Hirai. J Macromol Sci Chem A13:633, 1979.
8. M Ohtaki, N Toshima. Chem Lett 489, 1990.
9. K Esumi, A Suzaki, N Aihara, K Usui, K Torigoe. Langmuir 14:3157, 1998.
10. N Toshima, T Takahashi, H Hirai. Chem Lett 1245, 1985.
11. A Henglein. Chem Mater 10:444, 1998.
12a. H Hirai, H Wakabayashi, M Komiyama. Bull Chem Soc Jpn 59:367, 1986.
12b. H Hirai, H Wakabayashi, M Komiyama. Bull Chem Soc Jpn 59:545, 1986.
13a. H Bönnemann. Advanced Catalysts and Nanostructured Materials. W Moser, ed. New York: Academic Press, 1996, Chap. 6.
13b. H Bönnemann, W Brijoux, R Brinkmann, E Dinjus, T Joußen, B Korall. Angew Chem Int Ed Engl 30:1312, 1991.
13c. H Bönnemann, W Brijoux, R Brinkmann, R Fretzen, T Joussen, R Köppler, B Korall, P Neiteler, J Richter. J Mol Catal 86:129, 1994.
13d. H Bönnemann, GA Braun. Chem Eur J 3:1200, 1997.
14. T Teranishi, M Miyake. Hyomen 35:459, 1997.
15a. LD Rampino, FF Nord. J Am Chem Soc 63:2745, 1941.
15b. WP Dunworth, FF Nord. Adv Catal 6:125, 1954.
16. J Kiwi, M Grätzel. J Am Chem Soc 101:7214, 1979.
17a. JS Bradley, JM Millar, EW Hill, S Behal, B Chaudret, A Duteil. Faraday Discuss 92: 255, 1991.
17b. A Duteil, R Queau, B Chaudret, R Mazel, C Roucau, JS Bradley. Chem Mater 5:341, 1993.
18. LN Lewis, RJ Uriarte, N Lewis. J Catal 127:67, 1991.
19. MT Reetz, W Helbig. J Am Chem Soc 116:7401, 1994.
20. JH Sinfelt. J Catal 29:308, 1973.
21. JH Sinfelt. Acc Chem Res 20:134, 1987.
22. JH Sinfelt, GH Via, FW Lytle. J Chem Phys 72:4832, 1980.
23. JH Sinfelt, GH Via, FW Lytle. J Chem Phys 76:4832, 1982.
24. RS Miner, S Namba, J Turkevich. In: T Seiyama, K Tanabe, eds. Proc 7th Int Congr Catalysis. Tokyo: Kodansha, 1981, p 160.

25a. N Toshima, M Harada, T Yonezawa, K Kushihashi, K Asakura. J Phys Chem 95:7448, 1991.
25b. N Toshima, K Kushihashi, T Yonezawa, H Hirai. Chem Lett 1769, 1989.
25c. N Toshima, T Yonezawa, K Kushihashi. J Chem Soc Faraday Trans 89:2537, 1991.
26. D Richard, JW Couves, JM Thomas. Faraday Discuss Chem Soc 92:109, 1991.
27. B Xhao, N Toshima. Chem Express 5:721, 1990.
28a. N Toshima, Y Wang. Chem Lett 1611, 1993.
28b. N Toshima, Y Wang. Langmuir 10:4574, 1994.
28c. N Toshima, Y Wang. Adv Mater 6:245, 1994.
29a. N Toshima, P Lu. Chem Lett 729, 1996.
29b. P Lu, T Teranishi, K Asakura, M Miyake, N Toshima. J Phys Chem B 103:9673, 1999.
30. JS Bradley, EW Hill, C Klein, B Chaudret, A Cuteil. Chem Mater 5:254, 1993.
31. K Esumi, T Tano, K Torigoe, K Meguro. Chem Mater 2:564, 1990.
32. A Henglein. J Phys Chem 97:5457, 1993.
33. A Henglein, P Mulvaney, A Holzwarth, TE Sosebee, A Fojtik. Ber Bunsenges Phys Chem 96:2411, 1992.
34a. L Katsikas, M Gutierrex, A Henglein. J Phys Chem 100:11203, 1996.
34b. A Henglein, C Bransewicz. Chem Mater 9:2164, 1997.
35. A Harriman. J Chem Soc Chem Commun 24, 1990.
36. T Yonezawa, N Toshima. J Mol Catal 83:167, 1993.
37. TJ Schmid, M Noeske, HA Gasteiger, RJ Behm, P Britz, W Brijoux, H Bönnemann. Langmuir 13:2591, 1997.
38. LE Aleandri, H Bönnemann, DJ Johes, J Richter, J Roziere. J Mater Chem 5:749, 1995.
39. J Turkevich, G Kim. Science 169:873, 1970.
40. M Harada, K Asakura, N Toshima. J Phys Chem 87:5103, 1993.
41. G Schmid, H West, J-O Malm, J-O Bovin, C Grenthe. Chem Eur J 2:1099, 1996.
42. G Schmid, A Lehnert, J-O Malm, J-O Bovin. Angew Chem Int Ed Engl 30:874, 1991.
43. Y Degani, I Willner. J Chem Soc Perkin Trans 2:37, 1986.
44. Y Wang, N Toshima. J Phys Chem B 101:5301, 1997.
45. K Torigoe, K Esumi. Langmuir 9:1664, 1993.
46. K Torigoe, Y Nakajima, K Esumi. J Phys Chem 97:8304, 1993.
47. VM Deshpande, CS Narasimhan. J Mol Catal 53:L21, 1989.
48a. MT Reetez, W Helbig, SA Quaiser. Chem Mater 7:2227, 1995.
48b. MT Reetz, SA Quaiser. Angew Chem Int Ed Engl 34:2240, 1995.
49. T Imae. In: Chem Soc Jpn ed. Colloid Science IV. Tokyo: Tokyo Kagaku Dojin, 1996, Chap. 4.
50. T Yonezawa, T Tominaga, N Toshima. Langmuir 11:4601, 1995.
51. MT Reetz, W Helbig, SA Quaiser, U Stinming, N Breuer, R Vogal. Science 267:367, 1995.
52. Y Wang, H Liu. Polym Bull 25:139, 1991.
53. T Yonezawa, N Toshima. J Chem Soc Faraday Trans 91:4111, 1995.
54. K Furusawa. In: Chem Soc Jpn ed. Colloid Science IV. Tokyo: Tokyo Kagaku Dojin, 1996, Chap. 8.
55. H Hirai, H Chawanya, N Toshima. Reactive Polym 3:127, 1985.
56. PC Lee, D Meisel. J Phys Chem 86:3391, 1982.
57. O Siiman, A Lepp, M Kerker. Chem Phys Lett 100:163, 1983.

58. N Toshima. Macromol Symp 105:111, 1996.
59. N Toshima, T Yonezawa, M Harada, K Asakura, Y Iwasawa. Chem Lett 815, 1990.
60. N Toshima, M Harada, Y Yamazaki, K Asakura. J Phys Chem 96:9927, 1992.
61. M Harada, K Asakura, N Toshima. J Phys Chem 98:2653, 1994.
62. M Harada, K Asakura, Y Ueki, N Toshima. J Phys Chem 96:9730, 1992.
63. JS Bradley. In: G Schmid, ed. Clusters and Colloids. Weinheim: VCH, 1994, Chap. 6.
64. T Hashimoto, K Saijo, M Harada, N Toshima. J Chem Phys 109:5627, 1998.
65. N Toshima. Shokubai 40:536, 1998.
66. H Hirai, Y Nakao, N Toshima. J Macromol Sci Chem A12:1117, 1978.
67a. T Teranishi, M Hosoe, M Miyake. Adv Mater 9:65, 1997.
67b. T Teranishi, M Miyake. Chem Mater 10:594, 1998.
68. N Toshima, T Yonezawa. Makromol Chem Macromol Symp 59:281, 1992.
69. N Toshima. Supramol Sci 5:395, 1998.

9.2 POLYOL PROCESS

FERNAND FIÉVET
Université Paris 7–Denis Diderot, Paris, France

9.2.1 Introduction

Finely divided, nonagglomerated metal particles with controlled shape and size in the micrometer and submicrometer ranges find extensive applications in various fields, especially in the technology of advanced materials. Among the various methods for preparing powders, those that allow the formation of particles with well-defined morphology are chemical methods. They involve, in most cases, a step of precipitation from a homogeneous solution conducted under kinetically controlled conditions (1,2). Thus, fine metallic particles with a controlled morphology are synthesized either by reduction of dissolved metallic compounds and direct metal precipitation from a solution, or by reduction of a solid precursor obtained through precipitation with the desired final morphology, which is retained during the reduction (3). In the latter case, the precursor is an oxide precipitated in water by "thermal hydrolysis" (4). Among the former methods, the polyol process, in which the liquid polyol acts both as the solvent of the metallic precursor and as the reducing agent, is a suitable route to provide finely dispersed metals with well-defined morphological characteristics in order to fill the growing demand of such materials.

9.2.2 General Principles for the Preparation of Metal Powders by Reduction in Liquid Polyols

9.2.2.1 Reduction in Polyols

Various fine metallic powders can be prepared by the reduction of a suitable precursor in liquid polyols. Easily reducible metals such as the noble metals (Ru, Rh, Pd, Ag, Os, Ir, Pt, Au) and even less easily reducible metals (Co, Ni, Cu, Pb) can be obtained by precipitation in such organic media (5). Lately the process has been extended to the synthesis of iron powders (6) and to polymetallic powders: Ag–Pd (7), Co–Ni (8), Fe-based powders (6), Cu-based powders (9,10). Liquid α-diols such as 1,2-ethane diol (ethylene glycol, EG) or 1,2-propanediol (propylene glycol, PEG) are generally used, but diol ethers resulting from the condensation of α-diols, such as diethylene glycol (DEG), triethylene glycol (TEG), or tetraethylene glycol (TTEG), have been used as well. The starting compound may be rather soluble (nitrate, chloride, acetate) or it may be only slightly soluble (oxide, hydroxide) in the polyol. The solution or the suspension is stirred and heated to a given temperature, which can reach the boiling point of the polyol for less reducible metals; conversely,

Table 9.2.1 Relative Permittivity and Boiling Point Under Atmospheric Pressure of Some Polyols: Comparison with Water and Monoalcohols

	Water	1,2-Ethane diol	1,2-Propane diol	Diethylene glycol	Methanol	Ethanol	1-Octanol
ϵ_r	78.5	38	32	32	33	24	10
T_b (°C)	100	198	189	245	65	78.5	194

for easily reducible metals (e.g., Pd) the reaction can be carried out at temperatures as low as 0°C from a suitable precursor, e.g., $Pd(NO_3)_2$. The polyol acts simultaneously as the solvent, the reducing agent, the crystal growth medium for the metal particles, and in some cases as a complexing agent for the metallic cations.

Liquid polyols are interesting among nonaqueous solvents because, like water and monoalcohols, they are hydrogen-bonded liquids with a high value of relative permittivity (Table 9.2.1), and therefore they are able to dissolve to some extent ionic inorganic compounds. Moreover, reactions can be carried out in such solvents under atmospheric pressure up to 250°C, i.e., at a temperature range higher than in water or monoalcohols such as methanol or ethanol.

Polyols, like monoalcohols, are mild reducing agents that are able to reduce to the zero-valence state ions of the noble metals, copper, and some more electropositive metals such as cobalt or nickel. Polyols are not able to directly reduce Fe(II) or Fe(III) ions to the zero-valence state. Nevertheless, as exemplified later in this chapter, iron or iron-based particles can be obtained through the disproportionation of Fe(II) in polyols. Keeping the advantages of such reducing media to avoid the competitive oxidation into Fe(III), which is observed in water (11), the polyol is also used as a complexing agent to retain Fe(III) resulting from the disproportionation as dissolved complexes and hence to prevent the crystallization of magnetite.

Whatever the metal considered, the reduction reaction always proceeds via the solution rather than in the solid state. Indeed, the morphological characteristics of the various metal powders reported in the next section bring evidence, as discussed in Section 9.2.4, that the metal particles are formed by a reaction involving the following steps: 1) dissolution of the precursor; 2) reduction by the polyol of the dissolved species; and 3) nucleation and growth of the metal particles from the solution. Moreover, in some systems, the formation of a sparingly soluble intermediate phase is observed. The main feature of this process is that it is possible by acting upon the experimental conditions to control the kinetics of the precipitation and thus to obtain nonagglomerated metal particles with a well-defined shape, a controlled average size from micrometer to nanometer size range, and a narrow size distribution. In order to obtain particles with such well-defined morphological characteristics a general condition must be fulfilled: Nucleation and growth must be two completely separated steps (1,12).

9.2.2.2 Control of the Nucleation Step

Spontaneous nucleation occurs when the concentration of the metal generated by the reduction reaches a critical supersaturation level. If the metal is generated slowly and the nucleation rate is high enough, then the sudden nucleation almost immediately lowers the concentration below this critical nucleation level. Under these conditions the nucleation step is very short, and it is followed by the growth of the particles from the original nuclei as long as the metal is slowly generated.

In order to generate the metal slowly, different methods can be used, according to the metal considered. Starting from a very soluble precursor of an easily reducible metal [Ag (13), Au (14)], the reduction has to be carried out at a sufficiently low temperature. When various soluble precursors may be used it is possible to select the most suitable one [Pd (15); cf. Section 9.2.3.1]. In other cases [Co, Ni (8,16), or Cu (17)], it has been possible to control the concentration of the precursor species in solution, with these species being provided by the progressive dissolution of a sparingly solid phase acting as a kind of reservoir. The dissolution equilibrium regulates the release of these species, controls the supersaturation ratio, and then allows one to have a very short nucleation step. Nevertheless, it is possible to have an important yield of metal at the end of the growth despite the low supersaturation ratio. In some cases, to obtain more easily the separation between the nucleation and growth steps and a better control of the average size of the metal particles, homogeneous (spontaneous) nucleation can be replaced by heterogeneous nucleation by seeding the reactive medium with foreign nuclei obtained by adding a suitable nucleating agent (18).

9.2.2.3 Control of the Growth Step

Particle growth may proceed either by a stepwise addition of metal atoms or by coalescence of primary particles that form secondary larger particles. Coalescence of primary particles usually results in polydisperse secondary particles of various shapes. Thus, to obtain metal particles with well-defined morphological characteristics it is generally necessary to prevent the coalescence of the particles during their growth stage. In some cases [Co, Ni (8,16), Fe (6)] the polyol itself, acting as a protective agent, prevents the coalescence of the metal particles. In other cases [Cu (17), noble metals (7,13–15,19)], protective agents have to be used in order to produce steric stabilization. However, in particular cases [e.g., Fe–Ni alloys (8)], it is noteworthy that such a coalescence mechanism gives monodisperse, quasispherical secondary metal particles, as shown in Section 9.2.4.2.

9.2.3 Monodisperse Metal Powders: Preparation by the Polyol Process and Characterization

If it is possible to control the morphological characteristics of the metal particles obtained by reduction in liquid polyols by following the guidelines described in the previous section, the shape and size control of these particles still requires a specific approach for each metal, as shown in this section. Thus, we distinguish four cases regarding the redox potential of the metal:

- Noble metals, which are easily reduced to zero-valence state by polyols.
- Cobalt, nickel, or their alloys, which are more electropositive metals.
- Copper, which has an intermediate redox potential.
- Iron and iron-based powders, for which Fe^0 is obtained through disproportionation of Fe(II) rather than reduction by the polyol itself.

9.2.3.1 Noble Metals

Fine noble metals particles are used in the production of conductive inks and pastes to make electronic components by the thick film technique (20,21). Nonagglomerated particles in the micrometer or submicrometer size range are ideally required for this application (22). It has been shown over recent years that such powders can be made by the polyol process for various noble metals. Under uncontrolled conditions, these metals are obtained by reduction in polyols as polydisperse powders with particles of various shapes arising from chaotic growth by coalescence of tiny primary particles. The polyol itself is not effective enough as a protective agent to prevent the very strong tendency of small noble metals particles to coalesce. In order to obtain a monodisperse system, the growth step must be controlled by adding in the reacting medium a polymeric protective agent that adsorbs onto the metallic particles, ensuring their steric stabilization while they are growing. Polyvinylpyrrolidone (PVP) has been extensively used for this purpose. In general, it was shown that a critical PVP concentration is required to obtain monodisperse powders (13,14). Despite the use of PVP, noble metals powders prepared by the polyol process have a high purity as far as submicrometer-size particles are considered (Table 9.2.2). The precursors used are entirely soluble in the polyol, and no intermediate solid phase can be

Table 9.2.2 Chemical Analysis (wt%) of Some Noble Metal Powders Obtained by the Polyol Process

	Particle shape and size	C	O	N
Au	Quasi-spherical, sub-μm scale	<0.3		<0.2
Ag	Quasi-spherical, sub-μm scale	0.28	0.16	0.1
Ag	Rodlike, μm scale	0.26	0.16	0.09
Ag	Quasi-spherical, nm scale	4.59	2.32	0.83
Pd	Quasi-spherical, sub-μm scale	0.24	0.29	0.15
$Ag_{70}Pd_{30}$ alloy	Quasi-spherical, nm scale	5.3	4.9	1

Source: Ref. 19.

evidenced prior the precipitation of the metallic phases. A fast spontaneous nucleation step can be obtained, for instance in EG, at a temperature lower than the boiling point. In general it is not necessary to proceed through heterogeneous nucleation to achieve the separation of the nucleation and growth steps in order to get monodisperse particles. For various noble metals it has been possible to decrease the size of the particles from the submicrometer range to the nanometer range and then to get stable colloidal solutions by decreasing the precursor/polyol ratio (Fig. 9.2.1) (7,19,23,24). It can be inferred from the impurities content of these colloids (Table 9.2.2) that the particles are coated with residual organic species in spite of repeated washing; hence, air-dried powders are not pyrophoric, despite their very small particle size.

Gold. Polymer-protected gold powders have been synthesized by Silvert et al. (14,19) as follows: $HAuCl_4 \cdot 3H_2O$ (400 mg) dissolved in EG (5 mL) was added to a hot solution of PVP (5 to 12 g) in EG (70 mL). The reaction was carried out for 1 to 4 h, depending on the selected temperature (from the boiling point to 100°C). The vessel was protected to avoid light irradiation of the reacting solution. The powders were recovered by filtration under vacuum and washed with distilled water. The morphology of the particles appeared to be strongly dependent on the reaction temperature. At 100°C the nucleation rate is not high enough to ensure the separation of the nucleation and growth steps, so the system formed by particles or various shape is polydisperse (Fig. 9.2.2a). This separation was achieved in the temperature range 120–150°C as the nucleation rate increased; then monodisperse particles with a regular and polyhedral shape were obtained (Fig. 9.2.2b and c). Their mean diameter decreased from 0.40 μm (relative standard deviation $\sigma = 20\%$) to 0.24 μm ($\sigma = 19\%$) when the temperature was increased from 120°C to 150°C. Owing to the enhancement of the nucleation step with temperature, a larger number of nuclei are produced. Since the initial precursor concentration is the same for all experiments, the final particles are then smaller at high temperature. However, at 170°C the system became polydisperse (Fig. 9.2.2d), because of the PVP concentration being not high enough to ensure the steric stabilization of all the particles grown from the numerous nuclei formed at this temperature.

By a slightly different procedure, Seshadri et al. (25) obtained monodisperse Au particles in the submicrometer size range from reduction of $HAuCl_4$ in a EG-DEG mixture (pH $= 4.5$) in the presence of PVP.

Silver. Silver particles have been obtained generally from silver nitrate (13,23) or carbonate (26) in EG. Nonagglomerated quasi-spherical monodisperse silver particles have been obtained in the size range from 0.37 to 0.81 μm (Fig. 9.2.3a) (13). The experimental procedure was as follows: Silver nitrate dissolved in EG was added to hot ethylene glycol ($AgNO_3$/EG molar ratio $= 2.5 \times 10^{-2}$). Then a solution of PVP in EG was added dropwise by means of a peristaltic pump. The reaction was carried out for 15 min at a constant temperature in the range 160–180°C to ensure a complete reduction. The PVP/$AgNO_3$ ratio has to be optimized for a given reaction

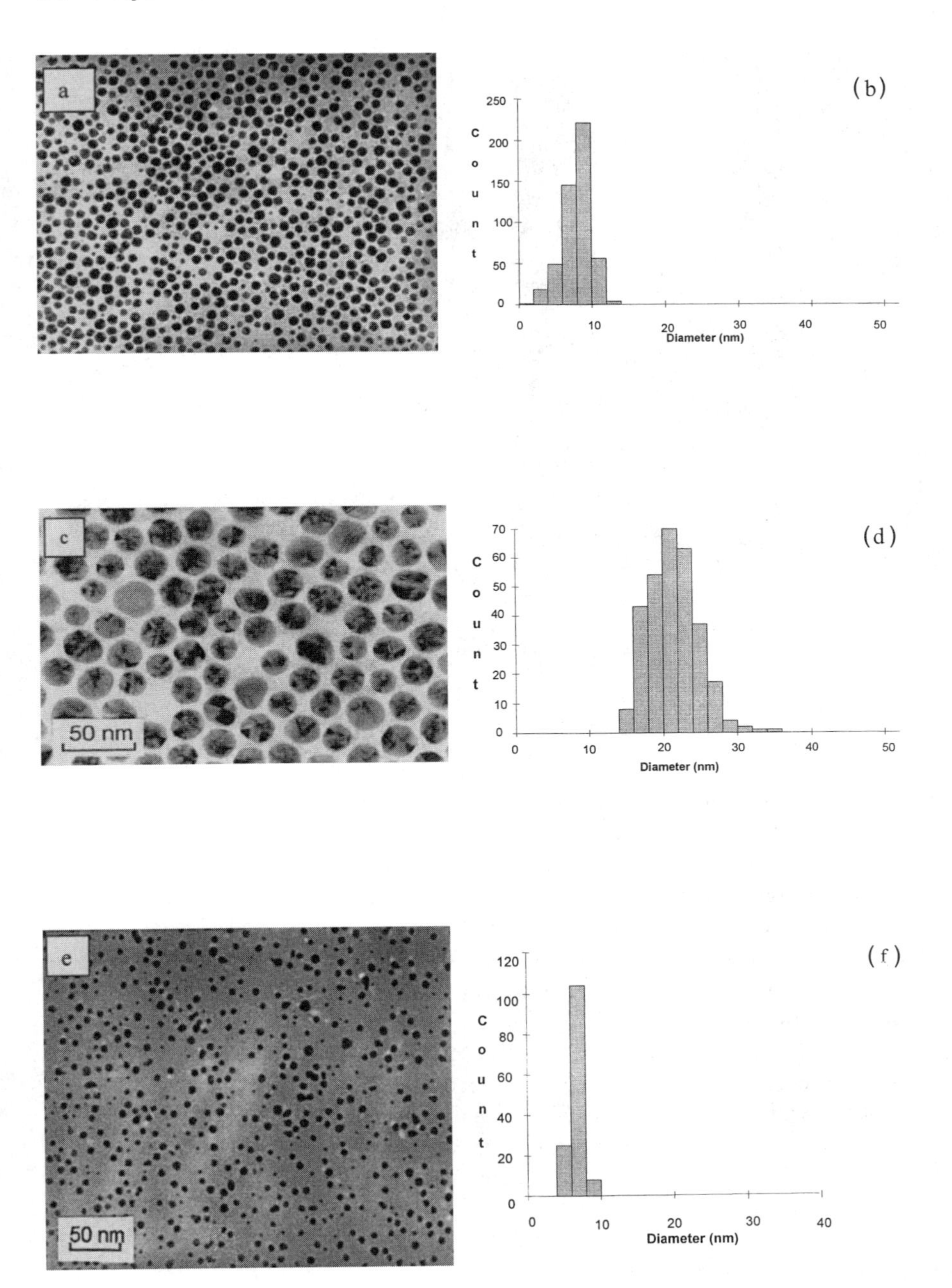

Fig. 9.2.1 TEM images and particle size distribution of monodisperse colloidal dispersions of precious metals: (a, b) Au, (c, d) Ag, (e, f) Ag–Pd alloy. (a, b from Ref. 27; c, d from Ref. 23; e, f from Ref. 7.)

Fig. 9.2.2 SEM images of gold powders obtained by reduction of $HAuCl_4 \cdot 3H_2O$ in ethylene glycol in presence of PVP at different temperatures: (a) 100°C, (b) 120°C, (c) 150°C, (d) 170°C. (From Ref. 14.)

temperature in order to obtain monodisperse systems (Table 9.2.3). The mean diameter was found to increase with temperature.

Nonagglomerated rodlike monodisperse silver particles have been obtained by heterogeneous nucleation (Fig. 9.2.3b) (13). Hexachloroplatinic acid was used as a nucleative agent with a critical Pt/Ag molar ratio equal to 10^{-3}. It was dissolved in EG and added to ethylene glycol at 160°C prior to the addition of both silver nitrate and PVP solutions. In situ reduction of this nucleating agent yields tiny platinum particles, which provide the nuclei for the deposition of metallic silver. This drastic change of the shape of the silver particles when spontaneous nucleation is replaced by heterogeneous nucleation has been tentatively explained by the selective adsorption of the protective agent (PVP) on a particular plane or edge of the nuclei, with the further deposition of metallic silver occurring on those planes not covered by the PVP. Thus, PVP acts in this system as both a protective agent and a crystal habit modifier. Moreover, the thickness of the rodlike silver particles was found to change when the amount of added PVP was varied.

Polymer-protected, monodisperse, nanoscale silver particles (Fig. 9.2.1c and d) have been obtained through spontaneous nucleation by the polyol process as follows (23). PVP (1–25 g) and $AgNO_3$ (50–3200 mg) were dissolved in EG (75 mL) at room temperature. Then the solution was heated up to 120°C at a constant

Fig. 9.2.3 SEM images of monodisperse silver powders obtained by reduction of $AgNO_3$ in ethylene glycol in the presence of PVP: (a) quasi-spherical particles obtained by spontaneous nucleation (d_m = 0.64 μm, σ = 0.13 μm); (b) rodlike particles obtained by heterogeneous nucleation using H_2PtCl_6 as nucleating agent (particle dimensions 3 μm long and 0.5 μm thick). (From Ref. 13.)

Table 9.2.3 Influence of Temperature and of the Relative Amount of the Protective Agent on the Size and Size Distribution of Silver Powders Obtained in Ethylene Glycol

T (°C)	PVP/$AgNO_3$ (wt%)	Mean diameter (μm)	Standard deviation (μm)
160	1/1	0.37	0.09
170	1.2/1	0.64	0.13
180	1.6/1	0.81	0.13

Source: Ref. 13.

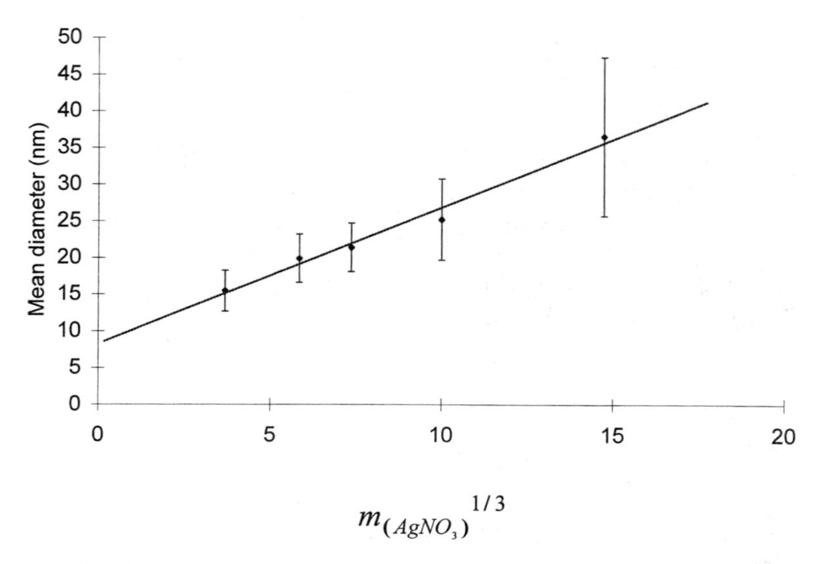

Fig. 9.2.4 Mean diameter of silver particle versus the cubic root of the precursor mass. (From Ref. 23.)

rate (1°C min^{-1}) and the reaction was carried out for 1 h at this temperature; after cooling, the silver colloids can be separated from the polyol after the addition of a large amount of acetone followed by centrifugation and can be eventually redispersed in ethanol. The concentration of PVP has to be optimized for each precursor concentration in order to have the narrowest particle size distribution. The mean particle size was found to increase from 15 to 36 nm with increasing precursor concentration, with a linear variation being observed when this mean diameter is plotted versus the cubic root of the precursor mass (Fig. 9.2.4). This linear variation can be explained assuming the following hypotheses:

- During the short nucleation step the number of nuclei formed at a given temperature does not depend on the amount of $AgNO_3$ introduced.
- During the subsequent growth step each nuclei gives rise to one nonporous final particle without any coalescence between the nuclei or the growing particles.

Consequently,

$$m_{Ag} = n \times 4/3 \times \pi(d/2)^3 \times \rho_{Ag}$$

where m_{Ag} is the mass of silver recovered, d the particle diameter, ρ_{Ag} the silver density, and n the number of silver particles, which is equal to the number of silver nuclei formed and is therefore constant. As m_{Ag} is proportional to m_{AgNO3}, the mass of precursor used, $m_{AgNO3} = kd^3$, in good agreement with the observed relationship.

Palladium. Under uncontrolled conditions, ''palladium blacks'' made up of agglomerated fine particles (average size smaller than 0.1 μm) are produced using the polyol process. Whatever the palladium precursor in hot ethylene glycol, the precipitation of the metal is mainly governed by the nucleation step, which predominates over particle growth with the formation of a very large number of nuclei. Ducamp-Sanguesa et al. (15) succeeded in obtaining palladium particles in the submicrometer range, which are suitable for conductive inks used to make electronic devices. The palladium(II) tetrammine complex was used as a precursor. The nucleation step was limited because the reaction was conducted below 0°C. Nevertheless, at such a low temperature an auxiliary reducing agent (hydrazine) had to be used. The process, which was improved later by Silvert (27), is as follows: $Pd(NH_3)_4^{2+}$ is prepared by adding $PdCl_2$ (100 mg) to ammonia (32wt%, 5 mL). This solution is mixed with the polyol (75 mL), and then hydrazine monohydrate (0.1 mL) is added. After mixing, the reaction is carried out for 1 h at room temperature without stirring. Submicrometer-size monodisperse quasi-spherical particles were obtained (Fig. 9.2.5). The mean diameter, which depends on the polyol, was found to be 0.15 μm, 0.23 μm, and 0.11 μm for EG, DEG, and TEG, respectively. For a given polyol, as for nanoscale silver particles, the mean diameter could be controlled accurately because of its linear variation versus the cubic root of the precursor mass.

Ag–Pd Alloy Powders (7,27). Bimetallic colloids, namely, Ag–Pd and Au–Pt, can be obtained by the polyol process. The composition $Ag_{70}Pd_{30}$ is of particular interest to make the internal electrodes of multilayer ceramic capacitors (MLCC). Polymer-protected, monodisperse, nanoscale $Ag_{70}Pd_{30}$ particles have been obtained

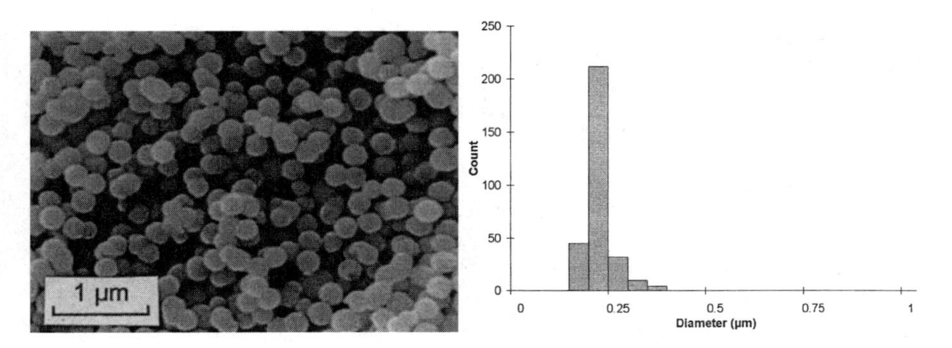

Fig. 9.2.5 SEM image and particle size distribution of a palladium powder obtained by reduction at room temperature of $Pd(NH_3)_4^{2+}$ by N_2H_4 in DEG (d_m = 0.23 μm, σ = 0.03 μm). (From Ref. 27.)

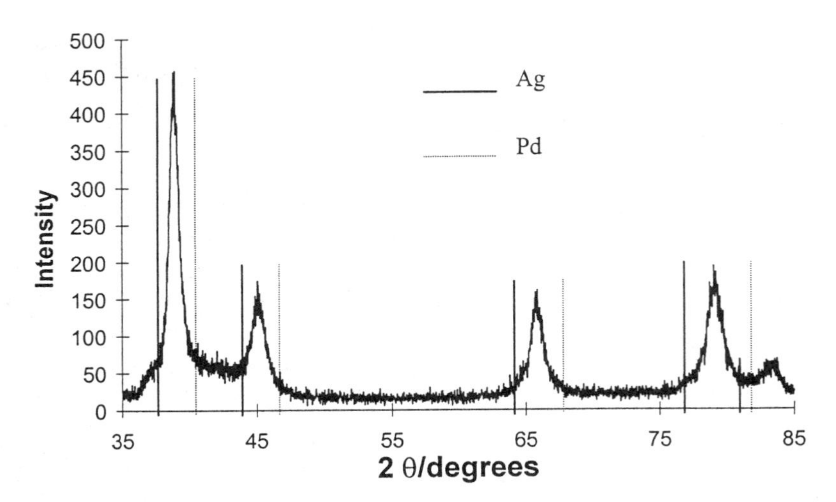

Fig. 9.2.6 X-ray diffraction pattern of PVP protected $Ag_{70}Pd_{30}$ alloy colloids synthesized in ethylene glycol at 120°C. (From Ref. 7.)

by reduction in EG (Fig. 9.2.1e and f). Nitrate salts of both metals are dissolved at room temperature in a solution of PVP in EG; then the solution is heated up to 120°C (1°C min^{-1}) and the reaction is carried out for 4 h. X-ray diffraction shows that these colloids are crystalline alloys as exemplified in Figure 9.2.6. As for silver colloids, the mean diameter can be varied in the nanometer size range by varying the concentration of the precursors.

For each noble metal it appears that reaction parameters such as the nature of the precursor, temperature, and time have to be optimized. Recently a study of the electrochemical reduction of noble metals species in EG was undertaken by Bonnet et al. (28). Better control of the experimental conditions leading to the preparation of monodisperse particles in the nanometer, submicrometer, and micrometer size range for various noble metals is expected from knowledge of the electrochemical fundamentals of the polyol process.

9.2.3.2 Cobalt, Nickel, and Alloys

In order to reduce quantitatively Co or Ni species to the zero-valence state the reaction must be carried out in boiling polyols (EG, PEG, mixture EG + DEG), i.e., at a temperature lying in the range 180–220°C. Owing to the low tendency of the metal particles to coalesce during the growth step, the addition of a polymer is not needed to ensure their steric stabilization. The metal precursor may be either very soluble in the polyol (acetate was used rather than chloride) or only slightly soluble (hydroxide, hydroxycarbonate).

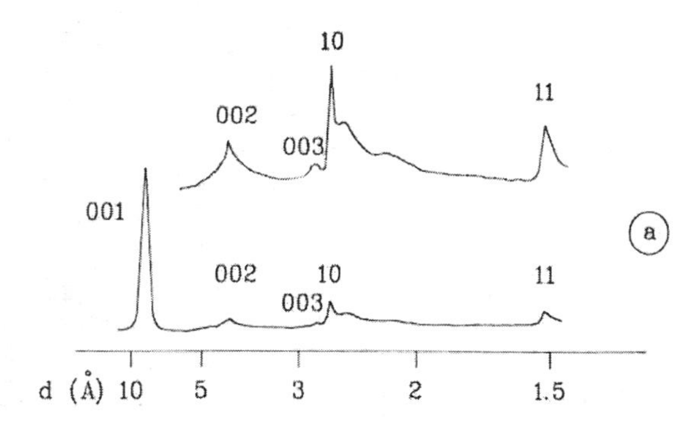

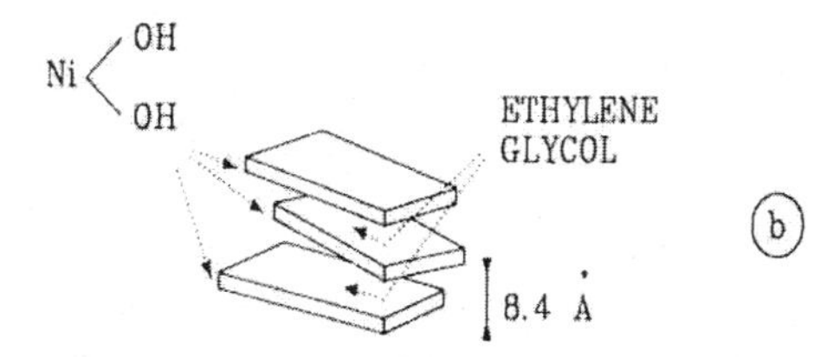

Fig. 9.2.7 Characterization of the intermediate phase obtained in the reduction of $Ni(OH)_2$ in ethylene glycol: (a) x-ray diffraction pattern, and (b) structural model. (From Ref. 29.)

Whatever the precursor, the formation of an intermediate solid phase was always observed. It can be inferred from X-ray diffraction (Fig. 9.2.7) and infrared spectroscopy that this intermediate phase shows a lamellar, incompletely ordered structure (turbostratic structure) built up with parallel and equidistant sheets like those involved in the lamellar structure of the well-crystallized hydroxides $Ni(OH)_2$ or $Co(OH)_2$,; these sheets are disoriented with intercalation of polyol molecules and partial substitution of hydroxide ions by alkoxy ions (29). The dissolution of this solid phase, which acts as a reservoir for the M(II) solvated species, controls the concentration of these species and then plays a significant role in the control of the nucleation of the metal particles and therefore of their final morphological characteristics. For instance, starting from cobalt or nickel hydroxide as precursor in ethylene glycol, the reaction proceeds according to the following scheme (8):

$$M(OH)_{2\,(solid)} \xrightarrow{\text{dissolution}} M^{2+}_{(solvated)} \xrightarrow{\text{reduction}} M^0 \xrightarrow[\text{growth}]{\text{nucleation}} M^0{}_n$$

$$(M = Ni, Co) \qquad \downarrow\uparrow$$

$$\phi_t \text{ solid phase}$$

(Metallic hydroxy-alkoxide with a turbostratic structure)

It is possible that this intermediate solid phase ϕ_t also plays a significant role to prevent the coalescence of the metal particles during their growth.

In order to obtain bimetallic particles with a good homogeneity in composition, according to this scheme, the intermediate phase ϕ_t must act as a reservoir for Co(II) and Ni(II) species as well. Moreover, this phase must be a double hydroxyethylene glycolate with a constant Co/Ni ratio in order to maintain such a ratio in the solution and therefore in the bimetallic particles growing from this solution. These conditions can only be fulfilled if the Co and Ni starting salts are soluble enough in the polyol and if the reduction occurs when almost all the Co(II) and Ni(II) species are involved in the intermediate solid phase ϕ_t. A mixture of acetates is readily soluble in the polyol and then, when the reaction medium is heated up to boil, a complete dissolution of these salts occurs before the precipitation of the phase ϕ_t; however, the precipitation rate of this intermediate solid phase is rather low due to the low concentration of hydroxide ions in the reaction medium, with those ions being provided by the acid–base equilibrium:

$$CH_3CO_2^- + H_2O \rightleftarrows CH_3CO_2H + OH^-$$

In these experimental conditions the precipitation of ϕ_t and the reduction of Co(II) and Ni(II) species occur simultaneously, and then ϕ_t is not able to act as a reservoir for these species. Thus, sodium hydroxide was added in a large excess to the reaction medium in the synthesis of CoNi monodisperse particles in order to provide an excess of hydroxide ions (8,30). This method can be extended to obtain nickel–noble metal alloys. Thus PVP-protected Ni–Pd colloids were obtained by Toshima et al. (31) for different Ni/Pd ratios.

A complete substitution of the spontaneous nucleation step by a heterogeneous one can be achieved easily in the synthesis of Co, Ni, or Co–Ni powders using a noble metal (Ag, Pd, Pt, . . .) salt as nucleating agent (18). Noble metal species are readily reduced by the polyol; hence metallic tiny particles are formed. These particles are preferential sites for the further growth of nickel or/and cobalt. Although micrometer-size particles are obtained by spontaneous nucleation, submicrometer- and nanometer-size ones are obtained through heterogeneous nucleation. Moreover, heterogeneous nucleation allows one to accurately control the mean diameter by varying the amount of nucleating agent, as shown later in this section.

Experimental Conditions of Synthesis

1. *Spontaneous nucleation.* Monodisperse micrometer-size Co or Ni particles have been obtained as follows (16,18): cobalt or nickel hydroxide was suspended in EG (250 mL); the molar ratio hydroxide/polyol was varied from 0.01 to 0.15. The suspension was stirred and brought to boil (195°C). The metal precipitation does not occur immediately when the boiling point is reached but only after an induction time. The water and the volatile products resulting from the oxidation of ethylene glycol were distilled off, while the main part of the polyol was refluxed. For Co the reduction was completed in a few hours. For nickel hydroxide it was difficult to achieve a complete reaction due to an incomplete dissolution of this hydroxide. This could be overcome by adding a small amount (a few milliliters) of

an aqueous solution of sulfuric acid (2.5 *M*) to make the dissolution of the aggregated $Ni(OH)_2$ particles easier.

According to the previous discussion, monodisperse micrometer-size Co_x-Ni_{1-x} particles have been obtained from cobalt and nickel acetate tetrahydrate with the Co/Ni ratio equal to the Co/Ni ratio of the final metal powder in the whole composition range (8). These salts were dissolved in EG (200 mL); sodium hydroxide was dissolved in an identical volume of polyol. The concentrations of the reactants were adjusted to get by mixing the two solutions, a metallic precursor concentration typically 0.2 mol L^{-1} and an NaOH concentration in the range $0.5 - 2$ mol L^{-1}. This solution was stirred and heated up to 195°C; as the temperature rose, a complete precipitation of the phase ϕ_t was achieved at 60°C, whereas the nucleation and subsequent growth of the metallic particles started at 150–160°C. Water and volatile organic products of the reaction were distilled off while the polyol was refluxed. After a few hours the metal was quantitatively precipitated from the solution and the metallic particles were recovered by centrifugation, washed with ethanol, and dried in air at 50°C.

2. *Heterogeneous nucleation.* Co or Ni powders have been synthesized in EG by pouring into the reaction medium, during the induction time, a small amount (10 mL) of a solution of $AgNO_3$ in EG, with the molar ratio Ag/M varying in the range 10^{-2}–10^{-5} (M = Co or Ni) (16). PVP-protected nanoscale Ni particles have also been made from $Ni(OH)_2$ in EG using Pd or Pt as nucleating agent (32).

Co_xNi_{1-x} particles have been obtained from cobalt(II) and nickel(II) acetate tetrahydrate dissolved with sodium hydroxide in PEG (30,33,34). The metal concentration $[Co^{2+}] + [Ni^{2+}]$ was fixed to 0.1 mol L^{-1}, with the sodium hydroxide concentration being 0.25 mol L^{-1}. Heterogeneous nucleation was achieved by adding a small amount of a solution of K_2PtCl_4 or $AgNO_3$ dissolved in a mixture of EG + DEG (1/1 vol), with the molar ratio $[PM]/[Co] + [Ni]$ varying in the range 10^{-6}–10^{-2} (PM = Pt or Ag). The solution containing all the reactants and the nucleating agent was slowly heated up to the boiling point under mechanical stirring and then maintained at this temperature during ~2 h.

Powders Characterization

1. *Morphological characteristics.* Co, Ni, and Co_xNi_{1-x} powders obtained in polyols either through spontaneous or heterogeneous nucleation are composed of equiaxed, nonagglomerated particles with a narrow size distribution. The standard deviation σ is generally lower than $0.2d_m$ (Fig. 9.2.8). The mean diameter of the micrometer-size particles obtained by spontaneous nucleation can be altered by modifying either the temperature or the precursor/polyol ratio. It decreases as the temperature increases. This trend that has been reported previously for gold particles can be explained in much the same way as earlier. When the precursor/polyol molar ratio is increased at a given temperature, the particle size significantly increases. That can be explained as done earlier for silver particles. The particle shape may also be modified in some extent: The application of an external magnetic field during

A

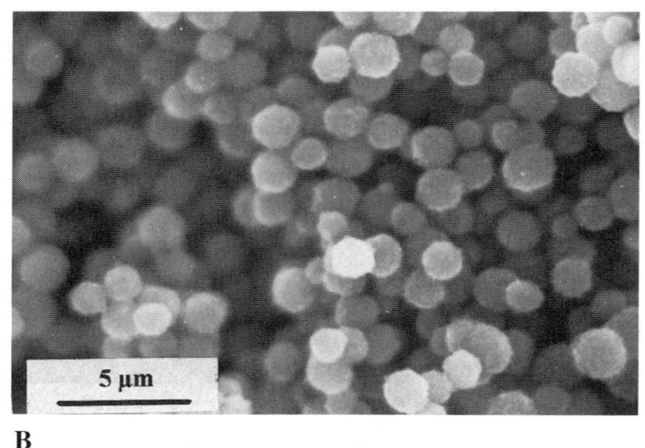

B

Fig. 9.2.8 SEM images of different powders in the CoNi system synthesized by reduction in polyols: (A) Co obtained by spontaneous nucleation, $d_m = 1.75$ μm, $\sigma = 0.14d_m$; (B) $Co_{20}Ni_{80}$ obtained by spontaneous nucleation, $d_m = 1.4$ μm, $\sigma = 0.15d_m$. (C) $Co_{20}Ni_{80}$ obtained by heterogeneous nucleation $d_m = 0.32$ μm, $\sigma = 0.14d_m$; (D) Ni obtained by heterogeneous nucleation, $d_m = 0.20$ μm, $\sigma = 0.30d_m$. (a, d from Ref. 18; b, c from Ref. 8.)

the precipitation leads to metallic filaments rather than spherical particles (Fig. 9.2.9). The magnetic field lines up during the early stages of their growth the small ferromagnetic particles that result from the spontaneous nucleation step. These tiny particles join together and lead to metallic filaments as subsequent growth proceeds. Thus, by varying different experimental factors acting upon the homogeneous nucleation and growth steps, it is possible to control to some extent the size and shape

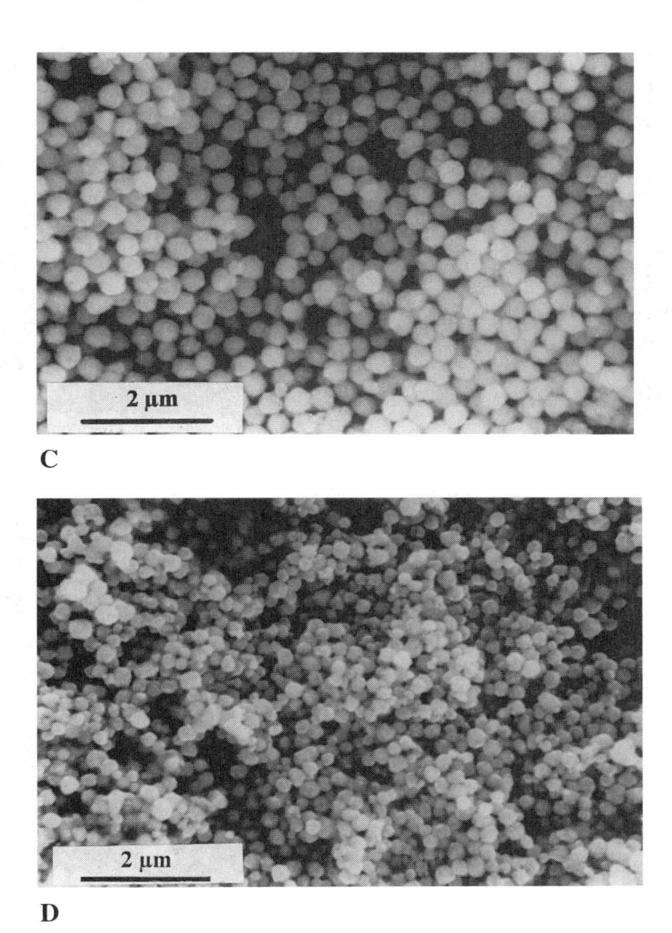

of the cobalt and nickel particles obtained by the polyol process through spontaneous nucleation. The diameter of the submicrometer-size particles obtained by heterogeneous nucleation is controlled by varying the amount of nucleative agent. As exemplified in Fig. 9.2.10 b and c for Co_xNi_{1-x} powders, a linear variation of d_m versus ([PM]/[Co] + [Ni])$^{-1/3}$ is observed. This relationship can be explained starting from similar hypotheses as those made earlier to explain the linear variation observed when the mean diameter of silver particles is plotted versus the cubic root of the precursor mass (cf. Section 9.2.3.1) (33). In contrast, when Ag is used as a nucleating agent, d_m is not found to be proportional to the ratio ([PM]/[Co + Ni])$^{-1/3}$ (Fig. 9.2.10d). This different behavior of Pt and Ag as nucleating agents and this discrepancy with the model observed for Ag can be tentatively explained by a stronger

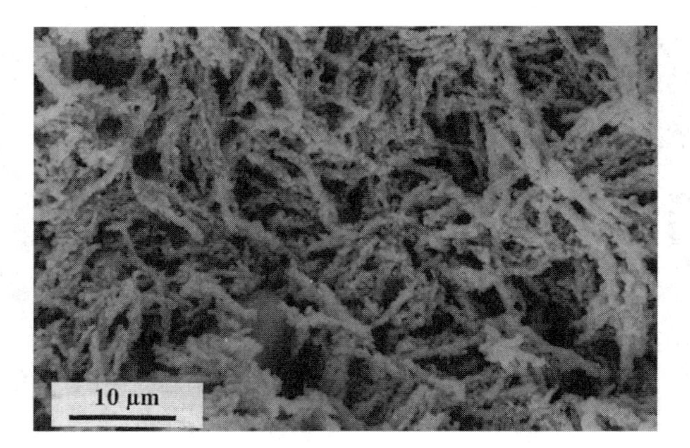

Fig. 9.2.9 SEM image of nickel filaments (10 μm × 0.5 μm) obtained from $Ni(OH)_2$ in EG at 205°C in a magnetic field. (From Ref. 16.)

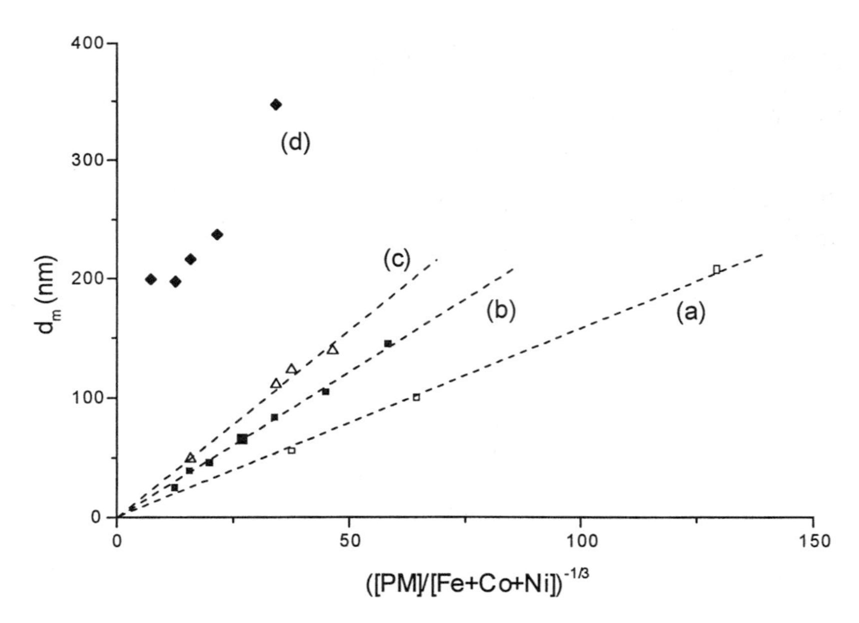

Fig. 9.2.10 Mean diameter of polymetallic particles prepared by heterogeneous nucleation versus ([PM]/[Fe] + [Co] + [Ni])$^{-1/3}$ for different compositions and different nucleating agent: (a) [Pt]/[$Fe_{0.13}(Co_{50}Ni_{50})_{0.87}$]; (b) [Pt]/[$Co_{80}Ni_{20}$]; (c) [Pt]/[$Co_{50}Ni_{50}$]; (d) [Ag]/[$Co_{80}Ni_{20}$]. (From Ref. 34.)

tendency of the silver nuclei to coalesce, specially for the highest silver concentrations; therefore the size of the noble metal particles acting as growth sites becomes concentration dependent. Nevertheless, even though the two different noble metals have a different behavior, they are quite complementary. For example, for $Co_{80}Ni_{20}$ (Fig. 9.2.10, b and d), Pt is used to synthesize the finest particles ($25\ \text{nm} < d_m < 150\ \text{nm}$), whereas Ag can be used to make larger ones ($200\ \text{nm} < d_m < 500\ \text{nm}$). It is worth noting that for the highest platinum concentration, nanoscale particles can be obtained as exemplified in Figure 9.2.11. The PVP-protected Ni–Pd particles

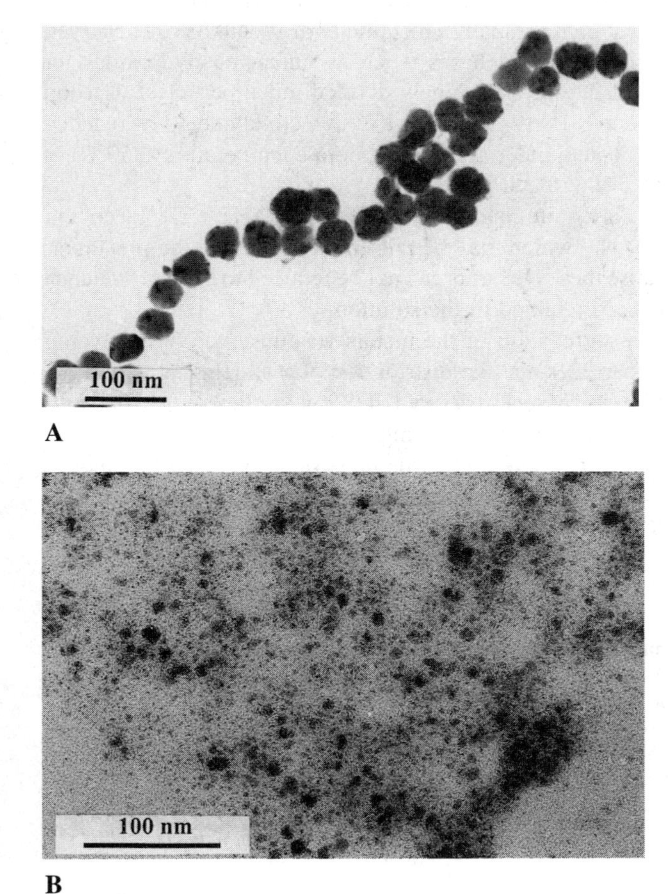

Fig. 9.2.11 TEM images of monodisperse colloidal dispersions prepared by heterogeneous nucleation with Pt as nucleating agent: (A) $Co_{35}Ni_{65}$, $d_m = 50$ nm, $\sigma = 0.10d_m$. (B) $Co_{80}Ni_{20}$, $d_m = 7$ nm, $\sigma = 0.26d_m$. (a from Ref. 34.)

obtained by Toshima (31) are even smaller (typical mean size ~2 nm), with a narrow size distribution and a low degree of aggregation.

2. *Chemical analysis.* As expected for powders synthesized in boiling organic solvents, the main impurities are carbon and oxygen. For an average particle size of ~200 nm, C and O contents were typically in the range 0.3–0.7 wt% and 1–2 wt% in the Co–Ni system, with the carbon and oxygen contents increasing steadily as the mean particle size decreased, then it may inferred that these impurities are mainly located at the surface of the particles. Owing to this impurity layer, which prevents a catastrophic oxidation of the particles in air, very fine powders prepared in polyols are not pyrophoric. It has been inferred from thermal analysis (30,32,33) that impurities can originate from organic adsorbed species (either polyol or degradation products), metallo-organic phases (such as metallic hydroxy-alkoxyde observed as intermediate phases), and inorganic phases (such as unreacted hydroxides, or oxides due to a superficial oxidation of the finely divided metal particles). Carbon and oxygen contents lower than 0.1 wt% and 0.2 wt% respectively could be reached after thermal treatments conducted under argon at moderate temperature (350°C) in order to avoid the sintering of the metal particles.

Other impurities levels are quite low as exemplified in Table 9.2.4 for cobalt. Impurities such as Ca, Fe, or Na, which may be present as cations in the precursor, are considerably lower because these elements can not be reduced to the zero-valence state by the polyol and then are retained in the solution.

For CoNi powders, no segregation of the metals was observed by analyzing the composition of isolated particles and the distribution of the elements within the particles by energy-dispersive x-ray spectroscopy (EDS); nevertheless, a concentration gradient for Ni and Co could be shown by fixing the nanoprobe (15 nm diameter for the higher magnification) at the core and at the edge on an isolated particle, as the core of the particles is always richer in Co and poorer in Ni than the edge. This

Table 9.2.4 Chemical Analysis of Two Cobalt Powders Obtained from the Same Parent Hydroxide by Solid–Gas Reduction and by the Polyol Process (Main Impurities/Cobalt Ratios in ppm by Weight)

	$Co(OH)_2$ (precursor)	Co (solid–gas reduction)	Co (polyol process)
Na	250	250	<10
Ca	150	150	<5
Fe	80	80	20
Ni	250	250	250
Si	150	150	130
S	150	150	<50
Cl	2000	500	60

composition gradient could also be seen on element distribution maps (Fig. 9.2.12) (8,30,33,34).

3. *Phase analysis and texture of the metal particles.* Over the whole composition range, whatever the particle diameter, a face-centered cubic (fcc) phase is always observed (Fig. 9.2.13) by x-ray diffraction (XRD) either as a single phase (Ni and Co_xNi_{1-x} with $x < 0.35$) or beside a hexagonal close-packed (hcp) phase with broad lines (Co and Co_xNi_{1-x} with $x \geq 0.35$). The lattice parameter of the fcc phase shows

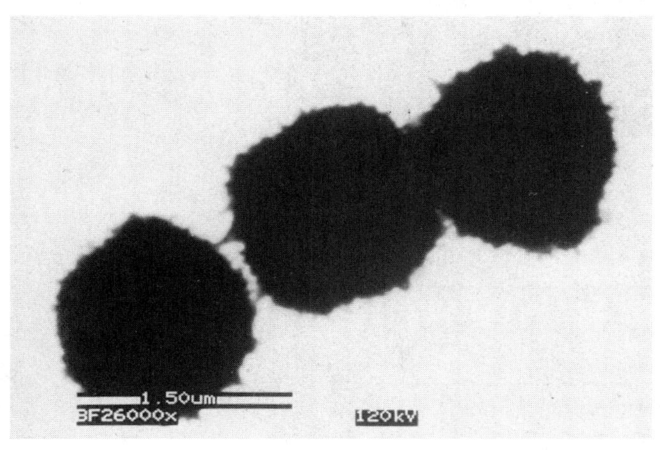

A

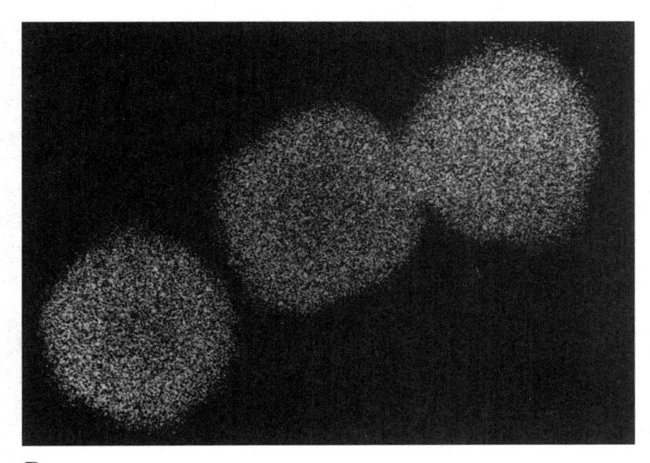

B

Fig. 9.2.12 Element distribution maps obtained by EDS with CoNi particles: (A) STEM image, (B) nickel, (C) cobalt. (From Ref. 8.)

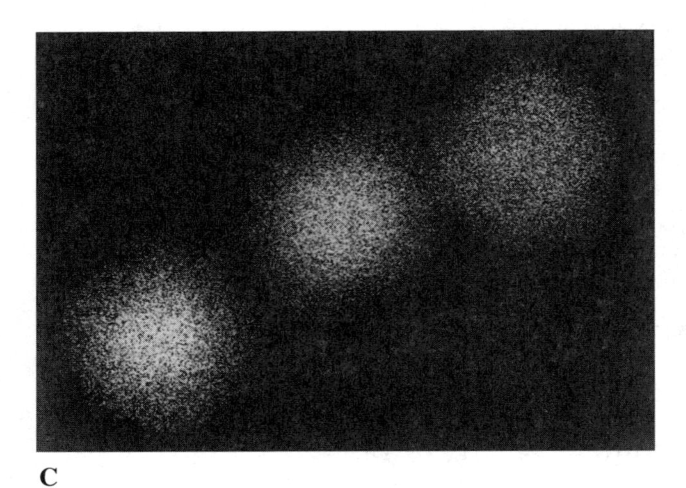

C

Fig. 9.2.12 Continued.

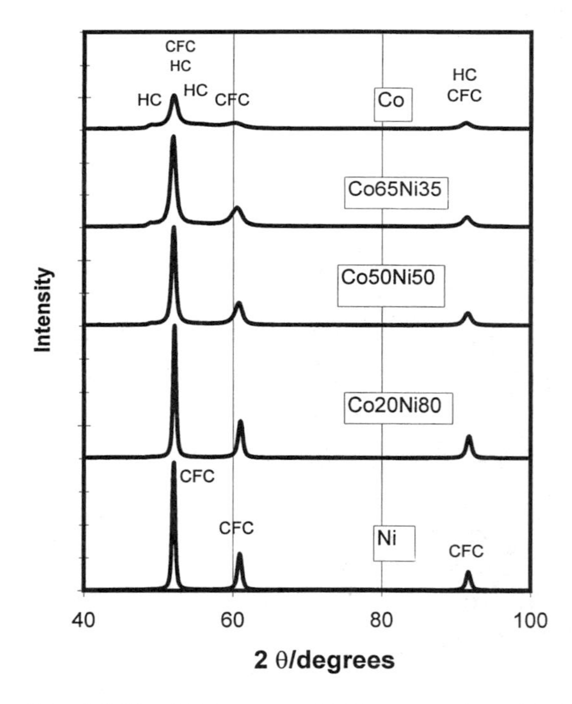

Fig. 9.2.13 XRD pattern of different powders with almost the same mean diameter (~200 nm) in the CoNi system.

a linear dependence against composition over the whole metal composition range, which provides evidence that this phase is a solid solution of the two metals (33). The occurrence of a hcp phase for high cobalt content is in qualitative agreement with the phase equilibrium diagram, with this hcp phase being expected for $x \geq 70$ instead of $x \geq 35$. The organic, metallo-organic, or inorganic phases present as impurities were not evidenced due to their low relative amount and/or their low crystallinity.

Transmission electron micrographs (TEM) of submicrometer-size particles show faceted particles, and selected area electron diffraction (SAED) patterns of isolated particles show that they are formed by a small number of crystallites (Fig. 9.2.14a). This result is consistent with the mean size of the crystallites, which can be inferred from the x-ray diffraction lines broadening analysis using a Williamson–Hall plot (35) in order to take into account the contribution of microstrains to the line broadening. Over the whole composition range, the mean crystallite size is in the range 40–60 nm for particles with a mean diameter in the range 200–300 nm (Table 9.2.5) (33).

The density of CoNi powder decreases steadily in the 8.5–7.5 g cm^{-3} range as the mean particle size decreases from 300 to 50 nm. The saturation magnetization follows the same trend. It has been inferred from those measurements that the particles are formed by a metallic, ferromagnetic, nonporous core whose density is close to the bulk value (8.9 g cm^{-3}). This core is coated by an impurity layer whose density and thickness are ~4 g cm^{-3} and 2 nm, respectively.[33]

Thus, monodisperse Co–Ni powders can be obtained by the polyol process in the whole composition range including pure Co or Ni powders. The particle size can be varied over a wide range, which can reach three orders of magnitude.

9.2.3.3 Copper

Copper powders can be obtained easily by reduction of copper salts or copper oxides using the polyol process. The reaction can be carried out in ethylene glycol at a lower temperature (150–195°C) than for cobalt or nickel. Starting from a very soluble precursor such as copper acetate or from slightly soluble copper(II) oxide, copper(I) oxide is obtained as an intermediate solid phase in both cases. Like noble metals, copper particles show a strong tendency to coalesce during the growth step, and then polydisperse particles with no definite shape are obtained under uncontrolled conditions. Copper powders composed of micrometer-size particles with well-defined morphological characteristics have been obtained by using a sparingly soluble precursor (CuO) (17). The solubility of this precursor and of the intermediate Cu_2O is enhanced and controlled by adding sodium hydroxide in the polyol. Hence, the dissolution step does not limit the overall reaction, and the reaction kinetics are governed by the rate of reduction of dissolved Cu(I) species into Cu and by the rate of nucleation and growth of the metallic phase. Moreover, a protective agent (D-sorbitol) has to be used in order to prevent the coalescence of the copper particles.

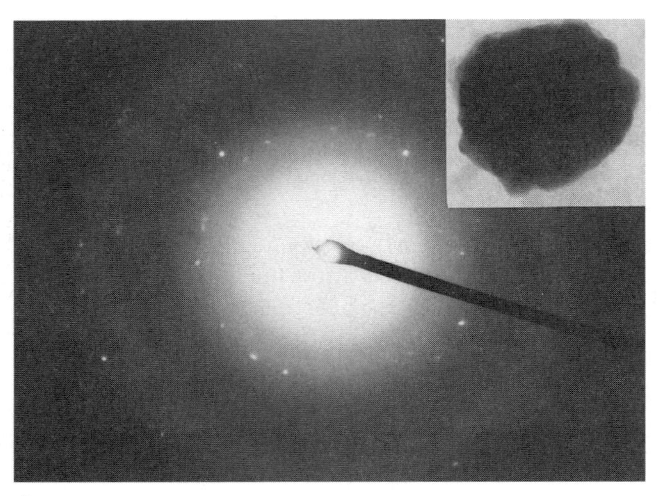

A

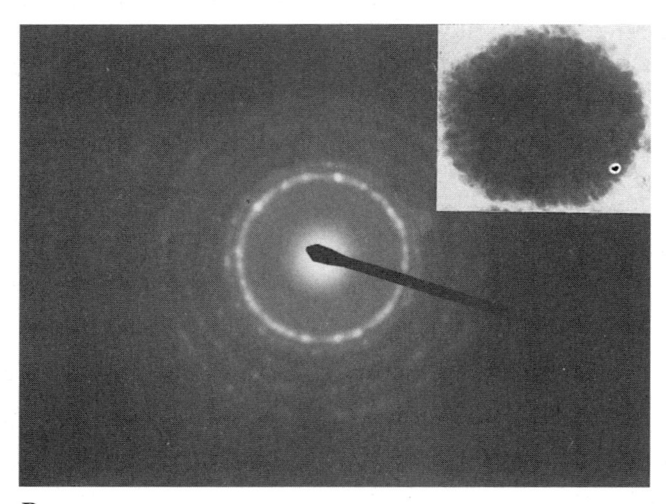

B

Fig. 9.2.14 TEM image and selected area electron diffraction (SAED) pattern of (A) a single $Co_{20}Ni_{80}$ particle, and (B) a single $Fe_{25}Ni_{75}$ particle. (From Ref. 8.)

Table 9.2.5 Mean Crystallite Size (D) and Microstrains Coefficient (η) Inferred from Williamson–Hall Plot for Co_xNi_{1-x} Powders with Almost the Same Particle Size (d_m)

	d_m (nm)	D (nm)	$\eta \times 10^3$
Ni	200	60	5
$Co_{20}Ni_{80}$	315	40	1.5
$Co_{50}Ni_{50}$	240	65	13
$Co_{80}Ni_{20}$	200	40	19
Co	—	50	26

The detailed experimental procedure was as follows (17): The reaction medium was prepared by dissolving NaOH (concentration range 10^{-3}–10^{-1} mol L^{-1}) in a solution of D-sorbitol in EG (molar ratio 0.06). CuO was dispersed in this solution (molar ratio CuO/EG, 0.07). The suspension was stirred and heated at a rate of 6°C min^{-1} up to a reaction temperature ranging from 150°C to 195°C. Typically the reaction time was 30 min at 195°C and 2 h at 175°C. Well-crystallized powders were obtained with a low impurity level ($C < 0.1$, $H < 0.1$, $O = 0.25$ wt%).

Under these experimental conditions the overall reaction appeared to be controlled by the nucleation and growth steps of the metallic particles, and therefore, nonagglomerated equiaxed copper particles with a narrow size distribution in the micrometer range were obtained. Moreover, by varying the NaOH concentration, the average diameter can be controlled in the range from 0.8 to 4 μm (Fig. 9.2.15), since NaOH enhances the rate of dissolution of the intermediate Cu_2O phase, then the rate of reduction of the dissolved Cu(I) species, and finally the number of nuclei formed during the nucleation step.

Nanoscale alloy materials composed of copper and a noble metal (Pd or Pt) have been obtained by Toshima et al. by reduction in EG (36). The method is characterized by the formation of a bimetallic hydroxide colloid in a first step carried out at room temperature by adding dropwise NaOH in a solution of the precursors and PVP in EG. This first step is designed in order to overcome the problem caused by the difference in redox potentials of the two metals. Then by refluxing this colloidal solution under a nitrogen flow, nonagglomerated, bimetallic, PVP-protected particles with a nanometer size range are obtained. Moreover, nanometer-size naked Cu–Pd particles have been synthesized by the same method (37). Materials composed of copper and a nonprecious metal have also been made by the polyol process. Hence nanocomposite powders of Co_xCu_{1-x} ($0.04 \leq x \leq 0.49$) have been synthesized by reduction of the metal acetates in EG (10). The powders were agglomerated and consisted of aggregates of nanoscale crystallites of copper and cobalt.

9.2.3.4 Iron and Iron-Based Powders

Polyols are mild reducing agents that do not allow Fe(II) or Fe(III) ions to reduce directly to the zero-valence state, with iron being more electropositive than nickel

A

B

Fig. 9.2.15 Effect of NaOH concentration on the copper powders obtained from the reduction of CuO in a sorbitol-ethylene glycol solution: (A) no NaOH; (B) [NaOH] $= 10^{-2}$ mol L^{-1}. (C) [NaOH] $= 5 \times 10^{-2}$ mol L^{-1}; (D) [NaOH] $= 10^{-1}$ mol L^{-1}. (From Ref. 17.)

or cobalt. Nevertheless, iron powders can be synthesized by disproportionation of ferrous hydroxide in such organic media. Polymetallic particles can be obtained as well by concomitant disproportionation of ferrous hydroxide and reduction of cobalt or/and nickel hydroxide. The disproportionation of Fe(II) hydroxide,

$$4Fe(OH)_2 \rightarrow Fe + Fe_3O_4 + 4H_2O$$

has been previously evidenced in aqueous medium (11), where it has been shown that it is favored by 1) alkaline conditions and 2) the presence in the medium of

C

D

weak complexing agents such as sugars or polyols, which leads to a poorly crystallized hydroxide (38). However, in water the formation yield of metallic iron remains very low due either to the competitive oxidation of the starting hydroxide by water or to the subsequent oxidation by water of the tiny metal iron particles formed at first by disproportionation of ferrous hydroxide. For these reasons, the disproportionation of $Fe(OH)_2$ in aqueous medium has never constituted a preparative route to iron metal particles. By substituting liquid polyols for water as reactive medium for the disproportionation of $Fe(OH)_2$, Viau et al. (6,30) showed that it was possible to take advantage of the reducing nature of the polyols in order to prevent undesirable oxidation reactions. It was also possible to take advantage of their complexing character in order to keep Fe(III) species in solution and hence to prevent the crystallization of magnetite.

Experimental Conditions of Synthesis

1. *Iron.* Metallic iron particles have been obtained by the disproportionation of Fe(II) hydroxide formed in situ by mixing Fe(II) chloride and sodium hydroxide in EG or PEG. The precipitation occurred in a temperature range as low as 80–100°C. In order to prevent the crystallization of magnetite and to retain unreacted Fe(II) species in solution, it was observed (6) that three conditions have to be fulfilled: 1) The initial concentration of Fe(II) species must be limited; 2) sodium hydroxide has to be used in large excess with respect to ferrous hydroxide stoechiometry; and 3) water must be distilled off the reaction medium as the reaction proceeds. However, under such conditions, iron particles were found to have no definite shape and a rather broad size distribution. Recently, iron powders with well-defined morphological characteristics were obtained by a two steps process. First, an iron(II) hydroxy-alkoxide was precipitated by boiling (15 min) a solution (200 mL) of $FeCl_2 \cdot 4H_2O$ (0.25 mol L^{-1}) and $CH_3CO_2Na \cdot 3H_2O$ (1.75 mol L^{-1}) in PEG. The suspension was cooled down to room temperature. Then a solution (60 ml) of NaOH (5.2 mol L^{-1}) in PEG was added. Second, the suspension was brought once more to a boil (30 min). In this alkaline medium the hydrolysis of the suspended hydroxy-alkoxide occurred progressively, yielding iron(II) hydroxide. This progressive formation of the hydroxide allowed control of the growth of the metallic phase, resulting in its further disproportionation. Iron particles could be obtained with nonnegligible yields (10–12%), which remain nevertheless low in comparison to theoretical yield of ferrous hydroxide disproportionation, i.e., 25%.

2. $Fe_z[Co_xNi_{(100-x)}]_{1-z}$ *powders.* Polymetallic powders were obtained by coprecipitation of Fe, Ni, and/or Co in EG or PEG (6,30,33,34). Mixed hydroxides were precipitated in situ by the addition of sodium hydroxide (1 mol L^{-1}) to a solution of tetrahydrated cobalt(II), nickel(II) acetate, and tetrahydrated iron(II) chloride in PEG ($[Fe^{2+}] + [Co^{2+}] + [Ni^{2+}] = 0.1$ mol L^{-1}). The precipitation could occur either by spontaneous or heterogeneous nucleation. In the latter case, platinum was used as nucleating agent with a molar ratio [Pt]/[Fe] + [Co] + [Ni] in the range 10^{-5}–10^{-7}. The solution was boiled while stirring for 2 h, with water and volatile organic products being distilled off. The precipitated metal powder was separated from the liquid by centrifugation, washed several times under ultrasonics with ethanol, water, and acetone, and finally dried under argon at 50°C. Cobalt and nickel were recovered almost quantitatively, whereas the iron yield was limited by disproportionation, and hence the composition of the powders could be varied in the ranges $0 \leq x \leq 100$ and $0 \leq z \leq 0.25$.

Powders Characterization

1. *Morphological characteristics.* Iron powders obtained by the two-steps procedure described earlier are formed of polyhedral, and even often cubic, nonagglomerated particles with a rather narrow size distribution and a mean size in the

range 0.2–0.7 μm (Fig. 9.2.16a). Fe–Ni and Fe–Co–Ni powders obtained by spontaneous nucleation are formed of monodisperse ($\sigma < 10\%\ d_m$), quasi-spherical, nonagglomerated particles with a mean size in the range 0.1–0.35 μm (Fig. 9.2.16b). Fe–Ni and Fe–Co–Ni powders obtained by heterogeneous nucleation had similar morphological characteristics except for their mean size, which was found in the range 50–200 nm (Fig. 9.2.16, c and d). The average diameter could be controlled

A

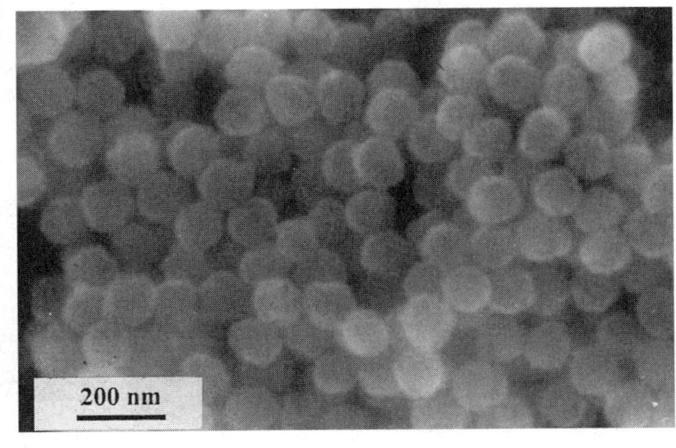

B

Fig. 9.2.16 EM images of different powders in the FeCoNi system synthesized by reduction in polyols: (A) Fe obtained by spontaneous nucleation, $d_m = 0.3$ μm; (B) $Fe_{0.13}[Co_{50}Ni_{50}]_{0.87}$ obtained by spontaneous nucleation, $d_m = 115$ nm, $\sigma = 0.08d_m$. (C) $Fe_{0.13}[Co_{20}Ni_{80}]_{0.87}$ obtained by heterogeneous nucleation, $d_m = 75$ nm, $\sigma = 0.08d_m$; (D) $Fe_{0.13}[Co_{50}Ni_{50}]_{0.87}$ obtained by heterogeneous nucleation, $d_m = 100$ nm, $\sigma = 0.09d_m$.

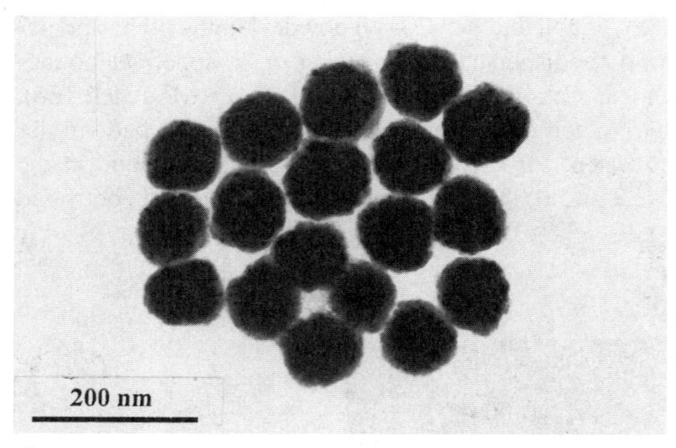

C

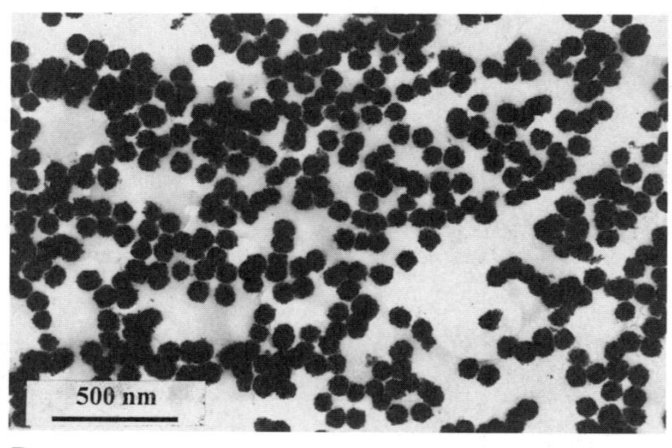

D

Fig. 9.2.16 Continued.

accurately by varying the amount of nucleating agent as for Co_xNi_{1-x} powders (Fig. 9.2.10a). In contrast, monodisperse FeCo powders have not been obtained by coprecipitation in liquid polyols; these powders always appeared as agglomerated submicrometer size particles.

2. *Chemical analysis.* Like Co–Ni powders, iron-based powders were found to have C and O contents that increase steadily as the mean particle size decreases; these contents were significantly higher in the FeCoNi system than in CoNi powders for a same particle size. For instance, C and O contents were found in the range 1.2–1.6 wt% and 3.5–4.5 wt%, respectively, for an average particle size of ~200 nm. C and O contents could be lowered to 0.3 and 2 wt%, respectively, after thermal treatment under argon.

For FeNi and FeCoNi powders, as for CoNi powders, no segregation of the metals was observed by EDS even if some local inhomogeneities within the particles could be shown (6,33). In contrast, for FeCo powders the segregation of the two metals was clearly evidenced. Some isolated particles did not contain any iron; the others were agglomerates formed by iron particles embedded in cobalt.

3. *Phase analysis and texture of the metal particles.* Iron powders are constituted of the α-Fe phase with a body-centered cubic (bcc) lattice, whereas Fe–Co powders appear as a mixture of three phases that are quite similar to those of pure metals (bcc for α-Fe and a mixture of hcp and fcc for cobalt) (6). In the Fe_zNi_{100-z} system, a single fcc phase is observed over the whole available composition range ($z \leq 25$) with a linear dependence of the lattice parameter versus z, which shows the existence of a fcc solid solution as already evidenced for the Co_xNi_{100-x} system (33). The XRD patterns of the $Fe_z[Co_xNi_{(100-x)}]_{1-z}$ powders depend on the composition: An fcc phase is always observed either as a single phase or as the main phase; a second hcp phase with weak and broad lines appears for a cobalt content $x \geq 35$; a third body-centered cubic (bcc) phase can be evidenced when $x > 80$.

The texture of monodisperse iron-based particles was inferred from TEM images, electron diffraction, x-ray line-broadening analysis, and density measurements (6,33). TEM images showed that the spherical particles are formed from the aggregation of smaller primary particles in the nanometer size range (Fig. 9.2.14b). The SAED pattern of an isolated particle showed the polycrystalline character of these particles as exemplified in Figure 9.2.14b. This feature is consistent with the mean size of the crystallites, which can be inferred using a Williamson–Hall plot. Mean crystallite sizes were found to be significantly smaller (10–25 nm) than those of Co–Ni powders (Tables 9.2.5 and 9.2.6), whereas the microstrain coefficients were significantly larger. Furthermore, density measurements evidenced low powder values with respect to bulk alloys (6,33). According to these various results, these particles could be described as porous submicrometer-size aggregates of nanoscale metallic particles that retain within the pores an appreciable amount of nonmetallic phases not removed by washing, which explains the high oxygen and carbon contents.

Table 9.2.6 Mean Crystallite Size (D) and Microstrains Coefficient (η) Inferred from Williamson–Hall Plot for Powders in the FeCoNi System

	d_m (nm)	D (nm)	$\eta \times 10^3$
Ni	200	60	5
$Fe_{13}Ni_{87}$	470	15	21
$Fe_{20}Ni_{80}$	—	24	55
$Fe_{0.13}[Co_{20}Ni_{80}]_{0.87}$	70	12	17

9.2.4 Particle Formation Mechanism

The morphological characteristics of the various metal powders that have been discussed in the previous section provide evidence that the reduction reaction involved in the polyol process is not a solid-solid reaction but occurs via the solution. This is obvious when the precursor is completely soluble; this is also true when the starting compound is only slightly soluble in the polyol or when the formation of a sparingly soluble intermediate phase is observed, because their is no similarity at all between the morphological characteristics of these solid phases and those of the metal powder obtained as the reaction is completed (18,29). Moreover, the narrow size distribution of the particles and the variation of their average diameter with the reaction temperature or with the precursor/polyol ratio as already exemplified for different systems are consistent with a mechanism of formation of the metal particles involving a short spontaneous nucleation step occurring as a burst followed by a progressive growth step. Further evidence was provided by the fact that it has been possible, for many systems, to replace this spontaneous nucleation step by an heterogeneous one though the addition of a nucleating agent. The variation of the mean diameter of the particles by varying the concentration of this nucleating agent is consistent with this mechanism of formation.

Moreover, according to this general schematic mechanism, a more detailed picture of the particle formation has been tentatively given in two particular cases: (1) synthesis of colloidal silver particles, and (2) formation of bimetallic ferromagnetic metal particles.

9.2.4.1 Colloidal Silver Particles

The development of the particle size during the synthesis of PVP-protected nanoscale silver particles carried out under the conditions described earlier has been followed

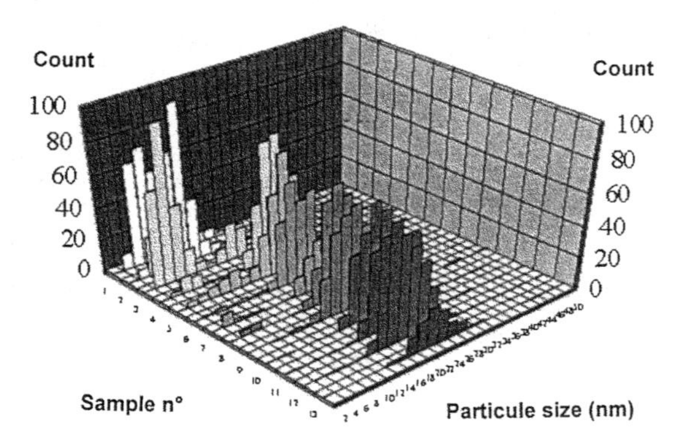

Fig. 9.2.17 Development of particle size during the formation in EG of monodisperse PVP-protected silver colloids. (From Ref. 39.)

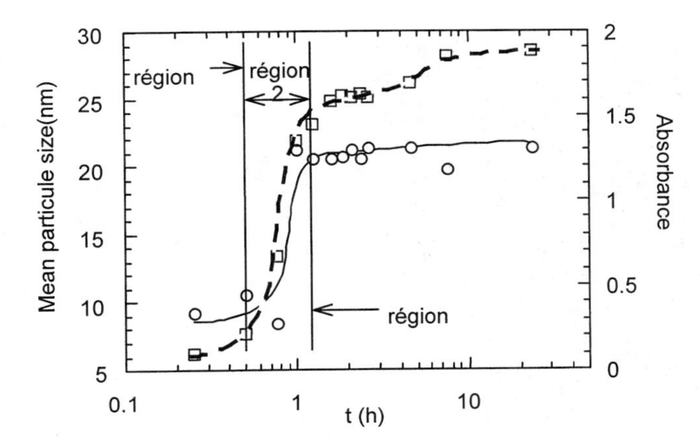

Fig. 9.2.18 Development of the mean particle size (○) and of the absorbance (□) (at 410 nm) during the formation in EG of monodisperse PVP-protected silver colloids. (From Ref. 39.)

by Silvert et al. (39). The system was heated up to 120°C (1°C min^{-1}) and then the reaction was carried out for 22 h, with samples being taken away during the temperature rise and at the temperature plateau as well. Each colloidal sample was characterized (Figs. 9.2.17 and 9.2.18) by scanning transmission microscopy (STEM), and the progress of the reduction of Ag^+ into Ag^0 was followed by measuring the absorbance (Fig. 9.2.18) of the suspension at a wavelength of 410 nm (absorption peak due to the resonance of surface plasmons in colloidal silver particles) (40). The formation of the particles appeared to proceed in three stages (Fig. 9.2.17). It was shown that the particles began to form in the early stage of the temperature rise and that they had a rather broad size distribution. A few particles already had the final mean size (21 nm), whereas most of the particles were smaller than 10 nm. Then, as the temperature rose and as the reduction went on, a very fast development of the particle size took place, which resulted in a shift of the mean size of the particles and a narrowing of their distribution size, with most of the particles having a diameter close to the final mean size with a few residual smaller particles (~6 nm). During the last part of the rise in temperature and during the beginning of the plateau, the number of these residual smaller particles decreased, and hence the standard deviation was found to decrease, whereas the mean diameter remained constant, and then a monodisperse colloid was finally obtained. These final particles appeared as twinned monocrystals with a polyhedral shape as exemplified in Figure 9.2.19. From this last result the authors were able to rule out the possibility of a growth mechanism of the large particles through the coalescence of smaller primary ones; polycrystals would be obtained with such a mechanism. The formation of a monodisperse colloid

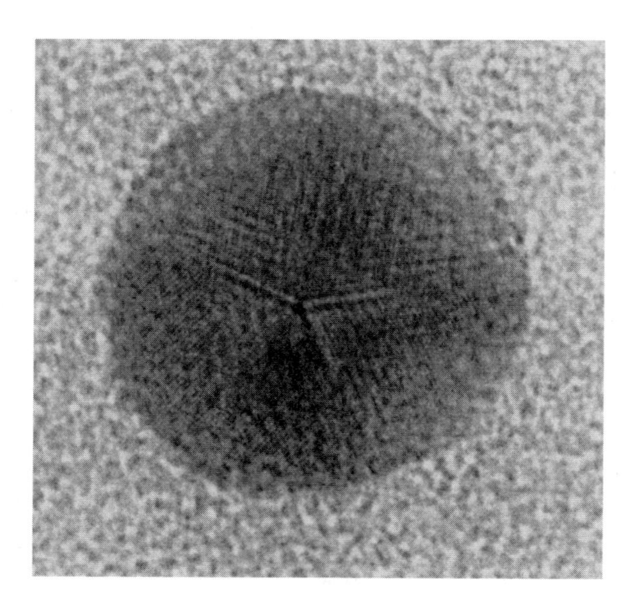

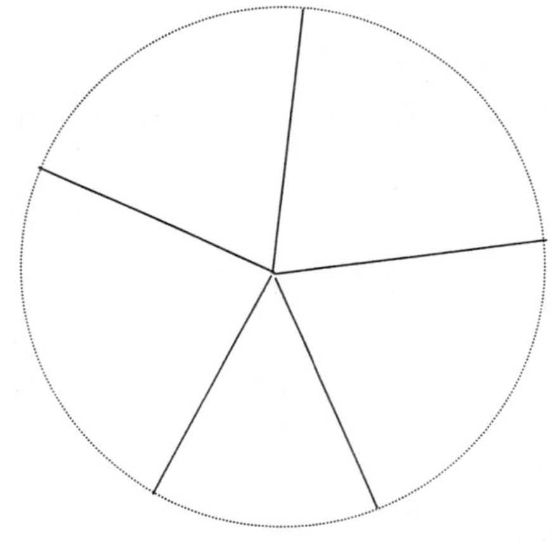

Fig. 9.2.19 High-resolution TEM image of a colloidal silver particle observed along [111] and its schematic model. (From Ref. 39.)

was tentatively explained by an Ostwald ripening mechanism: As the temperature increases, the smallest particles are no longer stable in the polyol and dissolve, contributing to the growth of the larger ones. This formation mechanism is rather unexpected for the formation of metal particles, and it needs to be ascertain by studies conducted under well-defined kinetic conditions, namely, isothermal conditions at temperatures lower than 120°C. Moreover, it must be stressed that PVP plays a complex role in such a synthesis of silver particles. Besides its role of protective agent, which has been tentatively explained by Yonezawa et al. (41) in the particular case of bimetallic clusters obtained in monoalcohol/water media, it has been shown that PVP plays a significant role as reducing agent when it is used in the polyol process (27,39).

9.2.4.2 Polymetallic Ferromagnetic Metal Particles

Despite the difference in the chemical reactions involved in the formation of cobalt or nickel on the one hand and in the formation of iron on the other hand in polyols, Co_xNi_{1-x}, Fe_zNi_{1-z}, and $Fe_z[Co_xNi_{(100-x)}]_{1-z}$ powders appeared as formed by quasi-spherical, nonagglomerated particles with a narrow size distribution and a good interparticular and intraparticular homogeneity in composition. However, a significant difference in the texture of Co_xNi_{1-x} particles obtained by heterogeneous nucleation on the one hand, and Fe_zNi_{1-z} and $Fe_z[Co_xNi_{(100-x)}]_{1-z}$ particles obtained either by spontaneous or heterogeneous nucleation on the other hand, has been evidenced by different techniques (TEM, SAED, x-ray line broadening, density measurements). The former particles are nonporous, well crystallized, formed by a small number of crystallites, and the density of the powder is close to the value expected for bulk samples. In contrast, Fe_zNi_{1-z} and $Fe_z[Co_xNi_{(100-x)}]_{1-z}$ particles are porous, polycrystalline, and formed by many nanoscale crystallites. These particles retain an important amount of nonmetallic solid phases in their pores and then exhibit a density that is much lower than the value expected for bulk samples. These different textural characteristics have been tentatively related to two distinct growth mechanisms. This growth may proceed either by a stepwise addition of atoms or by coalescence of primary particles, which form secondary larger particles. The textural characteristics of CoNi particles are consistent with the first type of mechanism, which can be described as follows. At first, the particles of the noble metal used as nucleating agent are formed; these particles are preferential sites for the further growth of nickel and cobalt. This growth, which occurs atom by atom, begins on every noble metal particle and goes on uniformly on all the surfaces of the growing particles. The growth rate can be limited either by the diffusion of the solvated species toward the metal–solution interface or by the interfacial reduction reaction itself. Such a mechanism is consistent with the possibility to control the mean diameter d_m of the particles by varying the amount of nucleating agent. Indeed, the linear relationship between d_m and $([PM]/[Co + Ni])^{-1/3}$ requires that each noble metal particle gives rise to one spherical, nonporous, final particle, and that there is no coalescence of these particles during the growth step. The composition homogeneity of the final metallic particles can be explained by the homogeneity of the intermediary solid phase, the dissolution of which controls the supersaturation ratio of both Co(II) and Ni(II) species. The reduction rate of these two species must also be quite similar. Indeed these rates must be slightly different in order to explain the small gradient of composition observed within the particles.

Obviously, the textural characteristics of Fe_zNi_{1-z} or $Fe_z[Co_xNi_{(100-x)}]_{1-z}$ particles obtained by precipitation in polyols are not consistent with this previous growth mechanism. These characteristics can be explained if these particles are considered as secondary ones made up by the coalescence of smaller primary particles. These primary particles are formed either by spontaneous nucleation and growth or by growth from noble metal particles acting as preferential growth sites. Then, during their growth, these nanoscale primary particles coalesce to form porous, polycrystalline secondary particles. Indeed, it is surprising to obtain monodisperse particles by a growth mechanism involving coalescence of primary particles. Nevertheless, it must be recalled that the formation of monodisperse particles by such a mechanism has been evidenced in several cases, all of them being related to the formation of oxides or hydroxides either in aqueous (42,43) or nonaqueous media (44). Nevertheless, it is not well understood why the final particles show such a narrow size distribution.

9.2.5 Conclusion

The precipitation of metals by reduction or disproportionation of metallic compounds in liquid polyols is a versatile method of preparation of metal powders in the micrometer to nanometer particle size range. If the reaction is carried out under kinetically controlled conditions, this process has potential in providing nonagglomerated metal particles with a well-defined shape and a narrow size distribution. Moreover, the substitution of the spontaneous nucleation step by heterogeneous nucleation, when it is possible, allows an accurate control of the mean size of the particles in the submicrometer and nanometer size range, whereas the use of a protective agent allows control of the particle growth of metals, like noble metals, that have a particularly high tendency to coalesce. The polyol process has potential to provide fine metal powders displaying properties that make them of interest for applications in different fields. Thus it has been successfully used to make finely divided cobalt powders, which have been of interest as an additive to make hard materials by powder metallurgy (45). A feasibility study to scale up this process to an industrial level has been conducted for cobalt. Furthermore, owing to their remarkable morphological characteristics and to the accurate control of their particle size, the polymetallic magnetic powders made by the polyol process have been used as model materials in order to study the influence of the particle size upon the microwave permeability (34,46,47). It will be possible by controlling the size and the composition of such magnetic inclusions to make granular materials with optimized absorption characteristics for microwave applications. In other respects, it has been shown that it is possible to control the growth of different noble metals in polyol and then to obtain nonagglomerated particles in the submicrometer size range (13–15). Such characteristics make these powders of interest for the production of conductive inks for the electronic industry. If the polyol process does not arouse a particular interest as a method of preparation of monometallic nanoscale colloids of noble metals, in contrast it appears as a valuable method to obtain bimetallic alloy colloids composed of two noble metals or of a noble metal and a metal such as Cu or Ni. Such bimetallic clusters showed promising characteristics in catalysis (31,36,37).

The polyol process has not been used only as a method of preparation of metal powders. As a mild process of reduction, it has been shown to be of interest for the preparation of supported (48) or intercalated (49) catalysts or as a method of metallization of various nonconductive surfaces (9,50).

REFERENCES TO SECTION 9.2

1. T Sugimoto. Adv Colloid Interface Sci 28:65, 1987.
2. M Haruta, B Delmon. J Chim Phys 83:859, 1988.
3. E Matijević. Faraday Discuss 92:229, 1991.
4. T Ishikawa, E Matijević. Langmuir 4:26, 1988.
5. F Fiévet, M Figlarz, J-P Lagier. Patents: Europe 0, 113, 281 (1987); USA 4, 539, 041 (1985); Finland 74, 416 (1998); Canada 123, 5910 (1988); Norway 163887 (1988); Japan 04, 024402 (1992).
6. G Viau, F Fiévet-Vincent, F Fiévet. J Mater Chem 6:1047, 1996, and French patent 2723015 (1996).
7. P-Y Silvert, V Vijayakrishnan, P Vibert, R Herrera-Urbina, K Tekaia-Elhsissen. Nanostruct Mater 7:611, 1996.
8. G Viau, F Fiévet-Vincent, F Fiévet. Solid State Ionics 84:259, 1996.
9. LK Kurihara, GM Chow, PE Schoen. Nanostruct Mater 5:607, 1995.
10. GM Chow, LK Kurihara, KM Kemmer, PE Schoen, WT Elam, A Ervin, S Keller, YD Zhang, J Budnick, T Ambrose. J Mater Res 10:1546, 1995.
11. FJ Shipko, DL Douglas. J Phys Chem 60:1519, 1956.
12. JTG Overbeek. Adv Colloid Interface Sci 15:251, 1982.
13. C Ducamp-Sanguesa, R Herrera-Urbina, M Figlarz. J Solid State Chem 100:272, 1992.
14. P-Y Silvert, K Tekaia-Elhsissen. Solid State Ionics 82:53, 1995.
15. C Ducamp-Sanguesa, R Herrera-Urbina, M Figlarz. Solid State Ionics 63–65:25, 1993.
16. F Fiévet, JP Lagier, M Figlarz. MRS Bull 14:29, 1989.
17. F Fiévet, F Fiévet-Vincent, J-P Lagier, B Dumont, M Figlarz. J Mater Chem 3:627, 1993.
18. F Fiévet, J-P Lagier, B Blin, B Beaudouin, M Figlarz. Solid State Ionics 32/33:198, 1989.
19. K Tekaia-Elhsissen, P-Y Silvert, S Lombard. Adv Powder Metal Particulate Mater 3: 109, 1996.
20. JR Larry, RM Rosenberg, RO Uhler. IEEE Trans Components Hybrids Manuf Technol 3:211, 1980.
21. G Fisher. Ceram Ind 120:80, 1983.
22. JG Pepin. J Mater Sci Mater Electron 2:34, 1991.
23. P-Y Silvert, R Herrera-Urbina, N Duvauchelle, V Vijayakrishnan, K Tekaia-Elhsissen. J Mater Chem 6:573, 1996.

24. K Tekaia-Elhsissen, F Bonet, S Grugeon, S Lambert, R Herreva-Urbina. J Mater Res 14:3707, 1999.
25. R Seshadri, CNR Rao. Mater Res Bull 29:795, 1994.
26. G Fischer, A Heller, G Dube. Mater Res Bull 24:1271, 1989.
27. P-Y Silvert. PhD thesis, Université de Picardie Jules Verne, Amiens, France, 1996.
28. F Bonet, C Guéry, D Guyomard, R Herrera-Urbina, K Tekaia-Elhsissen, J-M Tarascon. Solid State Ionics 126:337, 1999.
29. M Figlarz, F Fiévet, J-P Lagier. MRS Int Meeting on Advanced Materials, Tokyo 1988. MRS Conf Proc 3:125. Pittsburgh, PA, 1989.
30. G Viau. PhD thesis, Université Paris 7-Denis Diderot, France, 1995.
31. N Toshima, P Lu. Chem Lett 9:729, 1996.
32. MS Hedge, D Larcher, L Dupont, B Beaudouin, K Tekaia-Elhsissen, J-M Tarascon. Solid State Ionics 93:33, 1997.
33. P Toneguzzo. PhD thesis, Université Paris 7-Denis Diderot, France, 1997.
34. P Toneguzzo, G Viau, O Acher, F Fiévet-Vincent, F Fiévet. Adv Mater 10:1032, 1998.
35. GK Williamson, WH Hall. Acta Metall 1:22, 1953.
36. N Toshima, Y Wang. Langmuir 10, 4574, 1994; Adv Mater 6:245, 1994.
37. Y Wang, H Liu, N Toshima. J Phys Chem 100:19533, 1996.
38. GN Schrauzer, TD Guth. J Am Chem Soc 98:3508, 1976.
39. P-Y Silvert, R Herrera-Urbina, K Tekaia-Elhsissen. J Mater Chem 7:293, 1997.
40. JA Creighton, DG Eadon. J Chem Soc Faraday Trans 87:3881, 1991.
41. T Yonezawa, N Toshima. J Chem Soc Faraday Trans 91:4111, 1995.
42. M Ocaña, R Rodriguez-Clemente, CJ Serna. Adv Mater 7:212, 1995.
43. T Sugimoto, E Matijević. J Colloid Interface Sci 74:27, 1980.
44. D Jézéquel, J Guenot, N Jouini, F Fiévet. J Mater Res 10:77, 1995.
45. H Pastor, M Bonneau, J Pilot. Int Conf Adv Hard Materials Production, Bonn, Germany, 1992.
46. G Viau, F Ravel, O Acher, F Fiévet-Vincent, F Fiévet. J Appl Phys 76:6570, 1994.
47. G Viau, F Fiévet-Vincent, F Fiévet, P Toneguzzo, F Ravel, O Acher. J Appl Phys 81: 2749, 1997.
48. EA Sales, B Benhamida, V Caizergues, J-P Lagier, F Fiévet, F Bozon-Verduraz. Appl Catal A 172:273, 1998.
49. PB Malla, P Ravindranathan, S Komarneni, E Breval, R Roy. J Mater Chem 2:559, 1992.
50. GM Chow, LK Kurihara, D Ma, CR Feng, PE Schoen, LJ Martinez-Miranda. Appl Phys Lett 70:2315, 1997.

9.3 METAL PARTICLES MADE IN VARIOUS COLLOIDAL SELF-ASSEMBLIES: SYNTHESES AND PROPERTIES

MARIE-PAULE PILENI
Université Pierre et Marie Curie, Paris, France

9.3.1 Introduction

A key step in the control of mineralization employed by almost all organisms is the initial isolation of a space (1,2). Then, under controlled conditions, minerals are induced to form within the space. Filling up these spaces with amorphous minerals would appear to require a quite different strategy from that in filling spaces with crystalline material. The simplest way to fill a space with crystals is to create as high a local supersaturation as possible, and then induce nucleation or let the system spontaneously reach a lower energy state by crystallization, while at the same time removing the excess solvent. In terms of particle growth, a number of analogies between surfactant self-assemblies and natural media can be proposed. In both cases, this growth needs a supersaturated medium where the nucleation takes place.

Increasingly, chemists are contributing to the synthesis of advanced materials with enhanced or novel properties by using colloidal assemblies as templates. Colloid chemistry is particularly well suited to this objective with nanoparticles (3,4).

In solution, surfactant molecules self-assemble to form aggregates. At low concentrations the aggregates are generally globular micelles (5), but these micelles can grow on increasing surfactant concentration and/or upon addition of salt, alcohols etc. In this case, micelles grow to elongated, more or less flexible rodlike micelles (6–9), in agreement with theoretical predictions for micellization (10,11). The amphiphilic molecules spontaneously self-assemble to form highly flexible locally cylindrical aggregates with the average size reaching several micrometers (12,13). There are large fluctuations in details of the morphology (curvature of the film) of the mixture at a local scale. The contribution of the entropy of the folded film is predominant in the free energy of the solution, while the morphology has little influence. The interfacial curvature toward the water is, by convention, described as negative mean curvature and is known as type II or inverse curvature (14).

The fabrication of assemblies of perfect nanometer-scale crystallites (quantum crystal) identically replicated in unlimited quantities in such a state that they can be manipulated and understood as pure macromolecular substances is an ultimate challenge in modern materials research with outstanding fundamental and potential technological consequences. These potentialities are mainly due to the unusual dependence of the electronic properties on the particle size, either for metal (15–19) or semiconductor (3,20–29) particles, in the 1 to 10 nm range. The preparation and characterization of these nanomaterials have motivated a vast amount of work (30). One of the methods to control the particle size is the use of reverse micelles as microreactors (3,31). The achievements of an accurate control of the particle size, their stability, and a precisely controllable reactivity of small particles are required to allow attachment of the particles to the surface of a substrate or to other particles without leading to coalescence and hence losing their size-induced electronic properties. There are several reasons for forming films of inorganic particles attached to or embedded just under the surface. Moreover, the ability to assemble particles into well-defined two- and three-dimensional spatial configurations should produce advantageous properties such as new collective physical behavior. The synthesis of inorganic–organic supperlattices has been achieved using the multilayer casting of films (32). Asher et al. (33) have developed a method for creating both organic and inorganic submicrometer periodic materials. Recently, in our laboratory, spontaneous arrangements in either a monolayer organized in a hexagonal network or three-dimensional FCC arrangements of particles have been observed (34–38). Similar arrangements have been reported (39–43).

9.3.2 Control of the Particle Size

Reverse micelles are well known to be spherical water-in-oil droplets stabilized by a monolayer of surfactant. The phase diagram of the surfactant sodium bis(2-ethylhexyl) sulfosuccinate, called Na(AOT), with water and isooctane shows a very large domain of water in oil droplets and often forms reverse micelles (3,44). The water pool diameter is related to the water content, $w = [H_2O]/[AOT]$, of the droplet by (44): $D(\text{nm}) = 0.3w$. From the existing domain of water-in-oil droplets in the phase diagram, the droplet diameters vary from 0.5 nm to 18 nm. Reverse micelles are dynamic (45–48), and attractive interactions between droplets take place. The intermicellar potential decreases either by decreasing the number of carbon atoms of the bulk solvent or by increasing the number of droplets. This is due to the discrete nature of solvent molecules and is attributed to the appearance of depletion forces between two micelles (the solvent is driven off between the two droplets) (47). When the droplets are in contact, forming a dimer, they exchange their water contents. This exchange process is associated with the interface rigidity that corresponds to the bending elastic modulus of the interface (48). Hence, in collisions the droplets exchange their water contents and again form two independent droplets. This process has been used to make nanosized material by either chemical reduction of metallic ions or coprecipitation reactions. These various factors (water content, intermicellar potentials) control the size of the particles.

Syntheses in reverse micelles induce formation of nanoparticles dispersed in the solution. This can be followed by measuring the absorption spectra of the colloidal particles dispersed in the solvent. The average size of the particles can be determined by depositing a drop of this solution on a carbon grid and obtaining the transmission electron microscopy (TEM) patterns. Several types of material, such as CdS (3,21–29), CdSe (49), $Cd_yZn_{1-y}S$ (26,27), ZnS, $Cd_yMn_{1-y}S$ (28,29), and PbS for $(Ag^0)_n$ (22), $(Cu^0)_n$ (18,19,50), and Co_2B alloys (51), have been used. As an example, the synthesis of copper metal particles is described.

Reverse micelles are good candidates for making nanosized copper metal particles (18,19,50). To form this material, a functionalized surfactant such as copper bis(2-ethylhexyl) sulfosuccinate, $Cu(AOT)_2$, is needed. When this is replaced by copper sulfate (Cu^{2+}), oxidized copper instead of copper metal particles is formed. Furthermore, a few minutes after starting the reaction, the particles flocculate. Mixed micelles made of sodium and copper bis(2-ethylhexyl) sulfosuccinate are prepared. The water content, *w,* is defined as the ratio of water to the sum of Na(AOT) and $Cu(AOT)_2$ concentrations. The water content, $w = [H_2O]/\{[Na(AOT)] + [Cu(AOT)_2]\}$, is fixed at a given value. This solution is mixed with another Na(AOT) micellar solution in hydrazine, having the same water content. The copper bis(2-ethylhexyl) sulfosuccinate is reduced with the formation of copper metal particles, characterized by TEM, electron diffraction, and absorption spectroscopy. There is an increase in the particle size and a marked change in the absorption spectrum with appearance of a plasmon peak with increasing water content (18,19). As expected from expanded versions of Mie's theory (52–54), the absorption spectra vary with the particle sizes below 5 nm. The direct relationship between the particle size and the absorption spectrum allows obtaining a calibration curve relating the average particle diameter to its absorption (19).

9.3.3 Controlling the Shape of the Particles

Syntheses are performed either in self-assemblies having oil as the bulk phase and differing by their local arrangements or in dilute normal (oil-in-water) micelles.

9.3.3.1 Controlling the Shape by Using Water-in-Oil Self-Assemblies

Copper ions have been reduced in colloidal assemblies differing in their structures (55,56). In all cases, copper metal particles are obtained. Figure 9.3.1 shows the freeze-fracture electron microscopy (FFEM) for the various parts of the phase diagram. Their structures have been determined by SAXS, conductivity, FFEM, and by predictions of microstructures that require only notions of local curvature and local and global packing constraints.

The colloidal system consists of 5×10^{-2} *M* $Cu(AOT)_2$–isooctane–water. The colloidal structure is changed by increasing the amount of water in $Cu(AOT)_2$–isooctane solution (56,57). Syntheses are performed in various regions of the phase diagram (57) (Fig. 9.3.2).

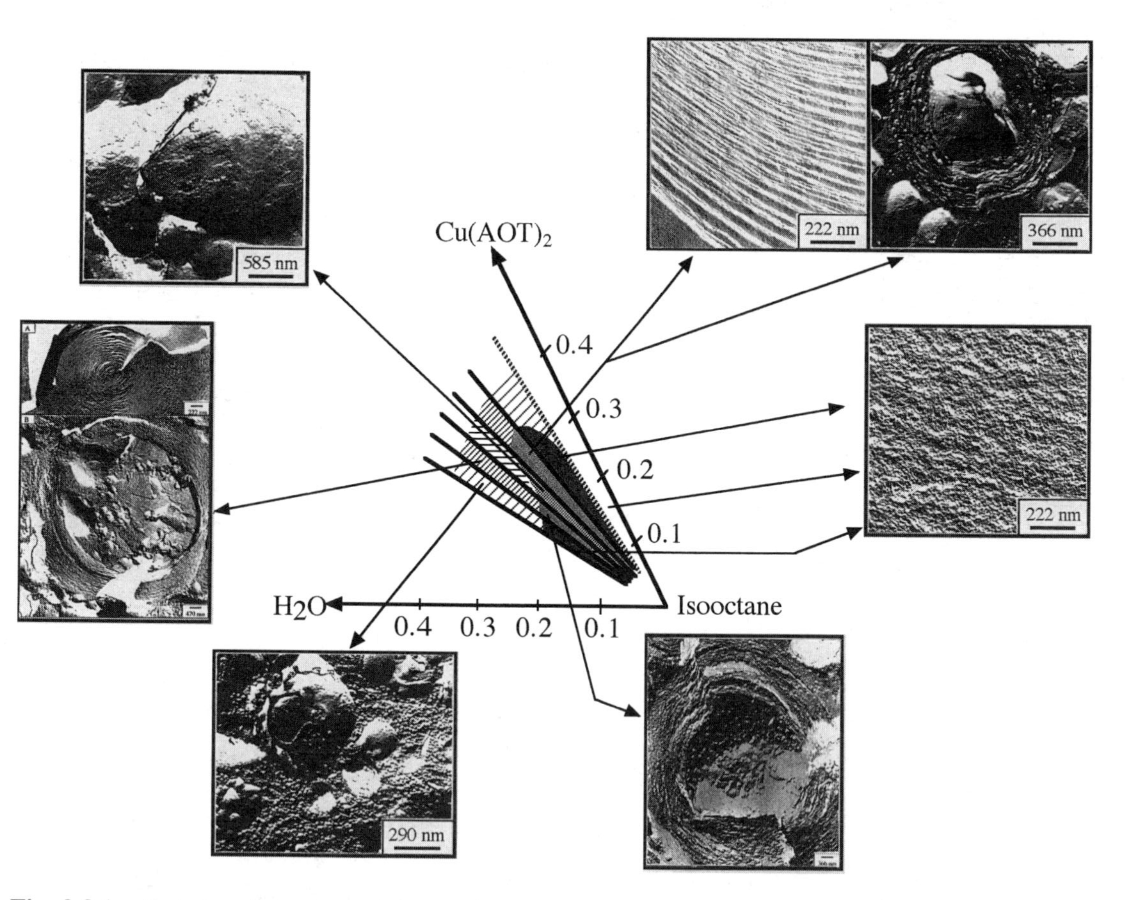

Fig. 9.3.1 Evolution of the phase diagram with the increasing in the water content, *w*, and keeping $Cu(AOT)_2$ concentration constant. $[Cu(AOT)_2] = 5 \times 10^{-2}$ *M*.

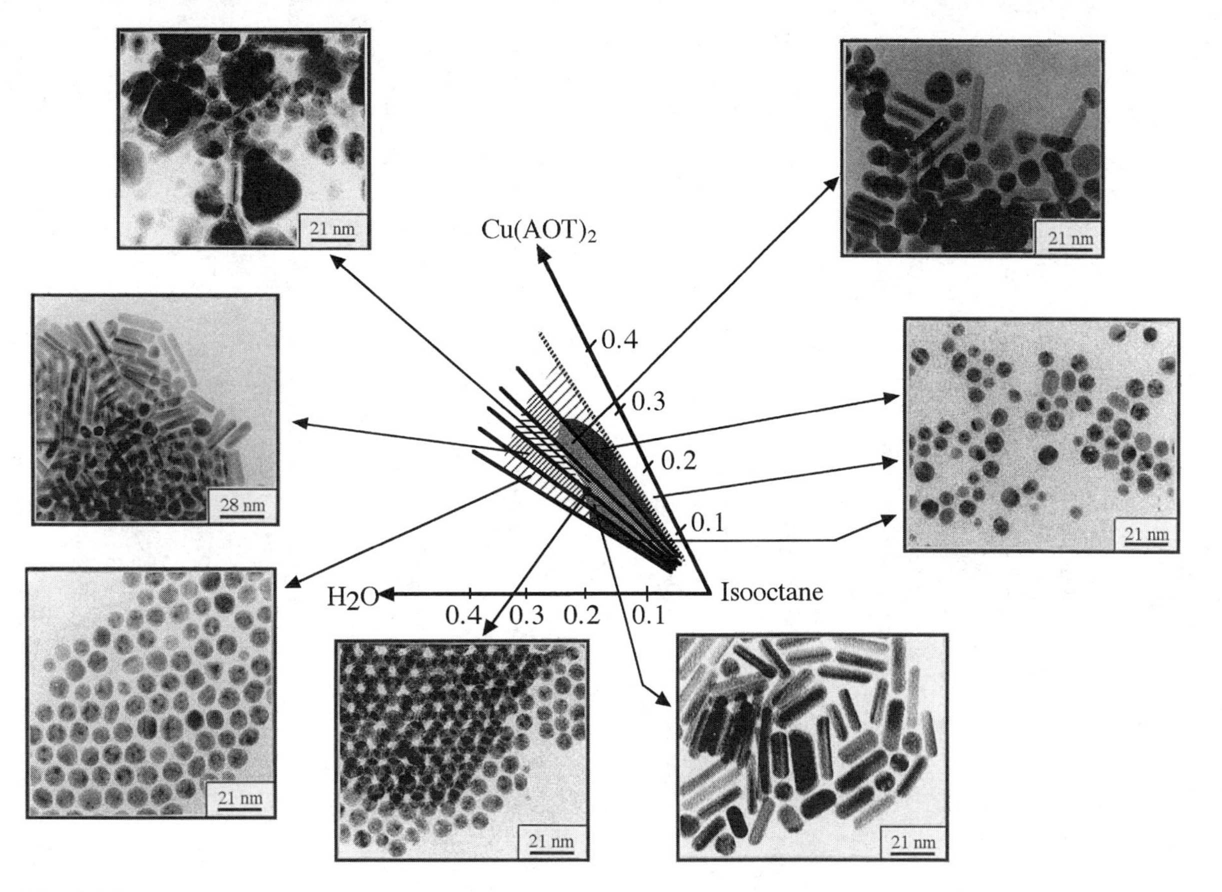

Fig. 9.3.2 Change in the copper metal shape with the increasing in the water content, *w*, of the colloidal assembly and keeping $Cu(AOT)_2$ concentration constant. $[Cu(AOT)_2] = 5 \times 10^{-2}$ *M*.

1. At low water content from $w = 2$ to 5.5, a homogeneous reverse micellar solution (the L_2 phase) is formed. In this range, the shape of the water droplets changes from spheres (below $w = 4$) to cylinders. At $w = 4$, the gyration radius has been determined by SAXS and found equal to 4 nm. Syntheses in isolated water-in-oil droplets show formation of a relatively small amount of copper metallic particles. Most of the particles are spherical (87%) with a low percentage (13%) of cylinders. The average size of spherical particles is characterized by a diameter of 12 nm with a size polydispersity of 14%.
2. An increase in the water content ($5.6 < w < 11$) destabilizes the solution, and the L_2 phase separates into a more concentrated reverse micellar solution (L_2*) and an almost pure isooctane phase. Structural studies indicate that L_2* is characterized by a bicontinuous network of cylinders with an increase in the number of connections with increasing w. The persistence length of the cylinders measured by SAXS does not change ($\langle l \rangle = 3$ nm) with increasing w. Syntheses at $w = 6$ show formation of a relatively large amount of copper metal cylinders (32%) in coexistence with 68% spheres. The average diameter of the spherical particles is 9.5 nm. The size distribution is rather large (27%). The average length over width ratio of the cylinder is found to be 3.5 with 40% polydispersity. The length and width of the cylinders are 22.6 ± 5.4 nm and 6.7 ± 1.4 nm, respectively.
3. At the phase transition ($w = 11$), an L_α phase made up of a mixture of planar lamellae and spherulites appears. Syntheses performed in this region of the phase diagram show formation of a very large amount of cylinders in coexistence with spheres. Very few large rods can be observed. Their sizes vary from 0.1 to 1 μm (58). In this region of the phase diagram, a slight increase in the number of cylinders is observed (38% of the particles are cylindrical whereas 62% are spherical). The average diameter of the spherical particles is 10.9 nm with 17% polydispersity in size distribution. The length and width of the cylinders are 25 ± 4 nm and 7.3 ± 1.4 nm, respectively. A slight increase in the size distribution is seen. The average length to width ratio of the cylinders is 3.7 with 44% polidispersity.
4. From $w = 15$ to $w = 19$, spherulites remain in equilibrium with isooctane. The spherulite size differs markedly (from 100 nm to 8000 nm). Syntheses in this phase region ($15 < w < 20$) show formation of particles having a higher polydispersity in size and in shape than those observed at low water content. As matter of fact, triangles, squares, cylinders, and spheres are observed.
5. At $w = 20$, an isotropic phase appears. By increasing the water content w from 20 to 29, the lamellar phase progressively disappears, giving rise to two phases consisting of isooctane and the isotropic phase. The latter is attributed to interdigitated reverse micelles. Syntheses at various water

contents from $w = 21$ to 29 induce formation of roughly 10% cylinders and 90% spheres. The percentage of cylinders decreases with increasing water content from 13 to 7%, whereas the cylinder length and width remain unchanged. These values are found to be 19.0 ± 2.5 nm and 6.7 ± 0.8 nm, respectively. The diameter of the spheres slightly decreases and the polydispersity increases with increasing water content from 9.7 ± 0.7 nm to 8.1 ± 1.2 nm at $w = 22$ and $w = 28$, respectively.

6. At $w = 30$, the isotropic phase remains in equilibrium with isooctane and is attributed to the L_2* phase, similar to that obtained at lower water content (in the range of $w = 5.6$ to 11). By increasing the water content from $w = 30$ to 35, the interconnected cylinders network is diluted with a decrease in the number of connections. Over all this water content range, formation of spherical and cylindrical copper metallic particles are observed. At $w = 34$, as at lower water content, cylindrical (42%) and spherical (58%) nanoparticles are observed. The average diameter of spherical particles observed in most of the cases is 9.5 ± 0.9 nm. The length and width of the cylinders are 19.8 ± 2.7 nm and 6.5 ± 0.8 nm, respectively.
7. At $w = 35$, an isotropic solution formed by water-in-oil droplets is obtained. On increasing the water content from $w = 35$ to 40, no drastic changes in the particle sizes and in the percentage of cylinders are observed. The TEM pattern obtained at $w = 40$. The particle size is 7.5 ± 0.6 nm (Fig. 9.3.2).

From these data it is concluded that the size, shape, and polydispersity of nanoparticles depend critically on the colloidal structure in which the synthesis is performed. This is well demonstrated when, by changing the water content, similar colloidal structures (reverse micelles or interconnected cylinders) are obtained:

1. Reverse micelles are formed below $w = 5.5$ and above $w = 35$. Most of the copper metal particles obtained in this region of the phase diagram are spherical. The percentage of spheres varies from 86% to 93%. At low water content ($w = 4$), the diameter of the particles is larger than that obtained at higher water content ($w > 35$). Hence, a slight decrease in the particle diameter with increasing water content is observed. In the reverse micellar regions, the length and width of the cylinders are larger at low water content and decrease with increasing w. Furthermore, the number of particles formed markedly increases with increasing water content. These differences could be mainly attributed to hydration of the head polar group.
2. Interconnected cylinders (L_2*) are formed in two water content range ($5.5 < w < 11$ and $30 < w < 35$). Syntheses in these two regions of the phase diagram show very strong correlation and similar data. Spherical and cylindrical particles are formed in both cases. No other particle shapes

have been observed. In both regions, the average diameter of spherical particles is the same (9.5 nm). Similarly, the size of the cylinders remains identical with an average length and width of 21.2 nm and 6.6 nm, respectively. The same average diameter and same ratio of cylinder axis ($\neq 3.3$) are observed at low ($5.5 < w < 11$) and high ($30 < w < 35$) water content.

Because of this strong similarity in various experimental conditions, this phenomenon is attributed to the structure of the colloid used as a template. This control of the particle shape could be explained as in nature (1,2) where the key step in the control of mineralization is the initial isolation of a space. As in the present case, a local supersaturation of reactants is needed to induce nucleation and let the system reach a lower energy state. In such a local supersaturation regime, the chiral molecules used to form the colloidal template [$Cu(AOT)_2$] impose an orientation of each reactant. The control of the crystal morphology could be due to the presence of high surfactant concentration (or compounds resulting from the chemical reaction), which specifically adsorbs on certain crystal faces. The same concept has been used to control the crystal morphology by the adsorption of impurities from solution onto specific crystal surfaces. Mechanisms of inhibition involved in the control of crystal morphology have been elucidated at the molecular level in studies on organic crystals and tailor-made inhibitors (59,60).

9.3.3.2 Controlling the Shape by Using Oil-in-Water Self-Assemblies

Copper dodecyl sulfate is made by mixing an aqueous solution of sodium dodecyl sulfate with copper sulfate, as described elsewhere (61). An aqueous solution of 0.1 *M* sodium docedylsulfate is mixed with 0.1 *M* copper sulfate. The solution is kept at 2°C, and the precipitate that appears is washed several times with a 0.1 *M* copper sulfate solution and recrystallized in distilled water. Copper dodecyl sulfate, $Cu(DS)_2$, forms micellar aggregates above the critical micellar concentration (cmc). The shape and size of these aggregates have been determined by SAXS and by light scattering and are found to be prolate ellipsoidal micelles with a hydrodynamic radius of 2.7 nm. The micellar size and shape are in good agreement with that observed with $Cd(DS)_2$ (62).

Copper(II) dodecyl sulfate, $Cu(DS)_2$, is reduced by sodium borohydride, $NaBH_4$, with an $NaBH_4/Cu(DS)_2$ ratio of 2, and copper metal aggregates are obtained for any of the experimental conditions described next. The synthesis is performed at the $Cu(DS)_2$ critical micellar concentration (cmc) of 1.2×10^{-3} *M*. TEM measurements made using a drop of this solution deposited on a carbon grid show formation of large domains of aggregates arranged in an interconnected network. Enhancement of this aggregate shows the interconnected network corresponds to a change in the particle shape and not to aggregation of strongly interacting small particles. Comparison of the absorption spectra for different shaped particles and spherical particles with 10 nm diameter shows a red shift in the plasmon peak when the shape of the copper metal particles changes from spheres to rods. Such changes in the absorption spectra of these particles can be related to those predicted at various r values, where r is defined as the ratio of the length to diameter of a cylinder (63–65). The plasmon peak due to the rod particles is centered at 570 nm. According to simulated absorption spectra (66), this peak corresponds to an r value of 2.5. From image analysis of the skeleton of the interconnected network corresponding to a plasmon peak centered at 570 nm, the average length of linear strands is 22 nm. The average minor diameter is 6.5 nm. Hence, the r value is 3.3. From the theoretical predictions, such an r value ($r = 3.5$) corresponds to a higher shift in the plasmon peak (620 nm) than that obtained (570 nm). This difference between r and the maximum of the plasmon peak is explained by the fact that the simulation is related to one size of cylindrical particles. In the present experiment, the particles are interconnected and not isolated cylinders. Furthermore, the polydispersity has to be taken into account. A qualitative shift in the maximum of the plasmon peak with the particle shape is in good agreement with those predicted. Thus, the shift and the ripening in the absorption band, compared to spherical and elongated particles, indicate that the interconnected network observed by TEM exists in solution. It is due to a change in the effective mean free path of conduction electrons and not to evaporation of the solutions.

9.3.4 Self-Organization in 2D and 3D Structures

The procedure to fabricate colloidal silver, $(Ag^0)_n$, spherical nanoparticles is similar to that already described (see Section 9.3.3): The $Cu(AOT)_2$ is replaced by the silver derivative. The relative concentration of Na(AOT), $Ag(AOT)_2$, and the reducing agent remain the same. Control of the particle size is obtained from 2 nm to 6 nm (67). To stabilize the particles and to prevent their growth, 1 μl/mL of pure dodecanethiol is added to the reverse micellar system containing the particles. This induces a selective reaction at the interface, with covalent attachment, between thio derivatives and silver atoms (68). The micellar solution is evaporated at 60°C, and a solid mixture of dodecanethiol-coated nanoparticles and surfactant is obtained. To remove the AOT and excess dodecanethiol surfactant, a large amount of ethanol is added and the particles are dried and dispersed in heptane. A slight size selection occurs, and the size distribution drops from 43% to 37%. The size distribution is reduced through the size selected precipitation (SSP) technique (38).

A drop of a dilute solution of 4.5-nm coated particles in hexane ($[(Ag)_n] = 2 \times 10^{-5}$ M) is deposited on a graphite surface. Figure 9.3.3A shows a large area of close-packed monolayers made of nanoparticles organized in a hexagonal network. The optical spectrum of the nanosized particles coated on graphite support is recorded in the reflectivity mode and is compared with the spectrum corresponding to free particles in hexane. The optical spectra are normalized to unity. The plasmon peak is shifted toward an energy lower than that obtained in solution (Fig. 9.3.3B).

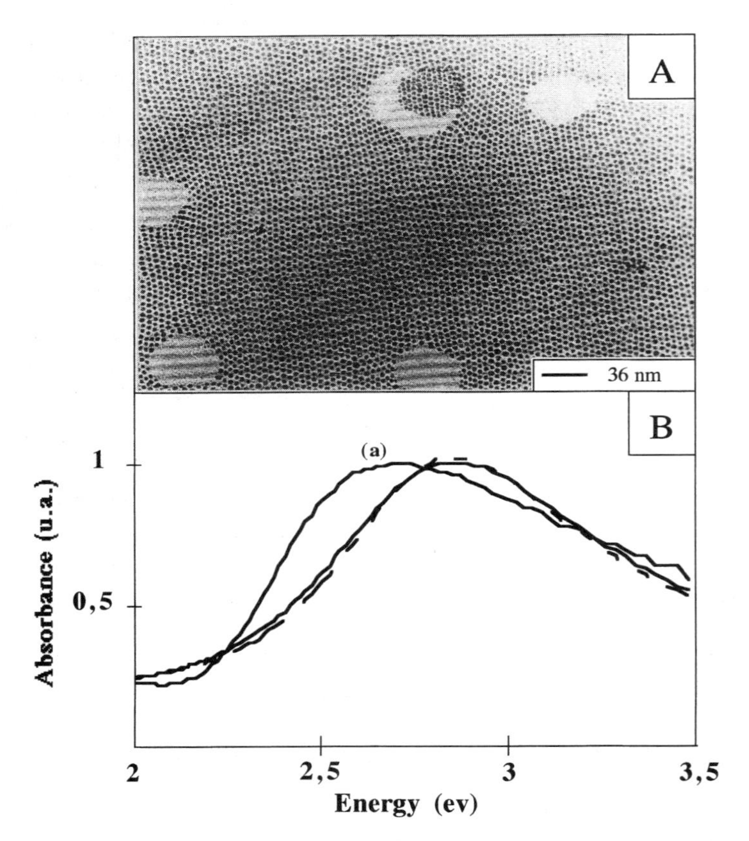

Fig. 9.3.3 (A) TEM micrograph of 4.5-nm silver particles deposited on graphite. (B) Absorption spectra of free 4.5-nm silver nanoparticles dispersed in hexane before (solid line) leaving a drop on the support, after washing the support with hexane (dashed line), and deposition on the support forming a 2D superlattice (a). ($[(Ag)_n] = 2 \times 10^{-5}$ *M*).

The coverage support is washed with hexane and the nanoparticles are redispersed in the solvent. The absorption spectrum of the latter solution is similar to that used to cover the support (free particles in hexane). This clearly indicates that the shift in the absorption spectrum of nanosized silver particles is due to their self-organization on the support. The bandwidth of the plasmon peak (1.3 eV) obtained after deposition is larger than that in solution (0.9 eV). This can be attributed to a change in the dielectric constant of the composite medium. Taking into account the variation of the simulated plasmon peak with the dielectric constant, it is concluded that the shift toward lower energy observed is due to an increase in the dielectric constant of the composite medium of the particles, which is the superposition of several factors (68), such as the spherical particles, the support, particle–particle interactions, and the air.

Similar behavior is observed for a diluted solution made of 5.2-nm and 6.2-nm nanosized particles.

When the particles are dispersed in hexane until complete evaporation of the solvent, large aggregates form. Figure 9.3.4A shows a rather high aggregate orienta-

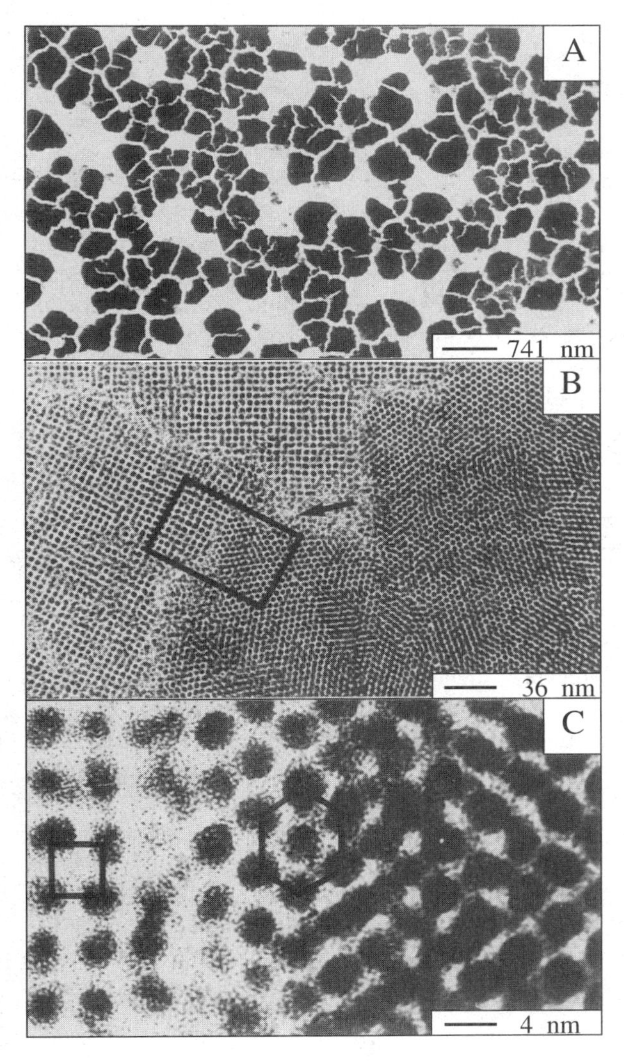

Fig. 9.3.4 Progressive enhancement of $(Ag^0)_n$ aggregates.

tion around a large hole or ring. The average distance between the oriented aggregates varies from 20 to 60 nm. The aggregate magnification confirms that it consists of $(Ag)_n$ nanoparticles and its average size is in the range of 0.03 to 0.55 mm^2. High magnification of one of these aggregates shows that the particles are arranged in two different symmetries. Figure 9.3.4B shows the formation of a polycrystal. Its magnification (Fig. 9.3.4C) shows either a hexagonal or a cubic arrangement of nanoparticles. The transition from one structure to another is abrupt, and there is a strong analogy with ''atomic'' polycrystals with a small grain called nanocrystals. Each domain or grain has a different orientation, clearly showing that the stacking of nanoparticles is periodic and not random. The ''pseudo-hexagonal'' structure corresponds to the stacking of a {110} plane of the fcc structure. A fourfold symmetry is observed in the same pattern (Fig. 9.3.4C), which is again characteristic of the stacking of {001} planes of the cubic structure. This cannot be found in the hexagonal structure. As a matter of fact, there is no direction in a perfect hexagonal compact structure for which the projected positions of the particles can take this configuration. This is confirmed by TEM experiments performed at various tilt angles, where it is always possible to find an orientation for which the stacking appears to be periodic. Hence, by tilting a sample having a pseudo-hexagonal structure, a fourfold symmetry is obtained. From these data, it is concluded that the large aggregates of silver particles are formed by stacking of monolayers in a face-centered cubic arrangement (38). Similar self-assemblies of silver sulfide (34–37) silver and gold nanoparticles have been obtained by using different experimental modes (39–43).

By increasing the concentration of 4.5-nm nanoparticles in hexane solution ($[(Ag)_n] = 4 \times 10^{-3}\,M$), the aggregates are so large that they cannot be investigated by TEM. By scattering electron microscopy, aggregates differing by their size and shape are observed (Fig. 9.3.5A). They are well faceted, as observed for the smaller aggregate by TEM. We choose to investigate one of these aggregates. Enhancement of one face of the aggregate shows well-defined defects in the crystal phase. The 60° tilt of aggregate confirms a three-dimensional (3D) structure. This permits us to estimate the size of the aggregate, which is 200 μm high and 100 μm large. To make sure that these aggregates are made of silver nanoparticles, EDX analyses were performed on the top of the aggregate and on the bottom (in the layers region). The analysis is made on a 0.786-μm^3 volume. The aggregate consists of silver, carbon, and sulfur atoms. This is fully consistent with a crystal made of silver nanosized particles coated by dodecanethiol. The ultraviolet (UV)-visible spectrum (Fig. 9.3.5B) of the aggregates just described shows a 0.25-eV shift toward lower energy of the plasmon peak with a slight decrease in the bandwidth (0.8 eV) compared to that observed in solution (0.9 eV). As described earlier for the monolayer, by washing the support, the particles are redispersed in hexane and the absorption spectrum remains similar to that of the colloidal solution used to make the self-assemblies.

For isolated particles, the increase in the dielectric constant induces a shift to the lower energy and an increase in the bandwidth of the plasmon peak. For particles

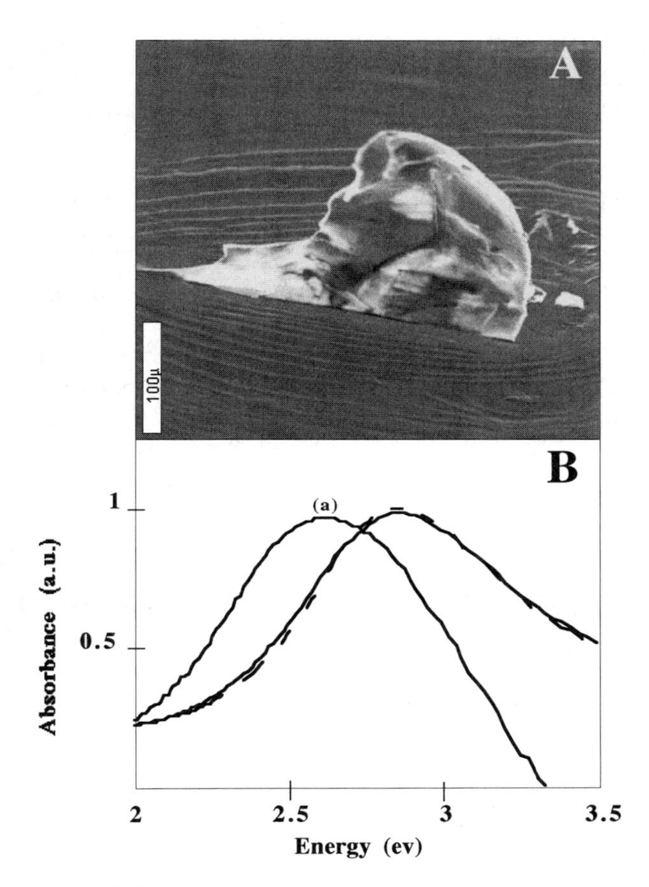

Fig. 9.3.5 (A) Scanning electron microscopy on concentrated solution of 4.5-nm silver particles ($[(Ag)_n] = 4 \times 10^{-3}$ *M*). Large aggregates on silver multilayers are present. (B) Absorption spectra of free 4.5-nm silver nanoparticles dispersed in hexane before (solid line) leaving a drop on the support, after washing the support with hexane (dashed line), and deposition on the support forming a 3D superlattice (a). ($[(Ag)_n] = 4 \times 10^{-3}$ *M*).

organized in an fcc structure, each silver nanoparticle is surrounded by 12 other particles, whereas in a monolayer it has 6 neighbors. So in 3D superlattices, the dielectric constant is the superposition of several factors, such as the external field, the dipole fields of all the other particles, and the surrounding due to the support. This induces an increase in the total dielectric constant, which, as for isolated particles, induces a shift toward the low energy of the plasmon peak. This is confirmed when Fig. 9.3.4 and 9.3.5 are compared: The shift of the plasmon peak of particles arranged in two-dimensional (2D) and 3D superlattice compared with the spectrum corresponding to free coated particles in hexane is 0.12 eV and 0.25 eV, respectively. Hence, the increase in the total dielectric constant induces an increase in the plasmon peak bandwidth. The bandwidth is of 0.8 eV, whereas for particles organized in 2D (Fig. 9.3.3) it is 1.3 eV. The value obtained for a 3D superlattice (0.8 eV) is close to that observed for free nanoparticles in hexane (0.9 eV).

These data show the effect of the medium dielectric constant of the particle when organized in 2D and 3D. The decrease in the bandwidth plasmon peak could be due to an increase in the mean free path conduction electrons of silver particles through a barrier of 2 nm. This is rather surprising, because the average distance between silver nanoparticles is 2 nm and we would not expect a tunneling electron effect through such large barrier. However, a recent paper published by Ung et al. (69) claims that a 1–2 nm distance between two metal surfaces is enough for tunneling of electrons across the double layers. Because we keep the same absorption spectrum and TEM pattern after washing the support, fusion between particles during the coverage can be excluded. No predictions have been given in the literature on the variation of the electromagnetic coupling of the particles when they are organized in 2D with a hexagonal network and 3D superlattice with a fcc structure. This does not permit us to definitively conclude a collective effect, due to the transport of the conduction electrons through the barrier due to the coating or due to changing in the dielectric medium.

Acknowledgments

I would like to thank very much my coworkers who have been in charge of these studies. Special thanks are due to A. Filakembo, Dr. I. Lisiecki, Dr. C. Petit, Dr. A. Taleb, and Dr. J. Tanori who participated in this work.

REFERENCES TO SECTION 9.3

1. L Addadi, S Weiner. Angew Chem Int Ed Engl 31:153, 1992.
2. S Mann. In: DW Bruce, D O'Hare, eds. Inorganic Materials. New York: J. Wiley & Sons, 1992.
3. MP Pileni. J Phys Chem 97:6961, 1993.
4. JH Fendler Chem Mater 8:1616, 1996.
5. C Tanford. The Hydrophobic Effect. New York: Wiley, 1973.
6. SJ Chen, DF Evans, BW Ninham, DJ Mitchell, FD Blum, S Pickup. J Phys Chem 90: 842, 1986.
7. DF Evans, DJ Mitchell, BW Ninham. J Phys Chem 90:2817, 1986.
8. IS Barnes, ST Hyde, BW Ninham, PJ Derian, M Drifford, TN Zemb. J Phys Chem 92: 2286, 1988.
9. G Porte, J Appell, Y Poggi. J Phys Chem 84:3105, 1980.
10. DJ Mitchell, BW Ninham. J Chem Soc Faraday Trans 2 77:601, 1981.
11. ST Hyde, S Anderson, K Larsoon, T Landn, SL Idin, Z Blum, B Ninham, eds. The Shape Language. New York: Elsevier, 1996.

12. JF Berret, DC Roux, G Porte, P Lindner. Europhys Lett 25:521, 1994.
13. JN Israelachvili, DJ Mitchell, BW Ninham. J Chem Soc Faraday Trans 2 72:1525, 1976.
14. MP Pileni, ed. Reactivity in Reverse Micelles. Amsterdam: Elsevier, 1989.
15. MP Pileni. N J Chem in press.
16. A Wokaun, JP Gordon, PF Liao. Phys Rev Lett 48:957, 1982.
17. C Petit, P Lixon, MP Pileni. J Phys Chem 97:12974, 1993.
18. I Lisiecki, MP Pileni. J Am Chem Soc 115:3887, 1993.
19. I Lisiecki, MP Pileni. J Phys Chem 99:5077–5082, 1995.
20. LE Brus. J Chem Phys 79:5566, 1983.
21. C Petit, MP Pileni. J Phys Chem 92:2282, 1988.
22. C Petit, P Lixon, MP Pileni, J Phys Chem 94:1598, 1990.
23. L Motte, C Petit, L Boulanger, MP Pileni, Langmuir 8:1049, 1992.
24. K Suriki, M Harada, A Shioi. J Chem Eng Jn 109:245, 1996.
25. T Hirai, H Sato, I Komasawa. Ind Eng Chem Res 33:3262, 1994.
26. J Cizeron, MP Pileni. J Phys Chem 99:17410, 1995.
27. J Cizeron, MP Pileni. J Phys Chem 101: 1997.
28. L Levy, JF Hochepied, MP Pileni. J Phys Chem 100:18322, 1996.
29. L Levy, N Feltin, D Ingert, and MP Pileni. J Phys Chem 1997, 101, in press
30. Nanostruct Mater Chem Mater 8(5), 1996.
31. MP Pileni. Langmuir 13:3266, 1997.
32. N Kimizuka, T Kunitake. Adv Mater 8:89, 1996.
33. AS Tse, Z Wu, SA Asher. Macromolecules 28:6533, 1995.
34. L Motte, F Billoudet, MP Pileni. In preparation.
35. L Motte, F Billoudet, MP Pileni. J Phys Chem 99:16425, 1995.
36. L Motte, F Billoudet, E Lacaze, MP Pileni. Adv Mater 8:1018, 1996.
37. L Motte, F Billoudet, E Lacaze, J Douin, MP Pileni. J Phys Chem 101:138, 1997.
38. A Taleb, C Petit, MP Pileni. Chem Mater 9:950, 1997.
39. RL Whetten, JT Khoury, MM Alvarez, S Murthy, I Vezmar, ZL Wang, CC Cleveland, WD Luedtke, U Landman. Adv Mater 8:429, 1996.
40. M Brust, D Bethell, DJ Schiffrin, CJ Kiely. Adv Mater 7:9071, 1995.
41. SA Harfenist, ZL Wang, MM Alvarez, I Vezmar, RL Whetten. J Phys Chem 100:13904, 1996.
42. JR Heath, CM Khobler, D Leff. J Phys Chem B 101:189, 1997.
43. SA Harfenist, ZL Wang, RL Whetten, I Vezmar, MM Alvarez. Adv Mater 9:817, 1997.
44. SA Safran, LA Turkevich, PA Pincus. J Phys Lett 45:L69, 1984.
45. TF Towey, A Khan-Lodl, BH Robinson. J Chem Soc Faraday Trans 2 86:3757–3762, 1990.
46. C Robertus, JGH Joosten, YK Levine. J Chem Phys 93(10):7293–7300, 1990.
47. G Cassin, JP Badiali, MP Pileni, J Phys Chem 99:12941–12946, 1995.
48. TK Jain, G Cassin, JP Badiali, MP Pileni. Langmuir 12:2408–2411, 1996.
49. ML Stigerwald, AP Alivisatos, JM Gibson, TD Harris, R Kortan, AJ Muller. J Am Chem Soc 110:3046, 1988.
50. I Lisiecki, M Borjling, L Motte, BW Ninham, MP Pileni. Langmuir 11:2385, 1995.
51. C Petit, MP Pileni. J Mag Mag Mater 166:82, 1997.
52. CV Fragstein, H Roemer. Z Physik 151:54, 1958.
53. H Roemer, CV Fragstein. Z Physik 163:27, 1961.

54. CV Fragstein, FJ Schoenes. Z Physik 198:477, 1967.
55. C Petit, P Lixon, MP Pileni. Langmuir 7:2620, 1991.
56. J Tanori, T Gulik, MP Pileni. Langmuir 1996.
57. J Tanori, MP Pileni. Langmuir 1996.
58. J Tanori, MP Pileni. Adv Mater 7:862, 1995.
59. I Weissbuch, L Addadi, Z Berkovitch-Yellin, E Gati, M Lahav, L Leiserowitz. Nature 310:161, 1984.
60. I Weissbuch, F Frolow, L Addadi, M Lahav, L Leiserowitz. J Am Chem Soc 112:7718, 1990.
61. Y Moroi, K Motomura, R Matuura. J Colloid Interface Sci 46:111, 1974.
62. C Petit, TK Jain, F Billoudet, MP Pileni. Langmuir 10:4446, 1994.
63. EJ Zeman, GC Schatz. J Phys Chem 91:634–643, 1987.
64. MP Cline, PW Barber, RK Chang. J Opt Soc Am B 3:15–21, 1986.
65. DS Wang, M Kerber. Phys Rev B 24:1777, 1981.
66. I Lisiecki, F Billoudet, MP Pileni. J Phys Chem 100:4160–4166, 1996.
67. CD Bain, EB Troughton, YT Tao, J Evall, GM Whitesides, RG Nuzzo. J Am Chem Soc 111:7155, 1989.
68. BW Ninham, RA Sammut. J Theor Biol 56:125, 1976.
69. T Ung, M Giersig, D Dunstan, P Mulvaney. Langmuir 13:1773, 1997.
70. SW Charles. J Mag Mag Mater 65:350, 1987.
71. N Moumen, MP Pileni. Chem Mater 8:1128, 1996.
72. N Moumen, MP Pileni. J Phys Chem 100:1867, 1996.
73. N Moumen, I Lisiecki, V Briois, MP Pileni. Supramol Sci 2:161, 1995.
74. N Moumen, P Bonville, MP Pileni. J Phys Chem 100:14410, 1996.
75. N Feltin, MP Pileni. Langmuir 13:3927, 1997.
76. E Matijevic. Chem Mater 5:412, 1993.

9.4 FORMATION IN GAS PHASES

KEISAKU KIMURA
Himeji Institute of Technology, Akou-gun, Hyogo, Japan

9.4.1 Introduction

In this chapter, we treat metallic fine particles whose size is less than micrometers; in many cases down to nanometers, produced by physical methods in the gas phase (aerosol technique). Physical methods have a great advantage for producing fine particles because of their versatility and universality for application to many sorts of substances, rather than chemical methods, although they have a weak point in size control and mass production. It should be emphasized that chemically clean surfaces can be obtained by a physical method without any sophisticated techniques. If chemical reaction takes place, the surfaces of metallic particles are generally covered with unknown by-products. It is difficult to remove these contaminating species, once they have occurred, to reach the desired purity level.

9.4.2 Overview of Metal Fine Particles in Gas Phase

Particles produced in the gas phase must be trapped in condensed media, such as on solid substrates or in liquids, in order to accumulate, stock, and handle them. The surface of newly formed metallic fine particles is very active and is impossible to keep clean in an ambient condition, including gold. The surface must be stabilized by virtue of appropriate surface stabilizers or passivated with controlled surface chemical reaction or protected by inert materials. Low-temperature technique is also applied to depress surface activity. Many nanoparticles are stabilized in a solid matrix such as an inert gas at cryogenic temperature. At the laboratory scale, there are many reports on physical properties of nanometer-sized metallic particles measured at low temperature. However, we have difficulty in handling particles if they are in a solid matrix or on a solid substrate, especially at cryogenic temperature. On the other hand, a dispersion system in fluids is good for handling, characterization, and advanced treatment of particles if the particles are stabilized.

The basic idea of the formation of nanometer-sized particles in the gas phase is a rapid quenching of supersaturated metal vapor by room-temperature or cold inert gas. Therefore the preparation is inherently processed in nonequilibium conditions. In Figure 9.4.1 is a schematic diagram of a typical aerosol process. At the extreme of a high-temperature furnace, vaporized metal atoms cooled down by an inert gas, forming an embryo or subnucleus within a very short time interval, a few microseconds, are traveling across a temperature gradient of thousands of kelvins per centimeter. In this supercritical region, numerous nuclei or embryos formed

Quenching time $\Delta t = \text{m s} \sim \mu\text{s}$

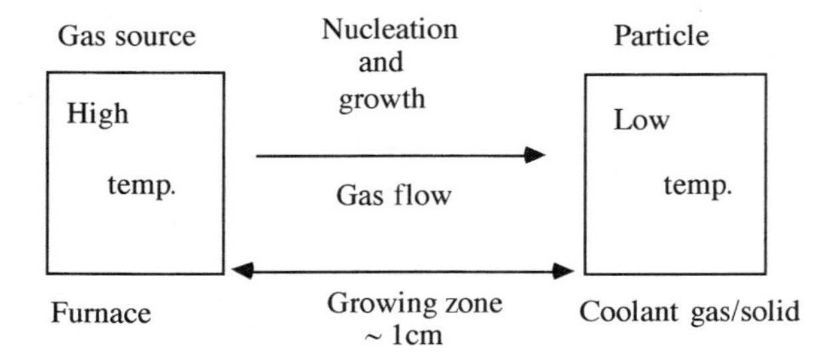

Temperature gradient $\Delta T = 1000 \sim 10{,}000\text{K/cm}$

Fig. 9.4.1 Schematic diagram of a particle growth in the aerosol technique. Nucleation proceeds in between two substrates with high and low temperature, the difference of which is several thousands of kelvins. High-temperature substrate is a heating element and low-temperature substrate is a kind of coolant such as gas, liquid, or solid substrate, depending on the operation mode. (From Ref. 1.)

coalesce with each other, forming the nucleation phase where stable nuclei formed from embryos give larger clusters and stop growing when they are traveling out of the growth zone where there is no further supply of atoms, embryos, and nuclei. Finally, particles are quenched by a room-temperature gas stream or by a cooled solid substrate. This is a rapid process and hence proceeds in a non-thermoequilibrium condition. The temperature difference between furnace and coolant is normally several hundred or a thousand kelvins, leading the gradient being 1000–10,000 K/cm. The resultant quenching rate is on the order of a million kelvins per second, nominally enough to quench thermodynamically unstable phase, such as a nanophase, high-temperature, quasi-stable phase and a quasi-crystal phase. Finally, these materials are collected as minute particles at room temperature.

9.4.3 Growth Mechanism in Gas Phase

We discuss here two major processes, the absorption growth process where particle nuclei grow up with absorption of metal atoms, and the coalescence growth process where particles grow up by collision of particle nuclei or clusters (2). The combination of these processes leads to two well-known distributions, the normal-like distribution and the lognormal distribution, depending on the growth condition. These distributions are frequently found for the size distribution of small particles. The general equation describing crystal growth in the aerosol phase is very complicated and is practically impossible to solve. Under a special condition, the equation is solvable and gives the two well-known distribution functions already mentioned. Hence, from the size distribution analysis, we can estimate the growth mechanism. In general, the lognormal function is found and results from a coalescence growth of the clusters. The normal distribution function is not often found in the size distribution. However, the absorption mechanism is effective in the very earliest stages of the crystal growth—that is to say, only applicable to minute particles whose size is less than 2 or 3 nm, the size is scarcely found in literature.

Absorption Growth Mechanism. Here we assume that a nucleus collides with a single atom at each time. This is equivalent to assuming that there are many free atoms in the gas phase and less opportunity for clusters to meet each other. This is a step-by-step reaction. The rate equation for *j*th step is then expressed as

$$\mathrm{u}_j + \mathrm{u} \rightarrow \mathrm{u}_{j+1} \tag{1}$$

where u stands for a metal atom and u_j is a cluster made of j atoms.

The other processes, such as

$$\mathrm{u}_i + \mathrm{u}_j \rightarrow \mathrm{u}_{i+j} \tag{2}$$

is neglected because $\mathrm{u} \gg \mathrm{u}_i, \mathrm{u}_j$. Assuming the shape of a particle being sphere, a distribution function in a special case (2) is derived and gives a skew distribution with a tail toward smaller size. On the other hand, a lognormal distribution mentioned later is a skew distribution with a tail on the larger size end. The combination of these distributions leads to an expression that is approximated by the well-known Gaussian distribution function, expressed by

$$f(d) = \frac{1}{\sqrt{2\pi}\,\sigma} \exp\left[\frac{(d - \overline{d}}{2\sigma^2}\right] \tag{3}$$

in which σ is a standard deviation, d is the diameter of a particle and $\overline{d}$ stands for the mean diameter. Note that $f(d)$ is symmetric against $d = \overline{d}$, which implies the existence of negative size and does not present a real size distribution. However, it is convenient to describe more-or-less symmetrically shaped size histograms using Eq. (3). It should be noted that this function is only applicable to the beginning period of a crystal growth.

Coalescence Growth Mechanism. Following the very early step of the growth represented by Eq. (1), many nuclei exist in the growth zone. Hence Eq. (2) would be a major step for the crystal growth. Since there are many nuclei and embryos with various sizes in the zone, u_j in Eq. (2) can be assumed to be a random variable. Due to mathematical statistics, the fraction of volume approaches a Gaussian after many coalescence steps (3). A lognormal distribution function is defined by

$$f_{\ln}(d) = \frac{1}{\sqrt{2\pi} \ln \sigma} \exp\left[-\frac{1}{2}\left(\frac{\ln d/\bar{d}}{\ln \sigma}\right)^2\right] \tag{4}$$

Here the symbols σ, d, and $\bar{d}$ have the same meaning as in Eq. (3). In the size histogram, the fraction of the number of particles per diameter interval is written as $\Delta n/n = f_{\ln}(d) \cdot \Delta d/d$.

We will show two examples on the analysis of the size distribution. Figure 9.4.2 stands for the size histogram of In particles produced by a gas flow cold-trap method in acetone (4). The best fit with the lognormal distribution is shown by a broken line in the figure. The fit is generally good, but it is difficult to achieve the best fit only from the appearance of fitting to the distribution curve. Instead, a graphical analysis is usually engaged. If the size histogram is a linear normal distribution, cumulation of the number of particles at a certain size follows the error function, and plotting of the points on the normal probability paper will give a straight line. Figure 9.4.3a is a normal-probability plot of the sample shown in Figure 9.4.2. It is clear that the size distribution does not fit with the normal distribution. On the other hand, a lognormal probability plot of the same sample shows a good straight line as seen in Figure 9.4.3b, indicating that the size distribution of this sample is of a lognormal. For sample number 250 and number of classes 29, the best fit was obtained with σ of 2.13 and $\bar{d}$ of 20 nm and is shown by the broken line in Figure 9.4.2. The correctness of the fitting was also checked by the χ^2 test. The significance level obtained was over 75%, showing the strong reliability of this fit, because the rejection level is normally set at 1 or 5%. The In sample produced by a matrix isolation method in acetone also gave a lognormal distribution with $\sigma = 1.8$ and $\bar{d} = 10.8$ nm (Fig. 9.4.4). These two examples are typical for the lognormal distribution and for the larger size particles.

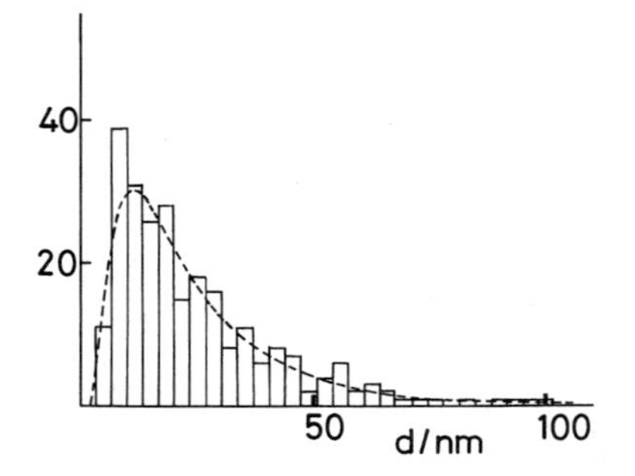

Fig. 9.4.2 Size histogram of In small particles produced by a gas flow–cold trap method in acetone. The pressure of He and Ar mixed gas was 1.3 kPa. The ordinate represents the number of samples at a given size interval. Total number of samples, 250. Broken line, calculated curve from the lognormal distribution with $\sigma = 2.13$ and $\bar{d} = 20$ nm. (From Ref. 4.)

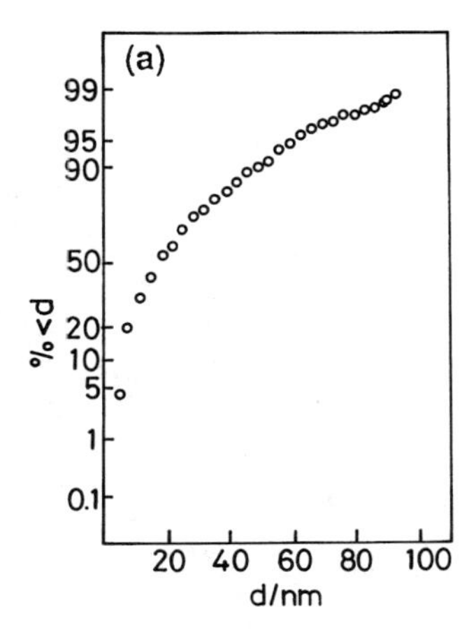

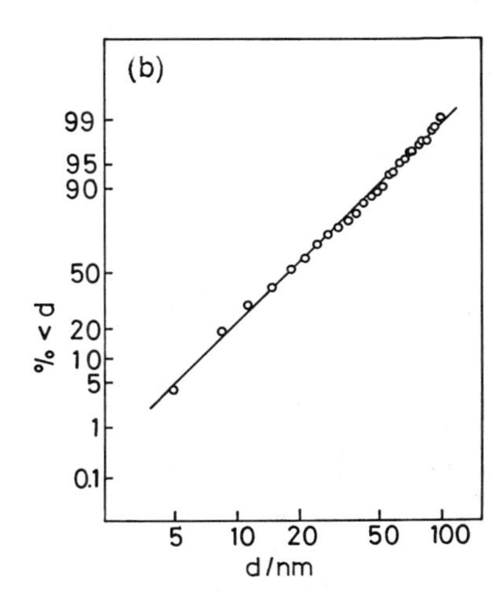

Fig. 9.4.3 Normal-probability plot (a) and lognormal probability plot (b) for In particles shown in Figure 9.4.2. The ordinate stands for the cumulative percent of particles, with diameters smaller than d on the abscissa. This follows an error function and should give a straight line if the plot obeys the correspondent distribution, as seen in case (b). (From Ref. 4.)

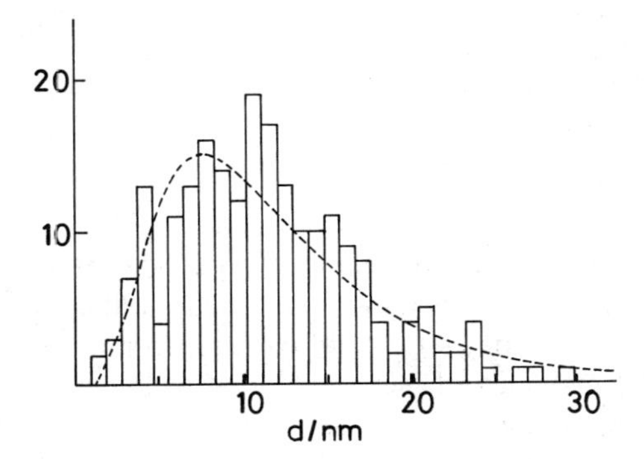

Fig. 9.4.4 Size histogram of In ultrafine particles produced by a matrix isolation method in acetone. Ordinate is the same as for Figure 9.4.2. Broken line is a calculated curve from the lognormal distribution with $\sigma = 1.8$ and $\bar{d} = 10.8$ nm. The χ^2 test gave a significance level of 30%. (From Ref. 4.)

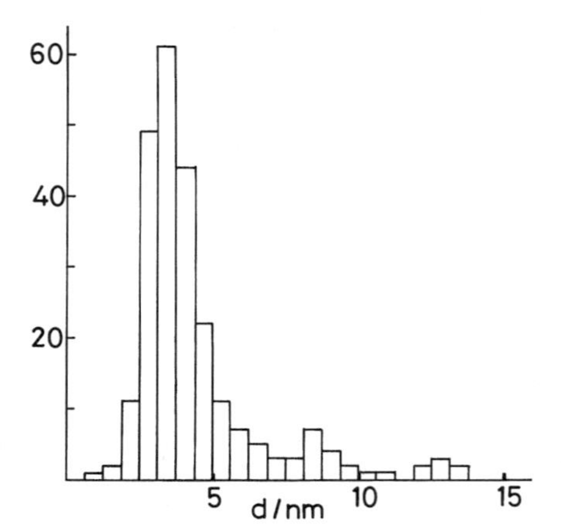

Fig. 9.4.5 Size histogram of In ultrafine particles produced by a gas flow–solution trap method in acetone. The gas pressure was 1.3 kPa pure He. The maximum height is at d = 3 nm. Ordinate represents the number of samples at a given size interval. No good was obtained for normal and lognormal probability plots. (From Ref. 4.)

For minute size samples, we expect the normal-like size distribution. Figure 9.4.5 shows the size histogram of In ultrafine particles prepared by the gas flow–solution trap method in acetone under He atmosphere. The maximum size distribution appeared at about 3 nm in the histogram, suggesting that the coalescence growth process is not a major process in this size region. Indeed, the size histogram shown in Fig. 9.4.5 has a tail at both smaller and larger size ends and does not fit with either the exact lognormal or the normal distribution. The long skew tail toward larger size may indicate that there occurs amalgamation of particles in the suspension. We discuss this in Section 9.4.6. Note that in all examples here, the particles are sampled from a suspension state—that is to say, the size distribution reflects that in a dispersion state and not in the gas phase.

9.4.4 Methods

As mentioned in the preceding section, to stabilize a particle, metallic particles must be produced in an inert gas atmosphere and then trapped in appropriate liquids to finely disperse them. Unless liquid is used, particles tend to coalesce, forming a larger particle or aggregate, the size of which exceeds several tens of nanometers as a powdered sample. First we introduce the normal gas evaporation technique to show the principle of aerosol method. Then several modifications are described to get dispersed metallic systems as a colloid.

Inert Gas Evaporation Technique. Inert gas evaporation technique is a familiar method in an aerosol production of ultrafine particles as already mentioned in Section

9.4.3. Gas evaporation using Ar for the preparation of various sort of metal fine powders was first reported by Kimoto et al. in 1963 (5). The production chamber of this method is basically the same as that of a vacuum sublimation chamber. A target material is heated in this chamber with several torr inert gas atmosphere. The nanometer-sized particles are easily formed in the chamber space. However, by this method, it is difficult to get genuine nanoparticles whose sizes are several nanometers. This is because of the radiation heating in a production chamber, resulting particle coalescence on the chamber wall or particle collector, as well as the direct particle contact in the deposited particle layer (powders). Therefore the size becomes several tens to hundreds of nanometers. Several ultrafine metallic powders are now commercially available, including Cu, Ag, Al, Ni, Co, Fe, and Au, with a size of several tens of nanometers.

Matrix Isolation Method. The radiation heating and coalescence of nanoparticles mentioned earlier can be avoided with the use of a cold substrate instead of using a room-temperature chamber wall. Application of cryogenic wall to nanoparticles was first reported by Wada and Ichikawa (6,7). Later this technique was modified for several applications and was widely used by many researchers (8–10). Figure 9.4.6 depicts an apparatus for the matrix isolation method. This is basically the same as that used for the vacuum sublimation in the cryogenic matrix method widely used in metal vapor deposition (11,12).

Organic vapor is sublimed through a solvent feeder into a glass Dewar cooled with liquid nitrogen (LN_2). After forming a cryogenic matrix film, a piece of metal is sublimed in the low-pressure inert-gas atmosphere, giving ultrafine particles in the gas phase. The nanocrystallites formed adhere to the surface of the frozen organics. Then the inert gas is again pumped off to high vacuum. Organic vapor is introduced again to form a cryogenic matrix. This procedure is repeated several times until enough amount particles accumulate in the cold matrix. Finally, the matrix is melted at room temperature, giving a colloidal suspension. Note that the total process is done in a rigorously anaerobic condition. Using this method with tetrahydrofuran as a dispersion medium, potassium and sodium ultrafine particles were formed. Deep blue color was perceived in these suspensions, a characteristic color of solvated electrons. The colloids were extremely unstable due to reactivity of alkali metals. Calcium colloid had a life of about 20 min and its optical spectrum can be recorded (10).

One can control the size of particles in a wide range from 1 μm to 1 nm by changing the pressure of ambient gas and the temperature of a crucible. Figure 9.4.7 shows how the size is varied by various combinations of species of gases and experimental conditions, taking Mg particles as an example (4). The largest particles were obtained from a conventional gas evaporation technique, while the smallest one is from a matrix isolation method using tetrahydrofuran as a supporting liquid in a helium atmosphere. The peak positions in the size histogram are at 2, 19, and 100 nm, respectively.

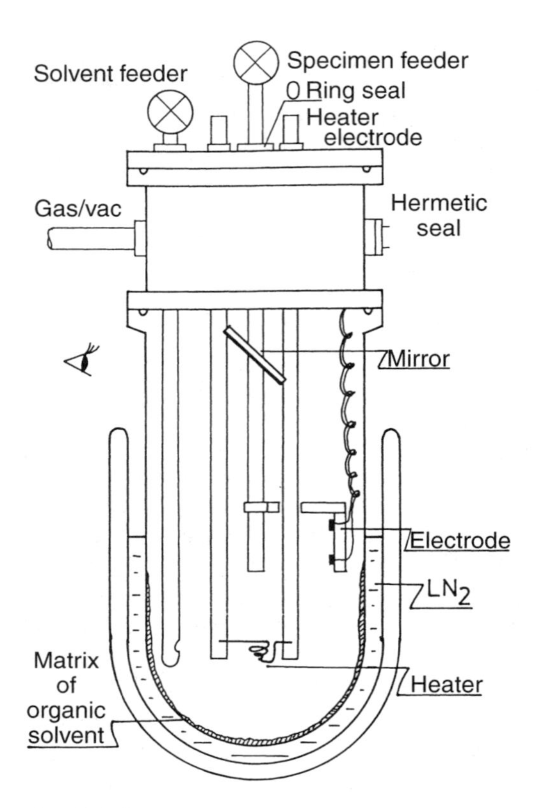

Fig. 9.4.6 Apparatus for the matrix isolation method. Organic liquids are sublimed into a Dewar vessel through a solvent feeder to form a cryogenic matrix on the Pyrex glass wall cooled with liquid nitrogen. Inert gas is introduced via gas inlet. A target material in a crucible is heated in a gas to form ultrafine particles, which are deposited on a cryogenic matrix. The processes are repeated several times until enough particles are accumulated on a cryogenic matrix. (From Ref. 10.)

Gas Flow–Cold Trap Method. The apparatus for the gas flow–cold trap method is shown in Figure 9.4.8. A small piece of metals is heated in a crucible set in a vacuum chamber made of stainless steel under a flow of helium or argon from several to several tens of torrs. The chamber was evacuated by a rotary pump prior to the introduction of gas. Under a constant flow of the gas, fine particles are formed in the gas stream just after sublimation at the crucible. A vapor of organic liquid, which was completely degassed in advance, was introduced through a Teflon valve in the carrier gas stream and then trapped in the glass container cooled with liquid nitrogen. The particles formed then are trapped in a glass container together with

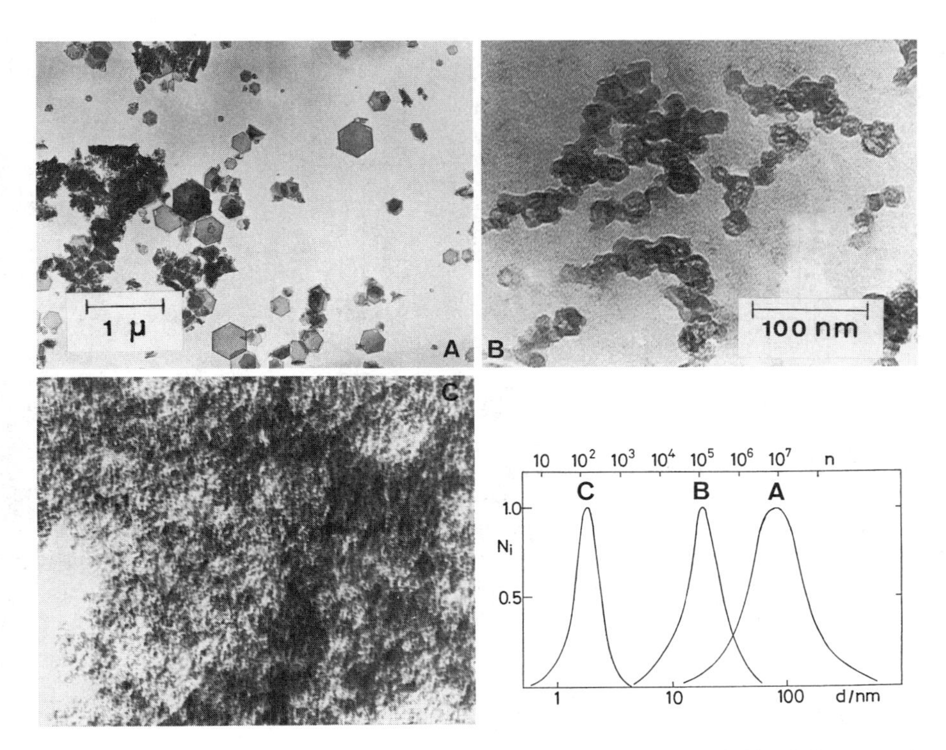

Fig. 9.4.7 Electron microscopic images of Mg small particles prepared by different methods and their size histograms. (A) Mg fine powders produced by a conventional gas-evaporation method with Ar at 4 kPa. (B) Mg fine particles produced by a matrix isolation method with Ar at 300 Pa in tetrahydrofuran. (C) Mg ultrafine particles produced by a matrix isolation method with He at 1.3 kPa in tetrahydrofuran. The scale bar for (C) is the same as for (B). Abscissa at top (n) is a rough estimate of the number of Mg atoms in a single particle whose diameter is represented by logarithm of diameter (nm) in the bottom scale. Ordinate (N_i) is a normalized number of particles in a unit size width. (From Ref. 4.)

organic vapor to form the cryogenic matrix. After warming at room temperature, metal sols are formed. We have made denser colloids by this method than by matrix isolation method. This has merit because the particle formation chamber is separated from the particle trapping chamber, which is free from the radiation heating of the furnace. However, separation of the chamber introduces a low particle yield.

Gas Flow–Solution Trap Method. Figure 9.4.9 shows an apparatus for a gas flow–solution trap method. Two kinds of gases, helium and argon, were introduced: He for the production of fine particles and Ar for the collection and transportation of particles to a trapping vessel. The procedure is the same as for the case of the

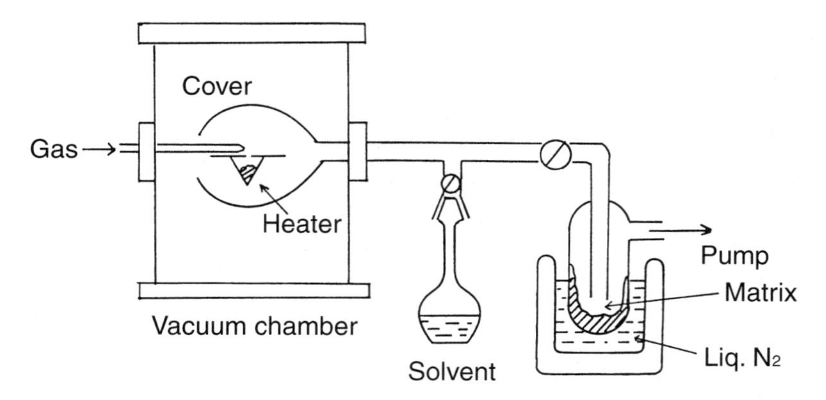

Fig. 9.4.8 Apparatus for the gas flow–cold trap method. One of the gas inlets is abbreviated in the figure. A cover in the figure is used to improve a collection yield. Without this, the particles produced just above a crucible are drifting in the chamber to deposit on the chamber wall. The end of the gas line is evacuated by a rotary pump so as to constantly flow the carrier gas. A solvent feeder is inserted on the gas line for the sublimation of organic liquid to deposit at the trapping apparatus cooled with liquid nitrogen. (From Ref. 10.)

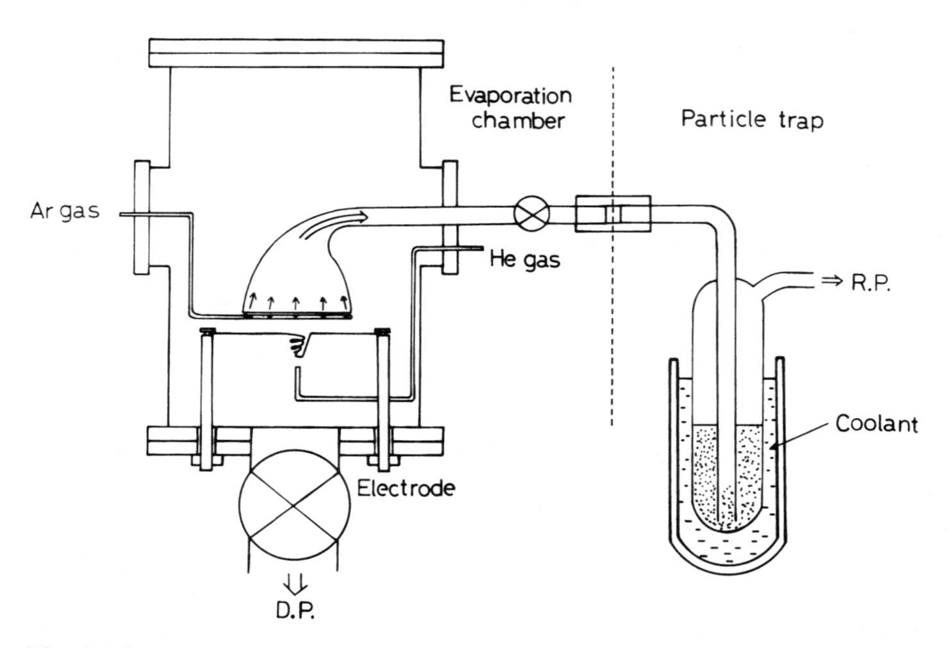

Fig. 9.4.9 Apparatus for the gas flow–solution trap method. Nanoparticles produced in the inert gas stream are trapped in a cooled liquid to form a colloid. (From Ref. 10.)

gas flow–cold trap method except that the cold trap is replaced by an organic liquid, which is cooled by cold ethanol in the temperature range -30 to $-40°C$ depending on the kind of solvent. On one hand, the matrix isolation method and the gas flow–cold trap method sometimes gave coagulated samples and did not give a finely dispersed suspension. On the other hand, the gas flow–solution trap method made reproducible, well-dispersed suspensions at the expense of sample dilution.

We point out here that the colloid prepared by these methods is very clean, because the carrier gas used is usually high-purity grade at six-nine, the chamber is once evacuated to depress the extent of contaminating oxygen and moisture, and the liquids themselves are always purified by sublimation process except for the solution trap method. To transfer the colloidal suspension after preparation, a specially designed stock bottle with a Luer-lock syringe is normally used in order to enable the operations under Ar flow to avoid unexpected air contamination. Therefore, we can carry the suspension liquid away from the production chamber without exposure to air, which means that the surface of colloidal metal is very clean if it does not react with suspension liquids.

9.4.5 Sublimation Technique and Apparatus

In order to sublime metals in an inert gas atmosphere, target substances must be heated up to their melting temperature to get enough vapor pressure for the condensation. So far, many heating devices have been developed, such as wire or plate resistance heating, contact resistance heating, laser irradiation, Ar arc plasma discharge, glow discharge, inert gas sputtering, electron beam heating, and microwave induction heating. However, due to the restrictions posed on an experimental setup, only a few of them are popular in preparation of ultrafine particles. We describe some of them next.

Resistance Heating. Three types of heating tools are frequently used; tungsten wire, boat, and crucible made of refractory materials. In the former two, target metals are directly put on the wires or boats. Special care should be taken for the reactivity of metals to filament materials, such as the case of aluminum on tungsten filament. In such a case, an alumina-coated crucible is used instead. The heating substances often used are tungsten for the boat and filament, and molybdenum, tantalum, boron nitride, and carbon for the boat. The power of the heater is controlled by the sectional area and the length of the filament and the applied electric power. A filament with desired resistance value can be easily hand made by adjusting the number of turns or can be purchased commercially. Crucibles often used are made of a tungsten heater covered with silica or alumina.

Plasma Discharge. The low-voltage and high-current-density process in plasma gas is often called arc plasma or hot plasma. A high-power ion beam formed in the plasma flame hits the target substances to heat them to several thousand kelvins, which is enough to melt all solid materials. Under the inert gas flow, the vaporized

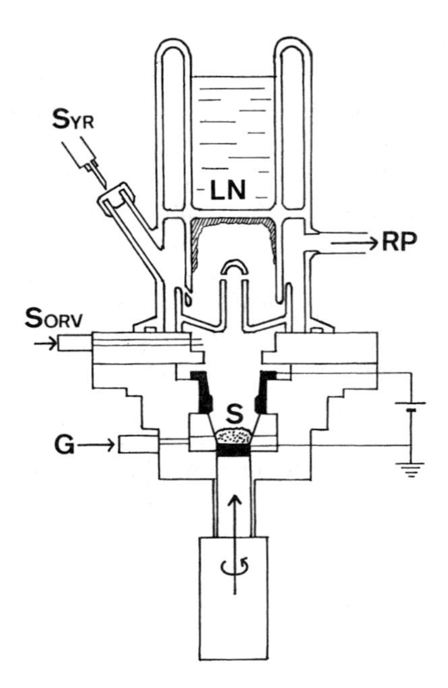

Fig. 9.4.10 Apparatus for the gas flow–arc plasma method. The apparatus is composed of two components. The upper part is a glass Dewar, which accumulates small particles in a cryogenic matrix on the trim cooled with liquid nitrogen (LN). Sorv, inlet of organic vapor; Syr, syringe for transferring produced colloids under anaerobic conditions; RP, rotary pump; S, target sample. Lower part is for plasma discharge. A BN furnace has gas inlets (G) and is specially designed for Ar gas to flow in screwed stream; hence the plasma is emitted in a jet flame due to a plasma pinch effect. The black parts are copper electrodes cooled by water. In order to maintain a constant spacing between the surface of sample and the upper electrode, the sample position can move vertically so that the current through the sample to the upper electrode is precisely controlled and constant. This is very important to produce powders with a narrow size distribution.

substances condense to form nanoparticles in a gas phase, which are carried away in gas stream from heating area. The shield of the plasma flame is very important to produce nanosized materials. There are many reports on the plasma discharge apparatus in a dry process (13,14). Figure 9.4.10 depicts a schematic diagram of the gas flow–arc plasma method with cold trap apparatus. We can arrange the system in the solution trap mode as well. The yield rate is very efficient, such as 10 g/h, at the expense of the size distribution, often mixed with micrometer-size droplets.

Sputtering Method. The high-voltage (around 1000 V) and low-current (micro-ampere) plasma process is called glow discharge. Under this plasma condition (cold plasma), solid materials are sputtered by the bombardment of accelerated positive ions. Normal sputtering takes place at a pressure less than 1 torr. However, in order to form nanoparticles in the gas phase, the pressure must enhance to the level of several torrs. We must devise an electric circuit to stabilize the current in such a high gas pressure. The production of nanoparticles utilizing a sputtering technique was first developed by Yatsuya et al. (15). Following this report, this technique was applied to the production of ultrafine particle of metals and semiconductor (16). In Figure 9.4.11 is shown the sputtering apparatus for the production of liquid suspension system. The merit of the sputtering mode is that it is a long-time stable operation, such as 1 h, and thus can be used for a source of nanoparticle beam. In addition, it can be used for refractory materials, including W, Mo, Pt, and Si. However, the production rate is very low, such as micrograms per minute per square centimeter per watt.

Size and Yield as a Function of Flow Rate and Gas Pressure. The effect of gas pressure on the size of particles prepared by the gas evaporation method was noticed at an early period of the development of this method (17). It was reported that to obtain the same size, the pressure of helium had to be about 10 times as large as that of argon, and the size was larger under high-pressure gas. More quantitative work was done by Yatsuya et al. (18) on the pressure dependence of the size of the Al fine particle. In these studies, particles were sampled in the gas phase.

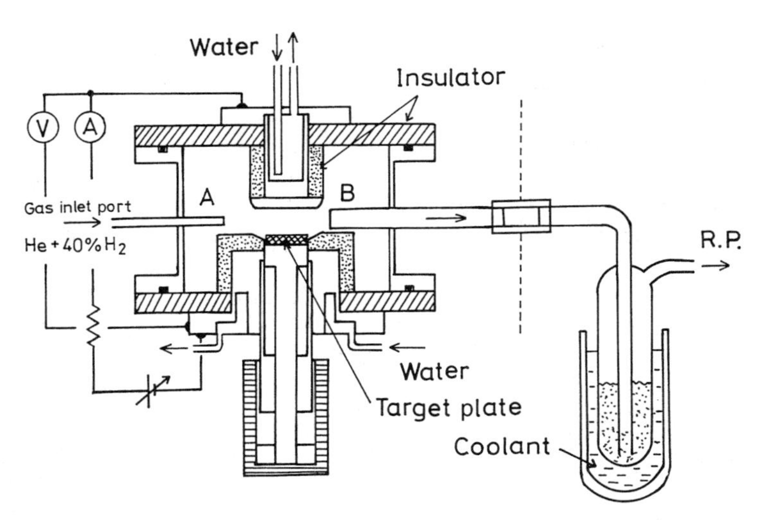

Fig. 9.4.11 Apparatus for the gas flow–sputtering method. A sputtering gas (mixture of He and 40% H_2) was supplied from nozzle A. Produced particles are flowed into the other port B and trapped by a solution trap apparatus cooled at approximately −50°C. (From Ref. 16.)

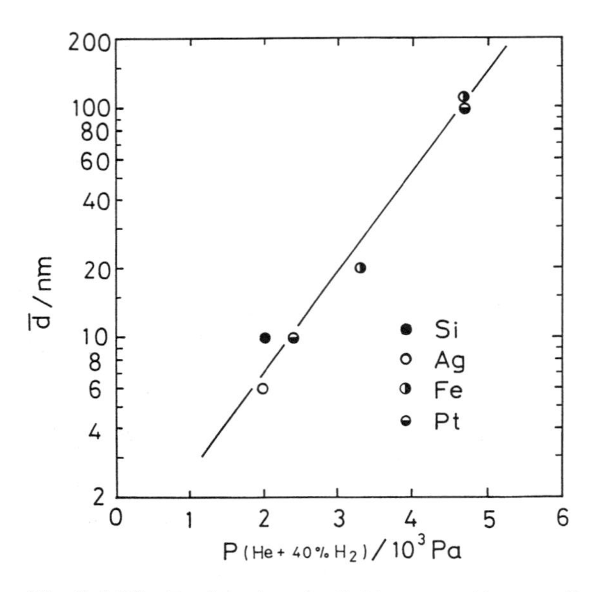

Fig. 9.4.12 Particle size of colloids prepared by a gas flow–sputtering method as a function of flowing gas pressure. Ethanol was used as a trapping medium. (From Ref. 16.)

What happens when particles are trapped in a liquid as a colloid? Figure 9.4.12 displays the size dependence of various kinds of particles (Si, Ag, Fe, Pt) on the ambient pressure made by the gas flow–solution trap method combined with a sputtering technique (16). We should note that the size of particles varies over a wide range, from several to a hundred nanometers, irrespective of metal species, by changing only the total pressure of flowing gas. The size was measured by the dynamic laser light scattering method; that is, the size was directly measured in a suspension state. Some samples were also measured by TEM observation, where the sample suspension was dropped on to a copper mesh with carbon film. The coincidence of both measurement suggests that particles are growing in the gas phase as mentioned in Section 9.4.2, and that liquid molecules effectively prohibit coalescence of particles in the suspension phase if the colloid concentration is not high.

The yield of particles in the gas flow method is generally low, because many particles produced in the chamber have tendency to deposit onto the inner wall of the chamber. The production yield as a function of gas flow rate was examined in case of the gas flow–solution trap method, and the result is shown in Figure 9.4.13 (19). Since the size is dependent on the total gas pressure, we maintained the pressure at 35 torr by mixing He and Ar gases. Total pressure and carrier gas supply rate were regulated by controlling the evacuation speed of pumping. The carrier Ar gas was supplied at the side of the crucible, and the weight of trapped Mg particles was measured. The yield was calculated in comparison to the loss of weight of the metals

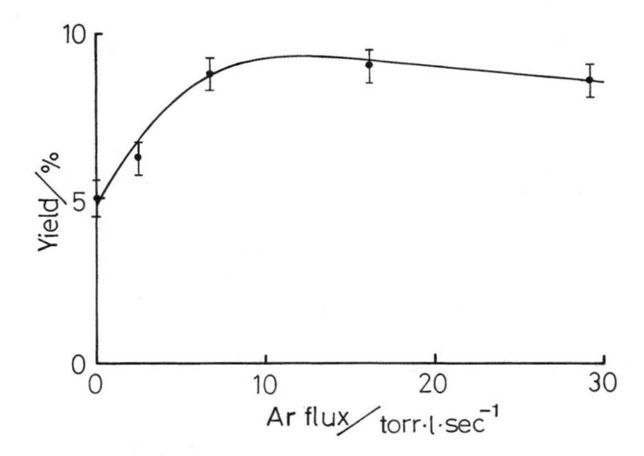

Fig. 9.4.13 Yield of metallic ultrafine particles as a function of Ar flow rate. The flow rate of He gas was maintained constant at 9.4 torr L/s. Total pressure of the chamber was 35 torr. (From Ref. 19.)

in the crucible with the recovered weight of particles in the glass container. At most, 10% of the sublimed metals was recovered as colloid particles by this method.

9.4.6 Examples of Ultrafine Metallic Particles

So far, we have prepared and tested many kinds of colloids, mainly in nonaqueous suspensions with combinations of metals or alloys as a dispersed phase and organic liquids as the dispersion media, without the use of any dispersing agents; these are listed in Table 9.4.1. We next give some examples of transmission electron micrographs of nanoparticles produced by an aerosol method. A sample for TEM measurement was obtained by dropping colloidal suspension onto a Cu mesh coated with an evaporated carbon film of 10 nm thickness. Many colloids were so unstable

Table 9.4.1 Colloids Produced by Gas Evaporation Method

Metals	Li[a], Be, Na[a], Mg, Al, K[a], Ca[a], Fe, Ni, Cu, Zn, Nb, Pd, Ag, In, Sn, Sb, Te, Pt, Au, Pb, BiCu, BiTe, PbTe
Media	Acetone, benzene, benzyl alcohol, butanol, 2-butanol, carbon tetrachloride, cetane, chlorocyclohexane, chloroform, cyclohexane, cyclohexanone, dichloroethane, dichloromethane, dioxane, ethanol, ether, ethyl acetate, hexane, methanol, 2-propanol, pyridine, tetrachloromethane, toluene, water, xylene

[a] Unstable in air. Particles must be kept on a solid substrate in vacuum or in strictly deaerated stable organic solvent.

in aerobic conditions that the suspensions were kept in an Ar atmosphere and the optical spectra were recorded under this condition. Other than this, the color of suspension has changed within a short time period.

Be. Figure 9.4.14 shows a TEM photograph of Be fine particles produced by a conventional gas evaporation technique. The size of the smallest particles is down to 5 nm.

Mg. We have already shown several Mg nanoparticles in Figure 9.4.7 produced by a conventional gas evaporation method and by a matrix isolation method. The color of the suspension was black for nanoparticles or dark grey for larger particles. Figure 9.4.15 shows a high-resolution TEM (HRTEM) image of Mg nanoparticles on a carbon-reinforced microgrid (20). Each particle seems to have a hexagonal crystal habit typical of hcp metals (21).

Al. Figure 9.4.16 shows A1 colloids produced by a gas flow–solution trap method. The color of the colloids was dark blue. The size of the majority of particles was around 30 nm. Note that the particle shape is spherical.

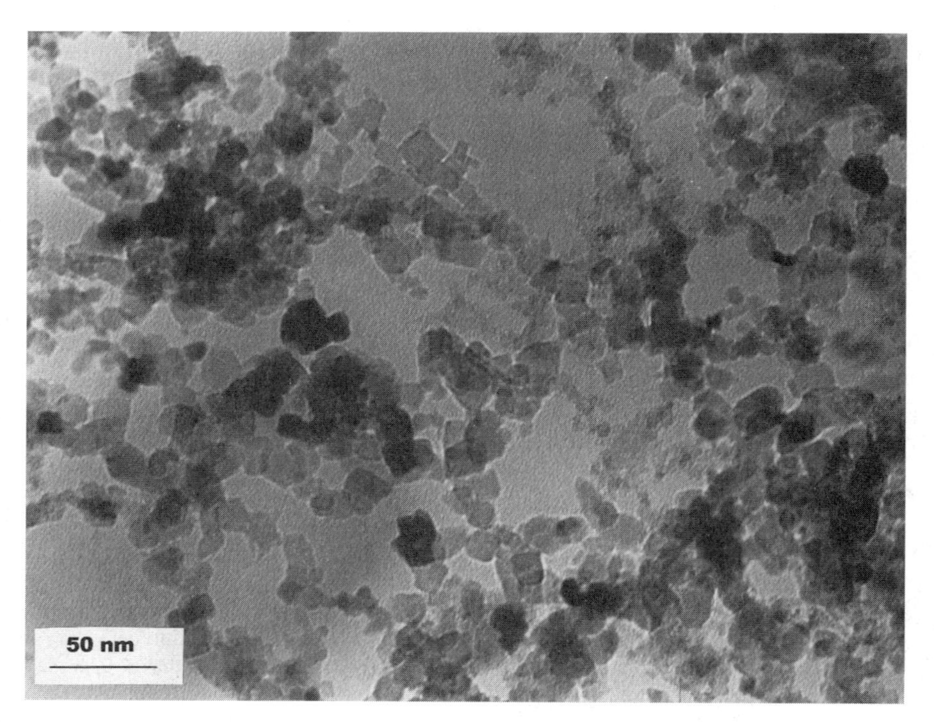

Fig. 9.4.14 Electron micrograph of Be fine particles produced by a conventional gas evaporation technique. Particles were sampled in gas phase.

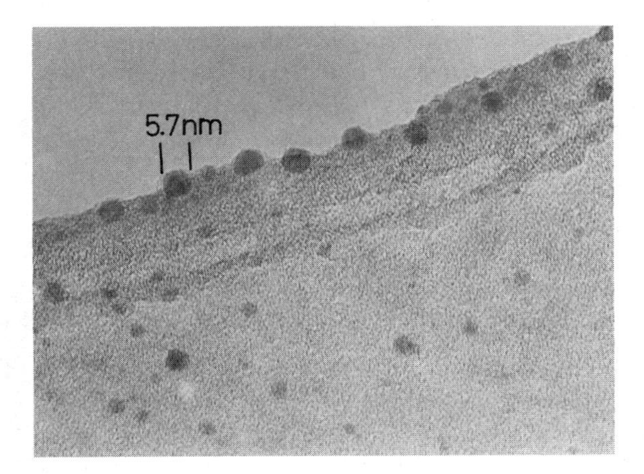

Fig. 9.4.15 High-resolution transmission electron micrograph (HRTEM) of Mg nanoparticles on a carbon-reinforced microgrid. Particles were prepared by a matrix isolation method with He gas. Each particle has a clear crystal habit, presumably hexagonal in shape. (From Ref. 20.)

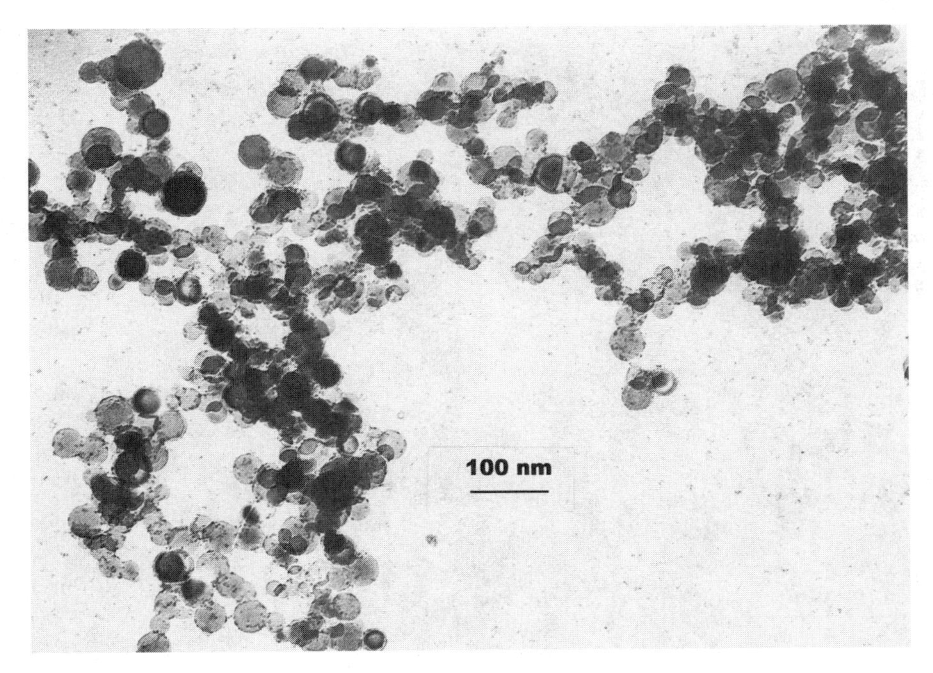

Fig. 9.4.16 Electron micrograph of Al ultrafine particles produced by a gas flow–solution trap method. (From Ref. 10.)

Cu. Figure 9.4.17 shows a photograph of Cu nanoparticles prepared by a gas flow–solution trap method. The color of the suspension was wine red just after preparation. This colloid was not stable in alcoholic solvent because of metal alkoxide formation (precisely discussed in Section 9.4.8) and its characteristic color disappeared.

Ag. If we pay special attention to evaporation conditions, as reported by Ichihashi (22), monodispersed fine particles are obtained. In Figure 9.4.18 are shown Ag particles thus obtained, many of which show icosahedron morphology, typical of multiply twinned metallic particles.

In. Figure 9.4.19 shows a TEM photograph of In ultrafine particles. The color of the colloid was light brown. The size distribution of this sample was rather broad, ranging from 3 to 80 nm in diameter and revealing bimodal size distribution. Larger particles may be produced by a coalescence growth in the liquid phase due to the low melting point of this material.

Sn. Figure 9.4.20 shows a photograph of Sn ultrafine particles produced by a gas flow–solution trap method. The diameter of many particles falls in the region of 15

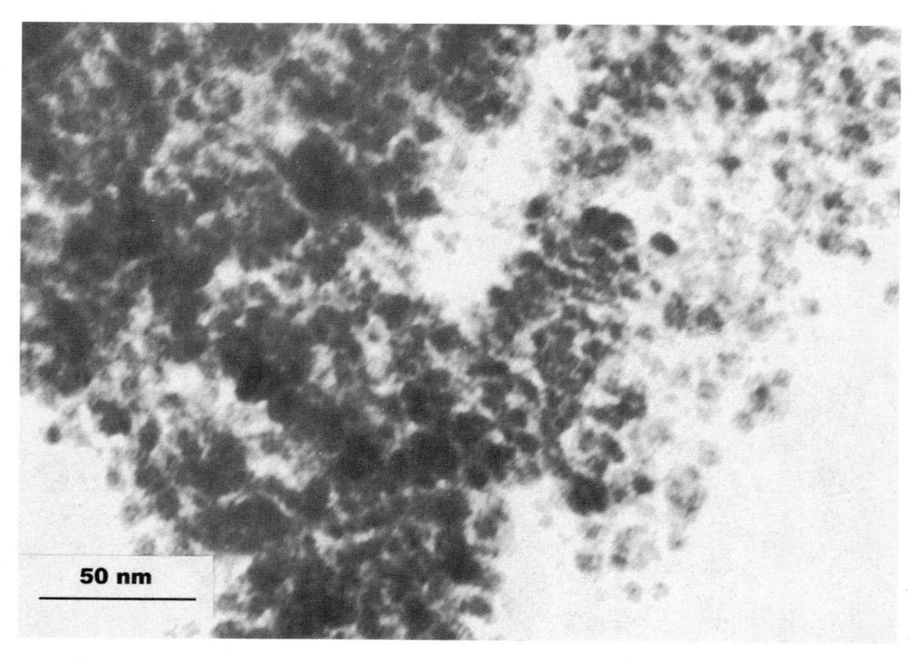

Fig. 9.4.17 Electron micrograph of Cu nanoparticles. The size of majority particles is near 5 nm. (From Ref. 10.)

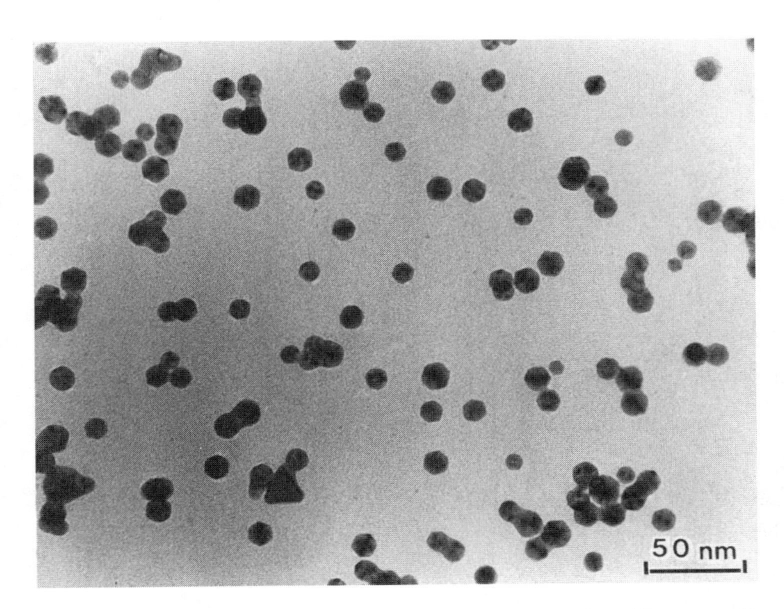

Fig. 9.4.18 Electron micrograph of Ag particles produced by a gas flow system. (From Ref. 22.)

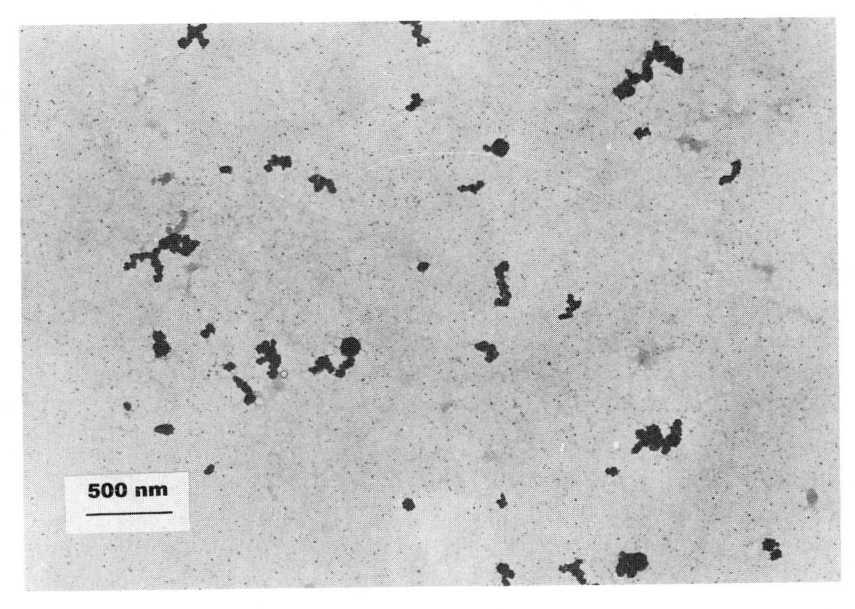

Fig. 9.4.19 Electron micrograph of In ultrafine particles produced by a gas flow–solution trap method. Note that there are large (several tens of nanometers) spherical particles with a background of very small nanoparticles scattered. (From Ref. 10.)

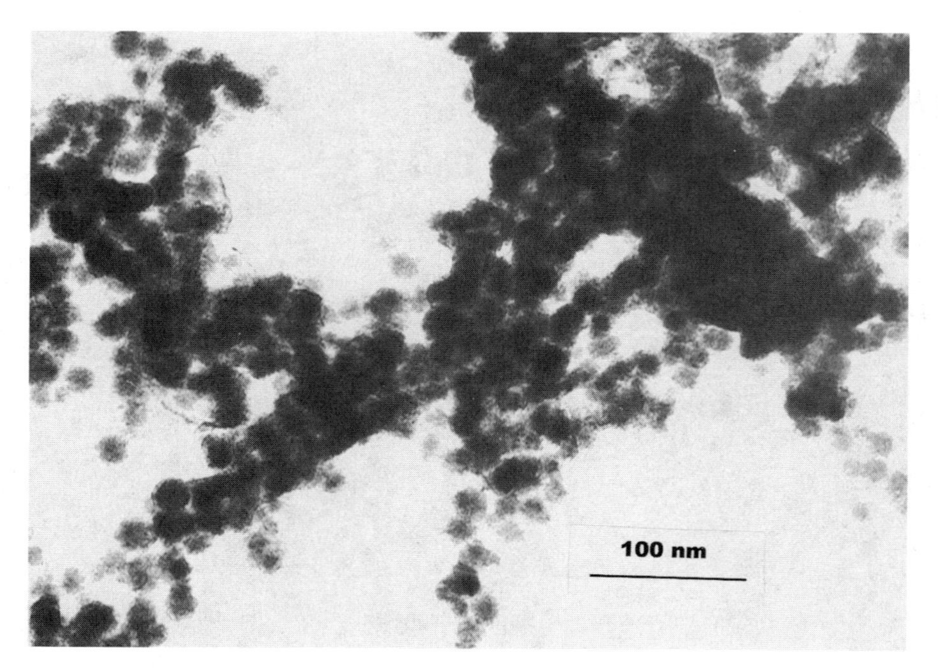

Fig. 9.4.20 Electron micrograph of Sn ultrafine particles produced by a gas flow–solution trap method. (From Ref. 10.)

nm. The particles seem to agglomerate to some extent. The color was dark brown. The optical spectrum of this sample is consistent with that of isolated particles in MgF_2 solid matrix.

Sb. Figure 9.4.21 shows a photograph of Sb ultrafine particles produced by a gas flow–solution trap method. The diameter of primary particles is in the range of 10–20 nm. The color of the suspension was dark grey. The chainlike structure found in the picture may be due to the amalgamation of the primary particles owing to the low melting point of Sb.

Pb. Figure 9.4.22 displays a photograph of Pb nanoparticles produced by a gas flow–solution trap method with ethanol as a dispersion medium. The diameter of many particles was less than 10 nm, and the color of the colloid was dark brown.

In terms of the common features of colloids just listed, nanoparticles with a clean surface have tendency to amalgamate when placed on a TEM sample mesh; that is evident in Figures 9.4.16 and 9.4.17. Furthermore, the particles often grow by collision (not by Ostwald ripening because metal ions are hard to dissolve in organic liquids) in the suspension state when the number density of particles in

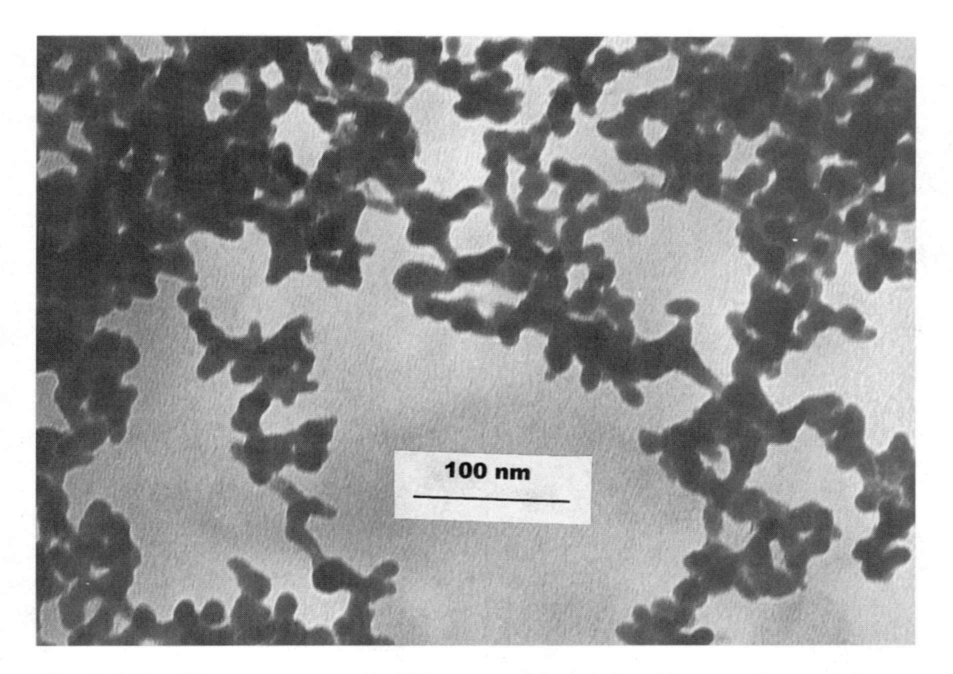

Fig. 9.4.21 Electron micrograph of Sb nanoparticles produced by a gas flow–solution trap method. Note that many particles are coalesced with each other, forming a fractal-like network structure.

suspension is very high. This is clearly seen in Figures 9.4.19, 9.4.20, and 9.4.21. We should emphasize that some of the metallic fine particles exemplified in this chapter with a large negative redox potential, such as alkali metals ($E_0' = -2.7$ to 2.9 V), Ca (−2.84 V), Mg (−2.37 V), Be (−1.85 V), and Al (−1.66 V), are unable to be prepared by the conventional chemical reduction process to form fine particles in a suspension state.

9.4.7 Colloidal Stability in Organic Liquids

In terms of dispersion and coagulation phenomena of colloidal suspension, it is known that there are two kinds of stability states in the colloids: the thermodynamically stable state and the metastable state. Examples of the former are surfactants, polymers, and biological materials in which the free energy is minimum in the dispersion state. On the other hand, in the latter case, a system is not thermodynamically stable but is held at the dispersion state due to an energy barrier. Therefore no further change is expected unless a large disturbance is added to the system. The important parameters are the energy of barrier height and that of disturbance. If the

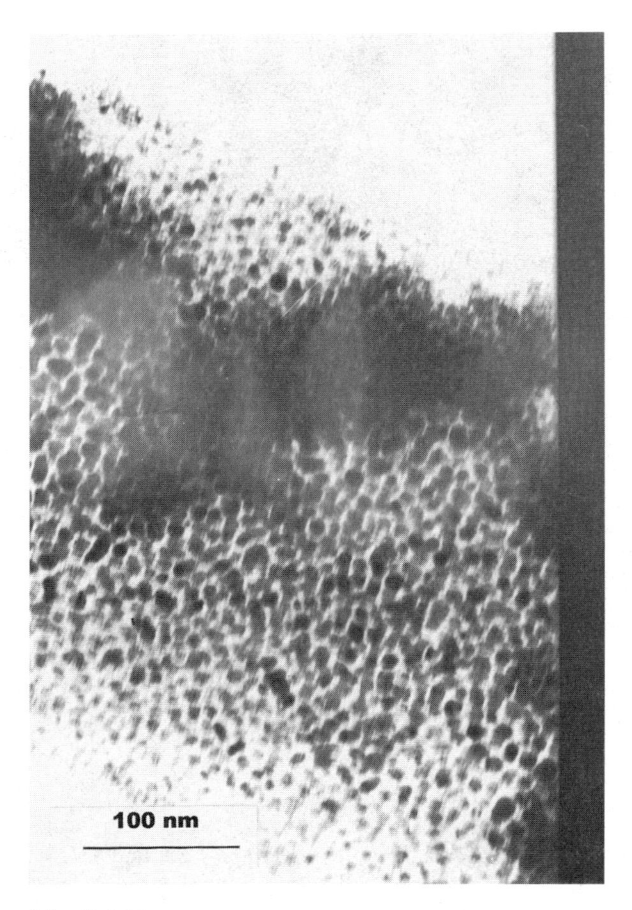

Fig. 9.4.22 Electron micrograph of Pb nanoparticles produced by a gas flow–solution trap method. (From Ref. 10.)

barrier height is not very high, even at the energy of kT (30 meV), a colloidal system becomes unstable, leading to gradual coagulation or degradation.

In aqueous suspension, the stability is discussed in reference to the DLVO (Deryaguin–Landau–Verway–Overbeek) theory. Within this framework, all solid substances have a tendency to coagulate due to their large van der Waals attractive force. The coulombic repulsive force among colloidal particles more or less prevents this tendency. These two opposite tendencies determine the stability of suspensions. What kind of parameters are concerned in the present nonaqueous system, for which little is known about the stability? This is an interest in this section.

Dispersion and Coagulation Diagram. As already listed in Table 9.4.1, we have made many suspension systems as functions of metal species and organics. During the production of these suspension systems, we have empirically noticed a characteristic metal dependence on the stability of colloids (23). It was difficult for Mg and Zn ultrafine particles to disperse in many organic liquids, while Cu, Ag, and Au are relatively easy to disperse. The dispersibility also depends on the kind of liquids used: the larger the polarity of the solvent, the more the metals can be dispersed. Although some sols were found to be very reactive toward solvents, as remarked in the previous section, a systematic behavior can be drawn out. Figure 9.4.23 displays a correlation diagram of the systematic combination of 11 liquids and 8 metal species. The liquids are classified according to their dielectric constants (ϵ_r is the relative dielectric constant taking that of vacuum, ϵ_0, as a reference) as an indicator of polarity of a molecule, while metals are distinguished by their electron

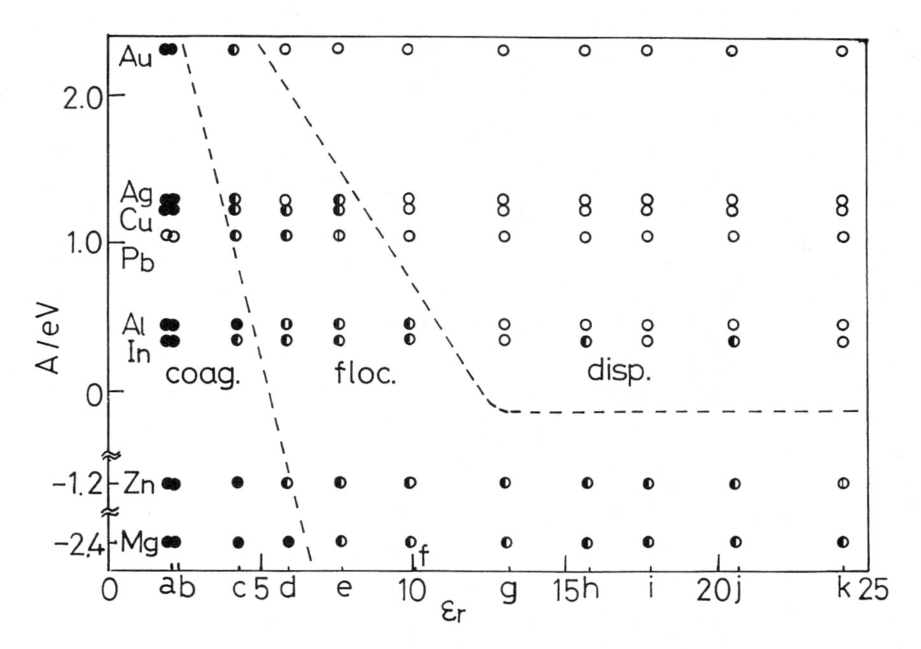

Fig. 9.4.23 Dispersibility of colloidal systems of a kind of metals in various organic liquids. ϵ_r, Relative dielectric constant of liquids; *A*, electron affinity; disp, dispersion (○); floc, flocculation (◐; upon stirring, the suspension becomes turbid then particles slowly sediment); coag, coagulation (●; immediately after stirring of the suspension, particles aggregate again to sediment). (⦶) Boundary between disp and floc; (◑) boundary between floc and coag. Broken lines divide each region. (a) Hexane, (b) benzene, (c) diethyl ether, (d) ethyl acetate, (e) tetrahydrofuran, (f) dichloroethane, (g) benzyl alcohol, (h) 2-butanol, (i) butanol, (j) acetone, (k) ethanol. (From Ref. 23.)

affinity (A) instead of redox potential E_0'. That is, the sols are more stable in liquids with high dielectric constants, irrespective of the metal species. It should be noted that a better correlation diagram was obtained with electron affinity, the atom- and molecule-related constant, rather than with physical constants for bulk-state parameters, such as work function. This finding strongly suggests that the electronic state of surface atoms of particles is responsible for this correlation, rather than the volume properties of particles.

This correlation diagram was supported by other observations. In hexane, all metal species tested coagulated to form sediment. However, we could redisperse these colloids again with the use of surfactants. The tendency of the dispersibility of the system of metal/surfactant/hexane was Au $>$ Sn $>$ A1 $>$ Mg, Zn. The last two still coagulated. Therefore, electron affinity is a good predictor of dispersibility even for the system containing surfactant. We should note that a surfactant molecule can interact only with surface atoms of a particle, not with inner metal atoms. This finding also suggests that the surface atoms are important for the dispersibility of sols in the present highly pure system. Although the diagram correlates well the dispersion and coagulation behavior of sols, this remains a macroscopic observation. It is more desirable to correlate this diagram with fundamental properties of metals, such as electronic state. However, this is a very difficult theme. Consider that the energy scale governing the stability of colloidal system is only 30 mV (kT value), in contrast to that of electronic state, several electron-volts. Hence we expect only a slight change to be detected in the electronic energy of a nanoparticle in the coagulation and dispersion process.

ESR Activity. A zinc nanoparticle is a special example in which we can use a spin probe method to trace a surface interaction with organic molecules through its electron spin resonance (ESR) signal (24). When a trace amount of O_2 is introduced on Zn small particles, the ESR signal due to the Zn^+ ion on the surface ZnO can be detected. This surface-localized ion can interact with the conduction electron of a metallic particle, leading to g-shift from its original value of $g = 1.96$ (the value when Zn^+ ion locates on the insulator). The g-value of conduction electron of Zn metal is 2.0033. Thus, this difference can be used as a indicator of the electronic change of a metallic particle, because at the nanometer scale, the metallic particle is no longer a normal bulk metal. This effect is known as a quantum size effect in solid-state physics. If Zn nanoparticles are firmly in contact with each other, the g-value may shift toward that of metals, and if it is isolated, the value approaches that of the free Zn^+ ion. Figure 9.4.24 displays the time course of the ESR spectra of Zn nanocolloids in acetone. Clear spectrum shift is evident from high magnetic field to low field as a function of sedimentation time, which is a measure of the extent of contact of particles. Initial g-value is considered to originate from an isolated particle, and low-field shift, indicating the particle is rich in metallic nature, is consistent with a tendency to coagulation. Through the contacting point, conduction electrons can move among particles, inducing more metallic properties than that of

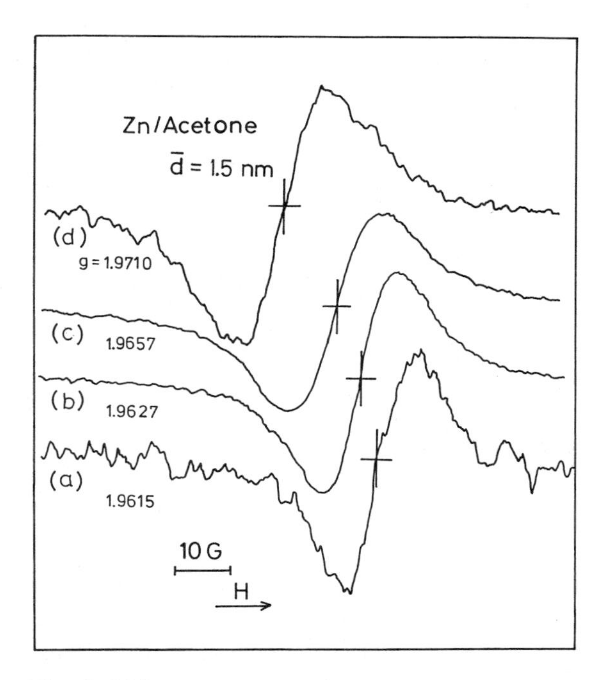

Fig. 9.4.24 Time course of the ESR spectra of Zn nanoparticles suspended in acetone. The diameter of particles was 1.5 nm: (a) immediately, (b) 30 min, (c) 80 min, and (d) 24 h after the sedimentation process started. (From Ref. 25.)

an isolated one. We also noticed in Fig. 9.4.24 that during the sedimentation process the line width was somewhat broadened. This is due to electron hopping among contacting particles and is regarded as a metallic tendency. At equilibrium, the *g*-value reached 1.971. From the *g*-value at equilibrium and those of free Zn^{+} and conduction electrons, calculation shows that the number of intimate contacting particles (this means the particle range of free electron hopping) is four or five. In this way, we can define what is contact on a microscopic basis. This time-dependent *g*-shift phenomenon was only perceived in a polar solvent and not in an apolar solvent such as hexane or benzene. In these liquids, the *g*-value revealed its equilibrium value around 1.97 just after preparation.

In Table 9.4.2 are shown the *g*-values of Zn colloid at equilibrium state as a function of organic species. Introduction of oxygen gas caused disappearance of the ESR signal, which is due to the oxidation of surface Zn^{+} to Zn^{2+}; the latter is ESR silent. This means that surface reaction (oxidation) takes place in the liquid phase. However, this reaction strongly depends on the kind of liquid, as seen in the figure. Moreover, we have found that there is a strong correlation between the behavior of

Table 9.4.2 ESR Characteristic of Zn Colloid as a Function of Organic Liquid

Liquid	*g*-Value	*d* (nm)	ϵ_r	O_2 Activity	Tendency of dispersion
Ethanol	1.9713	1.3	24.3	—	↑ Flocculation
Acetone	1.9711	1.5	20.7	—	
Butanol	1.9709	1.1	17.8	—	
Ethyl acetate	1.9683	0.9	6.0	—	
Ether	1.9711	1.1	4.3	+	↓ Coagulation
Toluene	1.9712	1.0	2.4	+	
Benzene	1.9712	1.2	2.3	+	
Hexane	1.9720	1.3	1.9	+	

Note: Oxygen activity: +, ESR signal disappeared upon addition of oxygen; —, ESR signal did not change upon addition of oxygen. *d,* Average diameter of particles; ϵ_r, dielectric constant. (From Ref. 23.)

colloids with dielectric constants and those with oxygen activity. That is, the presence and absence of oxygen activity relate to a difference in the accessibility of oxygen molecules to the surface spin of Zn particles. When the solvent is strongly bound to a metal particle, it is difficult for oxygen molecules to attack the surface. At the same time, the strong interaction between solvent and particles promotes good dispersibility. Accordingly, the dispersibility is a function of the direct activity of metal surface to liquid molecules. Since a rate constant of ESR quenching is an index of the O_2 reactivity, the lifetime of the surface Zn^+ ion should reflect the extent of the surface interaction. The longer the lifetime, the better is the dispersibility. In other words, we can quantitatively define the dispersibility of the colloids through the reactivity of oxygen toward the metal particles. This subject is treated more quantitatively in the latter section. These findings are peculiar to ultrafine metallic particles. Concerning the surface-related phenomenon, there is other kind of surface reaction with liquid molecules, which we discuss in the following section as well.

9.4.8 Reaction of Metals with Liquid Molecules

In the preceding section, we treated the surface reaction of Zn nanoparticles with oxygen. Here we mention the surface reaction of metallic particles with liquid molecules. We have found that alkali and alkaline earth metals are unstable in many polar organic solvents.

Copper Dissolution in Alcohol. Copper is believed to be stable to alcohol, at least in bulk, because copper still is used in distillation processes for the production of whisky. However, it was found that the color of copper sols prepared by the gas flow–solution trap method changed from the initial wine red to yellow within 30

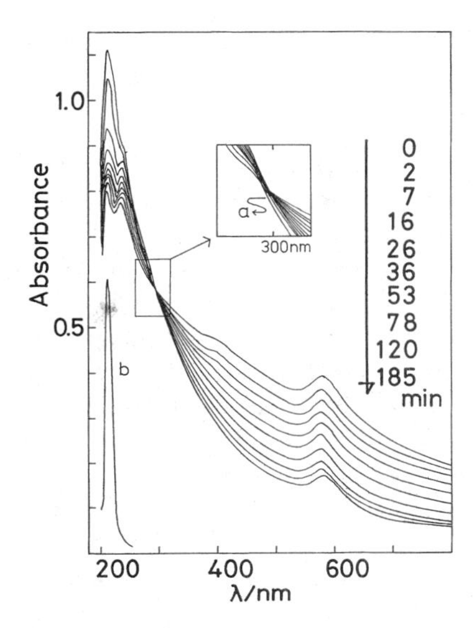

Fig. 9.4.25 Optical absorption spectra of copper colloid prepared by the gas flow–solution trap method as a function of time development. The numbers in the figure are the time after the preparation of the sample. The spectrum of sodium ethoxide in ethanol (authentic sample) is also shown in the same figure, marked by b. The insertion is the expansion of the region of the isosbestic point. The deviation from the isosbestic point at 10 h after the preparation of colloids is shown by a in the insert. (From Ref. 26.)

min under an aerobic condition within 1 day under an Ar atmosphere. Concomitant with this color change, the size of sols decreased with the passage of time from original 5.3 nm to 2.6 nm after 3 h. The color change coupled with the size decrease of copper particles is known as a kind of size effect. Figure 9.4.25 stands for this spectral change in the UV-visible region (26). The existence of an isosbestic point at 296 nm indicates that the reacting system contains two components mutually changeable. One component with a peak at 577 nm, which is characteristic in fine copper particles, gradually decreased as time passed. Concomitantly, a 213-nm absorption band increased. This peak was assigned to that from ethoxide ion when compared to the authentic sample. This peak is absent in the sol prepared in hexane as a liquid, in which no alkoxide formation is expected. This observation also confirms the earlier identification. The overall reaction suggested is

$$\text{Cu-UFP} + \text{ethanol} \longrightarrow \text{Cu-UFP}^{-}\text{-ethoxide} + \text{H}^{+} \tag{5}$$

where UFP stands for an ultrafine particle. With the progress of the reaction, the size of particles decreased further, and we have observed the peak at 577 nm shift to a longer wavelength. The alkoxide formation was also found in lead ultrafine particles in ethanol. Note that these findings are characteristics of ultrafine particles with an extremely clean surface. There was no change in the spectrum when copper powder was dropped into ethanol.

Reaction of Gold with Alcohol. One of the most stable elements, Au, was found to be active as nanoparticles. Figure 9.4.26 shows the time course of the optical absorption spectra of the Au nanoparticles in 2-propanol starting just after preparation. The peak position located around 530 nm characteristic of the plasmon band and the band width are strongly dependent on the dispersion state of colloids, such as isolated or coagulated. The sharpening of the initial broad plasmon band and the slight increase of the peak height as well as the color change from purple to reddish (this is coupled with the peak shift from longer wavelength to shorter one) suggest that the colloids have a tendency to be well isolated in suspension after a lapse of time. In fact, slightly aggregated particles just after preparation became isolated and were regarded as being in a nearly monodispersed state, having a diameter of 8 nm. The increase at 273 nm indicates the formation of acetone, which is also verified from nuclear magnetic resonance (NMR) measurement. All these behaviors are explained by the thermal kinetic effect due to DLVO theory. Acetone formation from 2-propanol indicates a reaction scheme of

$$\text{Au-UFP} + (CH_3)_2CHOH \rightarrow \text{Au-UFP}^{2-} + (CH_3)_2CO + 2H^+ \quad (6)$$

where Au-UFP^{2-} stands for the negatively charged Au ultrafine particles. During

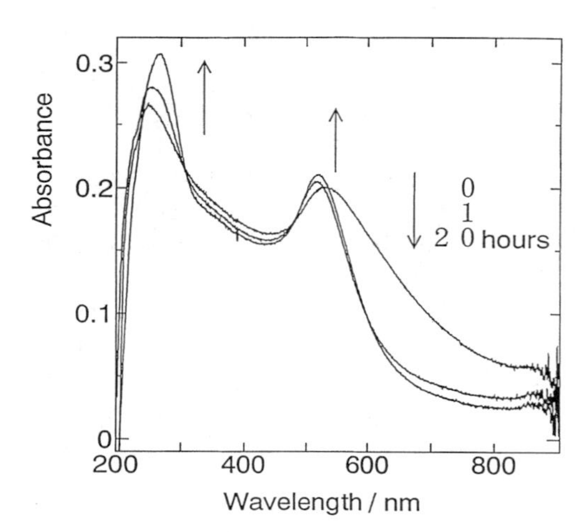

Fig. 9.4.26 Optical absorption spectra of Au-UFP/2-propanol suspension as a function of time. The numbers in the figure are the passage of time after the preparation of the sample. The sample colloids were prepared by a gas flow–solution trap method. (From Ref. 27.)

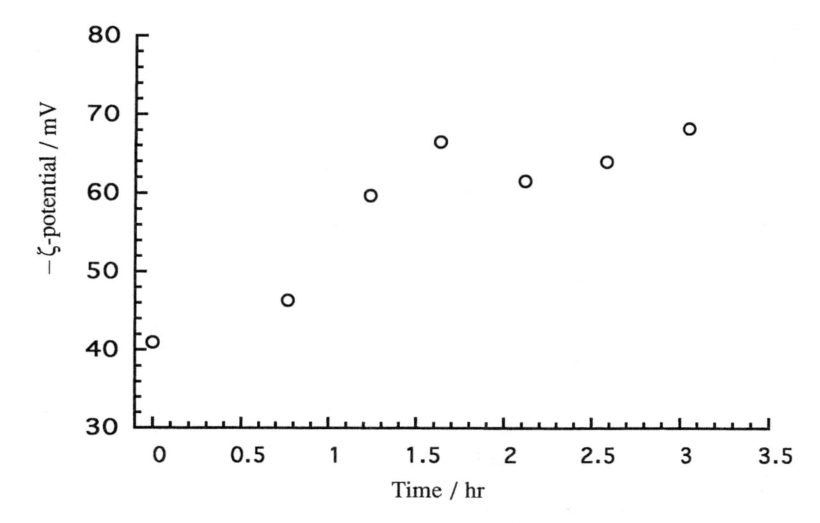

Fig. 9.4.27 Time development of ζ-potential of particles corresponding to the change in absorption spectra shown in Fig. 9.4.26. (From Ref. 27.)

this process, the charge of particles increased. Figure 9.4.27 presents the time course of the developing ζ-potential of particles corresponding to the process shown in Fig. 9.4.26. The charge of particles gradually increased concomitant with the color change of the colloid, indicating the gradual increase of a repulsive force. That is, increase of the absolute value of the ζ-potential stabilizes the redispersion state. Note that the charging process in this system is an electronic process and not the ionic process as found in normal colloidal systems. This system is basically free from any electrolyte and from air contamination.

Calculation of Coagulation Rate. Here we discuss an interaction potential of two charged particles in a liquid within a framework of DLVO theory. Following this theory, the overall interaction potential U_t of charged spherical particles of the same radius R and surface distance d is a sum of a coulombic repulsive force of charged particles and a van der Waals attractive force given by the equation (28):

$$U_t = \frac{4\pi\epsilon R(s-1)\psi_0^2}{s} \ln\left(1 + \frac{1}{s-1} e^{ikR(s-2)}\right) - \frac{A}{6}\left(\frac{2}{s^2-4} + \frac{2}{s^2} + \ln\frac{s^2-4}{s^2}\right) \tag{7}$$

in which s is the center distance scaled by R and equals $2 + d/R$, ψ_0 is a surface potential of a particle, k is the inverse of the electric double layer thickness, which is a function of formal charge of ions, ionic concentration, dielectric constant of

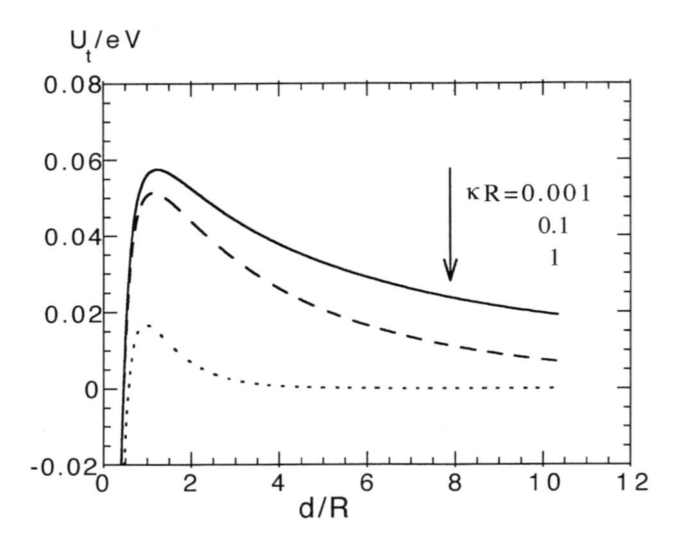

Fig. 9.4.28 Calculation curve for the interaction potential U_t. Curves for $kR = 1$, 0.1, and 0.001 (pure organic case), given $A = 2$ eV. (From Ref. 29.)

liquid, and temperature, and A is the effective Hamaker constant. Here no special interaction between the surface of a particle and a liquid molecule is taken into account. Figure 9.4.28 stands for the interparticle potential as a function of various kR such as 0.001, corresponding to very low ionic concentration, which is the case for organic liquids, and 1 for high ionic concentration, corresponding to aqueous solution using the Hamaker constant for gold, 2 eV (29). The potential barrier, 0.06 eV, prevents coagulation of particles. We should note that this value just falls in the experimentally observed value shown in Figure 9.4.27 where ζ-potential is in the region of -60 to -70 mV.

The stability ratio W is defined as the ratio of the rate of fast coagulation to that of slow coagulation and is given by

$$W = [\exp(V/kT)]/2kR \tag{8}$$

where k is the Boltzmann constant and V the potential barrier height. Substituting $V \approx 0.06$ eV, $kT \approx 0.025$ eV (room temperature), and $kR \approx 0.001$, we obtain $W \approx 10^4$, much less than the observed value of 10^7 (over several months). The discrepancy suggests that we must take into account the direct interaction between particle surface and a liquid molecule as discussed in the examples of Cu and Au surface reaction with liquids.

Interaction of Molecules with Particle Surface. We could not explain the observed stability of metal sols in alcohol over several months within the traditional framework of DLVO theory in which only particle charge repulsive force is taken into account as a stabilizing agent. The simplest approach to this problem is to take account of the special interaction between the particle surface and a liquid molecule. We must raise our attention to the tendency of oxygen activity shown in Table 9.4.2. The Zn^+ spin in apolar liquids is easily attacked by oxygen molecules, whereas in a polar liquid it is stable, suggesting that the spin is strongly guarded by organic molecules against oxygen in the latter case. The adsorption of solvent molecules to the surface of metals is a key issue in this system.

The lifetime τ of the adsorbed surface molecules is given as

$$\tau = \tau_0 \exp(-U_0/kT) \tag{9}$$

Here τ_0 is an average vibrational period in a surface potential well having a value of the order of subpicoseconds for a simple molecule, and U_0 is a depth of the potential, which strongly depends on the type of interaction. For strong interaction, τ exceeds 1 s or more, while in the case of no interaction, such as elastic collision, it is less than τ_0. If we assume the Lennard–Jones potential for the interaction potential and expand it at the equilibrium position r_0, τ_0 is given by relevant molecular parameters as

$$\tau_0 = \frac{\pi r_0}{3}\sqrt{\frac{M}{2\,|U_0|}} \tag{10}$$

where M stands for the molecular weight of a liquid molecule; r_0 is several angstroms in many simple organic liquids, and hence is set at 3 Å. For given molecules, we have calculated τ_0 and τ for several potential depths U_0. For ethanol, the typical hydrogen bond energy, 20 kJ/mol, and also the covalent bond energy relevant to the formation of ethoxide at the surface, 65 kJ/mol, were assumed. The result is displayed in Table 9.4.3. The large effect on τ appear solely from potential depth. This rough estimate suggests that the molecular interaction is crucial to the Zn surface, and the tendency found in τ is consistent with that found in Table 9.4.2 in which oxygen has much activity in an apolar solvent where oxygen has easy access

Table 9.4.3 Model Potential Parameters of Zn Surface to Organic Liquids

	Hexane	Benzene	Ethanol	
M	78	86	46	
$-U_0$ (kJ)	2	2	20	65
r_0 (Å)	3	3	3	3
τ_0 (ps)	1.4	2	0.3	0.3
τ (s)	3 p	4 p	1 n	60 m

Note: M, molecular weight; U_0, potential depth; r_0, distance between particle surface and a molecule, τ_0, vibrational period, τ, lifetime on the surface; p, ps; n, ns; m, ms.

to the surface. Our calculation also explains the finding in Fig. 9.4.3 as well as the discrepancy in the stability ratio W in Eq. (8).

Besides the DLVO theory, the stability and instability behaviors in organic liquids should have a close relationship to the contact angle of the substances in bulk. When the surface of a particle is not stable in a solvent, particles have a tendency to diminish their surface area that is exposed to liquids, suggesting coagulation of particles other than through the effect of the van der Waals attractive force. In this case, the affinity of the surface to the liquid is poor, and we observe a large contact angle if the particle is large and the liquid is a droplet. Figure 9.4.29 describes the relationship of wetting behavior of metal surface to liquids (case A) with the dispersion/coagulation behavior of metal ultrafine particles in a liquid suspension (case B). The left-hand side is a case of weak interaction, and the right-hand side is a case of strong interaction. In case A, the liquid is small, while in case B, the metal is small. Wetting (A) and dispersion (B) behavior are thus complementary. Therefore, the fact that Zn and Mg coagulate easily in apolar liquids suggests a hydrophilic nature of the surface.

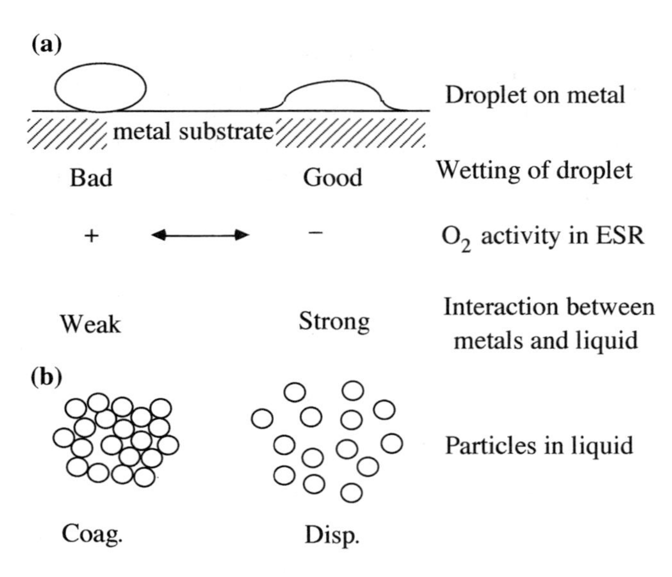

Fig. 9.4.29 Comparison of stability of metallic nanoparticles in bulk liquid with a droplet on a metal surface. (a) Wetting of a droplet on a metal surface. (b) Coagulation and dispersion of metallic particles in liquid. Figures on the left-hand side stand for weak interaction in case A causing coagulation in case B. Those on the right-hand side are a strong interaction between metal and liquid, suggesting good dispersion and good contact.

9.4.9 Control of Surface of Nanoparticles with Surfactant

In the preceding section, we emphasized that the surface interaction of metallic particles with liquid molecules is a very important parameter in the dispersion behavior of the sol. Since the surface nature is very sensitive to the surface modification, we can easily regulate it with the use of surfactant. As seen in Figure 9.4.23, almost all metallic particles cannot be dispersed in hexane as a suspension liquid. In this section, we show what kind of surfactant is effective in dispersion. The sample was prepared by the gas flow–cold trap method. We tested three surfactants, dimethyldi-

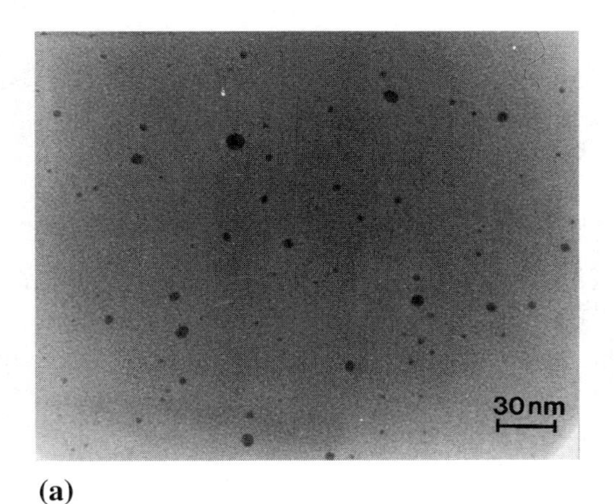

(a)

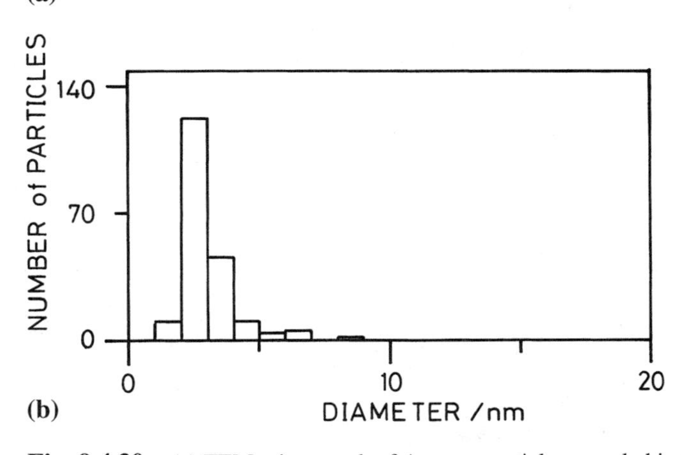

(b)

Fig. 9.4.30 (a) TEM micrograph of Au nanoparticles sampled in the gas phase. The exposure time of Cu sample grid to the gas stream was 1 s. (b) Size distribution of (a). (From Ref. 30.)

octadecylammonium chloride (abbreviated as DOAC) and (2-dodecylhexadecyl) trimethylammonium chloride (abbreviated as g-C28TAC) as cationic surfactants and sodium 1,2-bis(2-ethylhexyloxycarbonyl)ethanesulfonate (abbreviated as AOT) as an anionic surfactant.

First we show the TEM photograph and size histogram of primary particles sampled in gas phase in Figure 9.4.30. A number of particles of about 2.5 nm in diameter can be seen in the figure. The smallest one is less than 1 nm. From the histogram of this picture, a rather narrow size distribution was found. When these particles were introduced into hexane solution, strong coagulation took place, as seen in Figure 9.4.31. All particles were deposited within a few minutes. Original

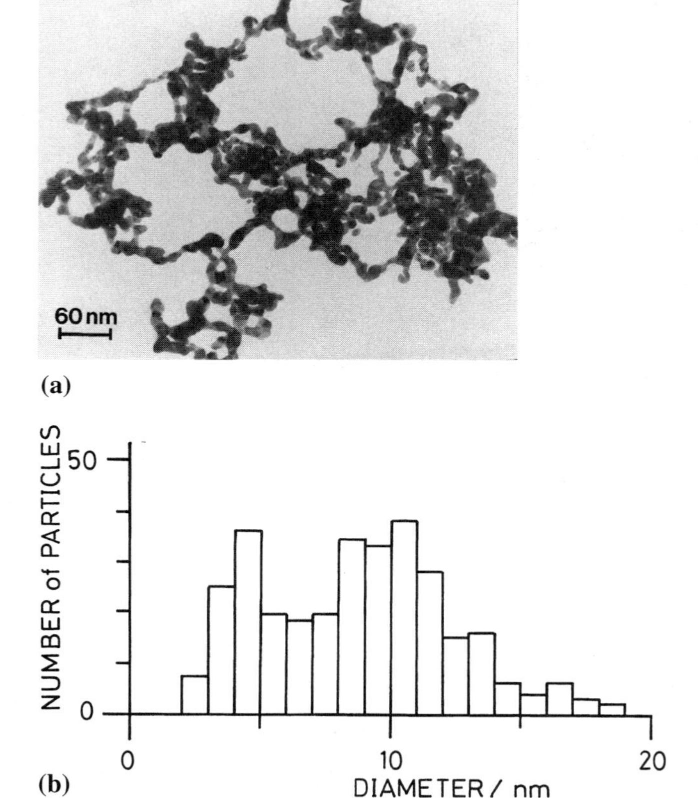

Fig. 9.4.31 (a) TEM micrograph of precipitates of Au nanoparticles deposited from hexane. (b) Size distribution of (a). (From Ref. 30.)

nanoparticles have come into contact with each other to form larger aggregates. The boundary of the primary particles seems to amalgamate and coalesce from the picture. A typical size distribution curve in Figure 9.4.31b shows a bimodal distribution. When we add DOAC solution in advance to the glass container before melting the matrix, the color of the suspension turned rose red and transparent, and the scattering light through the suspension was hardly observable in the visible region. A TEM picture of this sample is shown in Figure 9.4.32a together with its histogram. The peak position of this histogram is located in a position almost equal to those sampled in the gas phase. However, the width of size distribution became somewhat broader.

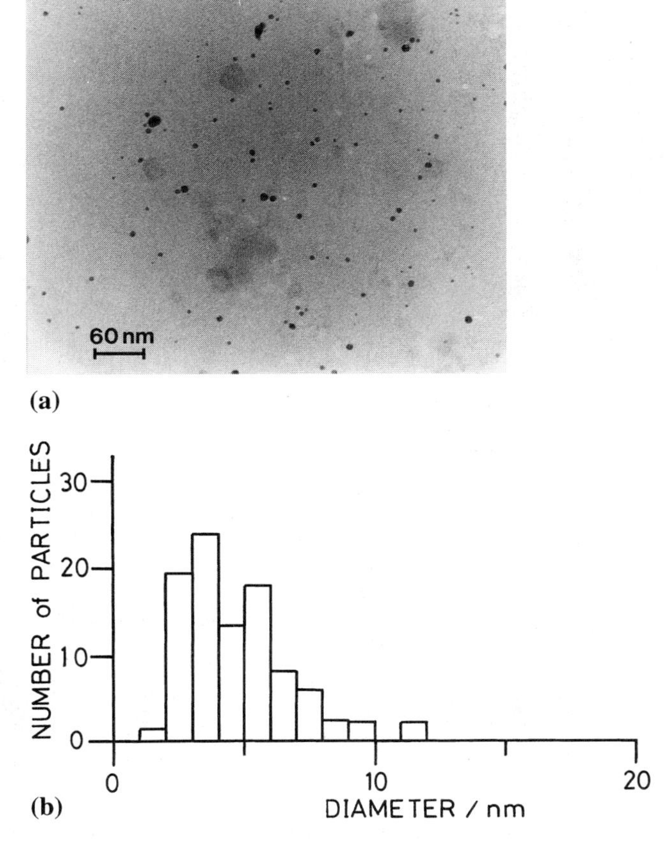

Fig. 9.4.32 (a) TEM micrograph of Au nanoparticles deposited from colloidal suspension of Au UFP/DOAC/hexane system. (b) Size distribution of (a). (From Ref. 30.)

It is impossible to completely depress coalescence growth in the suspension solely by using detergent.

The effect of surfactant on the dispersibility of Au nanoparticles depends on the sort of ionic type. When AOT was used instead of DOAC, the color of sols changed blue indicating the process of coagulation and the suspension was not stable. Within a few hours, the colloids fully precipitated. Figure 9.4.33 shows a TEM picture of this system sampled just after ultrasonic treatment to promote dispersion of sols. The particles in this picture are anisotropic and irregular in shape, and the size is larger than for the case of DOAC, as seen in the histogram. However, it is

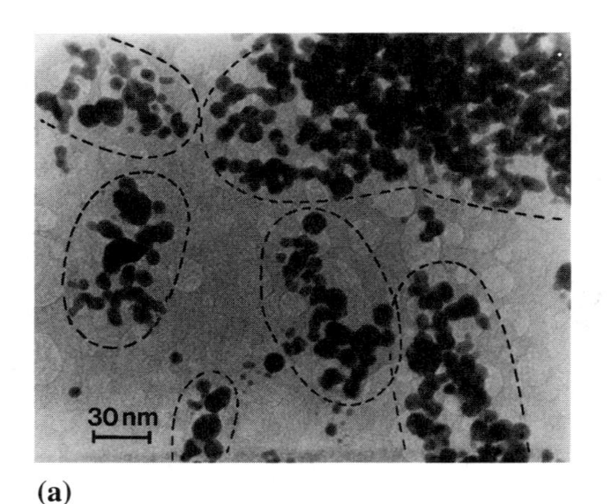

(a)

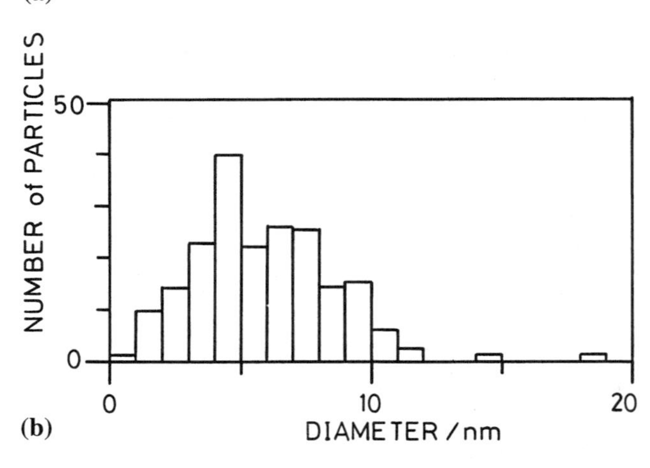

(b)

Fig. 9.4.33 (a) TEM micrograph of Au fine particles deposited from colloid suspension of Au/AOT/hexane system. The regions encircled by broken lines represent examples of the clustering unit of densely packed Au particles. (b) Size distribution of (a). (From Ref. 30.)

worth noting that coalescent growth stopped at some stage. Moreover, in the picture, we notice that the particles form a densely packed cluster. The difference of the activity of surfactant on the dispersion can be correlated with the type of charge of a detergent molecule, cation or anion. As already mentioned in Section 9.4.8, the charge of Au sols is negative. Thus it is reasonable to consider that a cationic molecule can easily attack the negatively charged surface. However, a detailed mechanism of this surface interaction remains to be solved.

This method has been applied to obtain greatly condensed particle assemblies using another type of cationic surfactant, g-C28TAC for Ag (31), BiCu, and BiTe (32) because of the low solubility of DOAC in hexane. It was found that redispersion of this colloids was possible after completely drying, and this process was reproducible (32). The maximum content of metal particles is determined by the solubility of detergent molecules. High-density, finely dispersed colloids was obtained using g-C28TAC for BiCu, for which the number density reached 10^{14} particles/cm^3 (33).

REFERENCES TO SECTION 9.4

1. K Kimura. Curr Topics Colloid Interface Sci 1:69, 1997.
2. G-H. Comsa. J Phys (Paris) 38:C2-185, 1977.

3a. CG Granqvist, RA Buhman. J Appl Phys 47:2200, 1976.

3b. CG Granqvist. J Phys (Paris) C2:147, 1977.

4. K Kimura. Bull Chem Soc Jpn 60:3093, 1987.
5. K Kimoto, Y Kamiya, M Nonoyama, R Uyeda. Jpn J Appl Phys 2:702, 1963.
6. N Wada, M Ichikawa. Jpn J Appl Phys 15:755, 1976.
7. N Wada. J Phys (Paris) C2:219, 1976.
8. H Abe, W Schulze, B Tesche. Chem Phys 47:95, 1980.
9. EC Honea, JS Kraus, JE Bower, MF Jarrold. Z Phys D26:141, 1993.
10. K Kimura, S Bandow. Bull Chem Soc Jpn 56:3578, 1983.
11. M Moskovits. Surf Sci 57:125, 1976.
12. GA Ozin. Angew Chem Int Ed 21:550, 1982.
13. N Wada. Jpn J Appl Phys 8:551, 1969.
14. H Ohya, T Ichihashi, N Wada. Jpn J Appl Phys 21:554, 1982.
15. S Yatsuya, T Kamakura, K Yamauchi, K Mihama. Jpn J Appl Phys 25:L42, 1986.
16. S Bandow, K Kimura. Nippon Kagaku Kaishi 1957, 1989.
17. N Wada. Jpn J Appl Phys 6:553, 1967.
18. S Yatsuya, S Kasukabe, R Uyeda. Jpn J Appl Phys 12:1675, 1973.
19. K Kimura, S Bandow. Nippon Kagaku Kaishi 6:922, 1984.
20. K Kimura. Mat Res Soc Symp Proc 272:193, 1992.
21. K Kimura. Z Phys D11:327, 1989.
22. T Ichihashi. Jpn J Appl Phys 25:1247, 1986.
23. N Satoh, S Bandow, K Kimura. J Colloid Interface Sci 131:161, 1989.
24. S Bandow, K Kimura. J Phys Soc Jpn 57:2805, 1988.
25. K Kimura. J Colloid Interface Sci 183:607, 1996.
26. K Kimura. Bull Chem Soc Jpn 57:1683, 1984.
27. Y Takeuchi, T Ida, K Kimura. J Phys Chem 101:1322, 1997.

28. RJ Hunter. Foundations of Colloid Science, Vol. 1. (Oxford: Clarendon Press, 1991, p. 182.
29. K Kimura. Surf Rev Lett 3:1219, 1996.
30. N Satoh, K Kimura. Bull Chem Soc Jpn 62:1758, 1989.
31. S Tohno, M Itoh. J Aerosol Sci 24:339, 1993.
32. S Tohno, M Itoh, K Kimura. Surf Rev Lett 3:59, 1996.
33. S Tohno, M Itoh, K Kimura. Surf Rev Lett 3:71, 1996.

10

Carbon Nanotubes

TAKASHI KYOTANI, AKIRA TOMITA, and YAHACHI SAITO

10.1 REACTION IN ANODIC ALUMINUM OXIDE FILMS

TAKASHI KYOTANI and AKIRA TOMITA
Institute for Chemical Reaction Science, Tohoku University, Sendai, Japan

10.1.1 Introduction

Carbon nanotubes are nanometer-wide needlelike cylindrical tubes of concentric graphitic carbon, and they have attracted much attention due to their unusual structure, their potential applications, and a fundamental interest in their properties. Carbon nanotubes were discovered in 1991 by Iijima (1) when he investigated the material deposited on the cathode for the arc-evaporation synthesis of fullerenes. A short time later, Ebbesen et al. found the optimum arc-evaporation conditions for the production of carbon nanotubes in bulk quantities (2). Apart from the arc discharge method, several other methods have been proposed, e.g., catalytic pyrolysis of hydrocarbons (3–7) and condensation of a laser-vaporized carbon-catalyst mixture (8). None of these methods, however, allow the accurate control of length, diameter, and thickness of carbon nanotubes. In 1995, Kyotani et al. and Parthasarathy et al. have prepared carbon nanotubes independently using uniform and straight channels of anodic aluminum oxide film as a template (9,10). The most striking feature of this method is to allow one to produce monodisperse carbon tubes with uniform length, diameter, and thickness. This chapter highlights how effectively uniform carbon nanotubes can be prepared by the template carbonization technique and how versatile this technique is for the production of novel one-dimensional carbon composites.

10.1.2 Formation of Anodic Aluminum Oxide Film

When aluminum plate is anodically oxidized in an acid electrolyte, a porous oxide layer is formed on one side of the plate, whose porosity consists of an array of parallel and straight channels with a uniform diameter. The cross section of a porous anodic film is illustrated in Fig. 10.1.1. The pore size and pore density can be controlled by changing the anodizing voltage, and the thickness of the oxide film is determined by the anodizing period, i.e., the amount of charge transferred. It was reported that films with pore size of 10–250 nm, pore densities of 10^{12}–10^{15} m^{-2}, and thickness of over 100 μm can be prepared (11). Although such anodic aluminum oxide films are sold commercially as a membrane filter, only a limited number of pore size is available. Kyotani et al. prepared anodic aluminum oxide film by anodic

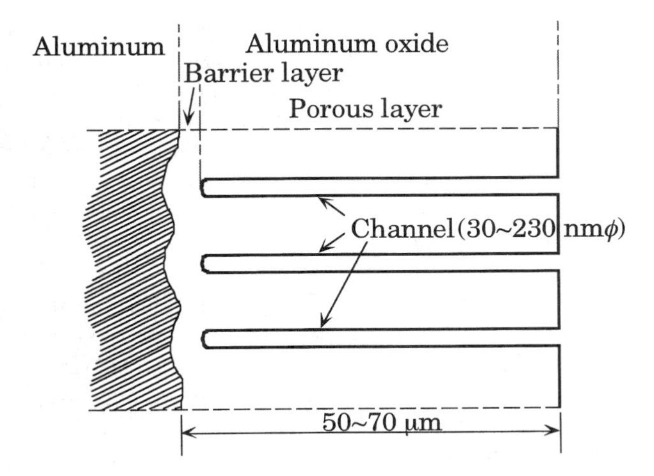

Fig. 10.1.1 Structure of anodic aluminum oxide film.

oxidation of an electropolished aluminum plate at a cell voltage of 20 V in 20 wt% of sulfuric acid at 0–5°C for 2 h (9,12). Following the electro-oxidation, the anodic oxide film was separated from the aluminum substrate by reversing polarity of the cell. Then an impervious layer (referred to as a barrier layer in Fig. 10.1.1) was etched by immersing the film in 20 wt% sulfuric acid for 1 h. The diameter and the thickness of this film were 15 mm and 75 μm, respectively, and the diameter of its straight pores was about 30 nm. Figure 10.1.2 shows the SEM (scanning electron microscope) photographs of the surface and the cross section of the anodic oxide

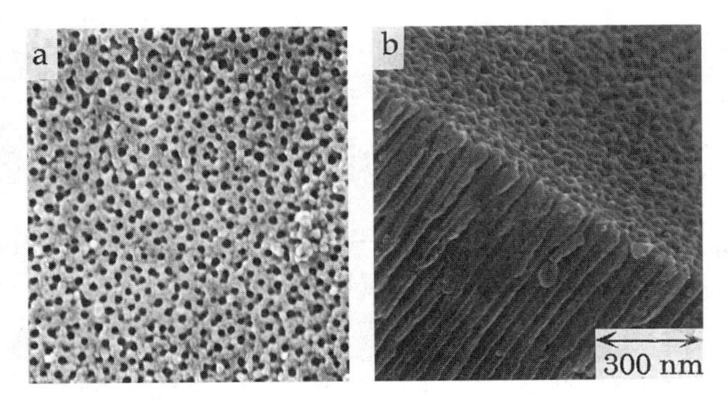

Fig. 10.1.2 SEM photographs of the anodic aluminum oxide film with 30 nm channels: (a) surface and (b) cross section. (From Ref. 12.)

films. Many openings with a uniform diameter of about 30 nm are observed on the surface. The cross-section view (b) indicates the presence of parallel and straight channels perpendicular to the film surface. The BET surface area of the anodic oxide film was determined as 23 m^2/g, and the area calculated by the geometrical dimension estimated from the SEM observation was 19 m^2/g. The pore distribution curve of this film had a sharp peak at about 30 nm pore diameter, which is consistent with the channel diameter from the scanning electron microscopy (SEM) observation (Fig. 10.1.2a). These findings imply the size and shape of the channels to be very uniform.

A variety of nanomaterials have been synthesized by many researchers using anodic aluminum oxide film as either a template or a host material: e.g., magnetic recording media (13,14), optical devices (15–18), metal nanohole arrays (19), and nanotubes or nanofibers of polymer, metal and metal oxide (20–24). No one, however, had tried to use anodic aluminum oxide film to produce carbon nanotubes before Kyotani et al. (9,12), Parthasarathy et al. (10) and Che et al. (25) prepared carbon tubes by either the pyrolytic carbon deposition on the film or the carbonization of organic polymer in the pore of the film. The following section describes the details of the template method for carbon nanotube production.

10.1.3 Carbonization of Organic Polymer in Straight Channels of Anodic Oxide Film

Kyotani et al. tried to prepare carbon tubes from organic polymer by using anodic aluminum oxide film (12). As a film, they employed a commercially available membrane filter of 25 mm diameter and 60 μm thick (Whatman Ltd., Anodisc 25), whose porosity consists of an array of parallel and straight channels with a diameter of about 230 nm. The films were dried at 150°C for 3 h under vacuum. After cooling down to room temperature, the films (five or six sheets) were impregnated under vacuum with furfuryl alcohol (100 cm^3) together with oxalic acid (0.25 g) as an acid catalyst. The mixture was stirred with the films for 3 days. The polymerization of furfuryl alcohol and its subsequent carbonization in the channels of the films were carried out by heating the impregnated films under N_2 flow up to 900°C at a rate of 5°C/min and holding for 3 h. The resultant carbon-anodic oxide composite films were washed with an excess amount of 46% HF solution at room temperature to dissolve the anodic aluminum oxide template. As a result, carbon was obtained as an insoluble fraction.

Figures 10.1.3a and 10.1.3b show the SEM and TEM (transmission electron microscope) photographs of the resultant carbon sample, respectively. The SEM photograph indicates the formation of tubular carbon, whose diameter is almost equal to the channel diameter (230 nm) of the anodic oxide film template. Each tube ramifies into several thin tubes near the end. This branchlike structure originates from a similar branching in pore structure of the commercial anodic oxide film near its surface. A more clear view of the structure of the carbon samples is given by

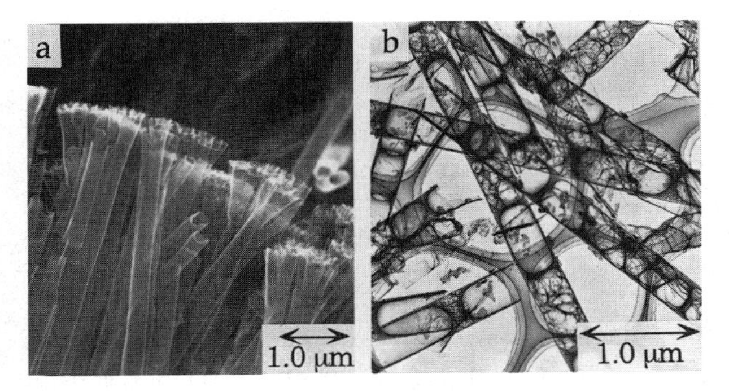

Fig. 10.1.3 Photographs of the carbon prepared from poly(furfuryl alcohol): (a) SEM image and (b) TEM image. (From Ref. 12.)

the TEM photographs (Fig. 10.1.3b), which exhibits the presence of voids and knots in the tubes. This is just like bamboo, although the authors do not have a clear explanation for the formation of such structure. It was confirmed that the carbonization in the one-dimensional channels gives the formation of one-dimensional carbon.

Martin and coworkers tried to prepare carbon tubes from the carbonization of polyacrylonitrile (PAN) in the channels of anodic oxide film (10). A commercially available film with a pore diameter of 260 nm was immersed in an aqueous acrylonitrile solution. After adding initiators, the polymerization was carried out at acidic conditions under N_2 flow at 40°C. The PAN formed during the reaction was deposited both on the pore walls and on both sides of the film. Then the film was taken from the polymerization bath, followed by polishing both faces of the film to remove the PAN deposited on the faces. The resultant PAN/alumina composite film was heat-treated at 250°C in air, and then it was heat-treated at 600°C under Ar flow for 30 min to carbonize the PAN. Finally, this sample was repeatedly rinsed in 1 *M* NaOH solution for the dissolution of the alumina film. The SEM observation of this sample indicated the formation of carbon tubes with about 50 μm long, which corresponds to the thickness of the template film. The inner structure of these tubes was not clear because TEM observation was not done. The authors claim that it is possible to control the wall thickness of the tubes with varying the polymerization period.

10.1.4 Formation of Carbon Nanotubes by Pyrolytic Carbon Deposition on Anodic Film

In addition to the impregnation method with furfuryl alcohol, Kyotani et al. attempted to deposit pyrolytic carbon on the inside of the straight channels of anodic oxide film in the following way (9,12). They used two types of anodic aluminum oxide films with different channel diameters (30 and 230 nm). Each anodic oxide film was placed on quartz wool in a vertical quartz reactor (ID 20 mm). The reactor temperature was increased to 800°C under N_2 flow and then propylene gas (2.5% in N_2) was passed through the reactor at a total flow rate of 200 cm^3 (STP)/min. The thermal decomposition of propylene in the uniform straight channels of the anodic oxide films results in carbon deposition on the channel walls. After a desired period, the reactor was cooled down to room temperature and the films were taken out of it. Then they were washed with HF as described earlier, and only carbon was left as an insoluble fraction. The formation process of carbon tubes is illustrated in Figure 10.1.4.

Figure 10.1.5 shows the SEM photographs of the carbon samples from the two types of films. These photographs reveal that in both cases the samples consist of only cylindrical tubes, and their outer diameter is almost the same as the channel diameter of the corresponding anodic oxide film. No other form of carbon was found in the microscopic observation. In the SEM photographs with low magnification (Fig. 10.1.5c), many bundles of the tubes can be observed, and the length of a whole tube in a bundle was almost the same as the thickness of the corresponding template film. The presence of many such bundles implies that most of the tubes are connected at the both ends of each tube, because the carbon deposition also took place on the external flat surface of the anodic film. However, some of the tubes were separated from the others during the stirring in the HF solution, as observed in the other pictures. It is noteworthy that the tubes from the anodic oxide film with the smaller

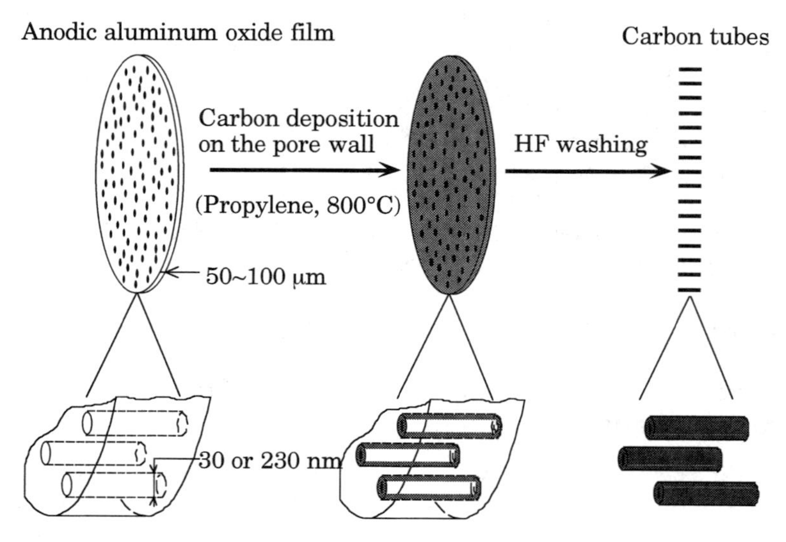

Fig. 10.1.4 Schematic drawing of the formation process of carbon tubes. (From Ref. 12.)

Fig. 10.1.5 SEM photographs of the carbon tubes prepared by carbon deposition of propylene: (a, b) carbon deposition period of 1 h on the anodic oxide film with 30-nm channels; (c, d) a period of 12 h on the film with 230-nm channels. (From Ref. 12.)

channels (Fig. 10.1.5b) look transparent under the SEM observation with an acceleration voltage of 15 kV, indicating that the wall thickness of these tubes is very thin.

Figure 10.1.6 shows the SEM photographs for the carbon tubes under different carbon deposition periods. The photographs, which were taken from the end of the tubes, show the open end structure of these carbon tubes. Furthermore, these images clearly demonstrate that the wall thickness of the tubes increases with an increase in deposition period. The wall thickness could be roughly estimated from the TEM photographs of these tubes (not shown here), where the thickness was in the ranges of 3–5, 40–45, and 60–80 nm for 1-, 6-, and 12-h deposition, respectively.

Electron diffraction analysis for these carbon tubes revealed that the tube wall consists of cylindrically stacked carbon layers. The lattice image for the carbon tubes with a diameter of 30 nm is shown in Figure 10.1.7, where at least four tubes cross each other. The thickness of the walls is about 10 nm, and consequently the carbon has a hollow with a diameter as small as 10 nm. Many small lines, which correspond to 002 lattice planes, are observed in the cross section of the walls for each tube.

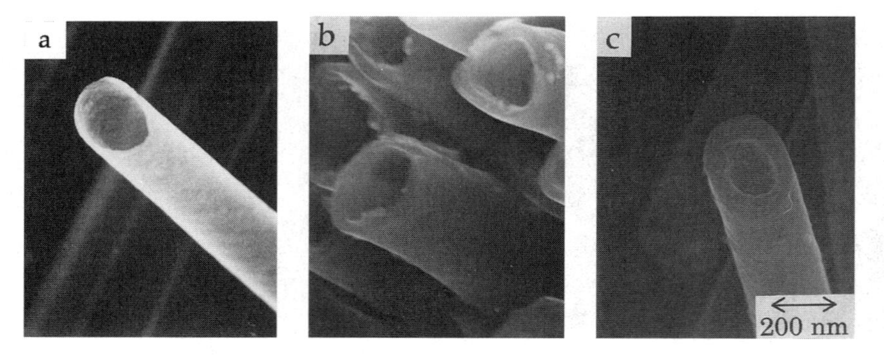

Fig. 10.1.6 SEM photographs of the carbon tubes prepared in the anodic oxide film (channel diameter 230 nm) with different carbon deposition periods: (a) 1 h; (b) 6 h; (c) 12 h. (From Ref. 12.)

This image demonstrates that the size of most 002 planes is less than 10 nm and they wrinkle to a great extent. This structure is far from the graphitic one. It should be noted that all the 002 planes are orientated toward the direction of carbon tube axis.

Che et al. recently demonstrated that not only propylene but also ethylene and pyrene can be used for the production of carbon tubes (25). They subjected anodic aluminum oxide template (channel diameter 200 nm) to either ethylene or pyrene gas stream at 900°C for 10 min and then the resultant carbon tubes were removed

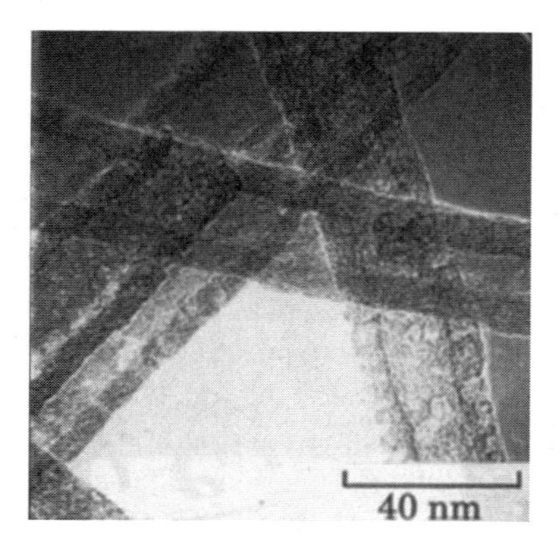

Fig. 10.1.7 High-resolution TEM image of the carbon tubes from the anodic oxide film with 30-nm channels with a carbon deposition period of 6 h. (From Ref. 12.)

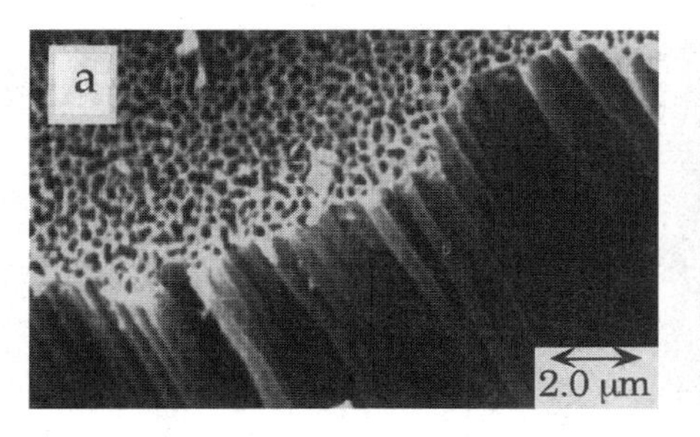

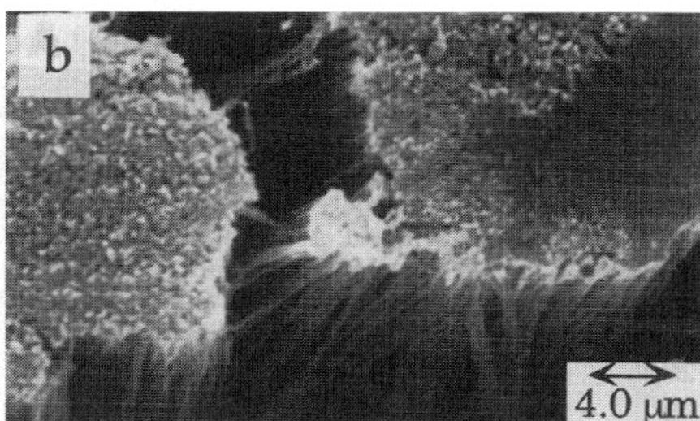

Fig. 10.1.8 SEM photographs of (a) the carbon nanotube and (b) the nanofibers prepared by ethylene CVD. (From Ref. 25.)

from the template with HF washing. The carbon tubes obtained in their method have almost the same feature as those in the ones just mentioned of Kyotani et al.—i.e., the same outer diameter as that of the channels of the template, and uniform carbon tubes with open ends. They found that a longer carbon vapor deposition (CVD) period (40 min) produces carbon nanofibers without clear hollows, instead of carbon nanotubes. The SEM photographs of carbon nanotube and nanofibers from ethylene are shown in Figure 10.1.8. Furthermore, using the template coated with Ni catalyst, they could synthesize highly crystallized carbon nanofibers at as low a temperature as approximately 500°C.

By the template technique using anodic oxide films and pyrolytic carbon deposition, one can prepare monodisperse carbon tubes. Since the length and the inner diameter of the channels in an anodic oxide film can easily be controlled by changing the anodic oxidation period and the current density during the oxidation, respectively, it is possible to control the length and the diameter of the carbon tubes. Furthermore, by changing the carbon deposition period, the wall thickness of the carbon tubes is controllable. This template method makes it possible to produce only carbon tubes that are not capped at both ends. Various features of the template method are summarized in Table 10.1.1 in comparison with the conventional arc-discharge method.

Table 10.1.1 Comparison of Carbon Nanotubes

Method	Template	Arc discharge
Products	Nanotube only	Mixture of various carbons
Tube size (length, diameter, thickness)	Monodisperse	Broad distribution
Size control	Easy	Very difficult
Tube wall	Stacking of small crystallites	Single crystal
Tube ends	Open	Closed

10.1.5 Metal Filling in Carbon Nanotubes Prepared by Arc-Discharge Method

As reviewed by Freemantle (26) and others (27,28), metal-filled carbon nanotubes could have a variety of industrial applications such as catalysts, electronic devices, an improved magnetic tape, and biosensors. Thus, the preparation and application of such metal-filled carbon nanotubes now becomes promising and challenging research. There have been several attempts to insert metal into the tubes prepared by an arc-discharge evaporation technique. These attempts can be classified into the following two methods: One is based on arc-evaporation of metal loaded carbon anode, and the other one is a two-step method. The former high-temperature method can produce mainly metal carbide-filled nanotubes in a single stage (29–33), but it suffers from a low yield of filled nanotubes, and it produces impurities such as encapsulated carbon clusters and soot. A higher yield of filled nanotubes can be achieved by the latter method, which consists of the opening of nanotube ends and the inclusion of metal into the opened tubes (34–38). Early work by Ajayan et al. (34) demonstrated capillary action to fill molten lead into the nanotubes, which were uncapped beforehand by oxidation with air. A similar approach has been applied by Ugarte et al. (38) using molten silver nitrate. Green and coworkers (36,37) have developed a wet chemical method, where nanotubes were treated in HNO_3 in the presence of soluble metal salt. As a result, the nanotubes got opened by the acid treatment and filled by the salt. Thus, the wet chemical method can be applied to more variety of materials than the arc-discharge method. A drawback of this chemical method is that some loading on the outer surface of nanotubes is unavoidable. For all the methods described here, it is still difficult to control the structure, shape, and size of encapsulated material in a hollow nanotube. On the other hand, only the template method makes it possible to prepare metal-filled uniform carbon nanotubes that are completely free from metal on the outer surface. The following section shows how such metal-filled carbon nanotubes can be prepared by the template method.

10.1.6 Formation of Pt-Filled Uniform Carbon Nanotubes

Kyotani et al. tried to prepare Pt-filled carbon nanotubes in the following way (39). They employed anodic aluminum oxide film whose pore size is about 30 nm and prepared carbon-deposited film by the CVD technique using propylene as described in the preceding section. The film was impregnated with an ethanol solution of hexachloroplatinic acid ($H_2PtCl_6{\cdot}6H_2O$) at room temperature for 3 h, and then ethanol was evaporated at 80°C under N_2 flow. The reduction of chloroplatinic acid in the channels was performed by the following two methods: (1) heat treatment at 500°C under H_2 flow, and (2) stirring in an excess of 0.1 *M* $NaBH_4$ aqueous solution at room temperature. After the reduction, $Pt/C/Al_2O_3$ composite film was washed with an excess of 46% HF solution at room temperature to dissolve the anodic

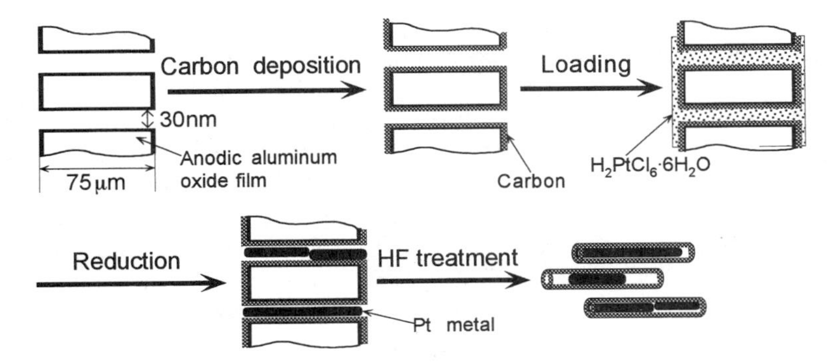

Fig. 10.1.9 Schematic diagram of the formation process of Pt/carbon nanotube composites.

aluminum oxide template. As a result, Pt metal/carbon nanotube composite was obtained as an insoluble fraction. The schematic diagram of the formation process is illustrated in Fig. 10.1.9.

Figure 10.1.10, a and b, shows the TEM bright-field images with different magnifications for the Pt/carbon tube composite prepared at 500°C. These two images exhibit the presence of uniform carbon nanotubes, and their outer diameter and

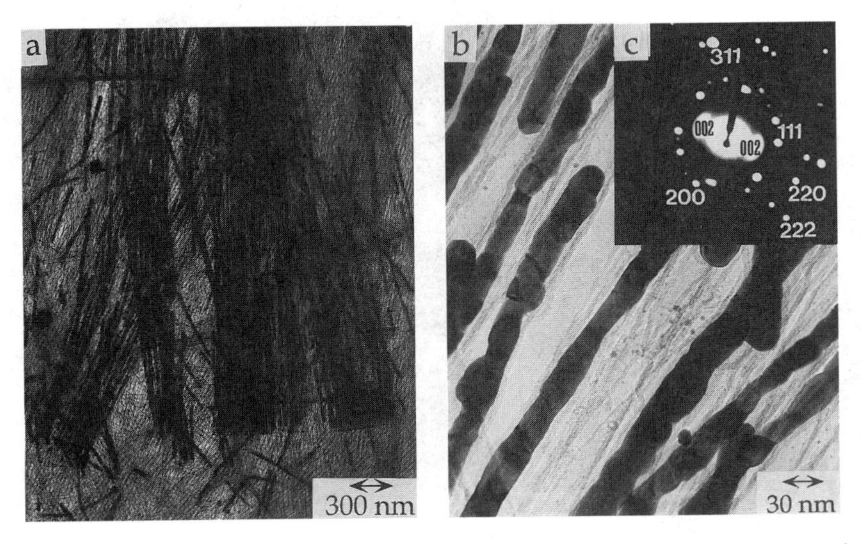

Fig. 10.1.10 (a, b) TEM photographs of the Pt/carbon nanotube composites reduced at 500°C; (c) electron diffraction pattern taken from TEM image (b). (From Ref. 39.)

wall thickness can be estimated from the latter figure to be 30 nm and about 5 nm, respectively. Although some of the tubes are empty, the others are filled with many rodlike materials. The low-magnification TEM image (Fig. 10.1.10a) indicates some of the nanorods are more than 1 μm in length. Their structure was investigated with electron diffraction and with x-ray powder diffraction (XRD). Figure 10.1.10c shows the electron diffraction pattern that was taken from the TEM image (Fig. 10.1.10b). The pattern presents the diffraction from Pt metal crystallites as clear spots, together with carbon 002 diffraction as a pair of strong arcs. Some of the spots from the fcc structure of Pt metal and carbon 002 arcs are labeled in the figure. This pattern confirms the dark materials to be Pt metal. The appearance of the diffraction for Pt as clear spots, not as rings, indicates the high crystallinity of Pt nanorods. The small number of these diffraction spots suggests the presence of only a few Pt crystallites in the area of Fig. 10.1.10b. The XRD analysis also confirmed the reduction of chloroplatinic acid to Pt metal. From the peak width of the Pt (111) diffraction, the average crystallite size was calculated to be about 30 nm.

The TEM photographs of the composite reduced by $NaBH_4$ solution at room temperature are displayed in Figure 10.1.11. Like the composites reduced at 500°C, Pt metal is observed only in the carbon tube hollows, and some tubes look to be completely filled with the metal. There is, however, some difference in microscopic feature between the two types of Pt metals prepared at different temperatures (Figs.

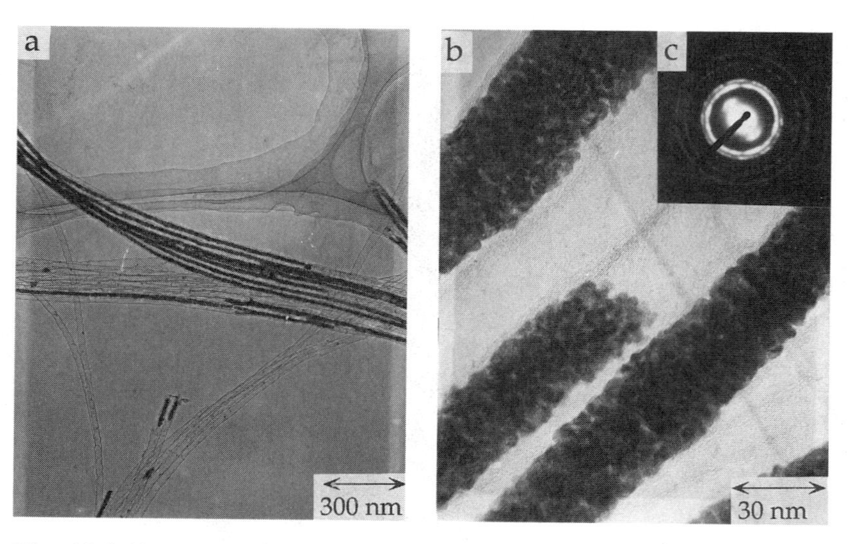

Fig. 10.1.11 (a, b) TEM photographs of the Pt/carbon nanotube composites reduced at room temperature; (c) electron diffraction pattern taken from TEM image (b). (From Ref. 39.)

10.1.10b and 10.1.11b). The Pt metal reduced at room temperature looks as if it consists of very fine particles with a size of 2–5 nm, which is in good agreement with the calculated value from the XRD analysis. The electron diffraction pattern of the Pt metal gives further information on its structure. The pattern (Fig. 10.1.11c) exhibits a set of diffraction rings from Pt metal (these rings can be labeled from the inner one as 111, 200, 220, and 311 of the fcc Pt metal), together with a pair of carbon 002 arcs. It is noteworthy that diffraction from Pt metal did not appear as clear spots, but as concentric rings, each of which consists of a large number of very small spots. This finding suggests that the metal reduced by $NaBH_4$ solution is comprised of many fine crystallites.

For the two types of the composites just described, Pt was loaded by the method of evaporation to dryness. On the other hand, when the impregnated film was removed from the chloroplatinic acid solution and then dried, less Pt was loaded in the carbon tubes. Figure 10.1.12 shows the TEM bright-field image of the composite prepared by this loading method, where the reduction of Pt was done with $NaBH_4$ solution at room temperature. The image exhibits very fine Pt particles in the hollows. Some Pt particles form agglomerates with each other. The sizes of Pt particles vary from 1 to 4 nm.

Che et al. also prepared Pt/Ru nanoparticle-filled carbon tubes with a diameter of 200 nm (40). They impregnated carbon-deposited film with a mixture of aqueous solutions of H_2PtCl_6 and $RuCl_3$. After drying in air, the metal compounds in the pores were reduced by H_2 flowing at 580°C for 3 h. Then the underlying alumina was dissolved away in HF solution. TEM observation of this sample revealed the presence of Pt/Ru nanoparticles (about 1.6 nm) dispersed on the inner wall of the tubes.

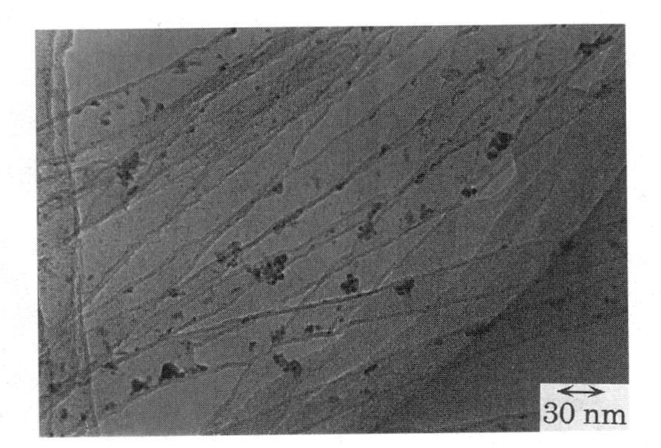

Fig. 10.1.12 TEM bright-field image of Pt nanoparticles in carbon nanotubes. (From Ref. 39.)

By applying the template technique, Kyotani et al. and Che et al. succeeded in preparing Pt and Pt/Ru metal-filled uniform carbon nanotubes in which the metal is present as either nanorods or nanoparticles. It should be noted that no metal was observed on the outside wall of the tubes. This is due to the preparation procedure, in which the metal precursor was loaded into the carbon-deposited alumina film before the dissolution of alumina by HF (see Fig. 10.1.9). Thus, there is no other space for metal to be loaded except in the channels.

10.1.7 Inclusion of Crystalline Iron Oxide Nanoparticles in Uniform Carbon Nanotubes

Pradhan et al. demonstrated that inclusion of iron oxide nanoparticles into template-synthesized carbon nanotubes was also possible when the MOCVD (metal organic chemical vapor deposition) technique was employed (41). A carbon-deposited film was subjected to MOCVD of ferrocene [$Fe(C_5H_5)_2$] in the following manner. Ferrocene was vaporized at 90 or 105°C (corresponding to a vapor pressure of 0.1 and 0.3 kPa, respectively) and then the vapor was introduced into the film in the quartz reactor with H_2 gas (50% in N_2) at a total flow rate of 100 cm^3 (STP)/min. The feed line was wrapped with heating tapes and maintained at a high temperature to avoid the condensation of ferrocene vapor. The following two series of MOCVD experiments were performed: one for 3 h with 0.3 kPa of ferrocene pressure at three different temperatures (350, 400, and 500°C) and the other for different time periods (0.5, 1, 3, 6, 12, 24 h) with 0.1 kPa at 400°C. After the MOCVD, the film was treated with 10 *M* NaOH solution at 150°C in an autoclave for 6 h in order to remove the anodic aluminum oxide template. As a result, metal/carbon nanotube composites were obtained as an insoluble fraction. In this case, HF washing was not applied for the alumina dissolution process because metallic component can be dissolved in HF.

Figure 10.1.13 shows the TEM bright-field images at different magnifications for the Fe/carbon tube composites prepared by the MOCVD at 400°C for 3 h with 0.3 kPa of ferrocene vapor. These images exhibit the presence of uniform carbon nanotubes with an outer diameter of 30 nm and a wall thickness of about 5 nm. Although some of the tubes are empty, the others contain many dark particles. It should be noted again that there is no metal deposited outside of the nanotubes. The high-magnification image (Fig. 10.1.13b) shows that the shape of some particles looks like a cube, implying the high crystallinity of these particles. The inset picture shows the SAD (electron diffraction for selected area) patterns taken from the area indicated by a circle in Figure 10.1.13a. The pattern presents sharp diffraction spots, which can be indexed to 111, 220, 311, 400, 511, and 440 reflections from cubic magnetite (Fe_3O_4). In addition to these spots, the authors could observe arcs and diffused rings from carbon 002, 10, and 11 reflections, which are too weak to recognize from this SAD picture. The appearance of the diffraction for Fe_3O_4 as clear spots, not as rings, indicates the high crystallinity of Fe_3O_4 nanoparticles. Some of

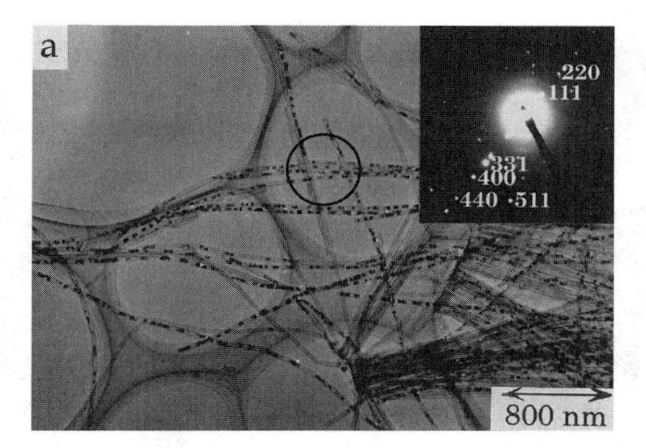

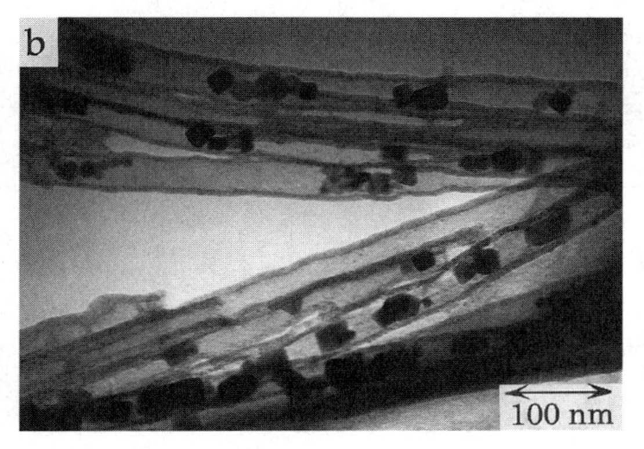

Fig. 10.1.13 TEM photographs of the Fe/carbon nanotube composites prepared at a MOCVD temperature of 400°C for 3 h with ferrocene vapor pressure of 0.3 kPa: (a) low and (b) high magnification images; the inset picture shows the corresponding SAD pattern for the area indicated by circle in image (a). (From Ref. 41.)

the particles observed in Fig. 10.1.13b must be single crystals. The maximum caliper diameter of a hundred particles in the image of Figure 10.1.13a was measured. It was found that their diameters range from 10 to 50 nm with a mean size of 24 nm and standard deviation of 7 nm. The reason why particles longer than the tube inner diameter (20 nm) were found is that many particles are not spherical, and the maximum caliper diameter was used as a measure of size.

The iron oxide was likely to be formed when the iron loaded carbon/alumina film was exposed to air and/or when the film was treated with an alkaline solution. In order to clarify this issue, the authors tried to characterize the iron-containing carbon nanotubes before the alkali treatment. After the MOCVD experiment, the iron-loaded carbon/Al_2O_3 film was taken out to the ambient air and broken into fine

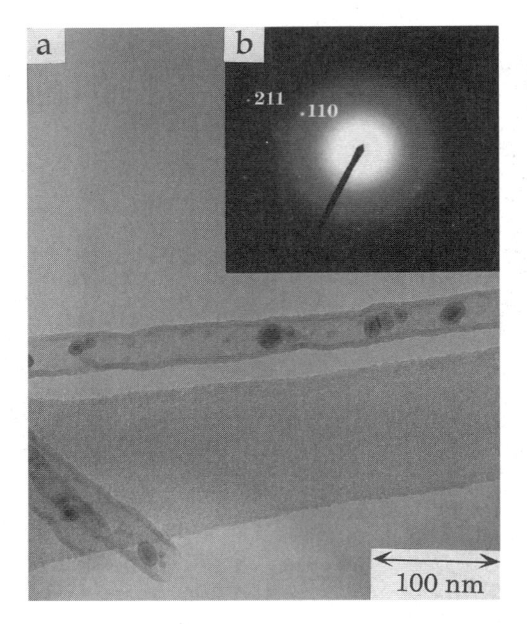

Fig. 10.1.14 TEM photographs of the Fe/carbon nanotube composites before the alkali treatment (MOCVD temperature, 400°C; MOCVD period, 24 h; ferrocene vapor pressure, 0.1 kPa: (a) a bright-field image and (b) SAD pattern for the nanotubes from the image (a). (From Ref. 41.)

pieces. TEM observation showed that some nanotubes stuck out of several broken Al_2O_3 pieces, and also some of the nanotubes were isolated from the film. Figure 10.1.14a exhibits the TEM image of one such isolated carbon nanotube containing nanoparticles. Most of these particles comprise a dark core surrounded by another type of substance, whereas such a dual structure was not observed in the case of very small particles. The corresponding SAD pattern (Fig. 10.1.14b) presents sharp clear spots, which were identified as the reflection from α-Fe (the spots can be indexed to 110, 211) and the diffused rings were from the carbon nanotube. Together with these clear spots, several very weak spots could be observed. All of the spots can be assigned to the reflection from Fe_3O_4. Taking the presence of two types of iron species (α-Fe and Fe_3O_4) into consideration, the authors could explain the formation of a dual structure in the following way. Initially, iron metal particles were deposited in the nanotube hollow by the MOCVD under H_2 flow. When these particles were exposed to air, their surface was oxidized to iron oxide (Fe_3O_4) to leave the metal (α-Fe) in the particle core, while very small metal particles were completely oxidized with the air exposure. Upon further exposure to air and/or subsequent alkali treatment, the remaining metal core was oxidized to Fe_3O_4 as observed in Fig. 10.1.13.

The MOCVD was done at the three different temperatures (350, 400, and 500°C) for 3 h with 0.3 kPa of ferrocene vapor. For all three cases, crystalline nanoparticles were observed, and their chemical form was identified to be Fe_3O_4 by the corresponding SAD patterns. It was found that the number of nanoparticles and their size increased with MOCVD temperature. Such increases in number and size reflected the weight gain of the carbon coated film; the gain was 3.4, 4.1, and 8.3 wt% at 350, 400, and 500°C, respectively. Assuming all the iron compounds to be Fe_3O_4, the authors could estimate that the iron oxide occupies 8, 10, and 20% of the total volume of the tube hollow when the MOCVD was performed for 3 h at 350, 400, and 500°C, respectively. At 500°C, although a few nanotubes remained empty, most of the nanotubes were filled, and some of them looked completely filled with the metal compound. The fraction of empty tubes in this case was quite low in comparison with the other low-temperature cases.

The effect of MOCVD period was also investigated at 400°C with 0.1 kPa of ferrocene vapor. When the MOCVD was carried out only for a short period, such as 0.5 h, most of the nanotubes were found to be empty. Only a few of them contained Fe_3O_4 nanoparticles, whose mean size was 5 nm with a standard deviation of 1 nm. With increasing MOCVD period, the size of nanoparticles gradually increased: 16 $\pm$ 5, 20 $\pm$ 9 and 22 $\pm$ 7 nm at 1, 6 and 12 h-MOCVD, respectively. The weight gain for 12 h of MOCVD was 9.3%, which corresponds to 22% filling of the total volume of the tube hollows.

In this work, Pradhan et al. demonstrated the encapsulation of crystalline Fe_3O_4 nanoparticles into the uniform carbon nanotubes by the MOCVD technique using ferrocene and that the size and number of such nanoparticles can be easily controlled by changing the MOCVD temperature or its period. At a proper MOCVD condition, Fe_3O_4 nanocrystals could be introduced into all of the nanotubes to different degrees, and more than 20% of the total volume of the tube hollow was filled with the nanoparticles.

10.1.8 Chemical Modification of Nanotube Inner Wall by Direct Fluorination

Taking into consideration that only the inner wall surface of carbon nanotubes is exposed to atmosphere in the stage of carbon-deposited alumina film, it would be possible to modify only the inner surface if the carbon-deposited alumina film is chemically treated. On the basis of this concept, Hattori et al. tried to fluorinate only the inner surface of carbon nanotubes (42). It is well known that fluorination is quite an effective way to introduce strong hydrophobicity to carbonaceous materials, and it perturbs the carbon π electron system (43,44). Thus, by the selective fluorination of nanotube's inner surface, it would be possible to produce carbon nanotubes whose inner surface is highly hydrophobic and electrically insulating while their outer surface is conducting. In this section, the details of the synthesis procedure and the results of characterization of such fluorinated carbon nanotubes are described.

A carbon-deposited film was prepared from the alumina film with 30-nm channels by the CVD technique using propylene. Fluorination was carried out by direct reaction of the film with dry fluorine gas (purity 99.7%). The film was placed in a nickel reactor and was allowed to react with 0.1 MPa of fluorine gas for 5 days at a predetermined temperature in the range of 50 to 200°C. Then the fluorinated carbon nanotubes were separated by dissolving the alumina film with HF. A schematic drawing of the fluorination process is given in Fig. 10.1.15.

From TEM observation of the resultant tubes, it was confirmed that their outer diameter is the same as the diameter of the channels of the alumina template. In order to check the chemical state of fluorine on carbon surface, X-ray photoelectron spectroscopy (XPS) measurement was performed for both the fluorinated carbon-deposited film and carbon nanotubes. Figure 10.1.16 shows the C1s XPS spectra of the carbon-deposited films that were fluorinated at the three different temperatures. The untreated carbon-deposited film had a single peak at 284.4 eV (Fig. 10.1.16a), which can be attributed to the sp^2 carbon atom of the carbon skeleton. Upon the fluorination reaction, two other peaks at 289.3 and 291.1 eV appeared. These are ascribed to the sp^3 carbon atom of the CF and CF_2 species, respectively. With increasing reaction temperature, the intensity of these fluorine-related peaks increased, while that of the carbon peak decreased. At 200°C, no peak was observed at 284.5 eV, but only the fluorine-related peaks remained, indicating that the surface of this carbon-deposited film is fully covered with covalent CF bonds. A large charging shift up to 4.3 eV observed in the XPS measurement with this 200°C-treated sample implies the surface to be electrically insulating. For every sample, the fluorine-related peaks disappeared with argon ion sputtering for 30 s, suggesting

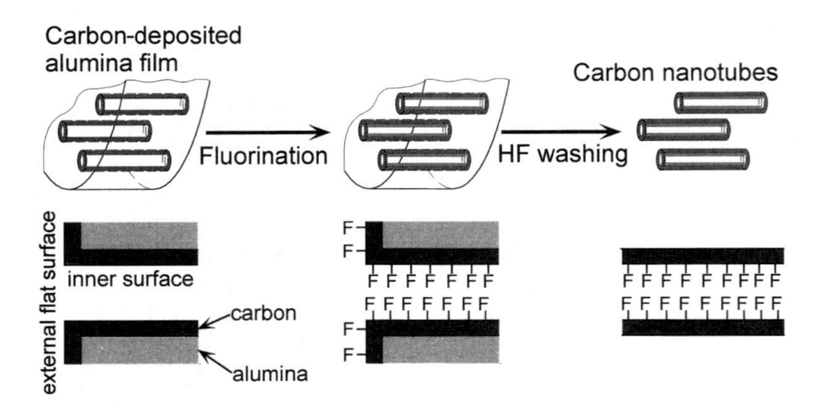

Fig. 10.1.15 Schematic diagram of the fluorination process of carbon nanotube.

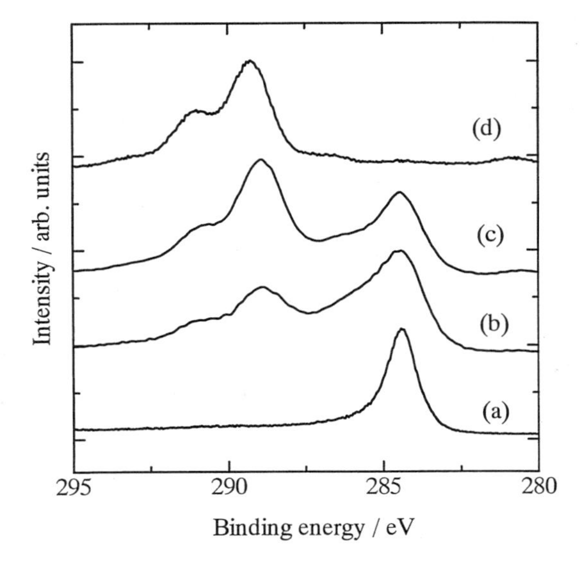

Fig. 10.1.16 C1s XPS spectra of the carbon deposited film: (a) original; (b, c, and d) the films fluorinated at 50, 100, and 200°C, respectively. (From Ref. 42.)

that fluorine atom are attached only on the surface. These XPS spectra mainly reflect the structure of the carbon on the external flat surface of the film, because the inner surface of the nanotubes could not be detected by XPS measurement. However, the chemical form of the inner surface can be expected to be the same as that of the external flat surface of the film. It can therefore be concluded that the inner surface of the nanotubes was completely fluorinated at 200°C.

In Figure 10.1.17, the C1s XPS spectrum of the carbon-deposited film fluorinated at 175°C is compared with that of the corresponding fluorinated tubes after the HF washing. The former spectrum and the latter one give information on the external flat surface of the fluorinated film and the outer surface of the fluorinated tube, respectively. In spectrum (a), the most intense component is the peak assigned to covalent CF bond. On the other hand, spectrum (b) exhibits a clear peak at 284.4 eV, which corresponds to the sp^2 carbon of the nanotube. These findings indicate selective fluorination of carbon tubes; the covalent CF bonds were formed almost exclusively on the inner surfaces of nanotubes, but the outer surfaces retained their sp^2 hybridization.

10.1.9 Potential Application of Template-Synthesized Carbon Nanotubes

As demonstrated in this chapter, the template technique using anodic oxide films makes it possible to control the length, diameter, and thickness of carbon tubes and

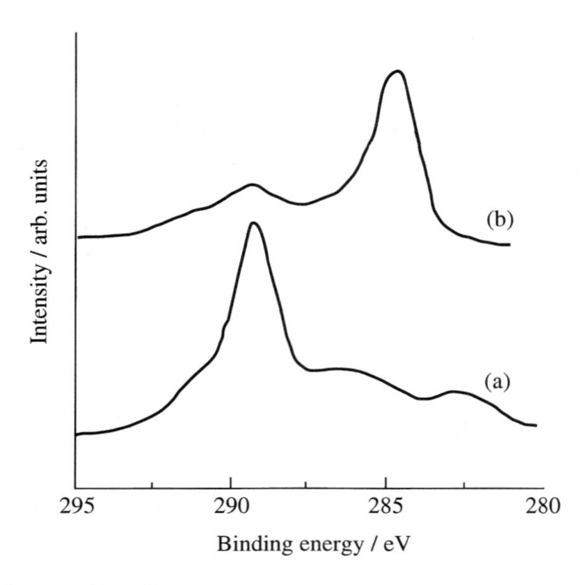

Fig. 10.1.17 C1s XPS spectra of (a) the fluorinated carbon deposited film and (b) the carbon nanotubes from the film (fluorination temperature 175°C). (From Ref. 42.)

to produce monodisperse carbon nanotubes. However, the carbon nanotubes prepared by this method have rather low crystallinity. This makes applications such as nanoelectrodevices or ''the ultimate carbon fibers'' difficult, because these rely on defect-free structure of nanotubes. However, for application to electrochemical devices such as lithium-ion batteries, this is not the case. Martin's group demonstrated that template-synthesized carbon nanotubes could be used as high-performance anode material in lithium-ion batteries (40). They loaded iron particles on the inner surface of carbon nanotubes in the stage of carbon-deposited film and then again carried out pyrolytic carbon deposition, which produce tinier tubes inside the larger tubes. It was found that the resultant ''tubes-within-tubes'' have a high Li^+ intercalation capacity in the lithium-ion battery.

Encapsulation of other material into carbon nanotubes would also open up a possibility for the applications to electrodevices. By applying the template method, perfect encapsulation of other material into carbon nanotubes became possible. No foreign material was observed on the outer surface of carbon nanotubes. The metal-filled uniform carbon nanotubes thus prepared can be regarded as a novel one-dimensional composite, which could have a variety of potential applications (e.g., novel catalyst for Pt metal-filled nanotubes, and magnetic nanodevice for Fe_3O_4-filled nanotubes). Furthermore, the template method enables selective chemical modification of the inner surface of carbon nanotubes. With this technique, carbon nanotubes whose outer and inner surfaces have different properties can be prepared, and unique adsorption behaviors and unique electrical properties can be expected for these carbon nanotubes with heterogeneous properties.

The template method will be able to produce various types of unique carbon nanotubes and one-dimensional carbon composites, and such unique materials would provide a variety of potential applications.

REFERENCES TO SECTION 10.1

1. S Iijima. Nature 354:56, 1991.
2. TW Ebbesen, PM Ajayan. Nature 358:220, 1992.
3. M Endo, K Takeuchi, S Igarashi, K Kobori, M Shiraishi, HW Kroto. J Phys Chem Solids 54:1841, 1993.
4. M Endo, K Takeuchi, K Kobori, K Takahashi, HW Kroto, A Sarka. Carbon 33:873, 1995.
5. V Ivanov, JB Nagy, P Lambin, A Lucas, XB Zhang, XF Zhang, D Bernaerts, G van Tendeloo, S Amelinkx, J van Landuyt. Chem Phys Lett 223:329, 1994.
6. WZ Li, SS Xie, LX Qian, BH Chang, BS Zou, WY Zhou, RA Zhao, G Wang. Science 274:1701, 1996.
7. M Terrones, N Grobert, J Olivares, JP Zhang, H Terrones, K Kordatos, WK Hsu, JP Hare, PD Townsend, K Prassides, AK Cheetham, HW Kroto, DRM Walton. Nature 388:52, 1997.
8. A Thess, R Lee, P Nikolaev, H Dai, P Petit, J Robert, C Xu, YH Lee, SG Kim, AG Rinzler, DT Colbert, GE Scuseria, D Tománek, JE Fischer, RE Smalley. Science 273: 483, 1996.
9. T Kyotani, L Tsai, A Tomita. Chem Mater 7:1427, 1995.
10. RV Parthasarathy, KLN Phani, CR Martin. Adv Mater 7:896, 1995.
11. RC Furneaux, WR Rigby, AP Davidson. Nature 337:147, 1989.
12. T Kyotani, L Tsai, A Tomita. Chem Mater 8:2109, 1996.
13. S Kawai, R Ueda. J Electrochem Soc 122:32, 1975.
14. D AlMawlawi, N Coombs, M Moskovits. J Appl Phys 70:4421, 1991.
15. CA Foss Jr, GL Hornyak, JA Stockert, CR Martin. J Phys Chem 96:7497, 1992.
16. CA Foss Jr, GL Hornyak, JA Stockert, CR Martin. Adv Mater 5:135, 1993.
17. CA Foss Jr, GL Hornyak, JA Stockert, CR Martin. J Phys Chem 98:2963, 1994.
18. GL Hornyyak, CJ Patrissi, CR Martin. J Phys Chem B 101:1548, 1997.
19. H Masuda, K Fukuda. Science 268:1466, 1995.
20. P Hoyer, N Baba, H Masuda. Appl Phys Lett 66:2700, 1995.
21. P Hoyer. Langmuir 12:1411, 1996.
22. BB Lakshmi, PK Dorhout, CR Martin. Chem Mater 9:857, 1997.
23. CJ Brumlik, CR Martin. J Am Chem Soc 113:3174, 1991.
24. W Liang, CR Martin. J Am Chem Soc 112:9666, 1990.
25. G Che, BB Lakshmi, CR Martin, ER Fisher. Chem Mater 10:260, 1998.
26. M Freemantle. Chem Eng News 74:62, 1996.
27. J Cook, J Sloan, MLH Green. Chem Ind 16:600, 1996.
28. TW Ebbesen. Phys Today 49:26, 1996.
29. M Liu, JM Cowley. Carbon 33:749, 1995.

30. Y Murakami, T Shibata, K Okuyama, T Arai, H Suematsu, Y Yoshida. Phys Chem Solids 54:1861, 1993.
31. S Subramoney, RS Ruoff, DC Lorents, B Chan, K Malhotra, MJ Dyer, K Parvin. Carbon 32:507, 1994.
32. Y Saito, T Yashikawa, M Okuda, N Fujimoto, K Sumiyama, K Suzuki, A Kasuya, Y Nishina. J Phys Chem Solids 54:1849, 1994.
33. C Guerret-Piecourt, Y Le Bouar, A Loiseau, H Pascard. Nature 372:761, 1994.
34. PM Ajayan, S Iijima. Nature 361:333, 1993.
35. BC SatishKumar, A Govindaraj, J Mofokeng, GN Subbanna, CNR Rao. J Phys B 66: 839, 1994.
36. SC Tsang, YK Chen, PJF Harris, MLH Green. Nature 372:159, 1994.
37. A Chu, J Cook, RJR Heeson, JL Hutchison, MLH Green, J Sloan. Chem Mater 8:2571, 1996.
38. U Ugarte, A Châtelain, WA de Heer. Science 274:1897, 1996.
39. T Kyotani, L Tsai, A Tomita. J Chem Soc Chem Commun 701, 1997.
40. G Che, BB Lakshmi, ER Fisher, CR Martin. Nature 393:346, 1998.
41. BK Pradhan, T Toba, T Kyotani, A Tomita. Chem Mater 10:2510, 1998.
42. Y Hattori, Y Watanabe, S Kawasaki, F Okino, BK Pradhan, T Kyotani, A Tomita, H Touhara. Carbon 37:1033, 1999.
43. G Li, K Kaneko, S Ozeki, F Okino, R Ishikawa, M Kanda, H Touhara. Langmuir 11: 716, 1995.
44. G Li, K Kaneko, F Okino, R Ishikawa, M Kanda, H Touhara. J Colloid Interface Sci 172:539, 1995.

10.2 PRODUCTION BY ARC DISCHARGE

YAHACHI SAITO
Mie University, Tsu, Japan

10.2.1 Introduction

Carbon nanotubes are attracting considerable attention because of their own novel physical properties and their promising application as field electron emitters (1–3), gas storage materials (4), probes of scanning tunneling microscopes (5), and so forth. Two types of carbon nanotubes are now available; one is single-wall nanotubes (SWNTs) and the other is multiwall nanotubes (MWNTs). SWNTs are made of one layer of graphene (a hexagonal network of carbon with monoatomic thickness) rolled up into a seamless cylinder (Fig. 10.2.1), while MWNTs are made of coaxially stacked cylindrical layers, ranging in number from 2 to about 40. Diameters of SWNTs are typically in a range between 1 and 5 nm, and those of MWNTs are between 4 and 50 nm. In 1991, Iijima (6) discovered MWNTs in a carbonaceous deposit that was left in a carbon arc reactor after collecting fullerene soot (soot containing fullerenes such as C_{60} and C_{70}) (7). Subsequently, SWNTs were discovered as an outgrowth of synthesis of carbon nanocapsules containing magnetic fine particles in 1993 (8–10). Catalytic metals have to be evaporated together with carbon to synthesize SWNTs; Iijima et al. (8), Bethune et al. (9), and Saito et al. (10) used iron, cobalt, and nickel as catalysts, respectively. MWNTs that have nearly perfect crystallinity are exclusively produced by carbon arc discharge, while SWNTs are synthesized not only by arc discharge but also by laser ablation.

The carbon arc is a versatile method to generate a wide variety of nanostructured carbon materials such as fullerenes (7,11), endohedral metallofullerenes (12–14), nanotubes (6,8–10,15), nanocapsules filled with metals (or carbides) (16–18), and so forth. Arc discharge is characterized by the high electric current

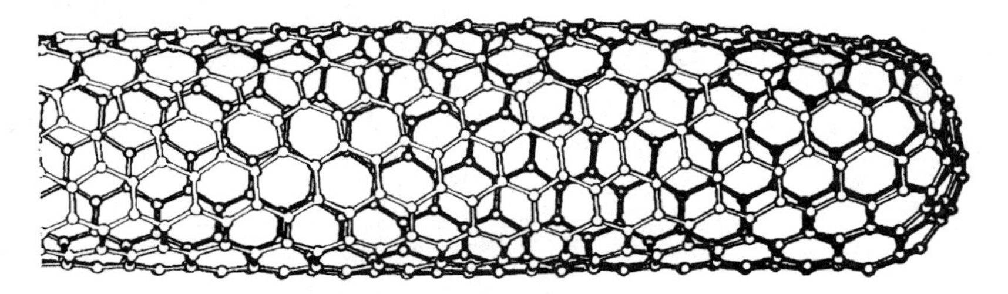

Fig. 10.2.1 Structure model of a single-wall carbon nanotube. (Courtesy of Prof. E. Osawa and Dr. M. Yoshida.)

and the high temperature of plasma, which exceeds 4000 K. Critical parameters that influence the yield of the products are the kind and the pressure of buffer gases, arc current, and metal catalysts. Transmission electron microscopy (TEM) is mainly employed for characterizing these nanoscale materials. In this chapter, structures and morphology of carbon nanotubes synthesized by arc discharge under various conditions are described.

10.2.2 Production Method

10.2.2.1 Carbon Arc Evaporator

The carbon arc apparatus is shown schematically in Figure 10.2.2. The reaction chamber, which is usually made of stainless steel, is connected to vacuum pumps and a gas supply line through valves. The front of the chamber has an observation window to enable monitoring the arc discharge (19). Electrodes, mounted on the end flanges, are supported horizontally. Alternatively, the electrodes can be fixed vertically (20). The electrodes and chamber walls can be cooled by water-cooling devices if necessary.

Power supplies commonly used for arc welding can be used as an electric power supply for generating arc discharge. Either direct current (DC) or alternative current (AC) arc can evaporate carbon electrodes. The DC mode is almost exclu-

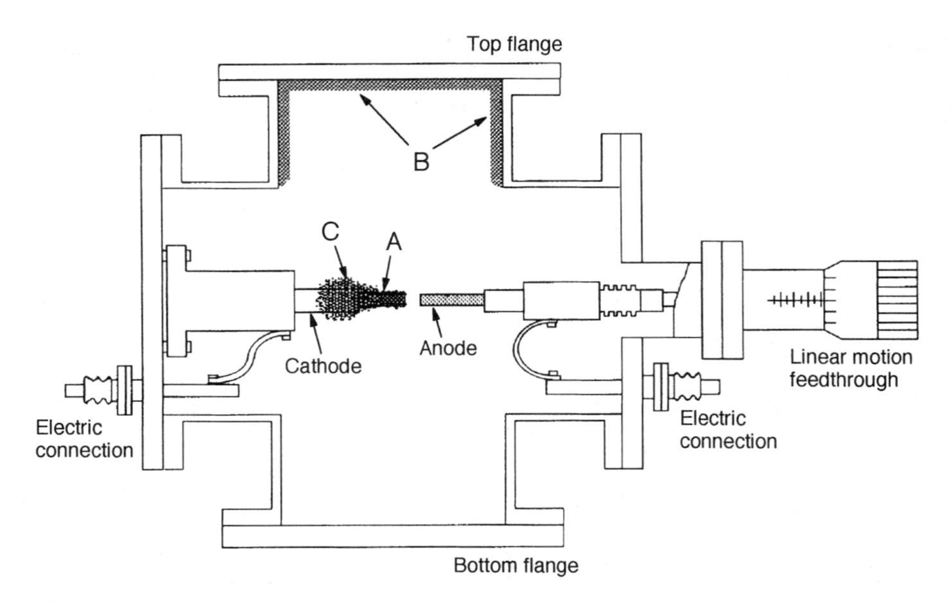

Fig. 10.2.2 Schematic drawing of a carbon arc apparatus. ''Cylindrical hard deposit'' (A) grown on the end of the cathode, ''chamber soot'' (B) deposited on the ceiling of the evaporator, and ''cathode soot'' (C) around the cathode surfaces are shown.

sively employed for synthesizing nanotubes because the DC arc is amenable and yields more nanotubes than the AC arc. Both the anode and the cathode are made of graphite rods (typically 6 mm in diameter and 50 mm in length for the anode, and 13 mm in diameter for the cathode). Graphite used to make electrodes is artificial graphite that is widely utilized for making crucibles, heaters and so on.

After evacuating (or purging) the chamber, inactive gas (usually helium) is introduced into it up to a desired pressure, and then arcing between the electrodes is started. Pressures of helium are usually in a range from 100 to 600 torr. The discharge current is typically 70 A for a 6-mm-diameter anode, and the voltage about 20 V. Since the anode surface is heated to a higher temperature (~4000 K) than the cathode surface (~3500 K) for the DC arc, the anode is consumed by evaporation. The position of the anode tip must be adjusted manually or under computer control (21,22) from outside the vacuum with a translation-motion feed-through in order to maintain the proper spacing (1 mm or less) between the electrodes. Approximately half of the evaporated carbon condenses on the end of the cathode and forms a cylindrical hard deposit (A in Fig. 10.2.2) containing MWNTs. The cylindrical deposit grows at a rate of about 1 to 2 mm/min. The other half of the evaporated material condenses in the gas phase and forms so-called soot containing fullerenes.

Pure carbon evaporation is commonly used for producing MWNTs as well as fullerenes. For synthesizing SWNTs, graphite anodes containing metal catalysts are evaporated; the method is described next.

10.2.2.2 Evaporation of Metal/Carbon Composites

When an anode contains an appropriate amount of metals (or metal oxide), novel carbon materials such as SWNTs, metallofullerenes, filled nanocapsules, ''bamboo''-shaped tubes (23), nanochains (10), and MWNTs filled with metal carbides (24,25) are formed. Especially SWNTs are now attracting a great deal of interest from researchers in physics and materials science, because exotic electronic properties that vary between semiconducting and metallic states depending on how a graphene sheet is rolled (i.e., diameter and helical pitch of a tube) are predicted theoretically (26–28) and because unique quantum effects are revealed experimentally (29,30).

In an early experiment, metal/graphite composite anodes were made of bored graphite rods that were packed with pressed mixtures of metal oxide powder, graphite powder, and pitch (12,13). The packed graphite rods had to be heated to about 1600°C for several hours under vacuum in order to cure the pitch. Subsequently the preparation of composite anodes was simplified by simply packing with a mixed powder of metal oxide (or metal) and carbon. Several kinds of composite rods, in which metal particles are uniformly dispersed in graphite, are now commercially available (31).

Vapor of the carbon and metal mixture condensed in a gas phase to form black soot, which finally deposited on the inner walls of the arc chamber and on a surface of the cathode. Hereafter we call the soot deposited on the chamber walls ''chamber soot'' (B in Fig. 10.2.2), and that sticking on the cathode surface ''cathode soot'' (C in Fig. 10.2.2). Structures and morphologies of carbon materials differ depending on where they grow; the details are described in the following sections.

10.2.3 Multiwall Nanotubes

A cylindrical hard deposit, which is grown on the tip of a cathode (A in Fig. 10.2.2), consists of two regions: an inner fibrous black core and an outer gray hard shell (15). The inner core had a columnar structure that was made up of bundles of nanotubes and flocks of polyhedral graphitic particles (32). On the other hand, the hard shell was made of stacked graphitic flakes.

Figure 10.2.3 shows a TEM image of MWNTs and carbon nanoparticles grown in the fibrous core of the cathode deposit. MWNTs consists of concentrically stacked layers, and hold cavities in their center. Lengths of MWNTs are over 1 μm. Both ends of an MWNT are capped by graphitic layers; the number of layers is the same as that for the sidewall of the tube. The caps have polyhedral shapes with sharp corners, at which pentagons are believed to be located (33). As for fullerenes, pentagons have to be introduced into the hexagon network for the graphene sheet to make a closure. Curvatures of a graphene sheet concentrate around the pentagons, so the locations where pentagons exist extrude like corners of a polyhedron, whereas other places without pentagons are flat. Tips of MWNTs show various shapes, depending on how six pentagons are located in the hexagon network.

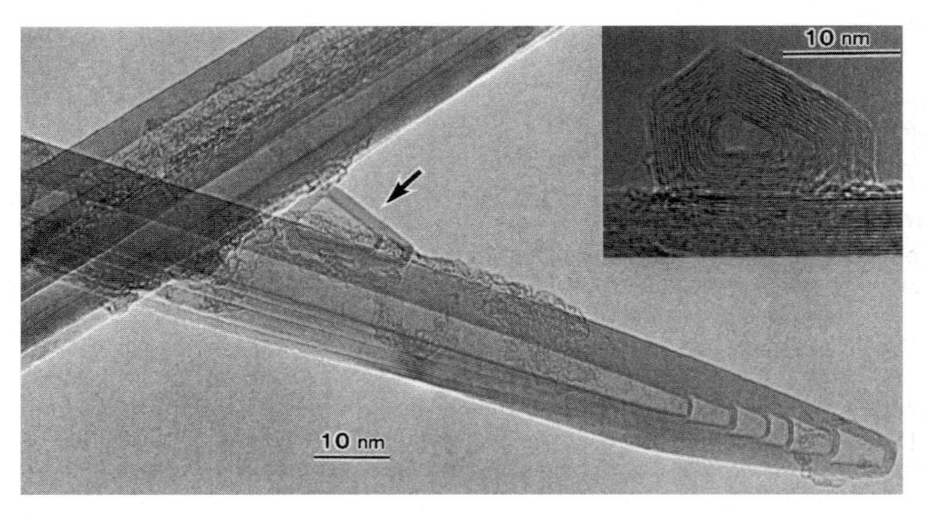

Fig. 10.2.3 TEM image of MWNTs grown in the cylindrical hard deposit. A carbon nanoparticle (also called nanopolyhedron) is indicated by an arrow. A high-magnification image in the inset shows another nanoparticle on a MWNT.

The interlayer spacings between rolled graphene sheets are close to those for planar graphite crystals (0.3354 nm), but strictly speaking the interlayer distance in MWNTs is by a few percent wider on average than that of the ideal graphite (34). The wide interplanar spacing is characteristic of turbostratic carbon, in which atomic positions are not registered between neighboring layers (35).

Since MWNTs produced by arc discharge have nearly perfect crystallinity of graphite and are very stiff [Young's modulus being 1.8 TPa on average (36)], almost all the MWNTs observed are straight. Elastically deformed (bent) ones are only rarely observed. Bent tubes are plastically deformed, with buckling on the concave side (37).

MWNTs recovered from the inner core of the cathode deposit contain graphite debris and nanoparticles [also called nanopolyhedra (32); see the inset of Fig. 10.2.3]. In order to obtain pure MWNT samples, these nanoparticles have to be removed. A few methods of purification, including an oxidation process, are reported (35–37).

10.2.4 Single-Wall Nanotubes

Metal catalysts are necessary to synthesize SWNTs of carbon for both the arc discharge (8–10) and laser vaporization methods (41). Typical catalysts for synthesizing SWNTs are listed in Table 10.2.1. The yield of SWNTs strongly depends not only

Table 10.2.1 Typical Catalysts for Synthesizing SWNTs

Metal catalysts	Methods and conditions	Density in soot[a]
Iron group		
Fe	Arc in Ar and CH_4	VL
Co	Arc in He/chamber soot	M
Ni	Arc in He/cathode soot	L
Fe–Ni	Arc in He	H
Co–Ni	Laser ablation in Ar at 1200°C	VH
Rare earths		
Y	Arc in He	M
La	Arc in He	M
Ce	Arc in He	M
Platinum group		
Rh	Arc in He	M
Ru–Pd	Arc in He	H
Rh–Pd	Arc in He	H
Rh–Pt	Arc in He	VH
Intergroup mixtures		
Ni–Y	Arc in He	VH
Ni–La	Arc in He	VH
Ni–Ce	Arc in He	VH

[a] Density of SWNTs in soot is graded into five classes: VH, very high (more than 20%); H, high (~20%); M, medium (several percent); L, low (~1%); and VL, very low (less than 1%).

on catalysts but also on buffer gases and growth places (i.e., chamber soot and cathode soot), as described later.

10.2.4.1 Elemental Metals as Catalysts

Iron-Group Metals. SWNTs were first discovered by using the iron-group metals (Fe, Co, and Ni) as catalysts. When iron alone is used as a catalyst, a CH_4/Ar mixture must be used as buffer to produce SWNTs (8). However, the yield of SWNTs from iron alone is very low.

Cobalt produces SWNTs relatively abundantly in helium gas (100–600 torr). The density of SWNTs in soot strongly depends on growth places; SWNTs are almost exclusively found in chamber soot. Figure 10.2.4 shows TEM images of SWNTs collected from the chamber soot. SWNTs are so long (more than 10 μm) that ends of tubes are rarely observed. As shown in the TEM pictures, SWNTs produced from Co catalysts are usually covered with amorphous layers.

Nickel shows fairly high activity in producing SWNTs in helium gas between 300 and 600 torr. The cathode soot was the major growth place of SWNTs for Ni, while the chamber soot rarely contained SWNTs (42), in striking contrast to the Co catalyst. In Figures 10.2.5 and 10.2.6 are shown TEM images of bundles of SWNTs produced from nickel. Capped ends of SWNTs are clearly observed in Figure 10.2.5. A cross section of a bundle of SWNTs, observed in Fig. 10.2.6, reveals that each SWNT has a circular cross section, and SWNTs in a bundle form a regular (two-dimensional) triangular lattice.

For Ni, it was also revealed that SWNTs grow radially from catalyst particles, which were either fcc Ni or carbide (Ni_3C). Figure 10.2.7 shows a TEM picture of bundles of SWNTs radially growing from a spherical Ni particle. The length of tubes was more than 1 μm for the majority of tubes.

Diameters of SWNTs are slightly dependent on the catalysts within the iron-group; the mode diameters (the most frequently observed diameters) were ~1.0 nm for Fe (3), 1.2–1.3 nm for Co (43), and ~1.2 nm for Ni (42).

Rare Earths. Among the series of rare-earth elements investigated (Sc, Y, and Ln = La through Lu except for Pm), lanthanum (La), cerium (Ce), and yttrium (Y) were the most active; abundant colonies of SWNTs growing radially from core catalytic particles were observed in both the chamber and the cathode soot. The fraction of SWNTs in soot obtained from these most effectual rare-earth catalysts (La, Ce and Y) is estimated to be a few percent by volume from visual inspection of TEM images (44). The length and diameter of tubes were typically 80 nm and 2 nm, respectively. Gadolinium (Gd), terbium (Tb), holmium (Ho), erbium (Er), and lutetium (Lu) showed intermediate activity. Praseodymium (Pr), neodymium (Nd), and dysprosium (Dy) showed weak activity. The remaining rare-earth elements including Sc did not produced any SWNTs. Except for Sc, correlation between the

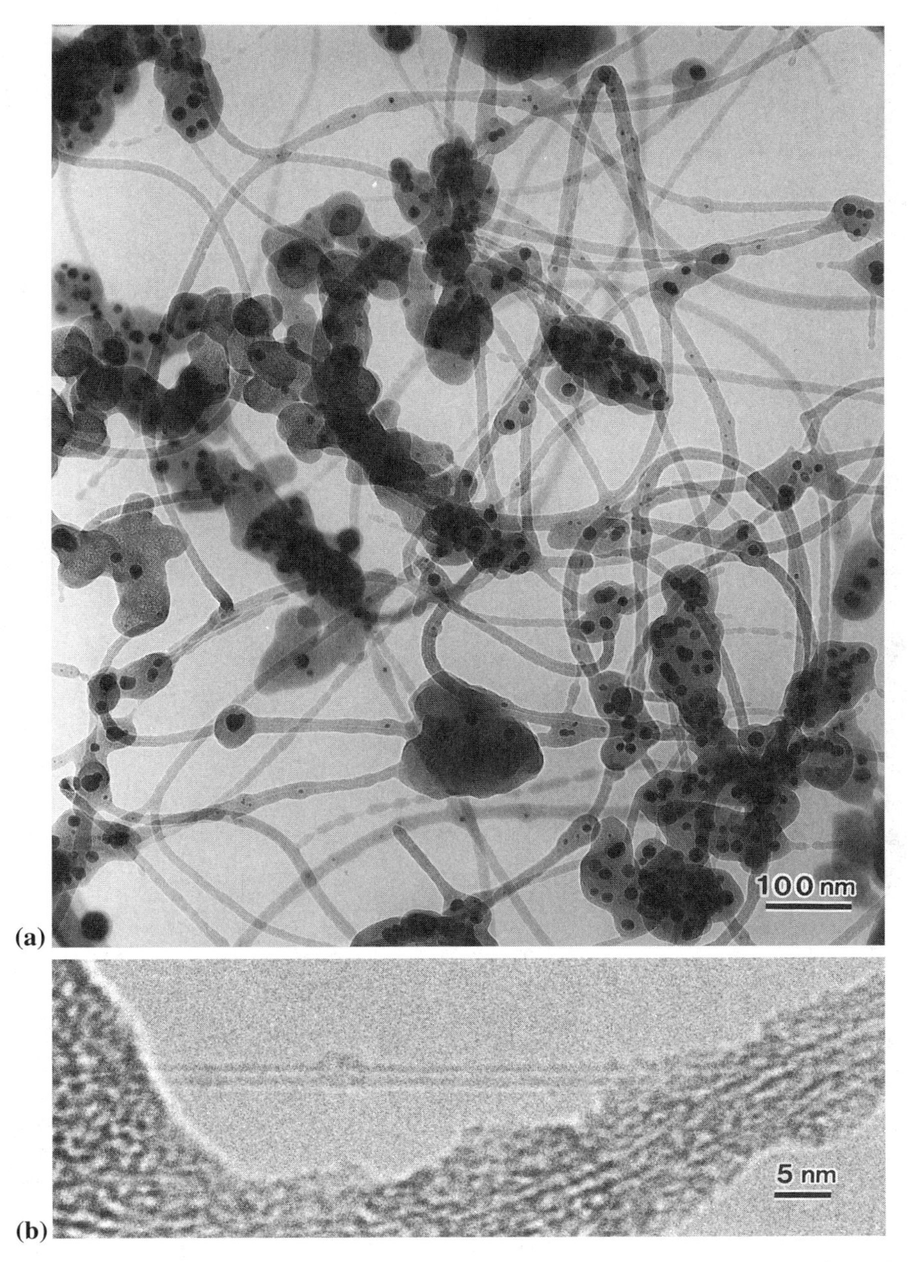

Fig. 10.2.4 TEM images of SWNTs in the chamber soot that was produced from Co catalysts at 600 torr of helium gas: (a) low-magnification image, and (b) high-magnification image revealing an SWNT covered with amorphous carbon.

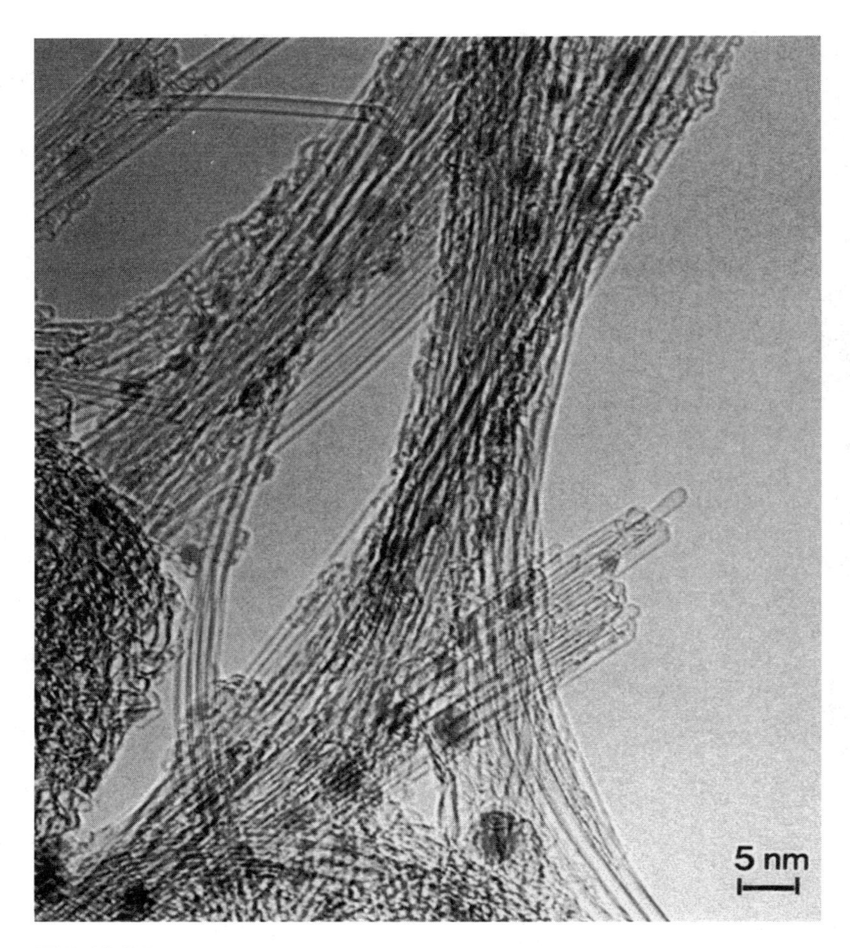

Fig. 10.2.5 TEM image of SWNTs produced from Ni, revealing that ends of SWNTs are capped.

activity for SWNT formation and the vapor pressure of the elements was found (42); nonvolatile metals (Y, La, Ce, Pr, Nd, Gd, Tb, Dy, Ho, Er, and Lu) showed the catalytic activity, while volatile metals (Sm, Eu, Tm, and Yb) did not.

Figure 10.2.8 shows a typical TEM image of SWNTs growing radially from La-containing core particles (45). SWNTs seem to be emerging directly from the surface of core particles (no graphitic layers are observed on the particle surface). The tips of SWNTs were capped and hollow inside. The growth pattern and the morphology of SWNTs are dubbed ''sea urchin'' (46). Similar growth morphology is observed for Gd (46) and Y (47) catalysts as well. However, for Gd and Y, core

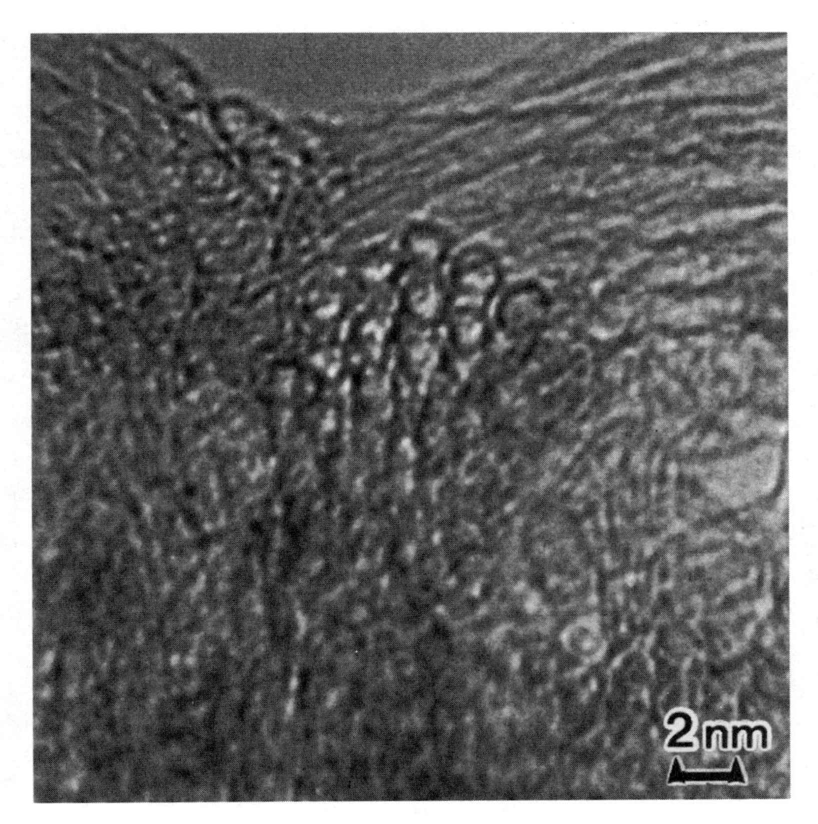

Fig. 10.2.6 TEM image showing a cross section of a bundle of SWNTs. When a curved bundle has a portion where the incident electron beam is parallel to the local axis of the bundle, a local cross section of the bundle can be viewed. (Reprinted with permission from Y Saito et al., Jpn J Appl Phys 33:L526, 1994.)

particles are covered with graphitic layers, over which SWNTs are emerging. The graphitic layers may have poisoned the catalytic activity, and stopped further growth of SWNTs, as suggested by Seraphin (47). The phases of core particles are rare-earth dicarbides (RC_2). Though rare-earth carbides are hygroscopic, this phase is preserved even in exposure to a humid air if the core particles are covered with graphitic layers.

The diameter of the SWNTs is typically 1.8–2.1 nm, thicker than the SWNTs produced from the iron-group metals (Fe, Co, Ni). On the other hand, a typical length of the tubes obtained from rare-earth catalysts is shorter (30 to 100 nm) than those obtained from the iron-group metals ($\sim$1 μm).

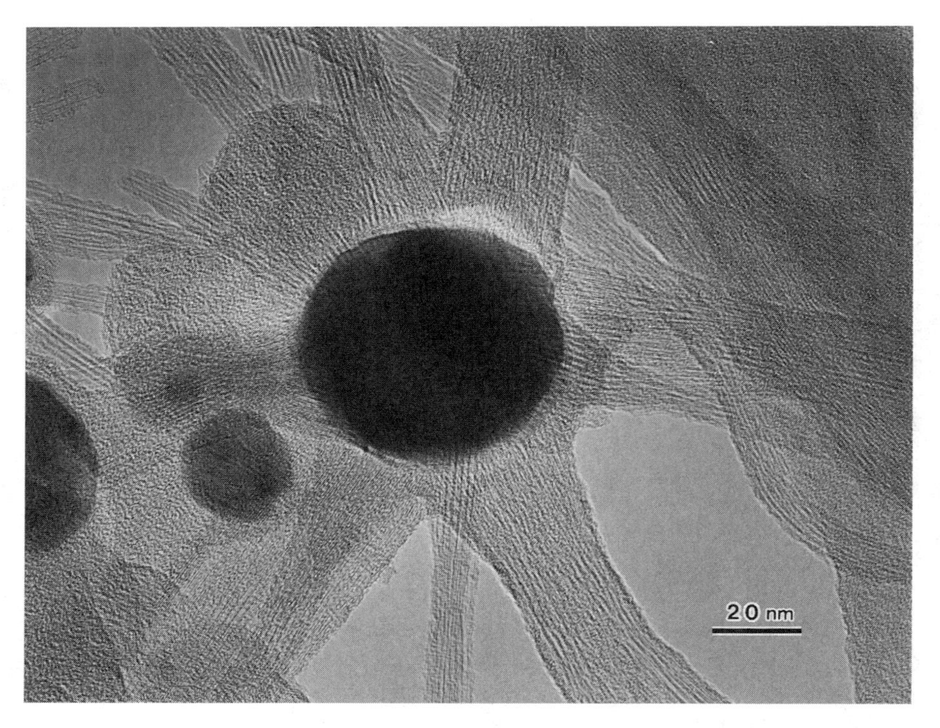

Fig. 10.2.7 TEM image of bundles of SWNTs radially growing from a spherical Ni particle.

Pt-Group Metals. Among the six metals (Ru, Rh, Pd, Os, Ir, and Pt) in the platinum group, rhodium (Rh), paradium (Pd), and platinum (Pt) assist the formation of SWNTs (48). Of the three metals, Rh is the most effective for producing SWNTs. Cathode soot produced from Rh contains SWNTs densely, whereas the density in SWNTs is comparable with that for rare-earth metal catalysts such as Y, La, and Ce.

Figure 10.2.9 shows a TEM picture of SWNTs growing from Rh particles. Diameter of the tubes ranged from 1.3 to 1.7 nm, and the length ranged from 10 to more than 200 nm. Several tens of SWNTs are extruding from an Rh particle; the growth morphology of the SWNTs is quite similar to the ''sea urchin,'' though the length of tubes is longer than that obtained from rare earths. Core Rh particles, with their diameter being 20–30 nm, were covered with graphitic layers, which were possibly formed after the radial growth of SWNTs.

For Pd catalyst, short SWNTs were radiating from large Pd particles (50 to 200 nm) (49). Bundles of SWNTs protrude radially and densely from the core parti-

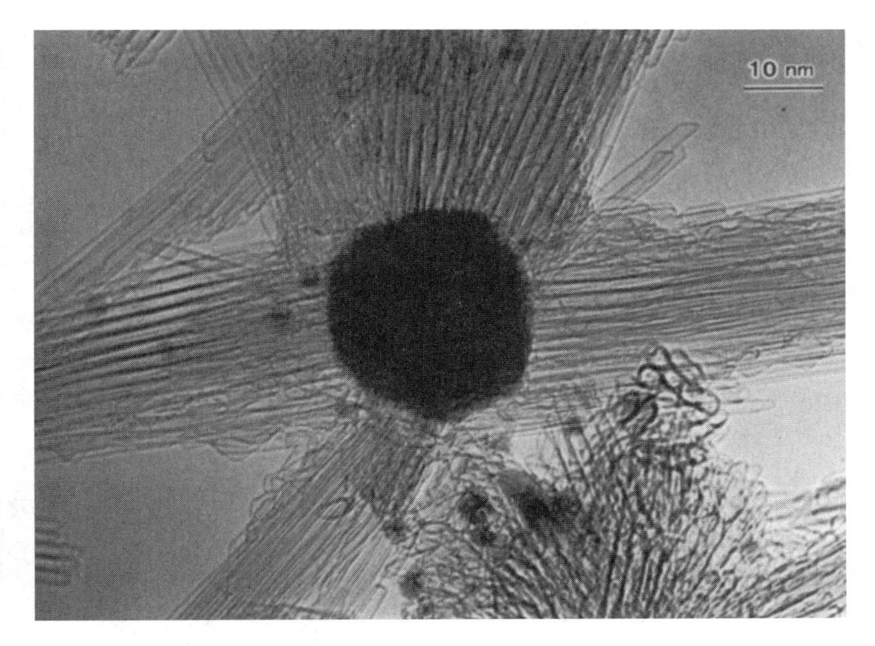

Fig. 10.2.8 SWNTs growing radially from an La compound particle.

cle. The length of the tubes was short (30 to 50 nm) compared to those grown from Rh, though the diameter (1.3–1.7 nm) was in the same range as that for Rh.

In contrast to Rh and Pd, a Pt particle emits only one or a few SWNTs (48). Moreover, the particle size of Pt, ~10 nm, is slightly smaller than that of Rh and much smaller than that of Pd. The diameter of SWNTs was typically 1.3–2.0 nm and sometimes ~3 nm, and the length of tubes ranged from 10 to 100 nm.

For Ru and Os, carbon globules comprised of single- or double-layered graphitic sheets curving in a complex fashion are formed (48). In the periphery of globules, cone-shaped protrusions are observed as shown in Figure 10.2.10. Measured apex angles of nanocones were concentrated around 20° and 35°. These values of angle agree well with the cone angles of 19.2° and 38.9° expected for carbon nanotubes with five and four pentagons in a tip, respectively. Together with the numerical agreement, contrast in the TEM image suggests that these nanocones are made of a single-layered graphitic sheet folded to the shape of a cone by inserting five or four pentagons into the network of hexagons.

10.2.4.2 Binary Mixtures as Catalysts

Some combinations of metals improve remarkably the yield of SWNTs, even though the respective metals show no or very low catalytic activity by themselves. Typical combinations showing such an effect are Fe–Ni, Rh–Pt, and Ni–Y.

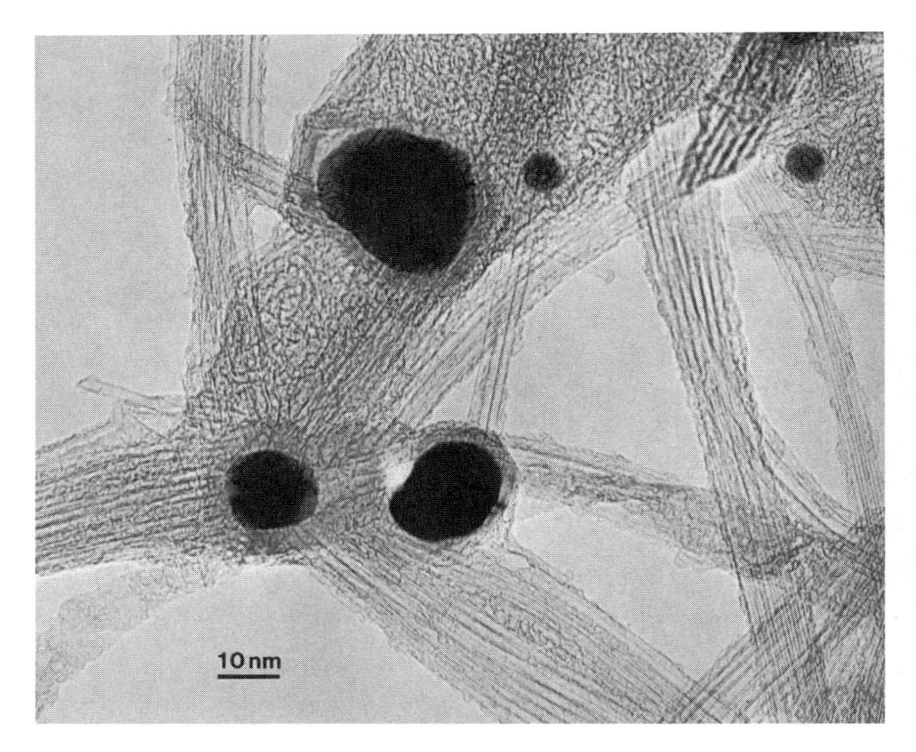

Fig. 10.2.9 TEM picture of SWNTs growing from Rh particles.

Mixtures of the Iron-Group Metals. It was first reported by Seraphin et al. (50) that a binary mixture of Fi and Ni yielded more abundant SWNTs than Fe or Ni alone catalyst. Saito et al. (51) showed that a mixture of Fe and Ni with the ratio of 1:1 (by weight) gave the highest yield of SWNTs, and deviation from the 1 to 1 composition reduced the yield. Approximately 10% of all the carbon in the raw soot (both the chamber and cathode soot) was incorporated into SWNTs at the highest yield. Diameters of SWNTs produced from Fe/Ni range from 0.9 to 1.4 nm, and the mode diameter is located in 1.1–1.2 nm.

A Co/Ni alloy is the next active catalyst among the binary combinations within the iron-group metals in the arc discharge method (51). Laser vaporization of metal/carbon composite in argon atmosphere at high temperature (1200°C) can also produce SWNTs (41). Guo et al. (41) reported that the Co/Ni alloy was the most effectual, with a yield of 50–90% in the laser ablation method.

Contrary to the other two alloys mentioned, Fe/Co was a poor catalyst for producing SWNTs by arc discharge.

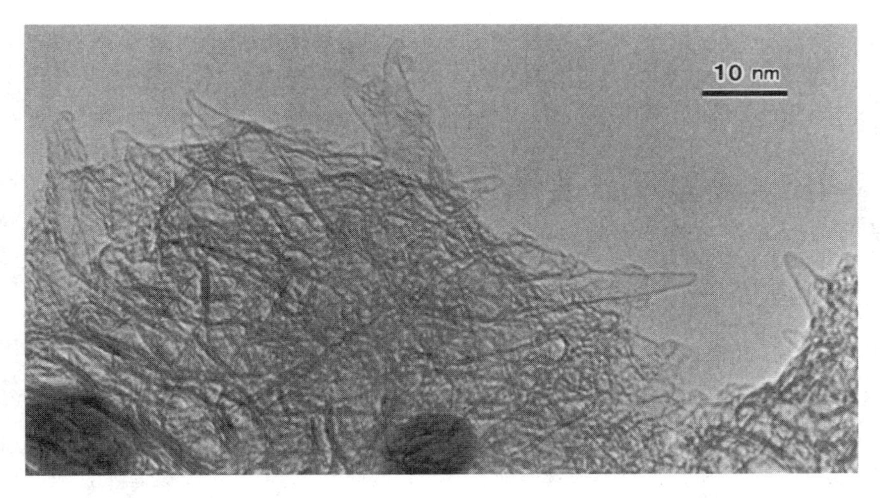

Fig. 10.2.10 TEM image of nanocones sticking up from carbon globules produced from Ru catalyst.

Mixtures of the Platinum-Group Metals. Rh–Pt mixture is the most effectual combination within the platinum group (52). The yield of SWNTs was strongly affected by the mixing ratio of Rh and Pt in a manner similar to the Fe–Ni system; i.e., the highest yield was obtained at a 1 to 1 weight ratio mixture. Helium pressure is another important factor for yielding SWNTs. TEM and Raman scattering spectroscopy suggested that the density of SWNTs reaches as high as 70–90% in the cathode soot prepared at helium 600 torr. A typical TEM picture of such raw soot is shown in Figure 10.2.11. Dark, spherical particles with diameters of 10–20 nm were Rh–Pt alloy with the fcc structure. The mean diameter of SWNTs was 1.28 nm and the diameter distribution was very narrow (standard deviation of only 0.07 nm). Bundles of long SWNTs (exceeding 10 μm in length) are entangled with each other, exhibiting ''highway junction''-like patterns. The growth morphology was observed to be similar for the Fe–Ni mixture.

The diameters of SWNTs depend on the pressure of helium gas in the reaction chamber. Thinner SWNTs were obtained at lower helium pressure, though the yield of SWNTs decreased. Thin SWNTs with diameters 0.7–1.3 nm were obtained at 50 torr of helium. Similar pressure dependence of the diameters was observed for other catalysts such as Fe–Ni.

Intergroup Mixtures. Effective catalysts for preparing abundant SWNTs can be made from mixtures between different groups of elements. Such an intergroup mixture was first reported by Kiang et al. (43), who added sulfur to cobalt. Not only an increase of the yield in SWNTs but also an enlargement of their diameters was

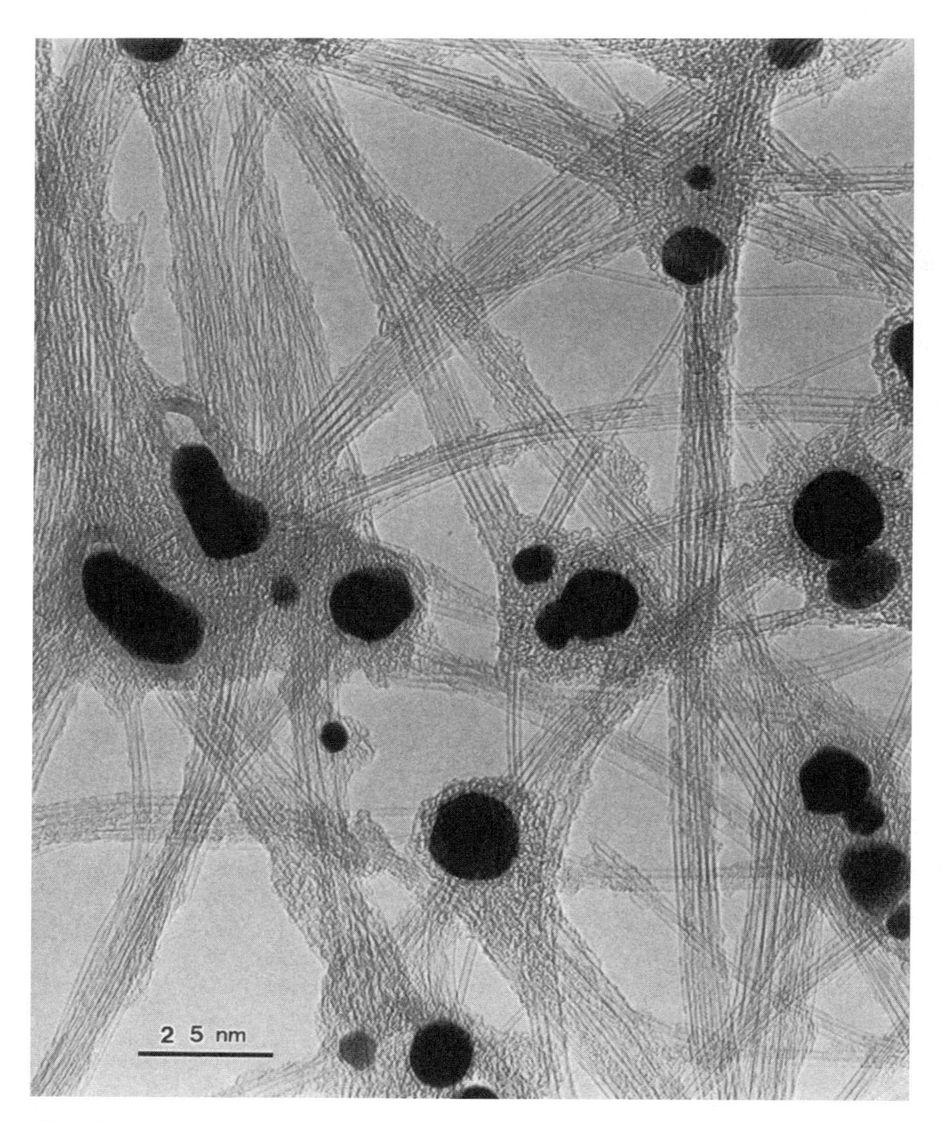

Fig. 10.2.11 TEM image of SWNTs synthesized using the mixed catalyst of Rh–50 wt% Pt.

brought about by the addition of sulfur. A much-pronounced increase in the yield of SWNTs was attained by a mixture of Ni and Y, which was found by Journet et al. (53). Other combinations of nickel and rare earths, e.g., Ni–La and Ni–Ce, also show marked enhancement in the SWNT formation, though the yield is slightly lower than that for Ni–Y. The Ni–Y binary mixture is the most effectual catalyst for producing abundant SWNTs by arc discharge among the catalysts studied so far.

10.2.5 Hypothetical Growth Mechanism

10.2.5.1 Multiwall Nanotubes

Several growth mechanisms of nanotubes have been proposed so far—e.g., ''open-end growth model'' (54), ''quasi-liquid tip model'' (29), and so forth. Here, we describe the quasi-liquid tip model for the growth of MWNTs.

A schematic illustration of the model is shown in Figure 10.2.12, together with that of polyhedral nanoparticles which grow as byproducts of MWNTs (see Fig. 10.2.3). An initial seed of an MWNT is the same as that of a polyhedral nanoparticle. Carbon neutrals [C, C_2 (19)] and ions (C^+) deposit and coagulate with each other to form atomic clusters and fine particles on a surface of the cathode. Structures of the particles at this stage may be amorphous with high fluidity (liquidlike) because of the high temperature (~3500 K) of the electrode surface and ion bombardment.

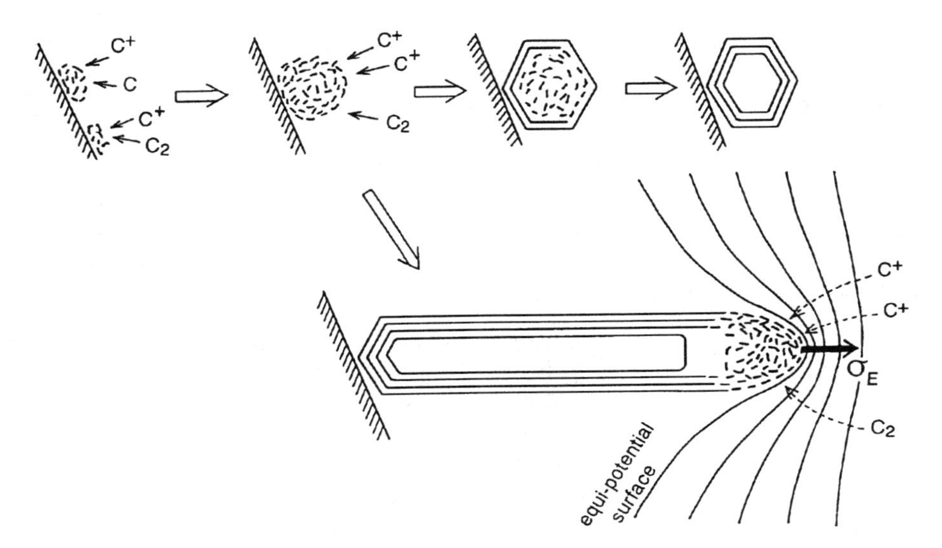

Fig. 10.2.12 Schematic drawings illustrating the ''quasi-liquid tip model'' for the growth of MWNTs (bottom) and nanoparticles (top).

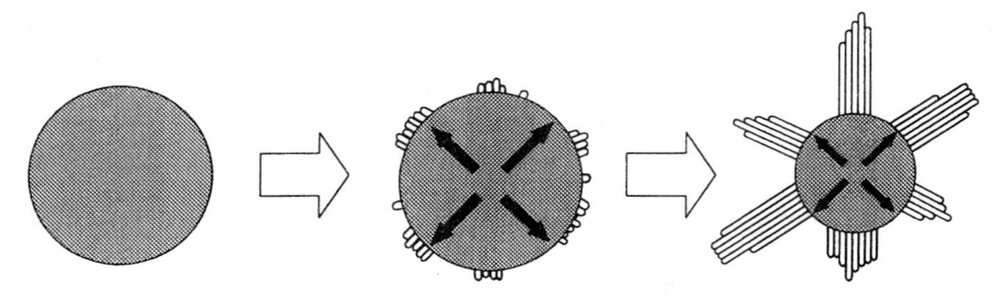

Fig. 10.2.13 Growth model of SWNTs based on the ''root growth'' on the surface of a catalytic particle.

There exists a strong electric field (10^4–10^5 V/cm) on the surface of a cathode. The driving force for the elongation is an electrostatic force (Maxwell tension) exerted on a liquidlike particle, which pulls the surface of the particle toward the direction of the field. Thus, the liquidlike particle deforms to an ellipsoidal form. The elongation of the particle brings about a further concentration of the electric field around the top of the particle, and accelerates the elongation. Since the tip of the elongated particle is exposed to ion bombardment, it remains in a high-fluidity phase. Solidification occurs from the bottom of the elongated particle, and proceeds with the increase of its length. Formation of graphitic sheets initiates from the surface, and the subsequent sheets are formed toward the interior. Not only carbon ions but also neutral atoms and molecules are attracted to the tip of a tube, owing to the polarization force (interaction between the inhomogeneous electric field and the induced dipole moment of molecules). Thus, the tip of a tube collects vapor and ions, and the tube continues to grow as long as carbon vapor and ions are supplied and the field is sustained stationary. Though the liquidlike phase at the tip is hypothetical in the present model, a recent molecular dynamic simulation of tip structures suggests the presence of quasi-liquid structures on a tip at high temperatures (55).

10.2.5.2 Single-Wall Nanotubes

A hypothetical growth process of SWNTs (42,45), based on the ''root growth'' on a surface of a catalytic particle, is illustrated in Fig. 10.2.13. When metal catalysts are evaporated together with carbon by arc discharge, carbon and metal atoms condense and form alloy (or mixed) particles in a gas phase. As the particles are cooled, carbon that dissolved in the particles segregates onto their surfaces because the solubility of carbon decreases with the decrease of temperature. Some singular surface structures or compositions in an atomic scale may catalyze the formation of SWNTs. After the nuclei of SWNTs are formed, carbon may be supplied from the core particle to the roots of SWNTs, and the tubes grow longer, maintaining hollow capped tips. Addition of carbon atoms (and C_2) from the gas phase to tips of the tubes may also help the growth.

REFERENCES TO SECTION 10.2

1. AG Rinzler, JH Hafner, P Nikolaev, L Lou, SG Kim, D Tomanek, P Nordlander, DT Colbert, RE Smalley. Science 269:1550, 1995.
2. WA de Heer, A Chatelain, D Ugarte. Science 270:1179, 1995.
3. Y Saito, S Uemura, K Hamaguchi. Jpn J Appl Phys 37:L346, 1998.
4. AC Dillon, KM Jones, TA Bekkedahl, CH Kiang, DS Bethune, MJ Haben. Nature 386: 377, 1997.
5. H Dai, JH Hafner, AG Rinzler, DT Colbert, RE Smalley. Nature 384:147, 1996.
6. S Iijima. Nature 354:56, 1991.
7. W Krätschmer, LD Lamb, K Fostiropoulos, DR Huffman. Nature 347:354, 1990.
8. S Iijima, T Ichihashi. Nature 363:603, 1993.
9. DS Bethune, CH Kiang, MS de Vries, G Gorman, R Savoy, J Vazquez, R Beyers. Nature 363:605, 1993.
10. Y Saito, T Yoshikawa, M Okuda, N Fujimoto, K Sumiyama, K Suzuki, A Kasuya, Y Nishina. J Phys Chem Solid 54:1849, 1993.
11. RE Haufler et al. J Phys Chem 94:8634, 1990.
12. Y Chai, T Guo, C Jin, RE Haufler, LPF Chibante, J Fure, L Wang, JM Alford, RE Smalley. J Phys Chem 95:7564, 1991.
13. H Shinohara, H Sato, M Ohkohchi, Y Ando, T Kodama, T Shida, T Kato, Y Saito. Nature 357:52, 1992.
14. DS Bethune, RD Johnson, JR Salem, MS de Vries, CS Yannoni. Nature 366:123, 1993.
15. TW Ebbesen, PM Ajayan. Nature 358:220, 1992.
16. M Tomita, Y Saito, T Hayashi. Jpn J Appl Phys 32:L280, 1993.
17. RS Ruoff, DC Lorents, B Chan, R Malhotra, S Subramoney. Science 259:346, 1993.
18. Y Saito. Carbon 33:979, 1995.
19. Y Saito, M Inagaki. Jpn J Appl Phys 32:L954, 1993.
20. Y Saito, J Ma, J Nakashima, M Masuda. Z Phys D 40:170, 1997.
21. DT Colbert et al., Science 266, 1218 1994.
22. H Lange, P Baranowski, A Huczko, P Byszewski. Rev Sci Instrum 68:3723, 1997.
23. Y Saito, T Yoshikawa. J Cryst Growth 134:154, 1993.
24. PM Ajayan et al. Phys Rev Lett 72:1722, 1994.
25. C Guerret-Piecourt, YL Bouar, A Loiseau, H Pascard. Nature 372:761, 1994.
26. R Saito, M Fujita, G Dresselhaus, MS Dresselhaus. Phys Rev B46:1804, 1992.
27. N Hamada, S Sawada, A Oshiyama. Phys Rev Lett 68:1579, 1992.
28. K Tanaka, K Okahara, M Okada, T Yamabe. Chem Phys Lett 191:469, 1992.

29a. SJ Tans, MH Devoret, H Dai, A Thess, RE Smalley, LJ Geerligs, C Dekker. Nature 386:474, 1997.

29b. SL Tans, ARN Verschueren, C Dekker. Nature 393:49, 1998.

30a. A Kasuya, Y Sasaki, Y Saito, K Thoji, Y Nishina. Phys Rev Lett 78:4434, 1997.

30b. AM Rao, E Ritcher, S Bandow, B Chase, PC Eklund, KA Williams, S Fang, KR Subbaswarmy, M Menon, A Thess, RE Smalley, G Dresselhaus, MS Dresselhaus. Science 275:187, 1997.

31. Toyo Carbon Company and Nihon Carbon Company are supplying metal/carbon composites according to customers' desire.
32. Y Saito, T Yoshikawa, M Inagaki, M Tomita, T Hayashi. Chem Phys Lett 204:277, 1993.
33. PM Ajayan, T Ichihashi, S Iijima. Chem Phys Lett 202:382, 1993.
34. Y Saito, T Yoshikawa, S Bandow, M Tomita, T Hayashi. Phys Rev B48:1907, 1993.
35. MS Dresselhaus, G Dresselhaus, K Sugihara, IL Spain, HA Goldberg. Graphite Fibers and Filaments. Berlin: Springer-Verlag, 1988, p 42.
36. MMJ Treacy, TW Ebbesen, JM Gibson. Nature 381:678, 1996.
37. JF Despres, E Daguerre, K Lofdi. Carbon 33:87, 1995.
38. TW Ebbesen, PM Ajayan, H Hiura, K Tanigaki. Nature 367:519, 1994.
39. H Hiura, TW Ebbesen, K Tanigaki. Adv Mater 7:275, 1995.
40. F Ikazaki, S Ohshima, K Uchida, Y Kuriki, H Hayakawa, M Yumura, K Takahashi, K Tojima. Carbon 32:1539, 1994.
41a. T Guo, P Nikolaev, A Thess, DT Colbert, RE Smalley. Chem Phys Lett 243:49, 1995.
41b. A Thess, R Lee, P Nikolaev, H Dai, P Petit, J Robert, C Xu, YH Lee, SG Kim, AG Rinzler, DT Colbert, GE Scuseria, D Tomanek, JE Fischer, RE Smalley. Science 273:483, 1996.
42. Y Saito, M Okuda, N Fujimoto, T Yoshikawa, M Tomita, T Hayashi. Jpn J Appl Phys 33:L526, 1994.
43. CH Kiang, WA Goddard III, R Beyers, JR Salem, DS Bethune. J Phys Chem 98:6612, 1994.
44. Y Saito, K Kawabata, M Okuda. J Phys Chem 99:16076, 1995.
45. Y Saito, M Okuda, M Tomita, T Hayashi. Chem Phys Lett 236:419, 1995.
46. S Subramoney, RS Ruoff, DC Lorents, R Malhotra. Nature 366:637, 1993.
47. D Zhou, S Seraphin, S Wang. Appl Phys Lett 65:1593, 1994.
48. Y Saito, K Nishikubo, K Kawabata, T Matsumoto. J Appl Phys 80:3062, 1996.
49. Y Saito, K Nishikubo. J Phys Chem Solids 57:243, 1996.
50. S Seraphin, D Zhou. Appl Phys Lett 64:2087, 1994.
51. Y Saito, M Okuda, T Koyama. Surface Rev Lett 3:863, 1996.
52. Y Saito, Y Tani, N Miyagawa, K Mitsushima, A Kasuya, Y Nishina. Chem Phys Lett 294:593, 1998.
53. C Journet, W Maser, P Bernier, A Loiseau, ML Chapelle, S Lefrant, P Deniard, R Lee, JE Fischer. Nature 388:756, 1997.
54. S Iijima, PM Ajayan, T Ichihashi. Phys Rev Lett 69:3100, 1992.
55. J-C Charlier, A de Vita, X Blase, R Car. Science 275:646, 1997.

11

Polymer Latices

HARUMA KAWAGUCHI

11.1 EMULSION POLYMERIZATION AND RELATED POLYMERIZATIONS

HARUMA KAWAGUCHI
Keio University, Yokohama, Japan

11.1.1 Introduction

Polymeric microspheres are produced by either of two methods. One starts from existing polymers and the other from monomers. In the first method, existing polymers in bulk or solution are converted to a particulate form by physical treatments such as crushing, crystallization, spray drying, solvent evaporation from emulsion, etc. In the second method, polymerization and particle formation are carried out in one step, so it is called ''particle-forming polymerization.'' Particle-forming polymerization includes various heterogeneous and homogeneous polymerizations such as emulsion polymerization, suspension polymerization, dispersion polymerization, and their modifications. Figure 11.1.1 shows the characteristic particle sizes formed by respective polymerizations.

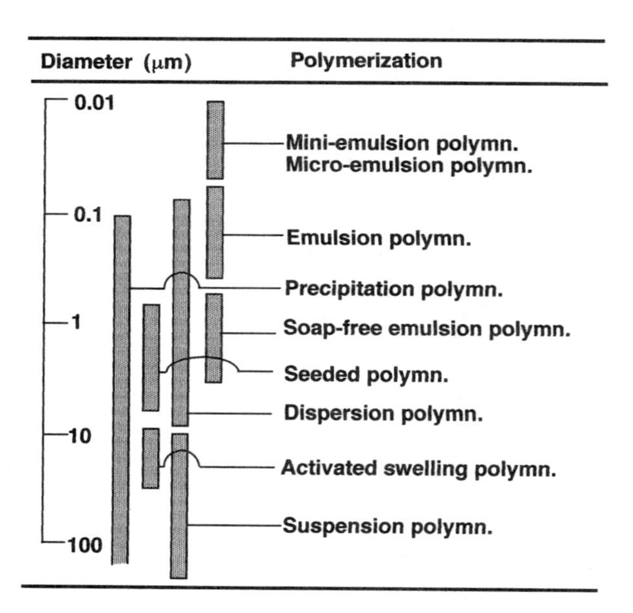

Fig. 11.1.1 Particle-forming or particle-modifying polymerizations and the size of particles prepared from them.

Suspension polymerization may be the most important particle-forming polymerization from an industrial viewpoint. The system is very simple, composed of monomer, initiator, stabilizer, and medium (water in most cases). The monomer droplets with dissolving initiator are dispersed in water and the stabilizer exists at the interface. But suspension polymerization is regarded as a kind of homogeneous polymerization because the polymerization occurs only in monomer droplets and water does not affect the polymerization. Water contributes only to dividing the polymerization locus into small droplets and absorbing the heat evolved by polymerization. On the contrary, in emulsion polymerization, which is another type of polymerization performed in water and as practically important as suspension polymerization, water affects the polymerization significantly. In this section, emulsion polymerization is first discussed, and then some modified emulsion polymerizations such as soap-free emulsion polymerization and micro and mini emulsion polymerizations are described.

11.1.2 Emulsion Polymerization

11.1.2.1 Emulsion Polymerization and Polymer Latex

''Latex'' originally meant the sap of the rubber plant and is a dispersion of particulate rubber. Emulsion polymerization produces a similar dispersion of synthetic rubber or polymers and was rapidly developed to obtain a substitute for natural rubber during World War II. Therefore the product of emulsion polymerization was first called ''polymer latex,'' but is now known simply as ''latex.'' Sometimes the product of emulsion polymerization is called ''polymer emulsion.'' But this terminology is incorrect for latices of solid polymer particles, because ''emulsion'' indicates liquid-in-liquid dispersion (1).

The emulsion polymerization system fundamentally includes hydrophobic monomer, water-soluble initiator, emulsifier, and water. Therefore, the initial system of emulsion polymerization is truly an emulsion.

The features of emulsion polymerization are:

1. High rate of polymerization, forming high-molecular-weight polymers
2. Volatile organic solvent free system
3. Effective control of the heat of polymerization
4. Easy handling of product due to its low viscosity

Features 2 to 4 are attributed to the aqueous medium. Emulsion polymerization forms submicrometer-sized particles, so-called latex particles. The particles are stabilized with ionic and/or noionic emulsifiers. The process to form submicrometer particles is very complicate because of the contribution of two phases, aqueous and oil, to particle. The mechanism of emulsion polymerization is described in the next section.

11.1.2.2 Harkins's Theory of Emulsion Polymerization

Harkins proposed the mechanism of emulsion polymerization in 1943 (2). His qualitative theory needs a few corrections but its basis has been and will be accepted. The theory is explained using Figure 11.1.2.

As mentioned earlier, the necessary components for emulsion polymerization are monomer, emulsifier, initiator, and water. The monomer is insoluble in water in principle, although a small amount of water-soluble monomer can be used as a comonomer in some practical cases. On the other hand, the initiator is usually water soluble. These relations mean that the loci for radical formation and radical polymerization are different, that is, in aqueous and organic phases, respectively. This might look irrational, but the unique advantages of emulsion polymerization lie in it. When radicals are formed from a water-soluble initiator in an aqueous medium in Figure 11.1.2, top, left, each radical undergoes one of the following reactions:

1. Addition reaction with monomers slightly dissolving in water
2. Diffusion into or collision with micelles
3. Diffusion into or collision with monomer droplets
4. Termination reaction in aqueous phase

The following experimental results in emulsion polymerization strongly suggested that reaction 2 is the most dominant and reaction 3 is excluded in general emulsion polymerizations due to the following reasons: the polymerization rate was small in the system at low emulsifier concentration, and the particles resulting from emulsion polymerization were much smaller than monomer droplets. The much smaller number and much smaller total surface area of droplets than of micelles also support the conclusion that monomer droplets have little chance to be entered by radicals because the small number and small surface area give the droplets little chance to collide with radicals and little chance to accept diffusing radicals, respectively, compared with micelles.

It is not reasonable to consider that radicals enter a micelle directly, because of their high solubility in water and strong repulsive force with polar groups on a micelle surface. In this situation, step 2 is believed to occur after the step 1; that is, the radical formed in aqueous phase must first react with some monomer molecules dissolved in water, although the polymerization is slow because of the low solubility of monomer in water. The oligomer radical increases hydrophobicity with increasing chain length and becomes amphiphilic like as an emulsifier molecule so that the oligomer radical can enter the micelle.

The radical that enters the micelle encounters monomers in it, and rapid polymerization takes place. Because the number of monomer molecules included in each micelle is 100 to 200, the monomers are used up at once. But monomer molecules are quickly supplemented from monomer droplets, which correspond to a reservoir of monomer, through the aqueous phase to maintain the equilibrium state between the inside of micelle and aqueous phase. Thus, the micelle in which polymerization

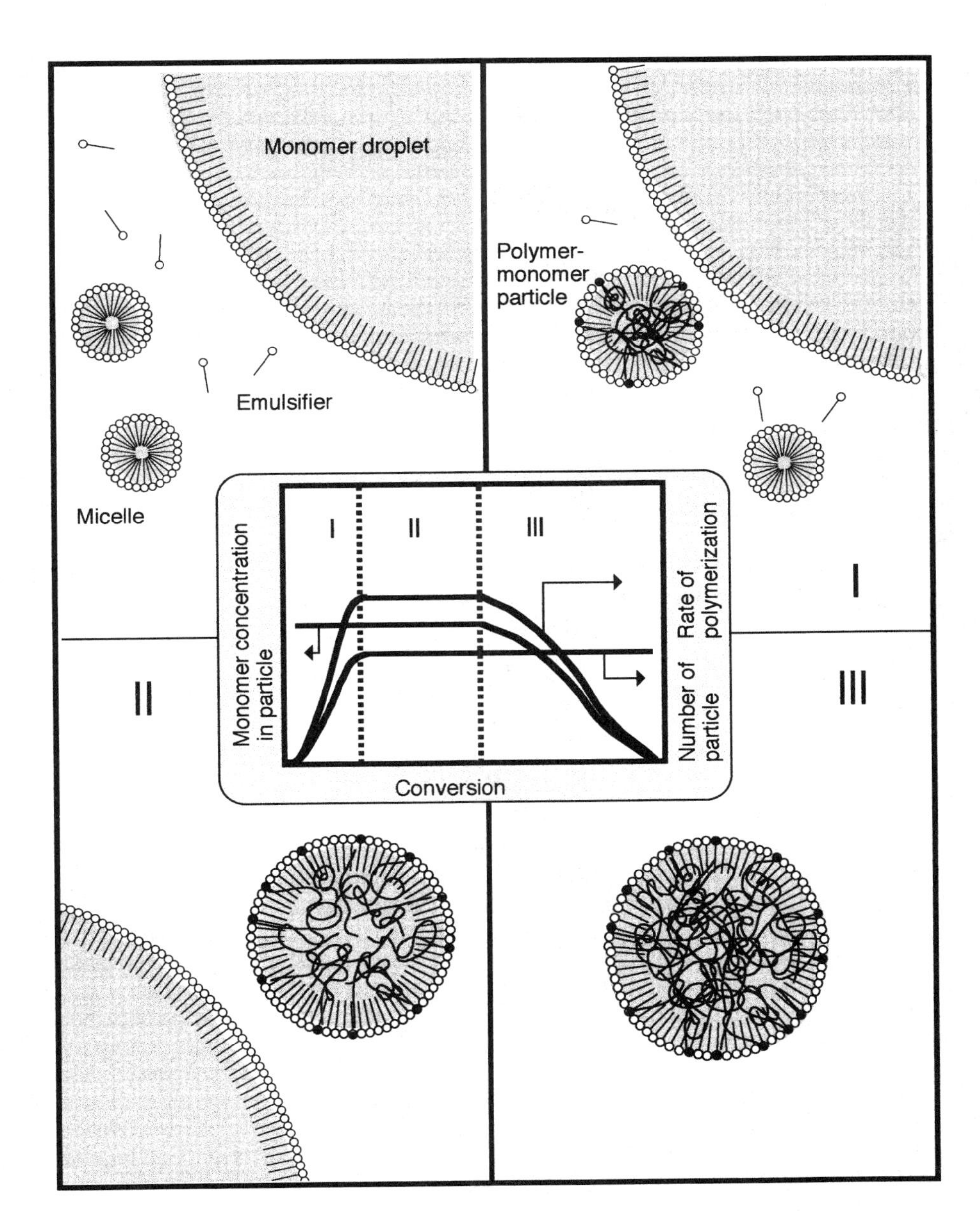

Fig. 11.1.2 Illustration of emulsion polymerization system at different steps: (I) particle-forming step; (II) steady step; (III) monomer-in-particle consuming step.

takes place grows up and loses its original shape, becoming a so-called polymer–monomer particle.

The increased surface of the growing polymer–monomer particle is stabilized by emulsifier molecules that are released from nonreacted micelles. When all of the micelles are used up by accepting radicals to start the polymerization or by breaking themselves to supply the emulsifier molecules to particle surfaces, the number of polymer–monomer particles is fixed. This is the end of the first stage. As long as the monomer droplet exists, the polymer–monomer particles at a constant number can hold the monomers at a constant concentration. This second stage of emulsion polymerization continues until the time when the droplets are used up. After this (third stage), the concentration of monomer in the particles decreases gradually with increasing conversion. Even in such nanosized particles, the Trommsdorff effect can take place if the viscosity exceeds a certain level.

11.1.2.3 Quantitative Analysis and Kinetics of Emulsion Polymerization by Smith and Ewart

In 1944 Smith and Ewart published a paper analyzing the Harkins model quantitatively (3). They considered three cases in terms of average number of radical (n) in each polymer–monomer particle.

Case 1: $n < 1$
Case 2: $n = 0.5$
Case 3: $n > 1$

Case 1, in which n is less than 1, is realized when termination in aqueous phase takes place appreciably so that the radical concentration in water is kept low, forcing the radical in particles to diffuse back into the aqueous phase (Fig. 11.1.3A) (4). Case 2, in which n is $\frac{1}{2}$, is found under the following condition (Fig. 11.1.3B): When one radical is polymerizing in a polymer–monomer particle and another radical enters the very polymer-monomer particle, the two radicals encounter in it and lead to termination reaction. That is, in each polymer–monomer particle, radicals entering

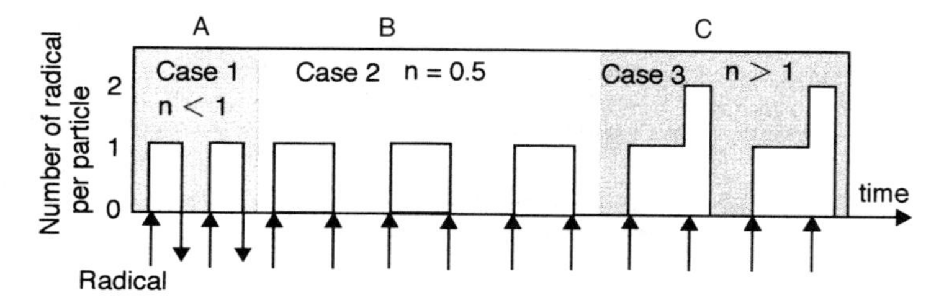

Fig. 11.1.3 Fate of radicals and average number of radicals in particle.

the particle in even order start the polymerization and those entering it in odd order stop the polymerization. Thus, half of the polymer–monomer particles are polymerizing and the other half are resting at each instance. Case 3 is the one in which multiple radicals coexist in each particle. Case 3 appears in particles of a diameter larger than 100 nm and/or of high viscosity in which the termination reaction between existing and entering radicals can not take place immediately, as shown in Figure 11.1.3C.

Smith and Ewart calculated the number of particles having been formed at the end of the first stage of polymerization. The number of particles is affected by the initiator decomposition rate (or radical formation rate) and total surface area of emulsifier to stabilize polymer–monomer particles. Smith and Ewart concluded that the number of particles is proportional to the 0.4 power of the initiator concentration and the 0.6 power of the emulsifier concentration, assuming that the surface area of total polymer–monomer particles is equal to the total surface area of emulsifier molecules when the last micelle disappears.

The rate of polymerization (R_p) in emulsion polymerization is expressed using n and N as

$$R_p = k_p\,[M\cdot]\,[M] = k_p n(N/N_A)\,[M]_p \tag{1}$$

where N_A is the Avogadro's number, $[M]_p$ is the monomer concentration in polymer–monomer particles, and $n(N/N_A)$ in Eq. (1) corresponds to the concentration of radical. In general emulsion polymerization, N is in the range of 10^{17} to 10^{18} particles/L and n is about 0.5 at the steady state. Therefore the radical concentration, $n(N/N_A)$, is about 10^{-6} mol/L. This value indicates the reason why emulsion polymerization runs faster than solution polymerization, in which the radical concentration can not exceed 10^{-8} mol/L because of the unavoidable termination reaction between two radicals close to each other. Emulsion polymerization is an exceptional system to maintain a high radical concentration by keeping radicals in isolated loci.

The typical emulsion polymerization of styrene was confirmed to proceed in Case 2. According to Eq. (1), the rate constant of propagation can be calculated by measuring the rate of polymerization (R_p) and the number of particles (N) and substituting 0.5 for n.

The degree of polymerization (Pn) of polymers obtained by emulsion polymerization is expressed as

$$\mathrm{Pn} = k_p\,[M]_p\tau = k_p[M]_p N/\rho \tag{2}$$

where τ is the interval of successive entries of radicals into polymer–monomer particle and ρ is the rate of radical formation. The relation between the rate of polymerization and the degree of polymerization of polymers is characteristic for emulsion polymerization. In general homopolymerization, the degree of polymerization is irreversibly proportional to the rate of polymerization. This is not the case in emulsion polymerization, as confirmed by Eqs. (1) and (2), which both contain

N. The increase in N leads to simultaneous increases in the rate of polymerization and the degree of polymerization.

11.1.2.4 Improvement and Development of the Harkins–Smith–Ewart Theory

Average Number of Radicals in Polymer–Monomer Particle. As mentioned in Section 11.1.2.3, the average number of radicals in a polymer–monomer particle in general emulsion polymerization is 0.5 due to the on–off mechanism. More detailed discussion was developed by Stockmayer (5), O'Toole (6), and later by Ugelstad et al. (7,8). They gave precise solutions for the equation of material balance in terms of number of radicals in each particle. The solution is schematically presented in Fig. 11.1.4. The figure is of log(average number of radical per particle) versus log($\alpha = \rho^{w}v/N^{w}k_{tp}$) where ρ^{w} is the rate of radical production in the water phase per unit volume of water, v is the volume of a polymer particle, N^{w} is the total number of particles per volume of water, and k_{tp} is the rate constant of mutual termination of radicals per particle. As shown in Figure 11.1.4, the duration for the system to keep the condition of $n = \frac{1}{2}$ is a function of m, which is

$$m = k_{d}v/k_{tp} \tag{3}$$

where k_{d} is the rate constant of desorption of radicals from a particle. The smaller m, due to smaller k_{d} and/or smaller particle size, causes a longer period of time to keep $n = \frac{1}{2}$. More water-soluble monomers have larger k_{d} values, and the n values for these monomers are less than $\frac{1}{2}$ in the initial stage of polymerization. A higher rate of termination in water results in the shift of curves to the right, that is, extends the period in which the particles have a smaller n.

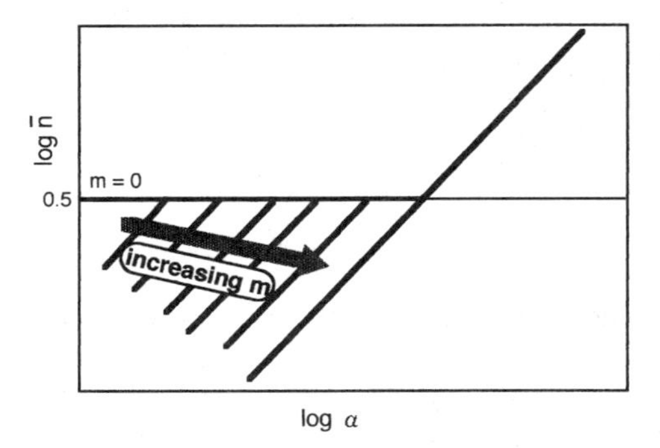

Fig. 11.1.4 Average number of radicals as a function of particle size, rate of radical release, and rate of termination in water.

Gardon employed new parameters in terms of the number of particles and checked the Harkins–Smith–Ewart theory qualitatively in 1963 (9).

Nucleation and Growth of Particles. Another argument was developed for particle nucleation because a significant number of particles were formed even in the system at a emulsifier concentration lower than the critical micelle concentration. The scheme in Figure 11.1.5 was proposed as a possible nucleation mechanism (10), in which growing radicals in aqueous phase lose their solubility in water because of increasing hydrophobicity with increasing chain length. The critical chain length (ccl) is, for example, 3 to 4 for styrene, about 12 for methyl methacylate, and more than 100 for vinyl acetate. When the chain length reaches the ccl, the radical precipitates. The precipitate itself or its aggregate is the nucleus of a particle. The microenvironment around the nucleus is hydrophobic enough to get the monomers together and to now attain a character similar to a monomer–polymer particle. This mechanism of particle formation is referred to as homogeneous nucleation.

Feeney et al. proposed a concept of precursor particle in which ultrafine particles named as precursor particles, of a diameter of a few nanometers, are first precipitated and then aggregate to become stable primary particles having a diameter of a

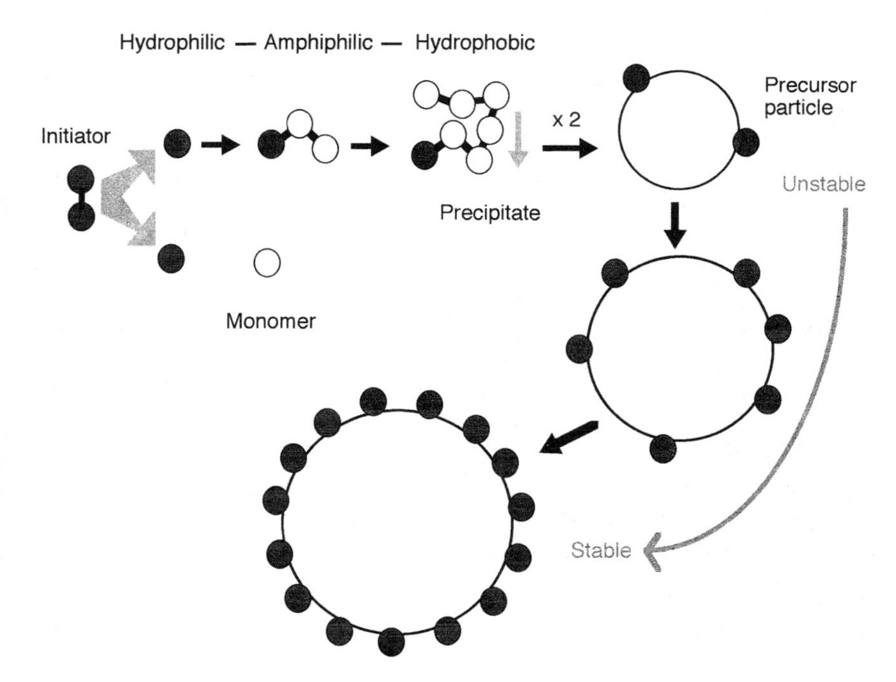

Fig. 11.1.5 Schematic representation of homogeneous nucleation.

few tens of nanometers. The basis of this idea was the distribution of particle size. If transformation of micelle to monomer–polymer particle is the main route for particle formation, the particle size distribution is expected to lean to the left. But the actual distribution leans to the right. This supports that the precursor-to-latex particle route is dominant.

Extent of aggregation of precursor particles depends on the qualities of polymer/monomer, initiator, and emulsifiers. The emulsifiers used for emulsion polymerization are divided into two classes, ionic and nonionic ones. The former class includes anionic, cationic, and amphoteric ones. The ionic emulsifiers stabilize particles by electrorepulsive force, but nonionic ones do so by steric stabilization effect, which generates repulsive force caused by hydrated emulsifiers on the particles. As the steric stabilization effect is less effective than the electrorepulsive force, precursor particles in a nonionic emulsifier system aggregate quickly and, as a result, are converted to a small number of particles. After establishing stable particles, a high rate of radical entry, which results from high rate of radical generation and/or small number of particles, is suggested to be the most important factor for the monodispersity of particles (11). In practice, a mixture of ionic and nonionic emulsifiers is often used to prepare monodisperse particles at a high rate.

The homogeneous nucleation theory may suggest that the dependence of particle number on the concentrations of emulsifier and initiator changes from monomer to monomer. In fact, z in Eq. (4) is 0.6 for a water-insoluble monomer such as styrene (as predicted by Smith–Ewert) but decreases with increasing solubility of monomer in water or increasing transfer of radicals out of particles (12):

$$N \propto [S]^{Z} \tag{4}$$

Locus of Polymerization. In all of the treatment of emulsion polymerization, it had been supposed that the monomer distributes homogeneously throughout the polymer–monomer particle. Williams refuted this supposition, claiming that the monomer concentration is high at the surface layer of polymer–monomer particle and consequently this is the main polymerization locus (13). The location of initiator fragments in the particle was investigated to find the solution on this argument. The study revealed that the location of initiator fragments did not directly relate to the polymerization locus but concerned the anchoring effect of the fragments. More polar fragments such as the sulfate group stayed more on the particle surface, but less polar fragments such as the amidine group were apt to be taken into the interior of particle due to a weaker anchoring effect.

The argument on the main polymerization locus continued for several years, until 1980, and closed without finding a clear conclusion.

Fate of Radicals Escaping from Polymer–Monomer Particles. The fate of a radical escaping from a polymer–monomer particle has been a hot topics and discussed quantitatively in the latest decade. An oligomer radical once entering a monomer–polymer particle seems to find it hard to escape from the particle because of its large size and extremely low water solubility. Instead, when a chain transfer reaction to a monomer molecule takes place in the particle, the resulting small-sized radical can escape the particle (14). The radical will restart polymerization, reenter a particle, or encounter another radical to perform a termination reaction. The escape of monomer radical to the aqueous medium makes the n smaller than 0.5 as mentioned earlier. The discrepancy of n from 0.5 also depends on the solubility of monomer, particle size, and the fate of escaping radical in aqueous medium. In aid of detailed discussion on this subject, oil-soluble initiators such as azo-bis-isobutyronitrile were used so that radicals are generated only in the monomer phase (15).

11.1.2.5 Novel Emulsion Polymerization

The emulsifier contributes a lot to emulsion polymerization, by providing the polymerization locus and stabilizing the formed particles. But it causes a severe disadvantage in the application of the resulting dispersion because emulsifier molecules are released from particles and behave as a kind of contaminant. One of the methods to cancel the disadvantage is to carry out the emulsion polymerization using a reactive emulsifier, which is taken into a polymer molecule and not released from the polymer (16). As the compound serves as a surfactant and monomer, it is named a ''surf-mer'' or polymerizable surfactant (17). One necessary condition for the surf-mer to achieve its aim is that it have low reactivity and polymerize at the later stage of polymerization. If not, the micelle of the surf-mer itself polymerizes first, so that the particle growth is suppressed due to stiffened layer of polymerized surf-mer. Allyl group-containing emulsifiers are desirable surf-mers in this sense. Table 11.1.1 shows a few surf-mers used so far. Some compounds that carry two functions of both initiator and surfactant, named ''ini-surfs,'' have a similar effect to surf-mers by being taken up into polymer chains.

11.1.2.6 Emulsion Copolymerization

In homogeneous copolymerization, the instantaneous composition of copolymer is decided only by monomer reactivity ratio. On the contrary, in emulsion copolymerization, the copolymer composition depends not only on the monomer reactivity ratio but also on the distribution of monomers between oil (polymer–monomer particles) and aqueous phases (18).

Artificial control of the monomer concentrations is possible by changing the monomer feed methods, which includes multishot, stage feed (19), and continuous feed. A multishot emulsion polymerization is expected to form multilayered particles if the monomers are chosen properly. When the layers are sufficiently thin, the particles exhibit unique thermal and mechanical properties. The stage feed system is shown in Figure 11.1.6. It makes it possible to produce particles having gradient composition of different monomer units.

Table 11.1.1 Some Polymerizable Surfactants

Surfactant	
$H_2C{=}CH{-}(CH_2)_8{-}\underset{\underset{\textstyle O}{\|}}{C}{-}O{-}CH_2{-}CH_2SO_3Na$	
$H_2C{=}CH{-}\underset{\underset{\textstyle O}{\|}}{C}{-}NH{-}(CH_2)_{10}COONa$	
$C_6H_4(CH{=}CH_2)(O{-}(CH_2)_{12}SO_3Na)$ (benzene ring bearing $CH{=}CH_2$ and, ortho to it, $O{-}(CH_2)_{12}SO_3Na$)	
$R{-}C_6H_3(SO_3Na){-}O{-}\underset{\underset{\textstyle O}{\|}}{C}{-}CH{=}CH_2$ (benzene ring bearing R, SO_3Na and $O{-}C(=O){-}CH{=}CH_2$)	$R = C_{10}H_{21}, C_{12}H_{23}$

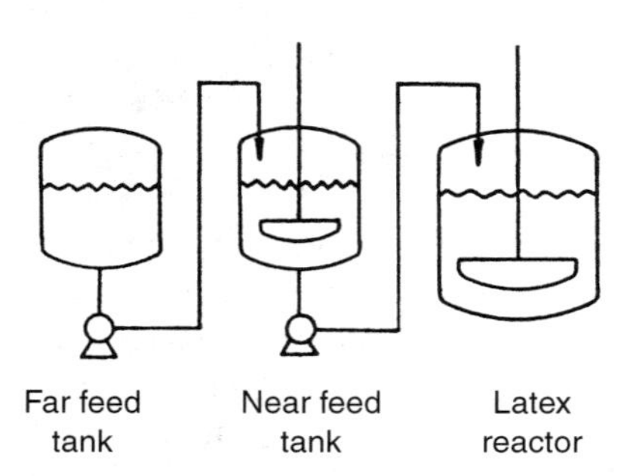

Fig. 11.1.6 Stage feed polymerization.

11.1.3 Soap-Free Emulsion Polymerization

Emulsifier is not a necessary component for emulsion polymerization if the following conditions are satisfied: The particles are formed by homogeneous nucleation mechanism, and the particles are stabilized by factor(s) other than emulsifier. As to the latter, the sulfate end group that is the residue of persulfate initiator serves for stabilization of dispersion via interparticle electrorepulsive force (20). When the stabilization mechanism works well, a small number of particles grow during polymerization without aggregation, keeping the size distribution narrow. Finally stable, monodisperse, anionic particles are obtained.

An alternative stabilization mechanism is steric stabilization by hydrophilic comonomer. When a hydrophobic monomer (e.g., styrene) and a hydrophilic monomer (e.g., acrylamide) are copolymerized in an emulsifier-free aqueous medium, particles have a graded distribution of monomer units along the radial direction. The hydrophilic component increases from the center to the surface of the particle, and the gradient depends on the difference in hydrophilicity of the two monomers. If the hydrophilicity of one monomer is too high, a large part of it does not contribute to particle formation but remains in water. As shown in Figure 11.1.7, acrylamide, a monomer having high water solubility, preferentially polymerizes in the aqueous phase in the emulsifier-free emulsion copolymerization of acrylamide and styrene (21). Occasional incorporation of styrene monomer in a growing radical makes it less soluble in water. Then it precipitates to form a nucleus of a particle. This homogeneous nucleation process finishes when a sufficient number of stable particles are prepared. After this, polymerization of styrene in existing particles becomes dominant, to consume styrene exclusively. Acrylamide remaining in the aqueous phase must wait for polymerization until the styrene is used up.

11.1.4 Miniemulsion Polymerization

Monomer droplets cannot be the source of particles in general emulsion polymerization because they have a much lower number and much less surface area than micelles so that they have no chance to accept the radicals from aqueous medium. But if such monomer droplets are made smaller to the level of micelles, the droplets get a chance to encounter and take in growing radicals to become the nuclei of particles. This can be realized by several methods—for example, by using a large amount of emulsifier, or using a hydrophobic alcohol such as $C_{16}H_{33}OH$ with emulsifiers (22). In the latter case, the alcohol and emulsifier form a liquid crystal-like layer, which stabilizes a wide interface area and, as a result, makes droplets that are smaller than 100 nm or even close to 10 nm. Differing from general emulsion, this emulsion is thermodynamically stable and havs a long lifetime. An emulsions including droplets of 50–100 nm and of less than 50 nm is commonly called a miniemulsion and a microemulsion, respectively. Polymerization in this system results in the formation of particles of almost the same number and same size of

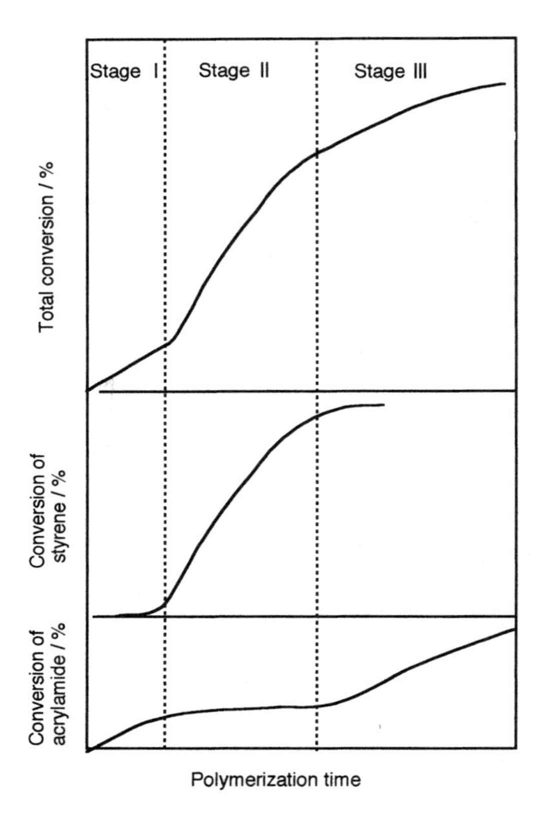

Fig. 11.1.7 Conversions of styrene and acrylamide in soap-free emulsion copolymerization.

monomer droplets. These polymerizations are referred to as mini- or microemulsion polymerizations. A typical recipe for miniemulsion polymerization is (23):

Styrene 8 mL
Water 25 mL
Sodium lauryl sulfate 10 m*M*
Cetyl alcohol 30 m*M*
Potassium persulfate 0.1–2.7 m*M*
Sodium hydrocarbonate 0.1–2.7 m*M*

One disadvantage of the miniemulsion polymerization is the low rate of polymerization. Even if almost the same number of droplets exist in the miniemulsion polymerization compared with the number of micelles in general emulsion polymer-

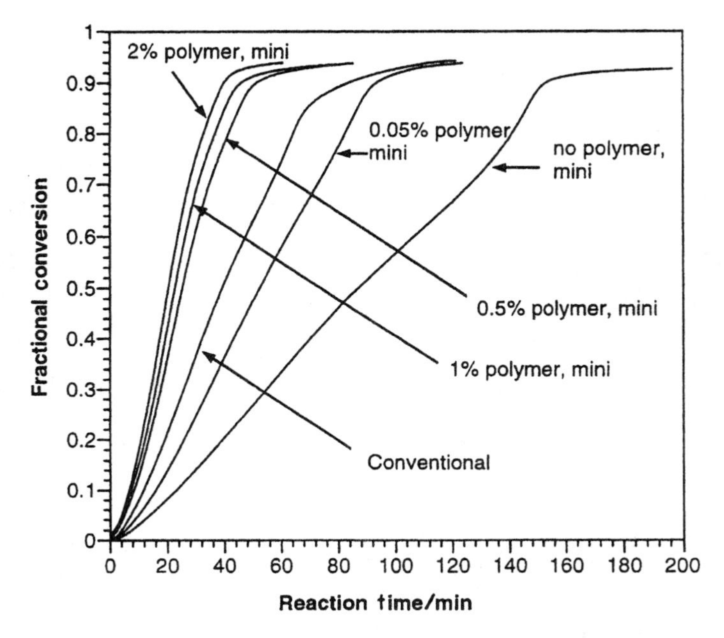

Fig. 11.1.8 Thermal analysis of polymerization course and number of particles.

ization, the polymerization rate of the former is lower than the latter by about one-third, as shown in Figure 11.1.8. This is because the firm liquid crystal-like surface layer over the monomer droplets rejects the smooth entry of radicals from aqueous medium. The disadvantage is overcome by adding a few polymer molecules in each monomer droplet before the polymerization. The polymer molecules are believed to disarrange the liquid crystal-like layer, to make it unsteady, and to help the radical entry. The effect of polymer addition on the rate of miniemulsion polymerization is clearly shown in Figure 11.1.8.

When chloro-octadecane was found to give the same result as a so-called cosurfactant, an argument arose in terms of the real role of this highly hydrophobic compound because it is not surface active and has no cooperation with surfactant. Taking account of these systems, the definition of miniemulsion polymerization will be revised to "the polymerization in which a water-insoluble compound in the dispersed phase retards or inhibits diffusion degradation of the emulsion."

11.1.5 Inverse Microemulsion Polymerization

Figure 11.1.9 shows the phase diagram of the ternary components of sodium bis(2-ethylhexyl)sulfosuccinate (AOT), toluene, and water (24). The dotted line indicates

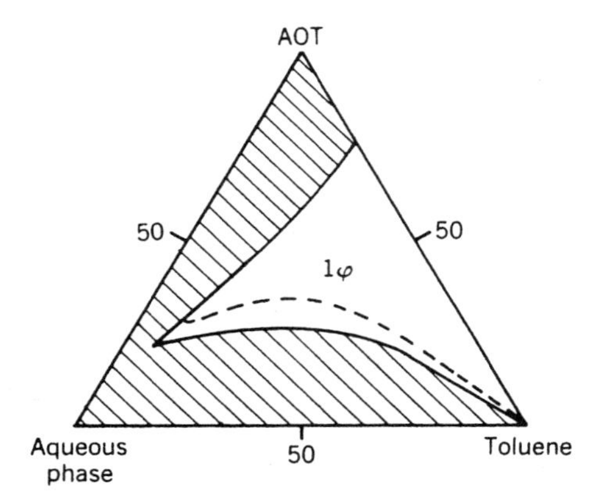

Fig. 11.1.9 Phase diagram of AOT/toluene/water (or water + acrylamide) system.

the boundary between microemulsion and emulsion phases. When water-soluble monomer, the same amount of acrylamide in the case of Figure 11.1.9, is added in water, the phase boundary in the diagram shifts to increase the microemulsion phase region as shown by the solid line. This means that the monomer contributes to the stabilization of the microemulsion by composing an interface layer with AOT. A similar stabilization effect is observed by using a long-chain alcohol. Microemulsion is carried out in such systems. Among various microemulsion polymerizations, inverse microemulsion polymerization of acrylamide has attracted attention because a versatile polyacrylamide of high molecular weight can be obtained quickly in a form easily handled. The emulsifier is chosen referring to the hydrophile–lipophile balance (HLB) of the emulsifier, which seems to be appropriate in the range of 4 to 6. Polymerization is usually started with a hydrophobic initiator such as azobisisobutyronitrile, or sometimes with a hydrophilic initiator such as persulfate (25). The polymerization results in the formation of particles with a diameter less than 30 nm. Each particle was found to be composed with one molecule or a few molecules of polyacrylamide. This kind of initiator did not seriously affected the rate of polymerization; that is, even an oil-soluble initiator enabled polymerization as fast as the hydrophilic initiator. This suggests strongly that the polymerization takes place mainly at the interlayer between the aqueous phase and oil phase. When toluene-soluble monomer—for example, styrene or methyl methacrylate—was added to the polymerization system, the polymerization of tolune-soluble monomer was delayed, compared with acrylamide, indicating that the polymerization in oil phase (toluene) was not dominant (25).

11.1.6 Concluding Remarks

Emulsion polymerization has a history of half a century in science and much more in industry. There seem to be few undissolved subjects in general emulsion polymerization of common monomers, although emulsion copolymerization includes some important problems to be investigated. Continuous emulsion polymerizations, which were not dealt in this chapter, should be also investigated more. Novel or unique emulsion polymerizations such as highly concentrated emulsion polymerization (26), multishot emulsion polymerization, emulsion polymerization using multifunctional reagents (surf-mer, ini-surf, etc.), and emulsion polymerization initiated by new initiators or initiating systems are now being developed. Even atom transfer radical polymerization has been recently applied to emulsion polymerization system (27). The related heterogeneous polymerizations have been also developed quickly. Among them, preparation of monodisperse microspheres composed of a single polymer molecule might be one of the most challengeable subjects. Reference 28 is strongly recommended to those who are going to study emulsion polymerization and related polymerizations in more details.

REFERENCES TO SECTION 11.1

1. DC Blackley. Polymer Latices. Science and Technology. London: Chapman & Hall, 1996.
2. WD Harkins. J Am Chem Soc 69:1428, 1947.
3. WV Smith, RH Ewart. J Chem Phys 16:592, 1948.
4. H Gerrens. Dechema Monogr 49:53, 1963.
5. WH Stockmayer. J Polym Sci 24:313, 1957.
6. JT O'Toole. J Appl Polym Sci 9:1291, 1965.
7. J Ugelstad, PC Mork, JO Aasen. J Polym Sci A-1 5:2281, 1967.
8. M Nomura, M Harada, W Eguchi, S Nagata. In: I Piirma, JL Gardon, eds. Emulsion Polymerization. ACS Symp Ser No 24, 1978, pp. 102–122.
9. JL Gardon. J Polym Sci A-1 6:623, 643, 665, 687, 1968.
10. RM Fitch, CH Tsai. In: RM Fitch, ed. Polymer Colloids. New York: Plenum, 1971, p 73.
11. PJ Feeney, DH Napper, RG Gilbert. Macromolecules 17:2520, 1984; J Colloid Interface Sci 118:493, 1987.
12. D Gershberg, AIChE Int Chem Eng Symp Ser 3:4, 1965.
13. MR Grancis, DJ Williams. J Polym Sci 8:2617, 1970.
14. BR Morrison, BS Casey, L Lacik, GL Leslie, DF Sangster, RG Gilbert, DH Napper. J Polym Sci Polym Chem 32:631, 1994.
15. JM Asua, VS Rodriguez, ED Sudol, MS El-Aasser. J Polym Sci Polym Chem Ed 27: 3569, 1989.
16. WD Winter, A Marien. Makromol Chem Rapid Commun 5:593, 1984.
17. A Guyot, K Tauer. Adv Polym Sci 111:45, 1994.
18. J Guillot. In: MS El-Aasser, RM Fitch, eds. Future Directions in Polymer Colloids. The Hague: Nijhof, 1987, pp. 65–77.
19. DR Bassett. In GW Poehlein, RH Ottewill, JW Goodwin, eds. Science and Technology of Polymer Colloids I. The Hague: Nijhoff, 1982, pp. 220–240.
20. A Kotera, K Furusawa, Y Takeda. Kolloid ZZ Polym 239:677, 1970.
21. H Kawaguchi, Y Sugi, Y Ohtsuka. J Appl Polym Chem 26:1637, 1981.

22. J Ugelstad, FK Hansen, S Lange. Makromol Chem 175, 1974.
23. CM Miller, ED Sudol, CA Silebi, MS El-Aasser. Macromolecules 28:2754, 1995.
24. F Candau, YS Leong, G Pouyet, S Candau. J Colloid Interface Sci 101:167, 1984.
25. J Balton. Makromol Chem Macromol Symp 53:289, 1992.
26. E Ruckenstein, KJ Kim. J Appl Polym Sci 36:907, 1988.
27. SAF Bon, M Bosveld, B Klumperman, AL German. Macromolecules 30:324, 1997.
28. PA Lovell, MS El-Aasser, eds. Emulsion Polymerization and Emulsion Polymers. New York: Wiley, 1997.

11.2 DISPERSION POLYMERIZATION

HARUMA KAWAGUCHI
Keio University, Yokohama, Japan

11.2.1 Introduction

Typical emulsion and suspension polymerizations are appropriate methods for preparing submicrometer and approximately 100 μm-ordered particles, respectively. They are suitable for mass production and are of little burden to environment due to their reaction in water, but have their own disadvantages in terms of the contamination of product with emulsifier, the wide distribution of size, etc. Dispersion polymerization overcomes these problems and forms monodisperse particles of micrometer order, which is a size just between those of particles produced by emulsion and those produced by suspension polymerization. The constitution of dispersion polymerization is quite different from those of emulsion and suspension polymerizations. The latter two polymerization systems are, roughly speaking, composed of multiple phases, that is, monomer phase(s) and medium. But dispersion polymerization starts from a homogeneous monomer solution. In addition, dispersion polymerization is normally a method to produce organic polymer particles in an organic solvent that is totally different from the other two methods where water is chosen as a continuous phase in general. Therefore, different considerations should be introduced to produce colloidally stable particle dispersion.

The history of dispersion polymerization begins during the 1950s (1). Until the 1970s, aliphatic hydrocarbons were used as a medium in almost every dispersion polymerization. The first generation of dispersion polymerization generated organic dispersions of submicrometer particles with high solid fractions and low viscosity, the so-called NAD (nonaqueous dispersion), and was employed for coating and industry use. Compared with this, dispersion polymerization of the 1980s should be called the second generation. It uses polar solvents as a medium, and the main purpose was to prepare micrometer-ordered monodisperse particles. The third generation is in the 1990s and aims for the same purpose with inventions of functional particles and consideration to the environment.

This section, discusses the mechanism of dispersion polymerization, role of components, control of particle size, and the application to new systems by tracing its history.

11.2.2 Dispersion Polymerization and Precipitation Polymerization

Precipitation polymerization is defined as polymerization starting from a homogeneous monomer solution in which the polymer is insoluble and forms coarse precipitates or coagulum. The polymerization of methyl methacrylate in dodecane is a good example. Even in a system like this, it is possible to prepare monodisperse particles by introducing a mechanism that resists the aggregation of polymers, that is, with the aid of dispersion stabilizer. Such polymerization is called dispersion polymerization. In the polymerization of methyl methacrylate in dodecane, a copolymer composed of the maleic acid ester of 12-hydroxystearic acid and methyl methacrylate (HSM) was used. Therefore, dispersion polymerization can be regarded as a special case of precipitation polymerization.

Arshady distinguished dispersion and precipitation polymerizations differently (2). His definition is shown in Table 11.2.1. In precipitation polymerization, the polymer precipitates from monomer solution and the monomer will not merge into the polymer. Therefore the locus of polymerization is in the solution, and the polymer continues to precipitate from it. The precipitation aggregates and finally becomes macroscopic particles or coagulums. On the other hand, in dispersion polymerization, as polymerization starts, a phase that is rich in polymer segregates from the solvent phase by phase separation. When the surface of the segregated polymer phase is stabilized, the polymer phase becomes a sphere (particle) and maintains a dispersion state. Polymerization after this occurs in both the continuous and polymer phases, and the contribution of each phase to polymerization depends on the partition of monomer between two phases. Arshady describes bulk polymerization of acrylonitrile as an example for the precipitation polymerization and copolymerization of butyl acrylate and acrylonitrile in the bulk for dispersion polymerization (3). Based on this interpretation, the polymerization of acrylamide (AAm) and methylene-bis-acrylamide (MB) in ethanol will belong to precipitation polymerization, and the methacrylic acid (MAc) added system, terpolymerization of AAm, MAc, and MB, would belong to the category of dispersion polymerization because the former formed coarse particles but the latter formed monodisperse particles (4). However, in either example there is no certain monomer composition that refers to the border of precipitation and dispersion polymerizations. Bulk polymerization of vinyl chlo-

Table 11.2.1 Comparison Between Precipitation and Dispersion Polymerizations Defined by Arshady

	Precipitation polymerization	Dispersion polymerization
Particle nuclei	Precipitate including no monomer	Phase-separated polymer containing monomer
Polymerization loci	1 (medium)	2 (medium and interior of particle)
Product	Coarse precipitate or coagulum	Stable particle with a smooth surface
Example	Bulk polymerization of acrylonitrile	Bulk polymerization of acrylonitrile and butyl acrylate

ride is usually regarded as a representative precipitation polymerization but, distinctively speaking, it does not completely satisfy Arshady's definition because the polymer adsorbs a small amount of monomer. As these examples show, there is not much meaning in considering a definite border between precipitation and dispersion polymerizations.

11.2.3 Formation and Propagation of Particles in Dispersion Polymerization

In Figure 11.2.1A the process of dispersion polymerization is shown focusing on the formation process of particles. Dispersion polymerization starts from a homogeneous solution, and when oligomeric radicals and polymer, formed in the monomer solution, do not have affinity for the medium, they become insoluble and precipitate. The precipitate is unstable and homoaggregates to become primary particles. Primary particles homoaggregate further until they become stable secondary particles. The mechanism to keep the particles stable depends on what type of stabilizer is used. The propagation processes from nuclei to primary particles and from primary to secondary ones does not have to be considered as discontinuous steps. However, it

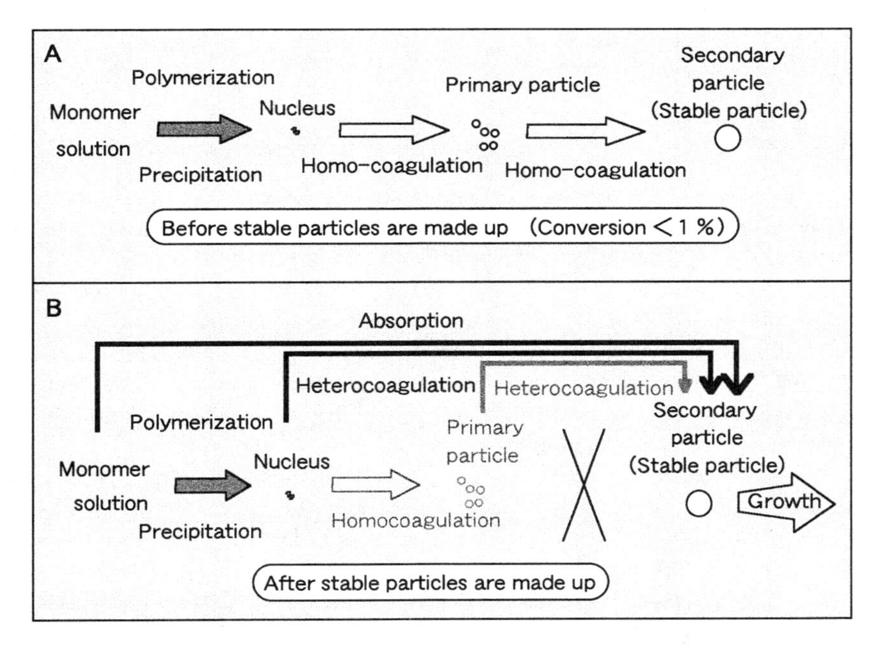

Fig. 11.2.1 Two steps to prepare monodisperse particles in dispersion polymerization: (A) before and (B) after stable particles are made up.

should be mentioned that the nuclei of several nanometers grow to become stable secondary particles of several tens of nanometers, by aggregation of several thousand nuclei. We cannot expect the formation of monodisperse particles unless the homoaggregation is fulfilled at the very beginning of polymerization where conversion is below 1%.

Figure 11.2.1B describes the propagation process of particles. When a sufficient amount of stabilized particles is formed, all the oligomeric radicals and polymer or primary particles are captured by existing stable particles before they precipitate and form new particles by themselves. This is what is called the growth process of particles by heteroaggregation.

In this way the formation of new particles is terminated and the particle number is maintained constant. After this, the enlargement of particles proceeds by (1) the capture of polymer, which is generated from solution, and (2) polymerization within themselves.

The balance of both terms affects the morphology and molecular weight distribution of the particles. If there is no unpredictable aggregation the particles grow together and the particle size distribution narrows down.

The number, size and distribution of the formed particles depend on the number of nuclei, the probability of particles becoming stable, and the aggregation of the particles. The amount and quality of the stabilizer used are related to these terms. Before discussing the qualitative analysis of number and size of the particles, we discuss the stabilizer, which has the greatest influence on the state of polymerization products.

11.2.4 Stabilizer

The dispersion polymerization system is composed of monomer, solvent, initiator, and stabilizer. The combination of monomer, solvent, and stabilizer is essential for particle preparation. That is to say, the stabilizer is chosen to meet the demand of the monomer and solvent. In any system, the stabilizer has affinity or cohesive strength for both the medium and the polymer particles. In a dispersion polymerization, the medium and polymer particles both are organic compounds. Therefore, it is not rational to rely on dispersion stabilization, which comes from the electrostatic repulsion force between particles. The stabilizer for dispersion polymerization that makes interfacial energy low must have affinity for particles due to the same quality and solvation at the surface of particles. It is desired that the stabilizer be a polymer that indicates a steric stabilization effect on the surface (5).

The types of stabilizers that are used fit into four classes: (a) homopolymer, (b) block copolymer, (c) macromonomer, and (d) others.

For (a), hydroxypropyl cellulose (HPC) (6–8), poly(vinyl pyrrolidone) (PVP) (9,10), poly(acrylic acid) (PAA) (9), and poly(dimethyl siloxane) (PDMS) (11) are usually employed. Ober et al. reported that the copolymers of isobutylene/isoprene and various methacrylates, which have weak polarity, are appropriate stabilizers for the dispersion polymerization in nonpolar solvents (6). For polar solvents, PAA, PVP, HPC, and polyvinylbutyral are used, and have polarity to some extent. Moreover, there are some reports that use ionic polymers as a dispersion stabilizer, containing ammonium or sulfonic acid groups, although their exact effect is not known. The homopolymer has nothing to do if it is just dissolved in the medium. As a widely accepted interpretation, homopolymer as a stabilizer works effectively when it is grafted and placed under restraint at the surface of particles.

For (b), a block copolymer, in which one side of the block has affinity to the solvent and the other block to the polymer particle, is the most reasonable stabilizer. Block copolymers of polystyrene/halogenated polybutadiene, polystyrene/polyethyleneglycol, and polystyrene/PDMS are examples of this type of stabilizer (12). When using a block copolymer, it is possible to provide appropriate amphiphilic and other surface properties by changing the block ratio. For example, when using a block copolymer of polystyrene/PDMS for polymerization of methyl methacrylate in hexane, the ratio of polystyrene/polydimehtylsiloxane should be below 4.4 (13). If the ratio is above 4.4, the block copolymer forms a stable micelle and will not function properly as a stabilizer.

For (c), a macromonomer that has a pendant group accustomed to the solvent is used as a comonomer in the dispersion polymerization of a monomer that composes the particle. The surface of resulting particles is covered with the pendant group and consequently stabilized by a steric stabilization effect (14,15). In this sense the macromonomer is a kind of stabilizer that shows its effect through polymerization, and it could be called as a "stabilizer formed in situ." A copolymer of macromonomer and particle-composing monomer, which joins the polymer particle, is much more effective for dispersion than a soluble stabilizer. With the dispersion polymerization of methyl methacrylate, which uses a macromonomer composed of an oligo-oxazoline pendant group, it is possible to cut the amount of stabilizer used to one-tenth or less compared to the oxazoline homopolymer stabilizer (16).

Of course, it is possible to first copolymerize the macromonomer and vinyl monomer and then use this as a stabilizer in dispersion polymerization. As mentioned earlier, HSM is a stabilizer copolymerized from a methyl methacrylate and a macromonomer that is a maleic ester of 12 hydroxystearic acid pentamer (17).

Macro azo-initiator can also possible be a "stabilizer formed in situ." For example, when a macro azo-initiator (I) composed of *n* chains of PDMS linked by *n* azo groups is used in the polymerization of methyl methacrylate in heptane, the initiator starts the polymerization and itself becomes one part of a block copolymer and contributes to the preparation of stable particles in the succeeding polymerization (18).

$$\left[\underset{CH_3}{\overset{CH_3}{\overset{|}{\underset{|}{C}}}}(CH_2)_2\underset{CH_3}{\overset{CH_3}{\overset{|}{\underset{|}{C}}}}N{=}N\underset{CH_3}{\overset{CH_3}{\overset{|}{\underset{|}{C}}}}(CH_2)_2CNH(CH_2)_3(\underset{CH_3}{\overset{CH_3}{\overset{|}{\underset{|}{Si}}}}O)_x\underset{CH_3}{\overset{CH_3}{\overset{|}{\underset{|}{Si}}}}(CH_2)_3NH \right]_n \qquad \text{(I)}$$

$$X \fallingdotseq 80, \quad n \fallingdotseq 7\text{–}8$$

Heptane is chosen as the medium for this polymerization because it is a good solvent for poly(dimethylsiloxane) and a nonsolvent for poly(methyl methacrylate).

11.2.5 Particle Size and the Control of Size Distribution

Solvent/polymer solubility is the dominating factor for the formation process of particle nuclei from solution. Here we show real examples and then discuss the overall control of size distribution.

11.2.5.1 The Solubility Parameter and the Control of Particle Size

Many reports have been published on the control of particle size by changing the composition of the solvent. The result obtained with the AAm/MAc/MB system is shown in Figure 11.2.2 (19). The particle size seems to increase almost linearly with solubility parameter in the range of 11 to 13. In the case of hydrophobic monomer

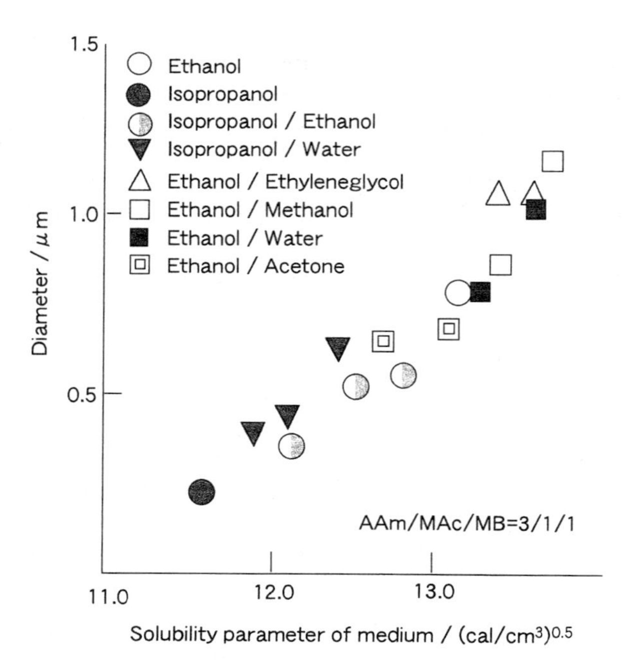

Fig. 11.2.2 Dependence of particle size on medium for precipitation polymerization of acrylamide, methacrylic acid, and methylenebisacrylamide.

systems, opposite results are obtained. Almog et al. reported that dispersion polymerization of styrene (St) in alcohol gave a constant decrease in diameter of the resulting spheres when changing the solvent from butanol to propanol, ethanol, and methanol—that is, when increasing the solubility parameter. If the difference of the solubility parameters of the dispersing medium and the polymer is chosen for the horizontal axis and the third power of diameter for the vertical axis, the relation between them is roughly proportional. Consequently, the smaller the affinity between the polymer particle and alcohol, the easier is the precipitation. Therefore a large number of nuclei will be formed, and finally a large number of small particles are obtained in such systems.

In a mixed solvent system, it is possible to control the particle size by changing the composition ratio. In a dispersion polymerization of methyl methacrylate in a mixed solvent of trimethylpentane and carbon tetrachloride, the size of the particle increases as the fraction of carbon tetrachloride, which has affinity to methyl methacrylate, increases.

When HPC is employed as a stabilizer and alcohol/methyl cellosolve as a solvent for the dispersion polymerization of styrene, particle size enlarges as the content of methyl cellosolve increases. Pentanol gives the largest size among a variety of alcohols examined (8). If the size of particles obtained in a variety of solvents is plotted down in a Hansen solvency map—a matrix of hydrogen bonding versus polarity parameter of solubility parameters—and the plots to draw a contour line of the same size particle are connected, the lines will gather as concentric circles and the coordinate of HPC will appears the greatest peak. We conclude from this that the size depends not on the solubility parameter of solvent and stabilizer itself, but upon the component of solubility parameter, and that the largest size is obtained when a solvent has values of polar and hydrogen parameter near to those of the HPC stabilizer. This index of particle size control would be consulted by those who examine new dispersion polymerization systems.

Before polymerization begins, the monomer is just a part of the medium, and the composition of the medium changes as the polymerization progresses in a dispersion polymerization. In consequence, the tendency for the polymer to precipitate changes and the number of particles is influenced. This is one factor that makes the control of particle size in dispersion polymerization difficult. In the case of dispersion copolymerization, the story gets much more complicated (6). According to the examination of a dispersion copolymerization styrene and butyl methacrylates, the size distribution broadens out unless the polymerization progresses under a controlled condition in terms of the ratio and amount of monomer feed, which are adjusted to give the least drift in the solubility parameter of the medium. It is reported that the drift of solubility parameters should be maintained below 0.7 to obtain monodisperse particles. To suppress the drift, divide feed and continuous charge of monomer are examples to overcome the situation.

11.2.5.2 Prediction of Particle Size

There are a lot of qualitative analysis reports on the parameters that affect the particle size in dispersion polymerization, but only a few are reported on quantitative analysis about the prediction of the size (20). Paine suggested a model to predict particle size in a dispersion polymerization that uses a type (a) stabilizer, that is, a medium-soluble homopolymer (21). In this case the premise of the model is that the graft polymer of the homopolymer, with the polymer from which the particle is formed, contributes as the stabilizer.

Here, how the number of particles are determined quantitatively by Paine is followed. In his model, two unstable nuclei or primary particles homoaggregate with a reaction rate constant of k_2 and the number of particles decrease with conversion by -1 power. The surface area (SFA), which is decided by the rates of formation of unstable nuclei or primary particles and of their homoaggregation, could be written as a function of conversion X as follows (the details are in Ref. 21):

$$\text{SFA} \propto k_2 X^{1/3} \tag{1}$$

The surface is stable if it is fully covered with stabilizer-grafting polymers—in other words, if Eq. (2) is valid in which Q_{min} is the minimum amount of graft polymer required for surface stabilization and r_s is the radius of gyration of the graft polymer.

$$Q_{\text{min}} = \text{SFA}/\pi r_s^2 \tag{2}$$

Therefore, as in Figure 11.2.3 (21), log Q_{min} versus log X has a straight line of slope 1/3. On the other hand, the amount of graft polymer formed (GP_{avail}) in the continuous phase by chain transfer reaction should be proportional to X. As shown in Fig. 11.2.3, Q_{min} and GP_{avail} lines will intersect at some point. This convergence point is called "A." When A is reached, further homoaggregation will not occur because newborn nuclei and primary particles are captured by existing stable particles. Another parameter, Q_{max}, is introduced. Q_{max} is the maximum amount of graft polymer to cover the surface completely with its most compact conformation. The conformations of graft polymer taking Q_{min} and Q_{max} are shown in Figure 11.2.3B (extended perpendicularly from the surface) and 11.2.3c (random coil). Q_{min} was estimated to be one-tenth of Q_{max} for HPC.

When the condition passes point A and enters field II, the conversion dependencies of SFA, Q_{min}, Q_{max}, and GP_{avail} change and the particle is enlarged by (1) the capture of polymer formed from the solution and (2) polymerization of the monomers that the particle itself absorbs; then log Q_{min} ($\propto$ log SFA) versus log X develops a slope of 2/3.

When the contribution of polymerization inside the particles increases, the rate of graft polymer formation decreases because it is due to the polymerization in solution, and the curve of log GP_{avail} versus log X has slope less than 1. But as long as the curve of log GP_{avail} versus log X remains in the Q_{max}–Q_{min} zone, the condition for the particles to remain monodisperse will be fulfilled. If the curve doesn't become

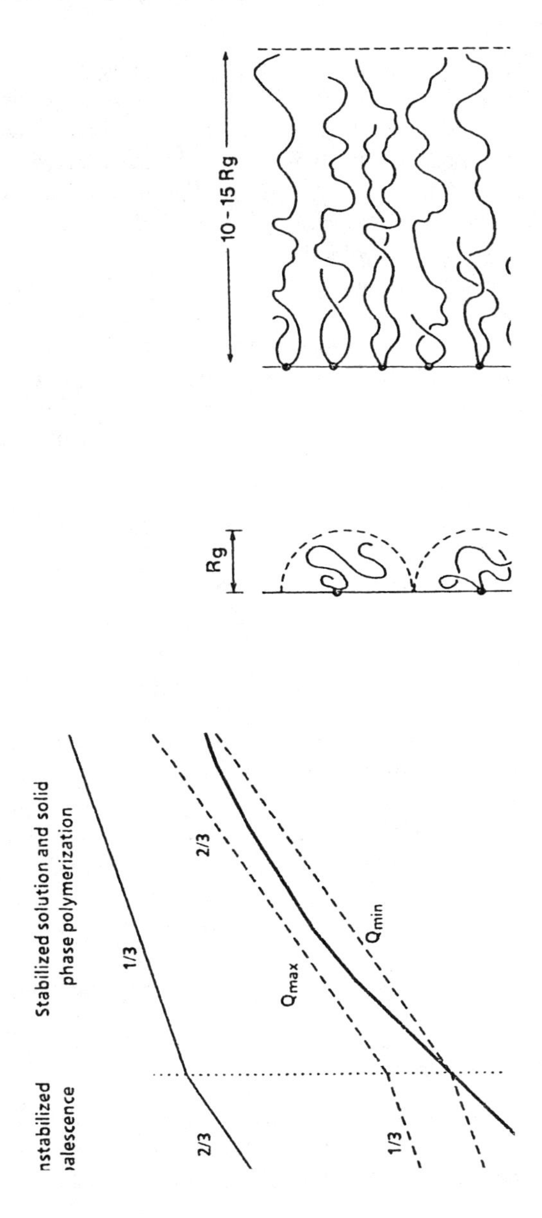

Fig. 11.2.3 Paine's model for particle formation and growth.

gentle enough and reaches Q_{max}, newborn graft polymer will not find any surface to adsorb and the excess graft polymer will serve to form new particles. This is one case of losing monodispersity. If the slope of GP_{avail} becomes too small and falls below Q_{min}, the existing stable particles become unstable and aggregation between particles progresses, forming macroscopic particles. This is the other deprivation of monodispersity. The ratio Q_{max}/Q_{min} is a value dependent on the flexibility of the polymer. It is reported that the GP_{avail} of PVP, which is flexible and has a larger Q_{max}/Q_{min} than HPC, is less susceptible to fall out of the Q_{max}–Q_{min} zone so it is easier to obtain monodisperse particles.

If it is assumed that all the graft polymer formed in the continuous phase is adsorbed by the particle and there is no change in number of particles after point A, the diameter of the monodisperse particles finally formed (d_f) could be calculated according to Eq. (3). The experimental values were reported to agree well with the calculation.

$$d_f = \left(\frac{1}{0.75A(\mathrm{Mw})}\right)^b \left(\frac{6[\mathrm{M}](\mathrm{Mm})}{\rho\pi NA}\right)^{2/3} \left(\frac{1}{(\mathrm{Cs})[\mathrm{S}]}\right)^{1/2} \left(\frac{k_2}{0.386k_p}\right)^{1/6} \left(\frac{10k_t}{fk_d[\mathrm{I}]}\right)^{1/12} \quad (3)$$

Here [M] is monomer concentration, [S] stabilizer concentration, [I] initiator concentration; A and b are constants in the function of radius of gyration $r_g = A(\mathrm{Mw})b$ (Mw is the molecular weight of the stabilizer); and Mm is the molecular weight of the monomer, ρ is density, and Cs is the chain transfer constant to the stabilizer.

The adsorbability of graft polymer to the particle is influenced by the solvent. For HPC, the graft polymer's tendency of adsorption to the particle is high in a strong polar solvent such as methanol and ethanol, and then a large amount of small particles that have the diameter assumed in Eq. (3) forms. On the other hand, if less polar solvents such as butanol or toluene/ethanol are used, the graft polymer is only adsorbed partially. Then Eq. (3) will not apply and large particles will be formed.

From a similar idea based on the surface area stabilized by a stabilizer, a presumption equation could also be made for dispersion polymerization that uses macromonomer, a precursor stabilizer, which has a pendant group as a solvating component (14).

11.2.6 New Dispersion Polymerization

The polymerization of the 1990s intended to invent functional particles and the development of polymerization techniques that accommodate with the environment. Some examples of dispersion polymerization of the third generation are introduced in this section.

11.2.6.1 Environmentally Friendly Dispersion Polymerization

The major drawback to dispersion polymerization is the use of VOC (volatile organic compounds). A proposal of organic solvent-free dispersion polymerization is made to decrease the load to the environment. Here two examples are taken up. One is the use of supercritical carbon dioxide (22) and the other is of water (23). In the first method, temperature and pressure are adjusted to a supercritical point of carbon dioxide. Carbon dioxide in the supercritical state has great monomer solvency and low viscosity and possesses an advantage of easy separation after particles form polymerization. However, soluble polymers, serving as a stabilizer in this dispersion polymerization, are limited to poly(dialkylsilicone) and fluorinated polymers. To stabilize poly(methyl methacrylate) precipitated from supercritical carbon dioxide, poly(methyl methacrylate-graft-fluorinated alkyl methacrylate) was used as a stabilizer. Azo-initiator containing polydimethylsiloxane fragments was used as an initiator, and has a higher initiator efficiency than in common solvents because termination reaction was suppressed due to the low viscosity of the medium.

Next, the precipitation or dispersion polymerizations of acrylamide derivatives in water is discussed. Acrylamide derivatives such as *N*-isopropylacrylamide (NIPAM) give a polymer having a lower critical solution temperature (LCST) in water. The isopropyl group in NIPAM is in hydrophobic hydration at low temperatures, but dehydrates and performs hydrophobic interaction at temperatures higher than the LCST. At the molecular level, coil–globule transitionlike conformational change occurs at LCST. The polymer is soluble in water at low temperatures but becomes insoluble at high temperatures. Therefore, if a monomer solution is prepared and polymerized at a temperature higher than the LCST of PNIPAM, a precipitation polymerization occurs as shown in Figure 11.2.4. Although PNIPAM becomes hydrophobic at high temperatures, it seems to retain a certain level of hydrophilicity to disperse the polymer precipitates stably in water. If it is a polymerization at low concentration, to say 5%, monodisperse particles form. The concentrated PNIPAM dispersion is transparent at low temperatures and becomes cloudy due to dehydration and aggregation at high temperatures. The PNIPAM particles exhibit unique temperature dependence for electrophoresis and protein adsorption (24,25).

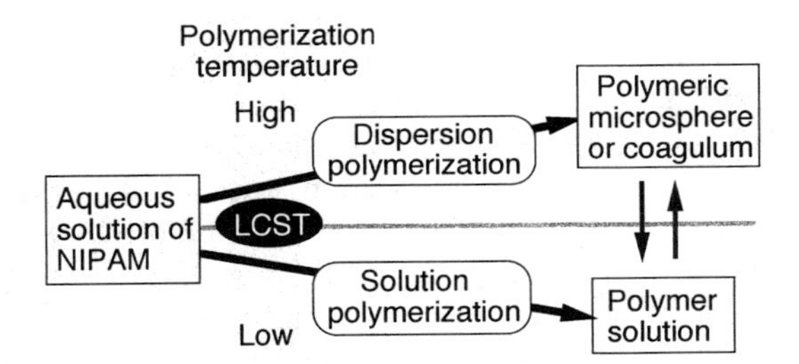

Fig. 11.2.4 Aqueous polymerization of *n*-isopropylacrylamide at different temperatures.

11.2.6.2 Gel Microspheres and Highly Cross-Linked Microspheres

Poly-AAm is insoluble in alcohols. As mentioned before, precipitation polymerization of AAm with cross-linking agents and dispersion polymerization where another monomer is added prepares hydrogel microspheres (26). The polymerization of *p*-nitrophenyl acrylate (NPA), MAc, and cross-linking agent in alcohol also prepares monodisperse hydrogel microspheres (27). NPA is an active ester, and the ester part of the particles could be easily modified. The amphoteric microspheres, prepared by the reaction between diamine and NPA units of the microspheres, reversibly expanded and shrank with changing pH. They contracted in a narrow pH range near their isoelectric point (IEP), although the pH for the most contraction and IEP do not necessarily coincide with each other if the cationic and anionic groups distribute unevenly in the particle. Not only the component distribution but also the cross-link distribution in the microsphere is usually uneven in this kind of polymerization. The AAm–MAc–methylenebisacrylamide terpolymer hydrogel microsphere of about 1 μm diameter was reported to contain highly cross-linked microdomains of about 30 nm diameter (26).

Dispersion polymerization, a method appropriate for micrometer-ordered monodisperse particles, is generally unsuitable for preparation of highly cross-linked particles, and in a dispersion polymerization where cross-linking agents are used in a large quantity, the resulting products are distorted or porous particles or aggregates. The formation of distorted particles is attributed to aggregation of particles during a certain stage of polymerization. Moderate cross-linking will make the particles stickier and facilitate aggregation; consequently, in order to prepare monodisperse highly crosslinked microspheres, it is recommended that one pass the sticky state promptly by increasing the cross-linking agent (28). The other way to obtain highly cross-linked polymer particles was reported to be dispersion polymerization in the presence of a small amount of oxygen. In this phenomenon it was considered likely that the oxygen radical induces the chain transfer to a type (a) stabilizer and promotes the formation of effective stabilizer (29).

11.2.6.3 Dispersion Polymerization of Nonvinyl Monomers

Here we discuss dispersion polymerizations that are not related to vinyl monomers and radical polymerization. The first one is the ring-opening polymerization of ϵ-caprolactone in dioxane–heptane (30). A graft copolymer, poly(dodecyl acrylate)-g-poly(ϵ-caprolactone), is used as a stabilizer. The polymerization proceeds via anionic or pseudoanionic mechanism initiated by diethylaluminum ethoxide or other catalysts. The size of poly(caprolactone) particles depends on the composition of stabilizer, ranging from 0.5 to 5 μm. Lactide was also polymerized in a similar way. Poly(caprolactone) and poly(lactide) particles with a narrow size distribution are expected to be applied as degradable carriers of drugs and bioactive compounds.

Monodisperse, optically active poly(lactide) microspheres are especially expected to attract much attention.

Electrically conducting polymer particles such as polypyrrole and polyaniline could also be prepared by dispersion polymerization in aqueous ethanol (31). The oxidation polymerization of pyrrole and aniline has been carried out at the electrode surfaces so far and formed a thin film of conducting polymer. On the other hand, polypyrrole precipitates as particles when an oxidizing reagent is added to a pyrrole dissolved ethanol solution, which contains a water-soluble stabilizer. In this way electrically conducting polymer particles are obtained and, in order to add more function to them, incorporation of functional groups, such as aldehyde to the surface, and silicone treatment were invented (32).

Next we consider the dispersion polymerization by polyaddition. In a typical method to prepare polyimide particles, polyamic acid solution is first obtained by coupling of pyromellitic dianhydride and oxy-dianiline, and then by heating the solution. The condensation reaction on heating causes crystallization of polyimide in a spherical form (Fig. 11.2.5, left) (33). However, on the contrary to this conventional method, polyamic acid microspheres could be obtained by dispersion polymerization if an appropriate medium is chosen (34). When a solvent that has a solubility parameter around 17 Mpa is used, submicrometer-sized monodisperse polyamic acid parti-

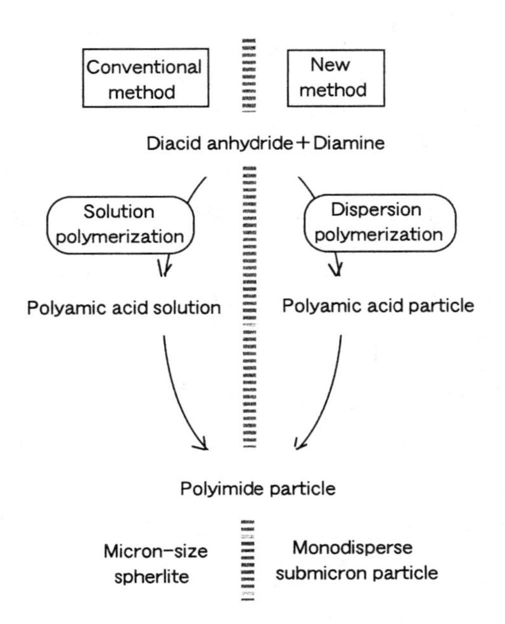

Fig. 11.2.5 Preparation of polyimide particles by different processes.

cles are prepared (Fig. 11.1.5, right). Polyimide particles are obtained by heating the polyamic acid particles.

11.2.7 Conclusion

An outline of dispersion polymerization from the first to third generations was explained in this chapter. However, the progress in this field is so rapid and vast that we must review the development constantly. Attention should be focused especially on the environmentally friendly dispersion polymerization, nonvinyl dispersion polymerization, and organic/inorganic composite dispersion polymerization (35).

REFERENCES TO SECTION 11.2

1. KE Barrett. Dispersion Polymerization in Organic Media. New York: John Wiley & Sons, 1975.
2. R Arshady. Colloid Polym Sci 270:717, 1992.
3. M Carenza, G Palma. Eur Polym J 31:41, 1985.
4. H Kawaguchi, Y Yamada, S Kataoka, Y Morita, Y Ohtsuka. Polym J 23:955, 1991.
5. U Genz, BD Aguanno, J Mewls, R Klein. Langmuir 10:2206, 1994.
6. CK Ober, KP Lok. Macromolecules 20:268, 1987.
7. AJ Paine. J Polym Sci Polym Chem 18:2485, 1990.
8. AJ Paine. J Colloid Interface Sci 138:157, 1990.
9. Y Almog, S Reich, M Levy. Br Polym J 131, 1982.
10. CM Tseng, YY Lu, MS El-Aasser, JW Vanderhoff. J Polym Sci Polym Chem 24:2995, 1986.
11. RH Pelton, A Osterroth, MA Brook. J Colloid Interface Sci 137:120, 1990.
12. C Winzor, Z Mrazek, MA Winnie, MD Croucher, G Riess. Eur Polym J 31(1):121, 1994.
13. JV Dawkins, G Taylor. Polymer 20:599, 1979.
14. S Kawaguchi, MA Winnik, K Ito. Macromolecules 28:1159, 1995.
15. S Kobayashi, H Uyama, SW Lee, Y Matsumoto. J Polym Sci Polym Chem 31:3133, 1993.
16. S Kobayashi, H Uyama. In: JC Salamone ed. Polymeric Materials Encyclopedia. New York: LRC Press, 1997, pp 4494–4500.
17. KEJ Barrett, HR Thomas. J Polym Sci A-1 7:2621, 1969.
18. K Nakamura, K Fujimoto, H Kawaguchi. Colloids Surf A Physicochem Eng Aspects 153:195, 1999.
19. H Kawaguchi, M Kashiwabara, N Yaguchi, F Hoshino, Y Ohtsuka. Polym J 20:903, 1988.
20. S Shen, ED Sudol, MS El-Aasser. J Polym Sci Polym Chem Ed 31:1393, 1993.
21. AJ Paine. Macromolecules 23:3109, 1990.
22. Y-L Hsiao, EE Maury, JM DeSimone, S Mawson, KP Johnston. Macromolecules 28: 8159, 1995.
23. RH Pelton, P Chibante. Colloid Surfaces 20:247, 1986.
24. H Kawaguchi, K Fujimoto, Y Mizuhara. Colloid Polym Sci 270:53, 1992.

25. H Kawaguchi. In: M Yalpani, ed. Biomedical Functions and Biotechnology of Natural and Artificial Polymers. Mount Prospect: ATL Press, 1996, pp 157–168.
26. Y Kamijo, K Fujimoto, H Kawaguchi, Y Yuguchi, H Urakawa, K Kajiwara. Polym J 28:309, 1996.
27. M Kashiwabara, K Fujimoto, H Kawaguchi. Colloid Polym Sci 273:339, 1995.
28. K Li, DH Stover. J Polym Sci Polym Chem Ed 31:3257, 1993.
29. M Hattori, ED Sudol, S El-Aasser. J Appl Polym Sci 50:2027, 1993.
30. S Sosnowski, M Gadzinowski, S Slomkwski. Macromolecules 29:4556, 1996.
31. SP Armes, JF Miller, BJ Vincent. J Colloid Interface Sci 118:410, 1987.
32. S Maeda, SP Armes. J Mater Chem 4:935–942, 1994.
33. Y Nagata, Y Ohnishi, T Kajiyama. Polym J 28:980, 1996.
34. A Okamura. Master's thesis, 1998, Keio University, Tokyo.
35. HH Stover, K Li. In: JC Salamone, ed. Polymeric Materials Encyclopedia. New York: CRC Press, 1997, pp 1900–1905.

12

Particles Modified in Surface Properties

KOHJI YOSHINAGA and HARUMA KAWAGUCHI

12.1 SURFACE MODIFICATION OF INORGANIC PARTICLES

KOHJI YOSHINAGA
Kyushu Institute of Technology, Kitakyushu, Japan

12.1.1 Introduction

Inorganic particles have intrinsic specific properties of magnetism, photochromism, conductivity, catalysis, and so on, which are practically utilized for many functional materials. In many cases, the particles have been incorporated into organic compounds or polymer materials in practical use. For instance, octadecyl group-grafted silica particles have been widely used for column packings as a stationary phase in high-performance liquid chromatography (HPLC) (1). Therefore, the chemical modification of inorganic particles is very important to give compatibility with organic materials in developments or designs of functional composite materials employing the particles (2,3). In this section, the chemical procedures, especially organic modifications of inorganic particles, and the properties of modified inorganic particles are described.

12.1.2 Functional Groups on Inorganic Particles

For the chemical modification of inorganic particles, it is required that the particles have a functional group that is a binding site of a modifier on the surface. In general, however, the intrinsic functional group existing on inorganic particles, mostly metal oxides, is only a hydroxyl group being available for the surface modification. For the determination of the surface hydroxyl group on the inorganic surface, many analytical methods have been proposed, but it is extremely difficult for chemical or spectroscopic determination to distinguish the hydroxyl group from adsorbed water molecules. The surface hydroxyl groups on typical particles—silica, titania, and ferrite—are shown in Table 12.1.1. Silica generally has three to four surface hydroxyl groups per square nanometer.

In chemical modifications of inorganic particles, the adsorbed water sometimes affects the reaction of the hydroxyl group with the modifier. According to Fripiat et al. (6), evacuation at 200°C and 10^{-7} mmHg for 24 h still leaves about 0.2 mmol/g adsorbed water on silica (Aerosil 200).

12.1.3 Binding of Organic Compounds

One conventional procedure for organic modification of inorganic particles is to utilize the reaction of the hydroxyl group with silane coupling agents. Many kinds

Table 12.1.1 Hydroxyl Groups on Inorganic Particles

Particles	Size (nm)	Surface area[a] (m^2/g)	Hydroxyl group mmol/g	Hydroxyl group groups/nm^2	Ref.
Silica					
Aerosil 200	12	200	1.19[b]	3.6	4
Aerosil 380	<7	380	2.16[b]	3.4	4
Silica gel-601 (Merk)	500–4000	460	2.54[b]	3.3	4
Titania	120	90–150	0.77[c]		5
Ferrite ($NiO \cdot ZnO \cdot Fe_2O_3$)	15	100	0.50[c]	3.0	5

[a] Determined by BET method.
[b] Determined by amount of dimethylsilyl [$Si(CH_3)_2$-H] group substituted for the hydroxyl hydrogen with Fourier Transform Infrared (FTIR) method.
[c] Determined by reaction of the hydroxyl group with triethylaluminum.

of the coupling agents are now commercially available. The general formula of the coupling agent is represented by $R\text{-}Si(R')_{3-n}X_n$ ($n = 2$ or 3). Ligand X is a hydrolyzable group of alkoxy, acyloxy, amino, or chloride, and R and R′ groups unreactive in the modification. Usually chlorosilane is not favorable, because of evolving hydrochloride in the reaction. The most popular agents are methoxy- or ethoxysilanes. The coupling reaction of the surface hydroxyl group with alkoxysilanes is generally proposed to proceed via the following steps, shown in Scheme 12.1.1: 1) hydrolysis and condensation of alkoxysilane to form oligosilanol, 2) adsorption of the oligosilanol onto the surface, and 3) condensation of the silanol with surface hydroxyl group to form the Si-O-M bond (7,8).

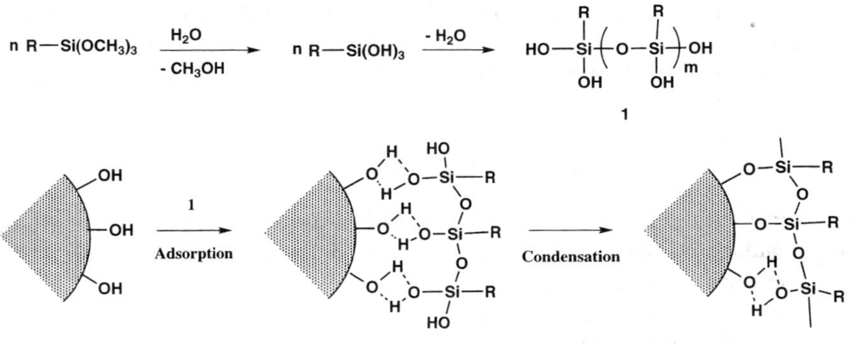

$$n\,R{-}Si(OCH_3)_3 \xrightarrow[-\,CH_3OH]{H_2O} n\,R{-}Si(OH)_3 \xrightarrow{-\,H_2O} HO{-}Si(R)(OH){-}\left(O{-}Si(R)(OH)\right)_m{-}OH \quad \mathbf{1}$$

Scheme 12.1.1 Surface modification with silane coupling agent.

In the alkoxysilanes, a trialkoxy compound is commonly more reactive, as compared with mono- or dialkoxysilane. The mono- or dialkoxysilane has lower affinity for the inorganic surface and is relatively less reactive for hydrolyzing to form the corresponding silanol. The reaction using the alkoxysilane gives stable binding via siloxane bonds on silica, titania, alumina, zirconia, tin oxide, and nickel oxide, but unstable bonds on boron oxide and ferric oxide. The alkoxysilane never forms stable binding to alkaline oxide or carbonate. Also, titanate coupling agents are commercially available for the chemical modification of silica, calcium carbonate, talc, etc. (9).

The silane coupling agents are also effective for converting the hydroxyl group on inorganic particles to various organic functional groups by the functionalized Y group in Reaction (1). The variety of the group is desirable for chemical modifications of the particles by using appropriate organic reaction in aqueous or organic solvent.

$$\bullet\!-OH + X_3Si-CH_2CH_2CH_2-Y \xrightarrow[-HX]{} \bullet\!-O-\overset{|}{\underset{|}{Si}}-CH_2CH_2CH_2Y \qquad (1)$$

X : $-Cl$, $-OCH_3$, $-OCH_2CH_3$

Y : $-NH_2$, $-NCO$, $-SH$, $-OCH_2CH\underset{O}{-}CH_2$ (epoxide), $-CH_2CH_2CH_2CH{=}CH_2$, $-OC(=O)-C(CH_3){=}CH_2$

12.1.4 Binding of Macromolecules

When the surface modification of inorganic particles is carried out using the silane coupling agent, the hydrophobicity or the organic characteristic features are added to the original inorganic properties. Incorporation of synthetic macromolecule characters into the particles is sometimes required. For instance, grafting of polymer onto inorganic particles has remarkably improved the stability against flocculation of colloidal dispersion in organic media and affinity for polymer matrix (2,3,10). Principally, there are the following methods for polymer grafting to inorganic surface via covalent bonds.

12.1.4.1 Polymerization from the Surface

Radical Polymerization. In this modification, a functional group like an azo or peroxy group, being able to initiate radical polymerization, must be introduced into the particle surface (Scheme 12.1.2).

Hamann et al. (11) have prepared phenylazo-bound silica via four-step reactions from the starting silica (Aerosil 200) and carried out the polymerizations of styrene, methyl methacrylate, acrylamide, acrylonitrile, acrylic acid, and 4-vinylpyri-

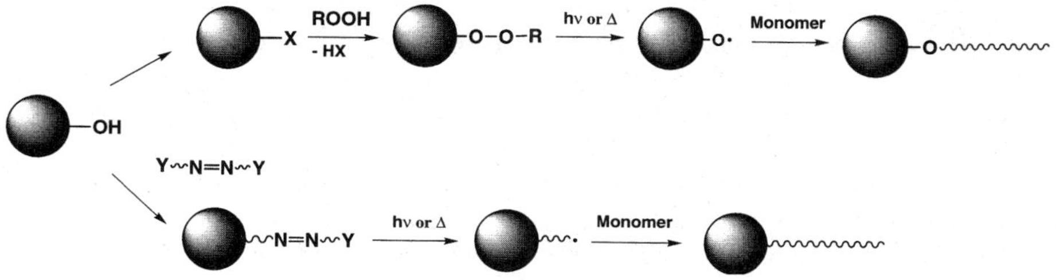

Scheme 12.1.2 Introductions of peroxy and azo groups on inorganic particles.

dine in organic solution, and then obtained polymer-grafted silica particles with amounts of grafted polymer of about 100–500 mg/g; see Reaction (2)).

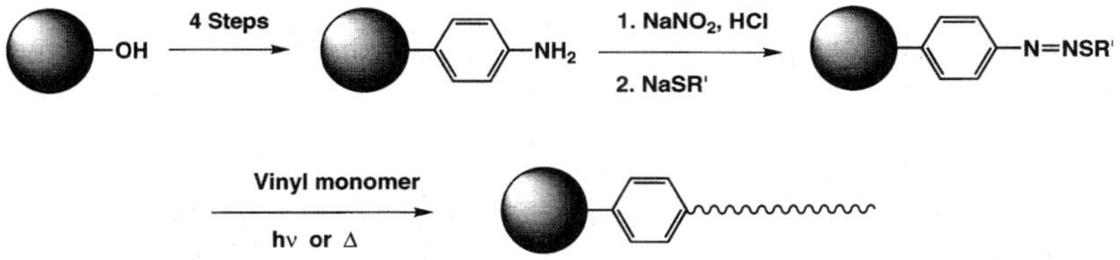

(2)

Tsubokawa et al. (12–14) have introduced radical sources of azo or peroxy groups by another methods, and successively conducted the radical polymerization of vinyl compounds, such as styrene or methyl methacrylate, to give polymer-grafted particles; see Reaction (3). The grafting by the radical polymerization of methyl methacrylate, initiated from a peroxy group introduced on silica, takes place at relatively high efficiency, compared with those from azo group-introduced particles.

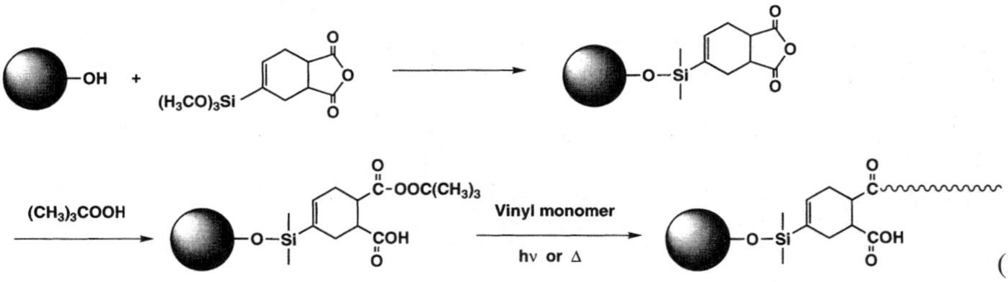

(3)

The introduction of the functional group, being capable to initiate the radical graft-polymerization, has been sometimes carried out by two- or three-step reactions using reactive reagents, such as butyllithium or thionylchloride. Therefore, some procedures in the functional group introduction are sometimes not applicable to the modification of colloid particles, because the reaction of colloid particles with reactive basic or acidic reagents mostly leads to quick coagulation among the particles. In this regard, in order to prevent coagulation of colloidal particles, the radical polymerization of styrene or methyl methacrylate in the presence of colloidal silica was carried out using 2,2′-azobis(amidinopropane) dihydrochloride as an initiator in tetrahydrofuran (15). In this case, adsorption of the initiator on the surface first takes place via electrostatic interaction, and then consecutive reaction of the radical species with monomer propagates the polymer chain; see Reaction (4).

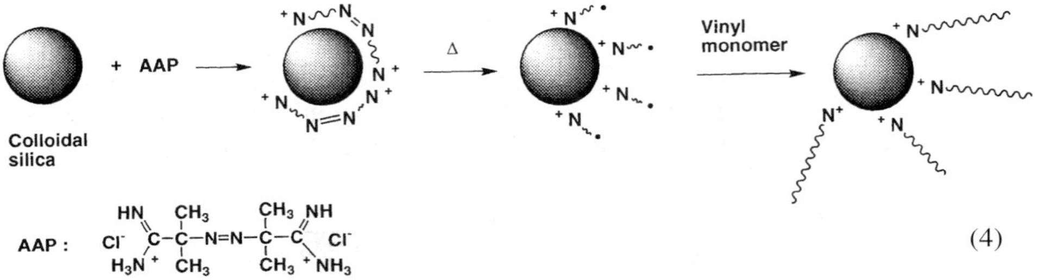

(4)

Radical polymerization in the presence of colloidal silica afforded polymer-coated silica with amount of 50–80 mg polymer per gram of particles without aggregation. Addition of divinylbenzene to the polymerization systems gave the cross-linked, polystyrene-coated silica particles, of which the coating polymer layer was never peeled off by washing with acetone or ethyl acetate. Furthermore, when the polymerizations of styrene were carried out with 3-mercaptopropionic acid or 3-mercaptopropylamine hydrochloride as a chain transfer reagent, the resulting polymer-coated silica had carboxyl or amino groups at 0.4–1.2 mmol/g or 2.4–2.8 mmol/g on the surface, respectively. However, it is ordinarily very difficult to control thickness of the polymer layer on colloidal silica in radical polymerization. Thus, in order to control the polymer layer, the polymerization of styrene in aqueous suspension has been carried out after the formation of hydrophobic layer on the surface through binding of a silane coupling agent having a long methylene chain or hydrophobic group (16).

Cationic Polymerization. The functional groups that enable initiation of cationic polymerization can be introduced on the inorganic surface. The introductions of acylium perchlorate [Reaction (5)], sulfonium or pyridinium salt, or active chloride

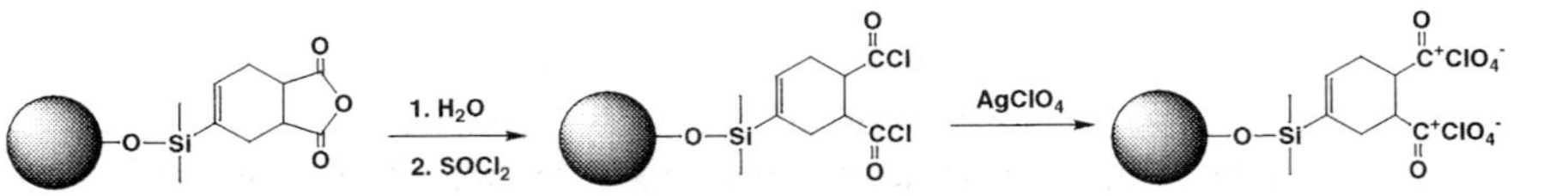

2

(5)

on silica or titania led to the cationic polymerization of styrene, cyclic ethers, cyclic acetals, and lactones [Reactions (6) and (7)] (17,18).

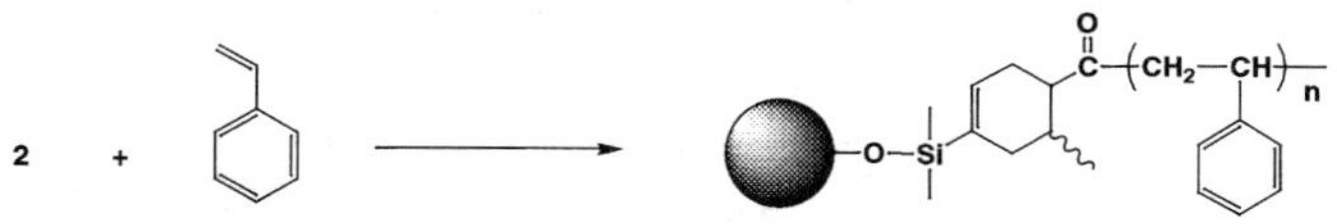

(6)

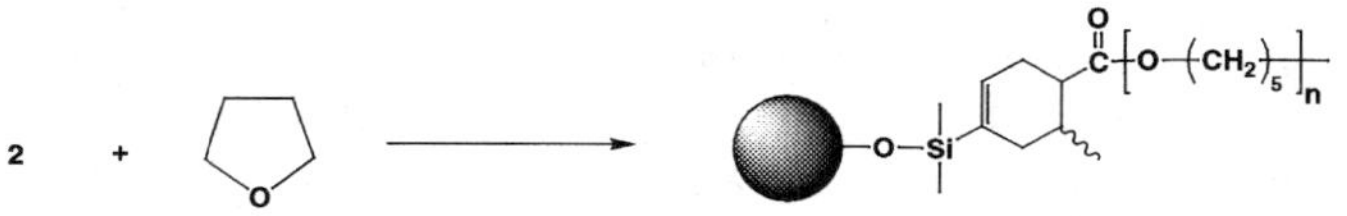

(7)

Recently, Spange et al. (19,20) have successfully achieved cationic graft polymerizations of vinyl ethers, vinyl furan, and cyclopentadiene onto silica, initiated by a stable ion pair formed from silanol and arylmethyl halide, such as di(*p*-methoxyphenyl)methyl chloride. The grafting of the polymer onto silica is proposed to take place via the propagation based on olefin insertion to a cation center being in a rapid equilibrium with the ion pair, as shown in Scheme 12.1.3.

According to Tsubokawa et al. (18), the molecular weight of grafted polystyrene from radical polymerization at the surface of silica is considerably higher than that from cationic polymerization.

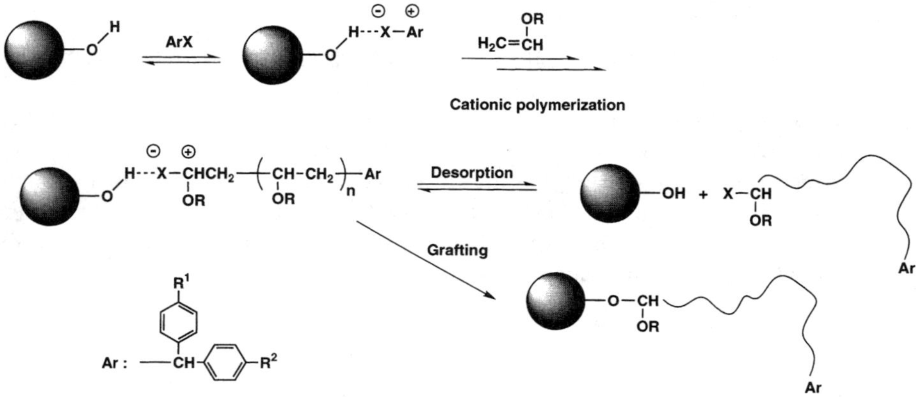

Scheme 12.1.3 Cationic polymerization from an ion pair formed from silanol and diaryl halide on silica surface.

Anionic Polymerization. Anionic grafting polymerization initiated on the inorganic particles is also possible to give composite particles. The amino group incorporated into the surface of silica caused anionic polymerization of *N*-carboxy-α-amino acid anhydride to give poly(amino acid) grafting [Reaction (8)] (21).

$$\text{●–C}_6\text{H}_4\text{–NH}_2 + n\ \text{O=C–CH(R)–NH–C(=O)–O (ring)} \xrightarrow{-\text{CO}_2} \text{●–C}_6\text{H}_4\text{–NH}\text{–}\left(\text{C(=O)–CH(R)–NH}\right)_n\text{–H} \qquad (8)$$

The potassium carboxylate group on inorganic particles also gave copolymerization of epoxides and acyl anhydrides on the surface through anionic polymerization (22). Moreover, the treatment of silica or glass fiber with butyllithium gives a lithium silanolate group, which enables initiation of anionic polymerization of styrene or methyl methacrylate (23). However, since it is ordinarily very difficult to remove completely the adsorbed water on the inorganic surface, it is scarcely possible to accomplish the anionic living graft polymerization via consecutive propagating of carboanion species from the inorganic surface.

Plasma Polymerization. Plasma-induced polymerization (24) of vinyl monomer from inorganic particles is also employed for polymer grafting. The conventional reactors for liquid-phase polymerization of vinyl compounds after generation of plasma on inorganic particles or powders have been recently invented by Ikeda et al. (25). Haraguchi et al. (26) have also prepared polymer-modified silica by plasma-induced polymerization of glycidyl methacrylate.

12.1.4.2 Reaction of Chemically Active Polymer with the Surface Functional Group

The reaction of a surface-active group with chemically active polymer is also effective for the modification of inorganic particles, as shown in Scheme 12.1.4. In this case, it is common that the particles must be activated to be incorporated into grafted macromolecules and the binding reactions must be carried out in the heterogeneous phase reactions. The procedure has the advantage of bindings of molecular weight-controlled polymers, which can be separately synthesized by a living polymerization, in spite of difficulty of bindings of high-molecular-weight polymers, because of steric hindrance of binding sites on the surface.

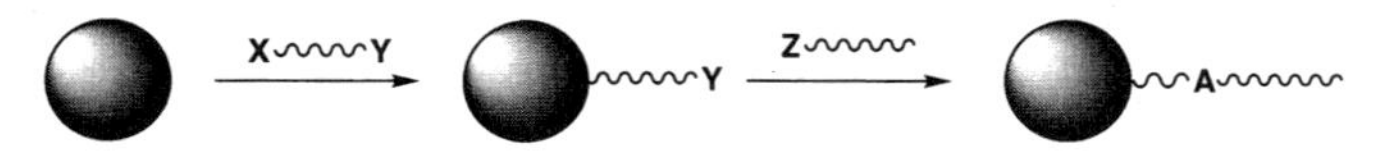

Scheme 12.1.4 Reaction of inorganic particle with chemically active polymer.

The reaction of hydroxyl group on inorganic particles with reactive polymers having isocyanate (27,28), alkoxysilane (29,30), and chlorosilane (29) groups efficiently gave the polymer-grafted particles [Reactions (9) and (10)].

(particle)—OH + NCO∿∿∿Polymer ⟶ (particle)—OC(=O)NH∿∿∿Polymer (9)

(particle)—OH + X–Si∿∿∿Polymer $\xrightarrow{\text{- HX}}$ (particle)—O–Si∿∿∿Polymer (10)

X : —Cl , —OCH_3

On the contrary, the reactive groups, isocyanate, amino, and acid anhydride, introduced on inorganic particles via the reactions with respective silane coupling agents, can also give covalent-bond binding to chemically active macromolecules capped with hydroxyl (31), carboxyl (32), and amino (33) groups, respectively. For alumina, heating of the suspension with hydroxyl-ended polyoxyethylene gave the polymer-grafted particles (34).

As described in a previous section, the reactions for the modification, including the introduction of the functional group, are sometimes unapplicable to those of colloidal particles, because of inducing flocculation or coagulation. For instance, the reaction of colloidal silica with polymeric chlorosilane immediately gives rise to flocculation to sedimentate due to making the suspension acidify and ζ-potential of the particles lower. In this respect, the reactions of trimethoxysilyl- or triethoxysilyl-terminated macromolecules with the surface hydroxyl group of inorganic particles are useful and convenient surface modification procedures. The trimethoxysilyl- or triethoxysilyl-terminated macromolecules can be easily synthesized by the radical polymerization in the presence of 3-mercaptopropyltrimethoxysilane as a chain transfer reagent (35), or by the coupling reaction of carboxyl-terminated prepolymer with 3-aminopropyltriethoxysilane using a condensation reagent, such as *N,N′*-dicyclohexylcarbodiimide (36). The modifications using the macromolecule silane coupling agents are effective not only for inorganic powders having hydroxyl group on the surface, but also for colloidal particles, such as silica or titania without aggregation.

Modifications of monodisperse colloidal silica, of 10 or 500 nm in diameter, were carried out using trialkoxysilane-terminated polymer in a low polar solvent, such as acetone or 1,2-dimethoxyethane, without coagulation during the coupling reaction (35,37–42). In this modification, the hydrophobic polymer can be efficiently bound to hydrophilic colloidal silica surface. The reaction mechanism of the binding

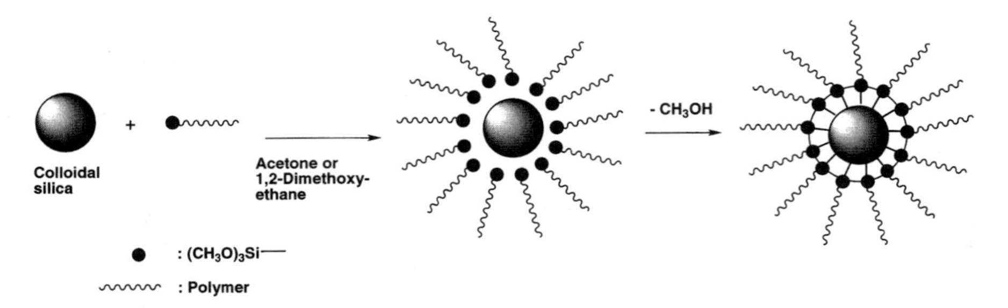

Scheme 12.1.5 Modification of colloidal silica ultrafine particles with polymeric silane coupling agents in low polar solvent.

reaction is proposed as follows (Scheme 12.1.5): (1) the polymeric silane coupler is adsorbed on the surface of colloidal silica through arraying of the agent on the surface in low polar solvent, and (2) the condensation between alkoxy group and the silanol on silica takes place to produce alcohol, ethanol, or methanol. The condensation is usually promoted by addition of a base, such as triethylamine.

In this modification, the solvent must be chosen to make the polymeric silane coupler tightly arrayed on the surface so as to prevent the aggregation of the particles due to the steric interaction among the polymer chains, because the adsorption of the coupler on the silica usually makes the ζ-potential of the surface lower. Interestingly, the reaction of the polymeric silane coupler with active ultrafine colloidal silica particles, of less than 50 nm in diameter, in acetone gave quantitative bindings of the polymer (39). The polymer modification of colloidal silica with maleic anhydride–styrene copolymer silane coupling agent gave composite particles dispersible in low polar solvent, and then led to the introduction of a carboxyl or amino group on the surface, as shown in Scheme 12.1.6. Thus, the maleic anhydride group in the grafted copolymer has an important role as a functional group, which can bind the secondary polymer or form the molecule having another active group. Further, it is possible to cross-link the grafted polymer chains on the particle surface with diisocyanate, like putting a net around the particle to prevent withdrawal of the polymer by hydrolysis of siloxane bonds (42).

The living polymerization gives the molecular-weight-controlled and monodispersed polymer. In general, living polymer species are usually active enough to react quickly with the surface hydroxyl group, and the reaction of the species with inorganic particles gives the composite grafted molecular-weight-controlled polymer chains. Horn et al. (43) have prepared polystyrene-grafted silica by the termination of polystyrene anionic living species with chlorinated silica surface to form the Si-C bond.

Tsubokawa et al. (44) have reported the grafting of polymeric cationic species of vinyl butyl ether or 2-methyl-2-oxazoline onto aminated silica particles via the termination. However, it is ordinarily difficult to apply the surface termination reac-

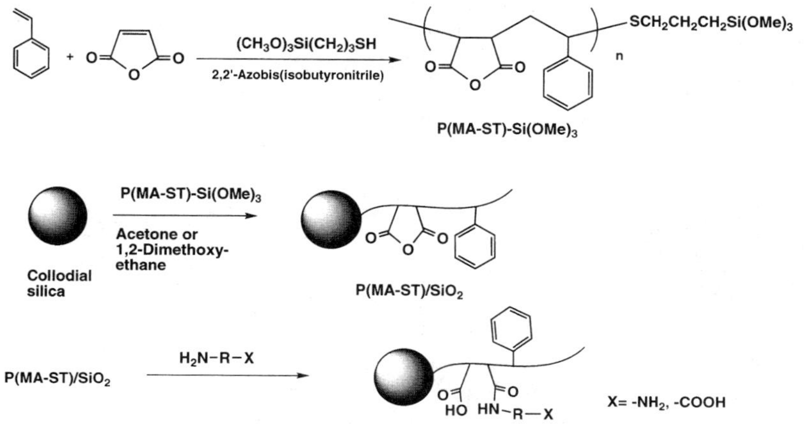

Scheme 12.1.6 Synthesis of trimethoxysilyl-terminated poly(maleic anhydride–styrene) and modification of colloidal silica with the polymer.

tion of reactive polymeric living species to the modification of colloidal particles, because the quenching of the living species with dispersion medium or adsorbed water competitively takes place much faster than the termination with the surface functional group on the particles.

12.1.4.3 Vapor-Phase Modification

Chemical vapor deposition (CVD) is also employed for the modification of inorganic particles. Fukui et al. (45,46) have conducted the modification of inorganic particles by contacting with 2,4,6,8-tetramethylcyclotetrasiloxane at 80°C by CVD method. In the modification of silica or titania, cross-linking polymerization took place on the surface to give hydrophobic composite particles via Type I and IIa shown in Scheme 12.1.7, while deposition onto mica or kaolin of the siloxane induced the ring-opening polymerization with molecular weight about 10^6 on the surface via Type IIa.

The CVD method can form a polymethylsiloxane layer of 0.6–0.8 nm on the inorganic particles (Table 12.1.2). Especially, the polysiloxane-modified titanium oxide has been used for the additives to cosmetics. Moreover, residual Si-H group grafted on the particles reacted with unsaturated organic compounds in the presence of Speier catalyst (47), H_2PtCl_6, to be functionalized by organic pendant group [Reaction (11)].

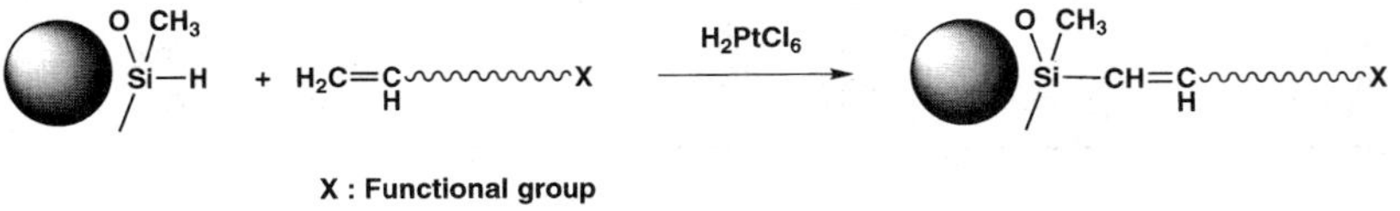

$$\text{Si–H} + H_2C{=}CH\text{∼∼∼X} \xrightarrow{H_2PtCl_6} \text{Si–CH}{=}CH\text{∼∼∼X} \qquad (11)$$

X : Functional group

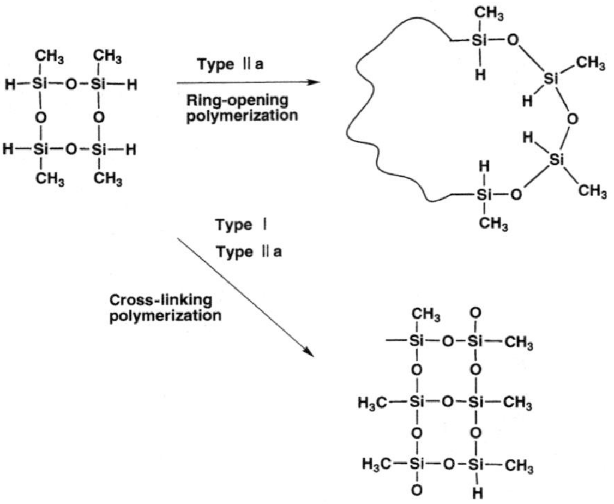

Type IIa
Ring-opening polymerization

Type I
Type IIa
Cross-linking polymerization

Scheme 12.1.7 Polymerization of polysiloxane in the modification of inorganic particles by a CVD method using 2,4,6,8-tetramethylcyclotetrasiloxane. (From Ref. 46.)

12.1.4.4 Properties of Polymer-Grafted Inorganic Particles

Polymer grafting onto inorganic particles ordinarily brings about extremely increased dispersibility in organic solvent, which is a good solvent for the polymer. However, the polymer-grafted particles lose dispersibility in a poor solvent, to give aggregation or precipitation. Polymer chains grafted onto the particles usually expand in a good solvent to prevent aggregation via steric interaction among polymer chains, but the chains shrink in a poor solvent, leading to aggregation due to decreasing of the steric

Table 12.1.2 Polymethylsiloxane Layer Deposited on Metal Oxides

Metal oxide	Weight increase (mg/g)	Specific surface area (m^2/g)	Thickness of layer (nm)
Silica	230	335.0	0.69
Titanium oxide	34	46.8	0.73
Zinc oxide	4	7.6	0.5
Iron oxide (black)	4	4.8	0.8
Iron oxide (red)	3	5.4	0.6

Source: Ref. 46.

repulsion. In this regard, the effects of grafting polymer chains onto silica composite particles, prepared by two-step modification, on the dispersibility in ethyl acetate–methanol cosolvent have been quantitatively investigated (40,48). The modification was first carried out using poly(maleic anhydride–styrene) to prevent aggregation and to make the binding site, and then the binding of amino-terminated secondary polymer, polymethacrylates, polyacrylates, and polystyrene to the composite particles was conducted in acetone [Reaction (12)].

P(MA-ST) / SiO_2 $\xrightarrow{H_2N\text{–Polymer}}$ (12)

H_2N–Polymer: $-(CH_2-CH(C_6H_5))_n-S(CH_2)_3NH_2$, $-(CH_2-C(CH_3)(COOCH_3))_n-S(CH_2)_3NH_2$, $-(CH_2-CH(COOCH_3))_n-S(CH_2)_3NH_2$

The surface polarity is found to be dependent on chain length of the secondary polymer, rather than on the number of binding polymer chains (40). From electron spin resonance (ESR) studies using spin-labeled poly(methyl methacrylate)- and polystyrene-grafted silica, the order parameter, *S*, which is an index of polymer chain flexibility in the solution, decreased with content of ethyl acetate in cosolvent, and increased with hexane content (48), as shown in Fig. 12.1.1. This result indicates that the secondary polymer chains are expanding in a good solvent-rich solution to make the particles disperse, and shrinking in a poor solvent-rich solution to aggregate via interparticle nonpolar interaction. Moreover, the ζ-potential of polymer-grafted silica can be also controlled by binding of amino-capped polyelectrolytes, polystyrenes having pendant ammonium or phosphonium groups, to the poly(maleic anhydride–styrene)-grafted silica in the range from -110 mV to $+65$ mV in ethanol (49).

The introduction of a temperature-response polymer onto inorganic particles leads to control of reversible dispersibility by changing temperature. The temperature-responsive silica particles were prepared by modification of monodisperse colloidal silica with iron(II)-bipyridyl branched poly(*N*-isopropylacrylamide) (50). The particles grafted with iron(II) complex were found to be self-aggregated above the low critical solution temperature (34°C) of the polymer, as shown in Figure 12.2.2. It has been proposed that aggregation of the composite particles takes place due to ligand exchange among the particles.

Ichimura et al. (51) have prepared silica particles showing photocontrolled reversible dispersibility in organic solvent. The silica particles incorporated spiropyran photochromic units, which showed reversible photoisomerization, were floccu-

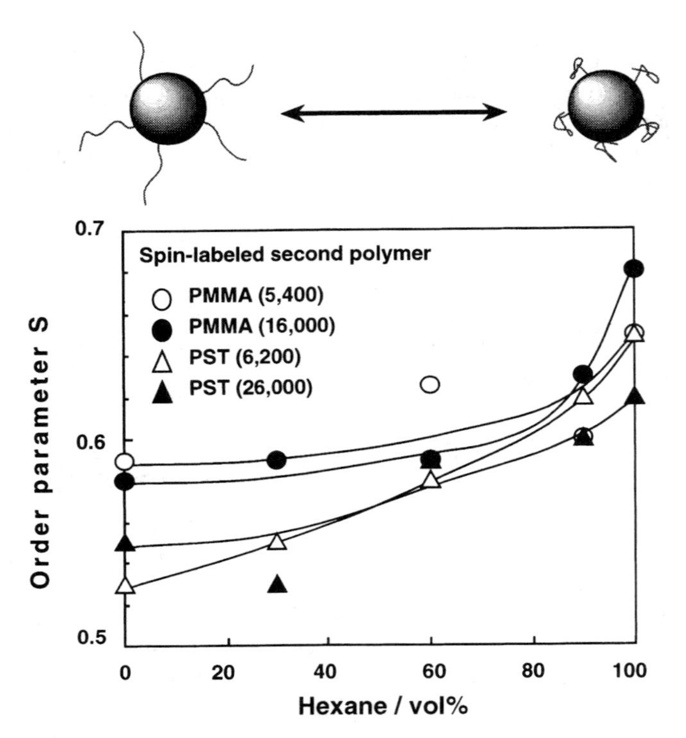

Fig. 12.1.1 Dependence of order parameter, *S*, of the secondary polymer, spin-labeled poly(methyl methacrylate) (PMMA) and polystyrene (PST), bound to poly(maleic anhydride–styrene)-grafted silica on hexane content in ethyl acetate–hexane cosolvent. Numbers in parentheses are number average molecular weight of the secondary polymer. (From Ref. 48.)

lated by ultraviolet (UV) irradiation to cause sedimentation, and were subsequently redispersed by visible light irradiation (Fig. 12.1.3).

Polymer-grafted inorganic particles have been also applied as biomaterial carriers. Shimomura et al. (52) have investigated the immobilization of glucoseoxidase on poly(acrylic acid)- or poly(acrylamide)-grafted silica or magnetite particles, in order to separate the enzyme from the reaction system for reusing. Also, maleic anhydride–styrene or –methyl methacrylate copolymer-grafted silica particles have been observed to efficiently immobilize bovine serum albumin (53,54), as shown in Figure 12.1.4. It is suggested that the protein-bound composite is applicable to immunological assay. Moreover, immobilization of enzymes on the silica grafted poly(maleic anhydride–styrene) markedly lowered their activities, in spite of effective immobilization. In this case, addition of hydrophilic secondary polymer of poly

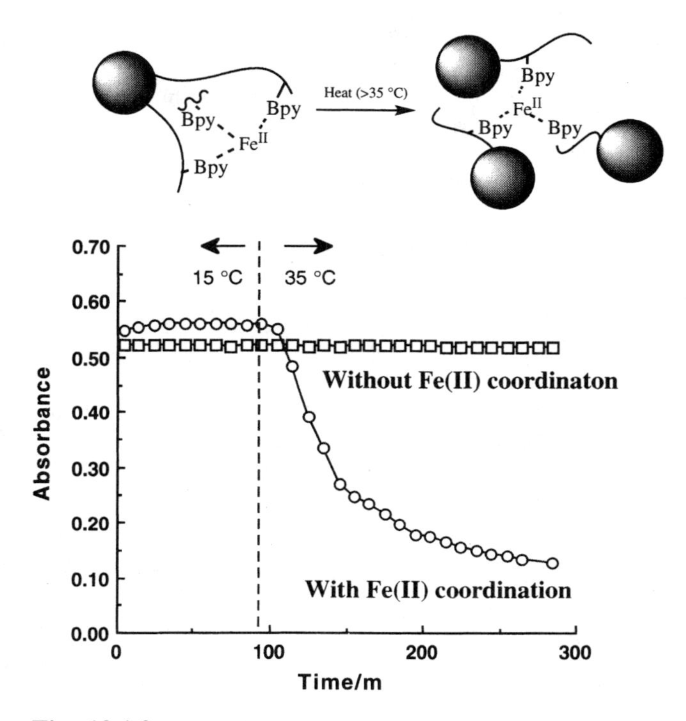

Fig. 12.1.2 Absorbance changes of the suspension (0.1 wt%) containing monodisperse colloidal silica modified with iron(II)-bipyridyl-branched poly(*N*-isopropylacrylamide). (From Ref. 50.)

(acrylic acid) to the polymer-grafted silica was found to increase the activity of α-chymotrypsin to attain 60% activity of the native enzyme (54).

12.1.5 Polymer Encapsulation of Inorganic Particles

In general, inorganic particles show hydrophilicity and have charge on the surface arising from the surface hydroxyl group. Therefore, if a surface-active compound having a polymerizable group could be adsorbed onto the inorganic particles via electrostatic interaction, consecutive polymerization would efficiently enable one to encapsulate the particles by polymer. The surface-active monomer has amphiphilic character, coming from a hydrophilic moiety and a hydrophobic tail group, in addition to a polymerizable group, such as a vinyl group. Therefore, the monomer can be adsorbed onto inorganic solid surfaces from aqueous solution to form a monolayer or bilayer at saturation of adsorption. Polymer-encapsulated inorganic particles can be obtained through the polymerization of the adsorbed surface-active monomer by heating or photoirradiation (55–60), as shown in Figure 12.1.5.

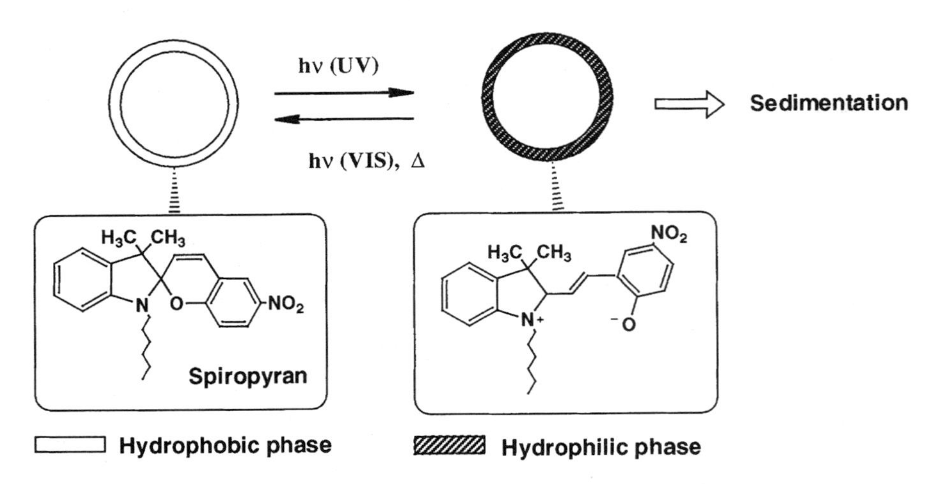

Fig. 12.1.3 Polarity change of colloidal silica surface by light stimulation. (From Ref. 51.)

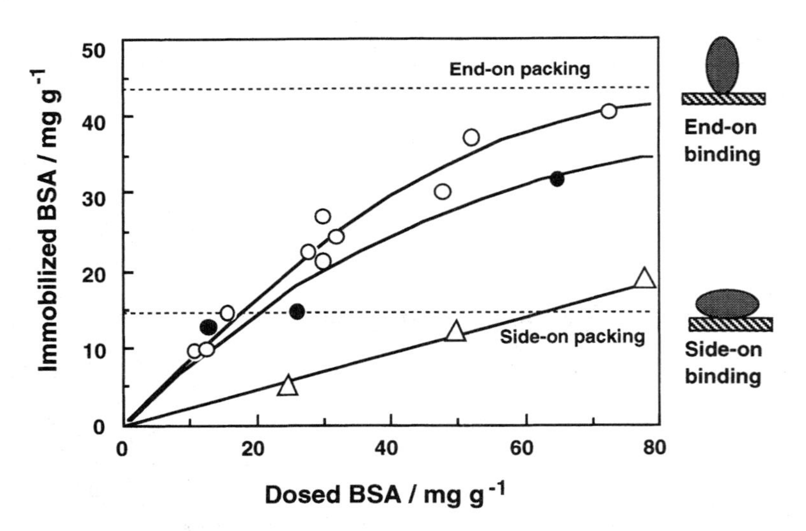

Fig. 12.1.4 Immobilization of bovine serum albumin on colloidal silica grafted with poly (maleic anhydride–styrene) (○), poly(maleic anhydride–methyl methacrylate) (●), and poly (ethylene glycol)–poly(maleic anhydride–styrene) (△). (From Ref. 42.)

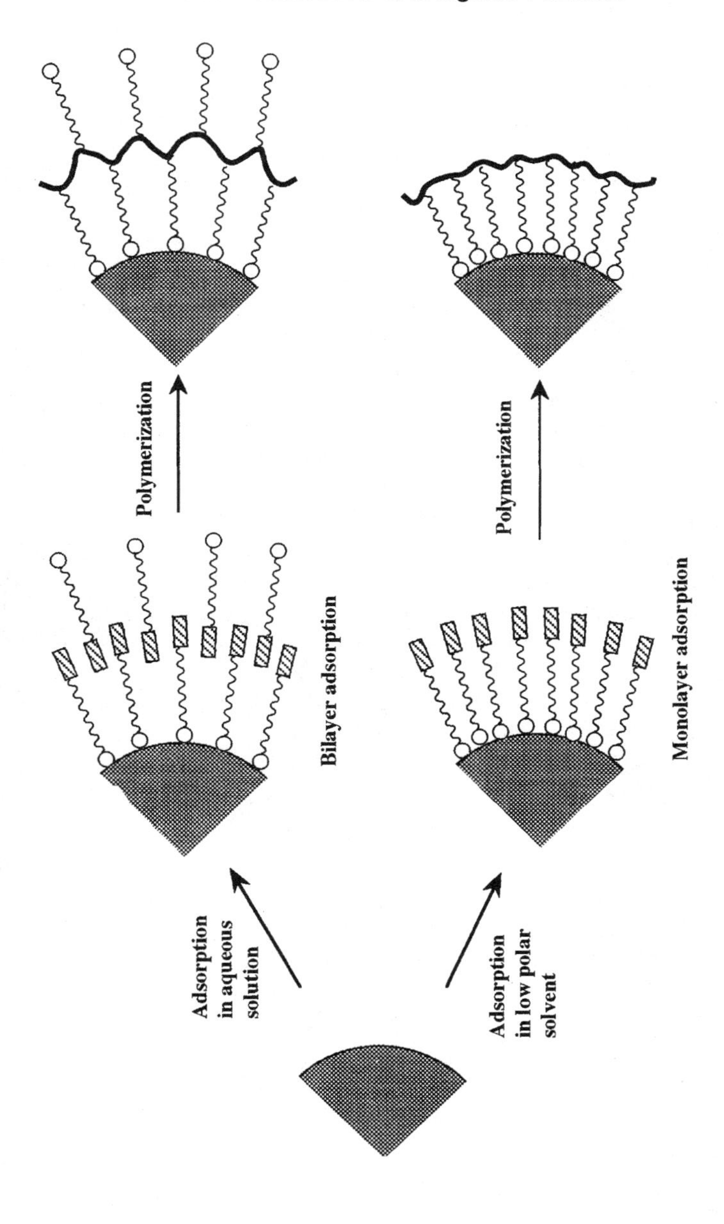

Fig. 12.1.5 Polymer encapsulation of inorganic particles by polymerization of adsorbed surface active monomer.

Polymer encapsulation of porous silica gel particles was achieved by the polymerization of a cationic surface-active monomer, *N,N*-dimethyl-2-methacryloyloxyethyleneammonium bromide, in the presence of the particles using 2,2′-azobis(isobutyramidine) dihydrochloride as an initiator (55,56). In this case, the adsorbed surface-active monomer was observed to be polymerized much faster than the micelle formed by the monomer in an aqueous phase. The enhancement of the polymerization rate might be related to a higher degree of orientation of the surface-active monomer adsorbed on the particles. Such a phenomenon has been also found in the polymerization of an anionic surface-active monomer, sodium 10-undecenyl sulfate, in the presence of alumina in aqueous solution (60). Moreover, addition of styrene to the monomer adsorbed onto alumina in aqueous solution gave copolymer-encapsulated particles. The polymerization of ionic surface-active monomers adsorbed onto an inorganic surface has been found to improve the dispersion stability of the particles in aqueous media (57). Furthermore, polymer-encapsulated silica particles were shown to exhibit good compatibility with a polymer matrix (56). On the other hand, successful polymer encapsulation of colloidal silica particles has been accomplished by the polymerization of cationic surface-active monomer in low polar organic media (61). The cationic surface-active monomer was adsorbed onto negatively charged colloidal silica particles in tetrahydrofuran or chloroform via electrostatic interaction. Successive heating of the organic suspension led to polymerization without an initiator, that is, spontaneous polymerization (62,63), of the particles. It is suggested that this is a promising procedure in surface modifications of inorganic colloidal particles having lesser amounts of surface hydroxyl group.

12.1.6 Polymer-Protected Nanoparticles

Nanometer-sized particles of metals and semiconductors have been intensively investigated in recent years, because of their size-dependence properties, quantum size effects (64), and the possibility of arranging them in microassemblies. The nanometer-sized particles have generally high surface energy, so that the particles must be stabilized in terms of preventing agglomeration by adsorption of polymer or surfactant. In most cases, nanometer-sized metal colloids are prepared by the reduction of metal ion in the aqueous solution containing water-soluble polymer or surfactant, as shown in Scheme 12.1.8.

Scheme 12.1.8 Formation of nanometer-sized metal particles protected by polymer.

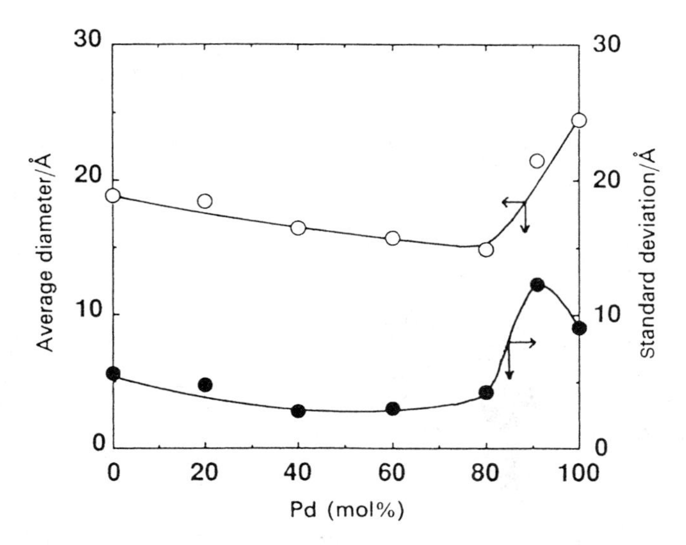

Fig. 12.1.6 Average diameter (○) and standard deviation (●) of palladium–platinum bimetallic clusters as a function of the palladium content. (From Ref. 68.)

The nanometer-sized palladium colloids were prepared by refluxing an alcohol–water solution of palladium ions in the presence of soluble polymer, poly(vinyl alcohol) (65) and poly(*N*-vinyl-2-pyrrolidone) (66), and the particles were shown to exhibit highly catalytic selectivity in the hydrogenation of dienes to monoenes (67). Interestingly, palladium–platinum bimetallic clusters, of 2–3 nm size, obtained by refluxing an alcohol–water solution of $PdCl_2$ and H_2PtCl_6 in the presence of poly(*N*-vinyl-2-pyrrolidone), exhibited highly catalytic activity (68). It has been shown that the cluster particle size is independent of respective metal ion concentration in the solution of less than 80 mol% Pd (Fig. 12.1.6), and the cluster particle showing highest catalytic activity has a Pd-surrounded Pt core structure (Fig. 12.1.7).

Stable palladium colloid was also prepared by the in situ reduction of $PdCl_2$ in the presence of protective water-soluble polymers and cationic polymer electrolytes. The palladium particles were in the nanometer size range and showed high catalytic activity in the hydrogenation of cyclohexene (69). Nanometer-sized, hydrophobic and oleate-stabilized silver organosols were obtained by reduction of $AgNO_3$ with $NaBH_4$ and the successive solvent exchanging method (70). The silver colloids capped by oleate are highly stable, and can be separated from the solution by evaporation. The particles can be stored in dried powder form and redispersed in various organic solvents. Recently, Akashi et al. (71) reported that poly(*N*-isopropylacrylamide)-protected platinum nanoparticles in 1–3 nm size, prepared by reduction of an alcohol–water solution of H_2PtCl_6, showed a different temperature dependence of

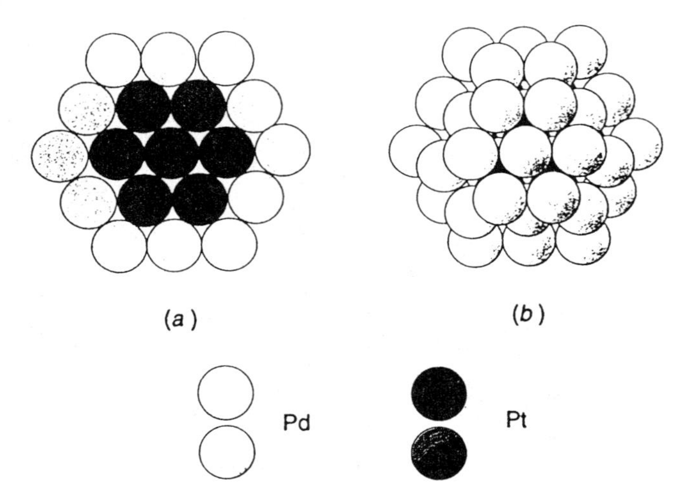

Fig. 12.1.7 Illustration of the (a) cross-section and (b) three-dimensional structure of the Pd-surrounded Pt-core model of Pd–Pt (4:1). (From Ref. 68.)

activity in the hydrogenation of allyl alcohol below the lower critical solution temperature (34°C) of the polymer from that over the critical temperature. Ziolkowski et al. (72) have also found out that TiO_2 nanoparticles capped with poly(styrenesulfonate) prevent the photodegradation of phenosafranin dye.

REFERENCES TO SECTION 12.1

1. LR Synyder, JJ Kirkland, JL Glyajch. Practical HLPC Method Development, 2nd ed. New York: Wiley-Interscience, 1997, Chap. 5.
2. R Raible, K Hamann. Adv Colloid Interface Sci 13:65, 1980.
3. N Tsubokawa. Prog Polym Sci 17:417, 1992.
4. K Yoshinaga, H Yoshida, Y Yamamoto, K Takakura, M Komatsu. J Colloid Interfaces Sci 153:207, 1992.
5. N Tsubokawa, A Kogure. J Polym Sci A Polym Chem 29:697, 1991.
6. JJ Fripiat, J Uytterhoven. J Phys Chem 66:800, 1962.
7. N Nishiyama, R Shick, H Ishida. J Colloid Interface Sci 143:146, 1991.
8. FD Blum, W Meesiri, H-J Kang, J Gambogi. J Adhesion Sci Technol 5:479, 1991.
9. SJ Monte, G Surgerman. Plast Rubber Mater Appl 117, 1978.
10. U Deschler, D Kleinschmit, P Panster. Angew Chem Int Ed Engl 25:236, 1986.
11. N Fery, R Laible, K Hamann. Angew Makromol Chem 34:81, 1973.
12. N Tsubokawa, A Kogure, K Maruyama, M Shimomura. Polym J 22:827, 1990.
13. N Tsubokawa, H Ishida. J Polym Sci A Polym Chem 30:2241, 1992.
14. N Tsubokawa, Y Shirai, H Tsuchida, S Handa. J Polym Sci A Polym Sci 32:2327, 1994.

15. K Yoshinaga, T Yokoyama, Y Sugawa, H Karakawa, N Enomoto, H Nishida, M Komatsu. Polym Bull 28:663, 1992.
16. K Yoshinaga, M Iwasaki, M Teramoto, H Karakawa. Polym Polym Composites 4:163, 1996.
17. N Tsubokawa, A Kogure. Polym J 25:83, 1993.
18. N Tsubokawa, H Ishida, K Hashimoto. Polym Bull 31:457, 1993.
19. S Spange, U Eismann, S Hohne, E Langhammer. Macromol Symp 126:223, 1997.
20. S Spange, and E Langhammer. Makromol Chem Phys 198:431, 1997.
21. E Dietz, N Fery, K Hamann. Angew Makromol Chem 35:115, 1974.
22. N Tsubokawa, A Kogure, Y Sone. J Polym Sci Polym Chem 28:1923, 1990.
23. H Kazama, Y Tezuka, K Imai. Polym Bull 21:31, 1989.
24. Y Osada, AT Bell, M Shen. J Polym Sci Polym Lett Ed 16:309, 1978.
25. Y Iriyama, S Ikeda. Polym J 26:109, 1994.
26. T Haraguchi, K Yamada, S Ide, C Hatanaka, Y Hatada, C Kajiyama. Polym Prep Jpn 44:2762, 1995.
27. K Burger, B Vincent. Eur Polym J 16:1017, 1980.
28. N Tsubokawa, K Maruyama, Y Sone, M Shimomura. Colloid Polym Sci 267:511, 1989.
29. KP Krenkler, R Laible, K Hamann. Angew Macromol Chem 53:101, 1976.
30. Y Chujou, E Ihara, T Saegusa. Macromolecules 22:2040, 1989.
31. R Yosomiya, K Horimoto, T Suzuki. J Appl Polym Sci 29:671, 1984.
32. S Nagae, Y Suda, Y Inaki, K Teramoto. J Polym Sci Polym Chem Ed 27:2593, 1989.
33. N Tsubokawa, A Kogure. J Polym Sci A Polym Chem 29:697, 1991.
34. ML Green, WE Rhine, P Calvert, HK Bowen. J Mater Sci Lett 12:1425, 1993.
35. K Yoshinaga, K Nakanishi. Composite Interfaces 2:95, 1994.
36. K Yoshinaga, R Horie, F Saigoh, T Kito, N Enomoto, H Nishida, M Komatsu. Polym Adv Technol 3:91, 1992.
37. K Yoshinaga, Y Hidaka. Polym J 26:1070, 1994.
38. K Yoshinaga, K Nakanishi, Y Hidaka, H Karakawa. Composite Interfaces 3:231, 1995.
39. K Yoshinaga, K Sueishi, H Karakawa. Polym Adv Technol 7:53, 1995.
40. K Yoshinaga, M Teramoto. Bull Chem Soc Jpn 69:2667, 1996.
41. A Kondo, T Urabe, K Yoshinaga. Colloid Surfaces A 109:129, 1996.
42. K Yoshinaga, T Kobayashi. Colloid Polym Sci 275:1115, 1997.
43. J Horn, R Hoene, K Hamann. Makromol Chem Suppl 1:329, 1975.
44. N Tsubokawa, S Yoshikawa. J Polym Sci A Polym Chem 33:581, 1995.
45. H Fukui, T Ogawa, M Nakano, M Yamaguchi, Y Kaneda. In: H Ishida, ed. Controlled Interphases in Composite Materials. New York, Elsevier Science, 1990, p 469.
46. H Fukui. Hyoumen 32:131, 1994.
47. FC Whitemore, FW Pietrusaza, LH Sommer. J Am Chem Soc 69:2108, 1947.
48. K Yoshinaga, J Shimada, H Nishida, M Komatsu. J Colloid Interface Sci 214:180, 1999.
49. T Kobayashi, H Karakawam, K Yoshinaga. Polym Prepr Jpn 46:4121, 1997.
50. K Yoshinaga, Y Sasao. Chem Lett 1111, 1997.
51. M Ueda, H-B Kim, K Ichimura. J Mater Chem 4:883, 1994.
52. M Shimomura, H Kikuchi, H Matsumoto, T Yamauchi, S Miyauchi. Polym J 27:974, 1995.
53. K Yoshinaga, K Kondo, A Kondo. Polym J 27:98, 1995.
54. K Yoshinaga, K Kondo, A Kondo. Colloid Polym Sci 275:220, 1997.

55. K Nagai, Y Ohishi, K Ishiyama, K Kuramoto. J Appl Polym Sci 38:2183, 1989.
56. K Nagai, H Kataoka, N Kuramoto. Kobunshi Ronbunshu 50:263, 1993.
57. K Esumi, N Watanabe, K Meguro. Langmuir 5:1420, 1989.
58. K Esumi, N Watanabe, K Meguro. Langmuir 7:1775, 1991.
59. K Esumi, T Nakao, S Ito. J Colloid Interface Sci 156:256, 1993.
60. DT Glatzhofer, G Cho, CL Lai, EA O'Rear, BM Fung. Langmuir 9:2949, 1993.
61. Y Yasuda, K Rindo, R Tsushima, S Aoki. Macromol Chem 194:1893, 1993.
62. K Nagai. Trends Polym Sci 4:122, 1996.
63. K Yoshinaga, F Nakashima, T Nishi. Colloid Polym Sci 277:136, 1999.
64. R Kudo. J Phys Soc Jpn 17:975, 1962.
65. H Hirai, Y Nakao, N Toshima. J Macromol Sci Chem A12:1117, 1978.
66. H Hirai, Y Nakao, N Toshima. J Macromol Sci Chem A13:727, 1979.
67. H Hirai, H Chawanya, N Toshima. Reactive Polym 3:127, 1985.
68. N Toshima, T Yonezawa, K Kushihashi. J Chem Soc Faraday Trans 89:2537, 1993.
69. ABR Mayer, JE Mark. Macromol Rep A33:451, 1996.
70. W Wang, S Efrima, O Regev. Langmuir 14:602, 1996.
71. C-W Chen, M Akashi. Langmuir 13:6465, 1997.
72. L Ziolkowski, K Vinodgopal, RV Kamat. Langmuir 13:3124, 1997.

12.2 SURFACE MODIFICATION OF POLYMER PARTICLES

HARUMA KAWAGUCHI
Keio University, Yokohama, Japan

12.2.1 Introduction

In most practical uses of polymeric particles, their surfaces play a very important role by taking part in interfacial interactions such as recognition, adsorption, catalytic reactions, etc. When we want to use polymer particles, we first check whether the chemical and physical structures of the surfaces meet the purpose. If some of them do not satisfy the criteria, we may seek other particles or try to modify the existing particles. This chapter mainly deals with the modification of surface of existing particles. In addition to chemical modification of particle surfaces, modification of the morphology of particles is also described.

The surface of polymer particles obtained by emulsion polymerization is occupied by emulsifier molecules, initiator fragments, and hydrophilic comonomer units. Therefore, desirable design of the surface ought to be done by choosing the emulsifier, initiator, and comonomer. Some of them are employed in aiming for postreaction at the surface to convert it into a functional one. When any change is necessary on the particle surface, modification of surface can be done by the following means:

1. Chemical reaction
2. Irreversible adsorption
3. Seeded polymerization

The strategy, methodology, and examples are presented next.

12.2.2 Chemical Modification

Emulsion polymerization of styrene using persulfate results in the formation of sulfate-group-carrying anionic particles if the pH for polymerization is kept not too low, that is, higher than 4. The sulfate group can be converted to a hydroxyl group when the dispersion is treated with acid. If the polymerization is carried out at a pH lower than 4, a significant fraction of the sulfate radicals and surface sulfates are hydrolyzed to hydroxyl radicals and hydroxy groups, respectively, during the polymerization (1). Carboxyl groups are also formed on the particle surface via oxidation caused by persulfate or other oxidizing reagents. Strong-acid-carrying particles can be thus reformed to weak-acid-carrying ones, and their electrophoretic mobility and dispersion stability can be controlled by the extent of reaction.

Soap-free emulsion polymerization of styrene and acrylamide forms monodisperse polystyrene core/polyacrylamide shell particles. The surface of the particle is completely covered with hydrophilic polyacrylamide. The amide groups can be converted to carboxylic acid via hydrolysis, and to amino groups via the Hofmann reaction using NaOCl and NaOH. Because the latter reaction is performed at a high pH, hydrolysis of the amide is accompanied by forming a carboxyl group, and, consequently, the resulting particle is an amphoteric one, whose isoelectric point ranges from 4 to 9, depending on the Hofmann reaction condition—that is, the reaction temperature, reaction time, and NaOCl/amide ratio as shown in Figure 12.2.1 (2). The amount of ionic groups formed by the Hofmann reaction increased with increasing NaOCl/amide, and the reaction at a lower temperature resulted in a higher ratio of amine to carboxyl groups on the particle. The reaction at higher temperatures formed more ionic groups released from the surface of particles. The Mannich reaction on the poly(styrene–coacrylamide) gives tertiary amine-containing particles. Thus, a variety of functional particles can be produced without changing particle size and distribution by the already mentioned modifications of single mother particles.

Treatment of polystyrene particles with dialdehyde caused the introduction of aldehyde groups onto the particles (3). The aldehydes are used as binding sites for functional compounds. Carboxyl groups on particles are also in common use as a

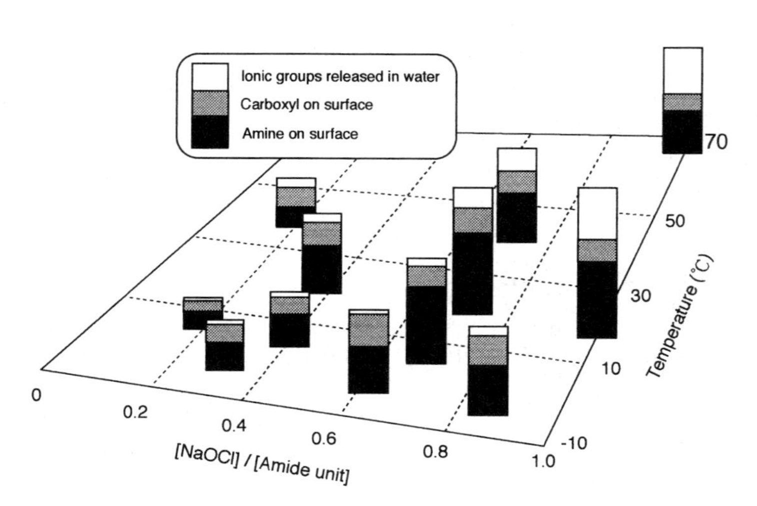

Fig. 12.2.1 Formation of ionogenic groups by the Hofmann reaction under different conditions. Full length of bar, total amount of ionogenic groups formed; solid part, amount of carboxyl group on surface; dotted part, amount of amine group on surface; open part, amount of ionogenic groups released to water.

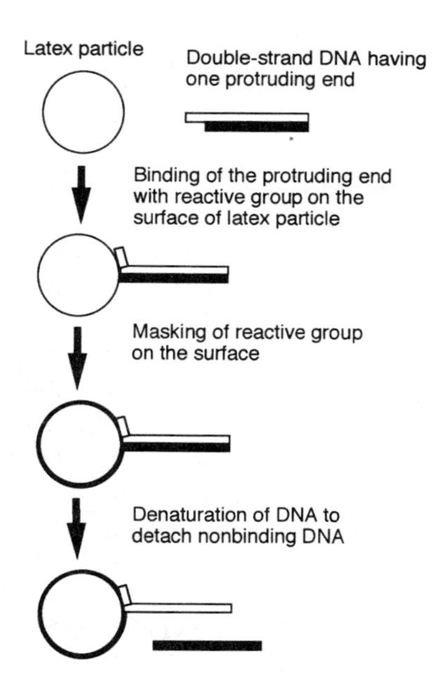

Fig. 12.2.2 Immobilization of single-strand DNA onto glycidyl-group-carrying particle at its chain end.

binding site. Before use, they are activated by carbodiimide and then reacted with amine in/on biomolecules or synthetic functional compounds.

Another possible binding site is epoxy or glycidyl groups, which easily react with amines (4) and thiols. Amino-group-containing molecules are often immobilized on particles through a coupling reaction with on-surface glycidyl groups, which are pendant groups of glycidyl methacylate (GMA). The emulsion or dispersion polymerizations of GMA must be performed carefully because glycidyl groups are cleaved by peroxide initiator during polymerization (5). But they are safe in the azo-compound-initiating system. GMA is a convenient monomer since its polymer forms a surface that does not suffer from nonspecific adsorption of proteins but has a binding site for immobilization of functional compounds. Therefore, polystyrene core/poly-GMA shell particles have been used for bioseparator and latex diagnosis. As to the immobilization of DNA, single-strand DNA can be attached onto poly-GMA particles at the chain end without including train- or loop-type attachment, as shown in Figure 12.2.2 (6). The procedure is;

1. Double-strand DNAs having one protruding and one blunt end are bound to a particle by the reaction between DNA amines and on-particle glycidyl groups.

2. Remaining glycidyl groups are masked.
3. Double-strand DNA is denatured to single-strand DNA. Exposed amines along DNA chains do not find reactive groups on the particle surface.

Several other methods are proposed, such as use of coulomb force between anionic DNA and cationic particle (7), use of biospecific affinity between biotin-labeled DNA and avidin-carrying particle (8), etc. Other methods for immobilization of DNA on the particle include reactions using cyanogen bromide, glutaraldehyde, and cyanuric chloride (9).

Thiol groups (-SH) can be introduced onto particle by the reaction of dithiothreitol (DTT) with on-surface glycidyl groups. The reaction accompanies the introduction of disulfide groups (-SS-) onto the particle surface, and the ratio of -SH to -SS- is dependent on the DTT concentration, reaction temperature, and reaction time. DTT-carrying particles have been used as a chaperone-like device, that is, to refold denatured proteins (10). Deactivated RNase, which has eight cysteines and so four -SS- groups in it, was refolded and recovered its activity when treated with DTT-carrying particles.

Here, active ester-carrying particles are introduced. Precipitation polymerization of acrylamide, methacrylic acid, and methylenebisacrylamide in ethanol results in the formation of monodisperse hydrogel particles with diameters in the range of 0.5 to 3 μm. Similar particles are obtained by replacing acrylamide with *p*-nitrophenyl acrylate (NPA) (11). The resulting particle (NPA particle) is a reactive one whose active ester can be converted to various functional groups as shown in Fig. 12.2.3. For example, amphoteric particles were obtained by treating the NPA particles with diamine. The volume of the particle became minimum at the pH at which amine and carboxyl groups neutralized each other. The pH did not necessarily coincide with the isoelectric point (iep) of the particle because the iep depends on the ratio of amine and carboxylic groups on the surface layer of particles, but the distribution of ionic groups in the modified NPA particle was not even.

Succinimide-group-carrying latex particles are another versatile reactive support (12). The fraction of active succinimide groups on the particle surface was measured to be about 1% of succinimide monomer charged for the heterogeneous copolymerization with diethylene glycol methacrylate.

A cationic latex particle of 60 nm diameter was modified with cobalt phthalocyaninetetrasulfonate (CoPcTs) via an ionic bond (13). The immobilized CoPcTs exhibited extremely high catalytic activity in autoxidation of 1-decanethiol.

Spacers are sometimes attached on particles prior to immobilization of functional components in order to eliminate steric hindrance from the functional components and make them mobile. Spacer molecules have to have a proper chain length and affinity with the dispersant but not with the particle. The latter condition is necessary to prevent the train conformation of spacer on the particle surface. Ethyleneglycol diglycidyl ether satisfies the prerequisite in water and is attached to amine-carrying particles.

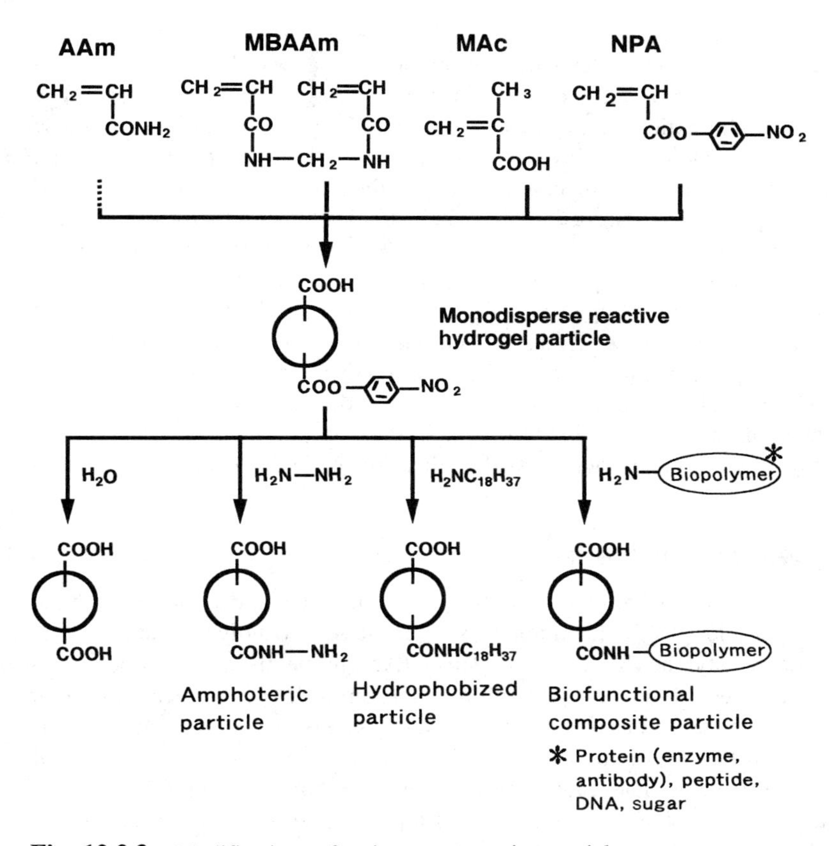

Fig. 12.2.3 Modifications of active ester-carrying particle.

12.2.3 Irreversible Adsorption

Dissolved polymer molecules can be adsorbed by polymer particles via electrostatic attractive force or hydrophobic interaction. When polyelectrolyte is adsorbed on an opposite-charge particle, the polymer molecules usually have a loop-and-tail conformation and, as a result, inversion of charge occurs. For example, sulfate-carrying particles behave as cationic ones after they adsorb poly(lysine). Then poly(-styrene sulfonate) can be adsorbed on such cationic particles and reinvert the charge of particles to anionic (14). Okubo et al. pointed out that the alternate adsorption of cationic and anionic polymers formed a piled layer of polyelectrolytes on the particle, but the increment of adsorbed layer thickness was much less than expected. This was attributed to synchronized piling of two oppositely charged polyelectrolytes (15).

Latex particles have been applied to agglutination test to detect a disease or disorder. Such latex particles carry an antibody or antigen, which catches its complementary antigen or antibody, respectively, relating with some diseases. The first latex for agglutination testing was proposed by Singer using polyvinyltoluene particles as a carrier (16). Poly(vinyltoluene) particles, highly hydrophobic polymers, adsorb antibody strongly and irreversibly. Hydrophilic polymers such as albumin and poly(ethylene glycol) are adsorbed on the same particles after antibody sensitization (immobilization). This is a necessary process in order to mask the polystyrene surface and prevent nonspecific adsorption on the particles.

In order to construct functional microspheres by modification of the surface with adsorbed proteins, e.g., enzymes and antibodies, the conformation and orientation of adsorbed proteins must be controlled to keep them as active as free proteins. If hydrophilic particles are used as a carrier, they hardly suffer nonspecific adsorption, but even antibody cannot be adsorbed. In this case, antibody is immobilized on the particles by chemical reactions such as those mentioned in the previous section (9).

12.2.4 Seeded Polymerization

Seeded polymerization is defined as ''polymerization of post-added monomer in the presence of particles.'' A conventional example of seeded polymerization is the polymerization of styrene (St) and acryronitrile (AN) in the dispersion of polybutadiene (PB) particles. In this polymerization, a portion of St and AN penetrates into the PB particle and the other portion stays in the aqueous phase. The former graft-polymerize with PB or polymerize by themselves in PB particle. The progress of polymerization in the particle leads to phase inversion to form a ''salami'' structured particle in which PB islands are dispersed in a poly(St-co-AN) sea. St and AN in aqueous phase may polymerize to form independent particles.

The already mentioned definition of seeded polymerization is in the broad sense of the word. Seeded polymerization in the narrow sense is ''polymerization after swelling the seed particle with monomer.'' Figure 12.2.4 shows what the seeded polymerization can form. Making larger particles is one of popular uses of seeded polymerization, but it does not belong to surface modification. Therefore, making larger particles is not discussed here.

12.2.4.1 Change of Surface Chemistry

Seeded polymerization using a slight amount of monomer leads to the surface modification without changing particle size. The resulting particles are a kind of core–shell particles or, more exactly, core–skin particles (Fig. 12.2.4C). Seeded polymerization of sugar-units-containing styrene derivative on polystyrene seed particle was carried out to obtain latex particles covered with sugar units (17). A necessary condition for this is that the monomer is more hydrophilic than the seed polymer. If not, the monomer permeates into the seed particle and only a small fraction remains on the

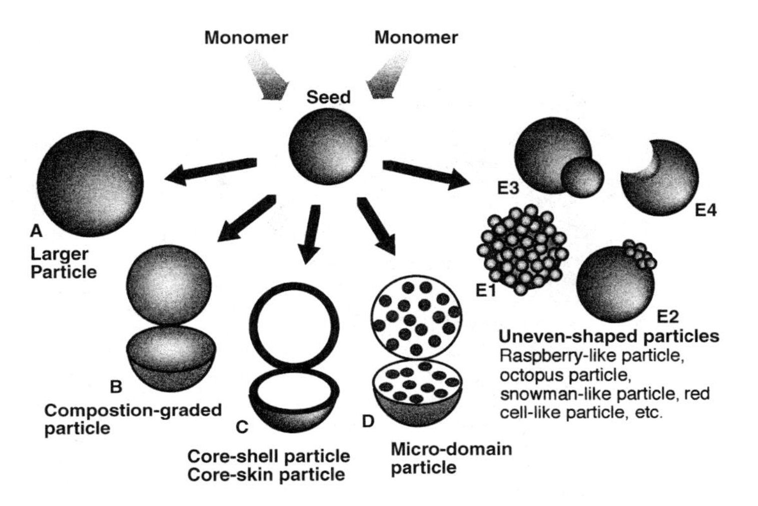

Fig. 12.2.4 Particles modified by seeded polymerization.

surface to cover it. On the other hand, if the monomer is too hydrophilic, the polymer is formed in the aqueous phase and does not contribute to surface modification. Monomers with moderate hydrophilicity form a particle having radially gradual composition (Fig. 12.2.4B). Such particles may be formed by one-shot (emulsifier-free) emulsion copolymerization. If it can, the emulsion copolymerization seems to be more operative than seeded polymerization. One of the advantages of the seeded method is that a series of particles with the same size but different functional group densities can be obtained by using the same seed particles.

A reactive surfactant shown next (RS) was used as a comonomer in a seeded polymerization. RS was easily adsorbed on seed particles due to their amphiphilicity. If dialkyl fumarate was preabsorbed in the particle, the polymerization proceeded quickly and resulted in the formation of skin layer of RS–fumarate copolymer. Because the vinyl group in RS is an allyl type, RS in aqueous phase hardly polymerizes and no water-soluble homopolymer was formed. The active ester group of RS on the skin layer was used for the preparation of functional microspheres (18).

$$CH_2{=}CH(CH_2)_6COO\text{-}Ph\text{-}\overset{\displaystyle CH_3}{\underset{\displaystyle CH_3}{\overset{|}{\underset{|}{S^+}}}}\ CH_3SO_4^-$$

12.2.4.2 Change of Interior Morphology

After monomer molecules permeate into seed particles, subsequent polymerization often results in phase separation because different kinds of polymers have little

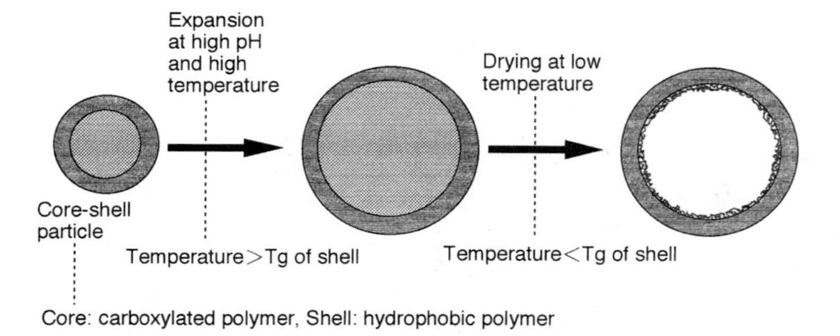

Fig. 12.2.5 Preparation of a hollow particle.

solubility in each other, even if monomer and seed polymer are compatible. The phase separation in the seed particles may result in microdomain structure or inverse core–shell particles (19). A typical example was described as ''salami'' in the previous section.

A kind of core–shell particle was employed for the preparation of a hollow particle (20). The procedure is shown in Figure 12.2.5. The starting material for the hollow particle is composed of carboxylic polymer core/poly(methacrylate) shell. The polymer particles are dispersed in alkaline solution at a temperature above the glass transition temperature (T_g) of the shell polymer. This procedure makes the particle highly swollen due to dissociation of carboxylic groups. Then the temperature is dropped down to a temperature much lower than the T_g and the particle is dried. The shell cannot deform during drying, and vaporization of water in the particle results in the formation of a hollow particle. The particles are light and have unique optical properties.

12.2.4.3 Uneven-Shaped Particle Formation

Phase-separated domains do not necessarily stay in the interior of particle. The domains may possibly appear on the surface and cause the formation of particles with nonsmooth surfaces. Especially when the seed particle is sufficiently cross-linked, severe phase separation takes place during seeded polymerization and brings about the formation of uneven-shaped particles. This happens due to squeezing out of newly formed polymers from cross-linked seed particles in the high-temperature polymerization. The driving force for squeezing is entropic elastics of network structure, which works when warmed, in addition to entropy loss accompanying polymerization (21). The force to suppress the projection is the interfacial energies between newly formed and existing polymers. If the newly formed polymer dislikes the seed polymer, the projection becomes appreciable. The uneven-shaped particles reported so far are raspberry-like particles [Fig. 12.2.4 E1: seed, poly(butyl acrylate) (BA);

monomer (M), St] (22), octopus-like particles [Fig. 12.2.4 E2: seed, St–BA–methacrylic acid terpolymer; M, St–BA] (23), snowman-like particles [Fig. 12.2.4 E3: seed, PSt; M, methyl methacrylate] (21), ice-cream-like particles (21), red-cell-like particles (Fig. 12.2.4 E4) (24), etc. A variety of morphology can be formed even if the kinds of seed particle and monomer are decided. The morphology strongly depends on the kind of initiator and the method for monomer charge, as well as the cross-link density of seed. The method of monomer charge affected the degree of swelling and monomer distribution in the seed particles.

St and divinylbenzene (DVB) were polymerized in a dispersion of acrylamide–methacrylic acid–methylenebisacrylamide terpolymer particles (25). Fine polystyrene particles were formed in/on each seed terpolymer particle. The former was smaller by about one-twentieth than the latter. The distribution of polystyrene particles depended on the cross-link density. Different amounts of St and DVB were charged in the seeded polymerization, and the resulting composite particles were used for protein adsorption measurement to assess the hydrophobicity of the particle surface. The adsorbed amount was almost proportional to the amount of St and DVB charged. In contrast, cells were less stimulated by the 5% St-containing particle than by the 0% St-containing one, that is, the seed particle. This phenomenon is attributed to selective protein adsorption on the 5% St-containing particle (26).

The red-cell-like particles in Fig. 4 E4 are prepared as follows (24);

First stage of polymerization: emulsion polymerization of styrene, hydroxyethyl methacrylate, and acrylic acid in the presence of isooctane.
Second stage of polymerization: seeded polymerization of already mentioned monomers and divinylbenzene.

In the second stage, cross-link formation triggers phase separation, and further cross-link fixes the unevenly localized and deformed shape of particle. Thus, the red-cell-like particles are prepared during polymerization and not formed during isooctane evaporation. The red-cell-like particles find practically use for controlling rheological and optical characteristics in the coating and paint industry.

12.2.5 Composite Particles

Some column-packed particles for chromatography are composed of silica particles covered with octadecyl chains to give them surface hydrophobicity. The alkyl groups are bound to the silica particles by use of coupling reagents. Carbon powders are sometimes masked with organic compounds for a special use—for example, to increase the dispersibility in organic solvents. The methodology for surface modification of carbon black with polymers was reviewed by Tsubokawa (27). The reactive surfactant described in the previous section was again used for the surface modification of carbon black.

Modification for inverse combination—that is, modification of polymer particle surface with inorganic materials—might meet more closely the scope of this section. One example is shown with the preparation of poly–St particles coated with magnetites. The composite particles were obtained by seeded polymerization of St in the presence of sodium oleate-modified magnetites powder and poly-St seed particles (28). Coating of dry inorganic powder on polymer particles in a hot blender, which is called a hybridizer, is another method to prepare composite particles.

An alternative method was developed to prepare magnetite-containing particles by modifying hydrogel particles (25). A proper amount of aqueous solution of ferric and ferrous chlorides was permeated into dry particles (about 1 μm diameter) of acrylamide–methacrylic acid–methylenebisacrylamide copolymer. When the particles were soaked in an ammonia solution, fine magnetites whose diameter was a few nanometers were precipitated in the hydrogel particles. The content of magnetites was 20% in weight. A large fraction of magnetites were distributed in the outermost layer of the particles but did not come out.

12.2.6 Others

Miscellaneous methods for surface modifications that were not taken up in the previous sections are presented in this section.

12.2.6.1 Growing Hair on the Surface

Surface grafting is another method to modify the particle surface. It mainly is useful in forming hairy particles. An example is shown of the preparation of particles bearing thermosensitive hairs. Glycidyl methacrylate (GMA)/St copolymer particles have reactive groups, glycidyl or glycol, on their surface. When the groups are reduced by ceric ions at room temperature, they offer radicals, which can be graft points for postadded monomer. Because no free radical is generated in the aqueous phase, polymerization proceeds exclusively on the surface if the chain transfer reaction is negligible. Poly(*N*-isopropylacrylamide) (PNIPAM) hair-carrying particles were thus prepared (29). They responded very sharply to temperature change compared with cross-linked PNIPAM-shell-carrying particles.

Hairy particles can be also prepared by several methods other than graft polymerization, as shown in Figure 12.2.6. This method has advantages in that it can offer a series of hairy particles having the same size core but different lengths and densities of hairs.

Some explanation could be added for the organized assembling in Fig. 12.2.6. The block copolymer is dissolved in a proper solvent. Two methods were proposed of preparing hairy nanoparticles from the solution;

1. A cast film was obtained in which an island/sea structure exists and the islands were cross-linked. Then they were dispersed in a good solvent.
2. Or, if a proper solvent is chosen, particulate assemblies of block copolymer can be obtained in it. They will be directly cross-linked. The length and density of hairs of particles were adjusted by postreactions. For example, long hairy particles could be converted to crew-cut particles.

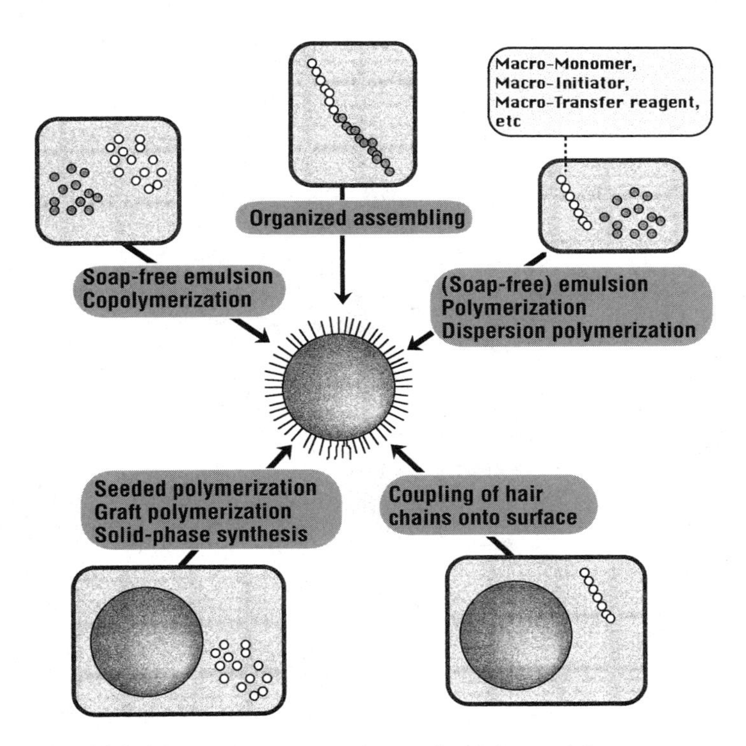

Fig. 12.2.6 Several methods of preparing hairy particles.

12.2.6.2 Dendrimers and Unsymmetric Particles

The dendrimer is a new type of cross-linked nanoparticle prepared by repetitive reactions to compose dendritic structure. There are two kinds of preparative methods, divergent (30) and convergent (31) methods, as shown in Figure 12.2.7. The dendrimer by the former method is constructed from the center of sphere to the outer layers, and therefore the surface chemistry is decided by the final reaction. When a dendrimer is formed from a central ammonia molecule, the forth-generation dendrimer of about 3 nm diameter has 48 functional groups at the surface. If a crowd of functional groups is not desirable on the outermost layer, they can be thinned down properly by replacing a two-functionality compound with a one-functionality one.

The convergent method takes an inverse way with the divergent method. The top surface is first prepared, and a coupling reaction of two semi-spheres is the final reaction. In this method, a unique unsymmetric nanoparticle can be obtained.

Unsymmetric particles have been prepared by several methods. Vapor deposition was used for their preparation. Gold (Au) was deposited first on the upper side

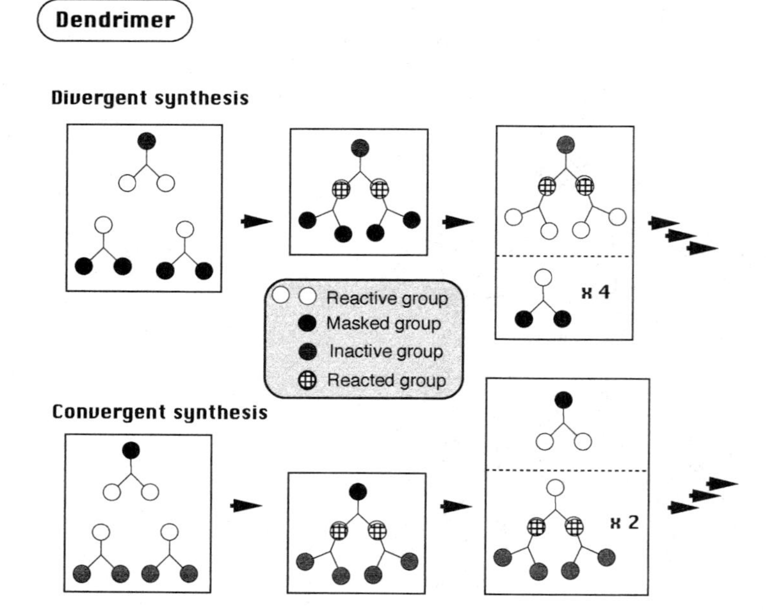

Fig. 12.2.7 Two routes to prepare dendrimers.

of particles arranged on a plate, and then thiol compounds were bound to the Au-covered portion of the surface (32). The various unsymmetric particles thus obtained can serve as probes that are sensitive to the gradient of the physicochemical microenvironment. Fujimoto et al. spread reactive microspheres over the surface of water and chemically modified the hemisphere facing the aqueous phase (33). When cationic nanospheres were loaded on the hemisphere of anionic particle, the composite particles were easily aligned in toluene under an electric field to exhibit electrorheology. If one side is hydrophilic and the other is hydrophobic, the particles can form a micellelike assembly.

12.2.7 Concluding Remarks

The surface of polymer particles is characterized by ionogenecity, hydrophilicity, softness, roughness, etc. The most prominent feature of the organic polymer particle, compared with inorganic materials, is that these characteristics can be easily modified. Various modifications were reviewed in this chapter to give a guide to those who intend to prepare functional polymer particles.

REFERENCES TO SECTION 12.2

1. SM Ahmed, MS El-Aasser, GH Pauli, GW Poehlein, JW Vanderhoff. J Colloid Interface Sci 73:388, 1980.
2. H Kawaguchi, H Hoshino, H Amagasa, Y Ohtsuka. J Colloid Interface Sci 97:465, 1984.
3. M Okubo, M Takahashi. Colloid Polym Sci 271:422, 1993.
4. E Zurkova, K Bouchal, D Zdenkova, Z Pelzbauer, F Svec, J Kalal. J Polym Sci Polym Chem Ed 21:2949, 1983.
5. T Nakamura, M Okubo, T Matsumoto. Kobunshi Ronbunshu 40:291, 1983.
6. Y Inomata, T Wada, H Handa, K Fujimoto, H Kawaguchi. J Biomater Sci Polym Ed 5:293, 1994.
7. T Delair, V Marguet, C Pichot, B Mandrand. Colloid Polym Sci 272:962, 1994.
8. M Uhlen. Nature 340:733, 1989.
9. H Kawaguchi. In: T Tsuruta et al., ed. Biomedical Applications of Polymeric Materials. Boca Raton, FL: CRC Press, 1993, pp 299–324.
10. H Shimizu, K Fujimoto, H Kawaguchi. Colloids Surf A Physicochem Eng Asp, 153: 421, 1999.
11. M Kashiwabara, K Fujimoto, H Kawaguchi. Colloid Polym Sci 273:339, 1995.
12. Y Morita, M Yoshida, M Asano, I Kaetsu. Colloid Polym Sci 265:916, 1987.
13. M Hassanein, WT Ford. Macromolecules 21:525, 1987.
14. S Sato, J Shirai, T Nashima, K Furusawa. Prepr 10th Polymeric Microspheres Symp, Fukui, November 1998, pp 143–144.
15. T Okubo, M Suda, S Morino, S Suzuki. Prepr 10th Polymeric Microspheres Symp Fukui, November 1998, pp 153–154.
16. JM Singer, CM Plotz. Am J Med 21:888, 1956.
17. MT Charreyre, P Boullanger, T Delair, B Mandrand, C Pichot. Colloid Polym Sci 271: 668, 1993.
18. K Takahashi, K Nagai. Polymer 37:1257, 1996.
19. S Lee, A Rudin. J Polym Sci Polym Chem 30:865, 1992.
20. Roam & Haas, US Patent 4,427,836, 1981.
21. YC Chen, V Dimonie, MS El-Aasser. J Appl Polym Chem 42:1049, 1991.
22. M Okubo, Y Katsuta, T Matsumoto. J Polym Sci Polym Lett 18:481, 1980.
23. M Okubo, K Kanaida, T Matsumoto. Colloid Polym Sci 265:876, 1987.
24. Mitsui Toatsu Kagaku, JP2-14222, 1990.
25. H Kawaguchi, K Fujimoto, Y Nakazawa, M Sakagawa, Y Ariyoshi, M Shidara, H Okazaki, Y Ebisawa. Colloids Surf A Physicochem Eng Aspects 109:147, 1996.
26. Y Urakami, Y Kasuya, K Fujimoto, M Miyamoto, H Kawaguchi. Colloids Surf B Biointerfaces 3:183, 1994.
27. N Tubokawa. Prog Polym Sci 17:417, 1992.
28. J Lee, M Senna. Colloids Polym Sci 273:76, 1995.
29. H Matsuoka, K Fujimoto, H Kawaguchi. Gels Polym Networks, in press.
30. DA Tomalia. Adv Mater 6:529, 1994.
31. JM Frechet. Science 263:1710, 1994.
32. H Takei, N Shimizu. Langmuir 13:1865, 1997.
33. K Fujimoto, K Nakahama, M Shidara, H Kawaguchi. Prepr 10th Polymeric Microspheres Symp, Fukui, November 1998, pp 121–124.

13

Particles of Specific Functions

MASATAKA OZAKI, MAMORU SENNA, MASUMI KOISHI, and
HIROTAKA HONDA

13.1 FORMATION OF MAGNETIC PARTICLES

MASATAKA OZAKI
Yokohama City University, Yokohama, Japan

13.1.1 Size of Magnetic Particles and Magnetic Properties

When a magnetic material such as magnetite or iron is placed in a magnetic field, it is magnetized, as shown in Figure 13.1.1, with the change in the applied magnetic field strength. Bulk material or powder is not magnetized originally unless it is exposed to a magnetic field, since the original directions of the magnetic dipoles in magnetic domains of bulk material or in powder are random. The magnetic moment of each domain results from the spontaneous alignment of electron spins due to superexchange interactions between electrons. Thus, the magnetization of magnetic solid is brought about by the orientation of magnetic dipoles. As the magnetic field increases, the degree of orientation of the dipole increases and saturation magnetization is attained at the perfect orientation of the dipoles. Once the magnetic material is magnetized, even though the magnetic field has decreased to zero, the material will have a certain magnetization, the so-called residual or remanent magnetization designated by M_r. A magnetic field of opposite direction is required to bring the magnetization to zero. This magnetic field is called coercive force, frequently desig-

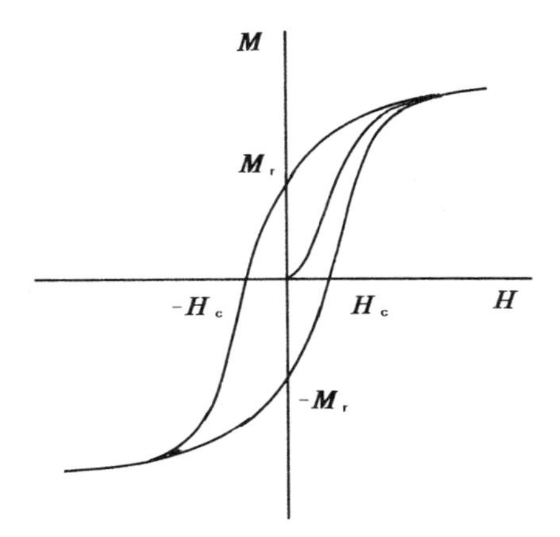

Fig. 13.1.1 Magnetization curve.

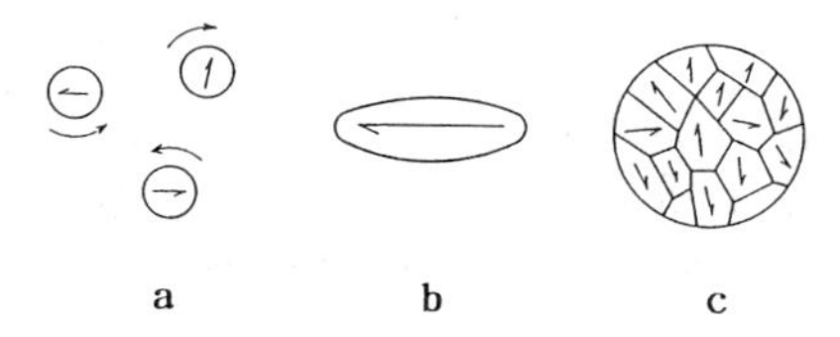

Fig. 13.1.2 Schematic picture of domain structures of magnetic particles: (a) superparamagnetic particles; (b) single-domain particle; (c) multidomain particle.

nated by H_c. Coercive force is essential for magnetic recording media as well as the remanent magnetization. If we remove the magnetic field with H_c, the magnetization of the material remains to some extent. A negative magnetic field slightly stronger than H_c is required to bring the magnetization to zero. This magnetic field is called remanent coercive force. The coercive force and the remanant coercive force are extrinsic, while the saturation magnetization is intrinsic.

As the size of the particle decreases, the number of the magnetic domains in the particle decreases, changing the domain structure from multidomains to a single domain as illustrated in Figure 13.1.2. Generally, the H_c of multidomain particles is smaller than that of single-domain particles, since the rotation of the magnetic moment in the former particles occurs easily from domain walls. The coercive force of a single-domain particle is determined by magnetocrystalline anisotropy together with the shape anisotropy. The H_c due to the shape anisotropy increases with the increase of the aspect ratio if the particle size remains the same. Therefore, elongated single-domain particles are preferentially employed for magnetic recording media. As we show later in this chapter, the particles available for magnetic media must be single magnetic domain particles having high saturation magnetization and proper coercive force.

If the particle becomes sufficiently small, the magnetic moment shows no preferential orientation due to the thermal agitation exhibiting superparamagnetic property. Such particles have very small coercive force and are not useful for the magnetic recording material. Thus, small magnetic particles having superparamagnetic nature are constituents of magnetic fluids.

In a stable dispersion of fine magnetic particles, particles are free to rotate. The magnetic field H, applied to a magnetic particle having a permanent magnetic moment m, will cause orientation along the direction of the field. The orientation may be disturbed by thermal agitation. The degree of the orientation can be related by the Langevin equation,

$$m_{av} = m(\coth \alpha - 1/\alpha) \tag{1}$$

where m_{av} is the average magnetic moment and $\alpha = mH/kT$, k being the Boltzmann constant and T the absolute temperature. Therefore the magnetization of a stable dispersion of magnetic particles can be given by

$$M = nm_{av} \tag{2}$$

where n is the number of the particles per volume. Thus, the dispersion of magnetic particles will show a magnetization curve similar to that of paramagnetic material.

The attractive magnetic energy, $U_A{}^{Mag}$, between two magnetic dipoles is expressed by

$$U_A^{Mag} = \frac{1}{4\pi\mu_o r^3}\,[\boldsymbol{m}_1 \cdot \boldsymbol{m}_2 - 3(\boldsymbol{m}_1 \cdot \boldsymbol{r}/\mathrm{r})(\boldsymbol{m}_2 \cdot \boldsymbol{r}/\mathrm{r})] \tag{3}$$

where $\boldsymbol{m}_1$ and $\boldsymbol{m}_2$ are the magnetic dipole moments of the particles, $\boldsymbol{r}$ is the vector joining the centers of the two dipoles, and μ_o is the magnetic permeability of the vacuum (1). The magnetic dipole moment, assuming a single-domain particle, is expressed as $m = I_0 v$ with I_0 being the magnetization per unit volume of the particle and v its volume. When two dipoles, $\boldsymbol{m}_1$ and $\boldsymbol{m}_2$, are oriented in the same direction on the same line, Eq. (3) can be written as

$$U_0 = -\frac{m_2 m_2}{2\pi\mu_o r^3} \tag{4}$$

where U_0 is the minimal potential energy of the system. The magnetic interaction energy increases with the second power of its volume, i.e., proportional to the sixth power of their radius. Therefore, the magnetic interaction energy increases rapidly as the size of the particle increases. The interaction between particles of strong magnetic material such as iron or magnetite is very strong, and stable colloidal dispersion is difficult to obtain unless the particle size is small, which constitutes superparamagnetic particles. The magnetic interaction energy between those particles will be reduced since the dipole moment in such particles has no preference for a particular orientation, due to free rotation of the magnetic moment by the thermal agitation. Accordingly, the average interaction energy $\langle U \rangle$ is expressed in

$$\langle \mathrm{U} \rangle / kT = -x^2/6 + 7x^2/1800 + 1.54 \times 10^{-4} x^6 \tag{5}$$

by Scholten and Tjaden (2) for small values of x, where $x = U_0/kT$. The first term of this equation gives the Keesom type interaction energy, $-U_0{}^2/12kT$. For large negative values of x,

$$\langle U \rangle / kT = x + 2 - 4x^{-1}/3 + 56x^{-2}/9 \tag{6}$$

This is identical to Eq. (4) in the limiting solution of the strong interaction.

The magnetic interaction energy increases if the interaction works between multiparticles. The multi-interaction becomes important in a concentrated dispersion (3).

13.1.2 Principles of the Formation of Monodispersed Magnetic Particles

In using fine magnetic particles, particle size is the most important parameter, as well as other qualities such as crystallinity and composition, since the magnetic properties of the particles are strongly influenced by them. Therefore, it is advantageous that the particles formed are uniform in their size and shape, i.e., monodisperse or of narrow size distribution in desired size ranges. The preparation method of fine particles having a narrow size distribution has been investigated by colloid chemists for a long time (4). The work by LaMer gave us some basic principles for the formation of monodispersed particles (5). According to the theory, to produce colloidal particles having a narrow size distribution, the nucleation and growth process must carefully be controlled. The conditions necessary for the formation of magnetic particles are essentially the same as for nonmagnetic particles, but some special precautions are necessary because of strong magnetic interactions among the particles. In producing monodispersed particles, the essential parameters are: 1) separation of nucleation process from growing process, 2) protection of particles from aggregation, 3) controlled supply of precursor material, and 4) temperature and pH in the solution. These parameters are in intimate relations to each other, and sometimes it is difficult to separate them. Thus, concentration of starting material, reaction temperature, and pH in the solution must be optimized. Dispersion reagents such as surfactants and polymers must be carefully chosen. Continuous supply of precursor material is sometimes made by controlled decomposition of other material. Transformation from nonmagnetic particles to magnetic particles is utilized to prepare particles for magnetic recording media. The details of the mechanism for the production of monodisperse fine particles are described by Sugimoto (6).

13.1.3 Nanometer-Scale Magnetic Particles and Magnetic Fluids

The properties of nanometer-scale magnetic particles differ from those of larger particles or differ from those of the same bulk material. Nanometer-scale particles may have some characteristic behavior due to quantum effects. Therefore the production of nanometer-scale magnetic particles having narrow size distribution has significant importance for scientific purposes as well as practical use for magnetic fluids. The difficulty in producing nanometer-scale particles is that these particles have a tendency to aggregate in order to minimize the surface energy.

It has been known that small magnetite particles can be prepared by precipitation from an aqueous solution containing ferric and ferrous ions in the following reaction:

$$2Fe^{3+} + Fe^{2+} + 8OH^- \rightarrow Fe_3O_4 + 4H_2O$$

The magnetite particles so produced were stabilized by protective colloids such as gelatin for use in magnetic domain observations. These precipitated magnetite particles are used for production of magnetic fluids, as we describe later.

Thomas et al. succeeded in forming fine cobalt particles with the sizes 2–30 nm, having narrow size distributions, by decomposing cobalt carbonyl compounds in the presence of suitable surface-active reagents (7). By the same procedure, iron and nickel particles were prepared. Papirer et al. studied the mechanism of the decomposition of a toluene solution of $Co_2(CO)_8$ precisely in the presence of a surface-active reagent, and found that at least two factors are responsible for the formation of particles having an extremely narrow size distribution: the division of the system into microreactors, and a diffusion-controlled growth mechanism of the individual particles (8).

Platonova et al. reported a preparation method of Co nanoparticles having good dispersibility using block copolymer (polystyrene–poly-4-vinyl piridine) micells where Co was generated by the reduction of micells loaded with $CoCl_2$ and by thermal decomposition of $Co_2(CO)_8$ in micellar solutions of the block copolymers (9).

Konno et al. prepared magnetite particles by using microemulsion method (10). The procedure was the mixing of water/isooctane or water/cyclohexane microemulsion with aqueous $FeCl_3$ and aqueous NH_3, followed by addition of aqueous $FeCl_2$ with vigorous stirring. In this procedure, Aerosol OT was necessary as a surfactant in order to solubilize adequate amounts of $FeCl_2$ in the hydrocarbon used. In using the microemulsion method, the size of the magnetite particle could be controlled by adjusting the size of water droplets in the microemulsion and the contents of ferric and ferrous ions.

Co nanocrystallines were also prepared by Pileni et al. using reverse micelles as a microreactor system. They examined magnetic properties of the superlattices formed from the Co crystallines and found the two-dimensional hexagonal networks of the self-organized Co particles showed collective magnetic properties of the bulk material (11). A variety of nanometer-scale magnetic particles have been developed by Pileni and her coworkers. Figure 13.1.3 shows an electron micrograph of nanometer-scale Co particles obtained by them (12).

Unilamellar vesicles have been used as a reactor for the synthesis of nanosmeter-scale magnetic particles (13,14). By adding alkaline solution to vesicles containing intravesicular solutions of Fe^{2+} and Fe^{3+}, the Fe^{2+}/Fe^{3+} resulted in the formation of membrane-bound discrete particles of different ion oxide particles. These results together with the particle formation in microemulsion are not only of interest in colloid chemistry but also have significance in mineralization in biosystems, such as magnetotactic bacteria, where particles are formed within enclosed organic compartments.

Magnetoferritin consists of maghemite particles of an average diameter of 10 nm encaged within a roughly spherical protein shell. It is attracting the interest of researchers because of its narrow size distribution with nanometer-scale sizes (15).

Small metal particles can also be obtained by vacuum evaporation in low pressure inert gases (16). Magnetic particles of metals such as iron, cobalt, nickel, and alloys of these metals can be prepared by this method. Though the amounts of particles obtainable by this method are limited, the particles are clean as compared with particles precipitated from solutions. They are mainly used for studies of physical properties of fine particles.

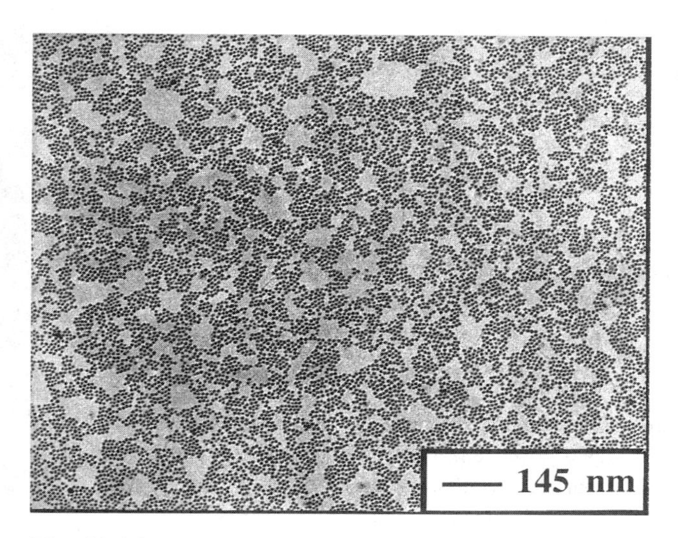

Fig. 13.1.3 Transmission electron micrograph of mondisperse nanometer-scale cobalt particles. (Courtesy of Dr. Pileni, Université Pierre et Marìe Curie.)

Coprecipitation has been used for the production of ferrite particles. According to Tang et al., $MnFeO_4$ particles of relatively small size (5 to 25 nm) can be obtained through the reaction

$$MnCl_3 + 2FeCl_3 + xNaOH \rightarrow MnFeO_4 + (x - 8)NaOH + 4H_2O + 8NaCl$$

for $x > 8$ ($[Me]/[OH^-] < 0.375$). For $Mn_xFe_{3-x}O_4$ ($0.2 < x < 0.7$), particles of bigger sizes up to 180 nm are produced from ferrous salts, where [Me] is the concentration of metal ions (17). In either case, the particle size appeared to be a unique function of the ratio of metal ion concentration to hydroxide concentration. Various ferrites can be obtained by similar reactions.

As stated earlier, historically, dispersions of small magnetic particles having superparamagnetic nature were used for the observation of magnetic domains. When a concentrated suspension of magnetic particles is placed in the gradient of a magnetic field, forces act on the particles and the magnetic interaction between them is enhanced. The magnetic field attracts particles, and the dispersing liquid moves together with the particles. Therefore the liquid behaves as if it is magnetized. Such a behavior is typical of a stable dispersion containing small magnetic particles. In 1960s after finding the use of concentrated dispersions of magnetic particles in space technology, a magnetic dispersion called ''magnetic fluid'' or ''ferro-fluid'' was introduced as a new material. Since then, it has been finding a variety of applications (18,19). Presently, magnetic fluids are employed in some industrial technologies

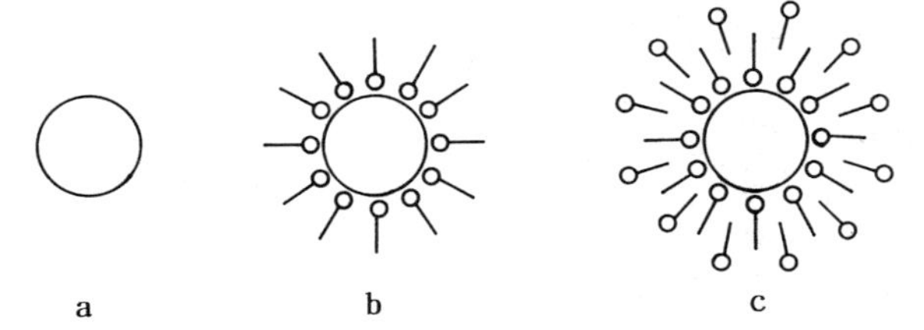

Fig. 13.1.4 Schematic picture of dispersion states of particles in magnetic fluids: (a) core particle; (b) oil-based magnetic fluid; (c) water-based magnetic fluid.

such as sealing of computer disk units. For example, lubricant oil in which magnetic particles are dispersed works as a good sealing material when it is suspended by a magnetic field. The magnetic particles remain in the magnetic field together with the lubricant oil, which works as a sealing material.

The first magnetic fluid was prepared by milling magnetic material in a nonpolar organic liquid in the presence of dispersion reagent. Concentrated dispersions of magnetic particles were also independently prepared by Shimoiizaka et al. using precipitated magnetite particles (20). According to their method, the magnetite particles were obtained by adding alkaline solution into an aqueous mixture solution containing Fe^{2+} and Fe^{3+} ions followed by adsorption of oleic acid. The magnetic particles covered by oleic acid were filtered and dispersed in an organic liquid to form a stable dispersion of magnetic particles. A schematic picture of the dispersed states of the particles in magnetic fluids is shown in Figure 13.1.4.

Since the magnetic fluid appeared, techniques for the production of variety of magnetic fluids including water-based and oil-based magnetic fluids were developed. Recently a number of magnetic fluids prepared by a variety of methods are on markets (19).

13.1.4 Magnetic Particles for Magnetic Recording Media

The original magnetic recorder invented by Poulsen one hundred years ago used ion wires. Extensive efforts have been made to produce better magnetic particles since particulate materials were employed for the magnetic recording system shortly before World War II (21–24). Originally, the use of magnetic particles was limited for audio tapes, but presently they are employed for many recording media, not only for audio and video recording systems but also for a variety of computer memories in the form of tapes, flexible and rigid disks, and cards. With the progress of these systems, magnetic particles having better recording performance have been pursued.

In order to ensure enough output signal strength and reliability of the storage of the signal, particulate magnetic recording materials are required to have 1) large saturation magnetization, 2) high coercivity, 3) good dispersibility and orientability, 4) narrow distributions in size and shape, and 5) high chemical and mechanical strength. For high-density recordings, the size of the particle must be small but not be too small, such as superparamagnetic particles. There are not so many materials that can satisfy all these demands.

After the finding that elongated maghemite (γ-ferric oxide, γ-Fe_2O_3) particles showed a superior magnetic property for recording performance, maghemite particles attracted the interest of recording tape manufacturers. After that, maghemite particles were the most popular material until cobalt-modified maghemite appeared. Recently, iron particles are employed for 8-mm videocassette tapes, still camera disks, high-quality audio tapes, and various digital recording systems (25). Much attention has been focused on fine barium ferrite particles for perpendicular recording materials as well as for use in videotapes to improve the recording property (26,27).

13.1.4.1 Acicular γ-Ferric Oxide Particles

Since γ-ferric oxide has cubic structure, it cannot be formed in elongated shape directly. Therefore, the particles are prepared from another kind of compound by solid-phase transformation. Though γ-ferric oxide can be prepared from either β-FeOOH or γ-FeOOH, most commercially available particles are produced from α-FeOOH (goethite) (21,28). The original shape and size of α-FeOOH particles are kept almost the same throughout the dehydration and reduction process followed by oxidation. For high-density recording, particles must be small, and particles are made to have a large aspect ratio to achieve large H_c. Therefore, the size and shape of the starting α-FeOOH particles must be carefully controlled.

First, seed particles of goethite are prepared by the reaction

$$Fe^{2+} + OH^- + \tfrac{1}{2}O_2 \rightarrow \alpha\text{-FeOOH}$$

The seed particles are then grown to desired sizes in the presence of iron and ferrous ions. Second, the goethite particles are dehydrated by heating, then reduced by hydrogen into magnetite, and finally oxidized into maghemite through the reactions

$$\alpha\text{-FeOOH} \xrightarrow[\sim 300^\circ C]{} \alpha\text{-Fe}^2O_3 \xrightarrow[300\text{–}400^\circ C]{} Fe_3O_4 \xrightarrow[\sim 250^\circ C]{} \gamma\text{-}Fe_2O_3$$

The temperatures and the length of the reduction and oxidation time are optimized and carefully controlled. To achieve good dispersibility and good orientability, surfaces of the particles are modified by chemical procedures after or prior to the reduction. The γ-ferric oxide particles on the market are 0.2–0.5 μm in length with aspect ratios of about 10. An electron micrograph of typical γ-ferric oxide particles is shown in Figure 13.1.5.

When α-FeOOH particles are used for the starting material, it is very difficult to avoid pore formation completely in the dehydration process, which necessitates

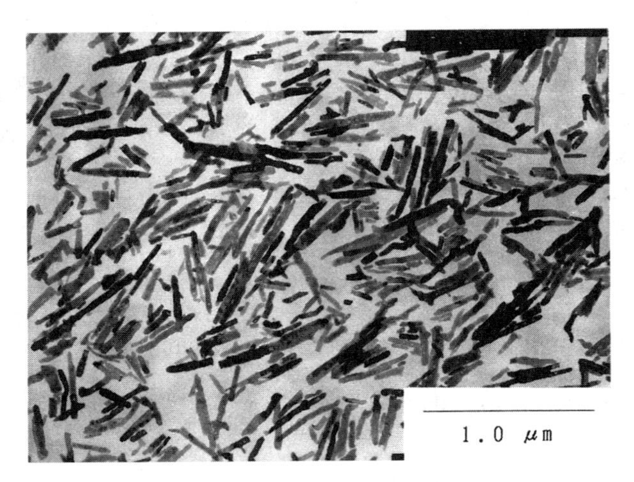

Fig. 13.1.5 Transmission electron micrograph of typical maghemite (γ-Fe_2O_3) particles. H_c = 360 Oe. (Courtesy of Dr. Horiishi of Toda Kogyo Ltd.)

heating at high temperatures to decrease pores. It was expected that nonporous particles could be formed if elongated hematite particles were used for the starting material. A successful method for the production of elongated hematite particles was developed by Matsumoto et al. using hydrothermal reactions (29) in the presence of phospho-organic compounds. The maghemite particles prepared from the hematite particles were poreless and were more ellipsoidal than other magnetic particles. According to Corradi et al., the good magnetic properties of the solids could be accounted for by the fact that the magnetic field lines in the particles were more parallel than in other particles (30). They called the particles ''non-polar particles (NP-particles).'' Maghemite particles having good magnetic properties could also be produced from spindle-type hematite particles prepared by forced hydrolysis of ferric chloride solutions in the presence of small amounts of phosphate ions (31). Maghemite particles of improved quality were also produced from acicular hematite particles prepared by doping tin in the hydrothermal transformation of precipitated $Fe(OH)_3$ (32).

13.1.4.2 Cobalt-Modified γ-Ferric Oxide Particles

According to the theory of magnetism, H_c of γ-ferric oxide particles larger than 500 Oe is obtainable, but the highest H_c of the solids reported is 450–470 Oe (33). It was very difficult to increase the H_c higher than 500 Oe. However, magnetic particles having higher H_c were required for high-recording-density recording media. It was well known that γ-ferric oxide particles containing cobalt had larger H_c. However, the temperature dependency of the H_c of the solids was large since most parts of

H_c of such γ-ferric oxide particles were due to the magnetocrystalline anisotropy rather than shape anisotropy. Great improvements of magnetic properties in H_c were achieved by forming a thin layer of cobalt oxide compounds on acicular γ-ferric oxide particles (34). The modification of γ-Fe_2O_3 particles was made by immersing the particles in a mixture solution containing Co^{2+} and Fe^{2+} followed by addition of alkaline solution. Nowadays, most VHS home videotapes and audio tapes of high bias position use this kind of particle.

13.1.4.3 Iron Particles

Since iron has a large saturation magnetization of 216 emu/g, it must be one of the best materials for magnetic recording media. Although iron particles were the first particulate materials employed for tapes, they were not used after the introduction of maghemite particles. This was partly because of the difficulty in overcoming corrosion with small iron particles. It did not take long until iron powders again attracted the interest of tape manufacturers as a material for a higher density recording system.

Though iron particles can be prepared by variety of methods, most of the particles on the market are produced from goethite through dehydration and reduction processes (25,35). Therefore, most processes for the productions are the same as the production processes of γ-ferric oxide particles. Dehydrated geothite particles are reduced into iron rather than magnetite, and surfaces of the particles are oxidized carefully to prevent further oxidation by forming a thin oxide layer (36). The saturation magnetization of iron powders is 60–70% of that of pure iron, but an H_c higher than 1000 Oe can easily be attained. Special procedures such as addition of silicone oil or Ag or Co compounds at the reduction process are effective in preventing mutual sintering of particles, leading to the powders' improved magnetic properties (37). Recently, Nikles et al. reported that iron particles of 70 nm in length with an aspect ratio of about 10 having H_c of 1500–2000 Oe can be prepared by reducing β-FeOOH particles suspended in liquid crystals using $NaBH_4$ as a reducing reagent at 0°C (38).

Elongated iron particles may be superior particulate materials for magnetic recording media, but the problem associated with corrosion has not been completely overcome for long storage time.

13.1.4.4 Barium Ferrite Particles

In the ordinary recording system, the magnetic medium is magnetized along the direction of the motion of the magnetic medium, which is called the longitudinal recording method. In an opposite method, Iwasaki proposed a different type of recording method called perpendicular recording for a high-density recording system, in which the magnetic medium is magnetized perpendicular to the direction of the motion of the magnetic medium (39). Therefore, the easy axis of the magnetic particles for this medium must be oriented perpendicular to the supporting film.

Platelet barium ferrite particles seemed a suitable material for this purpose since it has the easy axis perpendicular to the flat surface. Barium ferrite has widely been used for a precursor material of permanent magnets. However, the H_c of the particles was too large for the magnetic medium. Kubo et al. developed new platelike barium ferrite particles having the proper coercive force by a glass crystallization method in solid-phase reactions (40). New acicular barium ferrite particles doped with Co^{2+} and Ti^{4+} were prepared using α-FeOOH particles as the starting material by a conventional sintering method. Thus produced, $BaCo_xTi_xFe_{12-2x}O_{19}$ particles consisted of some grains with their *c* axis along the short axis of the acicular particle, having H_c of ~1500 Oe (41).

The perpendicular recording system was thought as a superior system for high-density magnetic recording, but in the last 10 years it has not received much attention. Recently, it is again attracting the interest of magnetic medium manufacturers for higher density magnetic recording systems.

Elongated magnetic particles are employed for the ordinary longitudinal recording system, as stated earlier. It was suggested by Lemke that the vertical components of the recorded signals play a important role as the recording density increases, that is, as the recording wavelength becomes shorter (42). Based on this principle, videotapes having improved recording performance were made by mixing small amounts of barium ferrite particles with cobalt-modified particles (27).

13.1.4.5 Other Magnetic Particles

Chromium dioxide particles were developed as a high-quality magnetic recording material (43) and were used until cobalt-modified γ-ferric oxide particles appeared on the market. The commercially available CrO_2 is produced by oxidation of Cr_2O_3 under high temperature and high pressure. The particles so produced are single crystals and acicular in shape with large aspect ratios and good magnetic properties. However, chromium dioxide particles are not widely used today because of high production cost in addition to toxicity. Although these particles are not used much in the world except for special purposes presently, the technical term "CrO_2 position" had been used for a period of time for the bias position together with "high" or "metal position" for audio recorders and tapes.

In 1972, Kim and Takahashi reported that $Fe_{16}N_2$ film was obtained by evaporating iron under a nitrogen atmosphere (44). This compound showed very large saturation magnetization of 2.8 μ_B per iron atom, which was larger than iron, where μ_B is the Bohr magneton. However, there had been some ambiguity about the presence of the compound until the single crystal of $Fe_{16}N_2$ was reported (45). By reducing dehydrated goethite particles in the presence of ammonia, iron nitride particles were obtained (46). They consisted of variety of compounds having different Fe:N ratios in addition to Fe_4N with high saturation magnetization of about 190 emu/g and large coercive force. Iron nitride seemed one of the desirable materials for the magnetic recording media; it is not used at present because of its unstability.

13.1.4.6 Coercive Force and Dispersibility of Magnetic Particles for Recording Media

The coercive force of a powder is an extrinsic property, which is influenced by many factors such as size and shape and packing density. Maghemite particles prepared from directly precipitated spindle-type hematite particles showed a tendency of increase in coercive force with decrease in particle size, and increase in their aspect ratios (31). However, a quantitative relationship between the coercive force and size and/or aspect ratio is still not available.

It is most desirable if measurements of the coercive force are carried out for individual particles, since particulate material contains particles of different sizes, although they are highly monodisperse in size and shape. A novel method was developed for the measurement of the rotation of magnetization of individual particles by Knowels, who applied a pulse magnetic field to a particle dispersed in a viscous liquid and observed under a microscope the rotation of the solid at a critical magnetic field (47). From the obtained remanent coercive force, he explained the rotation of the magnetization in the particle using fanning reversal mechanism. On the other hand, Aharoni showed that the rotation mechanism could be explained by a curling model using reported experimental data of individual particles (48). Recently, measurements of coercive force of individual particles were also made using Lorenz microscopy and atomic force microscopy (AFM) with maghemite particles prepared from directly precipitated spindle-type hematite particles (49,50). The isolated particles showed a very large H_c of about 1200 Oe; however, this was still somewhat below the value predicted by the coherent rotation model. From the measured spontaneous switching properties of the magnetization of single-domain maghemite particles, it was suggested that the dynamics of the reversal occur via a complex path in configuration space, and a complex theoretical approach is required to provide a correct description of thermally activated magnetization reversal even in a single-domain ferromagnetic particle.

Magnetic particles are embedded in a plastic binder for use as magnetic recording media. Therefore, particles must have good dispersibility in organic solvents, to have large squareness and smooth coated surfaces. However, it is especially difficult to disperse magnetic particles, as compared to nonmagnetic particles, due to strong magnetic interactions. Usually the surface of the magnetic particle is coated with a polymer to prevent aggregation. Inoue et al. investigated the effect of an epoxy resin adsorbed layer on the stability of γ-Fe_2O_3 particles dispersed in organic solvents and found that although both the height of the maximum and the depth of the secondary minimum in the total potential energy of the colloidal interaction strongly affected the dispersion stability, the former was more effective than the latter (51). The experimentally obtained surface roughness of the tape film produced by using polymer-resin-adsorbed magnetic particles increased with the depth of the calculated total potential energy minimum, rather than with the decrease in the height of the maximum energy.

Homola and Lorenz showed that the recording performance of rigid disks was considerably improved by using magnetic particles coated with small colloidal silica particles (52). The dispersion property of the coated particles were enhanced by the control of the separation distance between the magnetic particles, resulting in good particle orientability, which lead to improved recording performance.

13.1.5 Composite Magnetic Particles and Magnetic Particles in Magnetotactic Bacteria

In order to improve essential properties of particulate materials or to introduce new properties to them, a variety of composite magnetic particles have been developed. Introductions of magnetic materials can be performed by coating nonmagnetic materials with magnetic materials or by coating magnetic materials with nonmagnetic materials. Fine particles containing dispersed magnetic materials have also been developed. A schematic picture of composite particles is shown in Figure 13.1.6.

Iron microspheres covered with polystyrene were produced by depositing small latex particles on iron particles for making the core of a high-frequency transformer (53). Polymer particles coated with thin magnetic film were produced using a ferrite plating method developed by Abe and Tamura (54,55). Matijevic and coworkers developed many procedures for the preparation of well-defined, uniformly coated inorganic composite particles, including spindle-type iron particles coated with silica layers (56).

Small silica particles having cores of magnetic materials were obtained by first depositing a silica layer from sodium silicate on freshly prepared magnetite particles, followed by depositing a second silica layer from ethylmethoxysilane–ethanol solution (57).

Composite core–shell type microspheres were prepared by in situ heterogeneous polymerization on monodispersed seed latex particles suspended in an aqueous magnetite dispersion stabilized with sodium oleate (58).

The so-called ''dry mixing'' or ''dry blending'' method has been used for the modification of particles in the powder technology (59). In this method, surface modification of coarse particles is carried out by mixing fine particles and coarse particles with an auto ceramic mortar or with a centrifugal rotating mixer. This procedure can be applicable for the production of variety of composite magnetic particles.

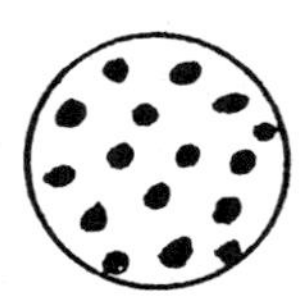

Fig. 13.1.6 Schematic picture of composite particles.

Hirano et al. reported that spherical carbon particles containing highly dispersed cementite (Fe_3C) particles were formed by heating copolymers of divinylbenze and vinylferrocene in a high-pressure bomb at 125 MPa and 650°C (60). The cementite particles thus formed could be transformed into α-Fe by further heating at 850°C for 6 h. When the same copolymers were heated with water, magnetite particles were formed. By using this technique, spherical carbon particles containing metal particles such as cobalt or cobalt alloys can also be prepared.

Inada et al. succeeded in combining an enzyme with synthesized magnetic particles, which made reuse of enzymes easy (61). Furthermore, magnetic polymeric microsphere particles are employed for labeling and separation of biocells (62). Recently many magnetic particles modified with specific proteins have been employed for extraction and cleaning of DNA (63).

It has been known that magnetic particles can be found in the bodies of some biosystems. In 1975, Blakemore found that some bacteria have magnetic particles in their body cells and navigate along geomagnetic fields using the magnetic particles (64). It is also believed that certain animals have the ability to detect magnetic fields (65). The magnetic particles obtained from the magnetotactic bacteria were confirmed, as well as crystallized magnetite particles with sizes of 50–100 nm (66). The particles found in some bacteria were cubiclike, having a narrow size distribution, and were aligned in a single or multiple chains, more or less parallel to the axis of the cell. A transmission electron micrograph (TEM) of a typical magnetotactic bacterium is shown in Figure 13.1.7. The magnetic moment of each particle is small,

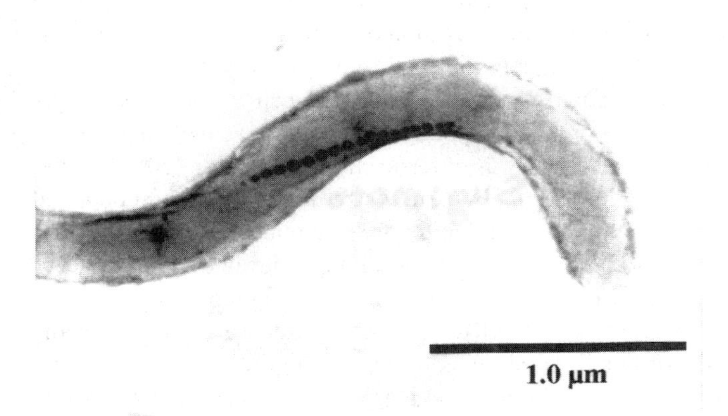

Fig. 13.1.7 Transmission electron micrograph of a magnetotactic bacterium. (Courtesy of Prof. Matsunaga of University of Tokyo Agriculture and Engeneering.)

but the total magnetic moments of the aligned particles are large enough to orient along the geomagnetic field. It is believed that magnetotactic bacteria navigate using these magnets as a direction detector toward north in the northern hemisphere and toward south in the southern hemisphere. The magnetite particles isolated from the bacteria showed magnetic properties similar to synthesized particles, and the aligned magnetic particles in their bodies were found to be a good model of the chain of spheres theory for the rotation of the magnetic moment of a magnetic particle (67). It was also found that the surfaces of the magnetic particles isolated from magnetotactic bacteria were covered with a strong organic membrane. Matsunaga et al. succeeded in immobilizing glucose oxidase and uricase on the organic membranes attached to magnetic particles (68). The enzymes bound on the particles were more active than enzymes combined on synthesized magnetic particles. Such particles could easily be separated from reactants solution. Introduction of the magnetic particles into bioparticles such as blood cells and microphages was also successfully done. Such small particles carrying magnetic particles can easily be moved to desired places by applying a magnetic field. Although the magnetic particles obtained from magnetotactic bacteria are useful, the amounts of magnetic particles obtainable from the bacteria are limited. In order to overcome this difficulty, gene techniques have been applied (69).

13.1.6 Weakly Magnetized Particles: Hematite

Hematite has been known as a material having a weak magnetic property called cant magnetism or parasitic ferromagnetism. The spontaneous magnetization of the solid is about 0.2 emu/g; it is less than 1/200 that of magnetite. Thus, hematite particles can be dispersed in an aqueous solution without aggregating strongly. Because of the weak magnetic property of hematite, little attention had been paid to it as a magnetic colloid. Although the magnetization of hematite is weak, dispersions of hematite particles show many magnetic properties, as we show later.

Hematite particles having narrow size distributions can be prepared in the form of spherical, cubic, spindle, or platelet types. Monodispersed spherical hematite particles were prepared by forced hydrolysis of ferric chloride solution (70). Spindle-type hematite particles having an extremely narrow size distribution were obtained by heating a ferric chloride solution containing small amounts of phosphate ions (71). The aspect ratio of the spindle-type particle could be controlled by the concentration of the phosphate ions. A transmission electron micrograph of the spindle-type particles is shown in Figure 13.1.8. Ocana et al. studied the mechanism of the formation of the ellipsoidal hematite particles by high-resolution electron microscopy and selected area electron diffraction and concluded that the ellipsoidal hematite particles are formed by ordered aggregation of smaller ellipsoids (72). Sugimoto et al. developed a new method for the production of spindle-type hematite particles by the so-called gel–sol method (73). Using their method, a large amount of particles

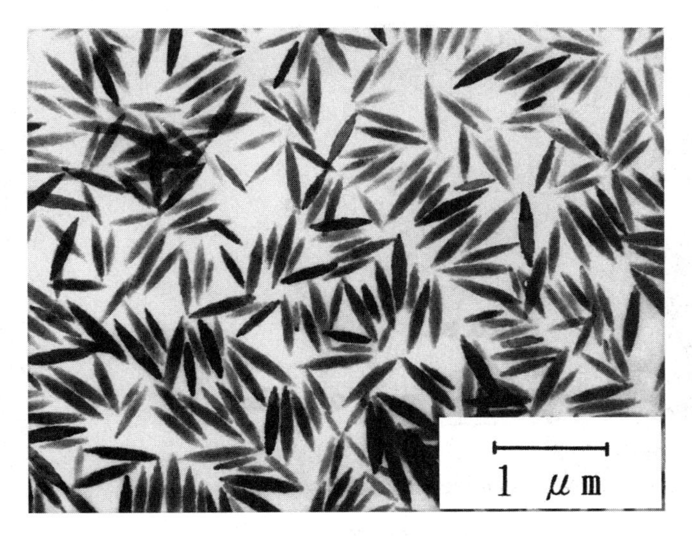

Fig. 13.1.8 Transmission electron micrograph of monodisperse spindle-type hematite particles.

can be produced from concentrated gels of ferric oxide in the presence of sulfate ions at high alkaline conditions.

Hematite particles are also produced by transformation from other particles in aqueous solutions. In this procedure, particles are recrystallized from other kinds of precipitates into final forms. Large cubiclike hematite particles were produced through conversion of previously deposited β-FeOOH (74) in an acidic solution of HCl at 100°C. Elongated hematite particles were produced in basic conditions by aging freshly precipitated ferric hydroxides at temperatures of 100–200°C in the presence of small amounts of organic compounds such as sulfonic acids or hydrocarboxyl acids or salts of these compounds as crystal growth control reagents (27). Platelike hematite particles were produced from α-FeOOH or $Fe(OH)_3$ in strong basic conditions at elevated temperatures (75,76). The platelets so produced were single crystals with (001) planes for the flat surfaces (77). There was a critical temperature for each alkaline concentration in the hydrothermal transformation (75).

The stability of hydrophobic colloids is well described by the well-known DLVO theory. If colloidal particles have magnetic moments, magnetic interactions must be taken into account in addition to the repulsive energy due to the double-layer interaction and the van der Waals attractive energy. Detailed discussions of the stability of the dispersion of magnetic particles are given by Scholten (3), who also discussed the effects of multi-interactions. Although stabilities of hematite dispersions were studied in the presence and absence of external magnetic fields, effects

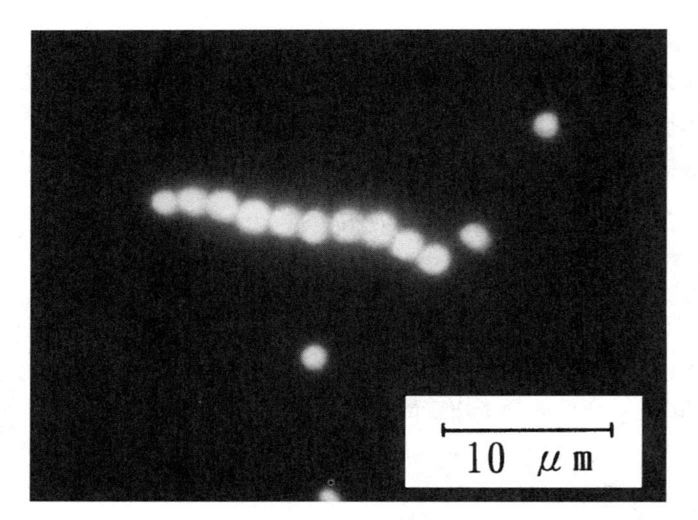

Fig. 13.1.9 Microphotograph of a chainlike agglomerate formed by cubiclike hematite particles in 10^{-5} mol/dm^3 aqueous solution of KCl.

of magnetic dipole moments due to the spontaneous magnetization of hematite had never been taken into account until loose agglomerations due to the magnetic property of hematite particles (78,79) were found. Figure 13.1.9 shows the loose and ordered chainlike agglomerate of cubiclike hematite particles observed under a reversed type metallurgical microscope in a dilute electrolyte solution. The loose agglomerate is not caused by the van der Waals attraction only; neither is it due to the application of a strong magnetic field. The agglomerate illustrated in Fig. 13.1.9 should depend on particle interactions based on the electrical double layer and magnetic property of the particle. As the microphotograph clearly show, there is a finite separation between individual particles in the agglomerated state. The particles also exhibited a limited freedom of vibration. The chain could easily be broken by applying a weak magnetic field vertical to the chain or by simple mechanical stirring. All these observations show that the attraction energy between agglomerated particles is rather weak, leading to the conclusion that the aggregation is due to interactions in the secondary minimum of colloidal interaction. In view of the magnetic property of hematite particles, a major contribution to the attraction energy may be made by the magnetic forces between permanent magnetic moments.

Elongated particles agglomerated with their long axes perpendicular to the external magnetic field, while platelike particles agglomerated with their flat surfaces parallel to the field. These facts suggest that the long axis of the spindlelike particles is a *c* axis and the flat surface of the platelets a *c* plane, since the magnetic moment of the hematite crystal is believed to be in a *c* plane above Morrin temperatures. These crystal habits agree with the results obtained by x-ray diffraction and selected area electron diffraction (73,77).

Chantrell et al. studied the structure of the agglomerates formed in a dispersion of a magnetic fluid in the presence and in the absence of an external magnetic field by computer simulation using a Monte Carlo method (80). According to their simulations, the agglomerates were straight under an external field but were randomly oriented when the field was absent. Formation of looplike agglomerates was also suggested. Similar looplike agglomerate were visually observed in a dispersion of coarse elongated hematite particles as shown in Figure 13.1.10 (81). In the absence of an external magnetic field, the chainlike agglomerate performed a snakelike motion. When the ends of the chain of the spindletype hematite particles happened to approach each other by Brownian motion, they made a bond to form a loop. The formation of the looplike agglomerates could be explained based on the theory of colloid stability by taking into account the magnetic interaction. The looplike structure observed with coarse spindle-type hematite particles may be essentially the same as suggested by Chantrell et al.

It should also be pointed out that the magnetic interaction between hematite particles with a diameter less than 200 nm is not so large as to cause aggregation at low ionic concentrations, but the magnetic moments are strong enough to orient the particle itself along applied external magnetic fields of moderate strength. Thus, changes in viscosity (82) and in electric conductivity (83) with applied magnetic fields are also observed.

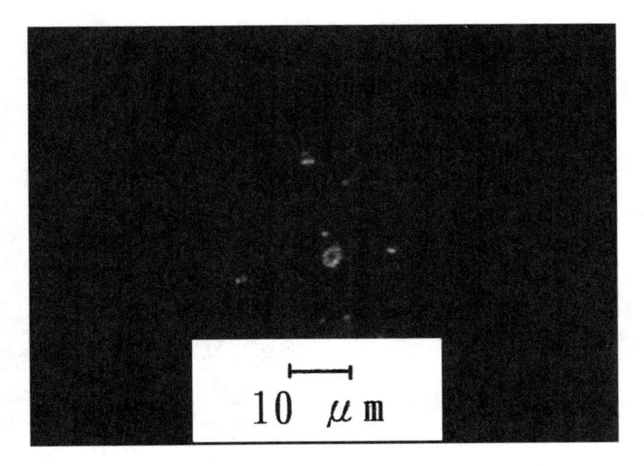

Fig. 13.1.10 Microphotograph of a looplike agglomerate formed by spindle-type hematite particles.

REFERENCES TO SECTION 13.1

1. S Chikazumi. Physics of Magnetism. New York: Wiley, 1965, p 5.
2. PC Scholten, DLA Tjaden. J Colloid Interface Sci 73:254, 1980.
3. PC Scholten. J Magn Magn Mater 39:99, 1983.
4. E Matijevic. Chem Mater 5:412, 1993.
5. VK LaMer, RH Dinegar. J Am Chem Soc 72:4847, 1950.
6. T Sugimoto. Adv Colloid Interface Sci 28:65, 1987.
7. JR Thomas. J Appl Phys 37:2914, 1966.
8. E Papirer, P Horny, H Balard, H Anthore, C Petipas, A Martinet. J Colloid Interface Sci 94:220, 1983.
9. OA Platonova, LM Bronstain, SP Solodvnikov, IM Yanovskaya, ES Obolonkova, PM Valetsky, E Wenz, M Antonietti. Colloid Polymer Sci 275:426, 1997.
10. M Gobe, K Konno, K Kandori, A Kitahara. J Colloid Interface Sci 93:293, 1983.
11. C Petit, A Taleb, MP Pileni. Adv Mater 10:259.12, 1998.
12. M Plileni et al. unpublished observations.
13. II Yaakob, AC Nunes, A Bose. J Colloid Interface Sci 171:73, 1995.
14. S Mann, JP Hannington. J Colloid Interface Sci 122:326, 1988.
15. RM Roshko, BM Moskowitz. J Mag Mag Mater 177–181:1461, 1998.
16. CG Granqvist, RA Buhrman. J Appl Phys 47:2200, 1976.
17. ZX Tang, CM Sorensen, KJ Klabunde, GC Hadjipanayis. J Colloid Interface Sci 145: 38, 1991.
18. RE Rosensweig. Sci Am 247:136, 1982.
19. S Taketomi, S Chikazumi. Tokyo: Jisei Ryutai, Nikkan Kogyo Shinbun, 1988, p 3.
20. T Satoh, S Higuchi, J Shmoiizaka. Abstr 19th Nat Meeting of Japan Chemical Society, 1966, p 293.
21. G Bate. In: DJ Craik, ed. Magnetic Oxides. New York: Wiley, 1975, p 689.
22. Y Imaoka, K Takada, T Hamabata, F Maruta. In: Ferrites: Proceeding of the ICF3, '80. Tokyo, 1981, pp 516–520.
23. H Hibst. J Magn Magn Mater 74:193, 1988.
24. G Bate. J Magn Magn Mater 100:413, 1991.
25. R Chubachi, N Tamagawa. IEEE Trans Magn Mag-20:45, 1984.
26. T Fugiwara, M Isshiki, Y Koike, T Oguchi. IEEE Trans Magn Mag-18:1200, 1982.
27. T Yashiro, Y Kikuchi, Y Matsubayashi, H Morizumi. IEEE Trans Magn Mag-23:100, 1987.
28. AH Morrish. In: HC Freyhard, ed. Crystals: Growth and Applications, Vol. 2. Berlin: Springer, 1979, p 172.
29. M Matsumoto, T Koga, K Fukai, S Nakatani. US Patent 4,202,871, 1980.
30. AR Corradi, SJ Andress, JE French, G Bottoni, D Candoflo, A Cecchetti, F Masoli. IEEE Trans Magn Mag-20:33, 1984.
31. M Ozaki, E Matijevic. J Colloid Interface Sci 107:199, 1985.
32. V Arndt. IEEE Trans Magn Mag-24:1796, 1988.
33. Y Yada, S Miyamoto, H Kawagoe. IEEE Trans Magn Mag-9:185, 1973.
34. S Umeki, S Saitoh, Y Imaoka. IEEE Trans Magn Mag-10:655, 1974.
35. S Asada. Nippon Kagaku Kaishi 1985:22, 1985.
36. S Asada. Nippon Kagaku Kaishi 1984:1372, 1984.
37. AA van der Giessen, CJ Klomp. IEEE Trans Magn Mag-5:317, 1969.

38. M Chen, DE Nickle. Abstr 216th ACS National Meeting MTLS No 066, 1998.
39. S Iwasaki. IEEE Trans Magn Mag-16:71, 1980.
40. O Kubo, T Ido, Y Hidehira. Toshiba Rev 43:897, 1988.
41. K Kakizaki, N Hiratuka. J Mag Soc Jpn 22(Suppl S1):129, 1998.
42. JU Lemke. IEEE Trans Magn Mag-15:1561, 1979.
43. HY Chen, DM Hiller, JE Hudson, CJA Westenbroek. IEEE Trans Magn Mag-20:24, 1984.
44. TK Kim, M Takahashi. J Appl Phys Lett 20:492, 1972.
45. M Komuro, Y Kozono, M Hanazono, Y Sugita. J Appl Phys 67(part A):5126, 1990.
46. T Tanaka, K Tagawa, A Tazaki. Nippon Kagaku Kaishi 1984:930, 1984.
47. JE Knowels. IEEE Trans Magn Mag-17:3008, 1981.
48. A Aharoni. IEEE Trans Magn Mag-22:149, 1986.
49. C Salling, R O'Barr, S Schultz, I McFadyen, M Ozaki. J Appl Phys 75:7989, 1994.
50. M Lederman, S Schultz, M Ozaki. Phys Rev Lett 73:1986, 1994.
51. H Inoue, H Fukke, H Katsumoto. IEEE Trans Magn 26:75, 1990.
52. AM Homola, MR Lorenz. IEEE Trans Magn Mag-22:716, 1986.
53. K Ochiai, H Horie, H Kamohara, M Morita. Nippon Kagaku Kaishi 1987:233, 1987.
54. K Ishikawa, M Ohishi, T Saitho, M Abe, Y Tamura. Abstr 6th Int Conf Ferrites '87, Tokyo, 1987, EB-04.
55. M Abe, Y Tamura. J Appl Phys 55:2614, 1984.
56. M Ohmori, E Matijevic. J Colloid Interface Sci 160:288, 1993.
57. AP Philipe, MPB Bruggen, C Pathmamanoharan. Langmuir 10:92, 1994.
58. M Lee, M Senna. Colloid Polym Sci 273:76, 1995.
59. K Ukita, M Kuroda, H Honda, M Koishi. Chem Pharm Bull 37:3367, 1989.
60. S Hirano, T Yogo, H Suzuki, S Naka. J Mater Sci 18:2811, 1983.
61. Y Inada, K Takahashi, T Yoshimoto, Y Kodera, A Matsushima, Y Saito. Trends Biotech 6:131, 1988.
62. RS Molday, SPS Yen, A Rembaum. Nature 268:437, 1977.
63. H Tajima. Nippon Oyo Jiki Gakkaishi 22:1010, 1998.
64. RP Blakemore. Science 190:377, 1975.
65. JL Kirshvink, JL Gould. Biosystems 13:181, 1981.
66. T Masuda, J Endo, N Osakabe, A Tonomura, T Arii. Nature 302:411, 1983.
67. BM Moskowitz, RB Frankel, PJ Flanders, RP Blakemore, BB Schwartz. J Magn Magn Mater 73:273, 1988.
68. T Matsunaga, S Kamiya. Appl Microbiol Biotechnol 26:328, 1987.
69. T Matsunaga, S Kamiya, N Tsujimura. In: U Häfeli et al., eds. Scientific and Clinical Applications of Magnetic Carriers. New York: Plenum Press, 1997, p 287.
70. E Matijevic, P Scheiner. J Colloid Interface Sci 63:509, 1978.
71. M Ozaki, S Krathovil, E Matijevic. J Colloid Interface Sci 102:145, 1984.
72. M Ocana, MP Molales, CJ Serna. J Colloid Interface Sci 171:85, 1995.
73. T Sugimoto, Y Wang, H Itoh, A Muramatsu. Colloids Surf A 134:265, 1998.
74. S Hamada, E Matijevic. J Colloid Interface Sci 84:274, 1987.
75. S Nobuoka, K Ado. Shikizai 60:265, 1987.
76. M Ozaki, N Ookoshi, E Matijevic. J Colloid Interface Sci 137:546, 1990.
77. D Shindo, BT Lee, Y Waseda, A Muramatsu, T Sugimoto. Mater Trans JIM 34:580, 1993.

78. M Ozaki, H Suzuki, K Takahashi, E Matijevic. J Colloid Interface Sci 113:76, 1988.
79. M Ozaki, T Egami, N Sugiyama, E Matijevic. J Colloid Interface Sci 126:212, 1988.
80. RW Chantrell, A Bradbury, J Poppelwell, SW Charles. J Phys D Appl Phys 13:L119, 1980.
81. M Ozaki, J Sanada, A Isobe. Colloid Surf 109:117, 1996.
82. M Ozaki, K Takamatu. Nippon Kagau Kaishi 1988:1960, 1988.
83. M Ozaki, K Nakata, E Matijevic. J Colloid Interface Sci 131:233, 1989.

13.2 FORMATION OF NANOCRYSTALLINE LUMINOUS MATERIALS

MAMORU SENNA
Keio University, Yokohama, Japan

13.2.1 Introduction

Small and well-defined fine particulate materials with the narrowest possible size distributions are desirable for vast numbers of industrial sectors. Tremendous efforts have been made toward nanoparticles, nanocrystals, or nanocomposites for these purposes. The concept of nanoparticles is quite different from area to area. In pharmaceutics, for instance, submicrometer particles are often categorized as nanoparticles. In the area of luminous materials, on the other hand, the definition of nanoparticles is much more strict and related to quantum size effects. Since they are associated with electron energy states, size criteria must be associated with a physically well-defined quantity such as the Bohr radius or mean free path of the electron. In these cases, nanoparticles are restricted to a single-nanometer regime, notably several nanometers in diameter, or at least a material geometry with one of their dimensions in a single-nanometer regime. It is particularly important and attractive that the band gap can be designed by controlling particle size. The threshold wavelength, λ_E, is calculated from the elementary model. In the case of CdS, the band gap is taken to be 2.53 eV. The calculated λ_E value for the bulk is then 490 nm. When the particle diameter decreases to 7.5 nm, 4.8 nm, or 2.1 nm, λ_E decreases to 478 nm, 444 nm, or 286 nm, respectively (1). By assuming that the exciton moves from site to site coherently and is delocalized throughout the whole cluster, the discrete energy band of an exciton can be constructed. The oscillator strength of an exciton per cluster increases linearly with the cluster volume at the low-temperature limit. This is called the giant oscillator strength effect (2).

Another important aspect is the proximity of particle layer and thin or thick films, whereby the border between thin and thick films remains ambiguous. Conventional processing of particle assemblies comprises molding of the particle mass, like dry compression or wet casting. In the case of nanoparticles, to form a film is much more rational than conventional powder processing. In a dry procedure, production of nanoparticles and film formation in most cases take place simultaneously. An assembly of nanoparticles can also be formed in a wet procedure by precipitation on the templates or even by simple dip or spin coating from a sol containing nanoparticles.

In any cases, we know that most of the cross-sectional view of such a film exhibits a rationally packed nanoparticles assembly with or without preferred orientation. Regulation of an assembly toward some organized manner has been widely studied in recent years in terms of self-organization. Many of the related techniques utilize organic templates like a Langmuir–Blodgett (LB) monolayer film of surfactants or fatty acid, or some biological regularities. The latter is called a biomimetic process.

The purpose of this chapter is to give a brief view of nanocrystalline (NC) phosphors and related luminous or optonic materials, including preparation technique and characterization. Special emphasis is laid on the ZnS:Mn nanocrystals modified by organic species like acrylic acid.

13.2.2 Nanocrystalline Phosphors

There have been a number of attempts to obtain NC phosphors or related optical materials, or, more generally stated, NC semiconductors. A common feature is to regulate and tailor a band gap. The basic idea of semiconductor crystallites as a class of large molecules came from the sense of controlling HOMO–LUMO separation of a molecule, which becomes the band gap in a solid (3). The method was correspondingly based on the effort toward the goal of pure isolated clusters, monodispersed at an atomic level. Surface derivatization, often referred as surface capping, was taken into account from the beginning. To obtain small particles by precipitation is not particularly difficult when an operation is made at low temperature in a diluted state. However, to prevent flocculation or agglomeration or to control the particle size distribution to be as narrow as possible is complicated and not easy. On top of that, there are different additional and critical demands, to bring surface states, concentration gradient, and atomic separation under control, particularly in the case of doped materials, which is much more difficult than to obtain well-dispersed small particles per se.

Preparative methods are broadly divided into two: a wet, colloid chemical route, and a dry, vacuum technological route. Many technologically attractive semiconductors for optical application are sulfides or selenides of various metals, notably transition metals, where coexisting oxides are regarded as contaminants. Efforts are made in common to maintain high chemical purity, including control of oxides in a chalcogenide, and phase homogeneity, i.e., a single-phased product with exact stoichiometry, without second phases.

To use templates or envelopes as a controlled reaction space was developed in the early 1980s, such as the use of inverse micelle technique (4). Another fundamental idea is to use the atomic periodicity of surfactant molecules by using them as surface ligands for sequential addition of anions and cations under the concept of semiconductive compounds like CdSe as a living polymer (3).

As for the state and morphology of the products, there are at least three categories of the state of NC semiconductive materials: those in the form of unsupported powders, particles dispersed in a matrix (supported NC particles), and those packed in a film. A comprehensive review is given in the book Nanoparticles and Nanostructured Films—Preparation, Characterization and Applications (5).

Control of near-surface composition with a proximate relationship with defect concentration is extremely important for the production of NC phosphors. Passivation of the near-surface region is one of the most promising way to get rid of the main shortcomings of NC phosphors. A well-regulated doping of the phosphors also reduces the negative effects of, among others, the nonradiative recombination of the exiton just created during activation of the phosphor.

A number of studies have therefore been devoted to the surface states of NC phosphors. Chen et al. (6) recently reported that the principal adsorption band of ZnS at 500 nm becomes more intense and exhibits a blue shift with decreasing particle size from 2.3 nm to 1.2 nm. They attribute these results to the surface states, which are size sensitive. By using the micelle-encapsulation technique, it is possible to link desired chemical species preferentially to the surface. This enables the control of surface concentration, surface structure, and isolation of molecular particles at the same time (7).

A general feature of doped semiconductive NC materials is described in a review paper authored by Bhargava, one of the founders of this area (8). It is generally accepted that solid particles are richer in crystallographical defects when their diameter becomes smaller. Quantum size effects associated with nanoparticles are very sensitive to the defects. In most cases, defects influence negatively the luminescent properties of phosphors. Most of these drawbacks of NC phosphors are attributed to the larger specific surface area, since defects tend to concentrate themselves in a near-surface region.

13.2.3 Size-Selective Precipitation from Molecular Precursors

It is not easy to keep chalcogenides completely free from oxides. In order to surmount this general difficulty, efforts are made to start from molecular compounds that contain a priori all the necessary elements for the final semiconductors. Murray et al. (9) tried to produce various II/IV semiconductive nanoparticles from molecular precursors like $[CH_3CdSe_2CN(C_2H_5)_2]_2$ dispersed in tri-*n*-octylphosphine (TOP), tri-*n*-octylphosphine oxide (TOPO), or their mixtures. They not only developed a method of nanocrystalline CdE (E = S, Se, and Te) preparation, but also proposed a technique of size-selective precipitation. At the same time, they emphasized and practiced the consistent surface derivatization that they called surface capping. Dispersion media serve as capping materials as well. After nanoparticles of cadmium chalcogenide were obtained, using standard airless procedures, and purified, they were redispersed in anhydrous butanol to obtain an optically clear solution. Methanol was then added until opalescence persisted upon stirring. Repeated centrifugation and the redispersion–precipitation procedure mentioned earlier resulted in the nar-

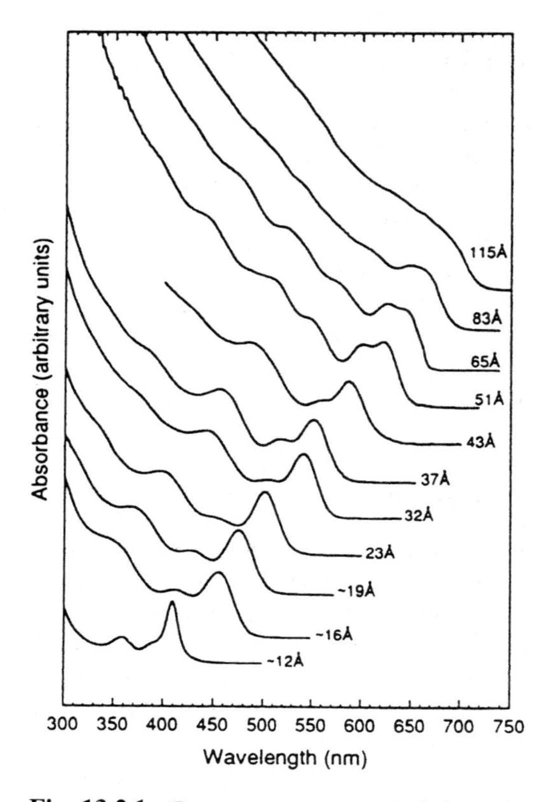

Fig. 13.2.1 Room-temperature optical absorption spectra of CdSe nanocrystals dispersed in hexane. (From Ref. 9.)

rowing of the size distribution as monitored by optical absorption spectra. Figure 13.2.1 demonstrates the room temperature optical absorption spectra of CdSe nanocrystals dispersed in hexane (9). Murray et al. also showed several transmission electron micrographs where a near monolayer of 5.1-nm CdSe crystallites exhibit short-range hexagonal close packing.

Trindade and O'Brien (10) also succeeded in the preparation of CdSe nanocrystals from a single molecular precursor in TOPO. Typical blue shift of the band edge of the absorption spectra and band maximum of the emission spectra are observed in a very well-defined manner. A further study by the same authors (11) demonstrated the synthesis of nanocomposites comprising CdS or CdSe and organic bridging ligands such as 2,2′-bipyrimidine (bpm), 4,4′-bipyridyl, or pyrazine. The authors emphasized the importance of appropriate surface capping to enhance the luminescence properties of the materials for technological applications. They further discussed the state of capping by TOPO and bpm.

13.2.4 ZnS:Mn–Acrylic Acid Nanocomposites

Industrial needs for brighter phosphors in the form of well-dispersed finer particles are ever increasing, for display with higher resolution. This became realistic after the reports of Bhargava et al. (12,13). In the authors' laboratory, ZnS:Mn was modified by acrylic acid (AA) to increase the photoluminescence (PL) intensity (14–18). In this section, methods and characteristic features of such nanocomposites are reproduced in detail. The mechanism of increasing PL intensity is also elucidated from various angles.

A detailed description of ZnS:Mn/AA nanocrystal is given elsewhere (17,18). The starting material is a mixed solution of Zn acetate and Mn acetate. The mixture was put into an Na_2S solution with a solvent comprised of equal volumes of methanol and water. Acrylic acid was added just after precipitation. After centrifugation, the precipitate was dried in air at 50°C for 24 h. Commercial powders with average particle radius 250 nm were also used for comparison.

Poly(acrylic acid) (PAA) with an average molecular weight of 250,000 was used to coat a dry tablet of ZnS:Mn, which was obtained by compressing the powder at 173 MPa. The amount and state of doped manganous ion, Mn(II), was examined by electron paramagnetic resonance spectroscopy (EPR). EPR enables us to gather various extra information with respect to the state of Mn(II).

All the synthesized particles are zincblende and are nearly spherical, with their average radii being 1.5 $\pm$ 0.3 nm, irrespective of AA concentration. Thus, the particles are in a single-nanometer regime and their radii are considerably smaller than the Bohr radius of ZnS, i.e., 2.5 nm.

As shown in Figure 13.2.2, the top peak of the excitation spectra of the ZnS:Mn without carboxylic acid prepared in this method shifts toward higher energy by 0.06 eV, to 3.67 eV (339 nm) from 3.61 eV (344 nm), for the particles prepared by a conventional method, with their average particle size being 0.5 mm. By preparing them in the presence of AA, however, a larger blue shift by 0.16 eV with respect to the commercial particles, to 3.77 eV (330 nm), is observed.

PL emission spectra exhibit characteristic bimodal curves, as shown in Figure 13.2.3, with two peaks at 2.13 eV (580 nm) and 2.88 eV (430 nm). The former is attributed to Mn(II) and the latter to the polymeric species of AA. The latter emission is not attributable to the additives of AA like the polymerization inhibitor. It is particularly noteworthy that the former peak, at 2.13 eV, increases with increasing the amount of Mn(II) concentration, at the cost of the latter peak at 2.88 eV, as shown in Figure 13.2.4. This suggests the possibility of energy transfer from the polymeric species to the active center, Mn(II), of the phosphor ZnS:Mn.

When a small amount of PAA with its average molecular weight 250,000 is dropped onto the ZnS:Mn tablet, an increase in the PL intensity was observed, not

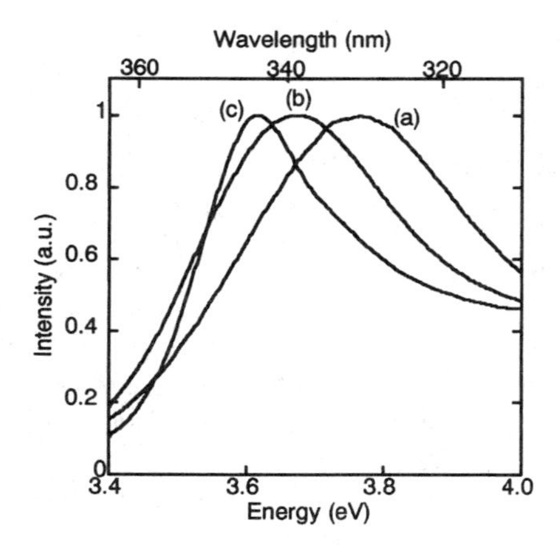

Fig. 13.2.2 Excitation spectra of ZnS:Mn: (a) prepared in the presence of AA, 0.72 mol, to 20 mmol ZnS:Mn; (b) without AA; and (c) ZnS:Mn particles with average diameter 0.5 μm.

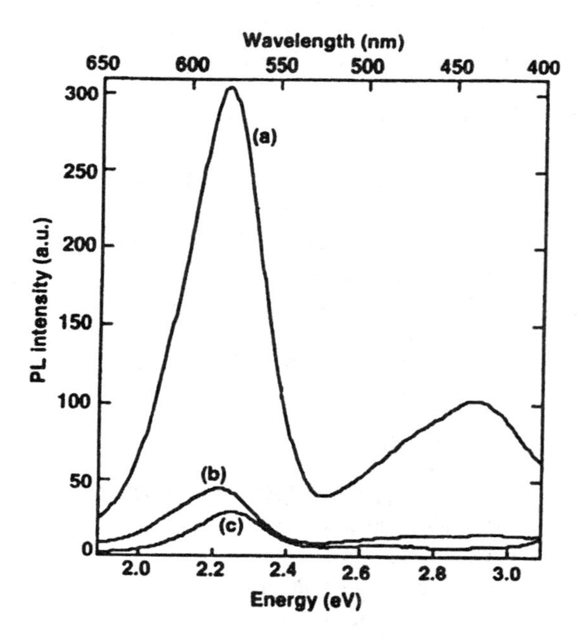

Fig. 13.2.3 PL emission spectra of ZnS:Mn: (a) prepared in the presence of AA, 0.72 mol, to 20 mmol ZnS:Mn; (b) without AA; and (c) ZnS:Mn particles with average diameter 0.5 μm.

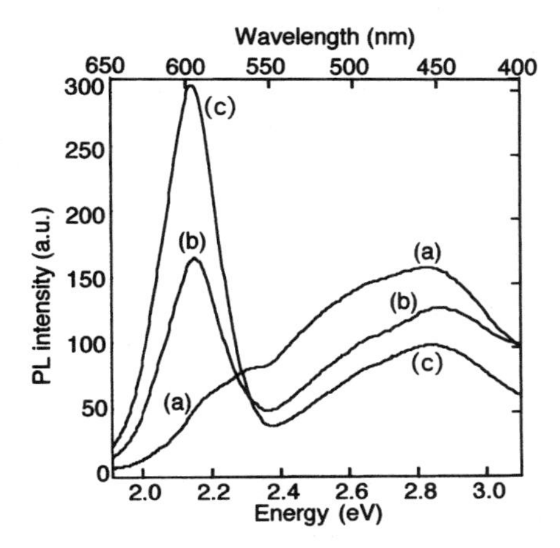

Fig. 13.2.4 PL emission spectra of ZnS:(Mn) prepared in the presence of AA, 0.72 mol, to 20 mmol ZnS:(Mn): (a) without Mn(II); (b) Mn(II) 0.4 mol%; and (c) Mn(II) 1.0 mol%.

only due to PAA itself, but also that due to Mn(II), i.e., at 2.14 eV. This is evidence of energy transfer from PAA to ZnS:Mn.

Absorption bands of Fourier transform infrared spectroscopy (FTIR) at 624 cm^{-1} and 1133 cm^{-1} are observed on the sample prepared without AA. Since they both are assigned to be ν_3 and ν_4 vibration of SO_4^{2-}, it is clear that the samples are oxidized to change from pure sulfide to partly oxidized sulfide. On the sample prepared with AA, however, these peaks are not observed. At the same time, two S $2p_{3/2}$ peaks due to S^{6+} and S^{2-} are observed at 168 and 162 eV, respectively, in the x-ray photoelectron (XPS) spectra. These suggest that the presence of AA prevents the phosphor from surface oxidation by air exposure and entitles a direct coordination of a sulfur atom to a carboxylic oxygen atom. After Ar^+ ion etching for 5 s, the peak at 168 eV decreased while that at 162 eV increased, indicating the chemical interaction between ZnS:Mn and AA being topochemically restricted to the near surface region.

Further information with respect to ZnS:Mn nanocomposite phosphors is available from EPR spectra. The ZnS:Mn/AA nanocrystal shows a typical sextet in X-band (9-GHz) ESR spectra, as shown by curve (a) in Figure 13.2.5. In contrast, the sample without AA exhibits the second component, signal II (Fig. 13.2.5b). From the sample with sulfur deficiency, ZnS(0.8):Mn, the relative intensity of signal II is higher (Fig. 13.2.5c). On the other hand, the g value, 2.0013, is significantly different from that of interstitial Mn(II) (g = 2.020) (19). From this, together with the value of the hyperfine structure constant (13), it is reasonable to assume that signal II is associated with Mn(II) with its coordination number lower than 3.

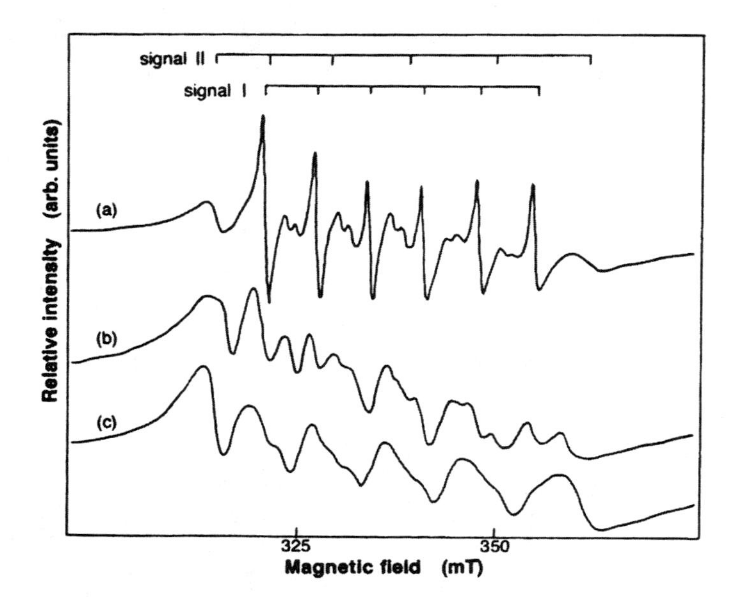

Fig. 13.2.5 EPR spectra of ZnS:Mn: (a) prepared in the presence of AA, 0.72 mol, to 20 mmol ZnS:Mn; (b) without AA; and (c) sulfur-deficient sample (Zn:S ratio 1:8) without AA.

Generally, however, Mn(II) in the interior of the solid can also have low coordination number in the neighborhood of vacancy or any similar structural defects. The signal I, on the other hand, is from the fully coordinated Mn(II), i.e., with coordination number 4. Decrease in the relative intensity of signal II by AA addition is obviously the result of capping of sulfur defects by AA.

The zero-field splitting constant of the sample ZnS:Mn/AA, obtained from X-band EPR spectra, is 9.1×10^{-3} cm^{-1}, which is significantly larger than that from the submicron particles, 2.0×10^{-3} cm^{-1}. This indicates the decrease in the ligand field symmetry by downsizing and addition of AA.

From the PL spectra of AA and its polymer, PAA, we concluded (13) that PAA exhibits a clear peak at 2.82 eV (440 nm), but monomeric AA does not. It is therefore clear that the 2.82 eV peak in Figure 13.2.3 is attributed to the polymerized AA. When the p electrons resonate between C═O and C═C, the latter being abundant in monomeric AA, the luminescence is severely suppressed. As a matter of fact, propionic acid, which does not contain C═C bonds, exhibits an appreciable PL signal at the same of position of PAA.

Aging the ZnS:Mn/AA nanocrystals at 80°C for different periods also brings about the increase in the intensity of PL peaks, both at 430 nm and 580 nm, as

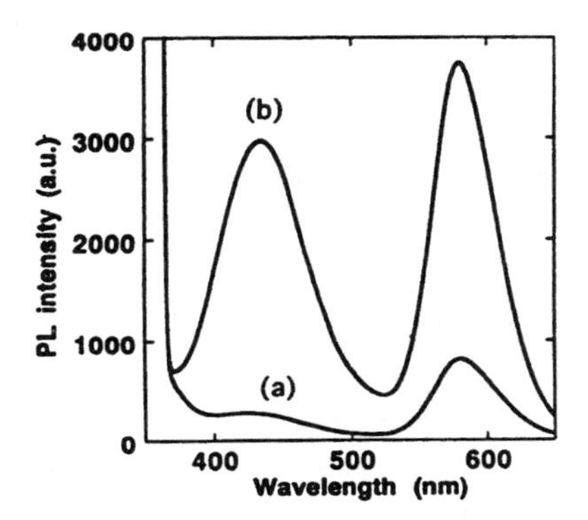

Fig. 13.2.6 PL emission spectra of ZnS:Mn prepared in the presence of AA, 0.72 mol, to 20 mmol ZnS:Mn: (a) without and (b) with subsequent heating at 80°C for 1 week.

shown in Figure 13.2.6. Progress of polymerization on aging was confirmed by the decrease of IR peak intensity due to C=C vibration. This is another indication of increased PL intensity on polymerization. Enhanced luminescence due to polymerization is likely to be associated with the change in the interaction of π electrons between vinyl and carbonyl groups.

Photoluminescence of ZnS:Mn occurs when the phosphor absorbs photon energy corresponding to the band gap of ZnS and relaxes to release the excess energy of the exciton (a pair of an s–p electron and a hole). Based on the selection rule of Laporte, the symmetrical field of 6-coordinated Mn(II) does not allow the d–d transition since it is not associated with the change in the parity. The 4-coordinated Mn(II), in contrast, allows a partial d–p hybridization, enabling the d–d transition.

Another important feature of the ZnS:Mn/AA nanocrystal is that PAA is excited by the photon, which simultaneously can excite ZnS with the same energy. This eases energy transfer from PAA to ZnS:Mn. At the same time, coordination of AA or PAA to ZnS:Mn increases the local electron density around the COO^- group due to the abstraction of some S^{2-} ions at the moment of oxidation to S^{6+} as we observed from IR and XPS spectra. The electrons concentrated near COO^- might contribute to enhance the quantum efficiency of the energy transfer from the s–p exciton of ZnS to the d band of Mn(II). All these hybrid effects involved in the ZnS:Mn/AA enhance the PL intensity as a whole.

13.2.5 Characterization of Semiconductor Clusters

Characterization of photonic nanoparticles is as important as its fabrication. Nanoparticles are often called clusters. While cluster size is determined in most cases by transmission electron microscopy, line broadening of the x-ray diffraction peaks, or absorption edge wavelength, Fischer et al. (20) applied high-performance liquid chromatography (HPLC) to determine the size distribution of CdS clusters. They assumed the plural of absorption maxima to be attributed to the structured size distribution; i.e., agglomerates have several preferential sizes under the concept of magic numbers. The principle was applied to monitor Ostwald ripening and photodissolution of CdS particles.

Detailed analysis of x-ray diffractometry reveals the local structure of semiconductor clusters. Bawendi et al. (21) applied the method to CdSe nanoclusters of 3.5–4.0 nm. They concluded that these clusters have a mixture of crystalline structures intermediate between zincblende and wurtzite, on the basis of detailed simulation studies taking the thermal fluctuation into account. Conventional diffractometry could overlook the coexistence of wurtzite component.

Chestnoy et al. (22) showed time dependence of the luminescence decay of CdS clusters of 2.2 nm and 3.8 nm as a function of wavelength and temperature. The decay over the entire band is in most cases not single-exponential, indicating the multicomponent lifetime of, e.g., 10-ns and 350-ns regimes. The luminescence lifetime increases by approximately three orders of magnitude when temperature decreases from 293 K to 4 K. The reciprocal of the lifetime is the rate constant, which is generally additive, i.e., $k = k_r + k_{nr}$, where k_r and k_{nr} are the radiative and nonradiative relaxation rate constants. With increasing temperature, the relative contribution of shorter lived nonradiative decay increases, making the overall rate constant smaller. Relationship between the overall decay time and temperature is shown in Figure 13.2.7. There are obviously two different regimes in the lifetime temperature dependence. Thermally activated radiationless decay occurs above 50 K, while the decay below 20 K is temperature independent. From these observations, Chestnoy et al. concluded that the carriers are very strongly coupled to lattice phonons. Furthermore, luminescence is attributed to a photogenerated, trapped electron tunneling to a preexisting, trapped hole, but the range of tunneling distances is almost independent of cluster size.

13.2.6 Colloidal Arrays for Photonics

To observe a lattice comprised of small particles instead of atoms is an old idea in the area of colloid science. However, with the development of synthesis techniques for various well-defined polymer microspheres with a very narrow size distribution, such an array of colloidal particles can now be applied for optical purposes. One of the exciting areas is the self-assembled mesoscopic spheres by which diffraction of light is regulated. A 200-μm-thick crystalline colloidal array (CCA) of monodispersed polymer microspheres can produce a sharp diffraction peak from its (111) plane at ~600 nm light (23,24). Optically nonlinear colloidal particles, like those containing an absorbing dye, are subjected to self-organization to make CCA. Remarkable nonlinear devices are fabricated (25). The frequency-doubling effect of yttrium–aluminum–garnet laser was observed from the CCA. Their optical proper-

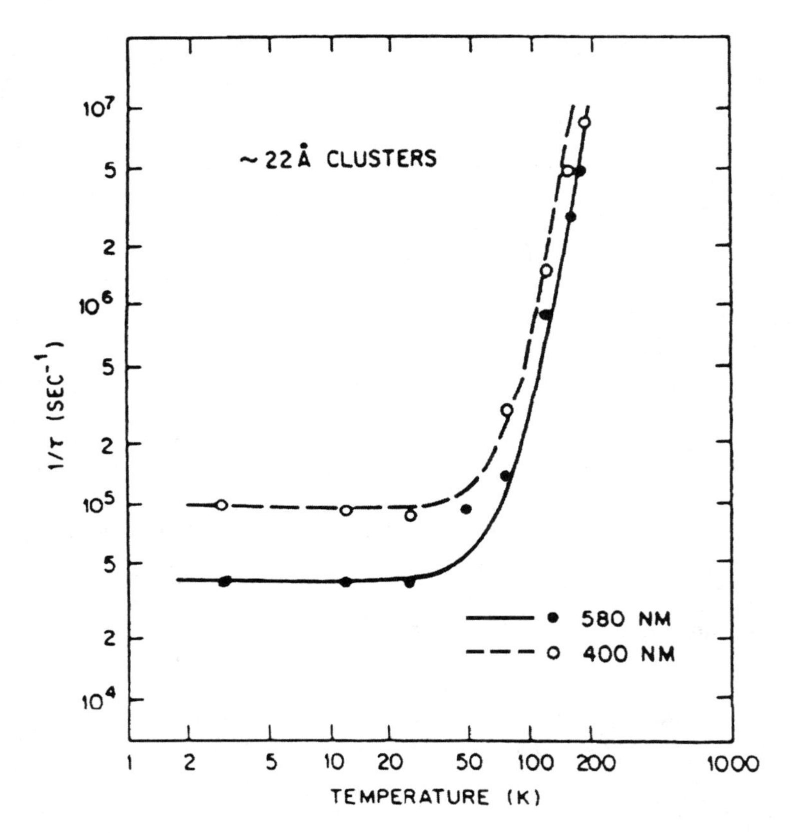

Fig. 13.2.7 Relationship between inverse lifetime and temperature. Calculated values based on the multiphonon decay are indicated by solid and broken lines. (From Ref. 22.)

ties can be regulated in many different ways, such as by adjusting the refractive indices of the medium relative to those of colloidal particles. As similar technique can be used to prepare thermally tunable photonic crystals (26,27). The main point of these unique materials is the change in the hydrophobicity of the colloidal particles with temperature. Some of the colloidal particles, like poly(*N*-isopropyl acrylamide), shrink from ~300 nm to ~100 nm when the temperature increases from 10°C to 40°C. The scattering power increases with temperature due to the increasing refractive index mismatch.

13.2.7 NC Composites in Thin Films

Once molecular complex precursors are successfully synthesized, they may be used for supported nanoparticles in the form of thin films. Hursthouse et al. (28) synthesized $MeZnS_2CNEt_2$, where Me and Et stand for CH_3^- and $C_2H_5^-$, respectively. The complex sublimes at below 150°C, 10^{-2} torr, and is suitable for thin film formation by conventional MOCVD at 450°C, 10^{-2}–10^{-3} torr. Other similar compounds for zinc and cadmium selenides were also synthesized by the same group (29).

To arrange NC or nanoparticles into an assembly is inevitable in order to utilize them for practical purposes. An assemblage in a finely regulated manner is always desired. It is therefore developed to control the interparticle separation. One of the most interesting methods, very rational but not too complicated, is the use of the thermal change in the state of surfactants adsorbed on the surface of nanoparticles. Kotov et al. (30) showed that the interparticle separation of titania NC of 2.0 ± 0.5 nm could be controlled from 0.9 nm to 0.2 nm depending on the heat treatment condition, from 90 min at 70°C to 120 min at 90°C.

When some NC particles are surface treated to make them hydrophobic by, e.g., silylation and floated on the surface of water, a nanoparticle layer can be made on a solid substrate by a method similar to Langmuir–Blodgett film formation. Interparticle separation is well controlled by choosing an appropriate surface pressure prior to making LB film, as discussed next.

13.2.8 Thin Films with Oriented Growth

As mentioned in the introduction, many of the thin films are understood as a densely packed assembly of nanoparticles. Among a large number of publications in this field, a summarizing review by Meldrum is recommended to obtain a general view (31). There are a large number of attempts to obtain oriented nanoparticles on the substrate. While molecular beam epitaxy (MBE) could bring about a single-crystal-like thin film (32), other techniques like metal–organic chemical vapor deposition (MOCVD) (33–35), metal organic vapor phase epitaxy (MOVPE) (36), or hydrogen plasma sputtering (HPS) (37) result in polycrystalline films.

In some of the special thin-film techniques, it is now really possible to make a stack of very well-defined layers by atomwise deposition. This can be best done in dry processing, usually in high vacuum. Atomwise deposition from solution, however, has also been attempted and seems partly successful. Furthermore, atomic layer epitaxy (38) enables control at a monatomic layer level.

Use of an organic monolayer can be either as prepared on the surface of a liquid or after being transferred to a solid substrate to make an LB monolayer. Use of such a molecular assembly as a stencil or a template has many advantages. Besides its relatively simple procedure for template preparation, the periodicity of the functional groups or particular ions can be regulated by appropriate additives and the surface area of the monolayer by using a known surface area–surface pressure relationship.

An outstanding example is the formation of PbS nanocrystals on monolayers of mixed arachidic acid and octadecylamine (39). By successive addition of octadecylamine to arachidic acid, the morphology of the product, PbS, changes from an equilateral triangles of 45 nm to square-shaped, 80-nm nanocrystals, passing through indented or right-angle triangles. Epitaxial orientations are determined in detail, i.e., from those on PbS {111} to {001} via those of {001} and {110}. This study was preceded by a very detailed TEM observation with respect to the orientation relationship between the substrate and the depositing nanocrystal (40).

13.2.9 Other Methods for Self-Assembled NC

A different aspect of preparation of an organized nanoparticles on a fluid is called as the rheotaxy technique. It is a well-established one to fabricate a well-crystallized film. Mobility of the atoms on the surface of the liquid substrate favors the aggregation of atoms in the growing films (41). In order to avoid a drawback of the rheotaxy, i.e., the negative effect of high surface tension of the substrate, a modification has been made by Romeo et al. (42,43), where they used substrates of elevated temperature close to but below their melting points. They prepared, e.g., ZnS:Mn thin films on some low-melting metals such as Pb, Bi or Bi-Sb alloy.

Self-assembled quantum dots were prepared by various further methods as well. Semiconductor technology has been well developed to fabricate well-organized quantum dots for device level. Although such a technology is not exclusively for luminous materials, the optical properties of such organized quantum dots are of significance in various aspects. One of the most developed technologies is the deposition of InGaAs quantum dots on GaAs single crystal. MBE is used commonly for these purposes.

Leonard et al. (44) reported self-organized islands of InAs on Ga As by MBE. The size of the islands ranged between 15 and 30 nm in a planer direction. The product exhibited a broad PL band at low temperatures, reflecting the fairly wide size distribution. In contrast, Nösel et al. prepared an InGaAs/GaAsB nanocrystalline array (45). The size of these strained InGaAs quantum disks can be controlled between 30 and 130 nm by changing the In content. The product can be used for a semiconductor laser. A comprehensive summary was given by Fendler and Tian (46).

It is well known that nucleation preferentially occurs at step edges, so that conventional MBE results in ridgelike alignment of quantum dots. When atomic hydrogen is introduced prior to the dot formation, however, quantum dots tend to distribute more uniformly on the surface (47,48). This is mainly attributed to the termination of the active dangling bonds at step edges by atomic hydrogen. It is also to be noted that the PL intensity from such quantum dots increases with atomic hydrogen pretreatment. This might be attributed to deactivation of the nonradiative center by hydrogen adsorption. Kawabe et al. also succeeded in preparing quantum dot arrays of InGaAs on GaAs(311)B substrate by conventional MBE (49).

The band gap of such a thin film cannot be determined easily by a conventional spectroscopic technique. Alperson et al. have developed two new techniques for these purposes: a photoelectrochemical photocurrent spectroscopy (50), and conductive spectroscopy (51). While photo current is measured as a function of illuminating wavelength, the latter current–voltage spectroscopy is done using a metallized atomic force microscope tip.

The $Si_{1-x}Ge_x/Si$ quantum well can be prepared by People et al. (52) or Kruck et al. (53). Their products can be applied to the quantum well infrared photodetectors. Similar intraband absorption in the mid-infrared can be achieved by the self-assembled InAs/GaAs as well (54,55).

Interesting optical materials have been prepared or designed by starting from a well-organized natural assembly of silica nanoparticles, i.e., opal (56). Particularly worth mentioning is the improvement of the emission efficiency and its directionality by modifying the opal structure (57). The photonic band gap (PBG) can be regulated by adding semiconductive materials like CdS (58). The refractive index of the silica assembly in an opal structure can be increased by coating silica nanoparticles with InP or TiO_2 (59). The top peak of the photoelectron spectrum of InP exhibits a blue shift by several tenths of an electron-volt with respect to the band edge of bulk InP and is consistent with that of 10-nm InP particles.

Many oxides exhibit photochromic effects, and the effects are enhanced when the oxides take the form of thin films. It is particularly interesting when WO_3 is subjected to ultraviolet (UV) irradiation (60). Depending on the intensity of UV irradiation, absorption of light with a wavelength higher than 500 nm increases. The effect is based on charge separation and subsequent trapping of charge carriers. It is therefore sensitive to quantum size effect.

Self-organization processes are now being seriously considered in the light of semiconductor patterning by combining with photolithographic technique. As a matter of fact, an atomic-step network can be organized by controlling the step motion during surface atom evaporation (61). For practical uses of this kind of technique, a wafer-scale control is necessary.

13.2.10 Summary

Some technologies for nanocrystalline photonic materials have been introduced, some in detail. One of the attractive points of nanocrystals in other fields of materials is that the resultant optical properties are very predictable. This enables us further to design new luminous materials.

Regulation of particle size and narrowing of its distribution are of common concern. Postclassification, like repeated centrifugation and redispersion, is only a method of second choice. Rational utilization of nuclei-growth phenomena is much more elegant. This can be done with and without the aid of substrate control. Controlled nuclei growth can be achieved via either homogeneous or heterogeneous nucleation. As for the alignment of nanoparticles, on the other hand, preparation of an appropriate substrate or template is decisive. Out of a number of methods for well-regulated periodicity of substrates, the monolayer of functional molecules on the liquid surface and the use of a well-prepared single crystal are used predominantly at present.

It is also important to emphasize that conventional consciousness of colloid or fine particle technology, like better dispersion and control of rheological properties of dipping solution, are not to be overlooked. The growth of nanoparticles in liquid phase is almost exactly regulated by the nuclei-growth theory as well as stability of lyophobic colloids suggested half a century ago.

REFERENCES TO SECTION 13.2

1. DY Godovsky. Adv Polym Sci 119:79, 1995.
2. E Hanamyra. Phys Rev B 38:1228, 1988.
3. ML Steigerwald, LE Brus. Acc Chem Res 23:183–188, 1990.
4. M Meyer, C Wallber, K Kurihara, JH Fendler. J Chem Soc Chem Commun 1984:90, 1984.
5. JH Fendler, ed. Nanoparticles and Nanostructured Films. Weinheim: Wiley-V CH, 1998.
6. W Chen, Z Wang, Z Line, L Lin. J Appl Phys 82:3111, 1997.
7. ML Steigerwald, AP Alivisatos, JM Gibson, TD Harris, R Kortan, AJ Myuller, AM Thayer, TM Dunkan, DC Douglass, LE Brus. J Am Chem Soc 110:3046, 1988.
8. RN Bhargava. J Luminescence 70:85, 1996.
9. CB Murray, JD Norris, MG Bawendi. J Am Chem Soc 115:8706, 1993.
10. T Trindade, P O'Brien. Adv Mater 8:161, 1996.
11. T Trindade, P O'Brien. Chem Mater 9:523, 1997.
12. RN Bhargava, D Gallagher, X Hong, A Nurmikko. Phys Rev Lett 72:416, 1994.
13. TA Kennedy, ER Glaser, PB Klein, RN Bhargava. Phys Rev B 52:14356, 1995.
14. I Yu, T Isobe, M Senna. J Phys Chem Solids 54:373, 1996.
15. I Yu, T Isobe, M Senna. J Soc Information Display 4:361, 1996.
16. T Isobe, T Igarashi, M Senna. Mater Res Soc Symp Proc 452. Pittsburgh: MRS, 1997, p 305.
17. T Igarashi, T Isobe, M Senna. Phys Rev B 56:6444, 1997.
18. M Senna, T Igarashi, M Konishi, T Isobe. Proc 4. Int Display Workshops, Inst Image Information and Television Eng Soc Information Display, Nagoya, Japan, 1997, p 613.
19. P Balaz, J Bastl, J Briancin, I Ebert, J Lipka, J Mater Sci 2(7):653, 1992.
20. C-H Fischer, H Weller, L Katsikas, A Henglein. Langmuir 5:429, 1989.
21. MG Bawendi, AR Kortan, ML Steigerwald, LE Brus. J Chem Phys 91:7282, 1989.
22. N Chestnoy, TD Harris, R Hull, LE Brus. J Phys Chem 90:3393, 1986.
23. D Thirumalai. J Phys Chem 93:5637, 1989.
24. PA Rundquist, P Photinos, S Jagannathan, SA Asher. J Chem Phys 91:4932, 1989.
25. G Pan, AS Tse, R Kesavomoorthy, SA Asher. J Am Chem Soc 120:6518, 1989.
26. Y Hirokawa, T Tanaka. J Chem Phys 81:6379, 1984.
27. XS Wu, AS Hoffman, P Yager. J Polym Sci Polym Chem A30:2121, 1992.
28. MB Hursthouse, MA Malik, M Motevalli, P O'Brien. Organometallics 10:730, 1991.
29. MB Hursthouse, MA Malik, M Motevalli, P O'Brien. Polyhedron 11:45, 1992.
30. NA Kotov, FC Meldrum, JH Fendler. J Phys Chem 98:8827, 1994.
31. FC Meldrum. In: JH Fendler, ed. Nanoparticles and Nanostructured Films. Weinheim: Wiley-VCH, 1998, p 23.

32. U Meirav, EB Foxman. Semicond Sci Technol 10:255, 1995.
33. K Hirabayashi, O Kogure. Jpn J Appl Phys 24:1484, 1983.
34. PJ Dean. Phys Stat Solidi a 81:625, 1984.
35. PJ Wright, B Cockayne. J Cryst Growth 59:148, 1982.
36. JH Fendler, FC Meldrum. Adv Mater 7:607, 1995.
37. H Sakama, M Ohmura, M Tonouchi, T Miyasato. Jpn J Appl Phys 32:1681, 1993.
38. D This, H Oppolzer, G Ebbinghaus, S Schild. J Cryst Growth 63:47, 1983.
39. J Yang, JH Fendler. J Phys Chem 99:5505, 1995.
40. J Yang, JH Fendler, TC Jao, T Laurion. Microsc Res Tech 27:402, 1994.
41. JQ Broughton, LV Woodcook. J Phys C 11:2743, 1978.
42. N Romeo. J Cryst Growth 52:692, 1981.
43. N Romeo. J Appl Phys 64:4762, 1988.
44. D Leonard, K Pond, PM Peroff. Phys Rev B50:11687, 1994.
45. R Nösel, J Temmyo, T Tamamura. Nature 369:131, 1994.
46. JH Fendler, Y Tian. In: JH Fendler, ed. Nanoparticles and Nanostructured Films. Weinheim: Wiley-VCH, 1998, p 429.
47. T Sugaya, M Kawage. Jpn J Appl Phys 30:L402, 1991.
48. YJ Chun, S Nakajima, Y Okada, M Kawabe. Physica B 227:299, 1996.
49. M Kawabe, YJ Chun, S Nakajima, K Akahane. Jpn J Appl Phys 36:4078, 1997.
50. B Alperson, S Cohen, Y Golan, I Rubinstein, G Hodes. In: E Pelizzeti, ed. NATO ASI Series 3, vol 12. Amsterdam: Kluwer, 1995, p 579.
51. B Alperson, S Cohen, I Rubinstein, G Hodes. Phys Rev B 52:R17017, 1995.
52. R People, JC Bean, SK Sputz, CG Bethea, LJ Peticolas. Thin Solid Films 222:120, 1992.
53. P Kruck, M Helm, T Fromherz, G Bauer, JF Nutzel, G Absteiger. Appl Phys Lett 69: 2785, 1997.
54. S Sauvage, P Boucaud, FH Julien, JM Gerard, V Thierry-Mieg. Appl Phys Lett 71: 2785, 1997.
55. S Sauvage, P Boucaud, JM Gerard, V Thierry-Mieg. Phys Rev B58:15, 1998.
56. SG Romanov, CM Sotomayer Torres. In: J Rarity, C Weisbuch, eds. Microcavities and Photonic Gandgaps. Amsterdam: Kluwer, 1996, p 275.
57. R Mayoral, J Requena, JS Moya, C Lopez, A Cintas, H Miguez, F Meseguer, L Vazquez, M Holgado, A Blanco. Adv Mater 9:257, 1997.
58. VN Bogomolov, SV Gaponenko, AM Kapaitonov, AV Prokoviev, AN Ponyavina, NI Silvanovich, SM Samoilovich. Appl Phys A 63:613, 1996.
59. SG Romanov, AV Fokin, VY Butko, NP Johnson, CM Sotomayo Torres, HM Yates, ME Pemble. Proc Int Conf Phys Semiconductors, World Sci, Singapore, 1996, p 3219.
60. S Hotchandani, I Bedja, RW Fessenden, PV Kamat. Langmuir 10:17, 1994.
61. T Ogino, Y Homma, H Hibino, Y Kobayashi, K Sumitomo, K Praghakaran, H Omi. J Surf Sci Soc Jpn 19(9):557, 1998.

13.3 FORMATION OF FINE COMPOSITES

MASUMI KOISHI and HIROTAKA HONDA
Science University of Tokyo, Hokkaido, Japan

13.3.1 Introduction

The physical or chemical modification of powder surfaces is a method for the preparation of highly functional and specific surface-characterized materials. High functionality and good characterization are important both in effective improvement and in new service applications of the original powder. Furthermore, the importance of the powder surface modification is also recognized from the point of view of being capable of improving and controlling the various powder properties, for example, dispersability, wettability, rheological properties, optical properties, electrical properties, electronic properties, reactivity, catalytic activity, and blendability.

The concept of microfabrication on powder surfaces can be used in many industrial fields, e.g., in pigments, printing inks, paints, foods, pharmaceuticals, detergents, cosmetics, dental materials, implant materials, copy toners, ceramics, cements, electrorheological materials, and metallurgy, etc. (1).

Throughout this section we refer to and discuss various dry process modification technologies for fine powders.

13.3.2 Dry Process Surface Modification

13.3.2.1 Mechanochemical Considerations of Powder Mixing and Coating of an Interactive Powder Mixture

The random mixing theory and mixing processes have been extensively explored (2). Randomization requires equally sized and weighted particles, with little or no surface effects, showing no cohesion or interparticle interaction, to achieve the best results; it cannot be applied to all practical mixing situations, especially where cohesive or interacting particles are mixed.

Thereafter, the dynamic mixing behaviors of fine cohesive particles adhered to the surface of a coarser excipient was discussed as considerable importance in the manufacture of solid pharmaceuticals (3). The term ''ordered mixing'' was given to this phenomenon by Hersey (4). An ''ordered mixture'' can be produced by a dry process, simple dry mixing of fine and coarse particles. When interparticle interactions, such as van der Waals and coulombic forces, exist between the two types of particles, the fine particle adheres to the surface of the coarse particle: that is, an ordered mixture spontaneously forms. As described earlier, ordered mixing does not require equally sized or weighted particles, but involves particle interaction, i.e., van der Waals forces, surface tension, frictional pressure, electrostatic charge, or any other forms of adhesion (5). Ordered mixtures are frequently more homogeneous than random mixtures; their standard deviations are unaffected by sample size. To study ordered mixing, considering that real mixing operations are a process of disordering, Egermann et al. (6) redefined ordered mixtures so as to feature a degree of homogeneity higher than that conforming to random mixtures. Instead of ''ordered mixture,'' they used the term ''interactive mixture'' for mixtures in which fine particles adhered to the carriers.

Many applications of interactive powder mixing (formerly called ''ordered powder mixing'') in the pharmaceutical field have been studied (7). That is, dry mixing methods have been developed to modify, coat, or hybridize particles with various materials. For example, granules and particles were coated with carnauba wax and magnesium stearate to develop sustained-release products (8). Drug-diluent hybrid powders in which drugs firmly adhere to the diluent surface were also prepared by the dry-mixing method. As far as durability of units is concerned, the hybrid powder seems to be more advantageous than the ordered mixture, and a surface-reforming system is considered the most effective of the dry mixing methods. The drug-diluent hybrid powder can be obtained by the dry impact blending method in a few minutes.

13.3.2.2 Hybridization of an Ordered Powder Mixture

Preparation of Ordered Powder Mixtures and Hybridization of the Mixtures. As already described, the concept of ordered mixing, in which one component consisting of fine particles adheres to a second component of coarser carrier particles, has been proposed as a theoretically ideal system capable of producing near-perfect homogeneous mixes. Such systems have previously been considered unsuitable for the production of homogeneous mixes because of the difference in particle size of the two components, which has been shown to produce segregation of the ingredients of random powder mix. However, an almost perfect homogeneous mix has been attempted, using triboelectrification as a novel mixing process. Figure 13.3.1 shows the three-dimensional system of fabrication technology. The formation of a perfect ordered mixture arises from just a surface modification of large particles by fine particle materials, depend on their mixing ratio and mutual triboelectrical interaction.

We developed a dry blending and an impact blending method (8) that has an impulsive force in addition to the interactive powder mixing procedure. It was clearly confirmed in previous works (8) that the dry impact blending procedure was capable of preparing composite or encapsulated-type particles; if inorganic fine particles were used as coating materials, the particles were fixed and embedded in the surface of the core particles, and if polymer or metallic fine particles were used as coating materials, the particles were partially melted for continuous film formation on the

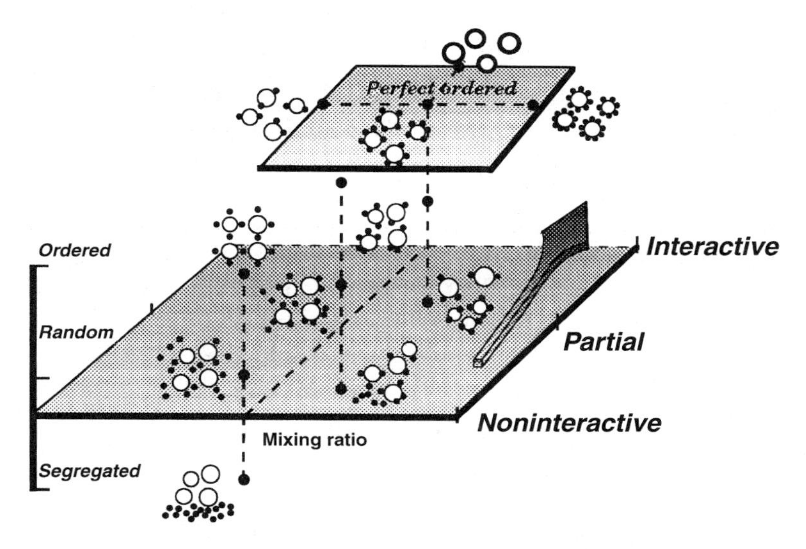

Fig. 13.3.1 Three-dimensional system of fabrication technology.

surface of the core particles. Furthermore, the randomized configuration of many fine particles on each core particle in the interactive mixing system was rearranged to an ordered state, giving a monolayer particle coated powder. Schematic preparation processes of ordered units and composite or encapsulated-type particles are shown in Fig. 13.3.2. As can be seen in Figure 13.3.2, the dry blending preparation method comprises a two-step blending process. The first step is a mechanical blending process (named dry blending) of the core and coating materials for the preparation of

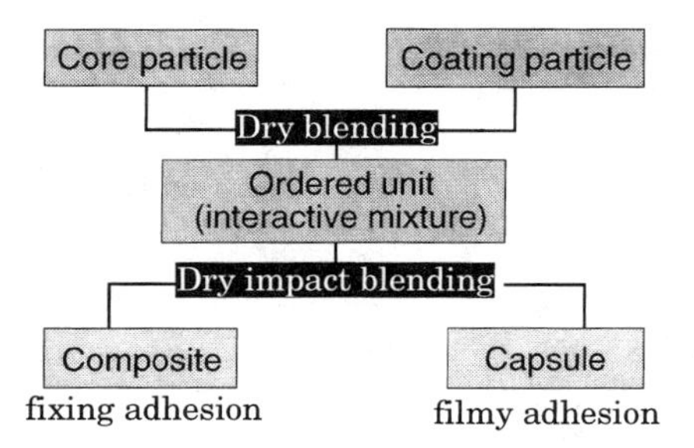

Fig. 13.3.2 Schematic preparation processes of ordered units and composite or encapsulated-type particles.

an interactive mixture formed by the adhesion of the fine coating particles on the surface of the coarse core particles. In general, during this stage, a centrifugal impeller rotating-type batch mixer was used (Mechanomil MM-10 type, Okada-seiko Co. Ltd., Tokyo). After the interactive mixture was prepared, the following coating process is responsible for the final preparation of the composite or encapsulated particles.

The second step is a mechanical impact blending process (termed dry impact blending) of the interactive mixture for the preparation of the composite or encapsulated particles. An impact-type hybridization machine with jacket was used (Hybridizer type-0, Nara Machinery Co. Ltd., Tokyo; the system is now patented). Figure 13.3.3 is a schematic diagram of the machine for producing the mechanical impacts. A thermometer was set in the circulation route to measure the inner atmospheric temperature of the machine.

The machine is surrounded by a jacket through which heat medium or coolant is circulated. Water was used as the coolant in these experiments. In this machine, powders (the interactive mixture) are guided through a feed chute into the center of the machine and blown off in a peripheral direction by the centrifugal force generated by the high-speed rotor. The dispersed powder particles hit striking pins rotating at 10,000–16,000 rev min^{-1}. Consequently, the powder receives mechanical impacts by these collisions on its surface. The powder reaching the periphery of the machine reenters the circulation route and returns to the center of the machine. This process is continually repeated. Since the powder particle (interactive mixture) repeatedly impacts on the surface, fine powder particles become attached or adhered to and arranged on the surface of the core particles. The time required for the circulation was quite short; the powder circulated at 300–600 rev min^{-1}, depending on the

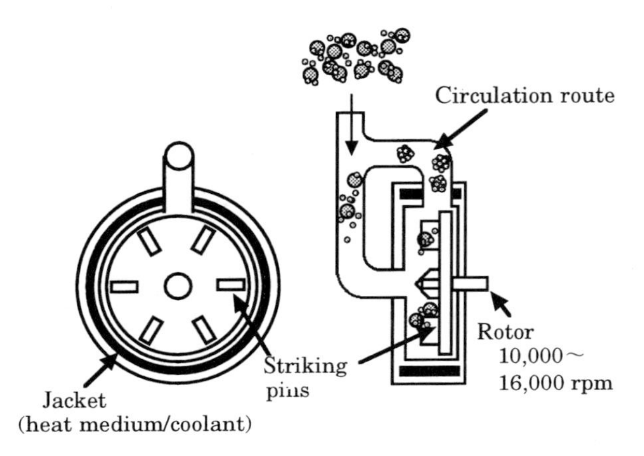

Fig. 13.3.3 Schematic diagram of the machine for producing the mechanical impacts.

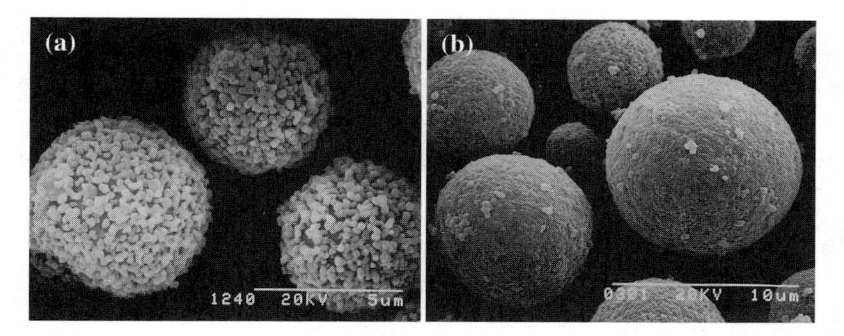

Fig. 13.3.4 Typical SEM photographs of the batch mixing (dry mixing) and the dry impact blending of titanium dioxide and nylon 12: (a) batch mixing of 30 wt% titanium dioxide (0.3 μm) and 70 wt% nylon 12 (5 μm); (b) dry impact blending of 30 wt% titanium dioxide (0.3 μm) and 70 wt% nylon 12 (3–10 μm).

revolution rate of the rotor. This means that powder particles experience a number of impacts for a short operating time.

Figure 13.3.4 shows typical SEM photographs of coated particles prepared by the dry blending or dry impact blending of titanium dioxide and nylon 12 particles.

As can be seen in Fig. 13.3.4b, many fine titanium dioxide particles are clearly fixed on the surface of each nylon 12 particle and are rearranged for monolayer particle adhesion by the dry impact blending treatment.

Generally, in the batch mixing of particulate matter, characterization of the resultant mixture is generally classified into two major groups: One involves only free-flowing particles and the other contains cohesive particles. Since the free-flowing mixture, the so-called noninteractive mixture, generally does not have a repulsive or an attractive interaction mechanism, the individual particle is able to move independently. However, the interactive mixture consisting of cohesive particles has some interparticulate bonding mechanism, and permits particles to move only with an associated unit of particles. During the first step of our experimental processes, an interactive mixture of titanium dioxide and nylon 12 particles was prepared by the dry blending method. However, many titanium dioxide particles randomly and in blocky manner adhere to the surface of nylon 12 particles in the interactive mixture, as shown in Fig. 13.3.4a. The geometrical morphology of the formed structure of adhered silica particles is very similar to that of the fractal structure of colloidal aggregates analyzed by a diffusion-limited aggregation model (9). Batch mixing over 60 min produces the fracture or deformation of the core particles and does not contribute to the preparation of a monolayer particle coated state. During the second step, 10 min of dry impact blending enables titanium dioxide randomly adhered on the surface of the nylon 12 particles to rearrange into an ordered state with some specific lattice structure, as observed in Fig. 13.3.4b.

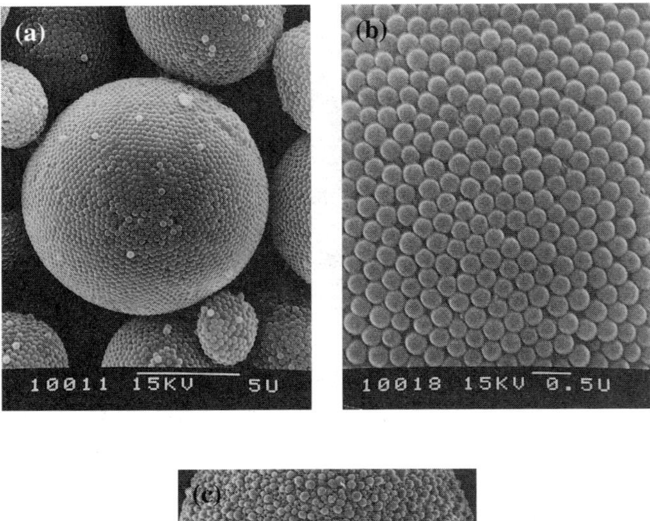

Fig. 13.3.5 Typical SEM photographs of the dry impact blending: (a) dry impact blending of silica (0.3 μm) and polyethylene (2–15 μm); (b) silica arrangement on polyethylene surface. (c) Pollen grain of *Clintonia udensis*. (From N Miyoshi, Okayama University of Science, with permission.)

Figure 13.3.5 shows typical SEM photographs of the silica-coated polyethylene particles prepared by dry impact blending utilizing mechanochemical treatment and pollen grain of *Clintonia udensis*.

The monolayer particle coated powder was formed by dry impact blending, when the ratio of the particle sizes (coating/core) was 0.3/3–12, as can be seen in Fig. 13.3.5a. For the combination of various sizes ratio, the arrangement of silica particles on the surface of the polyethylene always formed a monolayer particle coated powder. Furthermore, as shown in Fig. 13.3.5b, geometrical arrangement of the formed structure of adhered silica particles is observed as close packing. Judging from the photos, these morphologies are very similar situation to the surface appearance of a pollen grain of *Clintonia udensis*.

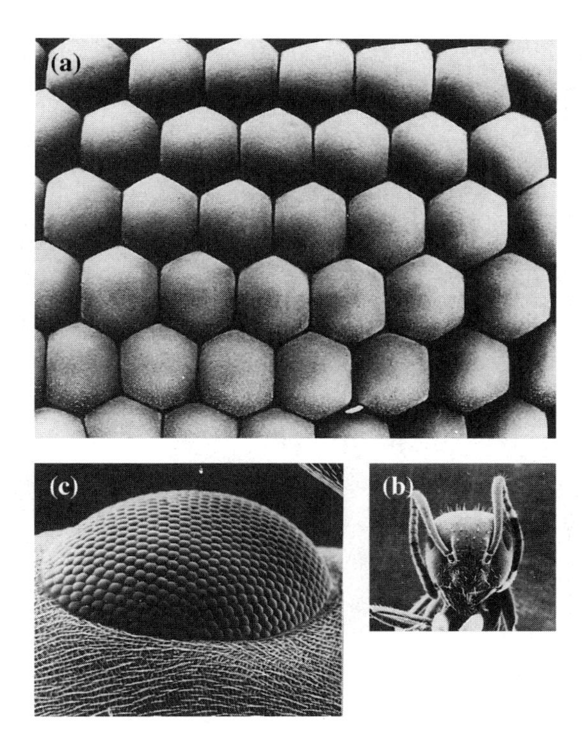

Fig. 13.3.6 Compound eye of insect *Lasius niger linne:* (a) arrangement of close packed ommatidia; (b) head composed of compound eye antennas and mouth; (c) ordered arrange ommatidia on the convex spherical surface (compound eye). (From O. Hojiro, with permission.)

Figure 13.3.6 shows a compound eye of the insect *Lasius niger linne.* The head composed of compound eye, antenna, and mouth is observed in Fig. 13.3.6c. Many ommatidia are ordered arrange on the convex spherical surface, as observed in Fig. 13.3.6b, and also are close-packed completely, as can be seen in Fig. 13.3.6a. Natural pollen grain and insect compound eye are the final approaching standard models for study of the preparation of composite and encapsulated powders.

Figure 13.3.7 shows scanning electron microscopy (SEM) photographs of the surface of the polyethylene particle after the silica particles were peeled off. The specimen was prepared in the following way. After the composite particles were potted in epoxy resin, the dried resin block was cut using a microtome to produce fine sections. The fracture surface appearance of the polyethylene was then observed under a microscope. The mean depth penetration into the surface of the core particles could be measured using the SEM photographs. Silica 0.3 μm in diameter was embedded in the surface of the polyethylene particles at a depth of 0.03 μm. In

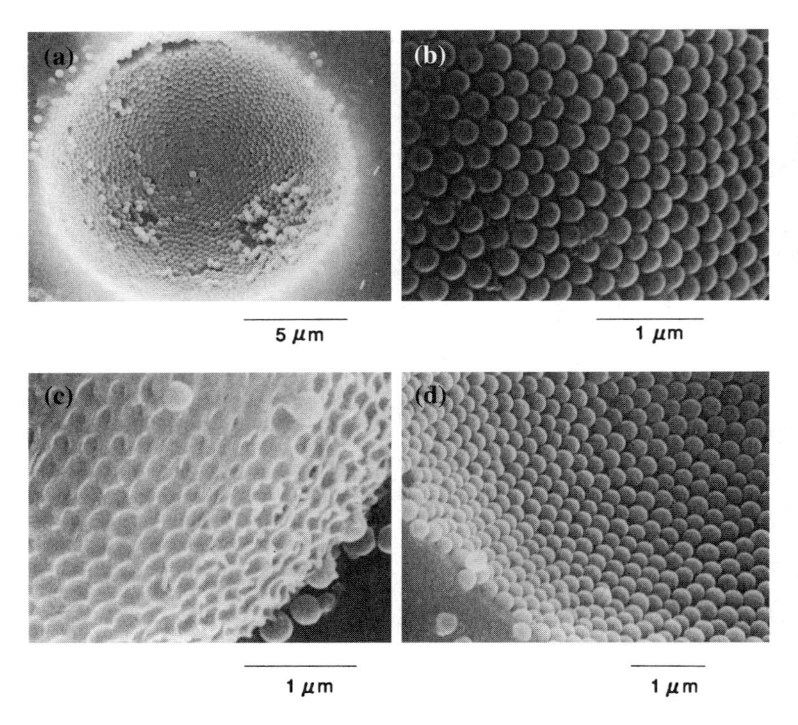

Fig. 13.3.7 Typical SEM photographs of a bare polyethylene surface and silica arrangements, embedded in epoxy resin after fracture: (a and d) detached silica (0.3 μm) arrangement after fracture; (b) silica (0.3 μm) arrangement before fracture; (c) bare polyethylene surface after fracture (10 μm).

these experiments, the depth of depression increased with the decrease in the size of the core particle. Also, the surface morphology of the composite particles was influenced by the combination of silica and polyethylene sizes. Good composite particles were formed, consistent with a decrease in the ratio of the size of the coating particle to that of the core particle. The particle size ratio of the core and coating particles was an important factor in the preparation of the monolayer particle coated powder. More basic discussion concerning the silica particles, deeply embedded in the surface of the polyethylene particles, has already been presented in a previous paper (10).

In contrast, the three- or two-dimensional morphologies of colloidal aggregates via Brownian particle trajectories show a fractal-like structure. One of the most prominent features of the surface deposits formed by the diffusion-limited aggregation mechanism is the formation of isolated treelike clusters (9). In our experiments, the surface morphology of the silica-coated polyethylene composite prepared by the dry impact blending method was a monolayer particle-coated structure. From thermophysical considerations, the random arrangement of coating particles on the surface of the core particles never spontaneously changes into an ordered arrangement. Therefore it is confirmed that the monolayer particle-coated structure is formed by the action of impulsive forces during the preparation. These SEM observations show that the particle size itself, the particle size ratio of the core and coating particles, and the action of impulsive forces are important factors in the effective preparation of a monolayer particle-coated powder.

Estimation of Binding Energy for the Preparation of a Monolayer Particle-Coated Powder. The main adhesion forces between particles are divided into five groups: 1) attraction forces (van der Waals interaction, electrostatic interaction, etc.); 2) physicochemical forces (mechanochemical interaction, chemical reaction, sintering, etc.); 3) interfacial forces and capillary pressure at the freely movable liquid (liquid bridge); 4) adhesion and cohesion forces not as a freely movable liquid (highly viscous bonding); and 5) form-closed adhesion (interlocking). The first group (1) is considered the main process to investigate the preparation of a monolayer particle-coated powder.

In this section, we calculate the energy for the preparation of monolayer particle-coated powder. Firstly, the energy is calculated on the assumption that the pure electrostatic interaction works mainly on adhesion between particles (11). Second, the energy is calculated on the basis of the assumption that the electrostatic and van der Waals interaction work together (12).

The model for the calculation on first assumption is schematically illustrated in Figure 13.3.8. The core particle having a radius a is coated by the coating particle with a radius b. The core has an electric negative charge $-Q$ at its center whereas the coating particles have positive charges $+q$. The positive charge is uniformly distributed over the surface of the core particle with a charge density $\sigma = Q/4\pi a^2$.

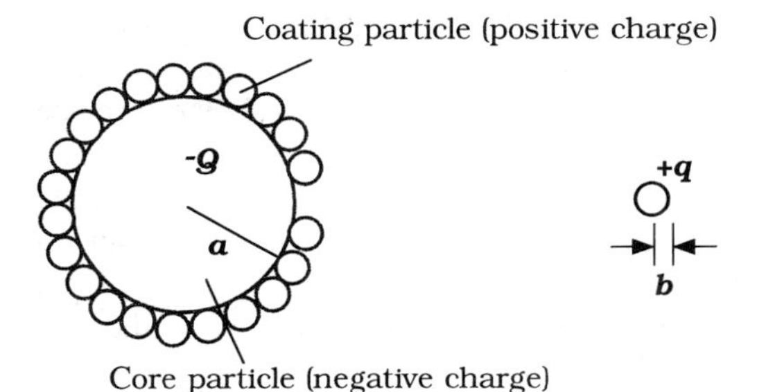

Fig. 13.3.8 Schematic model for calculation of the adhesion energy on the assumption that the electrostatic interaction is the main one.

It is considered that the energy needed to bring the final coating particle from infinity and adhere it to the core can be regarded as a binding energy. To estimate the binding energy, let us remove one coating particle from the monolayer particle-coated powder and make a circular hole in the sphere. We calculated the energy E done to bring one particle from infinity onto the circular hole in the sphere:

$$E = \frac{Q}{8\sqrt{2}\pi\epsilon_0 a}(1 - \cos\theta_0)^{3/2} \tag{1}$$

where θ_0 is the angle subtended by the surface of the circular hole as viewed from the center, and ϵ_0 is the dielectric constant of vacuum.

The assumption that b is much smaller than a gives a relation of $\sin\theta_0 = b/a$, allowing the following approximation:

$$\begin{aligned} 1 - \cos\theta_0 &= 1 - \sqrt{1 - \sin^2\theta_0} \\ &= 1 - \sqrt{1 - (b/a)^2} \\ &= \frac{b^2}{2a^2} \text{ for } (b/a)^2 \ll 1 \end{aligned} \tag{2}$$

Substituting Eq. (2) in Eq. (1), we get the final form for the mean binding energy of the adhesion as a function of a, b, and Q:

$$E = \frac{Q^2}{32\pi\epsilon_0 a}\left(\frac{b}{a}\right)^3 \tag{3}$$

If the adhesion mechanism between particles with different size is mainly subject to this attraction due to the electrostatic interaction, it is deduced that the increase in the ratio of core particle size to the coating particle size is advantageous for the formation of the monolayer particle coated powder.

Next, the energy of the monolayer particle coated system is calculated on the basis of the second assumption, that the electrostatic and van der Waals interaction work together. The system is schematically illustrated in Figure 13.3.9. As an outline of the monolayer particle-coated formation, a negatively charged spherical core particle with a radius a is uniformly coated on its surface by any number of positively charged spherical coating particles with radius b. D is the distance between centers of core and coating particle and is given by $D = a + b + l$. D' is the distance between centers of two neighboring coating particles, given by $D' = 2b + l'$; l and l' are the gap between the surfaces of the two spheres. When the number of coating particles in the system is N and each charge of core and coating particle is assumed to be $-Q$ and $+q$, respectively, the total positive charge is given by $Q' = Nq$.

The adhesive energy u_{tot} per one coating particle is then composed of the sum of four parts, i.e., $u_{e1,2}$, $u_{e2,2}$, $u_{v1,2}$, $u_{v2,2}$. The term $u_{e1,2}$ is a contribution from the electrostatic interaction between a core and each coating particle, and $u_{e2,2}$ is that between coating particles, $u_{v1,2}$ is a contribution from the van der Waals interaction between a core and each coating particle, and $u_{v2,2}$ is that between coating particles.

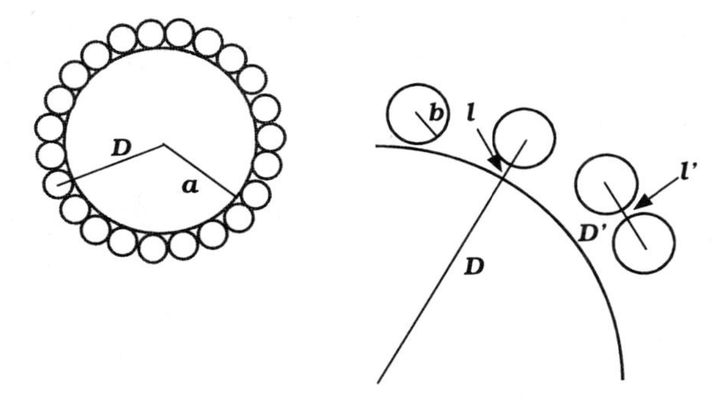

Fig. 13.3.9 Schematic model for calculation of the adhesion energy on the assumption that the electrostatic interaction and van der Waals interaction work together.

The energy due to the electrostatic interaction between a core and each coating particle is given by:

$$u_{e1,2} = \frac{1}{4\pi\epsilon_0}\frac{-Qq}{D} \tag{4}$$

The energy due to van der Waals interaction calculated by Hamaker is based on the microscopic theory of London–Heitler, and is expressed as:

$$u_{v1,2} = -\frac{A_{1,2}}{6}\left[\frac{2ab}{D^2-(a+b)^2} + \frac{2ab}{D^2-(a-b)^2} + \ln\frac{D^2-(a+b)^2}{D^2-(a-b)^2}\right] \tag{5}$$

where $A_{1,2}$ is the so-called Hamaker constant for two substances 1 and 2, namely, a core and the coating particles.

It is considered that the radius of the coating particles is very small in comparison with that of the core, so that coating particles are distributed on the surface of a core particles with some two-dimensional lattice structure. If n is the number of coating particles per unit area for an appropriate lattice structure on the surface of a core, the total number of coating particles is given by $N = 4\pi D^2 n$, and the charge per unit on the sphere becomes $\sigma_e = nq$, which in turn gives the total positive charge $Q' = 4\pi D^2\sigma_e = 4\pi D^2 nq$. Let dS_1 and dS_{12} then be surface elements at a distance r on the sphere of radius D for either the electrostatic or van der Waals interactions. The interaction energy per coating particle is generally written in the form

$$u_{i2,2} = \frac{1}{2N}\int u_i(r)dS_1dS_2 \tag{6}$$

where i denotes e or v, the factor $\frac{1}{2}$ occurs to avoid double counting for each pair of surface elements, and $u_i(r)$ is a kind of energy density depending on the type of interaction. The density usually depends only on the distance $r = 2D\sin(\theta/2)$, where θ is the angle subtended by the chord r as viewed from the center. Therefore the integration with respect to the other three angles except for angle θ is easily performed, and the equation becomes

$$u_{i2,2} = \frac{4\pi^2 D^4}{N}\int_{\theta_0}^{\pi} u_i\left(2D\sin\frac{\theta}{2}\right)\sin\theta d\theta \tag{7}$$

where a small angle θ_0 is defined by $\theta_0 = D'/D$. It is convenient for ease of performing the integration to change the integral variable from θ to t with the transformation $t = \sin(\theta/2)$, which reduces Eq. (7) to a simpler form:

$$u_{i2,2} = \frac{(4\pi D^2)^2}{N}\int_{\sin(\theta_0/2)}^{1} u_i(2Dt)tdt \tag{8}$$

In the case of electrostatic interaction, the energy density has the form

$$u_e(r) = \frac{1}{4\pi\epsilon_0}\frac{\sigma_e^2}{r} = \frac{1}{8\pi\epsilon_0 D}\frac{1}{t} \tag{9}$$

Inserting Eq. (9) into Eq. (8), and after integration has been completed, the energy is given by

$$u_{e2,2} = \frac{(4\pi D^2\sigma_e)^2}{8\pi\epsilon_0 DN}\left(1 - \sin\frac{\theta_0}{2}\right) \simeq \frac{(Q')^2}{8\pi\epsilon_0 DN}\,. \tag{10}$$

To obtain the final form of Eq. (10), θ_0 is set to be approximately zero because of the uniformity of positive charge on the sphere and the long-range nature of the electrostatic interaction.

In the case of van der Waals interaction, the energy density is given by the extension of Eq. (5) to the system with a particle density n on the sphere:

$$\begin{aligned} u_v(r) &= -\frac{n^2 A_{2,2}}{6}\left[\frac{2b^2}{r^2-(2b)^2} + \frac{2b^2}{r^2} + \ln\frac{r^2-(2b)^2}{r^2}\right] \\ &= -\frac{n^2 A_{2,2}}{12}\left(\frac{b^2}{D^2t^2-b^2} + \frac{b^2}{D^2t^2} + 2\ln\frac{D^2t^2-b^2}{D^2t^2}\right) \end{aligned} \tag{11}$$

where $A_{2,2}$ is the Hamaker constant for the coating particle material. By substituting Eq. (11) into Eq. (8) and further calculations, the following result for the interaction energy are derived:

$$u_{v2,2} = -\frac{NA_{2,2}}{12}\left[\left(1-\frac{b^2}{2D^2}\right)\ln\left(1-\frac{b^2}{D^2}\right) + \left(\sin^2\frac{\theta_0}{2}-\frac{b^2}{2D^2}\right)\ln\frac{\sin^2\frac{\theta_0}{2}}{\left|\sin^2\frac{\theta_0}{2}-\frac{b^2}{D^2}\right|}\right] \tag{12}$$

It should be noticed that the angle θ_0 cannot be set to zero in this case because of the singularity due to the short-range nature of the van der Waals interaction. Therefore the sum of Eqs. (4), (5), (10), and (12) gives the total interaction energy per coating particle, namely, the adhesive energy.

Let us now introduce the following dimensionless variables:

$$x = l/2a \quad y = b/a \quad z = l'/l \tag{13}$$

The final form of the adhesive energy within the present assumption is then

$$\begin{aligned} u_{tot} &= u_{e1,2} + u_{e2,2} + u_{v1,2} + u_{v2,2} \\ &= \frac{1}{8\pi\epsilon_0 a(1+y+2x)} q(Nq - 2Q) \\ &\quad - \frac{A_{1,2}}{12}\left(\frac{y}{x^2+xy+x} + \frac{y}{x^2+x+y+xy} + 2\ln\frac{x^2+x+xy}{x^2+xy+x+y}\right) \\ &\quad - \frac{NA_{2,2}}{12}\left\{\left[1 - \frac{y^2}{2(1+y+2x)^2}\right]\cdot\ln\left[1 - \frac{y^2}{(1+y+2x)^2}\right]\right. \\ &\quad + \left[\sin^2\frac{y+xz}{1+y+2x} - \frac{y^2}{2(1+y+2x)^2}\right] \\ &\quad \left.\cdot\ln\left[\frac{\sin^2\dfrac{y+xz}{1+y+2x}}{\left|\sin^2\dfrac{y+xz}{1+y+2x} - \dfrac{y^2}{(1+y+2x)^2}\right|}\right]\right\} \end{aligned} \tag{14}$$

The second line in Eq. (14) is a contribution from the electrostatic interaction; it can have a repulsive effect only when $Q' = Nq > 2Q$, whereas a contribution from the van der Waals interaction produces an attraction. The electrostatic interaction between particles with different electrical charges is not always advantageous for forming the monolayer particle-coating powder.

13.3.3 Application of Dry Process Surface Modification

13.3.3.1 Particle Shape Modification

In the preparation of particles of different shapes, various kinds of shape separation method have been investigated by means of inclined plates, horizontally rotating cylinders, and screens. On the other hand, a positive way to modify particle shape that was tried to make irregular-shaped particles spherical by utilizing a hybridizer. This was originally a mechanical method of dry impact blending to produce surface composite particles. As a result, it was found possible to adjust particle shape stepwise by setting the proper operating conditions (13). However, the following problems were pointed out to control the particle shape with various materials and to examine the shape modification process by this method: (1) Particle samples of relatively low ductility are suitable to prepare particles of different shape step-by-step, and (2) the shape evaluation method is not sufficient with an approximate ellipse equivalent to the perimeter and area of a projected particle outline. In a recent study, these points are improved by adopting their alternative and the particle shape modification process is analyzed on the basis of a plastic deformation model of particulate solids (14). The final conclusions of research results are brief summarized as the following: The particle shape modification process was investigated by the dry impact blending method. A simplified deformation model was proposed on the basis of the stress-strain relation. The fitness of the model was examined using the experimental results with stainless steel powder. As a result, the following were identified:

1. The model sufficiently conforms to the change in the shape index with longer treatment time, and it was proved that processing particulate materials of smaller ductility, the particle shape is more adjustable stepwise in terms of the shape index.
2. A bigger collision force due to high-speed impact was effective to produce more spherical particles, but it has a risk of crushing to generate fine particles.
3. The effect of powder concentration was not remarkable with stainless steel particles, but the model indicates the possibility of spherical deformation with materials of smaller strength.
4. The difference in the shape modification characteristics owing to the particle diameter was not recognized, as is the case with copper particles.
5. The unevenness coefficient of particle surface decreased with larger shape index and had a critical value. Therefore, the particle shape modification process was considered in such a way that the submicro uneven particle surface was smoothed at first and, in consequence, the macro shape was processed to be spherical.

Figure 13.3.10 shows the typical spheroidal deformation of stainless steel particles with treatment time. As compared with the result of copper particles, the rate of spheroidal deformation was lower. Thus, the stainless steel particles were easier for stepwise adjustment of the shape index with treatment time.

13.3.3.2 Sintering of Surface-Modified Composite

Hydroxyapatite (HAP), with basically the same crystal structure as Ca-deficient, carbonate-containing hydroxyapatite, is compatible with and reactive in a live human body. However, sintered HAP prepared by treating fine HAP particles under elevated temperature and pressure has insufficient mechanical properties, in particular fracture toughness, which greatly limits its commercial applicability. It is rarely implanted alone. On the other hand, zirconia, particularly partially stabilized zirconia (PSZ),

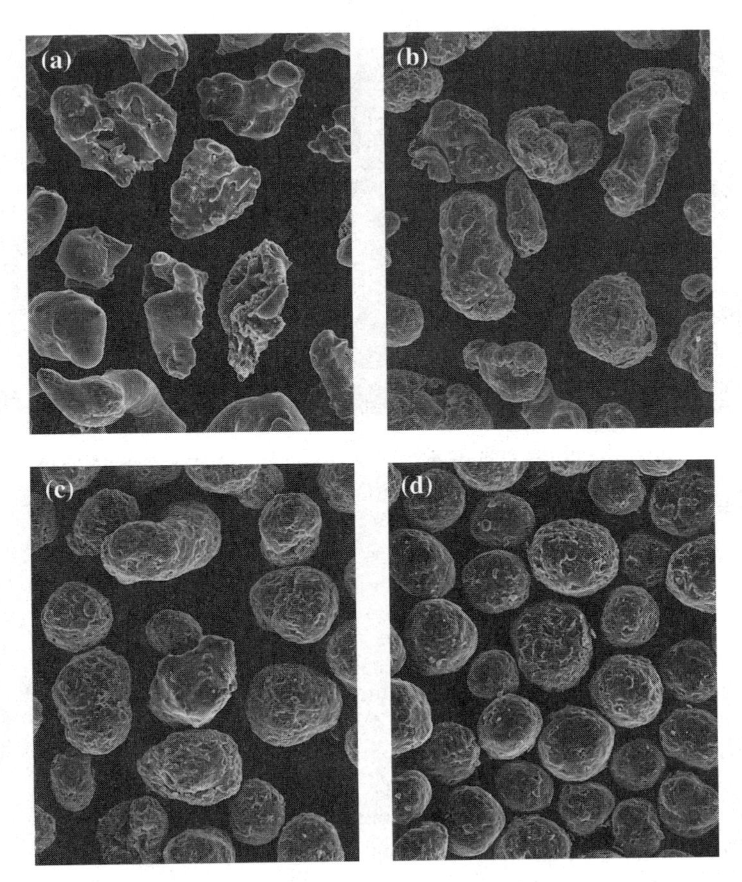

Fig. 13.3.10 SEM pictures of stainless steel particles prepared for different treatment times: (a) $t = 0$ s; (b) $t = 30$ s; (c) $t = 33$ s; (d) $t = 1200$ s. *Key:* 15–20 μm stainless steel particles were used. (From M Odani, Hokkaido Industrial Research Institute, and K Shinohara, Graduate School of Hokkaido University, with permission.).

has been attracting attention as an implantable material. It is harmless to a living body and has good mechanical properties—in particular, fracture toughness. However, it is inert and cannot be expected to show any reactivity in a living body. Our study focused on preparing materials with high reactivity in a living body and improving the fracture toughness by sintering a mixed compact of fine particles of HAP and PSZ (15). In consideration of the problem involved in the binary HAP/PSZ sinter, fine HAP particles were coated with the fine PSZ particles, using a dry impact blending method, and the composite particles were sintered in order to investigate the structures and the fracture toughness of the sinter.

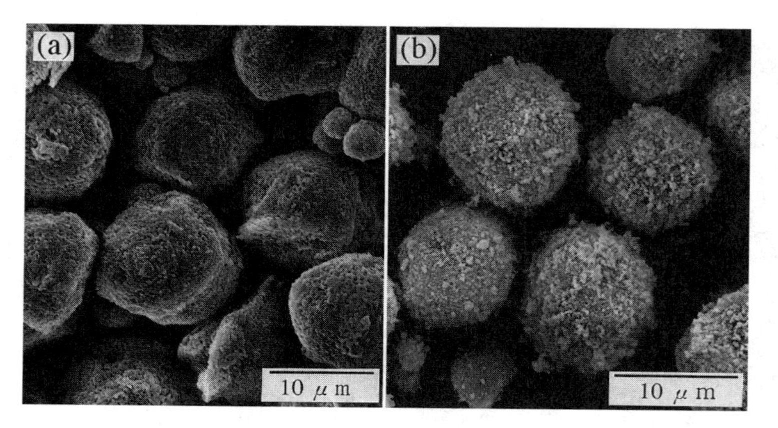

Fig. 13.3.11 SEM photographs of the spherical HAP particles and HAP/PSZ composite particles (HAP/PSZ weight ratio = 10/10): (a) spherical HAP particles; (b) HAP/PSZ composite particles. (From Ref. 15.)

Figure 13.3.11 presents SEM photographs of the commercial spherical HAP particles, heat-treated at 880°C for 2 h (a), and the HYB(hybridization system)-prepared HAP/PSZ composite particles (HAP/PSZ = 10/10 by weight) (b). The spherical HAP particles are agglomerated primary particles, with a number of irregularities on the surfaces. In contrast, the composite particles are HAP particles, completely coated with the PSZ particles. It is considered that the PSZ particles are sufficiently fine, at 0.12 μm average size, to fill the cavities in the HAP particles, when the latter particles are bombarded with the former. Figure 13.3.12 shows the cross sections of sintered bodies of HAP/PSZ composite particles. The sintered CIP (cold isostatic pressing)-prepared green compact consists of spherical HAP particles present independently in the PSZ phase, where the HAP particles nearly retain their original shapes and sizes, essentially without contacting each other. It is apparent that the PSZ particles fairly grow in the PSZ phase and there are a number of voids between them. The spherical HAP particles are cut at the sinter sections, which indicates that the HAP and PSZ particles are bonded tightly to each other. The HAP particles are present in the PSZ phase and do not contact each other; they are also in the HP (hot pressing)-prepared composite (Fig. 13.3.12b), where the HAP is represented by the dark portion and the PSZ phase by the networked white portion. The PSZ particles glow less than those in the CIP-prepared composite, mainly as a result of the lower sintering temperature (by 100°C) and the much shorter sintering time (one-twelfth). It is also observed that the number of large pore is smaller and the spherical HAP particles are cut at the sinter sections. As stated previously, we have used a new fabrication technique to prepare a HAP/PSZ composite powder.

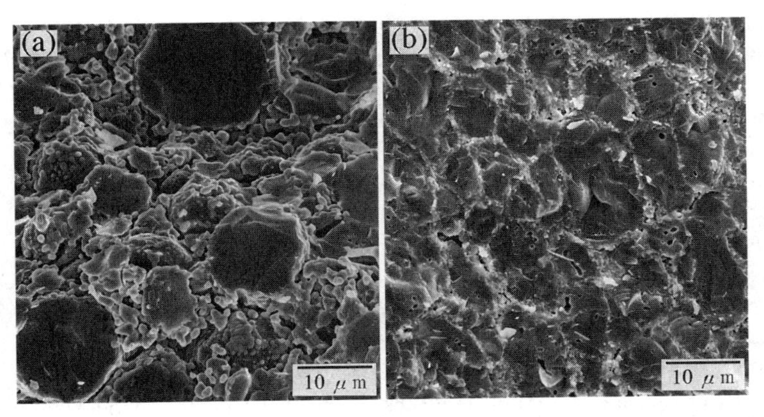

Fig. 13.3.12 SEM photographs of the cross sections of sintered bodies of HAP/PSZ composite particles (HAP/PSZ weight ratio = 10/5): (a) CIP-electric, and (b) HP. (From Ref. 15.)

From the experimental results concerning internal structures, phase transformation, and fracture toughness of the sinter, the following information can be summarized.

1. PSZ was present in the sinter as a continuous phase in a three-dimensional network structure, in which the HAP particles were distributed independently.
2. There were relatively large regions, consisting only of HAP in the sinter section, which strongly suggests the possibility of a hot-pressed composite to maintain excellent biocompatibility.
3. Tetragonal zirconia was partly transformed into the cubic phase during the sintering process of the UAP (uniaxially pressed) and CIP-prepared green compacts, and was accompanied by the formation of α-TPC. The transformation tends to accelerate as the PSZ quantity increases.
4. No phase transformation of zirconia was detected, using x-ray diffraction analysis, in the hot-pressed sinter at any of the HAP/PSZ ratios used in this study.
5. The sinter fracture toughness tends to increase as the PSZ quantity increases. The highest fracture toughness observed in this study was 2.8 $MPa/m^{1/2}$ in the hot-pressed sinter at HAP/PSZ = 10/10.

13.3.3.3 Surface Modification in the Field of Powder Technology

The dry impact blending method has recently been used in the investigation of surface modification techniques of particles in the field of powder technology. In the pharmaceutical field, many applications of interactive powder mixing (formerly called ''ordered powder mixing'') have been studied (8) as indicated earlier. Some papers (16) report 1) the coating and the encapsulation of an interactive powder mixture and its application to sustained release preparations, 2) characterization of a powder-coated microsponge prepared by dry blending method, 3) preparation of drug-diluent hybrid powders by dry processing, 4) drug dissolution from indometha-cin-starch hybrid powder prepared by the dry impact blending method, and 5) development of a sustained-release formulation of chlorpheniramine maleate using a powder-coated microsponge prepared by the dry impact blending method.

In the cosmetic field, a hybrid powder consisting of a spherical resin core, with its surface uniformely covered with either zinc oxide or aluminum chlorohydrate, was developed to quench offensive body odors (17). This hybridization technique enables us to overcome some of the aesthetic shortcomings that deodorizing actives possess, without sacrificing any deodorizing efficancy. Improvement in powder texture of zinc oxide and aluminum chlorohydrate when hybridized with spherical polyethylene powder is demonstrated by applying them directly to skin and also by measuring their coefficients of kinetic friction. The effect of hybridization on deodorizing efficacy was also investigated by headspace gas chromatography.

In the rodenticide field, a zinc phosphide microcapsule was prepared by dry impact blending (18). Zinc phosphide, a conventional rodenticide, was dispersed in a wax matrix by a dry impact blending method. The water resistance of zinc phosphide microcapsules prepared by this method was greatly improved; thus they are thought to be a useful rodenticide to distribute on the ground for field rats and mice. Moreover, the wax matrix is qualified as a slow-releasing preparation in the acid buffer solution. Therefore, this method is also useful for preparing rodenticides with chronic toxicity, such as warfarin or coumatet-ralyl. Furthermore, the taste repellency and lethal efficacy of the microcapsule were examined in mice and voles (*Microtus montebelli*). Stabilization of zinc phosphide was proven in the microcapsule without unfavorable influence on bait consumption and delay of the lethal period due to the slow-releasing property of the preparation.

In the cement field, fluidity of spherical cement and mechanism for creating high fluidity was studied (19). Spherical cement is cement in which the particle shape is round. The fundamental properties of spherical cement prepared by the dry impact blending method and the possibility of utilizing the concept of spherical cement to produce high-fluidity concrete, high-strength concrete, and high-durability concrete were discussed. The data and detail consideration are as following. The water/cementitious binder ratio of spherical cement concrete was decreased by 5–8%, and its unit weight of water was decreased by 14–30%, in the same level of fluidity as that of normal Portland cement concrete. The superplasticizer dosage could be reduced to a maximum 2/3. The mechanism for the creation of high fluidity was considered as follows.

1. The spherical shape is highly effective in increasing fluidity.
2. The particle size distribution, which is distributed over 3–40 μm, contributes to the high fluidity.
3. The adsorption of superplasticizer to spherical cement particle surface decreases by 40% because of a decrease of the specific surface area and localization of the intestitial phase with gypsum.
4. The initial heat evolution amount of spherical cement is smaller by 25% than that of normal Portland cement. This low activity at initial hydration contributes to the creation of high fluidity.

In high-performance liquid chromatographic (HPLC) column packings, application of nonporous silica ultramicrospheres was studied (10). Adsorption chromatography has been frequently used in the purification of proteins from biological materials and culture mediums. Crystalline hydroxyapatite (HA), which is a kind of calcium phosphate, is mainly employed as such an absorbant. HA has attracted notice also as an HPLC column packing for protein separations because chromatography using HA is demonstrated under mild conditions similar to physiological conditions in the body. We have already prepared HA-coated composite particles from orthodox fragile HA and polyethylene beads by the dry impact blending method. It demonstrates adsorption–desorption behavior of proteins similar to that of HA, and at the same time, it demonstrates satisfactory properties for an HPLC column packing (20). Furthermore, cattle bone powder (CBP) from natural resources was employed as a protein adsorbent instead of chemically synthesized hydroxyapatite (HA). Using CBP/40PE prepared from CBP and polyethylene beads (40 μm) by dry impact blending as an HPLC column packing, considerable correlation was observed between the elution concentrations of proteins and their p*I*. Such behavior was caused by the relatively large adsorption capacity for basic proteins. CBP/40PE could completely separate γ-globulin from bovine serum albumin (BSA) also as an open column chromatographic support, under relatively low concentration (20).

In the field of metallic powder applications, a method of plasma spray coating suitable for biomedical materials has been developed using titanium and calcium phosphate composite powder. By means of the mechanical shock process, the appropriate composite powder was prepared, and plasma sprayed on Ti substrate under a low-pressure argon atmosphere. A porous Ti coating layer was obtained in which the surface and the inside of the pores were covered thinly with hydroxyapatite. This surface coating is expected to show excellent bone ingrowth and fixation with bone (21).

Another application concerning the liquid-metal-cooled fast breeder reactors (LMFBR) was studied as development of advanced control rod materials for FBR (22). Fabrication tests and out-of-pile measurements were made of B4C/Cu cermet to obtain high-performance neutron absorber materials for LMFBR. A coating layer of Cu was formed on the surface of B4C/Cu powder, and then the coated B4C powder was hot pressed at 1050°C to form a B4C cerment. A high-density pellet of B4C/Cu cerment that contains about 70 vol% of B4C was successfully fabricated. The results of out-of-pile measurements indicated that the B4C/Cu cerment exhibited high thermal conductivity and excellent thermal shock resistance, which originated from the existence of metallic phase of Cu in B4C/Cu cerment (22).

For a combination of metallic and plastic powders, the preparation of metal capsule particles using a high-speed impact treatment was tried (23). Fine particles of gold, silver, or copper were blended with fine organic (nylon 12, polyethylene, polystyrene) or inorganic (glass bead, silica, fine iron) powders by hybridizer. Uniformity of metal layer thickness was identified by amount of metal component determined by XMA. It was recognized that uniform metal layers were formed on organic and on inorganic powders. The electric conductivity of epoxy molding consisting of 50% volume concentration of silver capsules was $2 \times 10^{-3}\ \Omega$ cm. EMI shielding effects of an acrylic resin layer consisting of 60% volume concentration of silver capsules are equal to EMI shielding effects of an acrylic resin layer containing the maximum permissible volume concentration of silver flake. The amount of silver in the silver capsules was half the amount of silver in the acrylic layer containing the maximum permissible volume concentration of silver flake.

On the other hand, wet process metal coating method was applied to the surface-modified nylon 12 particles (24). Nylon 12 particles modified by alumina or silica fine particles were given metallic coating by an electroless plating method.

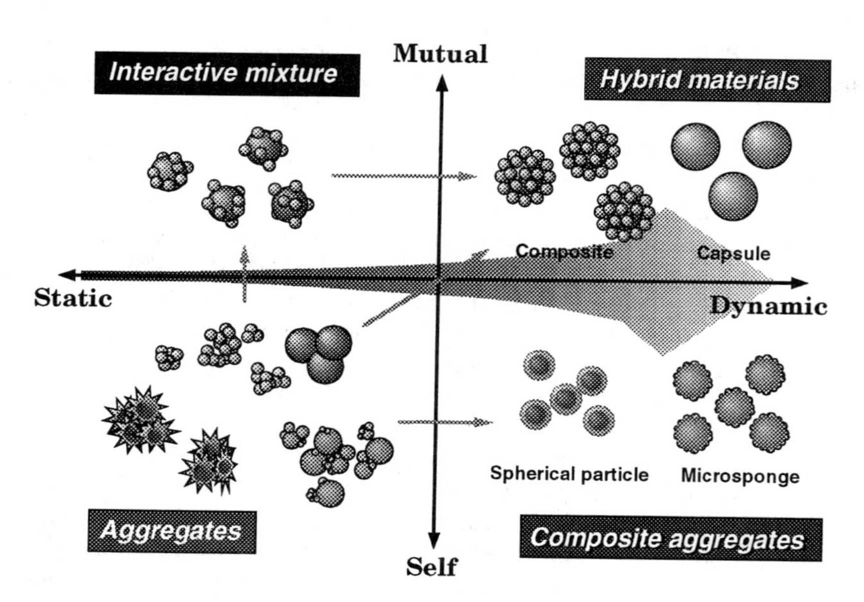

Fig. 13.3.13 Schematic drawing and a fabrication model of blending particles.

After fixing Pd catalyst on modified nylon 12 surfaces by various methods, they were coated by electroless Ni–P alloy plating. Alumina- or silica-modified nylon 12 was well wet by electroless plating liquid, and modified nylon 12 situations in a liquid were good dispersed. Nickel metal was deposited on silica or alumina surface. Finally, it was confirmed that the formation of metal layer was depended mainly on the method of Pd catalyst fixing.

As described already, many practical applications in dry process surface modification concern the interaction of particles. Figure 13.3.13 shows a schematic drawing and a fabrication model of blending particles. Under self or mutual harmonization during the dry impact blending treatment, final arranging and controlled composites can be obtained.

In a recent study, as the first trial case, a liquid injection technique was applied to a dry blending system (25). This introductory application concerns the study of the interaction of particles with injected liquid (solvent, polymer solution, colloidal solution, etc.) and will be reported elsewhere.

REFERENCES TO SECTION 13.3

1. M Koishi, ed. Biryushi Sekkei (Fine Particulate Design). Tokyo: Kogyo Chosakai, 1987, pp 32–35.
2a. PMC Lacey. Trans Instn Chem Eng 21:53–59, 1943.
2b. KR Poole, RF Taylor, GP Wall. Trans Instn Chem Eng 42:T305–T315, 1964.
2c. N Harnby. Chem Eng 214:CE270–CE271, 1967.
3. DN Travers, RC White. J Pharm Pharmacol 23:260S–261S, 1971.
4a. JA Hersey. J Pharm Sci 63:1960–1961, 1974.
4b. JA Hersey. Powder Technol 11:41–44, 1975.
5a. JN Staniforth, JE Rees. Powder Technol 30:255–256, 1981.
5b. JN Staniforth, JE Rees. J Pharm Pharmacol 34:69–76, 1982.
5c. JN Staniforth, JE Rees, FK Lai, JA Hersey. J Pharm Pharmacol 33:485–490, 1981.
5d. WJ Thiel, LT Nguyen. J Pharm Pharmacol 34:692–699, 1982.
5e. WJ Thiel, LT Nguyen. J Pharm Pharmacol 36:145–152, 1984.
6a. H Egermann. J Pharm Sci 74:999–1000, 1985.
6b. H Egermann. J Pharm Pharmacol 41:141–142, 1989.
6c. H Egermann, NA Orr. Powder Technol 36:117–118, 1983.
7. T Ishizaka, H Honda, Y Kikuchi, K Ono, T Katano, M Koishi. Powder Technol 41: 361–368, 1989.
8a. M Koishi, T Ishizaka. In: ST Hsieh, ed. Controlled Release Systems: Fabrication Technology, vol 1. Boca Raton, FL: CRC Press, 1988, pp 109–142.
8b. M Koishi, H Honda, T Ishizaka, T Matsuno, T Katano, K Ono. Chimicaoggi lugio-agosuto 43–45, 1987.
8c. T Ishizaka, H Honda, K Ikawa, N Kizu, K Yano, M Koishi. Chem Pharm Bull 36: 2562–2569, 1988.
8d. H Yoshizawa, M Koish. J Pharm Pharmacol 42:673–678, 1990.
8e. T Ishizaka, H Honda, M Koishi. J Pharm Pharmacol 45:770–774, 1993.
9. G Bushell, R Amal. J Colloid Interface Sci 205:459–469, 1998.

10. F Honda, H Honda, M Koishi. J Chromatogr 609:49–59, 1992.
11. H Honda, M Kimura, T Matsuno, M Koishi. Chimicaoggi-june 9:21–26, 1991.
12. H Honda, M Kimura, F Honda, T Matsuno, M Koishi. 1994. Colloids Surf A Physicochem Eng Aspects 82:117–128, 1994.
13. M Otani, T Uchiyama, H Minoshima, N Nakao, K Shinohara. J Soc Mater Eng Resources Jpn 7:35–45, 1994.
14a. M Otani, H Minoshima, T Ura, K Shinohara. Adv Powder Technol 7:291–303, 1996.
14b. M Otani, H Minoshima, T Uchiyama, K Shinohara, K Takayashiki, T Ura. J Soc Powder Technol 32:151–157, 1995.
14c. T Tanaka. Powder Sci Eng 21(4):39–45, 1989.
14d. K Watanabe. Powder Sci Eng 2(8):35–42, 1995.
15. T Matsuno, K Watanabe, K Ono, M Koishi. J Ceram Soc Jpn Int Ed. 104:942–945, 1996.
16a. K Ukita, M Kuroda, H Honda, M Koishi. Chem Pharm Bull 37:3367–3371, 1989.
16b. T Ishizaka, H Honda, Y Kikuchi, K Ono, T Katano, M Koishi. J Pharm Pharmacol 41:361–368, 1989.
16c. H Aritomi, Y Yamasaki, K Yamada, H Honda, M Koishi. Yakuzaigaku 56:49–56, 1996.
17a. F Kanda, T Nakane, M Matsuoka, K Tomita. J Soc Cosmet Chem 41:197–207, 1990.
17b. T Nakane, T Nanba, K Tomita. J Soc Cosmet Chem Jpn 25:161–170, 1991.
18a. F Nakaya, T Shimizu, T Tanikawa, Y Kikuchi, H Takahashi, M Koishi. Chimicaoggi December 27–31, 1991.
18b. F Nakaya, T Tanikawa, Y Kikuchi, H Takahashi, M Koishi. Material Technol (Zairyo Gijutsu) 13:154–159, 1995.
19a. I Tanaka, N Suzuki, Y Ono, M Koishi. Cem Concr Res 28:63–74, 1998.
19b. I Tanaka, M Koishi. Construction Building Materials 13:285–292, 1999.
19c. I Tanaka, N Suzuki, Y Ono, M Koishi. Cement Concrete Research 29:553–560, 1999.
20a. F Honda, H Honda, M Koishi. J Chromatogr A 696:19–30, 1995.
20b. F Honda, H Honda, M Koishi, T Matsuno. J Chromatogr A 813:21–33, 1998.
20c. F Honda, H Honda, M Koishi, T Matsuno. J Chromatogr A 775:13–27, 1997.
21a. S Oki, S Gohda, T Sohmura, H Kimura, T Kimura, K Ono, T Yoshida. Proc Int Thermal Spray Conf Exposition, Orlando, FL, 28 May–5 June 1992, pp 447–451.
21b. S Oki, S Gohda, T Kimura, K Ono. Proc Int Thermal Spray Conf Exposition, Orlando, FL, 28 May–5 June 1992, pp 381–386.
21c. S Oki, T Kimura, K Ono, S Gohda. J Jpn Soc Thermal Spray (Nippon Yosha Kyokaishi) 27(4):1–7, 1990.
22. T Maruyama, S Onose. Proc Int Symp Material Chemistry in Nuclear Environment, Tsukuba, 1996, pp 18–23.
23. Y Itou, A Kazama, Y Kawashima, N Segawa, M Tanaka. Mater Technol 8:287–295, 1990.
24. T Nakayama, Y Yamazaki. Mater Technol 11:208–215, 1993.
25a. K Hamada, M Ohwada. J Soc Powder Technol Jpn 35:447–450, 1998.
25b. K Hamada. Powder Sci Eng 30(4):63–69, 1998.

Index